Regional Conference on Materials and ASEAN Microscopy Conference (RCM & AMC 2017)

Journal of Physics: Conference Series Volume 1082

Penang, Malaysia
12 - 13 December 2017

Editors:

Mariatti Jaafar
Khairunisak Abdul Razak

ISBN: 978-1-5108-7299-8
ISSN: 1742-6588

Printed from e-media with permission by:

Curran Associates, Inc.
57 Morehouse Lane
Red Hook, NY 12571

Some format issues inherent in the e-media version may also appear in this print version.

Printed by Curran Associates, Inc. (2018)

For permission requests, please contact the Institute of Physics
at the address below.

Institute of Physics
Dirac House, Temple Back
Bristol BS1 6BE UK

Phone: 44 1 17 929 7481
Fax: 44 1 17 920 0979

techtracking@iop.org

Additional copies of this publication are available from:

Curran Associates, Inc.
57 Morehouse Lane
Red Hook, NY 12571 USA
Phone: 845-758-0400
Fax: 845-758-2633
Email: curran@proceedings.com
Web: www.proceedings.com

Table of contents

Volume 1082

Regional Conference on Materials and ASEAN Microscopy Conference 2017 (RCM & AMC 2017)

12–13 December 2017, Penang, Malaysia

Accepted papers received: 7 August 2018
Published online: 10 October 2018

Preface

Papers

IOP Conf. Series: Journal of Physics: Conf. Series **1082** (2018) 011001 doi:10.1088/1742-6596/1082/1/011001

PREFACE

This issue contains the proceedings of a conference entitled Regional Conference on Materials and ASEAN Microscopy Conference 2017 (RCM & AMC 2017) which was held in Penang, Malaysia from 12-13 December 2017. The conference was organized by the School of Materials and Mineral Resources Engineering (SMMRE), Universiti Sains Malaysia, and co-organized by Microscopy Society Malaysia (MSM), AUN/SEED-Net and JICA. With the conference theme of 'Creating sustainable society through innovation in Materials and Biological Sciences', this conference focused on sustainable development as one of the critical issues faced in the 21st century. The programme consisted of 20 invited talks, 77 oral presentations and 78 poster presentations from local and international speakers highlighting recent developments in each of these areas. All presenters were invited to contribute papers to the proceedings. All papers included in the conference proceedings were peer reviewed according to IOP Publishing standards.

The conference aimed to promote activities in various areas of materials, minerals and environment by providing a forum for exchange of ideas, presentation of technical achievements and discussion on their most recent innovations, trends, concerns, issues and challenges encountered with the solutions adopted towards the field of sustainable development. There were also participants from Life Sciences who presented in the ASEAN Microscopy Conference AMC 2017. The Conference was designed with oral and poster sessions and the topics related to Materials, Minerals, Environment and Life Science.

The success of RCM & AMC 2017 is a result of tremendous support and contribution from JICA, AUN/SEED-Net, MSM, industrial and sponsors, and of course, the enthusiastic response from all participants. The conference could not have happened without the commitment of the Organizing Committee, who helped in many ways to organize the conference. Finally, we are also very grateful to the contributors and reviewers for helping us to put together this proceeding.

IOP Conf. Series: Journal of Physics: Conf. Series **1082** (2018) 011002 doi:10.1088/1742-6596/1082/1/011002

Peer review statement

All papers published in this volume of *Journal of Physics: Conference Series* have been peer reviewed through processes administered by the proceedings Editors. Reviews were conducted by expert referees to the professional and scientific standards expected of a proceedings journal published by IOP Publishing.

Soil Buried Linear Low Density Polyethylene/Polyvinyl Alcohol/Kenaf Filler (Lldpe/Pvoh/Knf) Composites: Assessment of Tensile Properties and Morphological Feature

AL Pang[1*], I Hanafi[1,2], and A Abu Bakar[1]

[1]School of Materials and Mineral Resources Engineering, Universiti Sains Malaysia (USM), Engineering Campus, 14300 Nibong Tebal, Penang, Malaysia

[2]Cluster for Polymer Composites (CPC), Science and Engineering Research Centre (SERC), Universiti Sains Malaysia (USM), Engineering Campus, 14300 Nibong Tebal,
Penang, Malaysia

Email: pangailing87@yahoo.com

Abstract. The effect of soil burial duration on the tensile properties and morphological feature of linear low density polyethylene/polyvinyl alcohol/kenaf filler (LLDPE/PVOH/KNF) composites was investigated. A Thermo Haake internal mixer was used to fabricate LLDPE/PVOH/KNF composites with KNF content of 0, 10 and 40 parts per hundred resin (phr), respectively. The tensile properties (tensile strength, tensile modulus and elongation at break) and morphological feature of the composites were evaluated before and after 3 months and 6 months of soil burial period. The results revealed that the tensile properties decrease with prolonged period of soil burial. Field emission scanning electron microscopy (FESEM) micrographs showed presence of cracks and pores on composite samples after soil burial exposure.

1. Introduction

Plastics like polyethylene (PE), polypropylene (PE) and polystyrene (PS) are used worldwide in a variety of packaging industry due to its good mechanical properties, easy to process and reasonably priced [Sam and Ismail, 2014]. Nevertheless, the rapid growing in production of plastics has resulted negative impact to environment because of their resistant to degradation when buried in landfills for decades [Ismail et al. 2011]. In order to minimize environmental pollution, several efforts have been done in order to enhance the biodegradability of plastic products. For instance, some researchers blending the plastics with synthetic biodegradable polymers like polyvinyl alcohol (PVOH) [Ismail et al. 2009; Pang et al. 2015]. In addition, numerous researchers introduce natural resources (lignocellulosic materials) as cheaper alternative into non-degradable plastics [Sam and Ismail, 2014; Pang et al. 2016a].

Lignocellulosic materials like kenaf fiber have been extensively used as filler in plastic composites owing to their biodegradability, high stiffness and strength, as well as cost effective [Sarifuddin et al. 2013; Pang et al. 2016b]. The degradation of fiber-plastic composites depends on a few factors including the fiber loading, degradability of each component, and quality of the fiber-matrix interface [Abdul Khalil et al. 2010]. Generally, the addition of fiber increases the degradation of composites and could lead to a mechanically weakened polymer [Abdul Khalil et al. 2010; Ismail et al. 2011; Anankaphong et al. 2015].

This study aimed to investigate the degradation behavior of LLDPE/PVOH/KNF composites after exposed to soil burial test. The tensile test and surface morphology were measured to characterize the changes after soil burial exposure for a predetermined time.

2. Experimental Method

2.1 Materials

Linear low density polyethylene (LLDPE) and polyvinyl alcohol (PVOH) were purchased from PT. Lotte Chemical Titan Nusantara, Indonesia and Sigma-Aldrich Sdn. Bhd., Malaysia, respectively. Kenaf (KNF) was supplied by National Kenaf and Tobacco Board (LKTN), Kelantan, Malaysia. KNF was ground to particle size of ≤ 75 µm, prior to be used as filler in LLDPE/PVOH blend.

2.2 Preparation of composites

KNF was dried in oven at 80 °C for 24 hours prior to composites preparation. The weight ratio of LLDPE and PVOH was kept at 60:40 (by weight percent) for all loadings of KNF being studied such as 0, 10 and 40 parts per hundred resin (phr). LLDPE/PVOH/KNF composites were melt-blended with a rotor speed of 50 rpm at 150 °C, using a Thermo-Haake internal mixer (Model: R600/610). Subsequently, the composites were undergoing compression molding at 150 °C to form into a 1 mm thickness molded-sheet, using an electrically heated hydraulic press (GoTech Compression Machine, Model: KT-7014 A).

2.3 Soil burial test

A simple soil burial test was carried out to determine the degradation rate of composites and is conducted according to the method used by Sam et al. (2011) and Ismail et al. (2011). LLDPE/PVOH/KNF composites (dumbbell-shaped) were placed inside a polybag, covered by soil and was subjected to natural soil burial. The location for soil burial test was at Universiti Sains Malaysia with latitude 5°8'N and longitude 100°29'E. The effect of short-term response to soil burial test (3 and 6 months) was chosen, attributed to the present of biodegradable materials like KNF and PVOH. After 3 and 6 months of soil burial exposure, the composite samples were retrieved, cleaned and washed with distilled water to remove soil debris on the surface. Next, the samples were dried in oven (60 °C) until reaching a constant weight, and placed in desiccator.

2.4 Tensile test

The tensile test was conducted in accordance with ASTM D638 on the composite samples before and after soil burial exposure, using a universal testing machine (Model: Instron 3366). The test was carried out with a crosshead speed of 5 mm / min at 25 ± 3 °C. A total of five samples were tested to obtain the mean values of tensile properties such as tensile strength, tensile modulus and elongation at break.

2.5 Field emission scanning electron microscopy (FESEM)

Field emission scanning electron microscope (FESEM, Zeiss Supra-35VP) was employed to evaluate surface of specimens before and after subjected to soil burial test. Prior to surface examination, the specimens were sputter-coated with gold (thin layer) to avoid electrostatic charging and poor resolution.

3. Results and Discussion

3.1 Tensile properties

Table 1 shows the tensile properties (tensile strength, tensile modulus, elongation at break) of LLDPE/PVOH/KNF composites at different KNF loadings; before, after 3 and 6 months being buried in natural soil. As expected, the tensile strength, tensile modulus as well as elongation at break decreased with increasing soil burial duration. Furthermore, it is clearly presented in Table 2 that the composites exposed to 6 months soil burial exhibited lower tensile properties retention. The reduction in tensile properties and their retention over the 6 months soil burial exposure could be due to PVOH and KNF leached out from composites, resulting from the interaction with moisture. Subsequently, the composites start to lose their structural strength and thereby diminished its stress transfer ability. Moreover, the possibility of microorganism attack on composites surface could lead to further degradation, and in turn deteriorates the composites' tensile properties. Similar finding was reported by Ismail et al. (2011), who stated that the microbes attack on linear low density polyethylene/thermoplastic sago starch (LDPE/TPSS) resin may be responsible for the fine cracking and tearing that led to further degradation, and hence results in lower tensile properties. Meanwhile, according to Sam and Ismail (2014), degradation of composites in the soil environment was mainly attributed to microorganism consumption and moisture interaction.

Table 1. Tensile properties of LLDPE/PVOH/KNF composites at 0, 10, and 40 phr KNF loading, before, after 3 and 6 months soil burial.

KNF loading (phr)	Tensile strength (MPa)			Tensile modulus (MPa)			Elongation at break (%)		
	Control	After 3 months	After 6 months	Control	After 3 months	After 6 months	Control	After 3 months	After 6 months
0	9.3 ± 0.05	9.0 ± 0.19	8.5 ± 0.19	666.4 ± 11.06	552.9 ± 59.33	512.9 ± 40.38	7.5 ± 0.49	7.3 ± 0.21	6.6 ± 0.12
10	8.6 ± 0.17	8.3 ± 0.23	7.9 ± 0.22	796.5 ± 23.71	580.4 ± 66.81	559.5 ± 58.97	6.5 ± 0.56	6.2 ± 0.18	5.7 ± 0.11
40	7.1 ± 0.15	6.5 ± 0.26	6.2 ± 0.16	1169.0 ± 25.54	806.0 ± 13.85	675.2 ± 25.95	3.0 ± 0.13	2.6 ± 0.18	2.2 ± 0.24

Table 2. Retention of tensile properties for LLDPE/PVOH/KNF composites at 0, 10, and 40 phr KNF loading, after 3 and 6 months soil burial.

KNF loading (phr)	Retention for 3 months (%)			Retention for 6 months (%)		
	Tensile strength	Tensile modulus	Elongation at break	Tensile strength	Tensile modulus	Elongation at break
0	96.7	83.0	96.8	92.3	77.0	88.4
10	96.0	72.9	95.3	91.9	70.2	87.3
40	92.3	68.9	88.3	78.3	57.8	73.8

3.2 Morphological analysis

Figure 1 illustrates the FESEM micrographs of LLDPE/PVOH matrices and LLDPE/PVOH/40 phr KNF composites before, after 3 and 6 months soil burial exposure, respectively. For both LLDPE/PVOH matrices and LLDPE/PVOH/40 phr KNF composites, higher degradation effect was observed with increasing soil burial duration. Comparing Figure 1 (b and d) and (c and e), it is noticeable that the addition of KNF to LLDPE/PVOH matrices contributes to higher degradation rate after soil burial exposure. After 3 months soil burial exposure, there is presence of fine cracks on LLDPE/PVOH matrices surface (Figure 1c). However, more cracks and micro-pores are found on the surface of LLDPE/PVOH matrices after 6 months soil burial (Figure 1e). The pores on matrices surface possibly due to leaching of PVOH from the blend. Consequently, the matrices lose its strength with prolonged soil burial exposure.

On the other hand, Figure 1 (d and f) demonstrate the degradation surfaces of LLDPE/PVOH/40 phr KNF composites after 3 and 6 months soil burial. The composites surface shows the existence of large pores after 3 months and 6 months exposure. In Figure 1(f), it is visible that the large pores formed are deeper and KNF was seen exposed to the surface. The presence of large pores could possibly due to the leaching of both PVOH and KNF, due to the interaction with moisture in soil. These pores may act as stress concentration points and hence lead to the deterioration in tensile properties. Furthermore, it can also be a channel for moisture and/or microorganism to diffuse or enter into the polymer. This finding is in agreement with the lower tensile properties and its retention after prolonged soil burial exposure.

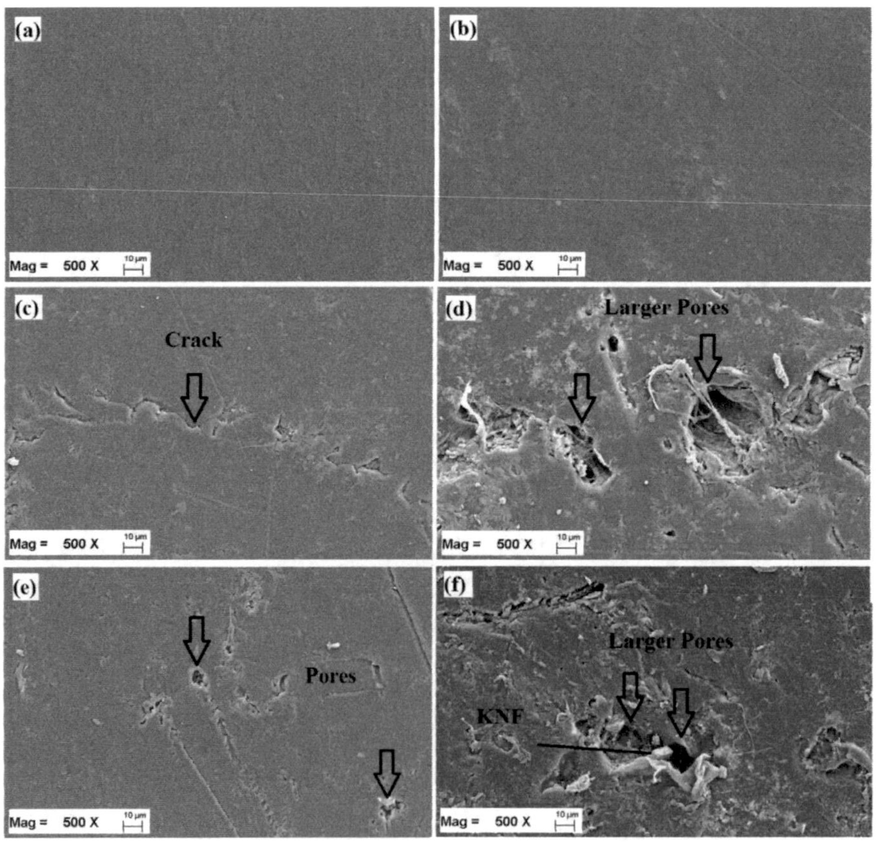

Figure 1. FESEM micrographs of LLDPE/PVOH matrices (a) before, (c) after 3 months, and (e) after 6 months soil burial; and LLDPE/PVOH/40 phr KNF composite (b) before, (d) after 3 months, and (f) after 6 months soil burial, respectively.

4. Conclusion

After subjected to soil burial exposure, the tensile properties (tensile strength, tensile modulus and elongation at break) of LLDPE/PVOH/KNF composites were decreased. As revealed from tensile results, greater reduction in tensile properties of LLDPE/PVOH/KNF composites with prolonged soil burial duration. Moreover, FESEM analysis shows formation of crack and pores on composite surface after soil burial exposure.

Acknowledgments

The authors would like to extend their appreciation to Universiti Sains Malaysia (USM) and National Kenaf and Tobacco Board (LKTN) for the RUC research grant (1001/PKT/8640014) and KNF supply for this work.

References

[1] Sam ST and Ismail H 2014 *Journal of Vinyl & Additive Technology* 20 42
[2] Tan Z, Yi Y, Wang H, Zhou W, Yang Y and Wang C 2016 *Applied Science* 6 147
[3] Ismail H, Majid RA and Taib RM 2011 *Pertanika Journal of Science & Technology* 19 189
[4] Ismail H, Ahmad Z, Nordin R and Rashid A 2009 *Polymer-Plastics Technology and Engineering* 48 1191
[5] Pang AL, Ismail H and Abu Bakar A 2015 *Biores* 10 7302
[6] Pang AL, Ismail H and Abu Bakar A 2016a *Procedia Chemistry 19* 505
[7] Muniyasamy S, Anstey A, Reddy MM, Misra M and Mohanty A 2013 *Journal of Renewable Materials* 1 253
[8] Sapuan SM, Pua FL, El-Shekeil YA, and AL-Oqla FM 2013 *Material & Design* 50 467
[9] Sarifuddin N, Ismail H and Ahmad Z 2013 *Journal of Physical Science* 24 97
[10] Pang AL, Ismail H and Abu Bakar A 2016b *Biores* 11 5889
[11] Abdul Khalil HPS, Bhat I and Sartika MY 2010 *Biores* 5 2278
[12] Anankaphong H, Pentrakoon D and Junkasem J 2015 *International Journal of Polymer Science* 2015 368341
[13] Sam ST, Ismail H and Ahmad Z 2011 *Polymer-Plastics Technology and Engineering* 50 851

Thermomechanical, Tensile and Morphological Properties of Allylic Monomer as Reactive Plasticizer and Reinforcement in Polystyrene/Polyphenylene Oxide Blends

Najamunirahafiza Najamulhak[1], Arjulizan Rusli[1*], Hanafi Ismail[1]

[1]School of Materials and Mineral Resources Engineering, Engineering Campus, Universiti Sains Malaysia, Nibong Tebal, Penang, Malaysia Sains Malaysia, Engineering Campus, Penang, Malaysia

*Email: arjulizan@usm.my

Abstract. The use of diallyl phthalate (DAOP) monomer as both reactive plasticizer and reinforcement in polystyrene/polyphenylene oxide (PS/PPO) blends was investigated. The blends of a fixed amount of PS (80%) with different contents of PPO-DAOP were prepared using an internal mixer and glass transition temperature (T_g), tensile and morphological properties of the blends were studied. T_g of the blends gradually reduced with the replacement of PPO with up to 4%DAOP might be due to the reduction of rigid component (PPO) but then plateau in the presence of 6%DAOP may be due to presence of higher content of cured DAOP. Tensile test conducted on the blends indicated that tensile strength and modulus increased with increasing DAOP content up to 4wt% and then slight reduced with further increased of DAOP with the value slightly higher than PS/PPO blend without DAOP suggesting that DAOP could act as reinforcement in PS/PPO blend. Morphological study of the blends showed single phase of 80PS/20PPO blend while blends with DAOP illustrated rough, two phase morphologies with DAOP particle size increased with increasing DAOP. However, in the presence of 6% DAOP, smoother surface was observed which translated into reduction of tensile strength and modulus of the blends.

1. Introduction

Engineering thermoplastics such as polyphenylene oxide (PPO), polyethersulfone and polyetherimide possess many desirable properties but difficult to be processed due to their rigid chain structure. They need to be processed at high temperature using special processing equipment in which they expose to degradation at that high temperature. One of the conventional methods to improve processibility of the thermoplastics is by blending with a more processable polymer. Commercially important application of the technique is blending of PPO with polystyrene (PS) producing segmentally miscible blend in all proportions due to favorable interaction between their aromatic rings with properties depending on the ratios of the components [1]. Addition of PS into PPO will ease processing by reducing glass transition temperature (T_g) (i.e. reduction of thermo-mechanical properties) with increasing proportion of PS [2].

Meanwhile, a few researchers reported the use of thermosetting monomers as reactive plasticizers in thermoplastics such as polyvinyl chloride [3], PPO [4] and polycarbonate [5]. In this application,

IOP Conf. Series: Journal of Physics: Conf. Series **1082** (2018) 012002 doi:10.1088/1742-6596/1082/1/012002

blending of small amount of crosslinkable monomers such as epoxies and allylic monomers reduce viscosity of the thermoplastic at the early stage of processing. Subsequently the monomer can be polymerized and will phase separated as dispersed particles in the thermoplastic matrix. The main advantage of this method is the improvement of processibility without or with a very slight reduction of thermo-mechanical properties throughout composition range [3,4]. In addition, there is possibility for the dispersed thermoset particles to act as toughener or reinforcement, depends on either the particles are rubbery or glassy. Most of the research conducted on improving processibility of thermoplastic in the presence of thermosets studied the effect of thermoset on viscosity and morphology of the blends. However, not much research has been done on the applicability of the thermoset as both reactive plasticizer and reinforcement/toughener in thermoplastic system. Therefore, this study aims to investigate the effect of different compositions of allylic monomers on the thermo-mechanical, tensile and morphological properties of PS/PPO blend.

2. Experimental

2.1 Materials

Two different thermoplastics were used in this research. Poly(2,5-dimethyl phenylene oxide) of polyphenylene oxide, PPO in powder form with a number and weight average molecular weights of 19,180 and 44,297 g/mol, respectively, was obtained from GE Plastics as C2000. Polystyrene (PS) in pellet form was obtained from Petrochemicals (Malaysia) Sdn. Bhd. Meanwhile, thermosetting monomer used was diallyl orthophthalate (DAOP) which was purchased from Sigma-Aldrich in the form of liquid. Dicumyl peroxide (DCP) or cumene peroxide, obtained from Sigma-Aldrich was used as an initiator to initiate chain growth polymerization of the allylic monomer. Both DAOP and DCP were used as received. The chemical structure of the thermoplastics, thermosetting monomer and initiator are shown in Table 1.

Table 1. Chemical structure of materials used

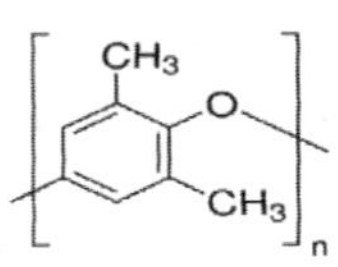	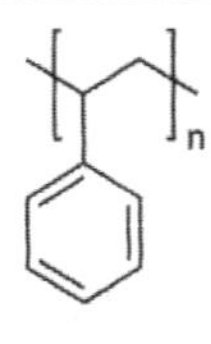	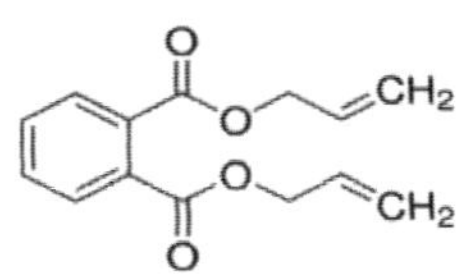	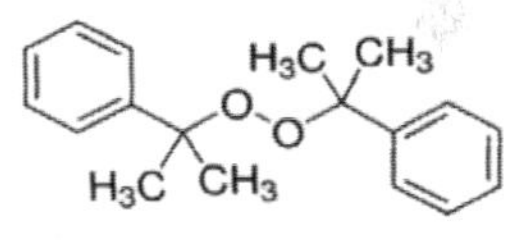
PPO	PS	DAOP	DCP

2.2 Melt blending of PS/PPO or PS/PPO/DAOP blends

Blending of PS and PPO with composition of 80 wt% PS and 20 wt% PPO was used as control sample. Before blending, the PPO powder and PS pellets were pre dried at 120C and 80 C, respectively for 12h. Then the PPO and PS were manually mixed and placed in an oven at 80°C for 2 hours to promote inter-diffusion of materials. After that, the sample was melt blended using Haake Polydrive Internal Mixer (Thermo Electron Corporation) at 220°C with 60 rpm mixing speed for a period of 15 minutes. The obtained blended sample was then compressed at 220°C with the pressure of 1000 psi for 5 minutes, using a compression moulding machine (Kao Tieh Go Tech Testing Machine) to obtain 1 mm thick plastic sheet.

Meanwhile, blends of 80 wt% PS with a total of 20 wt% of PPO and DAOP-DCP (formulation as shown in Table 2) were also prepared by melt blending. All the samples were melt blended using Haake Polydrive Internal Mixer at 200°C at 60 rpm and a total mixing time of 10 minutes. Before melt blending, a specified amount of PPO, DAOP and DCP were mixed at room temperature. The blending in internal mixer was initiated by firstly melted PS for 5 minutes and subsequently added quantified PPO/DAOP-DCP to the chamber. The mixing was then carried on for further 5 minutes. The obtained blended samples were then compressed at 200ºC and with the pressure of 1000 psi for 5 minutes in a compression moulding machine to obtain 1 mm thick plastic sheet. The samples were further cured at 150ºC for 3 hour.

Blends containing a reactive plasticizer and initiator are identified by the code wt%PS/wt%PPO/DAOP/DCP so that 80PS/18PPO/2DAOP-DCP represents the blend with 80 wt% PS and 18 wt% PPO and 2 wt% DAOP-DCP (the concentration of DCP in DAOP was fixed at 4 wt%). The compositions of PPO and DAOP-DCP were chosen based on the findings that uncured blends of PPO and DAOP were miscible at 70wt% and more of PPO content [4].

Table 2. Formulation of the blends

Blends	PS (wt%)	PPO (wt%)	DAOP-DCP (wt%)
80PS/20PPO	80	20	0
80PS/18PPO/2DAOP-DCP	80	18	2
80PS/16PPO/4DAOP-DCP	80	16	4
80PS/14PPO/6DAOP-DCP	80	14	6

2.3 Differential Scanning Calorimetry (DSC)

DSC studies were conducted using Mettler Toledo Star 1 DSC in order to determine the glass transition temperature T_g of the blends. The samples were obtained from sheet formed after compression mouding and curing of the sample. Samples of approximately 10 mg for each composition were analyzed using DSC machine by placing the samples in the aluminium pans. All samples were heated from 30ºC to 300ºC at the heating rate of 10ºC/min to flatten the samples and to remove thermal history of PPO/PS. Then the sample was cooled down from 300ºC to 30ºC at 10ºC/min by using liquid nitrogen at the rate of 50ml/min. The heating process was repeated with the same heating rate and heating procedure to determine the T_g of the blend.

2.4 Tensile test

Tensile tests for all samples were performed according to ASTM D882 using a testing machine (Instron Corporation, model 3366, USA) at 25°C and with the cross-head speed of 5 mm/min. Strip shaped specimens were cut from the sheets prepared from compression-moulding. At least five specimens of each sample were tested to determine the mean value of the results. Tensile strength, elongation at break and tensile modulus at 100% elongation were recorded.

2.5 Scanning Electron Microscopy (SEM)

The tensile fractured surface morphology of the blends was observed using Tabletop SEM Hitachi TM3000 with acceleration voltage of 5kV. The fractured samples were sputter coated with thin gold prior to examination using the SEM.

3. Results and discussion

3.1 DSC studies

The effect of incorporation of DAOP-DCP on the T_g of 80PS/20PPO was examined. Figure 1 shows the DSC curve of 80PS/20PPO blend and various blends of 80 wt% PS with a total of 20 wt% of PPO and DAOP-DCP compounds while T_g values are tabulated in Table 3.

For 80PS/20PPO blend, only single T_g is observed suggesting miscibility of the blend. The Tg observed was compared with T_g calculated from Fox equation [7]. PPO is reported to has a T_g of 250 [4] and 212 [6] which were obtained from Dynamic Mechanical Analyzer (DMA) and DSC, respectively while pure PS was observed to have a T_g of 90°C. The T_g obtained from DSC for the blends of 80PS/20PPO compares quite well with that calculated from Fox equation [7] which normally applicable for miscible blends and as well as for copolymers [8], assuming that no interaction occurs between PPO and PS. Meanwhile, in the presence of DAOP, single T_g was also observed with T_g slightly lower from the blends without DAOP. The presence of a single T_g is unexpected since it is believed that the thermosetting monomer will phase separate after cured thus could produce two T_g. This observation might be due to the presence of only a small amount of DAOP thus the second T_g could not be observed. On the other hand, the slight reduction of T_g in the presence of cured DAOP might be due to the reduction of PPO content (high T_g component) thus reducing the T_g of the blends. With the presence of 6wt% DAOP, T_g plateau at 94°C might be due to the presence of higher amount of cured DAOP which compensate the reduction of PPO.

Table 3. T_g of 80PS/20PPO blend and blends of 80 wt% PS with different compositions of PPO/DAOP-DCP

Compounds	Tg obtained from DSC (°C)	Tg predicted by Fox Equation (°C)
80PS/20PPO	100	109
80PS/18PPO/2DAOP-DCP	96	-
80PS/16PPO/4DAOP-DCP	94	-
80PS/14PPO/6DAOP-DCP	94	-

3.2 Tensile properties

The effects of various compositions of PPO/DAOP-DCP on tensile properties of 80PS/PPO/DAOP-DCP blends were studied. The tensile strength of 80PS/20PPO blend and blends of 80wt%PS with different amounts of reactive plasticizer (80PS/18PPO/2DAOP-DCP, 80PS/16PPO/4DAOP-DCP and 80PS/14PPO/6DAOP-DCP) are summarized in Figure 2. It can be observed that tensile strength of the blends increased with increasing amount of DAOP up to 4wt% DAOP but then slightly decreased with further increased of DAOP to 6wt%. In all compositions of PPO/DAOP-DCP blends, the value of tensile strength higher than 80PS/20PPO blend without reactive compatibilizer suggesting that DAOP could also act as reinforcement in 80PS/20PPO blends. It is believed that phase separated DAOP particles could cause higher force needed to break the blends thus improving tensile strength of the blends as normally found in polymer composite system [9]. However, in the presence of 6wt% DAOP, the tensile strength slightly reduced possibly because of high amount of DAOP could affect miscibility of PPO with PS or poor miscibility of PS and DAOP in the system with high DAOP content.

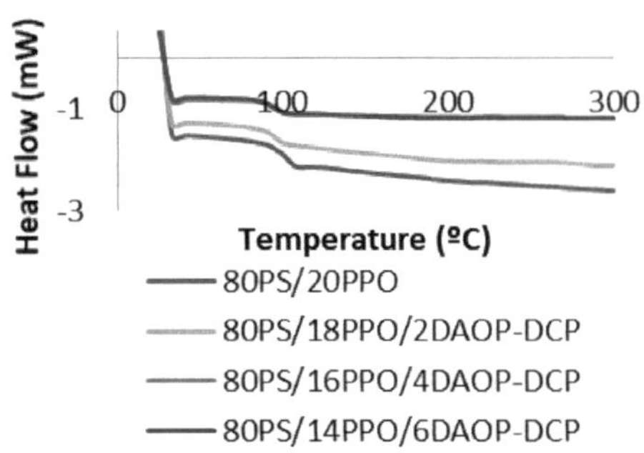

Figure 1. DSC curves of 80PS/20PPO and different compositions of reactive plasticized blends

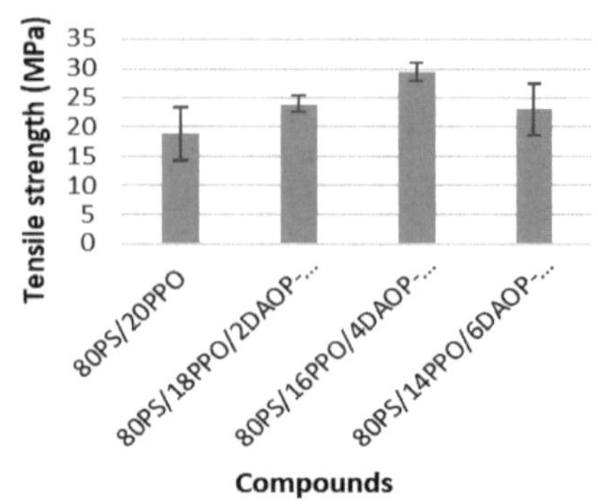

Figure 2. Tensile strength of 80PS/20PPO and various compositions of reactive plasticized blends

Figure 3 illustrates the modulus of 80PS/20PPO blend and different compositions of PPO/DAOP-DCP in 80PS/PPO/DAOP-DCP blends. It can be seen that tensile modulus slightly increased with increasing amount of DAOP up to 4wt% DAOP might be due to the presence of polyDAOP which disperses throughout the blends improves inter-chain friction thus decrease molecular mobility, as if the presence of filler in the blend [9]. However, in the presence of 6wt% DAOP, tensile modulus slightly reduced. The reason for that is less understood but possibly due to lesser miscibility among components.

Meanwhile, the presence of DAOP reduced the ability of the blends to elongate as shown by the significant reduction of elongation at break (Figure 4). The rigid DAOP might decrease deformability of the blends hence decrease the ability of the blends to elongate before rupture.

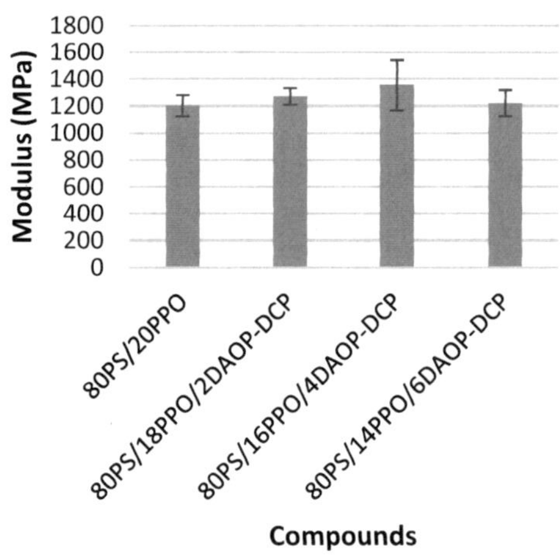

Figure 3. Modulus of 80PS/20PPO and various compositions of reactive plasticized blends

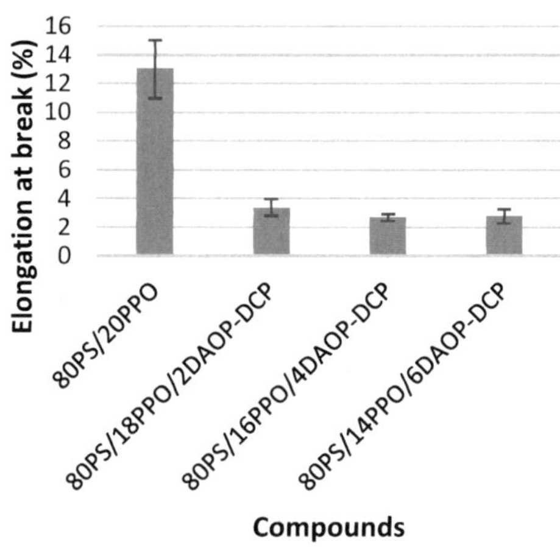

Figure 4. Elongation at break of 80PS/20PPO and various compositions of reactive plasticized blends

The tensile test results revealed that the optimum amount of DAOP that brought about improvement on the tensile strength and modulus is 4 wt%. These results show that DAOP can act as both reactive plasticizer and reinforcement in 80PS/PPO blend.

3.3 SEM

The tensile fractured surfaces of the blends were observed using SEM to evaluate the phase morphology of the blends. Evidence of phase separation and changes in the morphology with increasing DAOP content in the cured 80PS/PPO/DAOP-DCP blends are illustrated in Figure 5. For 80PS/20PPO blend without DAOP-DCP, single phase morphology was observed due to miscibility of PPO and PS. The presence of intermolecular interactions (π-π electron transfer) are said to be responsible for the miscibility of the blend system [2].

Meanwhile, in the presence of 4 wt%DAOP, a rough surface was observed with some indication of the presence of small dispersed poly (DAOP) particles in PS matrix. With a further increase of reactive plasticizer content to 6 wt%, smoother surface than the blend of 80PS/16PPO/4DAOP-DCP blends could be observed. In addition, bigger size of dispersed particles than that present in 80PS/16PPO/4DAOP-DCP was observed might be due to the present of more poly (DAOP) content in the former. In addition, smoother surface was observed in the blend of 80PS/14PPO/6DAOP-DCP which is believed to be responsible for the slight reduction of tensile strength and modulus of the blends compared with 80PS/16PPO/4DAOP-DCP blend.

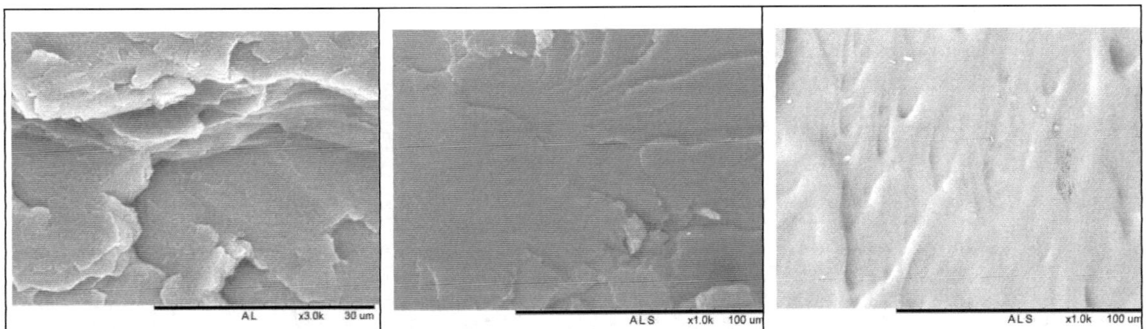

Miscible 80PS/20PPO	80PS/16PPO/4DAOP	80PS/14PPO/6DAOP

Figure 5. SEM images of 80PS/20PPO and various compositions of reactive plasticized blends

4. Conclusion

The replacement of PPO with DAOP in the blends of 80wt%PS slightly reduced the T_g of the blends. The presence of DAOP in 80PS/PPO blends improved the tensile strength and modulus up to 4wt% DAOP suggesting that DAOP could act as both reactive plasticizer and reinforcement in PS/PPO system. SEM micrograph indicated the presence of small particles in which DAOP phase separated in PS matrix.

Acknowledgement

The authors would like to thank research funding support Fundamental Research Grant Scheme (FRGS) (203/PBAHAN/6071281) awarded by Ministry of Higher Education of Malaysia.

References

[1] Cizek EP and Arbor A., U.S.P. Office, Editor. 1968 General Electric Company: United States
[2] Stoelting J, Karasz FE, and Macknight WJ 1970 *Polymer Engineering & Science* 10 133
[3] Liang G G, Cook WD, Sautereau HJ, Tcharkhtchi A 2009 *Polymer* 50 2655
[4] Rusli A, Cook W and Liang GG 2011 *European Polymer Journal* 47 1775
[5] Liang G G, Cook WD, Sautereau HJ, Tcharkhtchi A 2008 *European Polymer Journal* 44 366.
[6] Rusli A, Cook WD and Saito K 2013 *European Polymer Journal* 49 4072
[7] Fox TG 1956 Bulletin American Physical Society 1 123
[8] Brostow W, Chiu R, Kalogeras IM, Vassilikou-Dova A 2008 *Materials Letters* 62 3152
[9] Manson JA and Sperling LH, 1976 *Polymer blends and composites* (New York: Springer Science + Business Media

IOP Conf. Series: Journal of Physics: Conf. Series **1082** (2018) 012003 doi:10.1088/1742-6596/1082/1/012003

Formation of Titanium Hydride from the Reaction Between Magnesium Hydride and Titanium Tetrachloride

Sanjith Udayakumar[1], Atif Sadaqi[1], Najwa Ibrahim[1], M. N. Ahmad Fauzi[1], Sivakumar Ramakrishnan[1] and Sheikh Abdul Rezan[1*]

[1]School of Materials & Mineral Resources Engineering, Universiti Sains Malaysia, Engineering Campus, Nibong Tebal, Malaysia

*Email: srsheikh@usm.my

Abstract. The reaction between magnesium hydride (MgH_2) and titanium tetrachloride ($TiCl_4$) to produce titanium hydride (TiH_2) was carried out at varying temperature, time and Argon (Ar) gas flow rate. The reaction conditions were optimized for increasing the formation of TiH_2 by using the statistical design of experiments (DOE). The hydriding temperature was varied from 200 to 300 °C, while the reaction time was changed from 60 to 180 minutes. The flow rate of argon gas was controlled in the range of 20 to 60 mL.min^{-1} to minimize $TiCl_4$ dilution. The samples were characterized by X-ray diffraction (XRD), scanning electron microscopy (SEM), energy dispersive X-ray spectroscopy (EDX), Elemental hydrogen analyzer. From the experiments, the highest weight change (X_w) of the sample produced at 300 °C for 180 minutes with Ar flow rate of 20 mL.min^{-1} was 0.19 wt.% and the degree of hydriding (X_h) of 38.49% by elemental hydrogen analysis. The DOE analysis inferred that the reaction time was the most significant factor followed by the flow rate of Ar gas and reaction temperature. The SEM/EDX analysis indicated that the product was composed of MgH_2, $MgCl_2$, $TiCl_{2.3}$ and TiH_2.The product contained a minor phase of TiO_2 that could be due to seepage of ambient moisture into the reaction chamber.

1. Introduction

Titanium (Ti) is acknowledged for its strategic importance as it exhibits attractive properties such as low density, high strength and high structural efficiency for critical, high-performance aircraft, such as jet engine and airframe components [1, 2]. There were numerous processing methods to produce Ti metal that include the Kroll [3], TiRO [4], OS process[5] and oxycarbonitride process [6]. This transition element is being widely studied to be exploited in the energy sector as the next generation power source as it has high hydrogen storage affinity as TiH_2 and can be tailored to have different hydrogen concentration depending upon the nature of processing [7]. Though this metal hydride hold promise for clean energy storage, the cost of producing Ti by conventional ingot technology was considerably high due to costly batch manufacturing via the industry standard Kroll process and intricate processing steps [2]. Since there were numerous applications for $TiCl_4$, a titanium by-product from the Kroll process, a coordinated approach to utilize $TiCl_4$ in the production of TiH_2 was attractive [8]. Furthermore, a sustainable and low temperature method to produce $TiCl_4$ has been introduced [9-11]. In the pursuit to lower the production cost, a traditional powder metallurgy (PM) approach seemed viable and favourable route for cost effective fabrication of titanium alloys components [1]. The blended elemental powder metallurgy (BEPM) method used low cost starting blends of titanium powder and alloying powders which offers superior mechanical properties. The relative densities of different grades of titanium alloys must be greater than 98% of the theoretical densities in order to achieve a desired level of specific properties such as strength, ductility and creep resistance. However, the relative densities of the titanium alloys produced by powder metallurgy do not exceed 95% [1, 12]. In order to balance the achievable properties of titanium alloys without additional thermo-mechanical operations, the TiH_2 was used as the starting material instead of conventional Ti metal powder. TiH_2 was being produced by hydrogenation of low-grade titanium sponge, turnings, and other titanium materials until a revolutionary approach called the ADMA process was introduced by Institute for Metal Physics (Ukraine), Incorporated [13]. This process was developed and several improvements have been carried out to produce the low-cost, high-quality titanium hydride powder [14, 15]. In this study, to improve on the ADMA process, a laboratory scale work was conducted to develop a low cost, low temperature process for synthesizing TiH_2 powder by means of metallothermic (hydrides of alkali or alkali earth metals) reduction of $TiCl_4$. Previous work on converting titanium dioxide (TiO_2) to TiH_2 using MgH_2 was the starting point for this innovation [16].

2. Experimental Setup

Titanium hydride (TiH_2) is produced by the gas-solid hydriding reaction between an alkali earth metal hydride (i.e., MgH_2) and $TiCl_4$. The expected chemical reaction is as follows:

$$2MgH_2 + TiCl_{4(g)} = TiH_2 + 2MgCl_2 + H_{2(g)},\ \Delta G°(\text{kJ/mole}) = -0.0393*T - 488.42\ (\text{kJ/mol}) \quad (1)$$

The value for standard change in Gibbs free energy change ($\Delta G°$) indicated that the reaction was thermodynamically feasible and spontaneous at lower temperatures (<350°C) compared to the conventional ADMA process which takes place between 800-900°C [14]. Argon (Ar) gas was used as carrier gas in the reaction to facilitate the flow of $TiCl_4$ and dilute its concentration. A flow rate meter was used to control the flow of Ar gas that passed through the reaction chamber. The MgH_2 pellet was placed in an alumina crucible. The crucible with MgH_2 pellet was placed at the centre of the quartz tube at the desired temperature. A hot plate was used to evaporate the $TiCl_4$ liquid into gaseous state at 135°C. The vapour of $TiCl_4$ produced was carried by the Ar gas into the quartz tube for the reaction between MgH_2 and $TiCl_4$ to occur. To ensure the molar ratio of $TiCl_4/MgH_2$ was below 2, the partial pressure of $TiCl_4$ (p_{TiCl4}) vapour was estimated based on work done by Navarro et al. [17]. Residual exit gas was neutralized by NaOH into NaCl before excess Ar was released to ambient atmosphere. The percentage of TiH_2 formed was determined based on the gravimetric analysis of the reacted sample and final hydrogen content via elemental hydrogen analyzer (Perkin Elmer Series II CHNS/O analyzer, model 2400). The calculations for degree of hydriding was made assuming full conversion of $TiCl_4$ to TiH_2 with no other intermediate compounds formed. The study was carried out by varying three different factors for three distinct levels namely; furnace temperature of 200-300°C, reaction duration of 60-180 minutes and Ar gas flow rate of 20-60 ml.min^{-1}. The parameters were chosen based on the dehydriding temperature of MgH_2 while reaction time was based previous chlorination study on forming $TiCl_4$ [18, 19]. Design of Experiment (DOE), a statistical approach was used to study of interaction parameters with the measured responses so that the most significant parameters in the reaction can be identified. Besides this, all of the samples and reacted products were analysed by X-ray Diffraction using Cu-K$_\alpha$ radiation (XRD; Bruker D8-advance). XRD was used study the phases present in the reaction products and determine the extent of hydriding in the reaction. The surface morphology and the elemental composition of the sample was studied by Scanning Electron Microscopy equipped with Electron Dispersive X-ray Spectroscopy (SEM/EDX; Leo Supra 35VP). The degree of hydriding was measured using Elemental CHNS/O analyser by determining the hydrogen content present in the reactants before and after the hydriding process.

3. Result and Discussion

3.1 Characterization of Magnesium Hydride

The SEM/EDX analysis was performed to confirm the elements and ensure that the investigated MgH_2 powder was of high purity. The SEM images (**Figure 1**) show that the MgH_2 powder had an irregular flaky structure without any signs of particle agglomeration or oxide content [20]. Elemental analysis via EDX confirmed the presence of 100 wt.% of magnesium and hydrogen atoms were not detected due to instruments limitation to detect this element. The XRD pattern of the raw powder was shown in **Figure 2**. Since the powder was of high purity, the peaks of MgH_2 are very distinct [21, 22] and follow the reference pattern of ICSD 98-010-9357.

Element	Wt %	At%
Mg K	100.00	100.00

Figure1. SEM micrograph and elemental analysis of MgH_2 powder

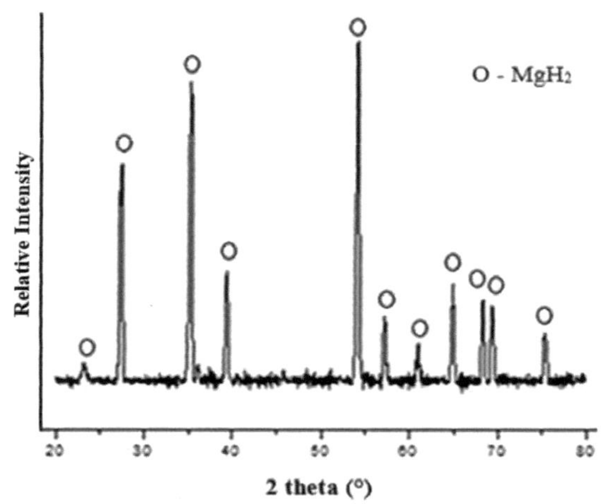

Figure 2. X-ray Diffractogram of raw MgH_2 powder

3.2 DOE Analysis

The results of the factorial design for the hydriding of TiH_2 are shown in the **Table 1**. The two responses chosen were percent of weight difference gained (wt.%) and degree of hydriding (X_h). From the experiments, the highest weight gain change (X_w) of the sample produced at 300 °C for 180 minutes with Ar flow rate of 20 mL.min^{-1} was **0.19 wt.%** and the X_h obtained to be **38.49 wt.%** (Run #4). Theoretical weight gained if full conversion of $TiCl_4$ to TiH_2 was **0.82 wt.%.** Model graph analysis using perturbation plots was also performed in order to interpret the experimental results based on factorial model. **Figure 3(a)** and **(b)** shows the perturbation plot for both responses which were X_w and X_h. From the latter figures shown, it can be observed that all the parameters namely temperature, time, and flow rate of Ar gas have a positive impact on X_w and X_h. This was corroborated by the positive gradient of the perturbation plot for all the factors. From the slopes of all three factors, the time factor showed the highest positive gradient which indicated the most significant parameter.

Table 1. Percent Weight difference and Degree of Hydriding for DOE Experimental Conditions

Standard Run	Temperature (°C)	Time (Minutes)	Flow rate of Argon Gas (ml.min^{-1})	Percentage of weight difference, X_w (wt.%)	Degree of hydriding, X_h (wt.%)
1	200	60	20	0.06	26.01
2	300	60	20	0.03	25.35
3	200	180	20	0.1	30.86
4	300	180	20	**0.19**	**38.49**
5	200	60	60	0.16	37.32
6	300	60	60	0.06	30.04
7	200	180	60	0.06	28.1
8	300	180	60	0.13	33.64
9	250	120	40	0.13	33.55
10	250	120	40	0.06	28.11
11	250	120	40	0.06	28.1

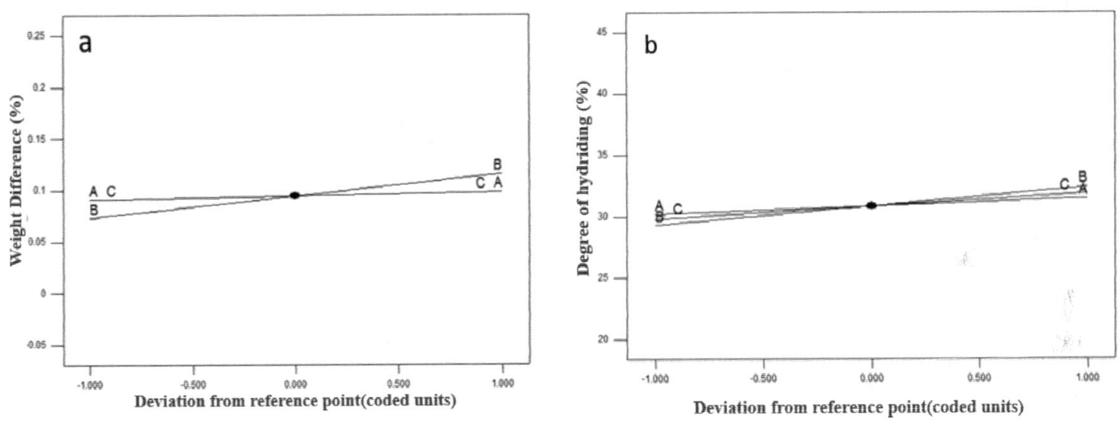

Figure 3. Perturbation plot of (a) Percentage of weight difference, X_w (b) Degree of hydriding, X_h
A = Temperature (°C), **B** = Time (Minutes), **C** = Flow rate of Argon Gas (ml/min)

3.3 Phase Characterization of Residual Product

The reaction products were characterized to identify and confirm the expected phases formed during the hydriding process. Of all the reactions listed in **Table 1** as per the factorial design, experiment designated standard run #4 showed the highest X_w. For this sample, the morphology, X-ray diffraction pattern was shown in **Figure 4**. The XRD of the product obtained in Run #4 (300°C, 180 minutes, 20 ml.min^{-1}) showed a lot of peaks corresponding to MgH_2 in spite of bearing the highest X_w. Since there was only a partial reaction with $TiCl_4$, residual peaks of MgH_2 (ICSD 98-010-9357) were observed. Few peaks of TiH_2 confirmed the presence of hydride formed during the reaction. Traces of $TiCl_{2.3}$ (ICSD 98-000-3839) were observed which is due to the low partial pressure of H_2 (p_{H2}) gas in the system since Ar gas was used. Furthermore, there was reaction between MgH_2 with $TiCl_4$ to form $TiCl_{2.3}$. From the DOE analysis, it was found that the use of Ar gas will not facilitate the full conversion of $TiCl_4$ to TiH_2 in the investigated range of parameters. Furthermore, due to the low p_{H2} in the reactor, formation of titanium sub-chlorides was favourable compared to TiH_2. A review of literature found most hydriding studies was done in H_2 gas atmosphere with p_{H2} close to 1 atmosphere or greater [1, 16, 24].

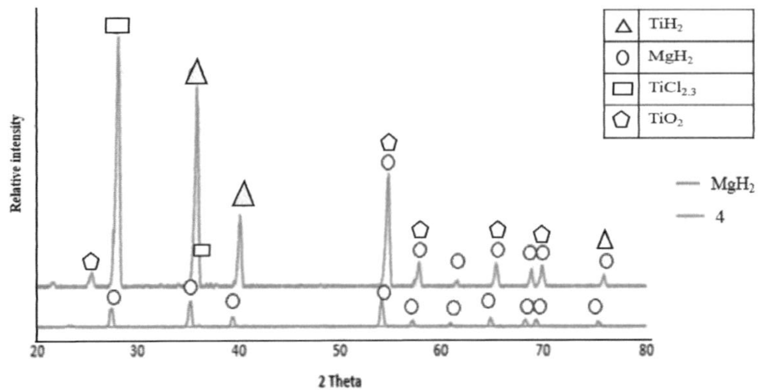

Figure 4. XRD pattern of reaction product (Run #4 - 300°C, 180 minutes, 20 ml.min^{-1})

4. Conclusions

The reaction parameters for MgH_2 with $TiCl_4$ gas to form TiH_2 was investigated and it was the first reported study by DOE. The study inferred that, the most significant parameter to affect the X_w and X_h was reaction time followed by Ar gas flow rate and reaction temperature. The highest X_w of 0.19 wt. % with X_h of 38.49% was achieved at 300°C for 180 minutes with 20 ml.min^{-1} of Ar flow rate. From the DOE statistical analysis, reaction time was identified as the most significant parameter on both responses measured, followed by flow rate of Ar gas and reaction temperature. The X_h was low in Ar atmosphere due to low p_{H2} that was more favourable for the formation of titanium sub-chlorides than TiH_2. This investigation has indicated a high p_{H2} close to 1 atmosphere was needed to make TiH_2 formation more favourable through this reaction.

Acknowledgements

The authors gratefully acknowledge the support from Universiti Sains Malaysia (USM) Fellowship Award. This work was financially supported by USM and Ministry of Higher Education (MOHE) of Malaysia through the following research grants: Fundamental Research Grant Scheme (FRGS) (No. 203/PBAHAN/6071364), USM Research University Individual (RUI) grant (No. 1001/PBAHAN/814273) and Nippon Sheet Glass Research Grant (NSGRG) (No. 304/PBAHAN/650360/N120). Special thanks to USM technician, Mr. Azrul for supporting the experimental work.

References

[1] Duz V, Matviychuk M, Klevtsov A, Moxson V 2017 *Metal Powder Report* 72 30
[2] Van Vuuren DS, Oosthuizen SJ, Heydenrych MD 2011 *Journal of the Southern African Institute of Mining and Metallurgy* 111 141
[3] Kroll W 1940 *Transactions of the Electrochemical Society* 78 35
[4] Doblin C, Chryss A, Monch A 2012 *Key Engineering Materials*: Trans Tech Publ.
[5] Ahmadi E, Yashima Y, Suzuki RO, Rezan SA. 2018 *Metallurgical and Materials Transactions B* 1-14
[6] Ahmadi E, Fauzi A, Hussin H, Baharun N, Ariffin KS, Rezan SA 2017 *International Journal of Minerals, Metallurgy, and Materials* 24 444
[7] Reilly JJ, Sandrock GD 1980 *Scientific American* 242 118
[8] Ivasishin O, Moxson V 2015 *Titanium Powder Metallurgy* 117-48.
[9] Ibrahim N, Ahmadi E, Rahman SA, Fauzi MNA, Rezan SA 2017 AIP Conference Proceedings 1805 040005

[10] Ahmadi E, Sereiratana E, Rezan SA, Yeoh F, Fauzi A, Zhang G 2016 *Materials Science Forum* Trans Tech Publication
[11] Ahmadi E, Rezan SA, Baharun N, Ramakrishnan S, Fauzi A, Zhang G 2017 *Metallurgical and Materials Transactions B* 48 2354
[12] Ivasishin O, Moxson V 2015 *Butterworth-Heinemann* 117-48
[13] Froes FH 1998 *JOM* 50 41
[14] Klevtsov A, Nikishin A, Shuvalov J, Moxson V, Duz V 2013 *Adma Products, Inc., assignee. United States patent US* 8 388,727
[15] Kasparov SA, Klevtsov AG, Cheprasov AI, Moxson VS, Duz VA 2011 *Adma Products, Inc., assignee. United States patent US* 8 007,562
[16] Fang ZZ, Middlemas S, Guo J, Fan P. A 2013 *Journal of the American Chemical Society* 135 18248
[17] Navarro RCS, Vasconcellos ET, Brocchi EA 2017 *Thermochimica Acta* 647 22
[18] Ahmadi E, Hamid SARBSA, Hussin HB, Baharun SRNB, Ariffin KSB, Fauzi M 2016 *International Journal of Jaipur National University* 5 11
[19] Ramakrishnan S, Kwok TC, Hamid SARSA 2016 *AIP Conference Proceedings*
[20] Itam Sulaiman N, Juahir N, Mustafa N, Halim Yap FA, Ismail M 2016 Dalton Transaction 45, 19380
[21] Phetsinorath S, Zou J-x, Zeng X-q, Sun H-q, Ding W-j *2012Transactions of Nonferrous Metals Society of China* 22 1849
[22] Yao M, Chen L, Rao G, Zou J, Zeng X, Peng 2013 J *Journal of Nanomaterials* 8
[23] Beattie SD, Setthanan U, McGrady GS 2011 *International Journal of Hydrogen Energy* 36 6014
[24] Zhang Y, Fang ZZ, Xia Y, Huang Z, Lefler H, Zhang T 2016 *Chemical Engineering Journal* 286 517

Synthesis of Porous Carbonate Apatite/Gelatin Scaffolds Via Freeze Drying Method

H. Noor Khairiyah Hanisah[1], R. Muhammad Ridhwan[1], and A. Nurazreena[1*]

[1]School of Materials and Mineral Resources Engineering, Engineering Campus, Universiti Sains Malaysia, 14300 Nibong Tebal, Pulau Pinang, Malaysia.

Email: nurazreena@usm.my

Abstract. Gelatin is widely used in medical field as scaffolding material. However, in order to encourage bone formation, carbonate apatite (CO_3Ap) which resembles the human bone composition is added into the gelatin scaffold. In this research, CO_3Ap/gelatin scaffold were fabricated via freeze drying method. CO_3Ap powder was obtained from precipitation reaction of calcium hydroxide and phosphoric acid accompanied by carbon dioxide gas as the carbonation source. The obtained powder were then analysed by X-ray diffraction (XRD), Fourier transformed infrared (FTIR) spectroscopy and carbon hydrogen nitrogen (CHN) test. The carbonate content by weight percentage is 12.8 wt. %. The mean particle size of the CO_3Ap powder was 7.76 μm. To fabricate the CO_3Ap/gelatin scaffold, the CO_3Ap powder were mixed with gelatin at various CO_3Ap/ gelatin ratio (15/5, 10/10) wt. % followed by casting and pre-freezing temperature of -10 °C, -20 °C and -40 °C before freeze dried. The highest compressive strength achieved was 7412.36 kPa for CO_3Ap/ gelatin with ratio (10/10) wt. %, freeze at -40 °C as the pre-freezing temperature. Field emission scanning electron microscope (FESEM) analysis shows that the CO_3Ap powder was distributed within the gelatin scaffold. As a conclusion, it was found that higher gelatin content and lower pre-freezing temperature increases the compressive strength of the CO_3Ap /gelatin scaffold.

1. Introduction

Tissue engineering field appears to be a promising approach for biological implants and prosthesis [1]. Generally, bone consists of collagen for the organic phase of bone and the inorganic phase are hydroxyapatite [2], while in some cases such as dental enamel; carbonate apatite became the main component for hard tissue [3]. Despite this, carbonate apatite is easier to dissolve compared to hydroxyapatite. The solubility will be increasing as the carbonate content increases [4] under acidic condition such as in the Howship's lacunae of osteoclast [5], plus with its bioresorbability characteristic [6-8] and it is believed to be an ideal bone replacement.

Although collagen is naturally found in the human bone, gelatin is used as the main component in this research due to its properties. As gelatin is readily assimilated by the body with it properties such as excellent biocompatibility, ease of processing and available at a low cost [9] however, pure gelatin scaffold are not satisfactory for hard tissue application as reported by J. Sundaram et al,. [10]. Therefore, a composite of gelatin and carbonate apatite are expected to be the ideal scaffold that possesses sufficient mechanical strength with good osteoconductivity and biodegradation [11].

2. Experimental study

2.1 Materials

The materials that were used in this research are calcium hydroxide ($Ca(OH)_2$); 98% purity, Fluka, UK and phosphoric acid (H_3PO_4); 15M, Merck, Germany and carbon dioxide, (CO_2); supply by Goldengas Industrial Supply, Malaysia while the raw material used in the fabrication of the scaffold were gelatin (food grade; supplied by Halagel), polyvinyl alcohol (PVA); Sigma Aldrich, molecular weight average of 14 000 and glutaraldehyde (GA); Merck, Germany.

2.2 Method

2.2.1 Synthesis of carbonate apatite (CO_3Ap) powder. Precipitation method was used in the preparation of carbonate apatite powder. The process started with the reaction of $Ca(OH)_2$ and H_3PO_4 alongside carbon dioxide gas as the carbonate source that flow into the reactor with 1 *l*/min . Initially, a solution of 125 ml of 2 M $Ca(OH)_2$ is added into the reactor that were continuously stirred at 400 rpm with 40 ±1 °C, then 75 ml of 2 M of H_3PO_4 solution is added slowly into the reactor as carbon dioxide gas is flowing through the reactor [12].

2.2.2 Characterization of carbonate apatite powder. The obtain powder were characterized using X-ray diffractometer (XRD) for phase identification and Fourier transform infrared spectroscopy (FTIR) for identification of the functional group presents. Elemental analyser; Carbon, Nitrogen, Oxygen (CHN) test was used to identify the carbonate content of the obtained powder, while the particle size of the powder were determined by Helos particle size analyser.

2.2.3 Preparation of porous carbonate apatite/gelatin scaffold via freeze drying method. Porous CO_3Ap/gelatin scaffold was fabricated by freeze drying method. Initially, 5 wt. % of gelatin were dissolved in 77.75 wt. % distilled water at the temperature of 60 °C. The gelatin solution is then stirred using mechanical stirrer at 300 rpm. Next, 15 wt. % of CO_3Ap powder and 2 wt. % of PVA is added slowly into the gelatin solution (to avoid bubble and coagulation) until homogenous mixture is form. Next, the mixture was put in the vacuum desiccator for 2 minutes to reduce the formation of air bubble. Then, the mixture is stirred again as 0.25 wt. % of GA is added. The mixture is then quickly poured into mold and allowed to freeze for 24 hours at different pre-freezing temperature; -10 °C, -20 °C and -40 °C. The solid specimen is then placed in freeze dryer for 24 hours. Similar steps are then repeated for the ratio (10/10) wt. % of CO_3Ap/gelatin. The microstructure of the specimen was observed using field emission scanning electron microscope (FESEM SUPRA 35VP ZEISS). The compressive strength of the scaffolds was measured using an Instron 3369 according to ASTM-F-451-95 standard at testing speed of 2 mm/min.

3. Results and discussion

3.1 Characterization of the synthesized CO_3Ap powder

Figure 1 shows the XRD pattern of hydroxyapatite (HA) as reference pattern and synthesized CO_3Ap powder. From these figure it can be observed that the peaks of CO_3Ap powder matches to the peaks of HA. However, the intensity of the peaks in CO_3Ap pattern are broader indicating low crystallinity apatite was obtained by precipitation method.

Carbonate content (wt. %) of the CO_3Ap powder calculated from the CHN analysis was determined to be about 12.8 wt. %. While, the mean particle size of the CO_3Ap powder was 7.76 μm; the particle size was measured by Helos particle size analyser.

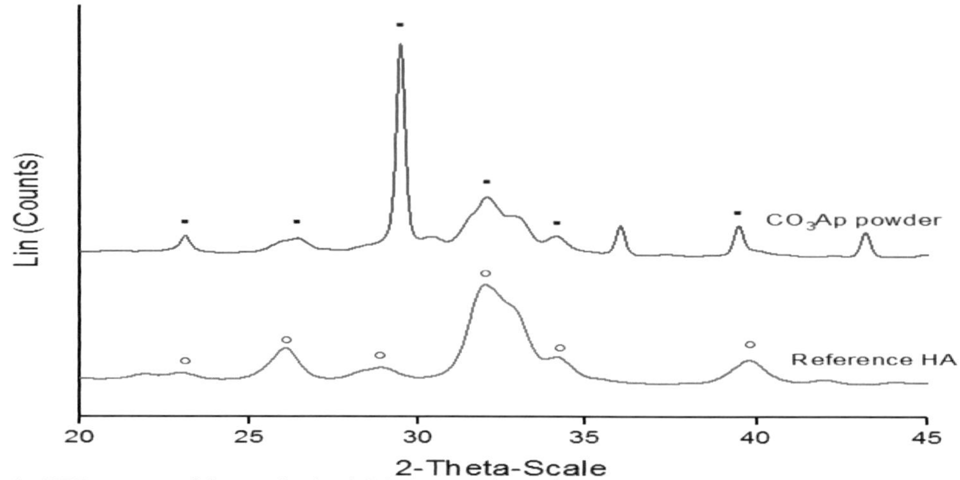

Figure 1. XRD pattern of the synthesized CO_3Ap powder and HA pattern as reference

Figure 2 illustrate the FTIR spectra of the obtained apatite powder proving that obtained powder is type-B carbonate apatite powder similar to the type of carbonate apatite that found in human bone. From this figure it can be observed an intense band at 1308.98 cm^{-1}, 1035.05 cm^{-1}, 602.95 -565.03 cm^{-1} similarly as reported by Mohamed K. R. et al. which corresponds to the PO_4 functional groups [13]. Absorption band assigned to C-O in CO_3^{2-} group were detected at 1636.96 - 1584.9 cm^{-1}, 1427.98 cm^{-1} and 874.81 cm^{-1}. This result is parallel with previous report by dos Santos et al., where stretching vibration band observed at 876 cm^{-1} and 1470 cm^{-1} [14]. In addition to this, Daitou, F. et al., also stated that carbonate ions can be identified at wave number of 1450 cm^{-1}, 1414 cm^{-1} and 874 cm^{-1} [15].

There are two types of carbonate ions exists in CO_3Ap structure [16]; at wavelength 1650 cm^{-1} to 1300 cm^{-1} are assigned to surface carbonate ions whereby the carbonate ions exist in the lattice of phosphate ions while the peak at 873 cm^{-1} is responsible for complete replacement the OH^- ions at the interface of growing crystal, as the hydroxyl ions are present at 3428.14 cm^{-1}.

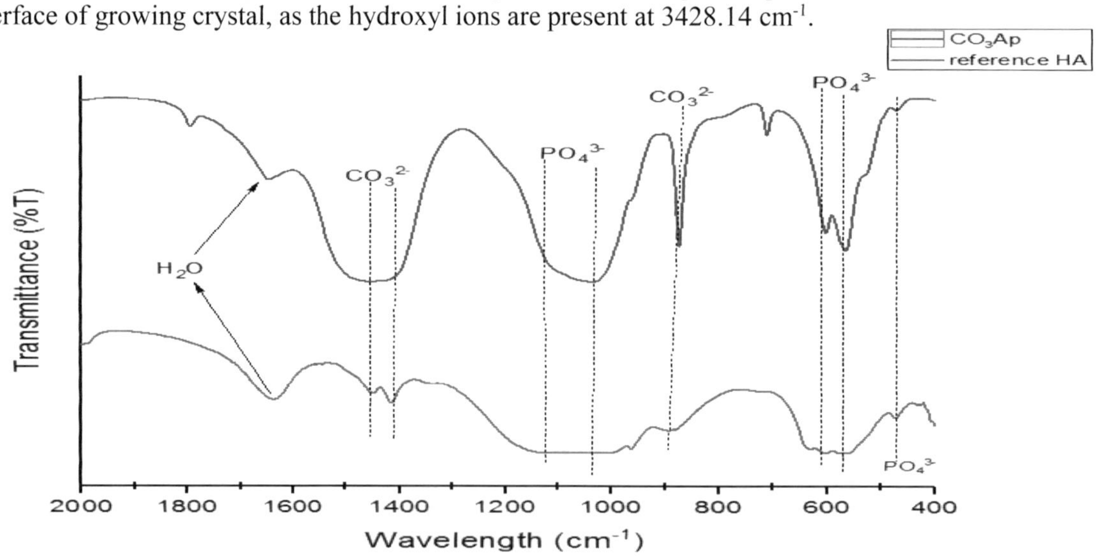

Figure 2. FTIR spectrum of synthesized carbonate apatite CO_3Ap and HA pattern as reference

IOP Conf. Series: Journal of Physics: Conf. Series **1082** (2018) 012004 doi:10.1088/1742-6596/1082/1/012004

3.2 Morphological analysis of porous CO_3Ap/gelatin scaffold

The morphology of the composite scaffold was examined by FESEM. The effect of CO_3Ap/gelatin ratio and pre-freezing temperature on the morphology of the CO_3Ap/gelatin scaffold are shown in Table 1. It can be observed that pores are presence throughout the specimens with different arrangement and size regardless of ratio and pre-freezing temperature. A good interconnectivity between the pores is achieved along with homogenous mixture between the gelatin and carbonate apatite with no coagulation present.

Having different pre-freezing temperature, i.e. -10 °C, -20 °C, -40 °C produced different pore size. For ratio of (15/5) wt. % the size obtained is 100-150 µm, 45-55 µm, and 200-400 µm. While for the ratio (10/10) wt. %, the size obtained is 80-270 µm, 125-150 µm, and 75-200 µm. These data correspond to the theory of pore size. As lower temperature is used, the smaller the pore size that will produced [17] but for the temperature -40 °C for the ratio (15/5) wt. % the pores appear to be largest among all this is due to the formation air bubble that trapped in the scaffold during casting process.

Table 1. Morphology of porous CO_3Ap/gelatin with ratio of gelation to CO_3Ap powder (15/5) and (10/10) at pre-freezing temperature of -10 °C, -20 °C, and -40 °C

Ratio of carbonate apatite powder to gelatin(wt. %)	Temperature (°C)		
	-10	-20	-40
15/5	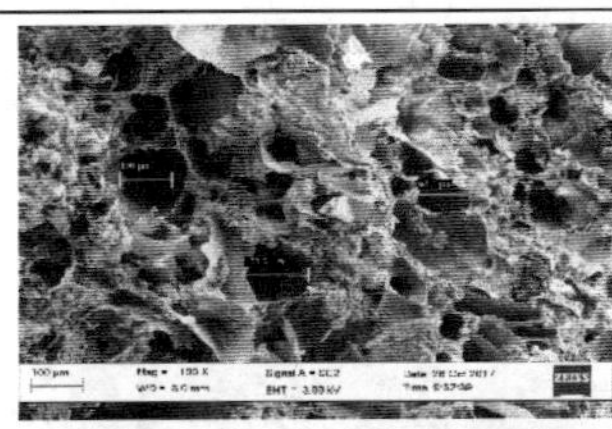	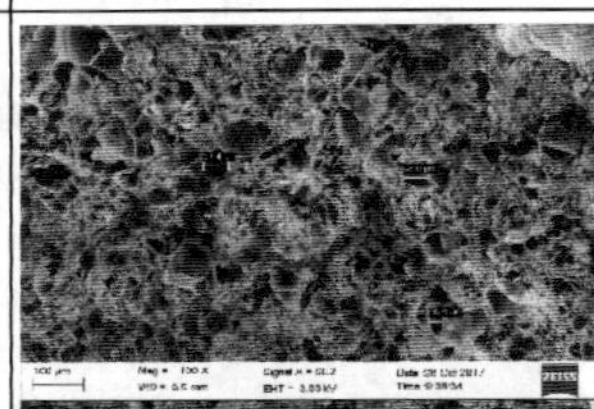	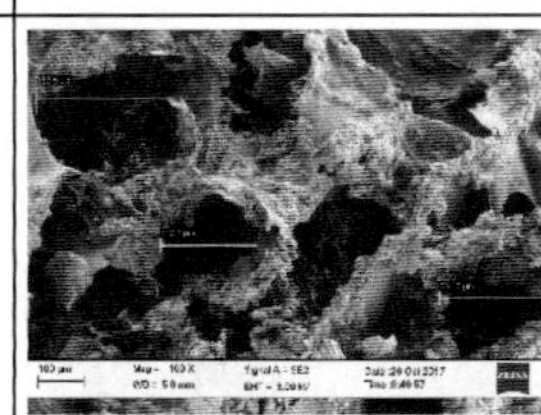
10/10	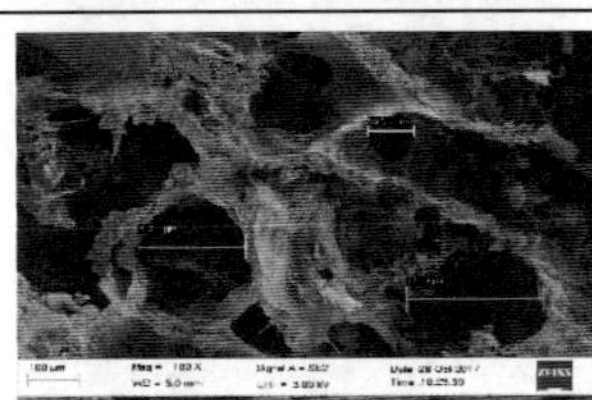	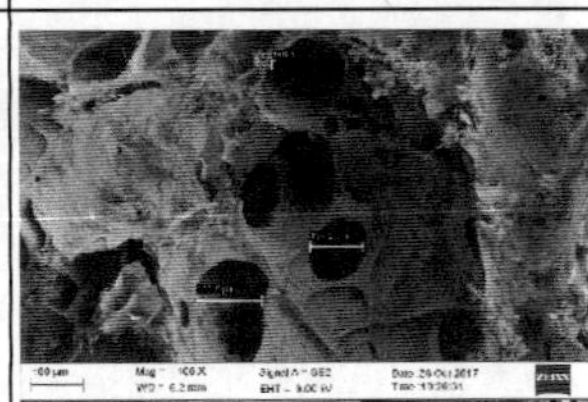	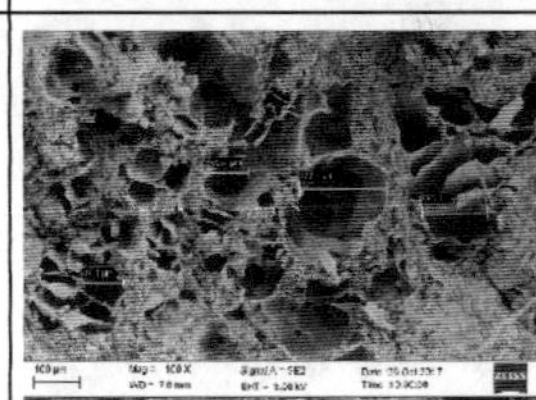

3.3 Compressive strength of porous CO_3Ap/gelatin scaffold

Figure 3 represent the maximum compressive strength of the scaffolds with different ratio of carbonate apatite to gelatin and different pre-freezing temperature used. From this figure it can be observed that, the higher the weight percentage of gelatin with lower pre-freezing temperature resulting higher compressive strength; as the highest compressive strength achieved is 7412.36 kPa for the ratio of (10/10) wt. %. Incorporation of ductile polymeric material will enhance the strength of the scaffold [18]. In this case, the stiffness of the carbonate apatite powders incorporated into the gelatin will increase the strength and toughness of the scaffold.

Theoretically, by controlling factors such as the freezing rate, freezing temperature and concentration will affect the porosity, pore size and pore morphology of the scaffolds [19] as freezing temperature is the most important steps in the microstructure formation as the formation of ice particles in the specimen will represent the final microstructure of the end product. Rapid cooling will results in small ice crystal and vice versa [20] as in can be seen in the pore formation in the scaffold in Table 1.

Smaller ice crystals formed as lower freezing temperature were used thus increasing the compressive strength. The ratio of (10\10) wt. % with -40 °C as pre-freezing temperature has the highest maximum strength due to more energy can be absorbed and distributed along the scaffold structure before failure occurred. However, for the temperature -40 °C for ratio of (15/5) wt. % the maximum compressive strength decreases dramatically due to the formation of non-uniform large size pores compared to the others.

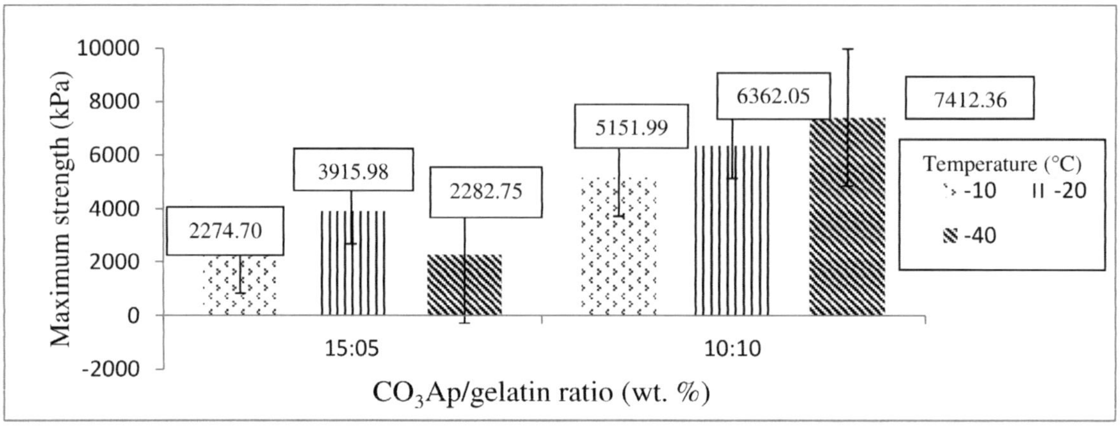

Figure 3. Compressive strength of porous CO_3Ap/gelatin scaffold containing different ratio of carbonate apatite to gelatin with different freezing temperature

4. Conclusion

As a conclusion, B-type CO_3Ap powder was obtained by the precipitation method of calcium hydroxide and phosphoric acid along with carbon dioxide gas as the carbonation source. The incorporation of CO_3Ap powder in the gelatin scaffold increases the compressive strength of the CO_3Ap/gelatin composites as the carbonate apatite powder induces stiffness while the gelatin introduced the ductility into the scaffold.

Acknowledgements

This work was supported by Trans Disciplinary Research Grant Scheme (TRGS) No: 203/PBahan/6761003.

References

[1] Langer R S and Vacantri, J P 1999 *Sci. Am.* 280 86
[2] Chen F Tang Q L, Zhu Y J, Wang K W, Zhang M L, Zhai W Y & Chang J, 2010 *Acta Biomaterialia* 6 3013
[3] Sonju Clasen A B and Ruyter IE 1997 *Adv. Dent. Res.* 11 523
[4] LeGeros RZ, Ming S 1983 *Caries Res* 17 419
[5] Gothlin G Osteoclast 1976 *Clin Orthop* 120 201
[6] Doi Y, Iwanaga H, Shibutani T, Moriwaki Y, Iwayama Y 1999 *J Biomed Mater Res* 47 424
[7] Barralet J, Akao M, Aoki H 2000 *J Biomed Mater Res* 49 176
[8] Landi E, Celotti G, Logroscino G, Tampieri A 2003 *J Euro Ceram Soc* 23 2931
[9] Rose J B, Pacelli S, Haj A J E, Dua H S, Hopkinson A White L J & Rose F R 2014 *Materials* 7 3106
[10] Sundaram J, Timothy D D, Rizhi W 2008 *Acta Biomaterialia* 4 932
[11] Bigi A, Panzavolta S, and Roveri N, 1988 *Biomaterials* 19 739

[12] Hasegawa T, Miwa M, Sakai Y, Niikura T, Lee S Y, Oe K, Iwakura T, Kurosaka M and Komori T 2010 *J Dent Res* 89 854
[13] Mohamed K R, Beherei H H and El-Rashidy Z M 2014 *J Adv Res* 5 201
[14] Dos Santos E A, Farina M, Soares G A and Anselme K 2009 *J Biomed Mater Res A* 89A, 510
[15] Daitou, F. (2010) *Dental Materials Journal* 3 303
[16] Rehman I & Bonfield W 1997 *Journal of materials science: Materials in medicine* 8 1
[17] Garg T, Chanana A, & Joshi R 2012 *Journal of Pharmacy* 2 072
[18] Bang L T, Kanji T, Melvin M, Kunio I, Radzali O 2013 *Ceram Int* 39 5631
[19] Harley B A, Leung J H, Silva E C C M, Gibson L J 2007 *Acta Biomater* 3 463
[20] Duarte A R C 2010 *Materials Science Forum* 636 22

IOP Conf. Series: Journal of Physics: Conf. Series **1082** (2018) 012005 doi:10.1088/1742-6596/1082/1/012005

Extraction of Nanohydroxyapatite from Waste Bovine Bone Using Alkaline Digestion Method

N. H. Adenan[1], I. Zainol[1*], N. A. Rahim[1] and C. N. A. Jaafar[2]

[1]Chemistry Department, Faculty of Science and Mathematics, Universiti Pendidikan Sultan Idris, 35900 Tanjong Malim, Perak, Malaysia

[2]Department of Mechanical and Manufacturing Engineering, Faculty of Engineering, Universiti Putra Malaysia, 43000 Serdang, Selangor, Malaysia.

*Email: ismail.zainol@fsmt.upsi.edu.my

Abstract. Bovine bones from food waste are discarded without any effort to reclaim its potential as calcium sources. This study reports the preparation and characterization of nano hydroxyapatite from bovine bone waste (BvHAp) collected from local restaurant using alkaline digestion method. Fourier transformed infrared spectroscopy (FTIR) and X-ray diffraction (XRD) analyses were used to verify the presence of hydroxyapatite in the bovine bone waste. Scanning electron microscopy (SEM) was also used to study the morphology of the apatite obtained. The SEM analysis revealed that the diameter size of nanoparticles BvHAp were around 20-30 nm. The energy dispersive X-ray (EDX) result indicated the calcium /phosphorous (Ca/P) ratio of obtained BvHAp after sintering at 1200 °C is 1.75 which is very closed to the theoretical value. The natural BvHAp produced has a potential source for biomaterials and food supplement applications.

1. Introduction

Many researchers have been reported production of natural hydroxyapatite (HAp) from bovine bone waste [1-2] as an alternative to synthetic HAp due to its cheaper price and easy availability of raw materials. However, most of the researchers have used fresh bone waste in their studies. None of the researcher has considered to recycle the bone waste from food origin such as bone soup, a local favorite dish for the production of HAp. It is estimated that within a year more than 2000 tons of waste bone from food origin is disposed especially from local halal restaurants [3]. Dumped bone waste without proper treatment would result in environmental hazards. Therefore, food origin bone waste from the local halal restaurant should be given more attention so that this precious halal bone source can be fully utilized without being a waste.

Conventional treatment methods to produce HAp from bovine bones have been reported by many researchers such as burning process [4], acid and alkaline digestion [5]. Alkaline digestion method however, offers some advantages such as direct formation of nanoparticles in alkaline solution, ease of preparation and low costs of processing [1, 6-7]. Thus, in this study alkaline digestion process followed by sintering at 1200 °C were applied to eliminate and dissolve the organic part present in the bovine bones waste from food origin. Alkaline method offer a huge amount of nanostructured HAp produced without the process of milling [8]. Thus this method will be more cost effective for big scale production.

2. Experimental study

2.1 Extraction of hydroxyapatite from waste bovine bones

A waste bovine bone (WBv) was obtained from a local restaurant in Kelantan, Malaysia and it was mainly a waste from bone soup dish. WBv was soaked in tap water and washed thoroughly in distilled water to remove dirty substances. The WBv was then deproteinized through washing with 0.1 M HCl and washed several times with distilled water to remove visible tissues on the bone surface and dried at 60 °C. A cleaned compact bone was cut into small pieces using band saw machine. The bone powder was heated in 5 M NaOH solution at 80 °C and stirred for 1 h. The treated bones was washed thoroughly with deionized water until the washing solution became neutral and then dried at 60 °C. The resulting WBv treated powder was then calcined in the electric furnace, at 1200 °C with 2 h holding time to produce nano BvHAp powder.

2.2 FTIR analysis

Fourier Transform Infrared (FTIR) was performed using FTIR Nicolet model 6700. Potassium bromide (KBr) was used in this analysis. About 0.003 g of samples was mixed with 0.3 g of powdered KBr (ratio of 1:10) and grinded together until become a fine powder using agate mortar. A transparent disk was formed by pressing a homogeneous mixture under 10000-15000 psi for a few minutes using hydraulic press. BvHAp disk samples were scanned for 32 times at spectral resolution of 4 cm^{-1} in the wavelength range between 400 and 4000 cm^{-1}. The location and intensity of the spectrum peak was determined using Omnic software.

2.3 Study of Surface Morphology by SEM and Ca/P molar ratio by EDX

SEM Hitachi SU 8020 UHR model was used to study the surface morphology and crystal size of the BvHAp samples. Firstly, the samples were coated with platinum using Quorum Q150R S cryocoater to give conductivity and protection from the beam damage. In this study, EDX spectrometer model Horiba was used to determine the percentage of elements Ca and P in BvHAp. The Ca/P molar ratios of BvHAp samples were obtained by using this analysis. Scanning transmission electron microscope (STEM) image was obtained to confirm the nano structure of BvHAp produced.

2.4 Phase identification and crystallinity analysis by XRD

Phase identification and crystallinity of BvHAp samples were analyzed using X-ray diffraction (XRD) model D8 Advance, Bruker AXS model, Germany. XRD source of Cu Kα radiation with a wavelength of 0.154 nm geometry was used in this analysis. The use of current and voltage X-ray tube was 40 kV and 40 mA. Data was taken from the diffraction angle 2θ between 10° to 70° with a scan speed of 2 $°Cmin^{-1}$ using a step size of 0.025. The phases were identified compared with the diffraction pattern that is indexed by file standard powder diffraction (ICDD) 00-009-0432 card for BvHAp hexagonal structure. BvHAp powder crystal size was calculated from measurements of the width of the XRD diffraction pattern (200) of the X-ray diffraction data.

3. Results and discussion

The average yield of BvHAp that produced from bovine bone after treatment using alkaline digestion process was 72 %. However after sintering at 1200 °C the yield of sintered BvHAp decreased into 58

% due to all the organic component was removed completely. This result was in agreement with previous study that showed the bovine bone consisting of mineral apatite between 41 to 84 % [8,9].

3.1 FTIR Analysis

FTIR spectra of WBv as treated bones were recorded in the range of 4,000–400 cm^{-1} as shown in Figure 1. In FTIR analysis, WBv showed the absorption band at 2967 - 2920 cm^{-1} and 2880 - 2850 cm^{-1} and these are related to the C-H band and those at 1653 - 1642 cm^{-1} and 1565 – 1541 cm^{-1} which are characteristic of amide I and amide II, respectively. The bands at 1455 – 1453 cm^{-1} and 1246 – 1239 cm^{-1} were identified as CH_2 and amide III bands. All these bands show the organic matter which identified as collagen present in the WBv sample. This observation has good agreement with the previous study by Sangkar et. al.[9] on fresh bovine bones. This result indicated that even though the bone was boiled in food preparation, the organic matters were not detached completely from the bones.

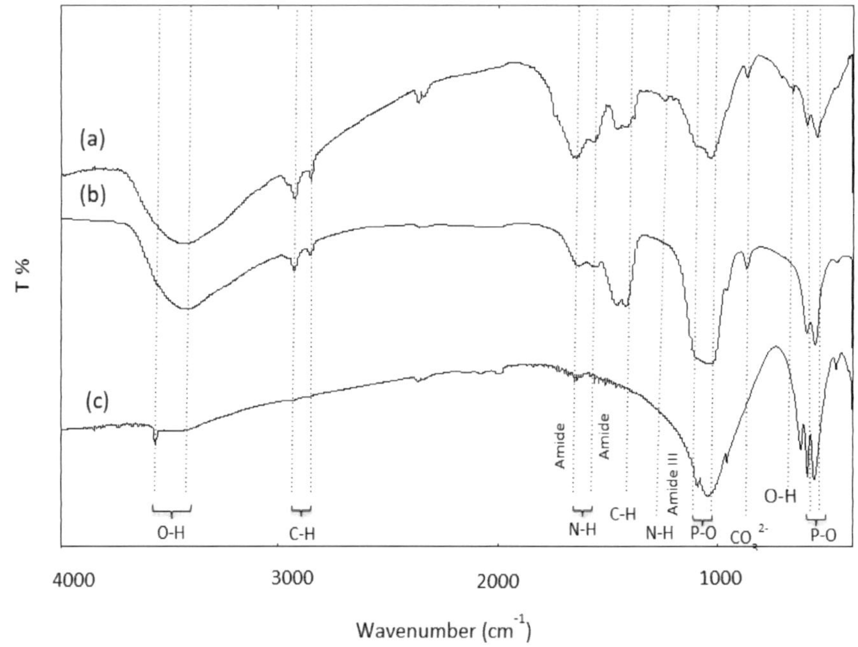

Figure 1. FTIR spectrum of (a) WBv (b) WBv threated with 5 M NaOH and (c) BvHAp sintered at 1200 °C.

The bands identified at range between 873 and 871 cm^{-1} indicated that the carbonat ion is present in the raw materials. For the WBv samples treated with 5 M NaOH, the bands at 1635 – 1628 cm^{-1}, 1558 – 1554 cm^{-1}, 1456 cm^{-1} and 873 cm^{-1} corresponded to amide I, amide II, C-H and CO_3^{2-} respectively. All this bands revealed that organic component and carbonate ion were not completely removed after been treated with 5M NaOH. However, it can be seen that the intensities of functional group associated to C-H and amide group (amide 1 and II) were decreased, this indicated that some of organic compound were dissolved in NaOH solution. C-H and amide III bands were totally disappeared in the case of treated WBv after sintering at 1200 °C for 2 hours. The present of three sharps peaks between 632 and 569 cm^{-1} revealed asymmetric bending vibration of P–O band which indicated high crystallinity of BvHAp sample [10]. The broad O-H peak at above 3500 cm^{-1} for WBv and alkaline treated WBv was changed to sharp O-H peaks for sintered WBv and this was due to the release of free H_2O in the unsintered samples and left O-H bonded groups for sintered WBv sample which is characteristic of BvHAp structure [11,12].

3.2 Surface morphology analysis

SEM and STEM micrographs of treated WBv with 5M NaOH are shown in Figures 2(a) and 2(b) respectively. The result in Figure 2(a) shows the treated WBv samples seen to be agglomerated on their morphology due to incomplete removal of organic component. However, STEM analysis of treated WBv shown in Figure 2(b) reveals the needle like nano particles with diameter was measured between 20 and 30 nm. The particle size obtained in this study is smaller than that reported by other researcher [13]. This was probably due to the fact that the 5M NaOH that used in this study resulting in the smaller particles size of the samples.

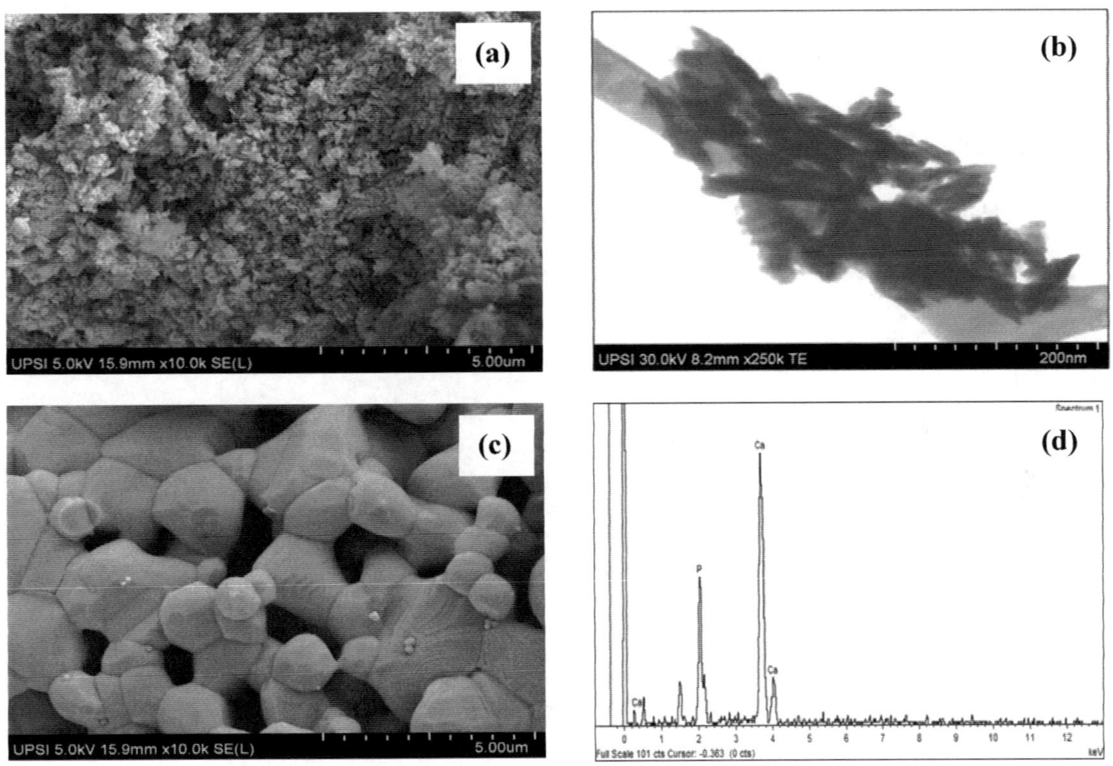

Figure 2. SEM micrographs of (a) treated WBv with 5 M NaOH and (b) STEM image of treated WBv (c) SEM micrographs of sintered WBv and (d) EDX spectrum of sintered BvHAp samples

The sintered WBv however showed homogenous BvHAp grain with inter connected porous structure (Figure 2(c)). Similar observation of SEM micrograph was also reported by other researchers [5,11]. EDX analysis of Ca and P of sintered WBv is shown in Figure 2(d). The Ca/P molar ratio of the BvHAp obtained was 1.75 similar to previous report using fresh bovine bones [14]. From this results, it can be deduced that the chemical composition of bones was not changed during food processing.

3. 3 XRD analysis

Figures 3 (a) and (b) show the XRD spectra of alkaline treated WBv and sintered BvHAp respectively. The broad XRD pattern in alkaline treated WBv is probably due to a lower crystallinity of the sample as a result of high content of organic compound present in the sample. The degree of crystallinity (Xc (%)) was calculated from XRD spectrum for alkaline treated WBv is 70.5 %. XRD pattern for sintered BvHAp is shown in Figure 3(b).

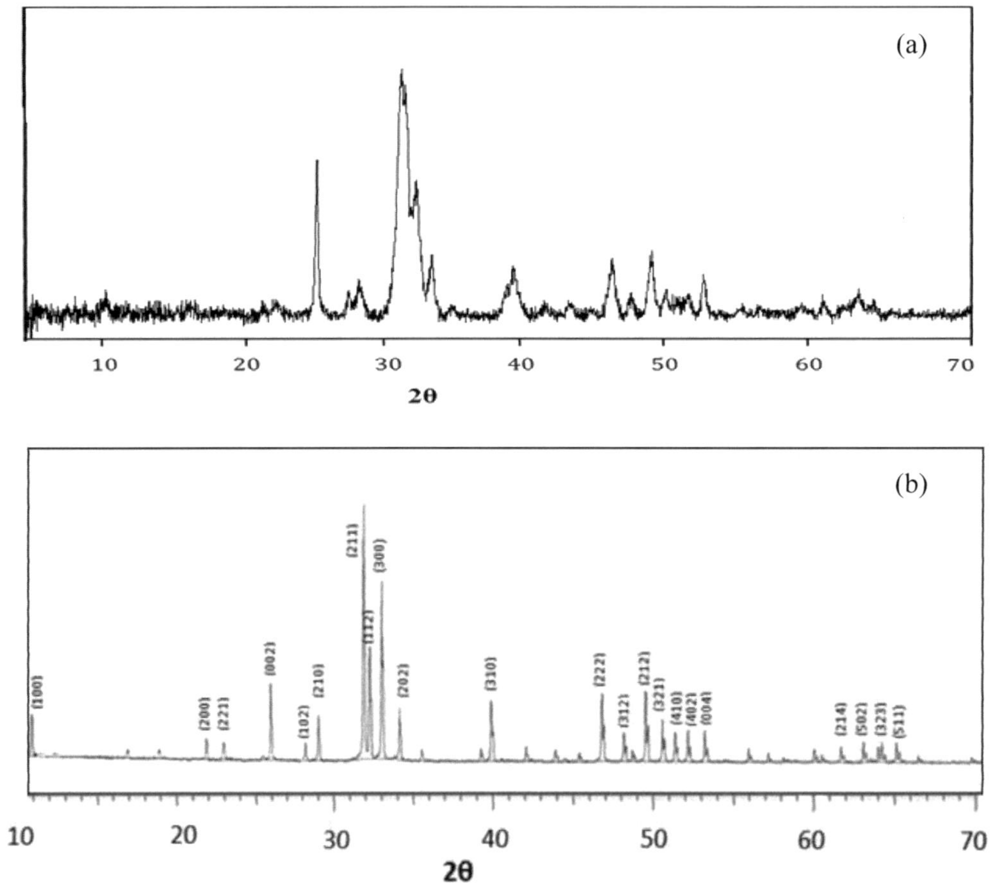

Figure 3. XRD spectrum of (a) alkaline treated WBv and (b) sintered BvHAp at 1200 °C

The result shows the sharp XRD spectrum match very well to standard HAp pattern (JCPDS card No. 09-0432). Three main peaks with the highest intensity were observed at 31.86 ° 33.04 ° and 32.53° associated with the planar hkl (211), (300) and (112) respectively. This is equivalent to the highest peak intensity reported by the previous study [12]. Some low intensity peaks were also observed in XRD pattern associated with the planar hkl (100) , (200), (2221), (002), (102), (210), (202), (310), (222), (312), (212), (321), (410), (402), (004), (214), (502), (323) and (511). The sharp intensity obtained for the sintered BvHAp at 1200 °C proved a complete crystallization occurred with high degree of crystallinity [15]. The calculated degree of crystallinity for this sample is 87.4%. Using Scherer equation [16], the crystallite size (τ) for treated WBv and sintered BvHAp was calculated as 55.69 and 83.96 respectively.

4. Conclusion

Based on this study, it can be concluded that nano particles of natural BvHAp was successfully extracted from waste bovine bone using alkaline digestion method. This method is simple and low cost for large scale production. Complete removal of organic matrix can be achieved via sintering process at 1200 °C. FTIR and XRD analysis results of BvHAp sintered materials confirmed the phase purity and high crystalline grade. From this study it is found that waste bovine bone from food origin has a very high potential to produce nano BvHAp for food supplement and other medical applications.

Acknowledgement

The authors would like to thanks to Universiti Pendidikan Sultan Idris for facilities support of this work and grateful to Mr Ridhuan from UKM, Malaysia for conducting XRD experiment.

References

[1] Barakat N A M, Khil M S, Omran A M, Sheikh F A and Kim H Y 2009 *Journal of Materials Processing Technology* 209 3408
[2] Masud R, Naznin A, Shaifur R, Hossen M J and Sikder M A *Int J Complement Alt Med* 2017 8 263
[3] Nor, AAMN, Mohamad, HR 2015 The Food and Fertilizer Technology Center (FFTC). (http://ap.fftc.agnet.org/ap_db.php?id=529&print=1, access at 10 April 2018.)
[4] Prabakaran K and Rajeswari S 2006 *Trends in Biomaterials and Artificial Organs* 20 20.
[5] Ooi CY, Hamdi M and Ramesh S 2007 Ceramics International 33 1171
[6] Nath N and Krishna P 2014 *Biomaterials* 37 433
[7] Kongsri S, Janpradit K, Buapa K, Techawongstien S and Chanthai S 2013 *Chemical Engineering Journal* 215 522
[8] Venkatesan J, Qian ZJ, Ryu B, Thomas NV and Kim SK 2011 *Biomedical Materials 6* 35003
[9] Sankar S, Sekar S, Mohan R, Rani S, Sundaraseelan J and Sastry TP 2008 *International Journal of Biological Macromolecules* 42 6
[10] Walters M, Leung YC, Blumenthal NC, LeGeros RZ and Konsker K 1990 *Journal of Inorganic Biochemistry* 39 193
[11]Baco S, Bambang L, Joseph N and Basri NF 2013 *Malaysian Journal of Fundamental and Applied Sciences* 220 224
[12] Balamurugan A, Michel J, Faur J, Benhayoune H, Wortham L, Sockalingum G, Balossier G 2006 *Ceramics - Silikaty* 50 27
[13]Mondal S, Bardhan R, Mondal B, Dey A, Mukhopadhyay SS, Roy S, Roy K 2012 *Bulletin of Materials Science* 35 683
[14] Bahrololoom ME, Javidi M, Javadpour S and Ma J 2009 *Journal of Ceramic Processing Research* 10 129
[15] Ramesh S, Aw KL, Tolouei R, Amiriyan M, Tan, CY, Hamdi M, Teng WD 2012 *Ceramics Internationa* 39 111
[16] Panda, NN., Pramanik, K., Sukla, LB 2013. *Bioprocess and Biosystems Engineering* 37 3 433

IOP Conf. Series: Journal of Physics: Conf. Series **1082** (2018) 012006 doi:10.1088/1742-6596/1082/1/012006

Effect of Silica Fillers on Mechanical Properties of Epoxy/Kenaf Composites

C. N. A. Jaafar[1]*, I. Zainol[2] and O. O. Aremu[1]

[1]Department of Mechanical and Manufacturing Engineering, Faculty of Engineering, Universiti Putra Malaysia, 43000 Serdang Selangor, Malaysia.

[2]Chemistry Department, Faculty of Science and Mathematics, Universiti Pendidikan Sultan Idris, 35900 Tanjong Malim, Perak, Malaysia.

Email: cnaiza@upm.edu.my

Abstract. The mechanical properties of epoxy/kenaf composites at various silica content has been investigated. Epoxy resin was mixed with silica filler ranging from 6.7 to 33.3 %w/w before applied into kenaf fibre mat (11.5 %w/w) to fabricate epoxy/silica/kenaf composites using hand layup method. The composites were precured at 80 °C for 2 hour and post cured at 110 °C for 1 hour. The mechanical properties of the composite were analysed using flexural and impact test, meanwhile the surface fracture of the composite was examined by means of scanning electron microscope (SEM). It is found that epoxy/kenaf composite with 20 %w/w silica content exhibited the best mechanical properties, with 3.1 kJ/m^2 of impact strength, 44.5 MPa of flexural strength and 2.7 GPa of flexural modulus. Its surface morphology showed the brittle fracture surface with pull out of fibre which indicated poor interaction between epoxy matrix and kenaf fibre.

1. Introduction

Epoxy resin is one of the most important commercial thermosetting polymer in multitude of industrial applications. It exhibits many desirable properties, such as low shrinkage, high thermal stability, high stiffness and excellent chemical resistance. However, epoxy is generally brittle properties because of its high density of cross-linked structure. Therefore, improving the toughness of epoxy resin is an important concerned in epoxy composites for better product performance [1]. Several types of rigid inorganic particulate fillers have been studied to improve the mechanical properties of epoxy composite, such as silica, alumina, nanoclay and others. Among the numerous fillers of epoxy, silica is the most commercially used as a filler in literature due to its significant improvement in flexural modulus, fracture toughness and impact strength in epoxy composite [2]. It has been reported that fracture toughness of epoxy resins can be significantly improved by silica modification [3].

Carbon fibres and glass fibres integrated in thermoset polymer are the traditional reinforced composite material. Natural fibre reinforced composites found to be an alternative solution for greener environment and thus received increasingly attention in recent years. Among the natural fibre, kenaf fibre is more attractive since it is available commercially, comparatively cheap and strong [4]. Therefore, implementation of kenaf as a reinforce materials would be an interesting to study in improving the mechanical properties of epoxy composites. Several researchers have studied on epoxy modification using different type of fillers to improve the mechanical properties of epoxy composites [1-3]. Marjetka (2013) has investigated the addition of silica improved the impact strength of epoxy from 6.4 kJ/m^2 to 8.9 kJ/m^2 [2]. Fiore et al. (2014) has proved that the used of multi-axial kenaf fibre mat increased the flexural modulus of epoxy from 3 to 7.82 GPa [5].

In this present work, polymer composites have been developed using kenaf fibre reinforced epoxy composites filled with different percentage of silica particles. The mechanical properties such as flexural strength, flexural modulus and impact strength have been investigated.

2. Experimental study

2.1 Materials

The epoxy resin used in this study was bisphenol A from Dow Chemical Pacific Singapore (DOW) whereas curing agent was cycloaliphatic amine (Jointmine 905-3S) from Epochemie. Nonwoven kenaf fibre mat was purchased from Kenaf Agro Vet Sdn Bhd, Malaysia. Silica powder with particle size 10 micron was bought from Maju Saintifik, Malaysia.

2.2 Preparation of epoxy/kenaf composites

Epoxy composites were prepared using hand lay-up method. The percentage weight of epoxy and kenaf mat were 88.5 % w/w and 11.5 % w/w respectively. The carbon steel mould (170 mm x 170 mm) was coated with mould release agent before placed with kenaf fibre mat. Epoxy resin was thoroughly mixed with curing agent before pouring into the mould to wet the kenaf fibre mat. The mould was then covered with top section and pressed with 1 kg load and heated at 80 °C for 2 hours. The curing was continued for another 1 hour at 110 °C for post curing process. The epoxy/kenaf composite sheet label as composite A was removed from the mould after cooled at room temperature and cut into specimen for mechanical testing.

2.3 Preparation of silica modified epoxy/kenaf composite

Epoxy resin was modified with silica filler by mechanical mixing process with silica content from 10, 15, 20, 25 and 30 %w/w for composites B, C, D, E and F respectively. The percentage weights of modified resin epoxy and kenaf mat were fixed at 88.5 % w/w and 11.5 % w/w respectively. The carbon steel mould (170 mm x 170 mm) was coated with mould release agent before placed with kenaf fibre mat. Similar to the above procedure, modified epoxy resin was thoroughly mixed with curing agent before pouring into the mould to wet the kenaf fibre mat. The mould was then covered with top section and pressed with 1 kg load and heated at 80 °C for 2 hours. The curing was continued for another 1 hour at 110 °C. The epoxy/kenaf composite sheet was removed from the mould after cooling at room temperature and cut into specimen for mechanical testing.

2.4 Mechanical properties

2.4.1 Flexural strength. Five samples from each epoxy composite were prepared for the test according to the ASTM D790 standard. Samples were cut using bandsaw machine and finished using milling machine. The samples were prepared in dimensions of 8 mm x 12.7 mm x 101 mm.

2.4.2 Impact properties. Charpy impact test was conducted according to ASTM D6110 standard. Five samples from each epoxy composites were prepared for the test. The samples were cut using bandsaw machine and finished using milling machine. The dimension of the sample was 5 mm x 12.7 mm x

63.5 mm. A v-notch was cut into the specimens using a sample notcher with an angle of 45°. The notched samples were left for 40 hours in room temperature before proceed for impact test in order to remove the stress concentration created by notching process.

3. Results and discussion

3.1 Flexural strength of epoxy/silica/kenaf composites

Figure 1 show the flexural strength of fully cured epoxy/kenaf composite (A) and epoxy/silica/kenaf composite labelled as B, C, D, E and F for 10, 15, 20, 25 and 30 %w/w silica content, respectively. Epoxy/kenaf composite (A) is able to withstand flexural strength up to 37.3 MPa as compared to 92.6 MPa in neat epoxy. The use of multi-axial kenaf fibre mat in composite was reported to lower flexural strength as compared to neat epoxy due to incompatibility of kenaf fibre and epoxy matrix which lead to poor interfacial bonding. Poor interaction between fibres, void formation and dispersion problem were also reported among the reason for lower flexural strength [5]. However, the present of silica fillers in the matrix composites B to F have improved the flexural strengths as compared to epoxy composite without silica (composite A). Composite C with 13 %w/w silica content show the highest flexural strength. Rigid particulate filler such as silica able to enhance the stiffness of the epoxy matrix [6].

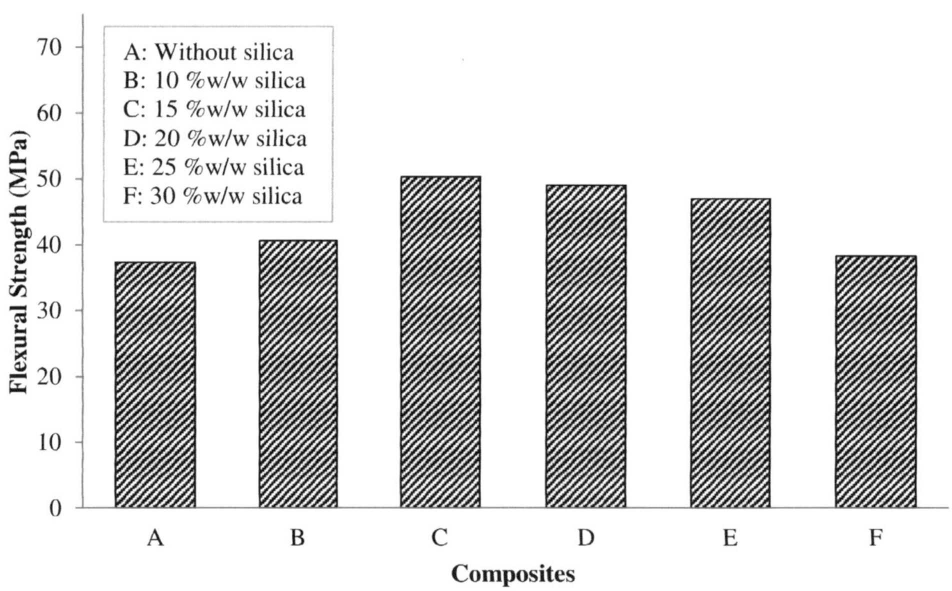

Figure 1. Flexural strength of epoxy/kenaf composites with different silica composition.

3.2 Flexural Modulus of epoxy/kenaf/silica composites

Figure 2 shows the flexural modulus of epoxy/kenaf composites (A) without silica and epoxy/silica/kenaf composite labelled as B, C, D, E and F. Similar pattern to the result of flexural strength was observed for flexural modulus. Epoxy/kenaf composites exhibits flexural modulus of 1.96 GPa as compared to 3.00 GPa as recorded for neat epoxy. This result is however not in agreement with the finding by Fiore *et al.* (2014), where epoxy composites with kenaf fibre mat should exhibits higher flexural modulus than neat epoxy [7]. This may because the fibres applied were not properly compacted and a lot of voids between the resin which may reduce the strength of the composites. Furthermore, the surface of fibre was not treated so it was not properly attached to the epoxy matrix. Therefore, the interfacial adhesion between fibre and epoxy resin is weak [8].

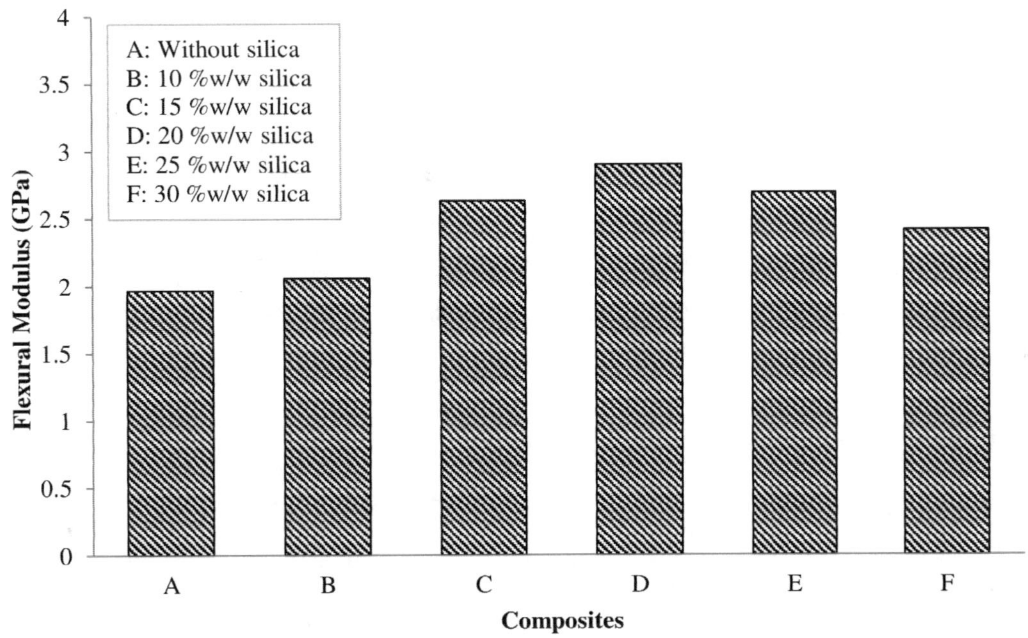

Figure 2. Flexural moduli of silica modified epoxy composites.

Addition of silica content has found increasing the flexural modulus as shown on composites B to D. The addition of 20 %w/w of silica in composites D denotes the highest flexural strength among the silica modified epoxy composites. However, there is no significant flexural strength improvement in composite E as compared to composite C. The results show that addition of silica above 20% w/w has reduced the flexural modulus. It is probably due to silica content overloaded and caused agglomeration formation in epoxy resin, which subsequently leads to poor stress transfer [9].

3.3 Impact strength of epoxy/kenaf/silica composites

Figure 3 shows the impact strength of epoxy/kenaf composite (A) as compared to epoxy/silica/kenaf composites (B to F) using charpy impact test. It is found that the impact strength of epoxy/kenaf composites is about 1.6 kJ/m^2 which is higher than the impact strength of neat epoxy (1.5 kJ/m^2). The impact strength was further increased as the silica content increased in the epoxy matrix of composites. Addition of continuous fibre of kenaf mat gave additional energy absorption during impact [10]. However, increasing silica content up to 15 %w/w gave significant improvement of impact strength by 93 %. The highest impact strength (3.1 kJ/m^2) was observed when 20 %w/w silica content was used in composites. The impact strength significantly reduced when the silica filler content above 20 %w/w.

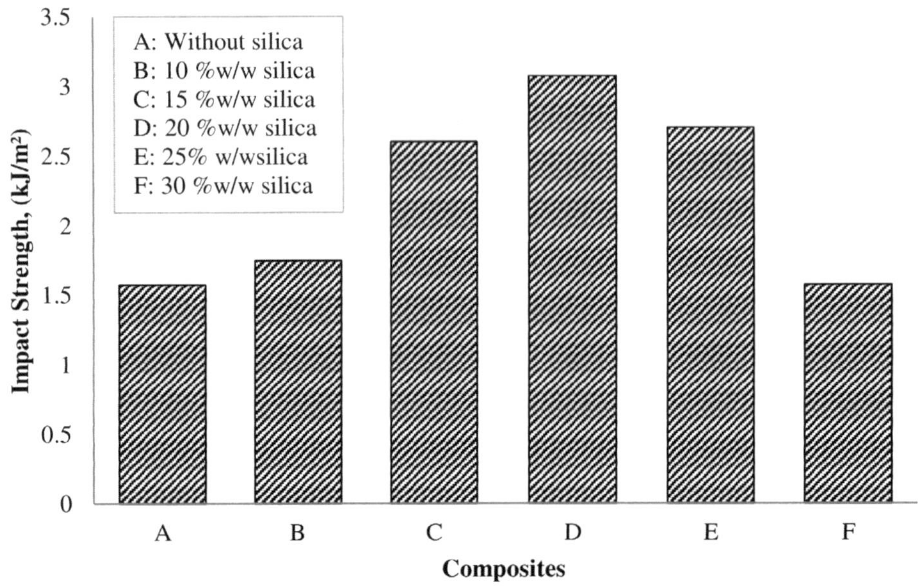

Figure 3. Impact strength of silica modified epoxy composites.

Overall results of mechanical properties proved that the addition of silica content improves the flexural strength, flexural modulus and impact strength of epoxy/kenaf composites. This is because the presence of silica creates plastic void growth around debonded particles, which may increase the difficulty of crack propagation during fracture [3]. In general, a relatively high loading of silica (up to 20 %w/w) is required to achieve reasonable toughening effect in epoxy composite, which would also increase the viscosity of the resin mixtures [11]. Composite D denotes the best impact strength and the best flexural modulus among silica modified epoxy composites. Above 20 %w/w of silica loading reduce the flexural strength, flexural modulus and impact strength of epoxy/kenaf composites. At higher filler loading disturb the matrix continuity leading to stress concentrations, which can act as micro-crack initiator and reduce the adhesion and energy absorption capacity of composites [6].

3.4 Morphology analysis of epoxy/kenaf/silica composites

Figure 4 reveals the morphology of fracture surface of composite D. Based on the SEM image, the cloudy spherical shape in matrix indicating the voids from silica particles were pulled out from the matrix of the composites. Silica particles at 20 %w/w was found well distributed in epoxy matrix, this it was in agreement with the findings of Xu *et al.* (2015) [12]. The dark circles in matrix indicated that the void have created by fibre pull-out from the matrix. The arrows in Figure 4(a) show the unbroken fibre pulled out from the opposite sample's surface. The clear pull out of the fibres indicate the poor interfacial adhesion of the kenaf fibres with the epoxy matrix. The existence of dark circles reflects to the failure of composite which is contributed by the poor interfacial adhesion between fibre and matrix, instead of the strength of the fibre itself [13]. Figure 4(b) shows the cross section of kenaf fibre embedded in the epoxy matrix. Small gaps were seen between the fibre and matrix (indicated by arrows), which supported that the interfacial bonding between kenaf fibre and epoxy resin was poor. The kenaf fibres found in the matrix were hollow in shape which hindered the resin to penetrate the bundle of the fibre as a result gave poor mechanical properties of composite [13].

High magnification image of fibre surface in epoxy composites is shown in Figure 4(c). It indicated that the smooth surface of kenaf fibre may limits the fibre and epoxy matrix interaction, thus decreased the mechanical properties [10]. A higher mechanical property of composites may be obtained if the surface of fibre is treated to become more hydrophobic for better interfacial adhesion with the epoxy matrix.

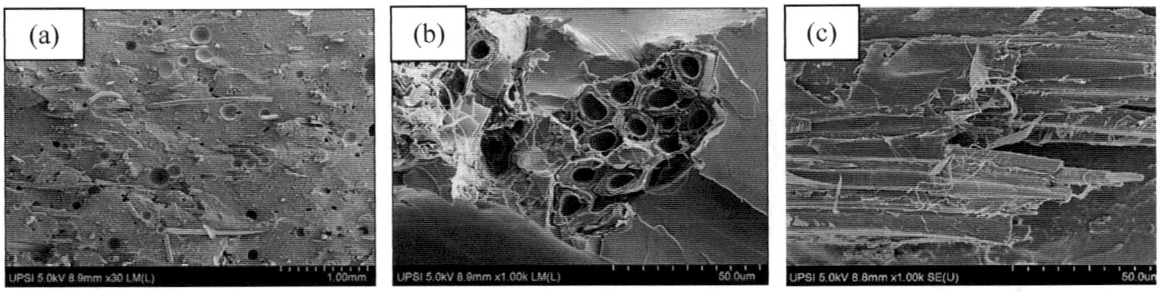

Figure 4. SEM images of (a) fracture surfaces composite E, (b) cross section of kenaf fibre and (c) fibre surface.

4. Conclusion

Overall results found that mechanical properties of epoxy composites has increased as the silica content increased up to 20 %w/w. Addition of 20 % w/w silica into epoxy composites have shown the highest impact strength i.e 3.0 kJ/m^2 and flexural modulus i.e 2.9 GPa whereas the flexural strength is 49 MPa. Beyond the addition of 20 %w/w silica content, epoxy composites exhibit reduction in flexural strength, modulus and impact strength. Surface fracture analysis using SEM has revealed that no interfacial interaction between kenaf and epoxy matrix.

References

[1] Payam S, Habibollah B, Azam R and Jalil M 2005 *Iranian Polymer Journal* 14 637
[2] Marjetka C 2013 *Materials and technology* 47 285
[3] Liang YL and Pearson RS 2010 *Polymer* 51 4880
[4] Saba N, Paridah MT and Jawaid M 2014 *Construction and Building Materials* 76 87
[5] Vijay C, AK Rajput and PK Bajpai 2017 Materials Today: Proceeding 4 9893
[6] Saiful I, Reza M and Hossein R 2013 Journal of Nanoscience 2013
[7] Fiore V, Di Bella G and Valenza A 2015 *Composite: Part B* 68 14
[8] Srisuwen S, Prasoetsopha N and Suppakarn N 2014 *Energy Procedia* 56 19
[9] Bajuri F, Mazlan N, Ishak MR and Imatomi J 2016. *Procedia Chemistry* 19 955
[10] Ramesh M 2016 *Progress in Material Science* 78 1
[11] Liu S, Fan X and He C 2016 *Composites Science and Technology* 123 132
[12] Xu T, Jia Z, Luo Y, Jia, D and Peng Z 2015 *Applied Surface Science* 328 306
[13] Yousif BF, Shalwan A, Chin CW and Ming KC 2012 *Materials & Design* 40 378

IOP Conf. Series: Journal of Physics: Conf. Series **1082** (2018) 012007 doi:10.1088/1742-6596/1082/1/012007

Fabrication and Characterization of Superhydrophobic Plasticized Polyvinyl Chloride Filled with Stearic Acid Coated Calcium Carbonate

S.N. Din[1,2], H.M. Akil[1], A. Rusli[1]

[1] School of Materials and Mineral Resources Engineering, Engineering Campus Universiti Sains Malaysia, Engineering Campus, Penang, Malaysia

[2]Faculty if Applied Sciences, Universiti Teknologi MARA, Perlis Branch, Arau Campus, 02600 Arau, Perlis, Malaysia

Email: sitinor432@perlis.uitm.edu.my

Abstract. Superhydrophobic surfaces are favourable for coating applications to protect various substrates from aggressive environments. In this work, the superhydrophobic coating materials was successfully prepared from plasticized polyvinyl chloride (pPVC) filled with stearic acid coated calcium carbonate particles ($CaCO_3$) by a simple approach. The thermal, surface properties and wettability of the materials were studied. Combination of microstructured surface roughness and low surface energy increases the superhydrophobicity of the pPVC surfaces. The results indicated that the increase in c-GCC fillers magnified the degree of roughness with the highest contact angle achieved at 167 ± 3°.

1. Introduction

The discovery of self-cleaning effect of the lotus leaves made by Barthlott and Neinhuis have inspired academics and industrial researchers in developing superhydrophobic surfaces[1]. The astonishing property of the lotus leaves was reported to be a result of the hierarchical surface topography composed of micro- and nano-structures with optimized geometry and the unique chemical composition of the epicuticular waxes[2]. A surface is said to be superhydrophobic when the water beads up with a contact angle of greater than 150°[3]. So far, there has been numerous attempts made to mimic the lotus leaf's multiscale topographic surfaces to achieve superhydrophobicity. However, multistep fabrication process, complicated surface treatments and the use of expensive materials in the formulation have limits their applications for a large scale production[4]. Furthermore, most of the fabricated superhydrophobic surfaces appeared to have low mechanical durability as they can be easily damaged by routine use, resulting in a decline of their superhydrophobicity [5]. The micro- and nano-particles filling was reported to be a promising method that would enhance their mechanical stability and durability[5].

Polyvinyl chloride (PVC), an intrinsically hydrophobic material, is one of the most widely used thermoplastics materials in various applications. The pPVC for instance, can be produced in the form of a suspension, known as plastisols, suitable for the production of foams, films, sheets, rotational casting and coatings. However, the addition of plasticizer in pPVC compound could actually reduce the water contact angle below than 90° hence decreasing its hydrophobicity[6]. The increase in hydrophobicity of pPVC compounds would therefore extend the long-term protection against corrosion, mechanical wear and fouling resistant [5]. In this research, we aim to fabricate the superhydrophobic pPVC by inexpensive ground calcium carbonate (GCC) particles filling. The presence of multiscale coated ground calcium carbonate (c-GCC) fillers would magnified the degree of surface roughness with the increase in water contact angle. The effects of coated c-GCC fillers on the thermal and surface properties of pPVC were evaluated. The surface morphology, topography and wettability of the samples were compared using field emission scanning electron microscopy (FESEM), atomic force Microscopy (AFM), and water contact angle measurement.

2. Experimental study

2.1 Materials

Emulsion grade PVC resin, EH250, with a K-value of 73, was kindly supplied by Kaneka Paste Polymers Sdn. Bhd. The c-GCC (C18H36O2) having an average particle size, d_{50} of 2.81 ± 0.3 µm was supplied by Omya Malaysia Sdn. Bhd. Diisononyl phthalate (DINP) plasticizer with a molecular weight of 418.62 g/mol and zinc oxide (ZnO) stabilizer were both purchased from Sigma-Aldrich. All chemicals were used as received without further purification.

2.2 Preparation of pPVC coating materials

The pPVC compounds as listed in Table 1, were pre-mixed using IKA mechanical stirrer with controlled stirring speed and a total mixing time of 15 minutes. Mixing was consequently completed using a high-shear homogenizer (Wiggen Hauser D-500) with a 10 000 rpm speed for 5 minutes. The pastes were aged for 2 weeks in a closed chamber at room temperature and were re-mixed for 5 minutes to reduce sedimentation prior to molding. The pastes were subsequently deaerated and poured into a metal mold of 1 mm thick, and heated in the oven at 190 °C for 12 minutes. All samples were stored for 48 hours to allow complete reformation of crystalline crosslinks in the PVC resin and were further rubbed with 500 grit sandpaper for 20 minutes, for surface analysis.

Table 1. Compositions of pPVC based coating materials.

Sample Codes		C0	C1	C2	C3	C4	C5
Components (phr)	PVC	100	100	100	100	100	100
	DINP	85	85	85	85	85	85
	ZnO	2	2	2	2	2	2
	c-GCC	-	10	30	50	70	100

2.3 Characterization

The effect c-GCC fillers on the thermal degradation of pPVC based coatings was quantified using a Mettler Toledo Thermogravimetric Analyzer (TGA) in a nitrogen atmosphere. The heating rate was set at 10 °C min^{-1} with heating range of between 30 °C and 900 °C. The surface morphological analysis of the samples that was sputter-coated with gold in vacuum conditions, conducted in top-view using the Field Emission Scanning Electron Microscope (FESEM, model Zeiss Supra 35VP) at an accelerating voltage of 10kV. Wettability of the samples were determined according to ASTM D7334 using the

Attension Theta Optical Tensiometer by KSV instruments. The measurement was conducted at an ambient temperature with approximately 5 µl of water droplet. The contact angles were obtained as an average values recorded after 10 s and the contact angle for each specimen was calculated from the average of five drops. The three-dimensional (3D) topography and surface roughness of the samples' surface were evaluated using the tapping mode of the SPA300HV Nanonavi Atomic Force Microscope (AFM). A minimum of five different locations were scanned for each sample, and the average roughness (R_a) value was calculated.

3. Results and Discussion

3.1 Thermal stability

PVC easily degraded at elevated processing temperature, giving off hydrochloric acid (HCl) from the polymer backbone that successively accelerates the degradation process[7]. It was proven that the addition of inorganic fillers such as $CaCO_3$ increases the thermal stability of PVC by scavenging the released HCl gas and further avoid its self-accelerating effect[8]. The thermal degradation of PVC based coating materials occurs through two main stages, as shown in Figure 1. The first stage (seen at 200-300 °C) can be linked to weight loss, which can be caused from the generation of HCl gas from the thermal decomposition of PVC, or from the release of carbon dioxide (CO_2) and water (H_2O) when $CaCO_3$ reacted with HCl, which is known as dehydrochlorination.

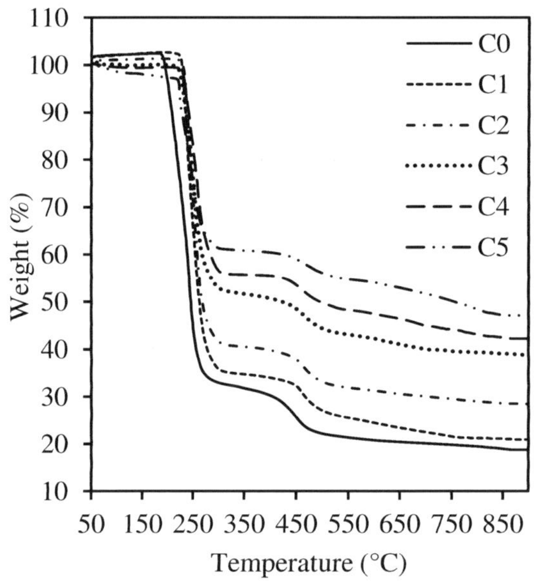

Figure 1. TGA curves of pPVC based coating materials filled with various filler loadings.

Weight loss that occurs at higher than 400 °C during the second stage of degradation can be attributed to the formation of H_2O and CO_2 when the dehydrochlorinated residues were thermally decomposed. Thermal degradation had significantly shifted to a higher temperature from 190 °C, up to more than 200 °C with increasing $CaCO_3$ filler loadings. The presence of higher GCC filler loadings can absorb more HCl gas compared to unfilled C0 samples. $CaCO_3$ micro particles also inhibited the fusion and entanglement of PVC molecules in the polymer compounds[9]. The increased in thermal stability of the coatings also evidence by the increased in char formation at 900 °C with increasing filler loadings. This condition suggests that only inorganic fillers were left in the samples at that temperature.

3.2 Observation of surface morphology

The change in morphology of the prepared coatings due to the increase in filler loadings is depicted in Figure 2. The characterization were made only on un-abraded samples to observe filler distribution throughout the samples' surfaces. Unfilled pPVC coating labelled as C0 sample exhibit smooth surfaces. It is clearly seen that c-GCC fillers progressively increases the roughness of pPVC based coatings. The coated fillers are also evenly distributed in the pPVC matrix without any agglomeration in all filled samples. It was reported that pPVC is slightly hydrophilic[10] and the introduction of surface roughness is therefore beneficial to enhance its hydrophobicity as depicted by Wenzel's law[3]. Addition of less than 50 phr c-GCC fillers (C1 – C3 samples) was unsatisfactory to increase the surface roughness of the coatings. Maximizing the filler loading up to 100 phr shows a significant modifications in the surface properties.

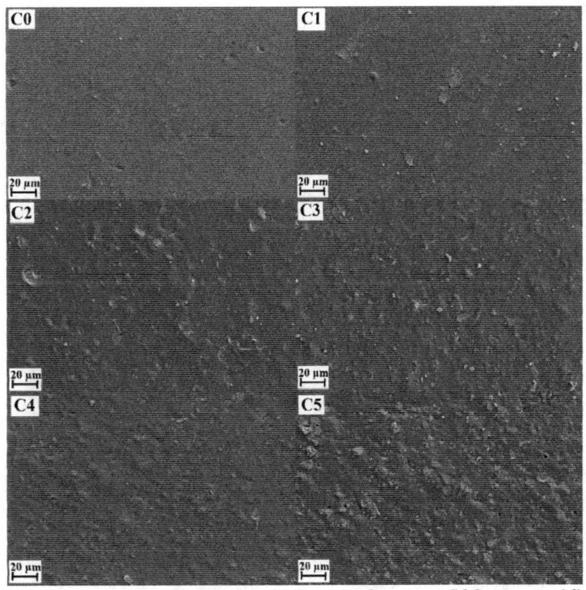

Figure 2. FESEM micrograph of un-abraded pPVC coating surfaces at 500× magnification.

3.3 Wettability

Hydrophobic property of the prepared coatings were studied by measuring water contact angles on both un-abraded and abraded surfaces. Wettability of the pPVC coating materials prepared in this research are tabulated in Figure 3. The water droplet that was placed on un-abraded, unfilled pPVC (C0) sample have shown to be partially wets the surface and shows contact angle of 77 ± 2°. The increase in filler loadings which slightly introduced roughness to the surfaces have steadily increased the hydrophobicity of up to only 106 ± 2°.

Although the use of c-GCC particles reduces the surface energy of the materials, it is however insufficient to increase wettability of the coatings. Hence, roughness of the surfaces were further enhanced with the help of sandpaper abrasion. For that reason, the contact angle of unfilled sample dramatically increases to 120 ± 5°. Superhydrophobic coating materials having water contact angle greater that 150° was achieved with the presence of 50 phr c-GCC fillers. The effect of abrasion along with the presence of high filler loadings further increases contact angle up to a maximum value of 167 ± 3° for C5 sample.

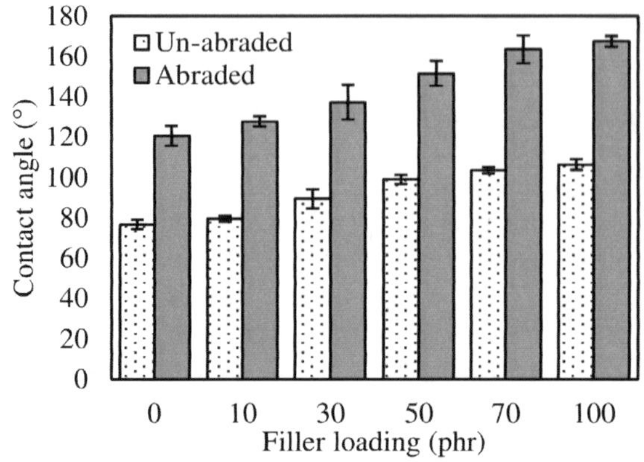

Figure 3. Water contact angle of un-abraded and abraded coating surface samples as a function of filler loading.

3.4 Surface roughness

The effect of surface roughness to wettability of the prepared coating materials were further witnessed using AFM with 100 × 100 µm². The 3D surface topography was presented in Figure 4 and the R_a was further charted in Figure 5. Abraded filled sample shows greater surface roughness compared to un-abraded coating surfaces as previously observed with FESEM.

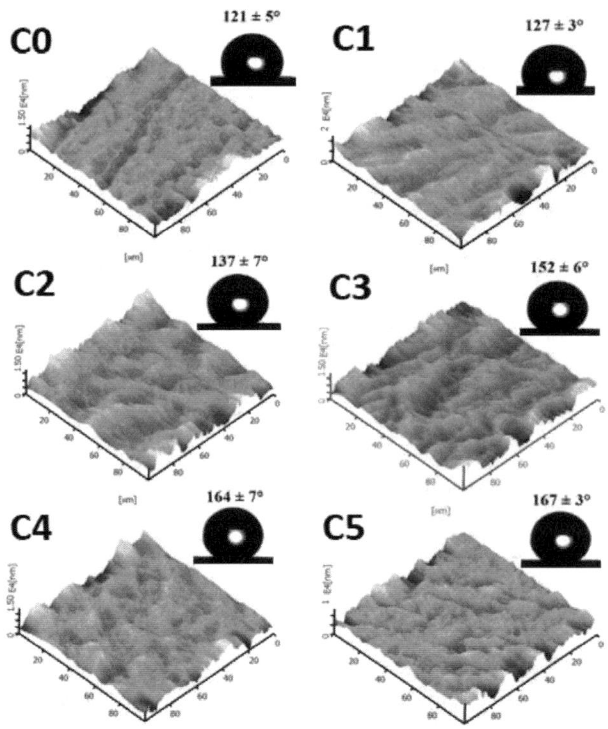

Figure 4. 3D topography and water contact angle of the abraded coating surface samples scanned using AFM.

As expected, the R_a of the surface was increased when the filler loadings were increased. There was also a remarkable increase in R_a due to the effect of sandpaper abrasion. Coating materials filled with 100 phr c-GCC fillers had resulted in the highest R_a in both un-abraded and abraded samples as they

IOP Conf. Series: Journal of Physics: Conf. Series **1082** (2018) 012007 doi:10.1088/1742-6596/1082/1/012007

introduced higher micro scale roughness to the surfaces. Accordingly the water droplet pins on roughness and the valleys are not penetrated due to the presence of air pockets entrapped between the roughness's[11]. The entrapped air is necessary to intensify the hydrophobicity of the solid surfaces[12].

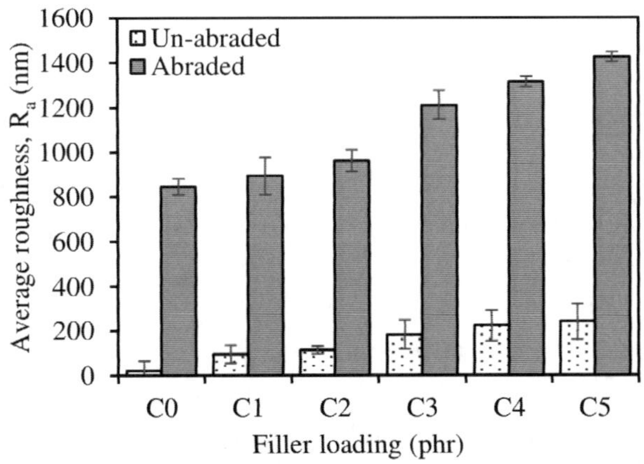

Figure 5. R_a of un-abraded and abraded coating surface samples as a function of filler loading.

4. Conclusion

In this study, we have fabricated superhydrophobic pPVC based coating materials from a versatile and simple method. The addition of c-GCC fillers in the prepared coating materials have steadily increased the thermal stability. The increase in fillers along with the effect of surface abrasion increases water CA to 167 ± 3°. The increase in filler loading had introduced higher surface roughness, as presented by AFM, with the introduction of higher amount of air pockets, which had decreased the adhesion of the water droplets to the surface.

References

[1] Barthlott W, Neinhuis C 1997 *Planta* 202 1

[2] Ensikat H J, Ditsche-Kuru P, Neinhuis C and Barthlott W 2011 *Beilstein Journal of Nanotechnology* 2 152

[3] Kang Y, Wang J, Yang G, Xiong X, Chen X, Yu L and Zhang P 2011 *Applied Surface Science* 258 1008

[4] Zhao N, Xu J, Xie Q, Weng L, Guo X, Zhang X and Shi L 2005 *Macromolecular Rapid Communication* 26 1075

[5] Qu M, Liu S, He J, Feng J, Yao Y, Ma X, Hou L and Liu X 2017 *Applied Surface Science* 410 299

[6] Määttä J, Koponen H -K, Kuisma R, Kymäläinen H -R, Pesonen-Leinonen E, Uusi-Rauva A, Hurme K -R, Sjöberg A -M, Suvanto M and Pakkanen T A 2007 *Applied Surface Science* 253 5003

[7] Folarin O M and Sadiku E R 2011 *International Journal of the Physical Sciences* 6 4323

[8] Karayildrim T, Yanik J, Yuksel M, Saglam M and Haussmann M 2005 *Journal of Polymers and the Environment* 13 365

[9] Chen C H O, Teng C C, Su S F, Wu W C and Yang C H 2006 *Journal of Polymer Science Part B: Polymer Physics* 44 451.

[10] Koponen H K, Saarikoski I, Korhonen T, Pääkkö M, Kuisma R, Pakkanen T T, Suvanto M and Pakkanen T A 2007 *Applied Surface Science* 253 5208

[11] Joneydi S, Khoddami A, Zadhoush A 2013 *Progress in Organic Coatings* 76 821

[12] Yuan Z, Chen H and Zhang J 2008 *Applied Surface Science* 254 1593

IOP Conf. Series: Journal of Physics: Conf. Series **1082** (2018) 012008 doi:10.1088/1742-6596/1082/1/012008

A Comparative Study on the Physiochemical Properties of Semi-Refined Carrageenan from Low Industrial Grade Seaweed and High-Grade Seaweed of Kappaphycus Alvarezii

S. Norhazariah [1], B. Azahari [2], R. Sivakumar [1] & A.R. Azura [1]

[1]School of Materials and Mineral Resources Engineering, Universiti Sains Malaysia, Engineering Campus, 14300 Nibong Tebal, Pulau Pinang, Malaysia.

[2]School of Industrial Technology, Universiti Sains Malaysia, 11800 Minden, Penang, Malaysia.

Email: srazura@usm.my

Abstract. A study on the physiochemical properties of semi-refined carrageenan (SRC) derived from low industrial grade seaweed (LIGS) were conducted and compared to semi-refined carrageenan from high-grade seaweed (HGS) of *Kappaphycus Alvarezii* and others researches. The morphology of SRC (LIGS and HGS) was investigated through scanning electron microscopy (SEM) and physiochemical properties of SRC through Fourier Transform Infrared (FTIR) analysis, particle size analyser, proximate composition (moisture, ash, acid insoluble ash, acid insoluble matter), sulphate and mineral content by inductively coupled plasma optical emission spectrometry (ICP-OES), colour, pH, viscosity and gel strength. SRC-HGS showed smaller particle size than SRC-LIGS from SEM and particle size analysis. SRC-LIGS exhibited lowest lightest, but highest yellowness, with L^* and b^* values reported 57.77 and 15.9, respectively. The gel strength of SRC-LIGS was lower than SRC-HGS. The SRC-LIGS show a comparable thermal properties, sulphur and mineral content, ash, acid insoluble ash and moisture content from SRC-HGS, which suggest that these SRC-LIGS have good potential as renewable resources materials.

1. Introduction

Seaweed has been utilized for ages as food, in cosmetics, pharmaceutical, fertilizer and nutraceutical [1–3]. The structural cell-wall polysaccharides usually contain an outer amorphous mucilage matrix, which is made up of linear sulphated galactan polymers such as carrageenan, agarans, and alginates as well as the inner rigid component which is cellulose fibril [4,5]. Thus hydrocolloids have been widely used as thickener, stabilizer or emulsifier [2,6]. *Kappaphycus Alvarezii* a kappa type carrageenan is one of red seaweed species, which is extensively farming in the east Sabah, Malaysia. The production of carrageenan divided into its purification process. There is refined carrageenan (RC) and semi-refined carrageenan (SRC).

However, production of RC is very costly and time consuming due to complex purification process during extraction, whereas SRC is much easier to prepare and cost-effective when compared to RC

[7,8]. The flourishing seaweed industry by Mini estate system driven by Department of Fishery, Sabah under Entry Point Project (EPP). This project identified seaweed as one of the high-value commodities in the Malaysia National Agro-Food-Policy (2011-2020). In 2013, production has achieved one metric ton per day and is targeted to produce 10 metric tons per day by 2020 [9]. Due to the high demand of RC and SRC in the market, there was an abundant leftover low-grade seaweed after processing thus RC and SRC from high-grade seaweed (HGS). There was no previous report of the production of SRC from low-grade industrial seaweed (LIGS). This study designed to introduce LIGS as a raw material in the production of SRC. The aim of this research to produce SRC from LIGS by alkaline treatment and compare with SRC from high-grade seaweed (HGS) in term of physiochemical and morphology properties.

2. Materials and methods

2.1 Preparation of SRC-LIGS and SRC-HGS

LIGS and HGS from species Kappaphycus Alvarezii obtained from a local supplier at the east coast of Sabah, Malaysia washed thoroughly with water to remove any residual salt, sand or other impurities. The cleaned LIGS and HGS then dried overnight in a laboratory oven at 45°C to remove any excess moisture.

For SRC, the dried LIGS and HGS treated with a hot alkaline solution of potassium hydroxide (KOH) at 70 °C for 30 min with KOH concentration of 6% w/v and seaweed to solution ratio (SSR) was maintain at 1:10. The treated LIGS and HGS was cooled down to room temperature and was washed with water several times to remove any excess KOH until it turned neutral (pH 7) before being dried in an oven (45°C) overnight. The dried SRC-LIGS and SRC-HGS were chopped, ground and sieved before kept in an airtight container to avoid further moisture absorption from the environment until further analysis.

2.2 Physical properties

2.2.1 Particle size analysis. The particle size was determined by using a Sympatec HELOS (Helium-Neon Laser Optical System, Sympatec Inc., Clausthal-Zellerfeld, Germany) laser diffraction instrument combined with RODOS dry particle disperser.

2.2.2 Density test. The density was measured using gas intrusion under a helium gas flow with an AccuPyc II 1340 pycnometer (Georgia, USA). Five sample was analysed at 27°C and the average value was computed.

2.2.3 pH. 1.5 % (w/v) LIGS and SRC solutions were prepared and cooled to room temperature (25 ± 0.5). pH meter (Eutech instruments, Singapore) was then used to measure the pH values of the LIGS or SRC solutions. This procedure was done in triplicate for each sample.

2.2.4 Colour. The surface colour of SRC samples was measured using Konica Minolta Chroma Meters CR-400 (Konica Minolta Sensing, Japan). The colorimeter previously calibrated measuring the colour parameters of standardized white glossy ceramic tiles. The colour was evaluated by means of CIELAB coordinate (L^*, a^*, and b^*) (CIE, 1976). Where L^* was whiteness ($L^*=0$) or brightness ($L^*=100$), a^* was redness ($a^*>0$) or greenness ($a^*<0$) and b^* was yellowness ($b^*>0$) or blueness ($b^*<0$). This procedure was carried out five times for each sample.

2.2.5 Viscosity. The SRC powder (1.5 g) was dissolved in 100 ml hot distilled water (± 90oC) and dispersed with high agitation using mechanical stirrer for 15-20 minutes. Next, the sample was allowed to stabilize in a water bath at 80-90 °C for 15 minutes to eliminate bubbles.

IOP Conf. Series: Journal of Physics: Conf. Series **1082** (2018) 012008 doi:10.1088/1742-6596/1082/1/012008

Viscosity was determined with Brookfield DV-II (Middleboro, USA) using spindle no 2 and speed of rotation for the viscometer was set at 50 rpm.

2.2.6 Gel strength. SRC solution was prepared by dissolving 1.5 g of SRC powder with 100 ml of hot distilled water at 90 °C with continuous string using mechanical stirrer for 20-30 min. Then, the sample was allowed to stabilize in a water bath at 80-90 °C for 15 minutes to eliminate bubbles. The viscous sample was poured into three containers with the same size (each of height 25 ml) moulded and left to cool to room temperature for 20-30 minutes before closed the container lid. The sample was stored for 24 h and maintain at room temperature (28°C) prior to analysis. TA-XT plus texture analyser single arm (Stable Micro Systems, England) was used to evaluate the compressibility of SRC product. A load cell of 5 kg was used where the maximum force was ± 5 kg. A fixture and an analytical probe were compressed into the sample with return distance 65 mm at return speed of 10 mm s-1 and contact force was 5g. Each run was replicated three times.

2.2.7 Scanning electron microscopy (SEM). The surface morphology of the SRC was observed under scanning electron microscopy (SEM), model ZEISS SUPRA 35 VP (Oberkochen, Germany) with an acceleration voltage of 5 kV.

2.3 Chemical analysis

2.3.1. Proximate analysis. The moisture content of SRC was determined using air-drying oven method (AOAC 934.01). As for ash, determination was carried out in the muffle furnace at 550 °C (AOAC 930.05). For acid-insoluble ash, hydrochloric acid was used for digestion prior to ash. For acid-insoluble matter, SRC samples were treated with sulphuric acid for an acid-insoluble matter determination.

2.3.2 Sulphur and mineral content determination. Approximately 2 g of dried SRC was weighed and placed in the 50 mL glass tube. About 20 mL concentrated sulphuric acid (H_2SO_4), was then pipetted into the glass tube and the vessel was gently shaken for 5 minutes by benchtop shaker SK-300 (Lab Companion, Minneapolis, USA) until stable with no observed bubbles formed. Next, the sample was left standing at room temperature for 2 hours. The solution was then filtered with glass fibre filter paper (90mm) (Toyo Roshi Kaisha, Ltd, Japan), to separate the leached sample. The concentration of the filtered solution was reduced by 100 times with deionized water and filtered again through an ashless Whatman filter. The solution was diluted according to the elements being analysed and was analysed by inductively coupled plasma optical emission spectrometry (Optima 4300 DV ICP-OES, Perkin Elmer, Sheldon, CT, USA).

2.4 Fourier Transform Infrared (FTIR) analysis

The functional group present in powder SRC were analysed by Fourier Transform Infrared (FTIR) analysis. Infrared spectra in attenuated total reflectance (ATR) accessory sampling device containing a diamond/ZnSe crystal. Spectra were scanned at room temperature over the wave range from 550 cm^{-1} to 4000 cm^{-1}, with a scan speed of 0.20 cm s^{-1} and 60 accumulations at a resolution of 4 cm^{-1} using Perkin Elmer Spectrum One (USA).

3. Results and discussion

3.1 Physical properties

Physical properties of SRC-LIGS and SRC-HGS are tabulated in Table 1. The particle size of SRC-LIGS was slightly larger compared to SRC-HGS. The morphology of SRC particle was analysed using

SEM (Figure 1). SRC-LIGS has rough an uneven surface, is irregularly shaped (flake-like) and is bigger. The small particles size was observed in almost the whole surface of SRC-LIGS. While SRC-HGS has a rough heterogenic surface, is the crooked and twisted shape and is smaller. This finding in accordance with the particle size analysis, whereas SRC-LIGS has larger particles size compared to SRC-HGS.

The weight of the material is the crucial factor in the material selection process, depending on its application. Density directly correlates to the weight. Both SRC-LIGS and SRC-HGS have same densities of 1.65 g cm^{-3}. Notwithstanding, the density of seaweed is lower than conventional glass fibre (2.5 g cm^{-3}) but slightly higher than other natural fibre such as banana fibre (1.35 g cm^{-3}) and kenaf (1.4 g cm^{-3}) [10]. The pH of both SRC is above 7 and in the range from 9.86 to 10.07. This finding was in agreement with JECFA (2006) [11], pH of SRC must in the range 8 to 11.

Calorimetry results showed that SRC-LIGS and SRC-HGS have similar value of yellowness b*and *a value more than 0. Thus indicates both SRC comes from red seaweed family. SRC-LIGS has lower *L value (57.77) compared to SRC-HGS (70.15), indicating that SRC-LIGS was slightly darker than SRC-HGS. This is might be due to the source of LIGS which from a low-grade seaweed in the market. The viscosity of SRC-LIGS was lower compare to SRC-HGS. Thus the high-grade seaweed might influence the viscosity of SRC solution. As the presence of free ions will influence the rate of viscosity [12], the SRC-HGS might have higher free ions that induce helical formation which give higher resistant to a shear [13]. Therefore, SRC-HGS exhibit higher viscosity compared to SRC-LIGS. Gel strength is the amount of force required to rupture the gel [14]. SRC-LIGS has lower gel strength (61.6 g cm^{-2}) as compared to SRC-HGS (73.26 g cm^{-2}). This might be due to higher degree aggregation and packing of SRC-HGS compared to SRC-LIGS [15].

Table 1. Physical properties of SRC-LIGS and SRC-HGS.

Sample	Particle size (µm)	Density (g cm^{-3})	pH	Colour			Viscosity (cP)	Gel strength (g/cm^2)
				*L	*a	*b		
SRC-LIGS	9.64	1.65 ± 0.00	10.07 ± 0.13	57.77 ± 1.63	+1.76 ± 0.08	+15.06 ± 0.12	50.93	61.6 ± 0.41
SRC-HGS	5.98	1.65 ± 0.00	9.86 ± 0.15	70.15 ± 2.39	+1.03 ± 0.27	+15.9 ± 0.38	93.87	73.26 ± 6.94

Figure 1. SEM morphology of (a) SRC-LIGS and (b) SRC-HGS particles at 300x magnification.

3.2 Chemical analysis

Seaweeds easily absorb moisture over time due to its hygroscopic properties. Moisture content for both SRC is 7.08 and 6.44 %. Thus met the FAO standard, the maximum moisture content allowed in SRC is 12 %. The international standard of ash levels in SRC is at least 15 %, but not more than 40 % [11]. Both SRC has the ash content meet the FAO specification. The acid insoluble ash (AIA) measurement is created to measure the amount of insoluble ash in diluted hydrochloric acid. The results show that AIA for both SRC has met the international standard, which must less than 1%. Acid-insoluble matter (AIM) carrageenan consists of a network of cellulose that is normally present in seaweed cell walls [7].

AIM for both SRC meet the FAO specification which less than 15 %. However, AIM for SRC-LIGS was higher compared to SRC-HGS. This might be due to the higher cellulose content and can be proven by the darker colour of SRC-LIGS than SRC-HGS. The sulphate in SRC is mainly derived from galactans in red algae and located at C-4 of the D-galactopyranose unit [2,16]. The lower sulphate content of SRC-LIGS is due to the lower grade of seaweed. Red seaweed (*K.Alvarezii*) contains high vitamins and minerals that are important in human nutrition. The mineral composition of *K.Alvarezii* is affected by the seasonal year, geographic area and water temperature [17,18].

From Table 2, the mineral content of SRC-HGS was higher compared to SRC-LIGS. The ash content paralleled the mineral contents. The ash contains SRC is made up from minerals such as sodium and potassium. Hence, the SRC-HGS with higher mineral content also had the higher ash content (22.97 %). Moreover, the mineral content had an extensive effect on the gel strength of SRC. As mention earlier, the gel strength of SRC-HGS was higher due to the degree of aggregation of SRC-HGS molecules. Thus aggregation induced by cation in SRC [13]. Apart from that, the higher mineral content of SRC-HGS also induces higher shear rate as potassium induce the formation of helical formation in SRC[13]. This concurrent with viscosity results which SRC-HGS exhibit higher viscosity.

Table 2. Chemical composition of SRC-LIGS and SRC-HGS.

Chemical composition	Moisture content	Ash	Acid-insoluble ash (%)	Acid-insoluble matter (%)	Sulphate (g/L)	Minerals (Mg/L)				
						Sodium, Na	Potassium, K	Calcium, Ca	Magnesium, Mg	Iron, Fe
SRC-LIGS	7.08 ± 1.15	19.39 ± 1.71	0.02 ± 0.02	10.75	322.1 ± 1.05	269.8 ± 2.56	3189 ± 91.9	605.4 ± 2.1	289.3 ± 2.74	24.3 ± 0.05
SRC-HGS	6.44 ± 1.79	22.97 ± 4.42	0.19 ± 0.00	8.52	388.2 ± 2.0	314.3 ± 4.28	4175 ± 107.3	1491 ± 6	453.3 ± 3.33	61.6 ± 0.28

3.3 *FTIR analysis*

The FTIR spectra for SRC-LIGS and SRC-HGS were almost identical (Figure 2). All spectra displayed an absorption band at around 1220 – 1230 cm^{-1} related to ester sulphate, 1035 to 137 cm^{-1} represent the glycosidic linkage, 920-930 cm^{-1} related to 3,6-anhydro-D-galactose and 845 cm-1 represent D-galactose-4-sulphate [19–21]. From this spectrum, it can be confirmed that the source of SRC-LIGS and SRC-HGS came from K.alvarezii species.

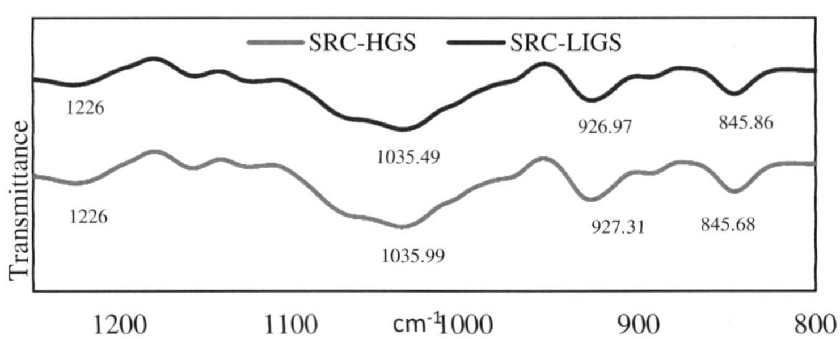

Figure 2. FTIR spectra of SRC-LIGS and SRC-HGS.

4. Conclusion

In this study, SRC from LIGS and HGS from Sabah were produced and the physicochemical properties were investigated and compared. The SRC-LIGS and SRC-HGS showed a significant difference in particle size, colour, particle morphology, viscosity, gel strength and chemical composition. These differences are possibly due to a geographical area, seasonal variation and variation in seaweed grade. Whereas, SRC-LIGS has a larger particle size and darker compared to SRC-HGS. The viscosity and gel strength of SRC-LIGS slightly lower from SRC-HGS. Moreover, the AIM content, Sulphur and mineral content of SRC-LIGS was lower compared to SRC-HGS. The AIA content for SRC-LIGS was higher from SRC-HGS. FTIR analysis showed that SRC-LIGS was from the same source as SRC-HGS. The characteristic shown by SRC-LIGS manifest that this seaweed also possesses a great potential as renewable resources in applicable applications in lower cost.

References

[1] Glicksman M, 1987 *Hydrobiologia* 15131
[2] Campo V L, Kawano D F, Silva D B Da and Carvalho I 2009 *Carbohydrate Polymer* 77 167
[3] Sjamsiah, Ramli N, Daik R, Yarmo M A and Ajdari Z 2014 *Journal Applied of Phycollogy* 26 1049
[4] Kumar K S, Ganesan K and Rao P V S 2008 *Food Chemistry* 107 289
[5] Arad S and Levy-Ontman O 2010 *Current Opinion in Biotechnology* 21 358
[6] McHugh D J 2003 *A Guide to the Seaweed Industry*
[7] Phillips G O 1995 *Gums Stabilisers Food Indudtries* 403
[8] Basmal J and Ikasari D 2014 *Squalen Bulletin of Marine & Fisjeries Postharvest & Biotechnology* 9 17
[9] Safari S 2014 *Food and Fertilizer Technology Center (FTCC) Agricultural Policy Articles* 1–4
[10] Wambua P, Ivens J and Verpoest I 2001 *13th International Conference Composite Materials Conferens Prociding* 1
[11] Joint FAO/WHO Expert Committee on Food Additives (JECFA) 2006 *Compendium of food additive specifications : 67th Meeting 2006.* (Food and Agriculture Organization of the United Nations)
[12] Asyakina L K and Dyshlyuk L S 2016 *Science Evolution* 1 11
[13] Morris E R, Doyle J, Giannouli P and K. Phillip 2002 *Gums and Stabilisers for the Food Industry 11* ed G O P Peter A Williams (Cambridge, UK: The Royal Society of Chemistry) 158
[14] Jain-Raina R, Babbar S B, Azevedo G, Torres M D, Sousa-Pinto I and Hilliou L 2015 *Asian Journal Biotechnology* 50 153
[15] Rees D A 1972 *Biochemistry Journal* 126 257
[16] Karthikaidevi G, Manivannan K, Thirumaran G, Anantharaman P and Balasubaramanian T 2009 *Global Journal Pharmacology* 3 107
[17] Krishnaiah D, Sarbatly R, Prasad D M R and Awang B 2008 *Asian Journal of Scientific Research* 1 166
[18] Kaehler S and Kennish R 1996 *Botanica Marina* 39 11
[19] Van de Velde F, Knutsen S H, Usov A I, Rollema H S and Cerezo A S 2002 *Trends Food Science Technology* 13 73
[20] Pereira L, Amado A M, Critchley A T, Van de Velde F and Ribeiro-Claro P J A 2009 *Food Hydrocolloids* 23 1903
[21] Pereira L, Gheda S F and Ribeiro-claro P J A 2013 *International Journal of Carbohydrate Chemistry* 2013 7

IOP Conf. Series: Journal of Physics: Conf. Series **1082** (2018) 012009 doi:10.1088/1742-6596/1082/1/012009

Optimisation of Formulation for Starch Modified Natural Rubber Composites by Using Response Surface Methodology

N. Mohamad[1], M. Mazliah[1], K.I. Karim[1], H. E. Ab Maulod[2], J. Abd Razak[1], A. Muchtar[3] and H. M. Mef'at[4]

[1]Advanced Manufacturing Centre, Fakulti Kejuruteraan Pembuatan, Universiti Teknikal Malaysia Melaka, Hang Tuah Jaya, 76100 Durian Tunggal, Melaka, Malaysia

[2]Fakulti Teknologi Kejuruteraan, Universiti Teknikal Malaysia Melaka, Hang Tuah Jaya, 76100 Durian Tunggal, Melaka, Malaysia

[3]Fakulti Kejuruteraan, Universiti Kebangsaaan Malaysia, 43600 Bangi, Selangor, Malaysia

[4]Sindutch Cable Manufacturer Sdn. Bhd., Lot 38, Alor Gajah Industrial Estate, 78000 Alor Gajah, Melaka, Malaysia

Email: noraiham@utem.edu.my

Abstract. Starch modified natural rubber vulcanisates were prepared by the melt compounding method using a two-roll mill at a constant starch loading of 20 phr. Four formulation parameters consisted of black filler, white filler, silane coupling agent (SCA) and glycerol content in the composites which were optimised using two-level factorial design via response surface methodology (RSM). The cure characteristics and rebound resilience were selected as the responses. The significance of the factors and interaction were analysed using ANOVA and the model's ability to represent the system was confirmed using the constant of determination, R^2 for a value above 0.90. The R^2 values for all responses were obtained in the range between 0.96 and 0.99 which are close to the union ($R^2 = 1$). The final combination of factors used to achieve the highest rebound resilience and cure characteristics were with a high black filler loading of 60 phr and high glycerol of 5 % loading. These findings were further supported by morphological the characteristics observed under scanning electron microscopy (SEM). Therefore, the chemical modification using both glycerol and SCA was proven to improve the cure characteristics and rebound resilience value through the enhancement of the interfacial interaction between the starch and natural rubber (NR) particles. It is expected that the findings in this study will be useful for rubber product manufacturers to explore the possibility of producing biodegradable products in the future.

1. Introduction

Recently, the use of organic fillers has attracted much interest given their low costs and relatively lightweight. Aside from being environmentally friendly, the composites produced display several mechanical property enhancements [1, 2]. Several cellulosic materials such as nut shells, bamboo, ground wood waste, cereal straw and white rice husk have been used as fillers for plastics [3] and elastomers [4]. Starch is one of the biopolymer substances that is naturally found in grains and vegetables and mostly consists of amylose and amylopectin. Further, starch is highly hydrophilic, and under strong inter and intramolecular hydrogen bond networks, and is therefore insoluble in most solvents. Several previous studies have explored starch reinforced elastomers [5-6]. In some studies, the incorporation of starch into polymer matrices has mainly focussed on phase separation. Notwithstanding, it has been widely acknowledged that good interfacial interaction between natural rubber (NR) matrices and starch, uniform particle distribution, and dispersion state of starch in the NR matrix are three vital parameters that strongly affect the mechanical properties of NR composites [7]. As a feasible solution, the modification of starch using a silane coupling agent (SCA) to enhance the interfacial interaction, and glycerol to assist in the plasticisation of starch-based NR compound appears to be a promising approach for further study.

The performance of NR composites is highly dependent on the formulation of the compound. Indeed, experimental work to optimise the formulation can be designed using suitable mathematical and statistical tools. The design of experiments (DOE) is a useful technique employed for multifactor experiments as it is less time-consuming and can detect the true optimum values of factors [8]. According to Mohamad et al. [9], it is possible to obtain quantitative relations for the effects of multi-process parameters on the properties of the NR composites via response surface methodology (RSM) in a minimum number of experiments. In this present study, RSM using a full factorial design was used to develop a mathematical correlation between formulations (effect of chemical modification) to produce NR composites with smaller sample size [10]. The mathematical models were applied in order to determine the feasible region of filler loading and the chemical modification of starch using SCA and the glycerol level to give NR composites with the desired cure characteristics and rebound resilience.

2. Experimental study

2.1 Materials

Table 1 displays the formulation used in this study. NR was purchased from Felda Global Ventures Holdings Bhd (FGV). Both white filler and black filler were purchased from Malaysian Rubber Board. The starch was purchased from Polyscientific Enterprise Sdn Bhd and other compounding ingredients such as sulphur, zinc oxide, stearic acid, were purchased from Systerm Classic Chemical Sdn. Bhd. The TMTD, SCA and glycerol were purchased from Aldrich Chemistry Sdn. Bhd., while Flexsys America, USA supplied 6PPD.

Table 1. Typical formulation

Materials	Compound (phr)[a]
Natural rubber (SMR 20)	100
Filler (white and black)	0,30,60
Zinc oxide	5.0
Stearic acid	2.0
Tetramethylthiuram disulfide (TMTD)	1.0
(1,3-dimethylbutyl)-N'-phenyl-p-phenylenediamine (6PPD)	1.0
Sulfur	2.5
Tapioca starch	20
Silane coupling agent (%)	0,2.5,5
Glycerol (%)	0,2.5,5

[a] Parts per hundred

2.2 Preparation and testing of NR composites

The compounding process was performed according to ASTM D 3184 [2] and carried out using a two-roll mill at a temperature of 50 °C. The NR was first masticated before adding all ingredients sequentially. Lastly, sulphur was added and mixed for around 1 min before dumped and left to cool to room temperature. The cure characteristics of the compound were obtained using a Monsanto moving die rheometer (MDR 2000) test at 150 °C. The rubber compounds were subsequently compression moulded at 150 °C using a hot press based on the respective cure time, t_{90} which is in accordance with ASTM D 2084.

2.2.1. Rebound Resilience Testing. The disk-shaped samples with a thickness of 12.5 ± 0.5 mm and diameter of 29 ± 0.5 mm were prepared according to the BS 903 PT A8 standard and tested using a DUNLOP Tripsometer at room temperature. At least seven samples were tested for every set of experiments to ensure a high confidence level of the results. The angle of rebound was recorded, and the average rebound resilience (%) was calculated using Eq. (1):

$$\text{Resilience, R} = \frac{1-\cos(angle\ of\ rebound)}{1-\cos(angle\ o\ fall)} \times 100 \qquad \text{(Eq. 1)}$$

2.2.2. Scanning Electron Microscopy. Scanning electron microscopy (SEM), model EVO-50 from Zeiss, USA, was used to analyse the fracture surface regarding the morphological properties and distribution of starch granules in the NR composites. Next, the samples were placed onto an aluminium stub and sputter coated with a thin layer of gold, of about 20 mm thickness before the scanning process in order to avoid the occurrence of electrostatic charging and poor resolution during the examination. The fracture surface was observed under secondary electrons at the magnification of 500x and 1000x. The quadratic method applied against the micrographs to analyse starch distribution in the NR composites and the skewness value was calculated applying Eq. (2):

$$\beta = \frac{q}{(q-1)(q-2)} \Sigma \left(\frac{N_{qi}-N_q^{min}}{\sigma}\right)^3 \qquad \text{(Eq. 2)}$$

2.3 Experimental design and analysis

The design of experiments (DOE) were carried out using Design Expert software (Statistics Made Easy, version 6.0.8, Stat-Ease, Inc., and Minneapolis, MN).

2.3.1. Two-level factorial design experiment. Independent variables in this study were black filler (X_1), white filler (X_2), glycerol (X_3), and SCA (X_4) with levels of variables as shown in Table 2. According to this design, there were 19 experiments with three replications at the centre point.

Table 2. Levels of Variables

Black filler (X_1;Phr)	White filler (X_2;Phr)	Glycerol (X_3;%)	SCA (X_4;%)
0 (-1)	0 (-1)	0 (-1)	0 (-1)
30 (0)	30 (0)	2.5 (0)	2.5 (0)
60 (+1)	60 (+1)	5 (+1)	5 (+1)

2.3.2. Analysis of variance (ANOVA). An appropriate polynomial relationship for the dependent variables (response) was obtained. The result of this design was used to fit a first-order polynomial as given in Eq. (3):

$$Y = \beta_0 + \beta_1 X_1 + \beta_2 X_2 + \beta_3 X_3 + \ldots + \beta_k X_k + \varepsilon \qquad \text{(Eq. 3)}$$

where Y is the predicted response, β are the coefficient values, X is the independent variables, and ε is a random error. In this study, $k = 4$ was used as there were four independent variables involved. From the RSM, a regression equation for the selected model for the response, Y was derived.

3. Results and discussion

In Table 3, the cure characteristics and rebound resilience results for each experiment are presented. In Table 4, the regression models for each response are presented. Accordingly, this mathematical relationship represents the quantitative effects of the independent variables and their interaction effects on the response. The positive values reflect the effects that lead to optimisation whereas negative values are the factors which provide an opposite effect on the response [11]. The R^2 values indicate the degree of agreement between the experimental results with those predicted by the model. The R^2 values for all responses are obtained in the range of 0.96 – 0.99 which are very close to the union ($R^2 = 1$). The model presented almost 100 % of the variation in the overall system. Therefore, this indicates that the regression model is accurate in describing and predicting the pattern of significance for each factor studied [9, 12].

Table 3. Cure Characterises and Rebound Resilience

Experiment	Scorch time (t_{s2})	Cure time (t_{90})	Rebound resilience (%)
1	0.57	1.17	36
2	0.78	1.19	34.15
3	0.57	1.11	37.54
4	0.86	1.36	32.18
5	0.37	0.96	11.6
6	0.69	1.44	51.46
7	0.49	1.08	27.72
8	0.57	1.17	54.68
9	1.02	1.65	24.44
10	0.62	1.04	51.06
11	0.71	1.5	29.24
12	0.4	1.05	46.92
13	0.5	0.91	23.52
14	1.46	2.25	48.5
15	0.54	1	31.2
16	0.49	1.15	53.2
17	0.69	1.12	37.92
18	1.25	2	39
19	0.46	1	39.1

Table 4. Regression Equation for different responses

Responses	R^2	Adjusted R^2	Equation of the models
Scorch time (t_{s2})	0.9979	0.9953	$Y_1 = 0.67 + 0.069\ X_1 - 0.086\ X_2 - 0.026\ X_3 + 0.052\ X_4 - 0.067\ X_1\ X_2 + 0.095\ X_1\ X_3 - 0.044\ X_1X_4 - 0.03\ X_2X_3 - 0.096\ X_2X_4 + 0.056\ X_3X_4 - 0.089\ X_1X_2X_3 + 0.11\ X_1X_3X_4$
Cure time (t_{90})	0.9971	0.9757	$Y_2 = 0.74 + 0.57\ X_1 - 0.03\ X_2 - 0.1\ X_3 + 0.014\ X_4 + 7.5 \times 10^{-3}\ X_1X_4 - 0.057\ X_2X_3 + 0.014\ X_2X_4 + 0.25\ X_3X_4 - 0.061\ X_1X_2X_3 + 0.047\ X_1X_2X_4 + 0.18\ X_1X_3X_4 + 0.076\ X_2X_3X_4 + 0.058\ X_1X_2X_3X_4$
Rebound resilience	0.9644	0.9244	$Y_2 = 37.09 + 9.43\ X_1 + 2\ X_2 + 0.65\ X_3 + 1.42\ X_4 - 1.77\ X_1X_2 + 4.79\ X_1X_3 + 1.98\ X_1X_4 - 0.052\ X_3X_4 - 4.46\ X_1X_3X_4$

IOP Conf. Series: Journal of Physics: Conf. Series **1082** (2018) 012009 doi:10.1088/1742-6596/1082/1/012009

3.1 Interaction between the variables for Cure Characteristics and Rebound Resilience Properties

Scorch time (t_{s2}) is the time offset during which a rubber compound is workable at a given temperature before 2 % of curing [13]. Figure 1(a) depicts the response surface for the variation in scorching time as a function of glycerol and black filler. According to the response surface plot, the t_{s2} increased as the black filler content increased from 0 to 60 phr, whereas, the t_{s2} value decreased as the glycerol content increased. This might be the suitable ratio between the fillers and glycerol which considerably delayed the scorch time, and improved the processing safety towards an efficient vulcanisation system [11]. Figure 1 (b) demonstrates the cure time versus black filler and glycerol in the 3D surface plot. The response surface plot shows a similar pattern to t_{s2}. Figure 1(c) demonstrates the rebound resilience versus black filler and glycerol in the 3D surface plot. Accordingly, it illustrates an almost identical pattern with the effect of both black filler and glycerol to t_{s2}. According to the response surface plot, the rebound resilience of the NR composites increases up to 51.96 % by an increase in the carbon black from 0 to 60 phr; as well as the increase in glycerol content. Therefore, according to these results, the maximum value of the rebound resilience is obtained at high black filler loading (60 phr) and high glycerol (5 %) loading.

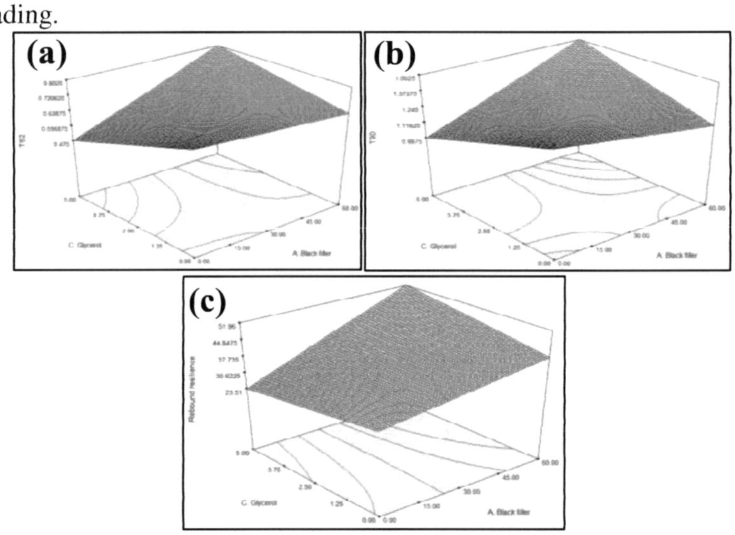

Figure 1. Response surface plot in (a) scorch time (t_{s2}), (b) cure time (t_{90}) and (c) rebound resilience

3.2 Morphological analysis

Figure 2 illustrates the SEM micrographs taken from the tensile fracture surfaces of the samples from Experiment 1 and Experiment 16. The starch particles appeared lighter and dispersed as spherical granules in the darker NR matrices. The granules were found to be uniformly distributed and dispersed in the range of 5 to 10 μm. At higher magnification (Figure 2(b)), the sample for Experiment 1 displayed larger starch granules with apparent interfacial gaps between the starch and natural rubber matrix. If compared with Figures 2(c) and (d) for Experiment 16, the starch granules appeared smaller and almost wholly embedded within the natural rubber matrices. Therefore, this represents that good compatibility, and good interaction between all components occurred in Experiment 16 composites. The dimension of the starch granules was strongly influenced by the treatment with 5 % glycerol and 5 % SCA. The reduction in particle size and improvement in the distribution of the starch particles in the NR matrices caused a remarkable development in the rebound resilience value. This is supported by the skewness value (β) obtained for a sample from Experiment 1 which was as high as 0.77 compared to ~ 0 for Experiment 16 thereby indicating the normal distribution of the starch particles [14].

4. Conclusion

The interaction between the formulation parameters established by the RSM is consistent with the morphological characteristics of the composites. The introduced chemical modification using both glycerol and SCA enhances the matrix-filler interfacial interaction and improves the cure characteristics as well as the rebound resilience value of the composites. In this study, the coefficient of determination (R^2) is sufficient to represent the model's fit between the experimental and predicted value obtained from the software. Hence, the generated regression model can be utilised to optimise the chemical modification contents as well as suitable filler loading for producing the starch modified NR composites with maximum rebound resilience and safe curing during processing. The final combination of factors used to achieve the highest rebound resilience and cure characteristics are at the highest black filler loading of 60 phr and the highest glycerol of 5 % loading.

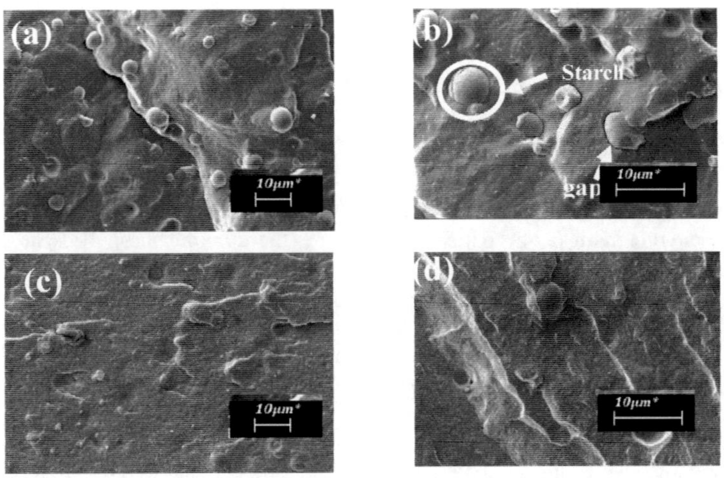

Figure 2. SEM micrographs of NR composites (a) Experiment 1 at 500x, (b) Experiment 1 at 1000x, (c) Experiment 16 at 500x and (d) Experiment 16 at 1000x magnification

Acknowledgement

I would like acknowledge Universiti Teknikal Malaysia Melaka for financial support through (PJP/2016/FKP/HI6/S01484).

References

[1] Mo J L, Wang Z G. Chen G X, Shao T M, Zhu M H, and Zhou ZR 2013 *Wear* 301 1
[2] Mazliah M, Mohamad N, Jeefferie A R, and Ab Maulod H E 2016 *Proceedings of Mechanical Engineering Research Day 2016*
[3] Nakason C, Kaesman A, Homsin S, and Kiatkamjornwong S 2001 *Journal Polymer. Science* 81
[4] Kahar A W M and Ismail H 2014 *Journal of Vinyl and Additive Technology*
[5] Kahar A W M, Ismail H, and Othman N K 2012 *Universiti Malaysia Perlis (UniMAP)*
[6] Konar B B, Roy S K, and Pariya T K 2010 *Journal of Macromolecular Science Part a-Pure and Applied Chemistry* 47 5
[7] Visakh P M 2017 Springer International Publishing: Switzerland
[8] Roshanida A R, Rosli M I, Mohd Ghazali M N, Ahmad Fauzi I, Osman H, Kamarulzaman K 2004 *Proc Biochem* 39 2053
[9] Mohamad N, Muchtar A, Ghazali M J, Mohd D H J, Azhari C H 2010 *Journal of Applied Polymer Science*
[10] Myers R H, Montgomery D C 2002 Wiley: New York
[11] Mohamad N, Yaakub J, Razak J E, Yaakob M Y and Muchtar A 2014 *Journal of Applied Polymer Science*
[12] Razak J E, Ahmad S H, Ratnam C T, Mahamood M A, Yaakub J and Mohamad N 2014 *AIP Conference Proceedings* 1614
[13] Annual Book of ASTM Standard; American Society for Testing of Materials: West Conshohocken, PA, 2001.
[14] Fernanda F and Gomes M I 2013 *Statistical Journal* 11 1

IOP Conf. Series: Journal of Physics: Conf. Series **1082** (2018) 012010 doi:10.1088/1742-6596/1082/1/012010

The Influence of Different Types of Rubber on Curing Behaviour and Dynamic Properties of Rubber Compound

M.S. Nafeesa[1], A.R.Azura[1*]

[1]Universiti Sains Malaysia, Engineering Campus, Penang, Malaysia

Email: nafesa.feesya.ms@gmail.com *srazura@usm.my

Abstract. The aim of the present work is to investigate the effect of the different types of rubber on cure characteristics. The paper also explores the dynamic properties of these rubbers. The curing behaviour of the rubber compound is the main consideration for the rubber moulding process maker, which govern the process and properties of the rubber product. The dynamic properties are another factor that needs to take into account as it is the important properties especially in designing an anti-vibration component. Three types of rubber were used includes natural rubber (NR), styrene butadiene rubber (SBR) and chloroprene rubber (CR). All rubber compounds were prepared based on semi-efficient (semi-EV) sulphur vulcanization system. These compounds were mixed by melt mixing on a two-roll-mill lab scale machine. The cure characteristics were analysed using the Moving Die Rheometer (MDR 2000). The results obtained show that SBR presented higher torque different (MH-ML) value followed by NR and CR. The hardness of the rubber compound showed a similar trend as torque different result. The dynamic properties; static stiffness, Ks and dynamic stiffness, Kd were tested using the Selvopulser Testing machine. For Kd value, the test was reported on the frequency of 100 Hz. The static stiffness, Ks shows a similar trend as the torque different and hardness results; SBR>NR>CR however, for dynamic stiffness, Kd value trend has shown significant changed to SBR>CR>NR. This is due to the bond breakage in a rubber chain also known as Payne effect. In conclusion, the Kd, torque different and hardness are correlated to the rubber molecular weight, meanwhile, for Ks is much more related to interaction of molecular structure in rubber compound (Payne effect).

1. Introduction

Rubber materials have been widely used in automotive engineering application nowadays. Rubber has found an indisputable role in many mechanical applications for example tires, seals, hoses and in automotive industries. Vibration is normally conceived as unwanted dynamic motions in mostly vehicle application and great effort is made reducing these motions and to increase vehicle component durability [1]. The rubber properties being generally soft while showing large inherent damping that make the material a perfect choice for use in vibration application. The viscoelastic behaviour of rubber material is the reasons for this usage.

The selecting the right rubber is very important as it is the main contribution of the elastic properties of end rubber products. The most commonly used rubber for the dynamic application is natural rubber (NR) due to its elastic properties. However, NR is known for poor ageing properties, chemical resistance and ozone resistance. Nowadays, the synthetic rubber has been also used in rubber automotive application such as Styrene Butadiene Rubber (SBR) due to its excellent abrasion resistance and Chloroprene Rubber (CR) due to its good ageing resistance, high ozone and weather resistance. Hence, the understanding of viscoelastic behaviour and rubber properties is essential for the efficient use and proper design of these materials for automotive engineering application [2] [3].

Although there were many studies on the effect of rubber types, those studies are only limited to the mechanical properties and fewer on static and dynamic stiffness. Besides that, the influences of stiffness test of rubber compound towards noise and vibration also become a concern. Thus, the present study focuses on the influence of different rubber types on curing characteristic and dynamic properties in term of static stiffness and dynamic stiffness based on NR, SBR and CR rubber compounds are discussed in this paper.

2. Experimental study

This work aimed to examine the influenced of different types of rubber; NR, SBR, and CR on the curing behaviour and dynamic properties for dynamic application of rubber products.

2.1 Material

There were three types of rubber used in this study; natural rubber (NR), styrene butadiene rubber (SBR) and Chloroprene rubber (CR). NR is Malaysian rubber (SMR 20 CV) supplied by Mardec Berhad. SBR (Nipol 1502) supplied by Centre West Industrial Supplies Sdn. Bhd., and CR (Bayprene 210) supplied by Imago Chemicals Sdn. Bhd. Table 1 shows the formulation used in this study. Other compounding ingredients such as zinc oxide, stearic acid, sulphur, N-cyclohexyl-2-benzothiazole sulfenamide (CBS) and tetramethylthiuram disulphide (TMTD) were purchased from Anchor Chemical Co. Compounding process was carried out in accordance with ASTM D 3184-89 using two-roll-mill.

Table 1. Rubber compound formulation

Material	Part per hundred rubber (phr)
Rubber *	100
Zinc oxide	5
Stearic acid	1
Sulphur	1.5
CBS	2
TMTD	1

* NR/SBR/CR

2.2 Cure characteristics

The cure characteristics were studied using the Mosanto Moving Die Rheometer (MDR 2000) according to the ASTM D 2084. About 4 grams of the respective compound samples were tested at the vulcanization temperature of 165°C.

2.3 Vulcanization process

Rubber button with respectively ~12 mm thickness was compression moulded at 165°C using hot press according to respective cure times, t_{90}, determined with the MDR 2000. For the vulcanization of static and dynamic stiffness, the dimension of the specimen was followed JIS K 6394 in the classification of N_2 type. The hardness test was carried out, followed by standard ASTM D 2240, type Shore A.

2.4 Static and Dynamic stiffness study

For the stiffness test (static and dynamic), the testing was conducted, followed by standard JIS K 6394 with N_2 type dimension specimen. Both tests were tested using a Servopulser testing machine at the test temperature of 23 ± 2 °C. The static stiffness test (Ks), first it undergoes the preload from 0 to 3.2 mm. The Ks value was measured at 1 mm to 2 mm (slope of force over displacement) under the compression mode. For dynamic stiffness test (Kd) also undergo the preloaded test before proceeding to the actual test. The Kd values were measured from sine swept frequency 0 Hz to 100 Hz using the amplitude of ± 0.05 mm under compression and tension mode and the reported value was at the frequency of 100 Hz.

3. Results and discussion

3.1 Cure characteristics and hardness

Figure 1 shows the torque different value (MH-ML) for NR, SBR and CR. It can be observed that SBR gave highest torque different value, followed by NR and CR. The trend might be due to the modulus and rigidity of each rubber. The SBR shows slightly higher than NR due to the pendant group that attaches to the backbone chain which is styrene bulky group. This bulky group of styrene will restrict the mobility of the polymeric chain by preventing the chains from sliding past each other easily, hence resulted in increasing the glass transition temperature thus increase the modulus [4].

In the comparison of synthetic rubber; SBR has higher in molecular weight which is ~158.244 g/mol compared to CR is ~ 88.534 g/mol. Based on Figure 1, the torque difference is about 50 % for SBR and CR. The higher molecular weight will increase the rubber modulus thus increase in torque different [5]. Even though NR has lower molecular weight than CR which is 68.12 g/mol, but the torque different is higher for NR. This might be due to the dirt contained in natural rubber which is produced from cup lump. This can be correlated to the Mooney of viscosity for each rubber which is 67 MU for SBR, 62 MU for NR and 49 MU for CR.

By referring Table 2, the scorch time (ts_2) for each rubber compound, SBR obtained higher ts2 value followed by a CR and NR. The ts_2 is known as the time from the beginning of the test to the time the torque has increased 2 units above ML value. It shows at which point the vulcanization (curing) actually starts. The vulcanization is a process that converts rubber or elastomer into the cross-linked polymer by forms bridges between individual polymer molecules when temperature is applied [6-7]. In this study, the vulcanization agent used is sulphur. The process starts with a polarized sulphur molecule or with a sulphur cation reaction with rubber polymer chain. The breakage of polymer bond is needed in order to form a sulphur cross-linked in the polymer chain with higher energy is required to break the higher bond energy. The presence of the styrene bulky group in SBR required higher energy to break the bond followed by CR which consists of C-Cl bond and C-C bond in NR. In this case, the similar temperature was used for vulcanization process and the rheometer. The longer time observed due to the time required to break higher bond energy. Thus, SBR showed longer scorch time and cure time tc_{90} (time of 90% cure).

Table 2. Cure characteristic for NR, SBR and CR rubber compounds

Cure characteristic	Type of rubber		
	NR	SBR	CR
ts_2, min	1.45	2.10	1.87
tc_{90}, min	1.92	3.07	2.92

IOP Conf. Series: Journal of Physics: Conf. Series **1082** (2018) 012010 doi:10.1088/1742-6596/1082/1/012010

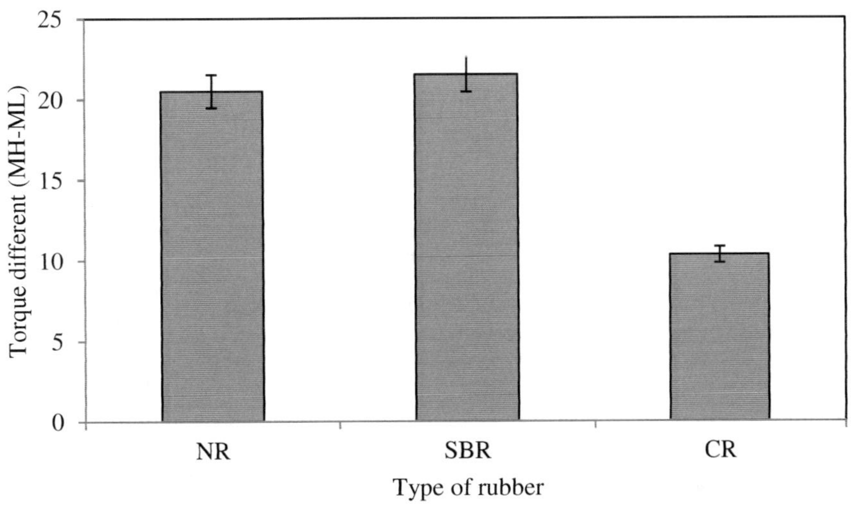

Figure 1. The torque different value for NR, SBR and CR

By referring Figure 2, its show the similar trend as torque different. As known that, hardness measure the resistance of material to deformation under compression load and dependent on ductility, elastic stiffness, plasticity, strain, strength, toughness, viscoelasticity and viscosity. When the compound is hard, it needs to obtain higher stress and lower strain in the stress-strain curve. Due to this, it will show the higher value of the slope (stiffness). The stiffness of the material is the amount needed to deform when force is applied. The higher force will give higher in hardness value. The torque different (MH-ML) can be correlated to the hardness and modulus [8]. Thus, the higher torque different material gives higher value in hardness.

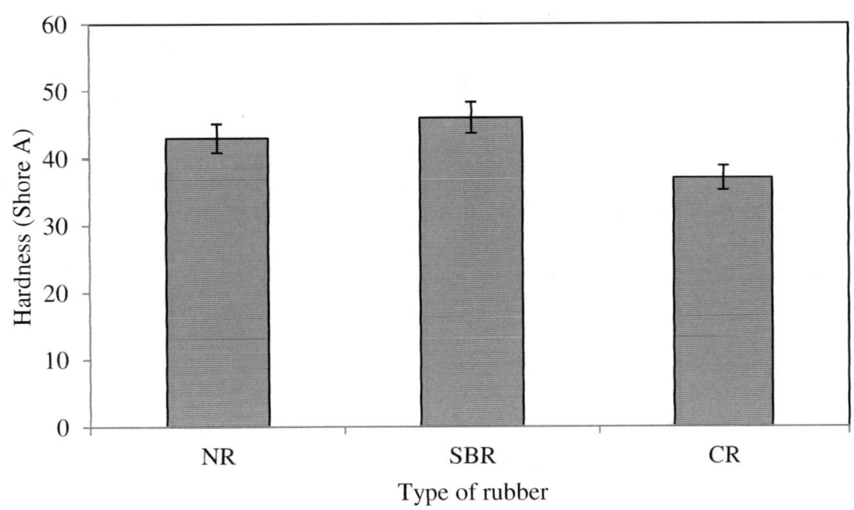

Figure 2. The hardness values for NR, SBR and CR

IOP Conf. Series: Journal of Physics: Conf. Series **1082** (2018) 012010 doi:10.1088/1742-6596/1082/1/012010

3.2 Static, Ks and dynamic, Kd stiffness study

Stiffness can be related to the rigidity of the rubber compound as its measure the force over deflection. The effect of different rubber on static, Ks and dynamic, Kd stiffness are shown in Figure 3. It can be observed that for static stiffness, Ks, the trend is clearly followed the hardness result. The higher value is observed for SBR followed by NR and CR. This might be due to the similar testing mode which is in compression mode for both tests. As discussed before, the bulky pendant group of styrene in SBR will increase the rubber modulus and rigidity, while for NR the linearity of the polyisoprene molecular structure gave higher Ks value compared to the CR. Besides that, the molecular weight of rubber also the main contribution for the rubber modulus [4]. Thus, the higher in molecular weight of SBR that consists of pendant bulky group gave higher in Ks value followed by NR and CR.

However, for dynamic stiffness, Kd test, (refer Figure 3) the result shows that the higher Kd value obtained for SBR followed by CR and NR. It can be observed that the Kd trend does not follow the Ks and hardness trend. This might be due to the different mode test of Ks and hardness compared to Kd test which is compression and tension mode (cyclic loading condition) test at the certain frequency. The dynamic stiffness calculation is based on the combination of viscous and elastic stiffness [9]. The dynamic stiffness is a particular feature of the stress-strain behaviour of rubber. This behaviour is also known as the Payne effect [10]. The effect is observed under cyclic loading conditions with small strain amplitude and is manifest as a dependence of the viscoelastic storage modulus on the amplitude of the applied strain.

Physically, the Payne effect can be attributed to deformation-induced changes in the materials microstructure for example to breakage and recovery of weak physical bond linkages. It depends on the filler content on the material and disappears for the unfilled elastomer. For this study, the rubber-rubber interaction only being considered because of the used of only unfilled rubber compounds.

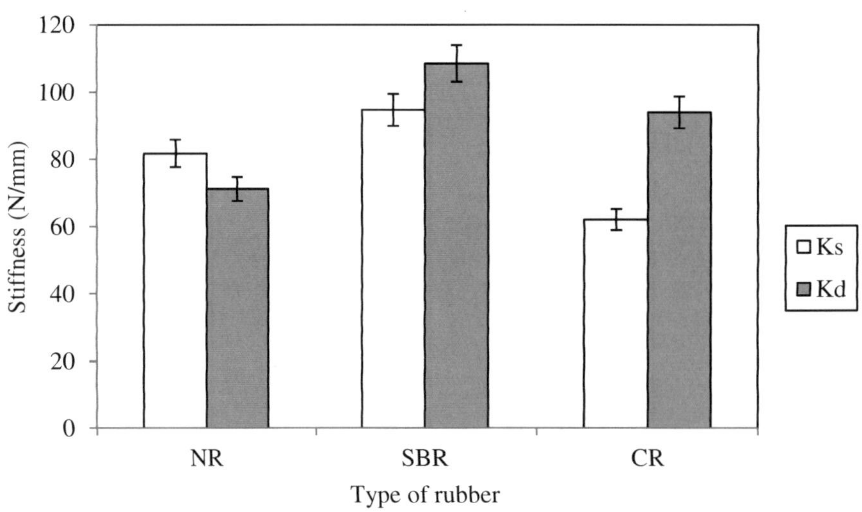

Figure 3. The effect of static, Ks and dynamic, Kd stiffness for each rubber

The higher of bond energy of the styrene bulky group attach at a rubber backbone chain in SBR required higher energy to break the bond and this will contribute the higher value of Kd. While for CR is due to the side chain of C-Cl with the bond energy of 339 kJ/mol. For NR, the polyisoprene structure only consists of carbon and hydrogen, which also similar to SBR and CR structure. Due to this bond energy in rubber chemical structure, the higher Kd value obtained for SBR is due to the styrene bulky group bond energy, followed by CR and NR. Thus, the stronger the bonding energy of rubber polymer chain, the higher the rubber stiffness.

4. Conclusion

In conclusion, the curing characteristic which is torque different (MH-ML) can be used to predict static stiffness, Ks value as it shows a similar trend. This also can be correlated to the hardness of rubber compound. While, for dynamic stiffness, Kd, the main contribution is the molecular structure of rubber as it is related to the viscoelasticity of rubber as stated in the Payne effect. In this study, Payne effect focuses on the rubber-rubber interaction. Higher the bond energy interaction in rubber molecular structure will need higher energy to break the bond structure. Higher energy can be referred to the force applied. By referring back to the stiffness calculation which is the force over deflection, thus it gives higher in Kd value.

References

[1] Sjoberg M 2002 *Dynamic Properties of Rubber Isolator*
[2] Cardillo R M 1964 *Journal of Applied Polymer Science* 8 53
[3] Hobaica E C 1980 *Journal of the Acoustical Society of America* 67 S23
[4] Fenoglio D J, Foster J J 1990 *Journal of Polymer Science Part A: Polymer Chemistry* 28 10
[5] Nicholson L M, Whitley K S, Gates T S, Hinkley J A, 2000 *NASA Langley Research Center,* Hampton VA 23681
[6] Flory P J 1953 *Principle of Polymer Chemistry* Cornell University
[7] Oae A 1977 *Organic Chemistry of Sulfur* Plenum Press, New York
[8] Ismail H, Nordin R, Noor A M *Polymer Testing* 565-569
[9] Frampton R C *The Dynamic Testing of Elastomer*
[10] Luo W, Hu X, Wang C, Li Q 2010 *International Journal of Mechanical Sciences* 52 168

IOP Conf. Series: Journal of Physics: Conf. Series **1082** (2018) 012011 doi:10.1088/1742-6596/1082/1/012011

The Effect of Alkali Treated Native Sago Starch Concentration on Mechanical Properties of Natural Rubber Latex Composite Films

M.F. Ab Rahman[1], S.R. Yahya[1], and A.R. Azura[1*]

[1] School of Materials and Mineral Resources Engineering, Universiti Sains Malaysia, Engineering Campus, 14300 Nibong Tebal, Pulau Pinang, Malaysia.

[*] Corresponding author: srazura@usm.my

Abstract. One of promising methods to improve starch dispersion and/or enhance interfacial strength between starch and rubber matrix is starch treatment using alkali. In this study, the effect of alkali treatment to prepare filler dispersion of native sago starch on mechanical properties of NR latex compounds was investigated. The alkali dispersed native sago starch was prepared with different percentages of total solid content (5, 10, 15 and 20%) at 10% w/w loading of NR latex compounds. Result showed the optimum mechanical properties of alkali dispersed NSS filled NR latex films were achieved at 10% TSC; with 29.74 MPa of tensile strength.

1. Introduction

In recent years, renewable resource fillers such as starch have gained major interest in polymeric composite materials. Starch is a non-toxic biopolymer, inexpensive material, abundant sources, naturally renewable polysaccharide and low environmental impact in which it is readily biodegradable in soil [1 – 2]. Starch has good potential for renewable fillers due to partly crystalline structures, offering interesting properties of polymer composites.

However, a finely dispersed starch in natural rubber (NR) matrix is difficult to achieve due to several limitations of native starch such as hydrophilic characteristic, large particle size and poor process ability [3 – 4]. These limitations lead to the poor mechanical properties of starch/NR latex composite. According to Liu et al. [5], a direct filling of native starch powders causes deteriorated mechanical properties of NR with decreased in tensile strength, tear strength and elongation at break.

Preparation of filler dispersion using alkali is a promising method to improve the distribution of native sago starch (NSS) in a polymer matrix, with proven improved the mechanical properties of NR latex films. A strong alkali is able to disrupt the molecular arrangement within the starch granules resulting from the disruption of intermolecular and intramolecular hydrogen bonding between the water and starch molecules, thus enhancing water solubility of starch [6 – 7].

This study investigates the effect of different total solid contents (TSC) percentages of alkali dispersed NSS (5, 10, 15 and 20%) at 10% w/w filler loading on mechanical properties of NR latex films. Preliminary studies on NSS dispersion have been conducted to characterize the state of dispersion.

2. Experimental

2.1 Materials

NR latex was purchased from Zarm Scientific and Supplies Sdn. Bhd. with the initial properties of dry rubber content (DRC) of 61.10%, TSC of 62.20%, mechanical stability time (MST) of 1213 s, and volatile fatty acid number (VFA no.) of 0.048. Meanwhile, sago starch was obtained from Sago Link Sdn. Bhd. with amylose and amylopectin content ranging from 22 – 31.7% and 68.3 – 78%; respectively. Other compounding ingredients such as diethyldithiocarbamate (ZDEC), zinc oxide (ZnO), potassium oleate, potassium hydroxide (KOH), antioxidant and sulphur were obtained from Bayer (Malaysia) Sdn. Bhd.

2.2 Preparation of alkali-based filler dispersion

The alkali treatment was carried out by mixing NSS with 100g of 5.35% KOH. The mixture was stirred at 200 rpm for 30 minutes before leaving it for two days under room temperature to gelatinise completely. All pH of filler dispersion was in the range of 12.60 – 12.80.

2.3 Preparation of NR latex compound and films

The latex compound was diluted to 50% of TSC with distilled water. The preparation of NR latex compound was carried out in accordance with a previous study [8]. All the ingredients were mixed in the beaker and were constantly stirred for two hours using Wise Stir HS-300 mechanical stirrer at stirring speed of 270 rpm under room temperature. The compound was pre-vulcanized at temperature of 70^0C in a water bath and until chloroform number 2 (degree of vulcanization of pre-vulcanized NR latex) has achieved. After that, the compound was left overnight under room temperature for maturation purposes. The aluminium plates were dipped into the maturated compound and cured in oven for 20 minutes at 100^0C. The films were stripped by the assistance of calcium carbonate powder.

2.4 Transmission Electron Microscope

The diluted dispersions containing alkali dispersed fillers were deposited onto glow discharged carbon coated grids and allowed to dry. These grids were analyzed using Philips CM 12 CRYO Transmission Electron Microscope (TEM).

2.5 Fourier Transform Infrared Spectrometry

The Fourier Transform Infrared Spectrometry (FTIR) spectrometer used in this study was Perkin Elmer, Model Spectrum One. Analysis was carried out under room temperature within a typical wave number range of 4000 cm^{-1} to 550 cm^{-1} under transmission mode.

2.6 Films Mechanical properties test

Tensile test was carried out according to ASTM D412; with the crosshead speed of 500 mm/min by using Instron Machine (Model 3366, USA). The films were cut into dumbbell shape and the recorded results are based on the average of 5 samples.

3. Results and discussion

3.1 Morphological analysis of filler dispersion

Figure 1 (a)-(d) show the states of NSS at different TSC percentages after dispersed in KOH, which were characterized by TEM. Referring to Figure 1 (a), it can be observed that the presence of alkali has destroyed the initial granular form of NSS. The alkali treatment of native starch basically caused starch gelatinization. Gelatinization is the swelling process of starch granule, followed by disruption of granule structures [9]. The distributions of small unreacted core particles which are less than 50 nm were detected in the 5% TSC of alkali dispersed NSS. At 10% TSC of alkali dispersed NSS (Figure 1 (b)), the gel network formation and the increased size of unreacted core particles of the NSS were observed. During alkali treatment, a majority of hydroxyl groups in the anhydrous glucose units are ionized at high pH and therefore, the agglomeration of starch chains and paste retrogradation can be retarded [7].

Starch gelatinization under strong alkali conditions is a heterogeneous fluid-particle reaction in which a liquid (water) reacts with a solid (NSS) to produce a paste, in the presence of an alkali (KOH). Strong alkali weaken molecular network within the starch granules by disrupting hydrogen bonds. This will allow further hydration and irreversible granule swelling which cause an increased in viscosity of the paste [10 – 11]. The starch gelatinization phenomenon was observed by the appearance of gel network, which held together by the formation of local regions of crystallinity that form physical rather than chemical crosslinks.

The ability of strong alkali at low concentration to leach out the amylose in the starch granules particular was reduced by the increase of TSC percentages of NSS where the enhancement of size of unreacted core particles was observed in Figure 1 (c)-(d). The unreacted core particles also agglomerated at 20% TSC of alkali dispersed NSS Figure 1 (d). The appearance of the crystalline region also became less significant which also indicating the increased in amorphous regions in the gel network at high TSC of alkali dispersed NSS. This result also supported by FTIR result (Figure 2).

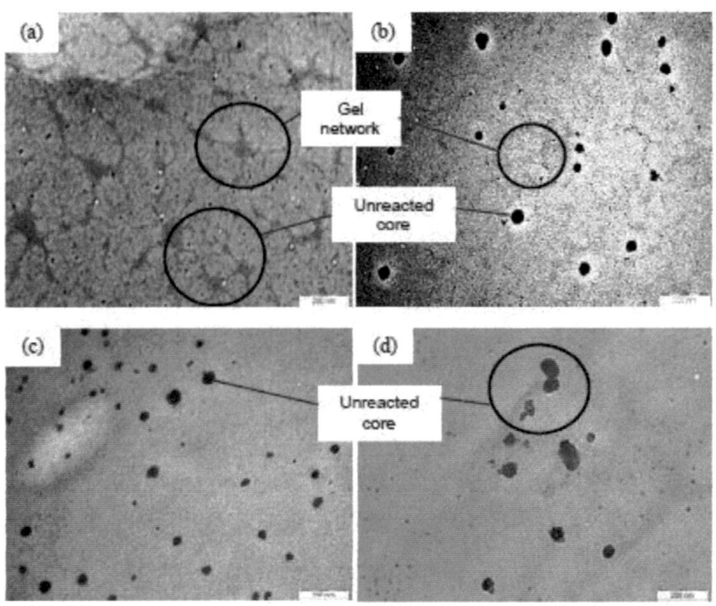

Figure 1. TEM characterization of alkali dispersed NSS at (a) 5% TSC, (b) 10% TSC, (c) 15% TSC and (d) 20% TSC (scale: 200 nm)

3.2 FTIR analysis

The changes in the chemical characteristics of NSS before and after dispersing in alkali medium at different TSC percentages were analysed by FTIR as given in Figure 2. After alkali treatment, all NSS at different TSC percentages showed the transmittance bands became stronger at 3420 – 3250 cm^{-1} and 1640 cm^{-1} due to the increased in O-H groups. As mentioned before, a strong alkali is able to interrupt the molecular arrangement within the starch granules to cause disruption of intermolecular and intramolecular hydrogen bonding between the water and starch molecules, thus enhancing water solubility of starch [7, 12]. The disappearance of bands at 2990 – 2850 cm^{-1}, 1417 cm^{-1}, 1338 cm^{-1} and 998 cm^{-1} were detected which indicated the alteration of CH_2, C-OH and C-O-C of alkali dispersed NSS. The appearance of a new band at 1020 cm^{-1} was observed and became stronger with the increased in TSC percentages of NSS after alkali treatment. Absorbance at 1020 cm^{-1} is assigned to the vibration of C-O-H deformation and is also related to the amorphous region in gelatinised starch [13 – 15].

Overall, 5% TSC of alkali dispersed NSS exhibited the significant disappearance of the main bands of glycosidic linkages which was indicated the large particle size of starch granules were easily destroyed by alkali at very low starch content. The bands at 1150 cm^{-1} and 1076 cm^{-1} were weakened as well as shifting to 1152 cm^{-1} and 1083 cm^{-1} after alkali treatment. The increased in TSC percentages gave insignificant changes on the bands at 1152 cm^{-1} and 1083 cm^{-1}. The transmittance bands at 1152 cm^{-1} and 1083 cm^{-1} are assigned as C-O, C-C and O-H bonds stretching of C-O-C glycosidic linkages [15– 16]. All filler dispersions for the bands at 860 cm^{-1}, 765 cm^{-1} and 709 cm^{-1} were disappeared and replaced by a broad band at 925-550 cm^{-1}. These results indicate the large particle size of starch granules were destroyed and changed into the gel network of alkali dispersed NSS.

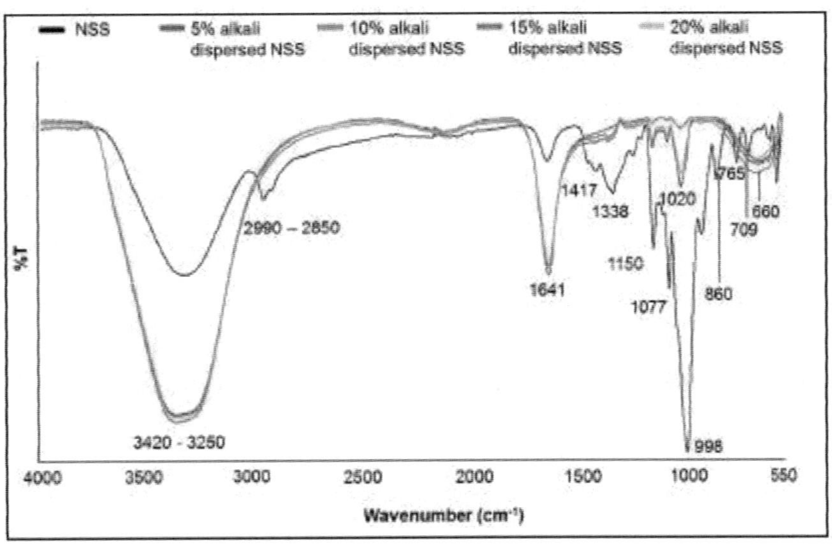

Figure 2. FTIR spectra of alkali dispersed NSS at different TSC percentages

3.3 Mechanical properties of NR latex films

In Figure 3, the stress-strain curves are presented in order to understand the influences of TSC percentages of alkali dispersed NSS on tensile properties of NR latex films. Stress-strain curves show that all curves are divided into three stages. The first stage showed the stress increased gradually at small strains. The second stage exhibited the stress increased at a uniform rate under moderate strains. At the third stage, the stress increases drastically at high strains. This stress behavior attributed to the strain-induced crystallization; endows these rubbers with good mechanical properties and resistance to crack growth [17].

At first stage, it can be observed that the addition of alkali dispersed NSS at different TSC percentages showed comparable stress of NR latex films before inclined from 150% up to 600% of strain compared to the unfilled NR latex films. This indicates that the addition of alkali dispersed NSS basically increases the modulus or stiffness at higher elongation. The addition of 5% TSC of alkali dispersed NSS lowered the tensile strength as well as elongation at break due to ineffective quantity as filler in NR matrix.

On the other hand, 10% TSC shows significant tensile strength improvement and higher elongation at break values. Further increase in TSC percentage, NSS exhibits significant deterioration in tensile strength due to the poor filler distribution of highly viscous filler dispersions in NR latex matrix.

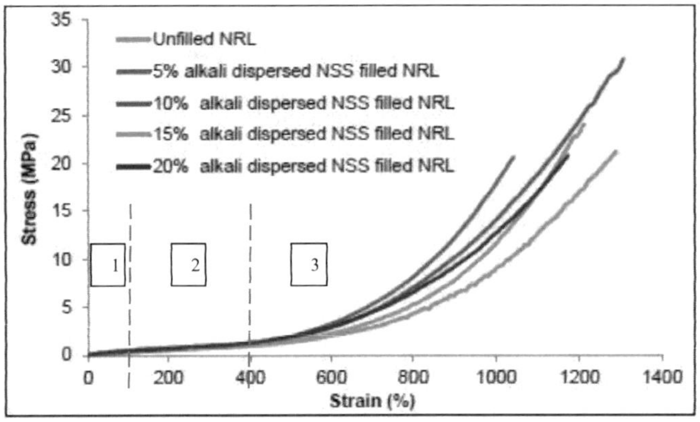

Figure 3. The stress-strain curves of alkali dispersed NSS filled NR latex films at different TSC percentages

4. Conclusion

In conclusion, low TSC of alkali dispersed NSS exhibits good dispersion state, whereas optimum mechanical properties of the NSS/NR latex films can be achieved at 10% of TSC.

Acknowledgements

This work was granted by a FRGS research grant (grant no: 203/PBAHAN/6071353) and laboratory equipment provided by the School of Materials and Mineral Resources Engineering, Universiti Sains Malaysia.

References

[1] Liu D, Wu Q, Chen H and Chang PR 2009 *Journal of Colloid and Interface Science* 339 117
[2] Chi H, Xu K, Wu X, Chen Q, Xue D, Song C, Zhang W and Wang P 2008 *Food Chemistry* 106 923
[3] Senna MM, Mohamed RM, Shehab-Eldin AN and El-Hamouly S 2012 *Journal of Industrial and Engineering Chemistry* 18 1654
[4] Wu YP, Ji MQ, Qi Q, Wang YQ and Zhan LQ 2004 *Macromolecular Rapid Communications* 25 565
[5] Liu C, Shao Y and Jia D 2008 *Polymer* 49 2176
[6] Chin SF, Azman A and Pang SC 2014 *Journal of Nanomaterials* 2014 1
[7] Han JA and Lim ST 2004 *Carbohydrate Polymers* 55(2) 193
[8] Afiq MM and Azura AR 2013 *Journal of Chemistry and Chemical Engineering* 7 137
[9] Nur A and Purwiyatno H 2010 *International Food Research Journal* 17 961
[10] Osoka EC, Njiribeako IA, Olebunne FL and Kamalu IC 2012 *International Journal of Science and Engineering Investigations* 1(9) 20

[11] Mohamed A, Jamilah B, Abbas KA, Abdul Rahman R and Roselina K 2008 *American Journal of Agricultural and Biological Sciences* 3(4) 639
[12] Chin SF, Azman A, Pang SC and Ng SM 2014 *Journal of Nanomaterials* 2014 1
[13] Mina J, Valadez-González A, Herrera-Franco P, Zuluaga F and Delvasto S 2011 *Dyna* 78 (166) 166
[14] Blaszczak W 2007 In: Yuryev V, Tomasik P, Bertoft E (eds) Starch: Achievements in Understanding of Structure and Functionality. Nova Science Publishers, Inc., New York, 179
[15] Liu Q, Charlet G, Yelle S and Arul J 2002 *Food Research International* 35 397
[16] Van Soest JJG, Tournois H, De Wit D, and Vliegenthat JJG 1995 *Carbohydrate Research* 279 201
[17] Nie Y, Gu Z, Wei Y, Hao T, and Zhou Z 2017 *Polymer Journal* 1

IOP Conf. Series: Journal of Physics: Conf. Series **1082** (2018) 012012 doi:10.1088/1742-6596/1082/1/012012

Effect of Glycerol Compositions on Protein Based Thermoplastic from Fish Waste

A. Siti Amirah[1] and K.I. Ku Marsilla[1*]

[1]School of Materials & Mineral Resources Engineering. Engineering Campus, Universiti Sains Malaysia,Seri Ampangan, 14300 Nibong Tebal, Pulau Pinang, Malaysia.

Email : ku_marsilla@usm.my

Abstract. The main objective of this work was to investigate the effect of glycerol at different compositions on physical properties, morphology and water absorption of protein based thermoplastic from fish waste. Fish waste containing fish head, bones and fins were collected from local fish product processing plant before further cooking, pressing, drying and milling. Compound formulation was done using twin screw extrusion and compression moulding. Samples at different glycerol compositions (Control, 20 pph and 35 pph) were characterized using Fourier Transform Infrared Spectroscopy (FTIR), Scanning Electron Microscopy (SEM) and water absorption. The addition of glycerol showed two broad bands between 1500-1700 cm^{-1} and 3000-3500 cm^{-1} which correlates with the amount of α-helix and β-sheets structure of the protein. The fracture surfaces at control changed from uneven surfaces with large particles to smooth surfaces. Plastic deformation was observed at control and 20 pph but was absence at high glycerol content. Increasing the glycerol content was found to influence the water resistance properties of fish protein by reducing the absorption capability due to amount of α-helix and β-sheets structure.

1. Introduction

Environmental pollution caused by petroleum-based plastics is a serious problem that becoming a threat to human life, animals and atmosphere. Their increasing use for one off or short-run applications has resulted in a great concern and interest in replacing petroleum based plastics with renewable biodegradable polymer materials. Natural biopolymers such as starch, cellulose, chitin, polylactic acids, soy protein, collagen, and gelatin have been successfully transformed into thin, transparent and flexible material of biodegradable films [1, 2]. Therefore, renewable raw materials such as protein extracted from agricultural, marine, animal or microbial sources can be degraded by the environment. For example, plant protein materials such as soy protein[3], potato protein [4], wheat gluten [5], and sunflower protein [6] have been studied among researchers. Besides that, animal protein such as blood meal [7], silk protein fiber [8], and fish myofibrillar proteins[9] also has been successfully being transform into thermoplastics. The advantages of biopolymer such as sustainable sources of raw materials and they are mainly environmental. Protein consists of repeating monomer similar to synthetic polymer but may contain up to 20 amino acids monomer that form polypeptide chain. Therefore, proteins are renewable, biodegradable/edible resource with great potentials to improve the quality and stability large range of agriculturl products.

However, there are some limitations because protein has lower mechanical properties as compared to synthetic materials. In previous study, waste blood has shown its ability to perform as thermoplastic polymers with high mechanical properties [10, 11]. It is important to control the material moisture content in order to ensure consistent material properties. In other research, fishmeal from Atlantic salmon has been successfully compression moulded and plasticized with glycerol [1]. Glycerol was added to either function as lubrication or increasing free volume. Moreover, glycerol has high boiling point and good stability and is regarded as one of the most efficient plasticizers for fish protein thermoplastics. Addition of plasticizer such as glycerol made soy protein thermoplastics and improved the flexibility by decreasing the tensile strength and modulus but increased its elongation [12]. Similar effect has also observed for sunflower protein that has been developed into thermo-moulded films [6]. During thermoplastic processing, protein molecule chains slide past each other and solidify to form a shape when the temperature is reduced in order to achieve flowing protein. The most common processing method for biopolymers is extruder. However, twin screw extruder offers more efficient mixing and conveying materials [10]. Some research of biopolymer uses internal mixer instead of conventional extruder [13]. In this study, fish waste from sardine is employed as raw material to evaluate whether this significant waste from fish processing industry can be converted into environmentally friendly bio plastics. This study also involves the influence of chemical additives and evaluation of a range of processing techniques such as compression moulding and extrusion.

2. Methodology

2.1 Materials

The fish waste used (Sardine) was supplied by Protigam Food Industries Sdn. Bhd (Butterworth). The fish waste is produced from fish products such as bones, heads, guts and fins. It has 46.4% of protein content (measured by Kjedahl method), 1% of fats and 3.29% of moisture. Glycerol and other chemicals including technical grade sodium dodecyl sulphate (SDS), analytical grade sodium sulphite (SS), urea, and glycerol were obtained from Sigma Aldrich.

2.2 Sample preparation

2.2.1 Fish Waste powder preparation. Fish waste was cooked at 90 °C for 20 min with a fish/water ratio of 1:1(w/w). A solid was separated by filtration and was pressed into a cake for further dried until it reach a constant weight at 90 °C and subsequently milled into a fine powder. Finally, the fish waste powder was sieved to an average particle size of 600 μm.

2.2.2 Preparation of Fish Protein Thermoplastic. Samples were prepared by dissolving urea, SDS and SS in water. The solution was heated until the temperature reached 50 °C followed by blending with fish waste powder in a high speed mixture for 5 min. Glycerol was added to the mixture and blended for another 3-4 min. The mixture was stored for at least 24 h prior to mixing process. Sample 1 was formulated without glycerol known as Control, and Sample 2 and 3 were formulated with varying amounts of glycerol as a plasticizer, and water as a secondary plasticizer, as shown below in Table 1. The samples were further processed using twin screw extruder.

IOP Conf. Series: Journal of Physics: Conf. Series **1082** (2018) 012012 doi:10.1088/1742-6596/1082/1/012012

Table 1. Fish protein with varying concentrations of plasticizer

No	Formulations composition	Sample 1(pph)	Sample 2(pph)	Sample 3(pph)
1	Fish Waste Powder	100	100	100
2	Urea	10	10	10
3	SS	3	3	3
4	SDS	3	33	3
5	Water	20	20	20
6	Glycerol	Control	20	35

2.2.3 Extrusion. Sino Alloy compact co-rotating twin screw extruder-PSM30 was used to extrude fish protein mixtures. The screw diameter was 31.2 mm with a length to diameter ratio of 40. The extruder was fitted with a single 10 mm circular die and the temperature processing from zone 1 until zone 9 as shown below in Figure 1. The extrusion is performed under screw speed 100 rpm and then the extrudate was crushed.

1	2	3	4	5	6	7	8	9
90 °C	95 °C	100 °C	100°C	110 °C	110 °C	115 °C	115 °C	120 °C

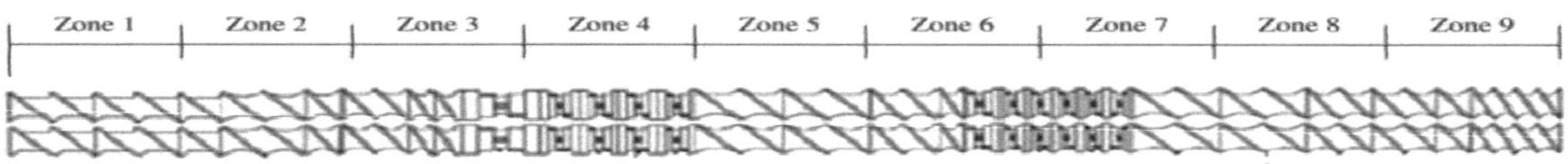

Figure 1. Twin-screw extruder. Screw and temperature profile [14]

2.2.4 Compression- Moulding. A hydraulic press (GoTech Test Model KT-7014 A) fitted with heating elements on the top and bottom plate was used for compression moulding. A mould was employed and a Teflon sheet was placed on the bottom of the mould to prevent sticking. Processing conditions such as temperature, preheat time, compression time, cooling time and pressure were optimized. Fish protein thermoplastics was generally pre-heat about 4 min, compressed time for 5 min, cooling time about 2 min each at 120 °C and pressure 9.7 Mpa. Then, the sample was straight to analysed. The samples was performed according to the ASTM D638-02a [15].

2.3 Characterizations

2.3.1 Fourier Transform Infrared Spectroscopy (FTIR). FTIR analysis was performed using a Perkin Elmer spectrophotometer, employing 16 scans at resolution from 4000 cm^{-1} to 400 cm^{-1}. The background was scanned prior to each sample scan and automatically subtracted from the spectrum by the FTIR software.

2.3.2 Morphology. The morphology of the thermoplastic fracture surface was investigated using scanning electron microscopy (SEM) (Hitachi S3000N) at an acceleration voltage of 5 kV. A small fragment of sample was placed on aluminium stubs with double sided tape. The stubs were coated with carbon paint and placed in an ion sputter which further coated the specimen with platinum prior to SEM analysis.

2.3.3 Water Absorption. The testing was performed according to ISO 62:2008 [16]. The water absorption capacity of each sample was studied by immersing a known mass of sample into distilled water at room temperature. The sample was periodically removed from the distilled water and reweighed until constant film mass was attained. The test was performed within 3 days in triplicate.

3. Results and discussion

3.1 Fourier Transform Infrared Spectroscopy (FTIR)

The FTIR spectrum of fish proteins for Control, 20 pph and 35 pph of glycerol are show in Figure 2. The appearance of a secondary peak at 3344 cm^{-1} suggests the formation of primary amine stretches. It was obvious that the peaks in the region 3100-3500 cm^{-1} shows increased in transmittance percentage and broadening which indicated the presence of NH-stretching with hydrogen bonding. The transmittance increased about 3% (20 pph) and 11% (35 pph) as compared with between peaks at control. These broadening were related to the possible extra interactions that arise between the plasticizer and the film structure. Generally, amide I (1600-1700 cm^{-1}) related to carbonyl group of proteins, amide II (1480-1575 cm^{-1}) related to vibrational groups N-H and stretching vibrations of C-N groups and Amide III (1229-1301 cm^{-1}) referred to vibration in plane C-N and N-H groups of bound amide[17]. Amides I band (highlighted in box) were found increased in absorbance and broadening about 1% and 3% when the addition of glycerol. There are 2 peaks of vibrational bands each sample in the Amide I (1600-1700 cm^{-1}) which could be associated with extended in β-sheets structure. This indicates that the interactions between polypeptide chains are located in β-structure regions and this is believed affected by temperature and concentration of molecules in forming hydrogen bonds interactions [18].

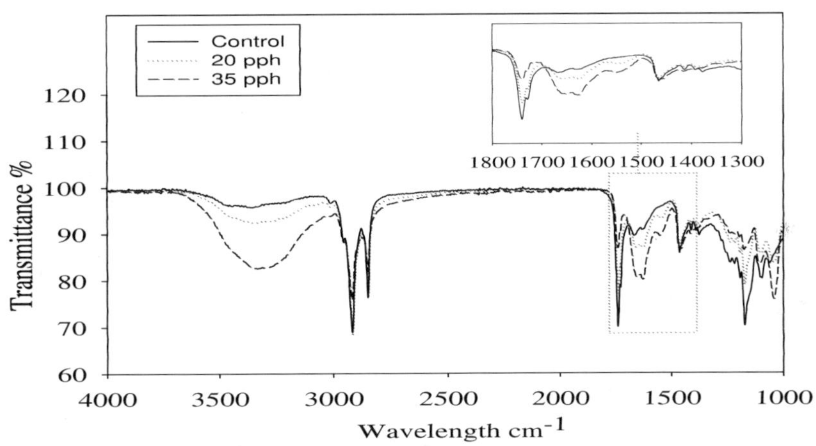

Figure 2. FTIR spectra of fish protein with different composition of glycerol (Control, 20 and 35 pph)

3.2 Scanning Electron Microscopy (SEM)

Figure 3 shows SEM images of fracture surface of fish protein thermoplastics with increasing composition of glycerol. It can be seen in Figure 3(a), at control shows an uneven surfaces with large particles and started to reduce when the addition of 20 pph of glycerol (Figure 3(b)). As the glycerol increased to 35 pph, an even and homogeneous surface without large particles has been observed in Figure 3(c). Although, the plastic deformation behaviour has been observed in Figure 3(a) and it shows more obvious plastic deformation when the increasing of the glycerol in Figure 3(b). This is because without plasticizer, the mixing was poor and brittle. In addition, the mixtures at control had a rough texture Figure 3(a) while the higher glycerol content had a smooth texture Figure 3(b). A material with plastic properties has good mechanical properties such as high in elongation and low in tensile strength. The addition of 35 pph of glycerol had a combination of both textures but with a higher proportion of smooth texture Figure 3(c). These results suggest that the increase of glycerol content in fish protein mixtures results in a better mixing between protein and plasticizer, which in turn, yields smaller particles and in a more smooth texture of the mixture. One possible explanation for this was that when plasticizer content was increased there were strong protein-protein interactions disrupted and the more partial unfolding such as loss of quaternary structure and consequently, the higher swelling capacity of the fish protein network.

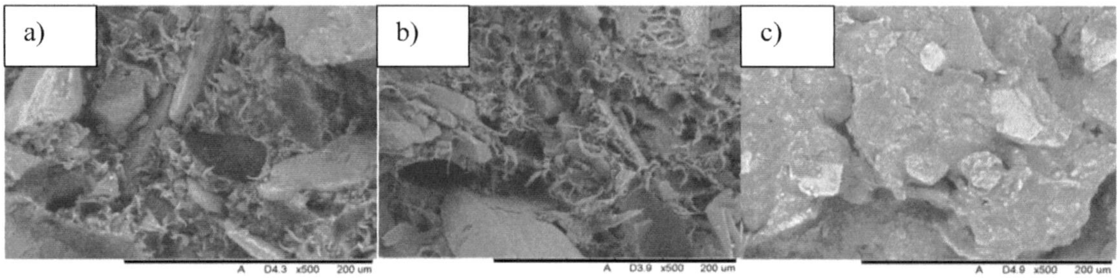

Figure 3. SEM images of fish protein with different composition of glycerol (a) Control, (b) 20 pph and (c) 35 pph

3.3 Water Absorption

Water absorption is a very important factor for many applications of bio-plastic products. The results of water absorption experiments, performed at the room temperature over 3 days are shown in Figure 4. It can be seen that the fish protein thermoplastic showed an increase of water absorption at the first stage. The absorption of all the thermoplastics was rapid during the initial 5 h followed by slower absorption rate until they reach equilibrium. This phenomenon may be due to the abundant active hydroxyl groups of glycerol which were vacant at the initial stage of absorption process. At control, the water absorption of fish protein thermoplastics has been improved to about 33 % at 72 h which is day 3. However, the addition of glycerol at 20 pph and 35 pph has reduced the water resistant of fish protein to about 28 % and 26% at day 3. It can be observed that the addition of plasticizers effectively decreased the swelling and water retaining capacity of fish protein thermoplastics. This could be explained by an increase in glycerol content caused the interactions with polypeptide chains became complex which lead to a less organised network.

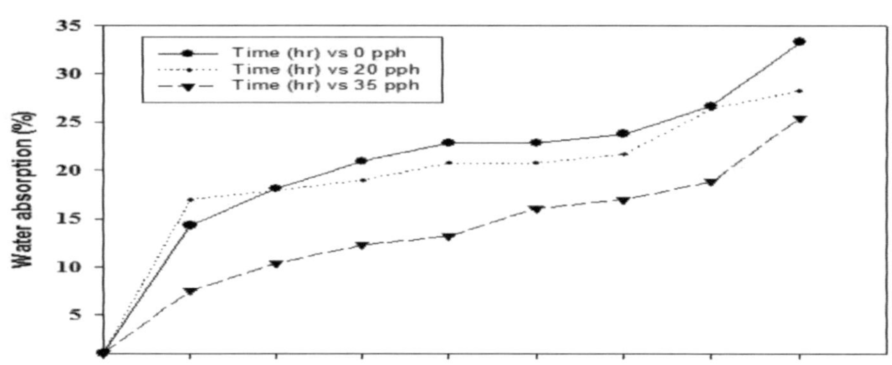

Figure 4. Water absorption of fish protein with different composition of glycerol (Control, 20 and 35 pph)

4. Conclusion

This study showed clearly with reference to the FTIR spectra, which fish proteins shown that intermolecular β-structures were induced by the additions of plasticizers. Plasticizers and heating are the main effect to significantly increase the intensity of the β-band. It shows that the interactions between polypeptide chains are located in β-structure regions and maintained by hydrogen bonds.

Without plasticizers, the fish protein films are brittle with many plastic deformation and uneven surfaces with large particles. Hence, the introduction of plasticizers helps to overcome brittleness and produces smooth surfaces with great mixing process. Moreover, increase in plasticizer effectively decreased the swelling and water retaining of fish protein thermoplastics.

Acknowledgement
The authors wish to acknowledge the Short Term Research Grant and Fundamental Research Grant Scheme (FRGS) for their valuable contributions to this study.

References

[1] J. Svenson, A. S. Walallavita, and C J R Verbeek 2013 *Waste and Biomass Valorization* 4 147
[2] P. Bergo and P J A Sobral 2007 *Food Hydrocolloids* 211285
[3] González A.and Alvarez Igarzabal C I 2013 *Food Hydrocolloids,* vol. 33, pp. 289-296, 2013/12/01/ 2013.
[4] Du Y, Chen F, Zhang Y, Rempel C, Thompson M R, and Liu Q 2015 *Journal of Applied Polymer Science,* 132
[5] Song Y and Zheng Q 2008 *Bioresource Technology* 99 7665
[6] Orliac O, Rouilly A, Silvestre F, and Rigal L 2003 *Industrial Crops and Products* 18 91
[7] Verbeek C J and van den Berg L E 2012 *Journal of Applied Polymer Science* 125
[8] Prachayawarakorn J and Hwansanoet W 2012 *Fibers and Polymers* 13 606
[9] Cuq B, Gontard N,. Cuq J L, and Guilbert S 1997 *Journal of Agricultural and Food Chemistry* 45 622
[10] Verbeek C J R, Lay M C, and Bier J M 2013 *CHEMECA 2013: Challenging Tomorrow*
[11] Marsilla K K and Verbeek C 2013 *International Journal of Chemical Engineering and Applications* 4 106
[12] Yang Y and Reddy N 2012 *Industrial crops and products* 36 116
[13] Felix M, Romero A, Martín-Alfonso J, and Guerrero A 2015 *Composites Part B: Engineering* 78 291
[14] Silva A, Ferreira Cipriano, Thomé da Asilva n H M da F, Cristina M, Rocha C G, Maria F, and Monteiro, G 2014 *Thermal, Rheological and Morphological Properties of Poly (Lactic Acid) (PLA) and Talc Composites* 24
[15] Standard test method for tensile properties of plastics (ASTM D638-02a) 2003 *American Society for testing and Materials In: Annual Book of ASTM standards* 7.01 46
[16] I 62:2008(E) 2008 Plastic- Determination of Water Absorption
[17] Kong J and Yu S 2007 *Acta Biochimica et Biophysica Sinica* 39 549
[18] Gueguen J, Viroben G, Noireaux P, Muriel Subirade 1998 *Journal of Industrial Crops and Products* 7149

Studies on the Influenced of Dextrin Filler on the Mechanical Properties of Carboxylated Nitrile Butadiene Rubber (Xnbr) Films

Daud, S[1], Afiq, M.M[1] and A.R. Azura 1*

[1]Universiti Sains Malaysia, Engineering Campus, Penang, Malaysia

Email: srazura@usm.my

Abstract. Dextrin used in this experiment is a product of acidified sago starch being heated at high temperature. The use of dextrin as filler in latex films is to produce environmental friendly products. The synthesized dextrin dispersion was prepared by mechanical (ball milling). Dextrin (DX) is compounded with carboxylated nitrile butadiene rubber (XNBR) latex. The DX/XNBR films were prepared by incorporating dextrin at filler loading of 0, 5, 10, 15 and 20 phr into XNBR latex. The dextrin was then mixed with XNBR compound as a filler and a film was prepared by coagulant dipping technique The effect of dextrin loadings on tensile properties of XNBR films were investigated. The results indicated that the tensile properties such as tensile strength, elongation at break, and crosslink density decreased with increasing filler loading. The changes of functional groups present in native sago, dextrin and DX/XNBR film were characterized using Fourier Transform Infra- Red (FTIR). The morphological properties of tensile fractured surfaces of the samples were observed using Scanning Electron Microscope (SEM).

1. Introduction

Carboxylated nitrile butadiene rubber (XNBR) latex is one of the largest contributors to world synthetic gloves. It is produced via emulsion polymerization technique where acrylonitrile, and 1, 3 butadiene are polymerized in an aqueous emulsion with presence of emulsifier (Lenko et al. 2013. Carboxylated nitrile butadiene rubber (XNBR) like other types of synthetic and natural latex, are not biodegradable due to nature of latex itself and other materials added to it such as conventional fillers to name a few carbon black, clay and silica. Degradation of XNBR mainly by means of physical processes such as thermal induction, photo-oxidation, ozonolysis, and radiation during the degradation process (Bhattacharjee et al. 1993). This causes serious threat to the environment. This results in a rise of awareness among researchers and people around the globe demanded for green-based products.

Demand for green based polymeric materials are increasing annually, due to increase awareness on environmental issue. Owing to this, researchers has found an alternative to conventional fillers used in latex and rubber industry by using lignocellulosic and polysaccharides natural fillers. The advantages of natural fillers include the biodegradability, renewable and environmental friendly (Blackburn, 2005). Natural fillers are also easily available, low density, cheap and non- abrasive (Muniandy et al., 2012). Thus, a blend of biodegradable fillers and synthetic polymer compound is seemed to be the best solution. The potential filler to be used is starch. In this experiment, dextrin is used as an alternative of sago starch. Dextrin is a product of acidified sago starch heated at high temperature until brownish colour powder obtained. Dextrin has low molecular weight and shorter chains compared to starch which undergo chain fragmentation through chemical and pyrolysis reaction (Bai et al., 2014). The low molecular weight of dextrin increase the chances of the XNBR films to be broken down easily upon disposal. Thus, make them more environmental friendly and greener option compared to conventional filler used in latex films. The objective of this research is to study the potential of using dextrin as filler in XNBR latex with the aim to study the effect of dextrin loading on the tensile properties of XNBR latex films.

2. Experimental study

2.1 Materials

Carboxylated nitrile butadiene rubber (XNBR) latex classified as XNBR X6322 was supplied by Synthomer (M) Sdn Bhd. Other chemical ingredients include zinc oxide, zinc diethyl dithiocarbamate (ZDEC), sulfur, and antioxidant were supplied by Farben Technique (M) Sdn Bhd. Sago starch was supplied by Sago Link (M) Sdn Bhd. The material was in white powder form granule size of 10-20 μm.

2.2 Preparation of Dextrination

The sago starch (100 g) was suspended in 150 ml water. The pH of the slurry was adjusted to 3.0 by 0.5 M HCl and stirred for 30 minutes. The sago starch then filtered using filter paper. The sago starch cake was dried in an oven at 40 for 24 hours. The starch then filtered and dried to form thick flakes with moisture content of approximately 50%. The starch flakes was dried in an oven at 40°C for 24 hours. The dried starch was grind, and strained through a sieve (100 mesh) and heated again in an oven at 170°C until the starch changed to darker colour (dextrin). The synthesized dextrin (50 g dry weight) then was dispersed in distilled water (45 mL), together with dispersing agent (5 g dry weight) to make dextrin dispersion. The compound was stirred for 1 hour and subjected to ball milling for 24 hours at 60 Hz.

2.3 Preparation of dextrin/carboxylated nitrile butadiene rubber (XNBR) latex films

The preparation of XNBR compounds was carried out by mixing all the ingredients in the beaker and was constantly stirred for one hour using Wise Stir HS-300 mechanical stirrer at stirring speed of 270 rpm at room temperature. The compounds were then pre-vulcanized for 1 hour at temperature of 80° C. Clean aluminium plate was dipped into coagulant (10% calcium nitrate) for 10 seconds and dried in convection oven for 15 minutes at temperature of 100°C. The plates was taken out and leave under at room temperature until the plate temperature reduced to 50-60°C. The plate then dipped into DX/XNBR latex compound for 17 seconds and cured in convection oven for 25 minutes at temperature of 100°C. The plate then cooled at room temperature for 24 hours before the films were stripped from the plate.

2.4 Tensile Properties of Dextrin/XNBR Films

Tensile test was conducted using Instron Machine according to ASTM 412-2006. AD/XNBR latex films was cut into dumbbell shape and thickness of each specimens was recorded.

2.5 Swelling Test

The swelling test was carried out according to ASTM D471 where a test piece weighing about 0.2 g was cut from the XNBR films. The films were immersed in 25 ml toluene and transferred into water bath and heated at 48 hours at 40 ° C. The test piece was taken out and blotted using filter paper and the swollen weight was immediately measured. The crosslink density of XNBR films [X] were calculated using Flory-Rehner equation.

$$\text{Flory Rehner Equation, } [X] = \frac{-[\ln(1-V_r) + V_r + \chi V_r^2]}{V_s(V_r^{1/3} - V_r/2}$$

2.6 Fourier Transform Infra- Red (FTIR)

The Fourier Transform Infrared Spectrometry (FTIR) spectrometer used in this study was Perkin Elmer, Model Spectrum One. Analysis was done under room temperature within a typical wave number range of 4000 cm^{-1} to 550 cm^{-1} under transmission mode.

2.7 Scanning Electron Microscopy (SEM)

Scanning electron microscope (SEM) is a tool for visualizing the surface of solid samples. The morphologies of fillers and XNBR films were observed using Hitachi Tabletop SEM model TM3000. The filler powder was sprinkled on the black tape mounted on aluminum stubs. The samples were sputter coated with thin layer of gold before analysis to avoid electrostatic charging during analysis.

3. Results and discussion

3.1 Swelling Test

Figure.1 show the swelling values of unfilled and dextrin-filled XNBR films at different filler loading. The incorporation of dextrin gradually increased in swelling percentage with increasing filler loading. The increased in swelling percentage is due to the hydrophilic dextrin was incompatible with hydrophobic XNBR resulted in the poor interfacial interaction between rubber and filler, thus increased the toluene uptake in the films (Lin et al. 2011). The increase in toluene intake indicates minimal interaction between filler and rubber as a result of a few crosslinks forming per rubber chains. The higher dextrin loading could also interrupt the formation of crosslinks between XNBR chains, thus reduce the mechanical properties of XNBR films.

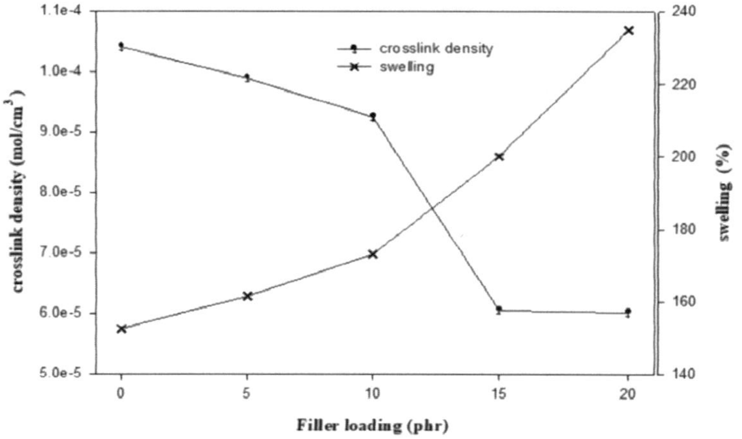

Figure 1. Crosslink density and swelling index of unfilled XNBR and dextrin filled XNBR

3.2 Mechanical Properties of Dextrin-filled XNBR films

The tensile strength of XNBR latex films attributed by three factors; the effectiveness of the dual crosslinking, the ability of ionic clusters to reform during stress and chain entanglement of XNBR macromolecules. The mechanical properties of the dextrin-filled XNBR latex films are shown in Table 1. The highest tensile strength of unfilled XNBR films ascribed the strength of the XNBR films. The presence of the different dextrin loadings showed decrement in tensile properties of XNBR films compared to unfilled XNBR films. The decreased in tensile strength was observed due to the breakage of sulfidic and ionic crosslinks (Blackley, 1997). The presence of polysaccharide molecules, the ionic crosslinks were hindered resulted from less sulfidic and ionic crosslinks. Dextrin with higher content of branched structures tend to reduce mechanical properties of rubber films compared to linear structure in latex films (Laurentin et al. 2003). The large particle size of dextrin also contributed to the decrement in tensile strength. Hence, the dextrin particle to become the stress concentrator during tensile test. The presence of weakness points on the films also restrict the ability of the films to elongate thus, reduce the elongation at break gradually with increasing dextrin loading as tabulated in Table 1. The moduli (modulus at 100% and modulus at 300 %) of XNBR films were decreased as increase in dextrin loading. Modulus is related to the crosslinks density of the rubber. Higher dextrin loadings prevent the formation of crosslinks in the XNBR rubber.

Table 1. Tensile properties of unfilled XNBR and dextrin-filled XNBR films

Samples	Tensile strength (MPa)	Elongation at break (%)	Modulus at 100% (MPa)	Modulus at 300% (MPa)
Unfilled XNBR	20.137	923.6	2.3305	4.549
5DX/XNBR	16.300	706.120	1.719	4.4905
10 DX/XNBR	14.633	663.920	1.691	3.5718
15DX/XNBR	11.380	550.100	1.514	3.3781
20DX/XNBR	10.680	517.617	1.359	2.5562

*DX/XNBR: dextrin/XNBR

3.3 Fourier Transform Infrared Analysis

Figure 2 indicates the functional groups shared by both native sago and dexrinized- sago. For starch, the peaks with representing its functional groups lies within major regions, 3600-3000 cm^{-1}, 1700-1600 cm^{-1}, and 1200-900 cm^{-1} which indicating the presence of –OH stretching, C-H stretch in starch main chain (CH_2) and CO stretching, respectively (Misman et al. 2013). The 3300-3310 cm^{-1} represents the OH group in starch and dextrin. The peaks within range 1630 cm^{-1} -1650 cm^{-1} were significant during dextrinization process as dextrin becomes more hygroscopic and soluble in water due to depolymerisation of the glucan as result of degradation by heat and acid (Yu et al. 2012). The crystalline region of starch can be detected within range 1047-1022 cm^{-1}. Whereas, the amorphous region can be seen within region 1550-1750 $cm-^{1}$ (Rajeev et al. 2010). It can be seen that the crystalline peak in the sago dextrin has diminished significantly and amorphous peak of dextrin appeared (Lin et al. 2011). The C-OH peak in carboxylic functionality can be found in both 1040 cm^{-1} unfilled XNBR and dextrin-filled XNBR films (Misman et al. 2016). There are two types of crosslinking exist in XNBR latex films. First one is the ionic crosslink that formed between carboxylic acid and zinc and the other one covalent bonding between sulphur and butadiene of XNBR. The absence of nitrile group band at 2238 cm^{-1} and presence of carbonyl group at band 1699 cm^{-1}, it can be assumed that the crosslinks take place mainly with –CN group even though the –COOH group also participate in the reaction. The incorporation of dextrin as filler does not affect crosslink process, it does not favour the reaction with –CN nor with –COOH due to unreactive group present on the dextrin surface (Ibarra et al. 2008).

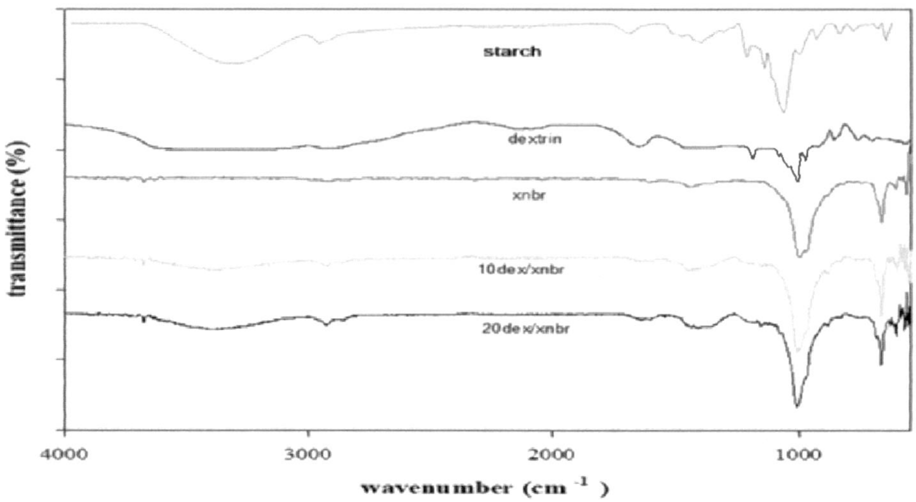

Figure 2. The Fourier Transform Infra-Red (FTIR) of sago starch, dextrin, unfilled XNBR and 10, 20 phr dextrin-filled XNBR films

3.4 Morphological Analysis

There is no obvious difference in the size and shape of starch granules detected between native sago starch and dextrinized sago starch after pyrolysis and acid treatment. The ball-milling method of preparing the dextrin dispersion did not efficiently reduce the particle size of the dextrin during preparations of dextrin dispersions. The only observation observed, that the structure of dextrinized sago starch particles are slightly cracked and damaged as pyroconversion leached out the crystalline ring growth consist of amylose content in native sago starch as shown in Fig.3 (a) and (b) (Karim et al, 2007; Dufresne, 2010). The SEM micrographs of tensile fractured surfaces of XNBR films are illustrated in Fig.3 (c) and (d). The presence of dextrin fillers sticking out of the fractured surface and formed holes indicate poor interfacial interaction between dextrin and XNBR due to polar characteristic of dextrin and non- polar characteristic of XNBR latex which lead to low tensile properties of XNBR films. Contrary, the unfilled XNBR (Fig. 3 (d)) has smooth fractured surface as no filler granule presents on the fractured surface.

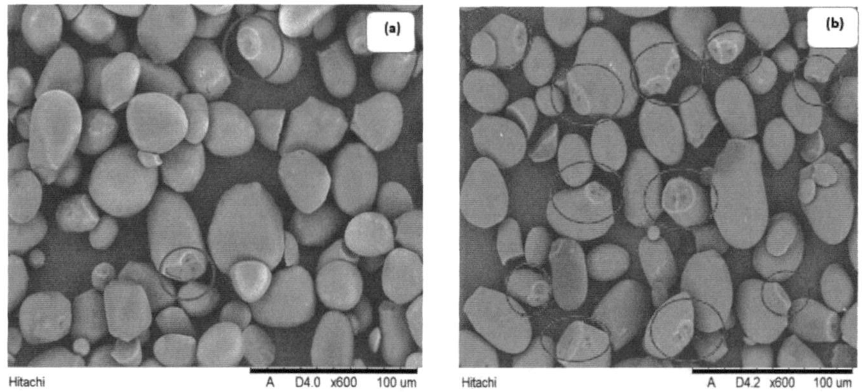

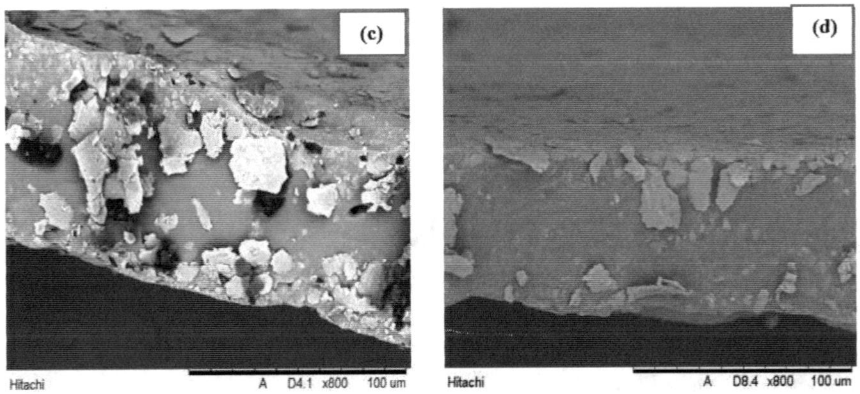

Figure 3. Morphology of (a) sago starch and (b) dexrinized sago (c) tensile fractured surface of 20 phr dextrin filled XNBR film (d) tensile fractured surface of unfilled XNBR film

4. Conclusion

In conclusion, the tensile properties of dextrin filled XNBR films, namely tensile strength, elongation at break, moduli 100 % and 300% decreased with increasing dextrin loading. The crosslink density of dextrin filled XNBR films also decreased as swelling index increased due to fewer formation of crosslinks hindered by presence of dextrin filler. The IR spectra showed that dextrin filler did not participate in crosslinking of rubber chains. The SEM showed that dextrin have poor adhesion with XNBR film due to hydrophilic nature of dextrin and hydrophobic nature of XNBR.

Acknowledgement

The authors would like to thank the Universiti Sains Malaysia research facilities and Ministry of Higher Education Malaysia for FRGS grant (grant no: 203/PBAHAN/6071353).

References

[1] Bai Y, Cai L, Doutch J, Gilbert E P and Shi YC 2014 *Journal of Agricultural and Food Chemistry* 4186 4194
[2] Misman M A and Azura A R 2013 *Journal of Chemistry and Chemical Engineering* 7 137
[3] Bhattacharjee S, Bhowmick A K, and Avasthi B 1993 *Elastomer Technology Handbook* 541
[4] Muniandy K, Ismail H and Othman N 2012 *Bioresources* 7 957
[5] Lenko D, Schlogl S, Temel A, Schaller R, Holzner A, and Kern W 2013 *Journal of Applied Polymer Science* 129 2735
[6] Dufresne, A 2010 *Molecules* 15 4111
[7] Blackley D 1997 *Polymer Latices* Vol 2 176
[8] Blackburn RS 2005 *Biodegradable and Sustainable Fibres* 271 279
[9] Laurentin A, Rdenas C. M, Ruales J, Peares E and Tovar, J 2003 *Journal of Agricultural and Food Chemistry* 51 5510
[10] Rajeev K R S, Ghanshyam L, Chauhan S, and Rakesh V 2010 *International Journal of Polymer Materials* 59 263
[11] Karim A A, Nadiha M Z, Chen F K, Phuah Y P, Chui Y M and Fazilah, A 2007 *Food Hydrocolloids* 22 1044
[12] Lin N, Huang J, Chang, P R, Anderson, D P and Yu J 2011 *Journal of Nanomaterials* 2011 1
[13] Misman M A, Azura, A R, Hamid, Z A A 2016 *Procedia Chemistry* 19 770
[14] Yu M, Yang R, Huang L, Cao X, Yang F, Liu D 2012 *Bioresources* 7 1802
[15] Ibarra L, Rodriguez A, and Mora-Barrantes I 2007 *Journal of Applied Polymer Science* 108 2197

IOP Conf. Series: Journal of Physics: Conf. Series **1082** (2018) 012014 doi:10.1088/1742-6596/1082/1/012014

Optimisation of Pre-structuring Process of Magnetorheological Elastomer Performance

S. Raa Khimi[1]*and N.T. Lai[1]

[1]School of Materials and Mineral Resources Enineering, Universiti Sains Malaysia, 14300 Penang, Malaysia

*Email: raakhimi@usm.my

Abstract. In this study, MREs based on natural rubber and carbonyl iron were prepared. The Taguchi method was utilised to study the effect of a number of dominating factors during the fabrication process, namely, pre-curing time, pre-curing temperature and applied magnetic field on the microstructure and tensile properties of MREs. Tensile properties were measured with a universal tensile testing machine. It was found that the pre-curing temperature had the greatest influence on ultimate tensile strength and elongation at break. Morphology characteristics of MREs were examined using scanning electron microscopy. The results revealed that isotropic MREs cured without an applied magnetic field had a homogenous carbonyl iron particle distribution in the rubber matrix, and that applying a magnetic field at elevated temperature allowed the particles to organise into chain-like columnar structure in anisotropic MREs.

1. Introduction

Rubber is one of the most commonly used materials for energy absorption and vibration suppression. However, rubber itself has low stiffness [1]. Many efforts have been put into modifying available rubber to improve its mechanical properties, such as inclusion of particle or fibre fillers [2-4]. In more recent research, magnetic particles have been included in rubber to create a magnetic rubber composite known as magnetorheological elastomers (MREs). MREs can be classified into two types: isotropic and anisotropic. Isotropic MREs have a uniform dispersion of magnetic particles in the rubber matrix. In anisotropic MREs, the particles form columnar structures, achievable by applying a magnetic field during curing of the matrix [5-8]. The addition of magnetic particles in a rubbery matrix allows more damping through magnetic interactions between neighbouring particles and particles-matrix interfacial damping. Moreover, MREs' modulus and stiffness can be changed during fabrication or in service by applying an external magnetic field. Most commonly used matrices for MREs include natural rubber, silicone rubber, polybutadiene, polyisobutylene, polyisoprene, and polyurethane rubber [5,9-13]. The magnetic particle choice for MREs is carbonyl iron [5-8].

The curing condition plays an important role in MRE performance. Pre-curing stage is essential for formation of magnetic particle alignment in the rubber matrix. In this process the magnetic particles are driven by the magnetic force and chain-like structures are formed in the rubber matrix. As far as the authors are aware, there is limited published work that study factors effecting the pre-structuring process. This work aims to study the effects of three predominant factors during pre-structure process on MREs, namely pre-curing time, pre-curing temperature and applied magnetic field [6,10] on the tensile strength of MREs. In this study, the Taguchi method is utilised to determine the optimum processing condition for MREs.

2. Taguchi Method

The Taguchi method, developed by Dr. Genichi Taguchi, proposes a simplified, standardised, consistent and cost-effective approach in design of experiments to study the influence of multiple factors on a product or a process.

In general, the Taguchi method includes the following steps: identification of the factors and their respective levels, design of experiments using an appropriate orthogonal array (OA), conducting of experiments, analysis of data and selection of the optimum level for each factor, confirmatory tests of the optimum condition.

The key steps in the Taguchi method are to identify the factors and levels, and to assign them to the appropriate OA. The Taguchi method allows for 18 standard orthogonal arrays, as mentioned in reference[14]. After the conditions for individual experiments, called trials, are specified using the OA, the trials are carried out, preferably in random order[14]. The S/N ratio for larger is better is then calculated for each factor, using the following equations:

$$\frac{S}{N} = -10log\frac{\sum_{i=0}^{n}\left(\frac{1}{{Y_{a_i}}^2}\right)}{n} \tag{1}$$

in which Y_{a_i} is the value of factor A at level a at the ith observation and n is the number of observations for factor A at level a. Another advantage of the S/N ratio is the determination of the trend of influence of the factors by calculating the average influence of a factor and plotting it against the corresponding level. The average factor effect is given by equation (2):

$$A_F = \frac{\sum_{i=0}^{n}\left(\frac{S}{N}\right)_i}{n} \tag{2}$$

in which A_F is the mean of the S/N ratios for factor F at level i, $(S/N)_i$ represents the S/N ratio and n the number of observations for factor F at level i. Moreover, the significance of each processing parameter is the difference between the maximum and minimum S/N ratio:

$$Significance\ of\ factor = \left(\frac{S}{N}\right)_{max} - \left(\frac{S}{N}\right)_{min} \tag{3}$$

S/N_{max} and S/N_{min} represents the maximum and minimum average S/N ratios, respectively, for a factor.

By using the Taguchi method, it is also possible to predict the optimum condition for the process. The optimum level for each factor is identified as the level at which the S/N ratio has the maximum value. However, the Taguchi method does not provide other statistically vital information, which can instead be obtained by performing an analysis of variance (ANOVA). Using ANOVA in the Taguchi method allows determination of sums of squares, degrees of freedom, percentage confidence level, and percentage contribution. Equations to calculate ANOVA parameters are discussed in reference [11].

3. Experimental

3.1 Materials

Compounding formulation used in this study is shown in Table 1. Natural rubber and all other chemicals in Table 1 were supplied by Zarm Scientific & Supplies Sdn. Bhd, Malaysia. Carbonyl iron particles (CIP) were purchased from Shanghai YY International Co, Ltd, China, and have a spherical shape and an average particle size of 4μm.

IOP Conf. Series: Journal of Physics: Conf. Series **1082** (2018) 012014 doi:10.1088/1742-6596/1082/1/012014

3.2 Design of experiments

In this study, three factors were selected to study: pre-curing time (t), pre-curing temperature (T) and applied magnetic field during pre-curing and curing (M), each with four levels as presented in Table 2. The smallest appropriate OA for three factors and four levels, L'16, was selected for the Taguchi method. The experimental and sample preparation was carried out according to L'16 OA as reported elsewhere [14]. The analysis of S/N ratios was subsequently used to evaluate the results. In this work, larger-is-better target for S/N ratios was chosen. ANOVA was performed with Statistica software to evaluate the percentage contribution of each factor.

Table 1. Compounding formulation

Materials	Functions	Loadings (phr)*
Natural rubber (SMR L)	Matrix	100
Zinc oxide	Activator	5
Stearic acid	Activator	1
Paraffin oil	Plasticizer	5
Carbonyl iron powder	Magnetic particles	70
IPPD	Antioxidant	2
CBS	Accelerator	2
TMTD	Accelerator	1
Sulphur	Crosslinking agent	1.5

*part per hundred rubber

Table 2. Experimental factors and their levels

Factors	Unit	Level 1	Level 2	Level 3	Level 4
Pre-curing time (t)	min	3	5	10	15
Pre-curing temperature (T)	°C	60	80	100	110
Magnetic field (M)	mT	0	100	150	165

3.3 Fabrication of MREs

The ingredients of the formulation in according to the OA and were compounded using a conventional laboratory two-roll mill according to ASTM D3184-07. The cure time at 150°C was determined by an Alpha MRD2000 rotational die rheometer according to D6204 - 15. Compounded rubber samples were placed in a 120mm × 65mm × 3mm mould and undergo a pre-structure process with their respective pre-curing parameters in Table 3. Isotropic samples were cured in a compression moulder at 150°C under a pressure of 10MPa. Anisotropic MREs were cured in a specially developed magnetic mould as shown in Figure 1. Each trial was repeated three times.

3.4 Characterisation

3.4.1 Morphology. The microstructures of MREs were observed using a Hitachi TM3000 scanning electron microscope. The surfaces were coated with a thin layer of gold prior to observation at accelerating voltage of 15kV.

3.4.2 Tensile test. Tensile test was carried out on dumbbell-shaped samples using an Instron universal tensile testing machine at a crosshead speed of 500mm/min according to ASTM D412-16.

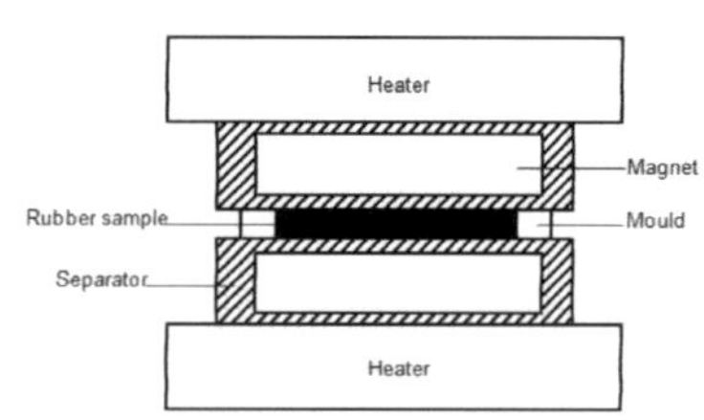

Figure 1. Schematic diagram of pre-structure process

4. Results and discussion

4.1 Morphology

Figure 2 shows SEM images of isotropic and anisotropic MRE samples. From the SEM images, it can be seen that isotropic MREs have a uniform distribution and dispersion of CIP with no obvious aggregation in the rubber matrix. For anisotropic MREs, clearly, as expected, the application of a magnetic field at elevated temperature allowed the CIP to form chain-like structures.

(a) (b)

Figure 2. SEM images of (a) isotropic MRE and (b) anisotropic MRE

4.2 Tensile testing

Table 3 shows the values of tensile strength and elongation at break, which were subsequently used to calculate the S/N ratios and ANOVA, each value representing an average from five samples. The highest value for each measurement is shown in bold.

Table 3. Tensile test results

Sample	Tensile strength (MPa)	Elongation at break (%)
1	12.60	**808.89**
2	10.37	642.22
3	12.53	710.56
4	11.02	619.44
5	2.44	285.56
6	12.03	752.22
7	12.56	710.56
8	11.10	643.89
9	2.95	345.56
10	8.96	643.33
11	11.09	766.11
12	9.13	582.78
13	12.54	742.22
14	11.16	623.89
15	10.45	681.11
16	**12.92**	726.11

Figure 3 shows the main effect plots of the S/N ratios for the ultimate tensile strength. The highest S/N ratios are observed at 15 min, 100°C and 0mT magnetic field during pre-curing. Table 4 shows ANOVA results for the ultimate tensile strength. It can be seen that the variability for each factor was tested at 99.9% confidence level. From the parameter significance, it can be seen that the most influential factor to form alignment of magnetic particles in the rubber matrix is temperature, followed by magnetic field and time. However, the percentage contribution is approximately the same for each factor. The minimal influence of all tested parameters on percentage contribution would appear to be due to tensile mode of loading during testing. This suggests the interaction between magnetic particles alignment is less efficient in a tensile mode, which is not surprising, given that in tension, it is largely only increasing the spacing between chains and the spacing within chain and breakdown of magnetic particles interaction between the neighboring particles is limited (see Figure 4), suggesting formation of magnetic particle alignment is less important in determining the ultimate tensile strength of MREs.

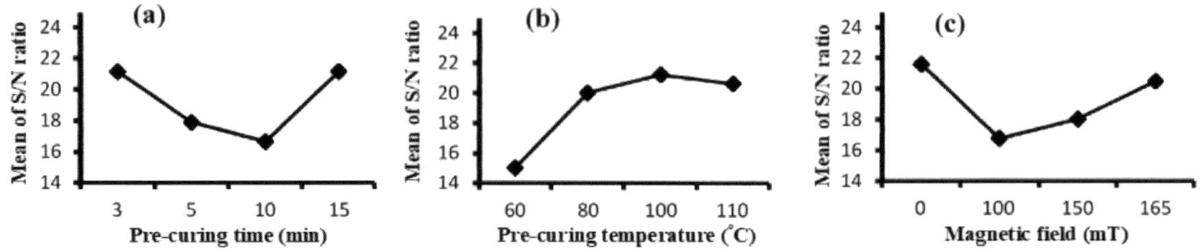

Figure 3. Main effect plots for S/N ratios of ultimate tensile strength: (a) effect of pre-curing time, (b) effect of pre-curing temperature and (c) effect of magnetic field during curing

Table 4. ANOVA Table and factor influence for ultimate tensile strength

Factor	Sum of Squares	Degree of freedom	Confidence Level (%)	Percentage Contribution (%)	Parameter Significance (S/N_{max}–S/N_{min})
Time (min)	115.522	3	99.99	21.31	1.500
Temperature (°C)	115.222	3	99.99	21.26	2.066
Magnetic field (mT)	119.963	3	99.99	22.13	1.602

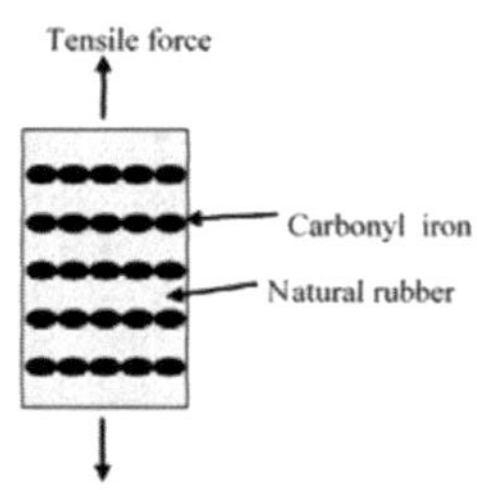

Figure 4. Influence of loading types on particle separation in tensile mode[11]

The main effect plots of the S/N ratios for elongation at break are presented in Figure 5. The highest S/N ratios are observed at 15 min, 100°C and 0 mT magnetic field during pre-curing. It is also noted that the suggested optimum conditions to obtain highest elongation at break are similar to those for the optimum conditions when assessing the ultimate tensile strength.

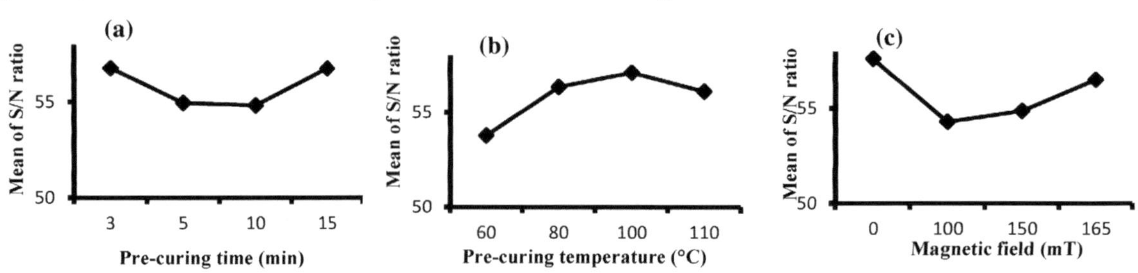

Figure 5. Main effect plots for S/N ratios of elongation at break: (a) effect of pre-curing time, (b) effect of pre-curing temperature and (c) effect of magnetic field during curing

Table 5. ANOVA Table and factor influence for elongation at break

Factor	Sum of Squares	Degree of freedom	Confidence Level (%)	Percentage Contribution (%)	Parameter Significance ($S/N_{max}-S/N_{min}$)
Time (min)	128579.265	3	99.99	12.97	0.648
Temperature (°C)	185832.806	3	99.99	18.74	3.333
Magnetic field (mT)	343874.089	3	99.99	34.68	3.330

Table 5 shows ANOVA results for the elongation at break. From the parameter significance, it can also be seen that the most influential factor to form alignment of magnetic particles in the rubber matrix is temperature, followed by magnetic field and time. From the levels of contribution, it can be seen that the most contributed factor to elongation at break value is magnetic field followed by temperature (34.68% and 18.74%, respectively). The influence of time was much less at 12.97%. Isotropic MREs was found to have better elongation at break when compared to anisotropic MREs. This can be explained by the better interfacial interaction in isotropic MREs, in which the particles are dispersed uniformly and the rubber chains coiled around the particles, thus allowed them to uncoil and better align along the stress direction. On the other hand, when anisotropic MREs underwent stretching, the force was in a perpendicular direction to the particle chains, and the rubber molecule chains were restrained by the particle chains and ruptured by the applied force. This is in agreement with similar finding reported by Wang et al[15].

5. Conclusion

In this study, MREs were fabricated based on natural rubber and carbonyl iron particles, and experiments were designed using the Taguchi method to evaluate the influence of three pre-curing time, pre-curing temperature and magnetic field during curing on the tensile properties of MREs. SEM images revealed that application of a magnetic field during the pre-structure and curing process produce columnar structures of magnetic particles in the rubber matrix, while samples fabricated without a magnetic field had a random dispersion of particles. Calculation of the S/N ratios for each factor implied that temperature had the greatest influence on both ultimate tensile strength and elongation at break. It was also found that both tensile strength and elongation at break for isotropic MREs are higher than those of anisotropic MREs, which could be due to the tensile loading mode. ANOVA results suggested that particle alignment caused little difference on tensile properties of MREs.

References

[1] Chung D D L 2001 *Journal of Materials Science* **36** 5733
[2] Chandra R, Singh S P and Gupta K 1999 *Composite Structures* **46** 41
[3] Prasertsri S and Rattanasom N 2011 *Polymer Testing* **30** 515
[4] Warn G P and Ryan K L 2012 *Buildings* **2** 300
[5] Li Y, Li J, Li W et al. 2014 *Smart materials and structures* **23** 123001
[6] Boczkowska A and Awietjan S 2012 Microstructure and Properties of Magnetorheological Elastomers *(Advanced Elastomers: Technology, Properties and Applications)* chapter 6 pp 147-180
[7] Wang X and Gordaninejad F 2008 Magnetorheological Materials and their Applications *(Intelligent Materials)* chapter 14 pp 362-408
[8] Li W, Zhang X and Du H 2013 Magnetorheological elastomers and their applications *(Advances in Elastomers I: Blends and Interpenetrating Networks)* chapter 12 pp 357-374
[9] Chen L, Gong X L, Jiang W Q et al. 2007 *Journal of Materials Science* **42** 5483
[10] Li J, Gong X, Xu Z et al. 2008 *International Journal of Materials Research* **99** 1358
[11] Shuib R K, Pickering K L and Mace B R 2015 *Journal of Applied Polymer Science* **132**
[12] Wu J, Gong X, Fan Y et al. 2010 *Smart materials and structures* **19** 105007
[13] Wang Y, Hu Y, Deng H et al. 2006 *Polymer Engineering & Science* **46** 264
[14] Roy R K 2010 *A Primer on the Taguchi Method*
[15] Wang Y, Zhang X, Oh J et al. 2015 *Smart materials and structures* **24** 095006

IOP Conf. Series: Journal of Physics: Conf. Series **1082** (2018) 012015 doi:10.1088/1742-6596/1082/1/012015

Flow Behavior Analysis of Emc in Molded Underfill (Muf) Encapsulation for Multi Flip-Chip Package

M.A. Azmi[1], M.K. Abdullah[1,*], M.Z. Abdullah[2],Z.M. Ariff[1],M.A. Ismail[2], M.S. Abdul Aziz[2]

[1]School of Materials and Mineral Resources Engineering, Universiti Sains Malaysia, Engineering Campus, 14300 Nibong Tebal, Penang, Malaysia

[2]School of Mechanical Engineering, Universiti Sains Malaysia, Engineering Campus, 14300 Nibong Tebal, Penang, Malaysia

Corresponding author: mkhalil@usm.my

Abstract. This paper present a numerical analysis of molded underfill (MUF) for multi flip chip package during encapsulation. Castro Macosko model is used in ANSYS fluent to describe the flow behavior of the EMC. The physical quantities such as viscosity, shear rate and velocity is plotted to analyzed the flow behavior of EMC during encapsulation. The EMC flow behavior at flip chip (FC) and free passage (FP) region in cavity are been compared. The viscosity profile of EMC is inversely proportional to the shear rate and velocity profile at FC and FP region The tendency of "frozen skin" layer occur in cavity package have been predicted with velocity profile during encapsulation.

1. Introduction

Transfer molding is used in MUF to fill the epoxy molding compound (EMC) in a mold during encapsulation process, a part of it is MUF. However, the behavior of EMC flow in the mold is very important to study since it will affect the performance and reliability of electronic package. There are many factors affect the EMC flow behavior and void formation during encapsulation process like transfer speed injection in transfer molding and effect of stacking [1-2].The EMC flow behavior at the free and chip areas in a mold are different each other [3].The possibilities of defect such as void, crack, etc occur at in the electronic package.

The flow behavior is dominated mainly by the viscosity and depends on the generalized shear rate in Generalized Newtonian Fluid (GNF) [4]. In addition, the physical quantity such as velocity is used to describe and analyze the EMC behavior during encapsulation process [5]. It is being more exacerbate, the

EMC is thermally cured during or after encapsulation process. For example, when the EMC fills the cavity and the molecules chain in EMC start to cross linking each other. As a consequences, the viscosity creases rapidly and the EMC transform into a gel, solidified and it becomes a highly cross-linked network [6-7].

In the present study, the numerical analysis of MUF encapsulation is carried out on multi flip chip package. The ANSYS Workbench 14.5 software is used to simulate the EMC flow and analysis EMC behavior during encapsulation process.

2. Mathematical model

The governing equations for continuity, momentum and energy for the three-dimensional mold filling are given as:

Continuity equation:

$$\frac{\partial \rho}{\partial t} + \nabla.(\rho \mathbf{V}) = 0 \tag{1}$$

Where $\mathbf{V} = (u, v, w)$ is the velocity vector.

Conservation of momentum in x , y ,and z- direction in an inertial:

$$\rho\left(\frac{\partial u}{\partial t} + \frac{\partial u}{\partial x} + \frac{\partial u}{\partial y} + \frac{\partial u}{\partial z}\right) = -\frac{\partial P}{\partial x} + \eta\left(\frac{\partial^2 u}{\partial x^2} + \frac{\partial^2 u}{\partial y^2} + \frac{\partial^2 u}{\partial z^2}\right) \tag{2}$$

$$\rho\left(\frac{\partial v}{\partial t} + \frac{\partial v}{\partial x} + \frac{\partial v}{\partial y} + \frac{\partial v}{\partial z}\right) = -\frac{\partial P}{\partial y} + \eta\left(\frac{\partial^2 v}{\partial x^2} + \frac{\partial^2 v}{\partial y^2} + \frac{\partial^2 v}{\partial z^2}\right) \tag{3}$$

$$\rho\left(\frac{\partial w}{\partial t} + \frac{\partial w}{\partial x} + \frac{\partial w}{\partial y} + \frac{\partial w}{\partial z}\right) = -\frac{\partial P}{\partial z} + \eta\left(\frac{\partial^2 w}{\partial x^2} + \frac{\partial^2 w}{\partial y^2} + \frac{\partial^2 w}{\partial z^2}\right) \tag{4}$$

where ρ is the fluid density, u, v and w is the fluid velocity components, P is the pressure respectively. The energy equation can be written as

$$\rho c_p\left(\frac{\partial T}{\partial t} + u\frac{\partial T}{\partial x} + v\frac{\partial T}{\partial y} + w\frac{\partial T}{\partial z}\right) = k\left(\frac{\partial^2 T}{\partial x^2} + \frac{\partial^2 T}{\partial y^2} + \frac{\partial^2 T}{\partial z^2}\right) + \eta\dot{\gamma} + \dot{\alpha}\Delta H \tag{5}$$

where T is the temperature, k is the thermal conductivity, c_p is the specific heat, η is the viscosity, $\dot{\gamma}$ is the shear rate, $\dot{\alpha}$ is the conversion rate and ΔH is the exothermic heat of polymerization. The EMC material is assumed to be a GNF and it was expressed by the Castro Macosko rheology model.

$$\eta(T,\dot{\gamma},\alpha)=\frac{\eta_o(T)}{1+\left(\frac{\eta_o\dot{\gamma}}{\tau^*}\right)^{1-n}}\left(\frac{\alpha_g}{\alpha_g-\alpha}\right)^{c_1+c_2\alpha} \quad (6)$$

with

$$\eta_o(T)=B\exp\left(\frac{T_b}{T}\right) \quad (7)$$

where η_o is the zero shear rate viscosity, τ^* is the parameter that describes the transition region between zero shear rate and power law region of the viscosity curve, B is exponential-fitted constant and T_b is the temperature fitted-constant. Meanwhile α is conversion of reaction, α_g is degree of conversion of gel point and c_1 and c_2 are fitting constants. This model will use in EMC material as rheology model, the present of curing reaction which is rate of conversion, α have change the viscosity and shear rate of material during encapsulation in simulation.

3. Simulation model and boundary condition

The cavity has 3 x 3 flip chip package in matrix array arrangement. The cavity have four inlet and seven outlet . The geometry of flip chip is summarized in Table 1 and its arrangement in the cavity is shown in Figure 1. The total tetrahedral element of meshing for cavity is 1,326,483. Hitachi CEL-9200 (LF) is defined as EMC material and its properties is summarized in Table 2. Castro Macosko model is used as rheology model of material with the rate of conversion (α=0.01). The cavity temperature (T_w) is set as 448 K and inlet temperature (T_{in}) set as 343 K. The inlet pressure during encapsulation process is 70 kg/cm^2[8].

Table 1. Geometry of flip chip

Flip chip parameter	Dimension (mm)
Substrate size	7 (W) x 7 (L) x 0.1 (H)
Solder bump diameter	0.45
Solder bump count	64
Bump standoff height	0.15
Pitch size	0.80

Table 2. Material Properties

Parameter	Value
ρ (kg/cm$^{3)}$	1856
c_p(J/kg.K)	1078
k (W/m.K)	0.74
T_b (K)	4810.3
B	0.042168
n	0.7773
τ^*(N/m^2)	0.0001

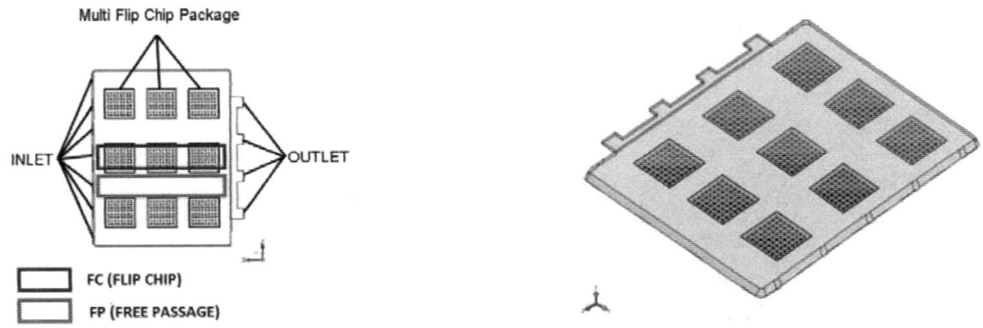

Figure 1. Multi flip chip package arrangement in a cavity package

4. Result and discussion

Figure 2 shows velocity versus Z-direction of EMC in cavity package at 5 s of filling stage during encapsulation. The graph was plotted half of the cavity, since another half is symmetry. The velocity of EMC is high at FP region and lower at FC region. When EMC flow through FC, the retardation flow occur due obstacle from flip chip package, it is caused the velocity of EMC drop at FC region. Besides that, the velocity of FP at middle area of cavity is higher than FP at side area. This is because the effect of inlet position at the middle area of cavity package. The shear stress also occurred between EMC and flip chip wall which contain substrate and solder bump.

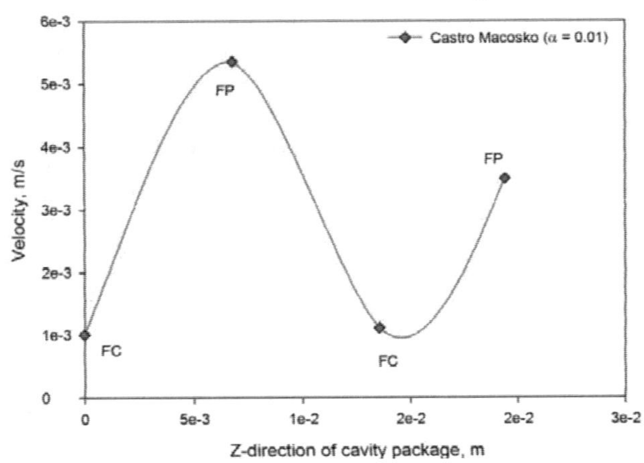

Figure 2. Velocity versus Z-direction of cavity package

Figure 3 shows the shear rate versus Z-direction of cavity package. The Shear rate of EMC at FC is lower than FP region. The shear rate of EMC reduced because the velocity of EMC drop due to obstacle from flip chip. When the EMC flow through flip chip, the effect of wall stress from substrate and solder bump have reduce the velocity and shear rate of EMC.

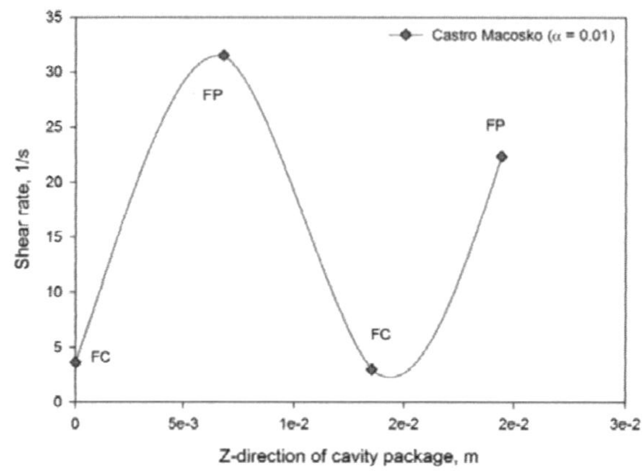

Figure 3. Shear rate versus Z-direction of cavity package

Figure 4 shows the viscosity versus Z-direction of cavity package for FP and FC during encapsulation process. From the figure, the viscosity of EMC is higher when it flow through the FC region while low through the FP region. The difference of these values is due to the flow obstruction by the flip chip package. The viscosity is inversely proportional to shear rate. When the shear rate of EMC is decreased, the viscosity of EMC will increase. In non-Newtonian fluid, the viscosity mostly dependent on shear rate. Besides that, the shear stress also occurred between EMC and flip chip. The velocity ,shear rate and viscosity of epoxy is an important factor in encapsulation process. Improper handling of these factors will contribute the imperfection of board package such as incomplete filling and void formation.

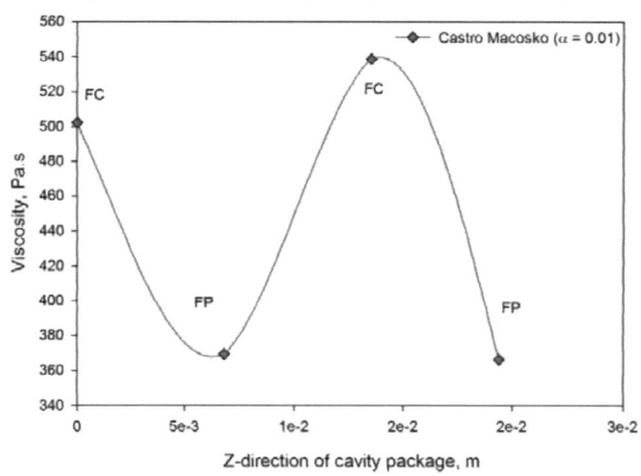

Figure 4. Viscosity versus Z-direction of cavity package

Figure 5 (a) shows the velocity profile for single multi flip chip package at y-direction of cavity package. The velocity profile of EMC before reach the flip chip is narrow with highest peak of velocity at the middle area of cavity package. When it reached at flip chip, the velocity profile have the curvature at the region (a gap between flip chip and cavity). The velocity of EMC is increased after the EMC flow past the flip chip. At flip chip, the velocity range is higher than velocity at free passage due to effect of obstacle from flip chip which it limit the EMC flow through the flip chip. As the results, the EMC flow will move toward top region of flip chip package and increase the velocity of EMC flow. The position of EMC flow which is far from the inlet can be seen start to cure from cavity wall, this call as the frozen skin layer effect, the increasing of frozen skin layer at cavity wall tend to increase the EMC flow velocity at the middle area of cavity package. The rectangle shape as shown in Figure 5 (b) shows the position of velocity profile along x-direction during encapsulation.

(a)

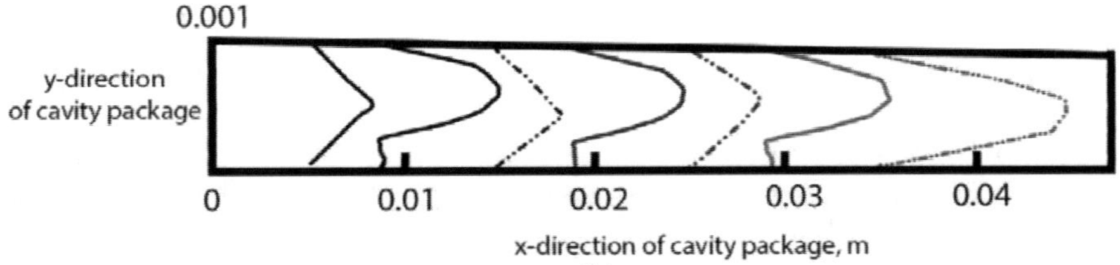

(b)

Figure 5. Velocity profile distribution of EMC at multi flip chip package.

5. Conclusion

The numerical investigation was been studied on molded underfill of multi flip chip package. The velocity, shear rate and viscosity profile have been plotted to analyze the flow behaviour of EMC in molded underfill of multi flip chip package during encapsulation. The viscosity profile of EMC is inversely proportional to the shear rate and velocity profile at FC and FP region. Furthermore, the velocity profile of EMC is used to describe the prediction of frozen skin layer during encapsulation in cavity package. The frozen skin layer tend to occur at the end of cavity package which is far from pressure inlet of cavity package.

References

[1] Hamid M F, Abdullah M K, Yusoff H, and Firdaus S M 2015 *Applied Mechanics and Materials* 786 318
[2] Abdullah M K, Abdullah M Z, Mujeebu M A,. Ariff Z M, and Ahmad K A 2010 *IEEE Trans Adv Packag* 33 438
[3] Abdullah M K, Abdullah M Z, Mujeebu M A, Gitano H, Ariff Z M, Razali R, and Ahmad K A 2010 *J Electron Packag Trans ASME* 132 0145021
[4] Wang H, Zhou H, Zhang Y, Li D, and Xu K 2011 *Comput. Fluids* 44 187
[5] Yang B, Ouyang J, Liu C, and Li Q 2010 *Chinese J Chem Eng* 18 600
[6] Lee M W, Khim J Y, Yoo M, Chung J, and Lee C H 2006 *Proc Electron Components Technol Conf* 2006 1029
[7] Gilleo K 1996 *Microelectronics International* 13 19
[8] Azmi M A, Abdullah M K, Abdullah M Z, Ariff Z M, Saad A A, Hamid M F, and Ismail M A 2017 *AIP Conf Proc* 1865

Improving the Flexural Properties of Abs/Muscovite Composites by Introducing Modified Muscovite

N.H. Che Ismail[1,2*], H.M. Akil[1]

[1]School of Materials and Mineral Resources Engineering, Engineering Campus, Universiti Sains Malaysia, Seri Ampangan, 14300 Nibong Tebal, Pulau Pinang, Malaysia

[2]Faculty of Applied Sciences, Universiti Teknologi MARA, Perlis Branch, Arau Campus, 02600 Arau, Perlis, Malaysia

Email: fiza.5856@gmail.com

Abstract. This paper reports the effect of unmodified and modified muscovite on the flexural properties of acrylonitrile–butadiene–styrene (ABS) composites. The composites were prepared by using melt compounding in an internal mixer. The morphologies of the composites were examined using field emission scanning electron microscopy (FESEM) whereas the structural characteristics of unmodified and modified muscovite were investigated by X-ray diffraction (XRD). The XRD results indicated that by modifying muscovite via two-stage ion exchange treatment, the basal spacing of muscovite was enlarged. The incorporation of treated muscovite (TM) in ABS matrix improved the flexural strength and modulus by 9% and 28%, respectively, compared to neat ABS strength of composites. The FESEM results showed that the fracture surface of ABS/TM composite indicated better interfacial adhesion between the filler and the matrix.

1. Introduction

The surge of layered silicates, mainly clay minerals incorporated with organic polymers, has recently attracted a great attention by researchers and industrial sectors mainly in the filling industry. The innovation of polymer-layered silicate was inspired by the Toyota research group in the early 1990[1]. They found that the addition of organoclay in minimal quantity showed a substantial improvement in various performances of polymer material[2, 3]. Among all the layered silicates used, those of smectite class, particularly montmorillonite (MMT), have been widely investigated due to the swelling behavior and intercalation properties. However, another important silicate group, mica, especially muscovite, has been studied to a lesser degree.

Muscovite, $KSi_3Al_3O_{10}(OH)_2$, is a type of 2:1 phyllosilicate. Muscovite has the highest layer charge density and homogeneous charge distribution, which makes this material ideal for reinforcement[3]. The extraordinary properties acquired by muscovite such as high aspect ratio, good dielectric properties, and high thermal stability make muscovite a potential alternative for MMT. However, the fabrication of muscovite with polymer matrices is very challenging due to the intrinsic incompatibility between hydrophilic clay and hydrophobic polymer matrix. Therefore, surface modification is required to render muscovite layers more compatible with polymer chains.

Acrylonitrile–butadiene–styrene (ABS) is an engineering thermoplastic that has been widely used in automotive hardware, pipe fitting, luggage, and appliance housing because of its outstanding hard and tough characteristics. In recent years, growing attention has been focused on ABS/layered silicate composite materials, commonly montmorillonite (MMT)[4–6], sepiolite[7], and bentonite[8] as the reinforcement filler. To the best of our knowledge, the properties of ABS/muscovite composites at various muscovite loadings have not been reported previously.

Therefore, this article aims to report on a new approach for the preparation of ABS/muscovite by melt mixing. The effect of untreated and treated muscovite on the flexural and morphological properties of ABS composites was investigated.

2. Experimental study

2.1 Materials

ABS pallets (Toyolac 700) from Toray Plastics (Malaysia) Sdn. Bhd. were used as the matrix. The inorganic filler, muscovite, was supplied by Lingshou County Xinfa Mineral Industry Co. Ltd. Figure 1 shows the particle size distribution of muscovite, showing that the mass median diameter (d50) was 21 μm. Lithium nitrate (LiNO3) and cetyltrimethylammonium bromide (CTAB) were supplied by Merck.

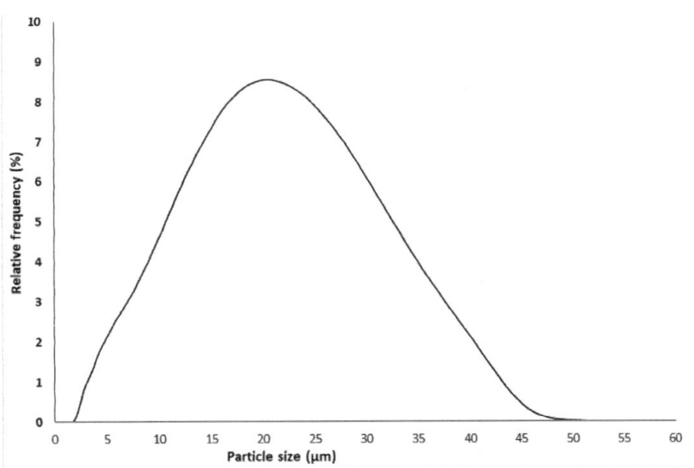

Figure 1. Cumulative particle size distribution of muscovite sample

2.2 Preparation of treated muscovite

The treated muscovite (TM), known as organomuscovite, was obtained with a two-stage ion exchange treatment.

First stage: Muscovite (5 g) was mechanically mixed with $LiNO_3$ (85 g) with a ratio of 1:17 of muscovite/$LiNO_3$. The mixture was heated in a furnace at 300 °C for 12 hr. Then, it was soaked in 200 ml of deionized water for 2 hr. The solution was vacuum filtered and the filtrate was dried in a vacuum oven at 110 °C for 12 hr. The product is known as Li-muscovite.

Second stage: Li-muscovite was added with an organic surfactant, CTAB, according to our previously reported paper[9]. After conducting the two-stage ion exchange treatment, it was found that the CTAB-modified muscovite produced an intercalated structure with the basal spacing of 2.79 nm. A similar organomodified-muscovite was used in this study.

2.3 Preparation of composite

Nanocomposite with three different filler loadings was prepared by melt mixing using an internal mixer (Polydrive Thermo Haake R600) at the temperature of 190 °C and rotor speed of 60 rpm. Both clays and polymers were dried for 12 hr at 80 °C prior to their use. The sequences of material addition were: ABS (3 min), followed by various clay loading concentrations, namely 1, 3, and 5% by weight of ABS with a total mixing time of 8 min. The compounded materials were then subjected to compression molding (using a Kao Tieh GoTech compression machine) into a 3 mm thick sheet, with the temperature set at 200 °C. The preheating time was set at 6 min, followed by hot compression for 3 min, and then cold-compressed for another 3 min. The detailed compositions of the compounds are listed in Table 1.

Table 1. Formulation of composite samples (wt%)

Sample code	ABS (wt%)	Untreated muscovite (UM) (wt%)	Treated muscovite (TM) (wt%)
Neat ABS	100	0	0
ABS/UM 1	99	1	0
ABS/UM 3	97	3	0
ABS/UM 5	95	5	0
ABS/TM 1	99	0	1
ABS/TM 3	97	0	3
ABS/TM 5	95	0	5

2.4 Characterization

X-ray diffraction (XRD) analysis was recorded on a Bruker D8 Advance diffractometer with a Cu target of k = 1.5405 A° at a generator voltage of 35 kV and a generator current of 30 mA. The diffraction patterns in the 2Θ range from 1° to 10° were collected using a step size of 0.05 and scan speed of 3°/s. The interlayer distance of organoclay in the composites was calculated using Bragg's equation of $2d\sin\Theta = n\lambda$. A flexural test was carried out in the three-point bending test using Lloyd tensile machine (Model 5533) according to ASTM D-790 at a strain rate of 5 mm/min. The specimen dimensions were 3 mm in thickness and 12 mm in width. The distance between the supports was 75 mm. Five samples were tested for each composition and their average values were reported. The observation of flexural fracture morphology of ABS and its nanocomposites was carried out using a scanning electron microscope (SEM) model ZEISS Supra 35 VP. The samples were mounted on aluminum stubs, sputter coated with a thin layer of gold prior to testing in order to avoid electrostatic charging, and observed at a magnification of 300 and 5000k×.

3. Results and discussion

3.1 X-Ray Diffraction

XRD is the best approach to study the changes in basal spacing and intercalation properties of clay in polymer matrix. Figure 2 displays the XRD patterns of muscovite, Li-Muscovite, and TM in the 2θ range of 1° to 10°. Muscovite showed a d(002) peak at 2θ = 8.87°, corresponding to an interlayer spacing of 0.99 nm. After the treatment with alkaline salt, Li-Muscovite exhibited a basal reflection peak at 2θ = 7.30°, corresponding to an interlayer spacing of 1.21 nm. After the two-stage ion exchange treatment, the d(002) plane diffraction peak of TM shifted to 3.17°, corresponding to a d-spacing of 2.79 nm. This observation indicated that modifying pristine muscovite with CTAB increased the interlayer spacing, which led to a shift in the diffraction peak towards a lower angle. A shift to lower angles suggests an increase in the interlayer spacing of nanoclay, which shows that the structure has intercalated.

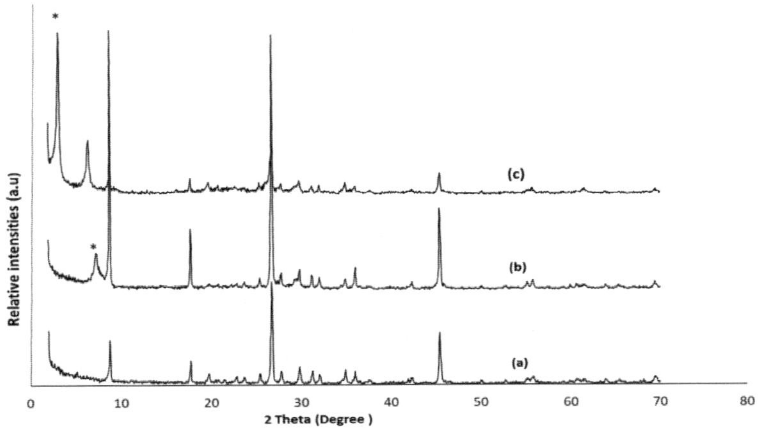

Figure 2. XRD diffraction patterns of a) muscovite, b) Li-Muscovite, and c) TM

3.2 Flexural testing

The results of the flexural strength and modulus of ABS/UM and ABS/TM composites following filler loading are illustrated in Figure 3. Regardless of the ion exchange treatment, the flexural strength of both untreated and treated muscovite-filled ABS showed a linearly increasing trend over neat ABS. Moreover, the flexural strength values of ABS/TM were higher than those of ABS/UM. The same observation was reported in the work conducted by E. Suryani [10]. When the clay loading was 5 wt%, the flexural strength increased from 78 MPa (control) to 85 MPa, corresponding to an improvement of 9%. The increment in flexural strength may also be due to the presence of silicate layer, which may improve the interfacial adhesion between the filler and the matrix.

Moreover, the addition of TM improved the dispersion and compatibility of hydrophilic clay and hydrophobic ABS, which facilitates effective stress transfer from matrix to clay. Therefore, it can be said that the incorporation of TM improved the compatibility with ABS matrix, which enables it to resist stress propagation when bending stress is applied to the samples. On the other hand, the flexural modulus of both ABS/UM and ABS/TM composites was observed to increase continuously with filler loading. Each system containing TM showed higher modulus increase than the ABS/muscovite system and neat ABS. As shown in Figure 3, the increase in flexural modulus from 2,687 MPa (control) to 3,435 MPa was achieved at 5 wt% loading of TM, corresponding to about 28% increment. This observation highlights the fact that the incorporation of organoclay, TM, into ABS can improve the mechanical properties of ABS composite.

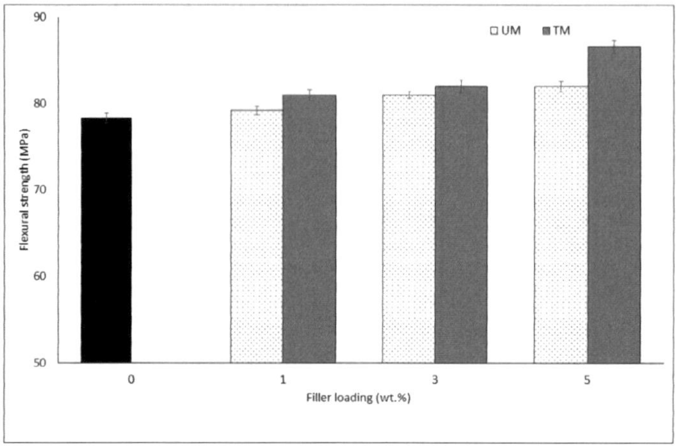

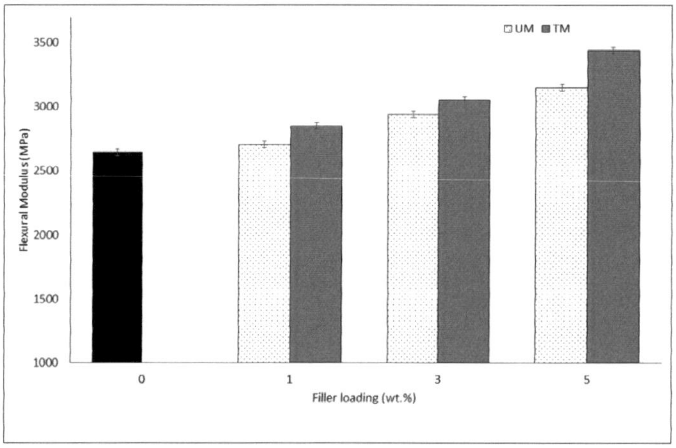

Figure 3: Variations of flexural strength and modulus for ABS/UM and ABS/TM composites

3.3 Morphological analysis

The fracture surface of neat ABS and ABS composites containing 5 wt% of filler is presented in Figure 4. From Figure 4 (a), it can be seen that neat ABS exhibited smooth fracture surface. Compared to ABS/UM composite, the micrograph revealed a slight increase of surface roughness after muscovite was embedded in ABS matrix. This probably suggests the inhomogeneity of filler dispersion, with some of the filler was found detached from the ABS matrix. Obviously, the occurrence of more voids and cavities are clearer, which creates sites for crack initiation and leads to localized stress concentration. The existence of gap signifies the weak and poor interfacial adhesion between the filler and the matrix, thus caused a detrimental effect on flexural properties.

On the other hand, the micrograph of ABS/TM as shown in Figure 4 (c) revealed a more homogenous dispersion of filler with reduced aggregation of the particles within matrix. In this case, TM has tight contact with the matrix, which improves compatibility, leading to good interfacial adhesion. Therefore, the improved interaction was subsequently validated with the results of flexural findings.

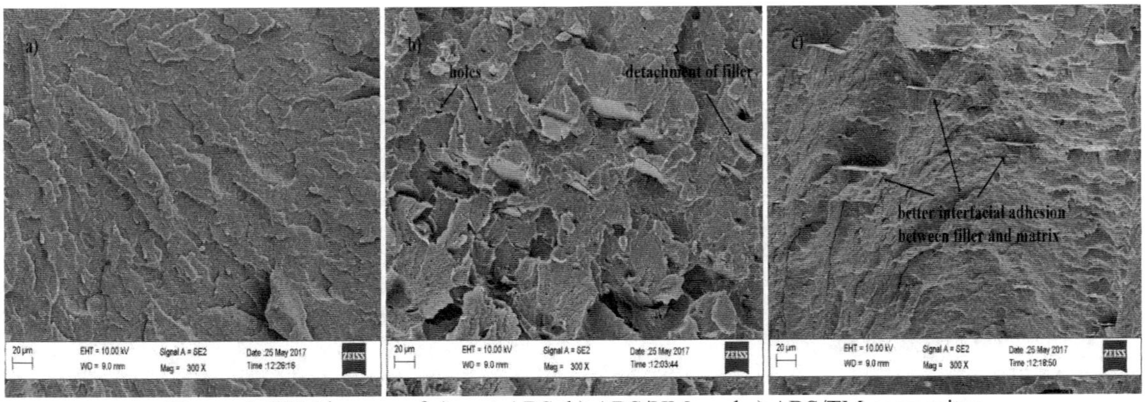

Figure 4: SEM images of a) neat ABS, b) ABS/UM, and c) ABS/TM composites

4. Conclusion

The flexural strength of ABS/UM and ABS/TM composites prepared by melt intercalation was investigated. Following the modification, intercalated muscovite was produced as viewed by XRD. The flexural strength of composites showed an increment of 9% in flexural strength and 28% in flexural modulus corresponding to neat ABS, which suggests that the flexural properties of composites could be enhanced by the addition of organoclay. In relation to this, it was found that ABS/TM nanocomposites displayed higher values of strength and modulus compared to ABS/muscovite composites. The improvement was contributed by the homogeneous dispersion of TM, which was provided by the two-step ion exchange treatment as indicated by FESEM.

References

[1] Kojima Y, Usuki A, Kawasumi M, 1993 *J Polym Sci Part A Polym Chem* 31 1755
[2] Fina A, Cuttica F, Camino G, 2012 *Polym Degrad Stab* 97 2619
[3] Maged A. Osman MP and UWS, 2003 *J Mater Chem* 13 2359
[4] Wang S, Hu Y, Zong R, 2004 *Appl Clay Sci* 25 49
[5] Choi YS, Xu M, Chung IJ, 2005 *Polymer* 46 531
[6] Xia Y, Jian X, Xia Y, 2007 *Polym Plast Technol Eng* 46 1173
[7] Basurto FC, García-López D, 2012 *Compos Part B Eng* 43 2222
[8] Lee HB, Lee HK, Shim MJ, Kim SW,1999 *Mater Chem Phys* 58 264
[9] Che Ismail NH, Ahmad Bakhtiar NSA, Md. Akil H, 2017 *Mater Chem Phys* 196 324
[10] Suryani E, Rashid A, Fadli M, 2011 *Appl Clay Sci* 52 295

Property Comparison Before and After Electron Beam Irradiation of Recycled Polypropylene/Peanut Shell Powder (Rpp/Psp) Composites

N.F. Zaaba[1], H. Ismail[1]

[1]School of Materials and Mineral Resources Engineering, Universiti Sains Malaysia, Engineering Campus, Penang, Malaysia

Email: fasehazaaba@usm.my

Abstract. This study investigated the properties of recycled polypropylene (RPP)/peanut shell powder (PSP) composites under the effect of electron beam (EB) irradiation. RPP/PSP composites were prepared by melt-mixing and compression moulding with 0 to 40 wt. % of PSP loading. The fabricated composites were then irradiated using 2.0 MeV EB accelerator at a fixed dose of 40 kGy. The properties of non-irradiated and irradiated composites were compared and characterized by tensile properties, Fourier transform infrared (FTIR), thermogravimetric analysis (TGA) and scanning electron microscopy (SEM). The results in tensile strength and tensile modulus of irradiated RPP/PSP composites show an increment while decrement in elongation at break. The thermal stability of irradiated composites was also improved compared to non-irradiated composites. The enhanced interfacial adhesion between components was evidenced by SEM morphology.

1. Introduction

Presently, owing to the growing demand for ambient and social queries, natural fillers are chosen as an alternative to substitute those damaging synthetic fillers In fact, due to low density and good mechanical resistance of natural fillers, the application of these materials as reinforcement in thermoplastic and thermosetting polymers are getting higher [1].

Though, like other natural plant based resources, natural fillers have high moisture absorption tendency and poor surface adhesion to hydrophobic polymers. Thus, it is not suitable for high temperature applications. In order to deal with the weak interfacial adhesion between natural fillers and polymer matrices, a number of filler surface modification techniques have been amplified to enhance the interfacial bond strength. Among all, the ionizing radiation is a process that capable to modify the properties of natural filler polymeric materials. It can be used as an alternative in the development of new polymeric materials by enhancing the interfacial adhesion between natural fillers and polymer matrix [[2].

Moreover, the main effects triggered by the ionizing radiation in polymers are the chain scission and appearance of crosslinking. Once the crosslinking is in high proportion, it tends to create three-dimensional nets. The procedure take place by the recombination of radicals formed by irradiation of polymer. In the crossed linking, the polymeric chains attach themselves via covalent bond caused by chemical joining or radiation. The crosslinked structure could reduce the interfacial tension and resulted in an improvement in material properties. [3].

Numerous studies have discovered that ionizing radiation have proficiently improve the material properties by introducing the cross linking and scission process throughout the polymer chains in multipurpose applications [4]. On the face of it, no attempt has been made to study the effect of EB irradiation on the physical, thermal, mechanical and morphological properties of recycled polypropylene (RPP)/ peanut shell powder (PSP) composites. It is expected that, it could induce the formation of crosslinking network and enhance overall properties of the composites.

2. Experimental study

2.1 Materials

Recycled polypropylene (RPP) with melt flow index of 30 g/10 min, density of 0.896 g/cm^3, and tensile modulus of 900 MPa was purchased from Zarm Scientific and Supplies, Penang, Malaysia. Peanut shell powder (PSP) was also supplied by Zarm Scientific and Supplies, Penang, Malaysia. Prior to composite fabrication, the peanut shells were crushed to produce particle sizes in the range of 70 to 250 µm. Then, the PSP was dried for 3 h at 70 °C using a vacuum oven.

2.2 Composite preparation

RPP/PSP composites were prepared using an internal mixer (Rheomixmixer, Haake, Model R600/610) at 180 °C and 50 rpm to obtain a homogeneous sample. The RPP was placed in the mixer and melted for 4 min before the PSP was added. The composites were mixed for another 8 min until the mixing torque stabilized. The total mixing time was 12 min for all samples. The processed samples were compression molded into a 1-mm thick sheet by an electrically heated hydraulic press (Compression Machine, Kao Tieh Go Tech) at temperature of 180 °C. Table 1 shows the formulation of the composites.

Table 1. Formulation of RPP/PSP composites

Composite	RPP (wt.%)	PSP (wt.%)
RPP	100	-
RPP + 10% PSP	90	10
RPP + 20% PSP	80	20
RPP + 30% PSP	70	30
RPP + 40% PSP	60	40

2.3 Electron beam irradiation

The composite sheet produced through melt-blending and compression-molding as mentioned earlier were then irradiated using an electron beam (EB) accelerator (NHV EPS, Model-3000) facility installed at the Malaysian Nuclear Agency, Bangi, Malaysia at a fixed dose of 40 kGy. All samples were irradiated at room temperature with accelerator energy of 2.0 MeV and beam current of 2mA.

2.4 Measurement of tensile properties

Tensile tests were carried out using a universal testing machine (Instron, Model 3366) according to ASTM D638. Dumbbell specimens of 1 mm thickness were cut from the compression molded sheets with a Wallace die cutter. A crosshead speed of 5 mm/min was used, and the tests were performed at 25 ± 3 °C.

2.5 Thermogravimetric analysis

A Perkin Elmer Pyris TGA analyser was used to determine the TGA of non-irradiated and irradiated RPP/PSP composites. About 10 mg of samples required in this analysis under a nitrogen atmosphere at 50°C to 600°C and 20°C/min heating rate.

2.6 Fourier transform infrared spectroscopy analysis

The functional groups and chemical characteristics of the composites were obtained by FTIR (transmission mode method) with a Perkin Elmer System 2000 at a resolution of 4 cm^{-1} in a spectral range of 4,000 to 550 cm^{-1}, using 32 scans per sample.

2.7 Morphology evaluation

The microstructure of the tensile fractured surfaces of the RPP/PSP composites were compared using a SEM ZEISS Supra, Model 35 VP operating at 5 kV. The samples were first sputter-coated with a thin layer of carbon, which provides an electron-transparent (low atomic number), conductive coating.

3. Results and Discussion

3.1 Tensile properties

Figures 1-3 demonstrate the results of tensile strength, elongation at break and tensile modulus of non-irradiated and irradiated RPP/PSP composites at different PSP loading, respectively. Obviously from Figure 1, the tensile strength of both non-irradiated and irradiated RPP/PSP composites decreased with an increase in PSP loading. The decline was due to the incompatibility between hydrophilic PSP and hydrophobic RPP. Owing to the different polarities, there was a weak interfacial adhesion between PSP and RPP. As comparing both composites, the irradiated composites showed a higher tensile strength than the non-irradiated composites. The increment in tensile strength can be related to the formation of crosslinking during the EB irradiation [5].

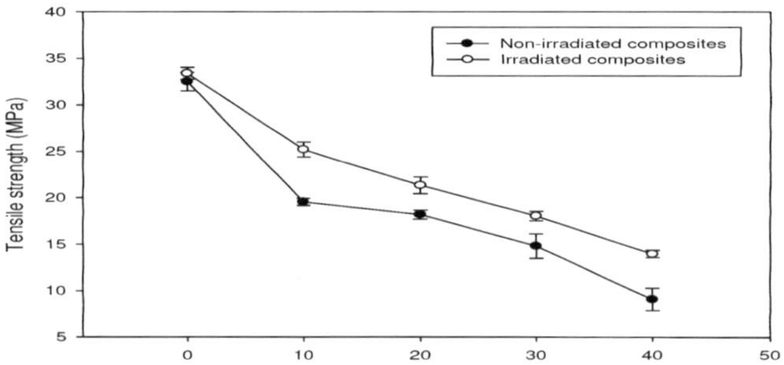

Figure 1. Tensile strength of non-irradiated and irradiated RPP/PSP composites at different PSP loading

Figure 2 shows the elongation at break of non-irradiated and irradiated RPP/PSP composites at different PSP loading. The elongation at break of both composites lessened with a rise in PSP loading. The lessening was due to the decrease in polymer chain mobility or deformability of a rigid interface between the PSP and the RPP matrix. Again, when comparing both composites, the irradiated RPP/PSP composites showed a lower elongation at break than the non-irradiated composites. The reflection is accredited to the existence of irradiation which had persuaded the crosslinking in the irradiated RPP/PSP composites [6].

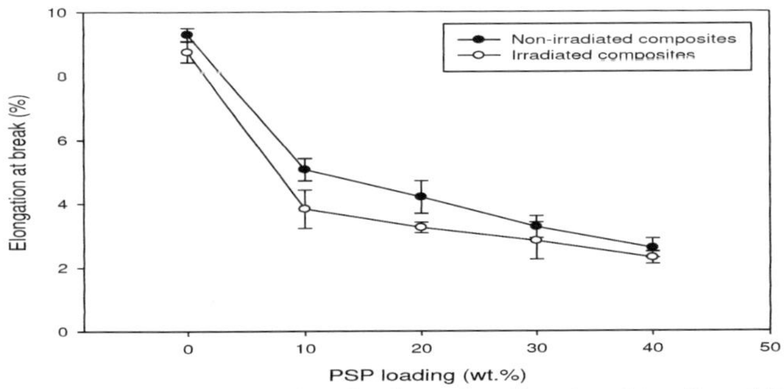

Figure 2. Elongation at break of non-irradiated and irradiated RPP/PSP composites at different PSP loading

Likewise, the tensile modulus of non-irradiated and irradiated RPP/PSP composites is also shown in Figure 3. Apparently, the presence of PSP was discovered to increase the tensile modulus of RPP/PSP composites. This can be accredited to the composite stiffness increases as the filler content increased. [7]. Conversely, when comparing both composites, the irradiated composites showed a higher tensile modulus than the non-irradiated composites. The formation of crosslink network structure upon exposure to irradiation caused improvement of interfacial adhesion between filler and matrix [8]. Eventually, leads to the enhancement of tensile modulus of irradiated composites.

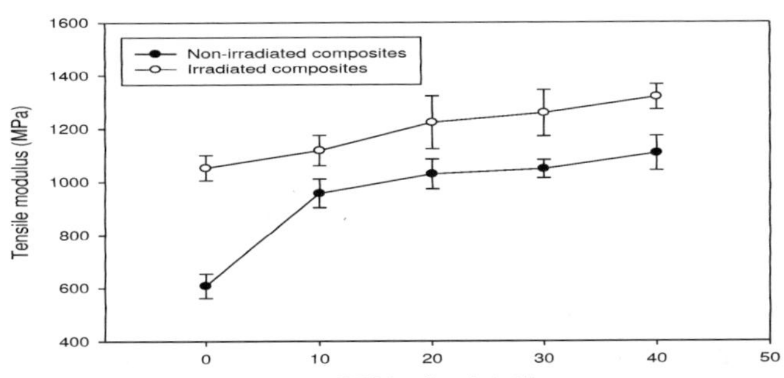

Figure 3. Tensile modulus of non-irradiated and irradiated RPP/PSP composites at different PSP loading

3.2 Thermogravimetric decomposition behaviours

As can be seen in TG thermograms in Figure 4, all composites experienced the three stages of weight losses; first region (130-170 °C), second region (200-400 °C), and third region (above 400 °C). Around 130-170 °C, the TGA result shows the removal of absorbed water in the composites [9]. Also, the decomposition temperature in the range 200-400 °C validates the thermal degradation of hemicellulose and cellulose while above 400 °C goes to the decomposition temperature of lignin and RPP [10].

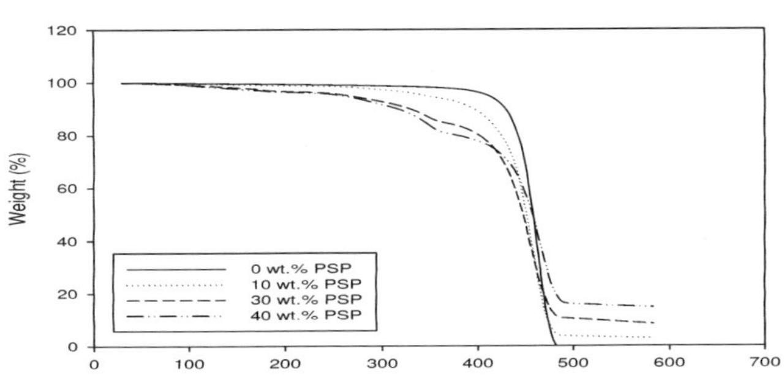

Figure 4. TG thermograms of irradiated RPP/PSP composites

3.3 Fourier transform infrared spectroscopy

Figure 5 exhibits the FTIR spectroscopy of non-irradiated and irradiated RPP/PSP composites at 10 wt.% PSP loading. The comparison of chemical change that might be occurred after being subjected to EB irradiation could be determined through this spectroscopy. After the EB irradiation, some changes in the absorption bands which were related to chemical reaction occurred in the blends can be observed in Figure 5(b). As can be seen in the figure, there was an increase in the intensity of the O-H stretching absorption bands at 3000-3600 cm^{-1}. This band was associated to the formation of oxidative products during the irradiation. This result is in agreement with Sam et al. in their study on LLDPE/soya powder blend [11]. In addition, there is a slightly increase and broad peaks existed at 1600-1800 cm^{-1} implied the formation of carbonyl group (like carboxylic acids at 1713 cm^{-1}, ketones at 1718 cm^{-1} and esters at 1741 cm^{-1}) whereas an oxidative product generated during irradiation [5]. Figure 6 shows the proposed crosslinking between propylene chains of RPP after being subjected to EB irradiation.

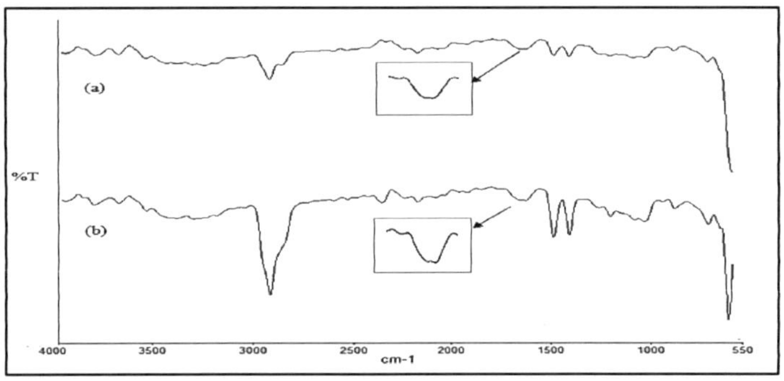

Figure 5. FTIR spectra of (a) non-irradiated and (b) irradiated RPP/PSP composites at 20 wt. % PSP loading and a zoom at C=O band

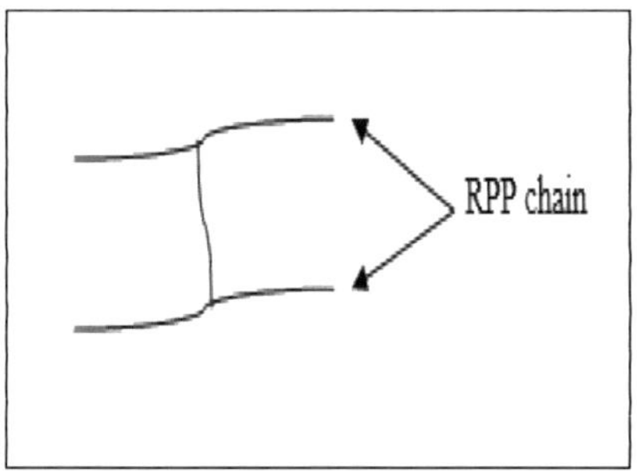

Figure 6. Proposed crosslinking between propylene chains of RPP

3.4 Morphological properties

Figure 7 demonstrates the morphology of RPP/PSP composites before and after being subjected to EB irradiation process. As shown in Figure 7(a), (b) and (c), for non-irradiated composites, the micrograph of tensile fractured surface presented a distinguish filler pulled out as well as filler tearing. As PSP loading increased, the presence of filler pulled out or remained loosely within the RPP matrix was more visible. More voids were also distinguished, indicated the places where the fillers were pulled out. Furthermore, the voids also reveal the poor adhesion of the filler to the matrix, where it acts as stress concentration points and contributes to the decrement of mechanical properties. Hence, lower tensile strength and elongation at break were obtained for non-irradiated composites.

While for irradiated composites, a considerable improvement in adhesion between filler and matrix can be observed in Figure 7(d), (e) and (f). The filler coated within the RPP matrix and the absence of the voids were an evidence of the improvement in adhesion mechanism [12]. The PSP filler surface is coated with RPP matrix indicated the good adhesion between them whereas the voids that being detected earlier are almost covered up after subjected to the EB irradiation. Therefore, the tensile properties of irradiated RPP/PSP composites were improved.

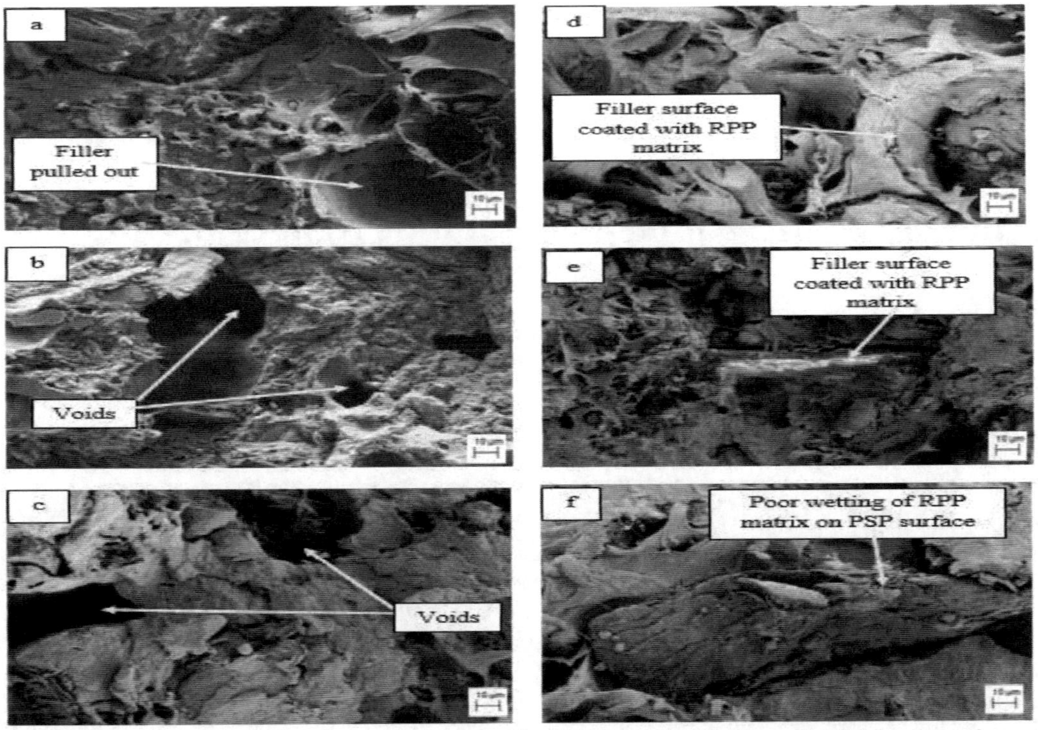

Figure 7. Tensile fractured surface of (a) 10 wt.%, (b) 20 wt.%, (c) 40 wt.% PSP loading of non-irradiated RPP/PSP composites and (d) 10 wt.%, (e) 20 wt.%, (f) 40 wt.% PSP loading of irradiated RPP/PSP composites

4. Conclusion

The overall results have shown that there is an improvement in the tensile strength and tensile modulus of irradiated RPP/PSP composites while decrement in elongation at break. The thermal stability of irradiated composites was also improved compared to non-irradiated composites. The enhanced interfacial adhesion between components of irradiated composites was evidenced by SEM morphology.

References

[1] Mohanty A K, Misra M, and Hinrichsen G 2000 *Macromol Mater Eng* 1 276.
[2] Ismail H, Galpaya D, and Ahmad Z 2010 *Journal of Vinyl Additives and Technology* 16 141-146.
[3] Nordin R, and Ismail H 2013 *International Journal of Engineering Research and Application* 3 1820.
[4] Choi H Y, Han S O, and Lee J S 2008 *Applied Surface Science* 255 2466.
[5] Murray K A, Kennedy J E, McEvoy B, Vrain O, Ryan D, and Cowman R 2013 *Nuclear Instruments and Methods in Physics Research B* 297 64-74.
[6] Sabet M, Hassan A, and Ratnam C T, *Polymer Bulletin* 68 2323.
[7] Ismail H, Abdullah A H, and Bakar A A 2011 *Journal of Vinyl and Additive Technology* 17 132.
[8] Ibrahim N A, Ahmad S N A, Yunus W M Z Y, and Dahlan K Z M 2009 *eXPRESS Polymer Letters* 3 226.
[9] Ma X, Yu J,and Kennedy J 2005 *Carbohydrate Polymers* 62 19-24.
[10] Liang J 2018 *Journal of Thermoplastic Composite Materials* 31 246-264.
[11] Sam S T, Ismail H, Ahmad Z, and Ratnam C T 2012 *Journal of Applied Polymer Science* 124 5220.
[12]Youssef H A, Ismail M R, Ali M A M, and Zahran A H 2009 *Journal of Elastomers and Plastics* 41, 245.

IOP Conf. Series: Journal of Physics: Conf. Series **1082** (2018) 012018 doi:10.1088/1742-6596/1082/1/012018

Properties of Graphene Filled Polyaniline Nanocomposite

H.M. Soe[1], Y.Z.N.Htwe[1], J.L.Lee[1] and M.Mariatti[1]

[1]School of Materials and Mineral Resources Engineering, Engineering Campus, Universiti Sains Malaysia, 14300 Nibong Tebal, Pulau Pinang, Malaysia

Email: mariatti@usm.my

Abstract. Graphene, a new carbonous material with two-dimensional (2D) nanostructure, has attracted great attention both in fundamental science and applied research. Polymer nanocomposites have been the subject of fundamental scientific interest due to their immense applicational potential and could be used in differential application. Among the conducting polymers, polyaniline (PANI) is the most promising material for electronic application due to its excellent capacity for energy storage, easy synthesis, high conductivity, and low cost. Graphene were applied in PANI to produce graphene filled PANI nanocomposite. In this study, PANI was synthesized by using interfacial polymerization. Graphene filled polyaniline nanocomposites were fabricated by sonochemical method. The morphology and properties of the composites were characterized by scanning electron microscopy (SEM), high resolution transition electron microscopy (HRTEM), fourier transform infrared (FTIR) spectroscopy and electrical conductivity test. Graphene filled PANI nanocomposites has great applicability in electronic, optical and magnetic materials. From the results of four point probe resistivity measurement, graphene filled PANI nanocomposite shows better electrical conductivity, 17.2831 S/cm which increases around 17 times compared to pure PANI, 1.1617 S/cm.

1. Introduction

Graphene is essentially a single layer of graphite in the form of a hexagonal lattice of carbon atoms bonded in the sp^2 configuration. Due to its ultrathin, two-dimensional, sp^2-bonded carbon nature and unprecedented properties, graphene have been intensively studied [1]. The combination of the nanomaterial with polymer is very attractive not only to reinforce polymer but also to introduce new electronic properties based on the morphological modification or electronic interaction between the two components. Depending on the nature of the components used and the method of preparation, significant differences in composite properties may be obtained. Conductive polymer composite have several advantages over their pure metal counterparts including lower cost, ease of manufacture, high flexibility, reduced weight, mechanical shock absorption ability, corrosion resistance and conductivity control [2].

Polyaniline is most likely chosen to be the conductive polymer backbone for graphene polymer composites because of its ease of synthesis, low cost, excellent environmental stability and many interesting chemical, electrical and optical properties [3]. Composite of graphene nanopowder (GNP) coated with PANI produced by in situ polymerization of aniline monomer was reported by Li et.al [4]. In this study, PANI was synthesized by using interfacial polymerization. The objectives of the present study are to synthesize PANI using surfactant and to investigate the properties of PANI nanocomposite produced using multilayer graphene.

2. Experimental study

2.1 Materials

The raw materials and chemical used in Polyaniline synthesis were aniline (ANI), ammonium persulfate (APS), hydrochloric acid (HCl) and sodium dodecylbenzene sulfonate (SDBS). Commercial multilayer of GNP used in graphene filled PANI nanocomposite. GNP and HCl were supplied by SkySpring Nanomaterials, Inc. and J.T. Baker and ANI, APS and SDBS were supplied by Sigma-Aldrich Sdn. Bhd.

2.2 Methodology

2.2.1 Synthesis of polyaniline. In the synthesis of PANI, 4 mmol (0.372 g) of ANI was dissolved in 20 mL HCl (1 M) and stirred with magnetic stirrer for one hour. About 4 mmol (0.913 g) of APS was dissolved in 20 mL HCl (1 M) and magnetically stirred for one hour. APS was used as an oxidant (initiator). Then, the APS solution was added into the ANI solution and was left at room temperature for 6 hours. Finally, the black green precipitate was collected and washed several times by acetone and distilled water. The pure PANI was then dried at ambient temperature for 24 hours.

Another set of PANI is prepared using the same method as above; the only difference is that surfactant with amount of 0.7 g of SDBS was added in the solution of APS and HCl. After both APS with SDBS solution and ANI solution was mixed, the colour of polymerization solution changed with polymerization step. In the initial stage, ANI/ SDBS/ APS solution was colourless with high transparency. As the polymerization proceeded, the colour of the solution changed from white through yellow, brown and finally to green, which indicates the formation of PANI emeraldine salt (ES).

2.2.2 Fabrication of Commercial multilayer graphene filled polyaniline nanocomposite. In the synthesis of commercial multilayer graphene filled PANI nanocomposite, GNP suspension was prepared by adding 0.5 g of graphene in the solution of HCl (100 mL). 2 mmol (0.7 g) of SDBS was added into 20 mL GNP suspension (5 g/L) and the mixture was ultrasonicated for 30 min. About 4 mmol (0.372 g) of ANI was rapidly poured into the above mixture and stirred for another one hour. Then, about 4 mmol (0.913 g) of APS, (the molar ratio of ANI/ APS is 1) which dissolved in 20 mL HCl (1 M) and magnetically stirred for one hour was added to the mixture drop by drop. The reaction was kept for 20 h with continuous stirring in the ice bath. Finally, the product was collected and washed repeatedly with distilled water, then dried in a vacuum oven at 80 °C for 24 h.

2.3 Characterization

In this study, PANI powder and commercial multilayer graphene (CMLG) filled PANI nanocomposite was characterized using a Perkin Elmer Spectrum One FTIR spectrometer by transmittance techniques. The surface topography, morphology of PANI, PANI nanocomposite samples were observed under FESEM model Zeiss Supra 35VP. HRTEM (Tecnai G2 20 s-twin, Fei.) was employed to monitor the multilayer graphene, surfactant-modified PANI and CMLG filled PANI composite powder morphology in terms of particle size and shape. A four point probe is a simple apparatus for measuring the resistivity of polymer samples and polymer composites sample.

3. Results and discussion

PANI was added with surfactant and CMLG filled PANI nanocomposites. Figure 1 shows the combination FTIR spectrum of PANI and PANI with SDBS nanocomposite. The vibrational bands of 850, 1150, 1300, 1500 and 1600 cm^{-1} are referred to formation of PANI while vibrational bands of 1000, 2900 and 3200 cm^{-1} are referred to incorporating of SDBS into PANI [6]. The peak at 850 cm^{-1} is attributed to the out-of-plane bending of C-H. The peak at 1150 cm^{-1} is due to quinonoid unit of doped PANI [7]. The peak at 1300 cm^{-1} is attributed to the C-N stretching mode. The main peaks at 1500 and 1600 cm^{-1} can be associated with C=N and C=C stretching vibrations of quinone and benzene rings, respectively [8]. The vibrational band at 1000 cm^{-1} is attributed to S=O bond, 2900 cm^{-1} which is assymmetric $-CH$ stretching and 3200 cm^{-1} which is symmetric $-CH_2-$ bond [9]. The presence of the above bands suggested the PANI was coated successfully on the surface of graphene.

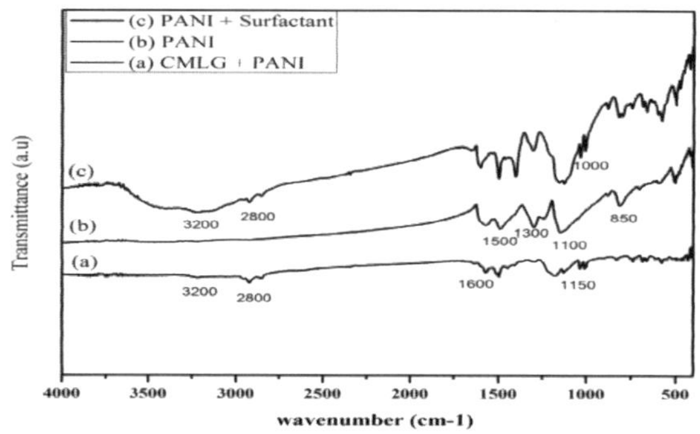

Figure 1. FTIR spectrum of the combination of (a) PANI with graphene, (b) PANI and (c) PANI with SDBS surfactant

The SEM images of PANI and PANI with SDBS surfactant show in Figure 2 and 3 respectively. In Figure 2, surfactant-free PANI shows a thicker and more aggregated structure compared to the surfactant-modified PANI (Figure 3). The particle size and surface morphology were dependent on surfactant because the surfactant was adsorbed physically to the growing polymer. SDBS surfactant consists of hydrophilic sulfonic acid and long hydrophobic alkyl chain [10]. The surfactant-modified PANI with hexagonal shaped is believed to have better performance compared to thicker, globular shaped surfactant free PANI, due to the high surface area and small diameter of hexagonal shaped PANI [11].

Figure 2. SEM images of PANI at magnifications of a) 5 KX, b) 10 KX and c) 30 KX

IOP Conf. Series: Journal of Physics: Conf. Series **1082** (2018) 012018 doi:10.1088/1742-6596/1082/1/012018

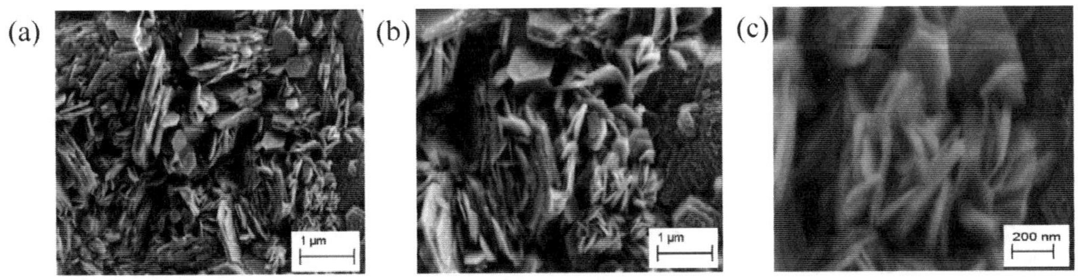

Figure 3. SEM images of PANI added with surfactant at magnifications of a) 5 KX, b) 10 KX and c) 30 KX

Figure 4 shows the SEM images of multilayer graphene filled PANI nanocomposite with different magnifications. It shows that PANI have been grown on graphene sheet's surface. There have some sharp corner and semi-transparent of graphene sheet between PANI. In some places embedded flakes of graphene in PANI matrix are seen, suggesting graphene interconnection with the polymer network and forming a highly conducting network.

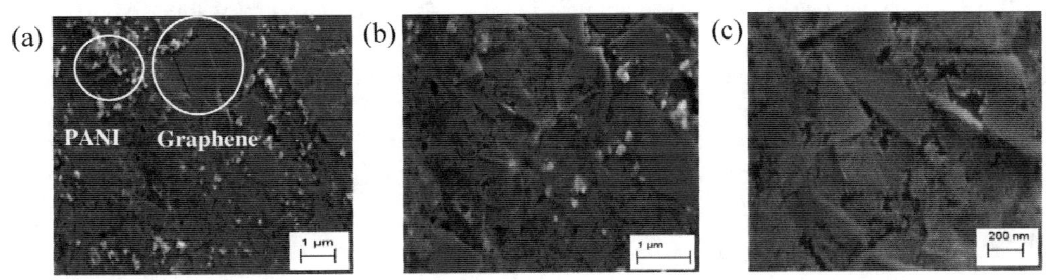

Figure 4. SEM images of CMLG filled PANI nanocomposite at magnifications of a) 5 KX, b) 10 KX and c) 30 KX

The morphology and structure of surfactant-modified PANI and CMLG filled PANI nanocomposite can be characterized by using HRTEM. Figure 5 shows HRTEM images of PANI added with surfactant, at 10 K and 50 K magnifications. PANI appears irregular, globular shaped in Figure 5. There is some agglomeration also observed in the images. Figure 6 shows HRTEM images of CMLG filled PANI nanocomposite. It shows that graphene sheets appear thin and semi-transparent while PANI appear globular shape in the surface. It is observed that graphene exists not only on the surface but also embedded in the PANI particles. These results confirm that the interaction between graphene and PANI produced a PANI coated graphene sheet. This suggests that the graphene sheets which has high specific surface area which could makes a strong connection with PANI when they are formed. Graphene participate directly in the polymerization of the monomer. Therefore, the properties of the resulting composite particles are strongly dependent on the surface area and thickness of the graphene [12].

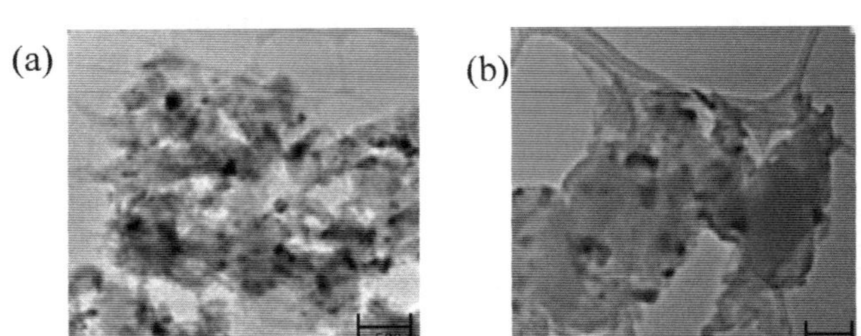

Figure 5. HRTEM images of PANI added with surfactant at magnifications of a) 10 K and b) 50K

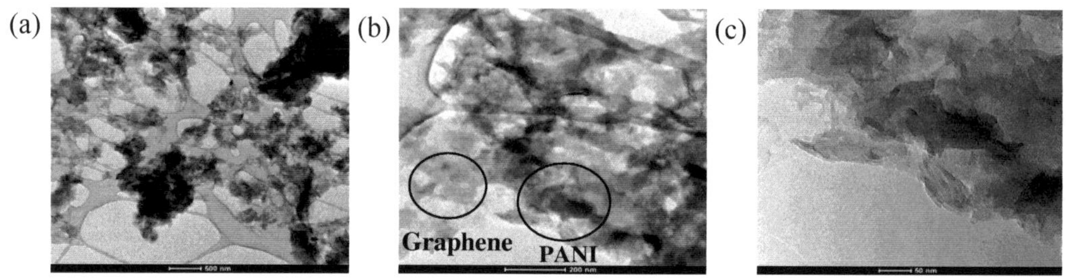

Figure 6. HRTEM images of CMLG filled PANI nanocomposite at magnifications of a) 10 K, b) 50 K and c) 100 K

The electrical conductivities of PANI and CMLG filled PANI nanocomposites are determined using a four point probe resistivity measurement system. Figure 7 shows the comparative bar diagram of PANI and CMLG filled PANI nanocomposites. Doping is the process by which PANI are exposed to charge transfer agent, SDBS through electrochemical oxidation or reduction. Doping of PANI creates new states (donor or acceptor states), which exist within the band gap and are energetically accessible to the pi electrons [13]. This process will increase the conductivity of the PANI because of the increased concentration of charge carriers [14]. In this figure, CMLG filled PANI nanocomposite has higher electrical conductivity than PANI. The increasing in conductivity is due to the high surface area of graphene sheets which could serve as effective percolative conducting bridges. Besides that, the PANI conjugated chains offer lots of conductive paths on the surface of graphene layer, which activates more electrons to participate in the directional movement. The PANI coated graphene layers may cause some interaction between PANI chains and graphene. The synergistic effect of these two parts leads to increase of the conductivity of the composites [15].

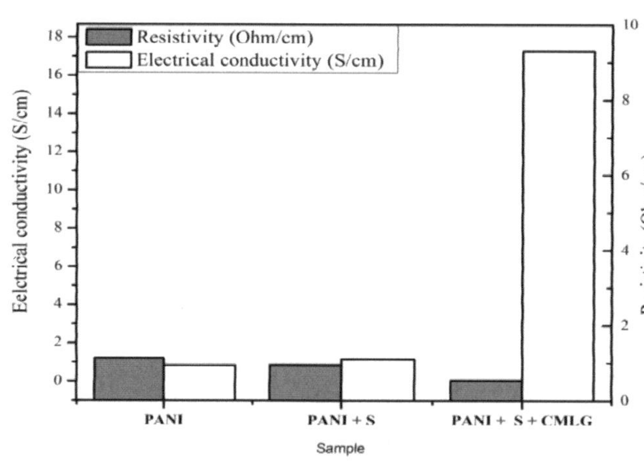

Figure 7. Electrical conductivities of PANI and CMLG filled PANI nanocomposites

4. Conclusion

Incorporation of SDBS which was used in polymerization modifies the structure of PANI. The HRTEM, SEM morphology and presence of the bands at 500, 850, 1000, 1150, 1300, 1500, 1600, 2900 and 3200 cm^{-1} in FTIR spectrum suggested that the PANI was coated successfully on the surface of graphene. From the results of four point probe resistivity measurement, CMLG filled PANI nanocomposite shows a better electrical conductivity compared to pure PANI. The electrical conductivity of surfactant-modified PANI shows a higher value which is 1.1617 S/cm compared to pure PANI which is 0.8319 S/cm. The incorporating of multilayer graphene into PANI increases the electrical conductivity of PANI.

Acknowledgements

The authors would like to express appreciation for the financial support from Fundamental Research Grant from Ministry of Education (account no. 6071284) and AUN/SEED-Net, as well as support from the School of Materials and Mineral Resources Engineering Campus, Universiti Sains Malaysia.

References

[1] Avouris P & Dimitrakopoulos C 2012 *Materials Today* 15 86
[2] Teh P, Ng H & Yeoh C 2011*Malaysian Polymer Journal* 6 98
[3] Goswami S, Maiti U, Maiti S, Nandy S, Mitra M & Chattopadhyay K 2011 *Carbon* 49 2245
[4] Li X, Song H, Zhang Y, Wang H, Du K, Li H, Yuan Y & Huang J 2012 *International Journal of Electrochemical Science* 7
[5] Wang L, Ye Y, Lu X, Wen Z, Li Z, Hou H & Song Y 2013 *Scientific reports* 3
[6] Shreepathi, S 2006 Dodecylbenzenesulfonic acid: a surfactant and dopant for the synthesis of processable polyaniline and its copolymers.
[7] Vera A R, Romero B H & Ahumada E 2003 *Journal of the Chilean Chemical Society* 48 35
[8] Abdulla H S & Abbo A I 2012 *Int J Electrochem Sci* 7 10666
[9] Kuczynska, A, Uygun A, Kaim A, Wilczura-Wachnik H, Yavuz A G & Aldissi M 2010 *Polymer International* 59 1650
[10] Han M G, Cho, S K, Oh S G & Im S S 2002 *Synthetic Metals* 126 53
[11] Weiller B H, Virji S, Baker C, Huang J, Li D & Kaner R B 2005 *Nanotech* 2 281
[12] Zhang S, Lin L, Deng H, Gao X, Bilotti E, Peijs T, Zhang Q & Fu Q 2012 *Express Polym Lett* 6 159
[13] Parvez K, Li R, Puniredd S R, Hernandez Y, Hinkel F, Wang S, Feng X & MüLlen K 2013 *ACS nano* 7 3598
[14] Stejskal J, Omastová M, Fedorova S, Prokeš J & Trchová M 2003 *Polymer* 44 1353
[15] Harish C, Sreeharsha V S, Santhosh C, Ramachandran R, Saranya M, Vanchinathan T M, Govardhan K & Grace A N 2013 *Advanced Science Engineering and Medicine* 5 140

IOP Conf. Series: Journal of Physics: Conf. Series **1082** (2018) 012019 doi:10.1088/1742-6596/1082/1/012019

Extraction and Characterization of Starch From Mango Seeds

N.A. Shahrim[1], N.Sarifuddin[1], H. Ismail[2]

[1] Department of Manufacturing and Materials Engineering, Kulliyyah of Engineering, International Islamic University Malaysia (IIUM), Kuala Lumpur, Malaysia

[2] School of Materials and Mineral Resources Engineering, Engineering Campus, Universiti Sains Malaysia (USM), Penang, Malaysia

Email: norshahida@iium.edu.my

Abstract. The mango (*Mangifera indica*) seed starch were extracted via distillation, alkaline, sedimentation and centrifugation method. The effects of extraction methods on its morphology, structural and physicochemical properties extracted were comparatively studied. Morphological characteristics like shape and size of the starch granules, display significant differences under scanning electron microscope (SEM). The extracted mango seed starch granules showed oval and irregular or cuboidal-shaped with smooth surface despite of different extraction method. Distillation method yields smallest (13.80 ± 2.90 μm) starch granules in size, followed by alkaline method (15.83 ± 3.65 μm), sedimentation method (15.97 ± 4.44 μm) and centrifugation method (19.91 ± 6.82 μm). X-ray diffraction pattern whereas, showed the isolated mango seed starch is an A-type starch, hence, corroborated to suggest the morphology of the starch granules by their size. Moreover, starch isolated via alkali method had highest starch yield of 91.72 % than starch isolated via distillation (81.87 %), sedimentation (74.82 %) and centrifugation (74.56 %), suggest that mango seed as one of the potential sources of starch.

1. Introduction

Starch is greatly in abundance and the main source of carbohydrate constituent for most plants [1]. The growing demand for starch in many applications including food, pharmaceutical, paper, adhesive and textile has rising attention in discovering new sources of this type of polysaccharide from other plants or fruits [2]. Many attempts have been focused on the use of starch obtained from various plant sources such as rice, wheat, corn and potato [3]. Yet, limited work available on the extraction of starch from fruit by-products such as seeds. Therefore, in this study, native starch is obtained from mango seed.

Mango (*Mangifera indica*) is a popular tropical fruit that is widely grown and cultivated in many tropical regions of the world including Malaysia, Thailand, and India [4]. Typically, in Malaysia, the mango seeds are discarded as wastes after the fruits are eaten. Thus, due to the availability and be a reliable source of starch, this type of agricultural waste could be used as value-added products [5]. Hence, with appropriate extraction method (via distillation, alkaline, sedimentation and centrifugation), this study is aiming to evaluate the morphological properties of starch extracted from mango seeds.

2. Experimental study

IOP Conf. Series: Journal of Physics: Conf. Series **1082** (2018) 012019 doi:10.1088/1742-6596/1082/1/012019

2.1 Sample collection and seed treatment

Ripe mangoes Lily cultivar seeds were collected from the local market. The seeds were washed with tap water to remove their impurities and dried in the oven at 70 °C overnight. Then, the dried seed's kernels were separated manually using a kitchen knife to recover the seeds and dried in the oven at 50 °C for 24 hours. The brown layer (seed coats) were removed manually to obtain the cotyledons that were used to prepare flour by grinding using a laboratory grinder. The obtained flour was sieved through sieve (500-50 μm mesh), packed and sealed in a plastic bag, then stored in a refrigerator (< 5 °C) until further subjected to starch extraction.

2.2 Extraction of mango seed starch via distillation method

The mango seed starch was isolated according to the procedure described by Noor et al. [5] with minor modifications. Firstly, 5 g of mango seed flour was added to 100 ml of distilled water with continuous stirring for 6 hours at room temperature. Next, a cloth bag (about 200 μm mesh) was used for filtering the slurry and the remaining residues were washed with distilled water for three times. The filtrate was mixed and precipitated overnight at 4 °C. Finally, the obtained starch was filtered and dried in the oven at 40 °C for 24 hours. The starch was ground with a mortar and pestle, packed in a sealed plastic bag and kept at room temperature until further use.

2.3 Extraction of mango seed starch via alkaline method

The mango seed starch was isolated using procedure mentioned by Noor et al. [5] with some modifications. Firstly, 5 g of mango seed flour was added to 0.5% concentration of sodium hydroxide (NaOH) with continuous stirring for 6 hours at room temperature. Next, the slurry was filtered using a cloth bag (about 200 μm mesh) and the remaining residues were washed with distilled water for three times. Then, the filtrate was mixed and precipitated overnight at 4 °C. Finally, the obtained starch was filtered and dried in the oven at 40 °C for 24 hours. The starch was ground with a mortar and pestle, packed in a sealed plastic bag and kept at room temperature until further use.

2.4 Extraction of mango seed starch via sedimentation method

The mango seed starch was isolated using Oates and Powell [6] method with minor modifications. Firstly, 50 g of mango seeds were steeped in water overnight, washed and ground in a kitchen blender. Next, the slurry was filtered through a cloth bag (about 200 μm mesh) and the filtrate was reserved to sediment the starch. Then, the starch was re-slurried in distilled water and sedimented three times. The final product was sedimented in 0.1 M sodium chloride (NaCl) and 1/10 volume of toluene for 3 hours. Finally, the obtained starch was thoroughly washed with distilled water and dried in the oven at 50 °C for 24 hours. The starch was ground with a mortar and pestle, packed in a sealed plastic bag and kept at room temperature until further use.

2.5 Extraction of mango seed starch via centrifugation method

The mango seed starch was isolated using Hassan et al. [4] method with some modifications. Firstly, 50 g of mango seeds were steeped in 0.16% aqueous solution of sodium hydrosulfide hydrate at 50 °C for 24 hours. Next, the solution was decanted and the seeds were ground in kitchen blender. The ground slurry was filtered through a cloth bag (about 200 μm mesh), washed thoroughly with distilled water and re-slurried in distilled water for 1 hour. Then, the supernatant was decanted from the filtrate and the settled starch layer was re-suspended in distilled water. Finally, the starch was centrifuged at 5000 rpm for 30 minutes and dried in the oven at 50 °C for 6 hours. The starch was ground with a mortar and pestle, packed in a sealed plastic bag and kept at room temperature until further use.

2.6 Scanning electron microscopy (SEM) analysis

The morphology of starch granules was observed using a scanning electron microscope (JEOL, Model JSM 5600 SEM). The samples were in powder form and sputter coated with palladium (Polaron SC515) and observed under the SEM at an accelerating voltage of 7 kV.

2.7 X-ray diffractometry (XRD) analysis

The structural properties of starch samples were examined using a wide-angle x-ray diffractometry (Rigaku, Model Ultima IV diffractometer). The samples were in powder form and scanned using Cu-Kα radiation (wavelength (λ) = 1.54056 Å) at 40 kV and 40 mA. The scanning region of the angles (2θ) was from 4° to 70° at a scanning speed of 0.02°/min.

2.8 Starch yield determination

The percentage yield of isolated starches was determined using equation (1) according to Noor et al. [5]:

$$\text{Starch Yield (\%)} = \frac{\text{Initial weight} - \text{Final weight}}{\text{Initial weight}} \times 100\% \qquad (1)$$

3. Results and discussion

3.1 Morphological properties

Table 1 shows the morphology of starch granules. The micrographs of four starch samples revealed the presence of starch granules at ×500 magnification, from small to large and oval to irregular or cuboidal shape with smooth surface. There is a little significant change was observed among the starch granules with different extraction method. Distillation and centrifugation method had majority of starch granules with oval shape, whereas sedimentation method had more irregular shape, but to a lesser extent than that of alkaline method. The average diameter of the starch granules ranges from 13.80 µm to 19.91 µm. The range of starch granules in size for distillation method (< 18 µm), followed by alkaline method (< 23 µm), sedimentation method (< 25 µm) and centrifugation method (< 37 µm). This indicates that the different starch extraction method had little impact on the granule shape and size of mango seed starch. The results revealed are consistent with those observed by Kaur et al. [7] for mango kernel starch granules with 15 µm to 21 µm in size.

Table 1. Micrographs of mango seed starch granules and size

Extraction method	SEM micrographs (magnification × 500)	Starch granule size (µm)	
		Range	Average
Distillation	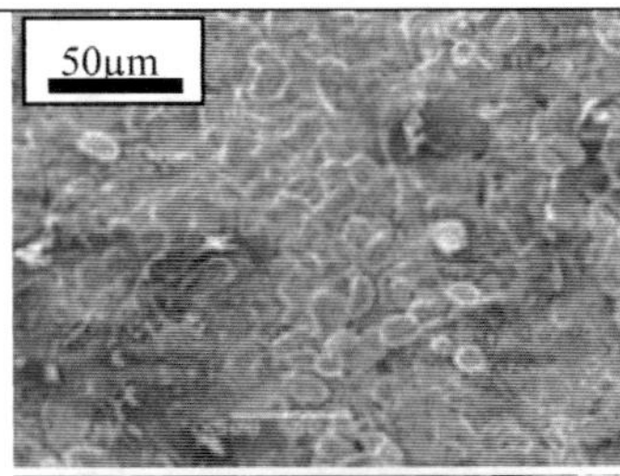	7.98 - 17.14	13.80 ± 2.90
Alkaline		10.20 - 22.30	15.83 ± 3.65
Sedimentation	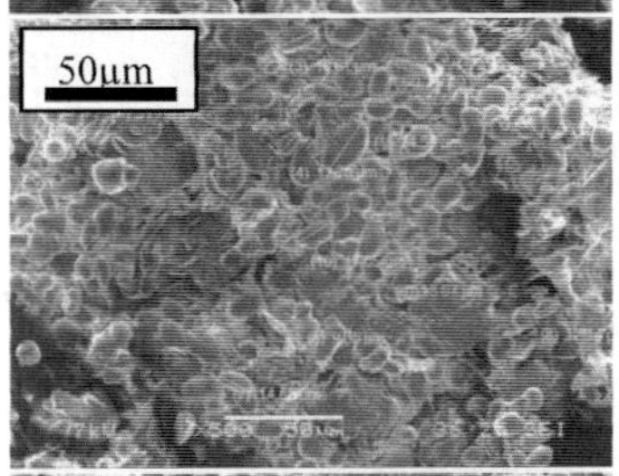	9.35 - 24.59	15.97 ± 4.44
Centrifugation	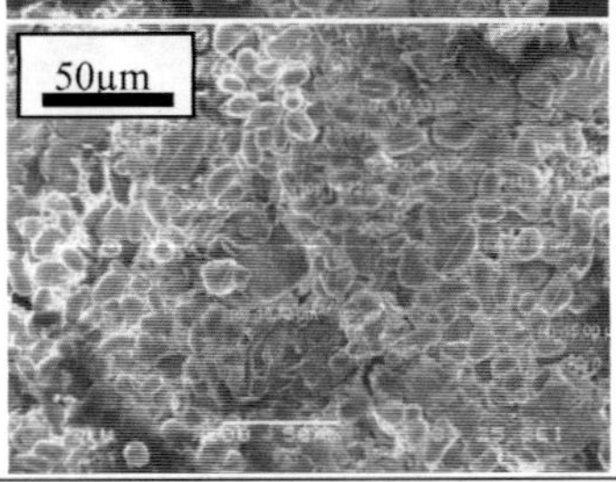	14.72 - 36.48	19.91 ± 6.82

3.2 X-ray diffraction analysis

The X-ray diffractograms of the four mango seed starch samples are presented in Figure 1. The broad x-ray diffraction peak pattern indicates the mango seed starch is in amorphous state similar to the jackfruit seed starch reported by Tulyathan et al. [8]. Furthermore, the mango seed starch exhibited strong diffraction peaks in Bragg angles 2θ at 18° and 23°. These diffraction patterns fit the A-type starch which is in agreement with tamarind kernel starch reported by Kaur and Singh [9]. The different starch extraction method had no influence on the type x-ray diffraction pattern; however, the crystallite size varies among them. The crystallite size was calculated using Scherrer equation [10], as presented in Table 2, shows significant reduction in crystallite size with respective diffraction peaks positioned at 2θ from centrifugation, sedimentation, alkaline and distillation method.

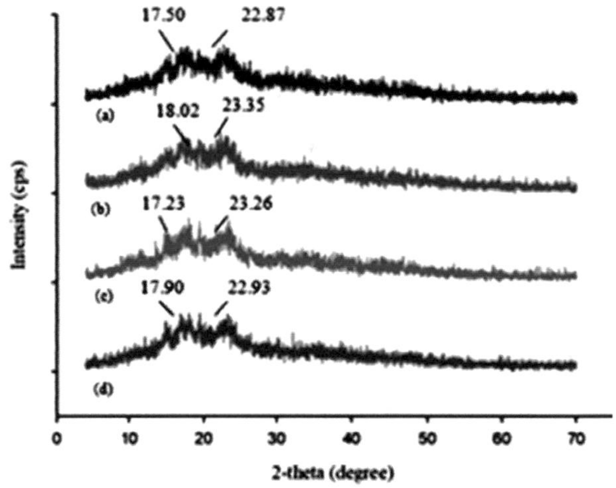

Figure 1. X-ray diffraction pattern of mango seed starch, (a) distillation, (b) alkaline, (c) sedimentation, and (d) centrifugation method

Table 2. XRD data for mango seed starch

Extraction Method	2θ (degree)	Crystallite size (nm)
Distillation	17.50	11.60
	22.87	33.00
Alkaline	18.02	13.80
	23.35	33.00
Sedimentation	17.23	16.5
	23.26	29.00
Centrifugation	17.90	14.90
	22.93	27.50

3.3 Starch yield

Figure 2 shows the percentage of starch yield of the mango seed with distillation, alkaline, sedimentation and centrifugation method were 81.87, 91.72, 74.82 and 74.56 %, respectively. These values were higher than that found by Noor et al. [5] who observed the yield of jackfruit seed starch was 50.69 % to 84.48 %, and yield of rice starch was 45.70 % to 65.00 % [12]. The high starch yield of the mango seeds compared to other sources of starch might be due to the granular size which allows the starch granule to be extracted easily [13]. Hence, the results suggest that the extracted mango seed as one of the potential sources of starch.

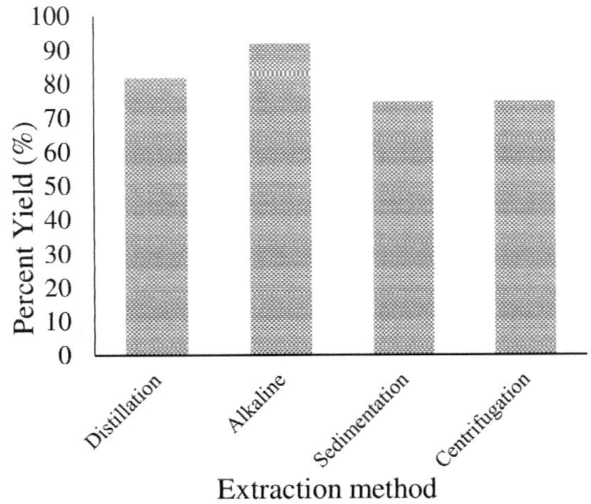

Figure 2. Percent yield of mango seed starch

4. Conclusion

This study has shown that, the starch was prepared from mango seed and extracted via different extraction methods which are distillation, alkaline, sedimentation and centrifugation method. The results showed significant differences in their properties. Investigation on morphology of starch granules revealed oval to irregular shape with small to large granule size. Moreover, the mango seeds revealed an A-type starch pattern and high starch yield. The information obtained in this work could be very useful for the mango seed as new source of starch. However, there are need to carry out more detailed investigation to characterize other properties, particularly proximate composition and physicochemical properties essential to applications as a potential biopolymer.

Acknowledgement

The authors would like to acknowledge the Research Management Centre, International Islamic University Malaysia for the research grant RIGS16-072-0236.

References

[1] Alcázar-Alay S C, and Meireles M A A 2015 *Food Science and Technology* 35 215
[2] Wang Y, Zhang L, Li X, and Gao W 2011 *Brazilian Archives of Biology and Technology* 54 243
[3] Singh N, Singh J, Kaur L, Sodhi N S, and Gill B S 2003 *Food Chemistry* 81 219
[4] Hassan L G, Muhammad A B, Aliyu R U, Idris Z M, Izuagoe T, Umar K J, and Sani N A 2013 *IOSR Journal of Applied Chemistry* 3 16
[5] Noor F, Rahman J, Mahomud S, Akter M S, Talukder A I, and Ahmed M 2014 *International Journal of Nutrition and Food Sciences* 3 347
[6] Oates C G, and Powell A D 1996 *Food Chemistry* 56 405
[7] Kaur M, Singh N, Singh K, and Singh H 2004 *Food Chemistry* 85 131
[8] Tulyathan V, Tananuwong K, Songjinda P, and Jaiboon N 2002 *ScienceAsia* 28 37
[9] Kaur M, and Singh S 2016 *International Journal of Food Properties*19 2432
[10] Bojari H, Malekzadeh A, Ghiasi M, Gholizadeh A, Azargohar R, and Kumar Dalai A 2013 *Crystal Research and Technology* 48 355
[11] Suely M, Samara F, De Albuquerque M, Rafaela I, Silva A, Silva D, Magnani M, and Queiroga V 2014 *Food Chemistry* 143 440
[12] Ashogbon A O, and Akintayo E T 2012 *International Food Research Journal* 19 665
[13] Izuagie T, Hassan L G, Uba A, Achor M, and Sahabi D M 2012 *Bayero Journal of Pure and Applied Sciences* 5 60

IOP Conf. Series: Journal of Physics: Conf. Series **1082** (2018) 012020 doi:10.1088/1742-6596/1082/1/012020

The Effect of Sintering Temperature on Silica Derived from Rice Husk Ash - Nickel Oxide (SiO_2-NiO) Foam Fabrication *via* Slurry Technique

Rizamarhaiza.M[1], S. Ahmad[1], H.A.Rahman[1], H.Taib[1]

[1]Department of Materials and Design Engineering,Faculty of Mechanical and Manufacturing Engineering,Universiti Tun Hussein Onn Malaysia,86400, Batu Pahat,Johor,Malaysia.

Email: hariati@uthm.edu.my

Abstract. Fabrication of ceramic foam has been an interesting study in the field of a porous material due to the excellent mechanical and physical properties. This study presents an approach for the fabrication of Silica (SiO_2) derived from rice husk ash (RHA) and Nickel Oxide (NiO) foams using the slurry technique, highlighting the sintering temperature affects. Polyvinyl alcohol (PVA) was used as binder and Polyurethane (PU) foam was applied as the space holder. The composition of SiO_2 applied in this study was 20wt%. The PU foams dipped into SiO_2-NiO slurry were dried and further sintered at three different sintering temperatures of 1050°C, 1150°C and 1250°C. The morphologies of SiO_2-NiO foams were observed by using the Scanning Electron Microscopy (SEM) and physical properties were determined by using Archimedes method for investigating the total porosity and bulk density. The identification of phases of SiO_2-NiO foams were analysed by using X-Ray Diffraction (XRD). The XRD analyses indicated that there were only SiO_2 and NiO present and no additional phases detected after sintering which implied the compatibility of SiO_2 derived from RHA and NiO even temperatures up to 1250°C. The density values of the SiO_2-NiO foams were found to increase with increasing of sintering temperature. The densities were found to be in the range 0.5239g/cm³ to 0.6210g/cm³ and the percentage of the foams porosity were in the range of 69.71% to 75.19% .Thus it is concluded that the slurry technique is found to be successful to fabricate the SiO_2 as derived from RHA and NiO foams. The sintering temperatures was found to affect the SiO_2-NiO foams in turns of the density and porosity of the foam.

1. Introduction

Ceramic foams are known as porous and brittle materials with pores of commonly open cell, closed cell and interconnect porosity[1–3].Ceramic foams have been widely applied in many fields including thermal insulation, catalyst support, filter and medical implant thanks to their cellular microstructure[4]. Open cell foams are applied in metal melt filtration and also as filter in diesel engine exhaust and industrial hot gases [5–8]. Moreover, the closed cells in ceramic foam usually used in wide range of application for thermal insulation and fire protection. It is known that the ceramic foam with high porosities up to 95% exhibit a reticulate (open pores)[6]. The porosity percentage, pore size distribution, and pore morphology will affect most the material's properties. As an example, the presence of cell walls influences both permeability and strength, while compositional purity will affect the chemical and oxidation resistance, as well as high-temperature creep, electrical resistivity and thermal properties[9].

Several techniques have been developed for the production of ceramic foams; which include replication technique, sacrificial templating and direct foaming[10–12]. The replication technique consists in the impregnation of a natural or a synthetic template polymer foam with a ceramic suspension [11]. Among these methods, the polymeric foam replication process is regarded as a promising one because it is cost-effective and easy to operate. This method involves i) soaking of PU foams with slurries containing ceramic particles and some appropriate binders, ii) removal of excess slurries until a thin ceramic coating forms over the struts of reticulated structure, and iii) drying and pyrolysis of the organic matter and pressure less sintering at designed temperatures[13].

Polyurethane (PU) foam used as a space holder or skeleton in fabrication of ceramic foam. Polyurethanes are the most versatile materials, which can be formed into rigid and flexible, foams, elastomers, adhesives, sealants, etc and are now widely used widely utilized as engineering materials in many industries [32]. PU foam are stiffness, hardness and density value was high. Among other advantage of using PU foam is a form of hollow structure and interconnect pores.

Silica (SiO_2) is an oxide of the element of silicon which is the second most abundant element found on earth. Silica can be obtain from natural resources including rice husk ash (RHA). Rice husk, a form of agricultural biomass, is generated in large quantities as a major by-product in the rice milling industry [14]. Rice husk ash (RHA) is released during rice processing; due to its high caloric value (16,720 kJ/kg). RHA has great potential as a thermal energy source. Due to its high silicon content, rice husk has become a source for preparation of elementary silicon and a number of silicon compounds , especially silica , silicon carbide and silicon nitride [15] . SiO_2 as derived from RHA usually constitutes three main phases which are quartz, cristobalite and tridymite depending on the firing temperature[16,17].

Silica is attractive as a catalyst support since has strong structural robustness, stable at elevated temperature and chemically inert. The application of silica in industry included of abrasive, building materials, fillers, electronics and water filtration [18].

Silica based foams are ideally suited for a wide variety of applications in light weight thermal protection systems. These foams are often fabricated by incorporation of air bubbles in an aqueous suspension of silica powders by direct foaming, which is well-recognized as a simple and effective process [19].

Porous silica have been proven invaluable for use as adsorbents, catalyst supports, and porous electrodes thanks to their low density, favorable thermal and mechanical stability, promising bioactivity, super paramagnetic, and chemical inertia[20].

Nickel is well known as non-noble catalyst materials with having large specific surface area and have gained tremendous attention in catalyst because they are active in hydrogenation, hydro treating, and steam-reforming reactions[21,22]. Nickel oxide (NiO) appears to be a promising alternative because of its high theoretical specific capacitance, low fabrication cost[23] and environmental impact [24].

In the present study, the feasibility of slurry technique as to fabricate the SiO_2 as derived from RHA and NiO foams was determined. The effect of sintering temperatures to the physical properties of the fabricated foams.

2. Materials and Method

In this study, the SiO_2 as derived from RHA local rice mill (Muar,Johor) with the average size of 63μm was used as a raw material. Slurry of SiO_2 (20wt%) and NiO (5wt%) composition was applied throughout the study.5wt% Polyvinyl alcohol (PVA) was used as ~~a~~ binder and Polyurethane (PU) polymeric sponge in cylindrical shape of 125mm x 260mm was applied as the template slurry of SiO_2-NiO mixtures was first prepared in an aqueous media. The dipped PU foams with SiO_2-NiO slurry were dried in the drying oven at 110˚C for 24 hours. The SiO_2-NiO foams were sintered at 1050˚C, 1150˚C and 1250˚C with ramp rate of 2˚C/min. The morphologies of the SiO_2-NiO foams was observed by using Scanning Electron Microscopy (SEM) (JEOL, JSM- 6700F, JAPAN). Phase analyses of the SiO_2-NiO foams were performed by using the X-Ray Diffraction technique (XRD) Bruker D8 Advance, German at the range of 10-100˚ (2θ).The density and porosity test was carried out using the Archimedes' method by Mettler Toledo Density Determination Kit model XS64 ,Switzerland to identify the bulk density and the total porosity of porous foams by using distilled water as the immersion medium.

3. Results and Discussion

3.1 Phase Identification of SiO_2-NiO Foams

Figure 1 shows XRD pattern SiO_2 and NiO foams sintered at temperatures of 1050°C, 1150°C and 1250°C. Both SiO_2 and NiO phases were identified in all green before sintering and after sintered foam. It was also clear that there was no secondary phases (apart from SiO_2 and NiO) identified in SiO_2-NiO foams sintered at all temperatures and there do not have any new phase will form after sintering process. The SiO_2 was identified as the phases of cristobalite (JCPDS 39-1425) and tridymite (JCPDS 42-1401) were present in green and all sintered foams. The peaks positions appearing at 2θ of 36° 43°, 62°, 76° and 80° were identified as NiO, as perfectly matched to JCPDS 47-1049 . The sharpness and the intensity of the peaks indicate the well crystalline nature of the prepared sample[25].

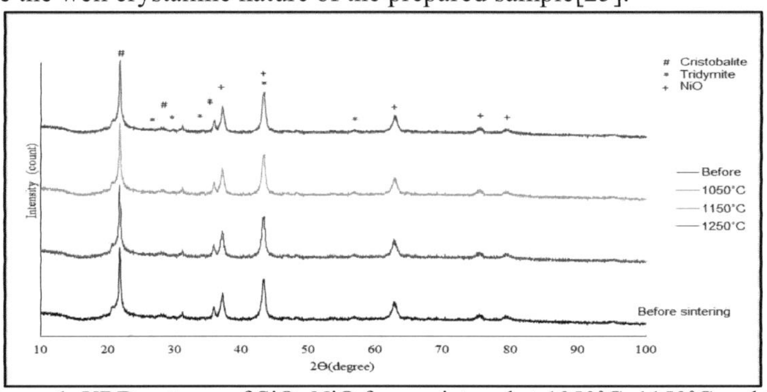

Figure 1. XRD patterns of SiO_2-NiO foams sintered at 1050˚C, 1150˚C and 1250˚C.

3.2 Morphologies of SiO_2-NiO Foams

Figure 2 shows the image of sintered SiO_2-NiO foams surface before and after sintering at 1050˚C, 1150˚C and 1250˚C.

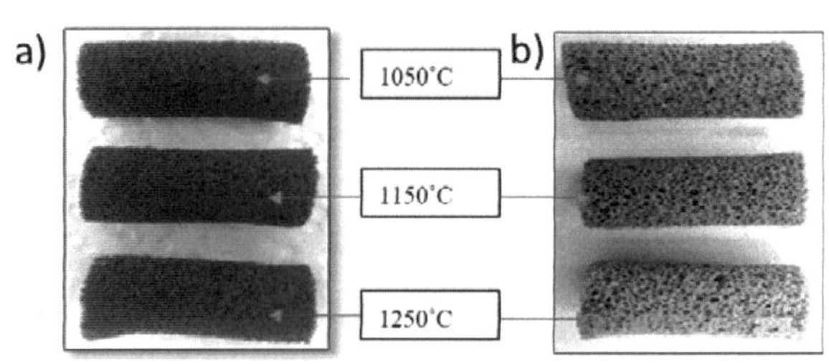

Figure 2. SiO_2-NiO foams a) before and b) after sintered 1050˚C, 1150˚C and 1250˚C.

From Figure 3, it is observed that all foam were black in colour due to raw SiO_2 and NiO before sintering due to the colour of the slurry. The change in colour of SiO_2-NiO foams was observed after sintering process from black to green colour. The colour changed from dark grey to green was due to the presence of NiO [26]. The changed in colour observed may be caused by oxidation of NiO during sintering[27].

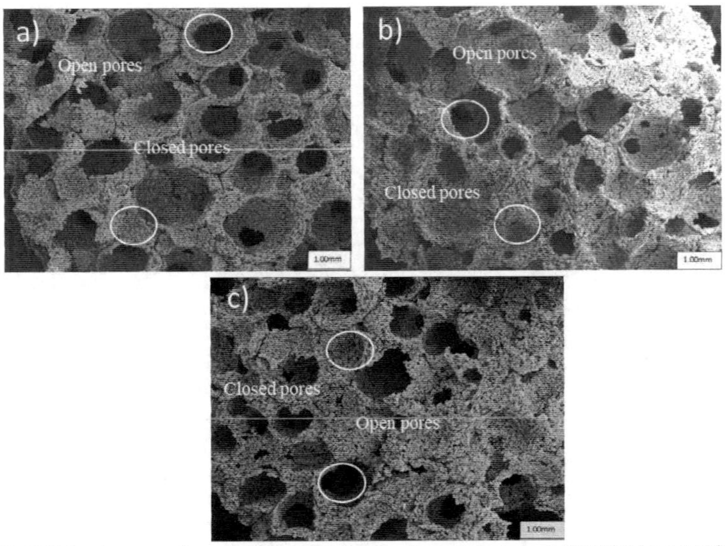

Figure 3. SEM micrographs of SiO_2-NiO foam after sintering a) 1050°C b) 1150°C c) 1250°C

As seen in Figure 3, all SiO_2-NiO foams fabricated shows appearance of the open and closed pores. The higher temperature would cause the gradual formation of the pores which yielded dense and pores with decreasing porosity.

3.3 Porosity and Density of SiO_2-NiO Foams

It was observed that good porosity percentages of more than 50% achieved in all samples indicated that the SiO_2-NiO foams were indeed successfully fabricated using the slurry technique. The properties of porous ceramic should consist of > 70 %[28,29] . The total porosity was found to decrease when the increasing of sintering temperature which can be seen in the Figure 4.

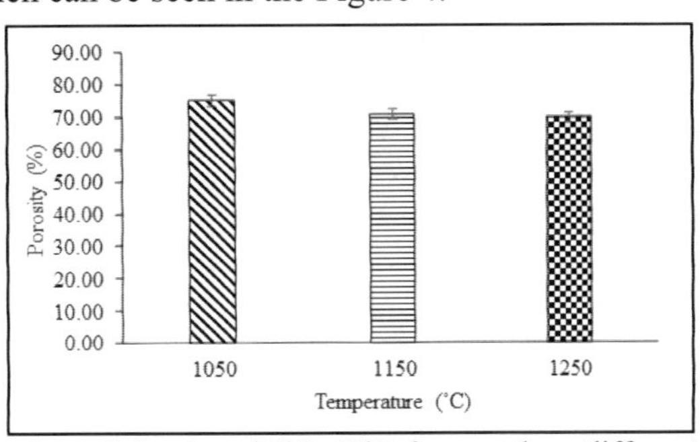

Figure 4. Porosity and density of SiO_2-NiO foam at three different sintering temperatures

Figure 4 shows porosity percentage SiO_2-NiO foams was 74% at 1050°C and decreased to 71.00% when temperature is increase to 1150°C and decreased to 69.00% when temperature change to 1250˚C. Thus high temperature causes the gradual occurrence formation of the pores where the microstructure will be denser thus with decreasing porosity [30].

Additionally, the density of SiO_2-NiO foams formed at 1050°C, (0.5239g/cm^3) were increased to 0.5547g/cm^3 at temperature 1150°C and further increased to 0.6210 g/cm^3 with sintering temperature of 1250˚C like in Figure 5. Thus, it is clear that when the higher sintering temperature is set, higher densities were achieved. This is may be due to the reducing pore size and decreasing number of porosity because of migration of the massive atoms towards the connective surfaces which promotes the sintering necks grown and the inter-particle spacing become shorten as the sintering temperature increase [31].

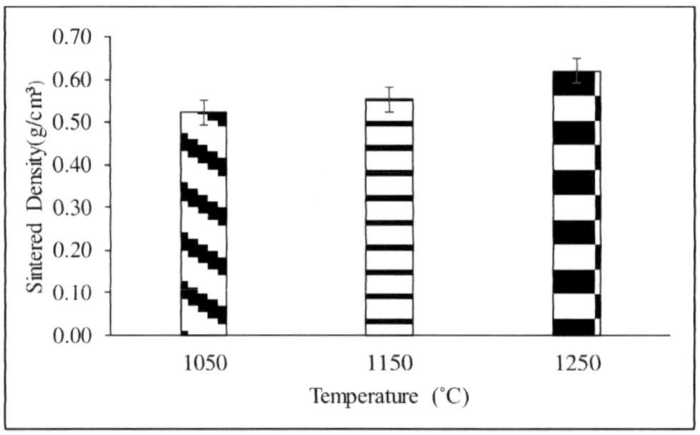

Figure 5. Density of SiO_2-NiO Foams

4. Conclusions

SiO_2-NiO foams were successfully fabricated by using the slurry technique at different sintering temperature. XRD analyses signifies that SiO_2 as derived from RHA is indeed compatible with NiO, as there were no additional phases detected in foams sintered at 1050˚C to 1250˚C. SEM micrographs showed formation of open pores and closed pores which supports the findings of decreasing porosity percentages and increasing density. Sintering at lower temperature of 1050˚C produced highest porosity (75.19%) in the SiO_2-NiO foam. However, although lower porosity (70.95% and 69.72%) were yielded from sintering temperatures of 1150˚C and 1250˚C percentage. The values are still in acceptable range for foams. Moreover the density of the SiO_2-NiO foams was found to increase proportionally with the sintering temperature at temperature 1250˚C to 0.6210g/cm^3.

Acknowledgements

The author gratefully acknowledges Universiti Tun Hussein Onn Malaysia (UTHM) and Ministry of Higher Education (KPM) for research sponsorship under FRGS Vot 1593 and KPT'S MyBrain 15 scholarship.

References

[1] J. Zhao, S. Shimai, G. Zhou, J. Zhang, and S. Wang 2017 *Colloids Surfaces A* 537 216
[2] E. Wichianrat, Y. Boonyongmaneerat, and S. Asavavisithchai 2012 *Trans. Nonferrous Met. Soc. China* 22 1674
[3] M. Parvanian and M. Panjepour *Mater. Des 2013* 49 834

IOP Conf. Series: Journal of Physics: Conf. Series **1082** (2018) 012020 doi:10.1088/1742-6596/1082/1/012020

[4] E. C. Hammel, .-R Ighodaro, and O. I. Okoli *Ceram. Int 2014* 40 15351
[5] Y. C. G. Hamimah Abd. Rahman *Int. Conf. Eng. Environ. 2007 (ICEE 2007)* 2
[6] J. Verma, R. Mitra, and M. Vijayakumar 2013 *J Eur Ceram Soc* 33 943
[7] B. N. M. S. Sharmiwati, R. M. Mizan 2014 *Int. J. Sci. Technol. Res* 3
[8] J. Luyten, S. Mullens, J. Cooymans, a. M. De Wilde, I. Thijs, and R. Kemps *J. Eur. Ceram. Soc.* 29 829
[9] S. Baharom, S. Ahmad, H. Taib, and R. Muda 2016 *AIP Conference Proceedings* 2016 0004.
[10] J. Locs, V. Zalite, L. Berzina-Cimdina, and M. Sokolova 2013 *J. Eur. Ceram. Soc.* 33 3437
[11] A. Celani, S. Blackburn, M. J. Simmons, L. M. Holt, and E. H. Stitt, "Formulation of Ceramic Foams: a New Class of Amphiphiles," 2017.
[12] B. Ma, Y. Li, G. Liu, and D. Liang *Ceram. Int.* 41 3237
[13] A. R. Studart, U. T. Gonzenbach, E. Tervoort, and L. J. Gauckler *J. Am. Ceram. Soc.* 89 1771
[14] P. Deshmukh, J. Bhatt, D. Peshwe, and S. Pathak 2012 *Trans. Indian Inst. Metm* 65 1 63
[15] V. Della, I. Kuhn, and D. Hotzam*Mater. Lett* 57 818
[16] S. Haastrup, D. Yu, and Y. Yue 2016 *J. Non. Cryst. Solids*n 450 42
[17] J. Zhu *et al 2017 Inorganica Chim. Acta*, 2017
[18] S. Mishra, R. Mitra, and M. Vijayakumar 2008 *J. Eur. Ceram. Soc* 28 1769
[19] S. Mishra, R. Mitra, and M. Vijayakumar 2009 *Mater. Lett* 63 2649
[20] W. Wan, Y. Feng, J. Yang, S. Xu, and T. Qiu 2015 *J. Eur. Ceram. Soc* 35 2163
[21] C. Li and Y. W. Chen 1995 *Thermochim. Acta* 256 457
[22] A. J. Majewski, J. Wood, and W. Bujalski 2016 *Int. J. Hydrogen Energy* 38 14531
[23] S. De, J. Zhang, R. Luque, and N. Yan 2016 *Energy Environ. Sci* 9 3314
[24] P. Cao, L. Wang, Y. Xu, Y. Fu, and X. Ma 2015 *Electrochim. Acta* 157 359
[25] Kemary M E L, NagyI N, El-Mehasseb 2013 *Mater. Sci. Semicond. Process* 16 1747
[26] F. Tietz, F. J. Dias, D. Simwonis, and D. Stöver 2000 *J. Eur. Ceram. Soc* 20 1023
[27] Higginson R L , Jackson C P, Murrell E L 2015 *Metadata Rec* 2 11
[28] Karageorgiou V and Kaplan D 2005 *Biomaterials* 26 5474
[29] E. Of, S. Temperature, O. The, P. Properties, and A. N. Foam, "Effect Of Sintering Temperature On The Physical Properties Of Titania- Alumina-Silver Nitrate Foam," vol. 466, pp. 877–880, 2014.
[30] B. Berger, M. B. Berger, and C. Cc, "The Southern African Institute of Mining and Metallurgy Refractories 2010 Conference The Importance And Testing Of Density / Porosity / Permeability / Pore Size For Refractories."
[31] Anon. n.d. "Review Of Polyurethane In Composites." Retrieved April 25, 2018

IOP Conf. Series: Journal of Physics: Conf. Series **1082** (2018) 012021 doi:10.1088/1742-6596/1082/1/012021

Synthesis and Characterization of Akermanite by Mechanical Milling and Subsequent Heat Treatment

Hossein Mohammadi, Yanny Marliana Baba Ismail, Khairul Anuar Bin Shariff, Ahmad-Fauzi Mohd Noor*

[1]School of Materials and Mineral Resources Engineering, Universiti Sains Malaysia, Engineering Campus, 14300, Nibong Tebal, Penang, Malaysia

*Email: srafauzi@usm.my

Abstract. In this study, nano-structured akermanite was synthesized by mechanical milling and mixing using calcium oxide, magnesium oxide, cobalt oxide and silicon dioxide in a planetary ball milling with the speed of 500 rpm for 3 h. The milled powder was pelletized and sintered at 1200°C subsequently were characterized by X-ray diffraction (XRD), transmission Fourier transform spectroscopy (FTIR) and field emission scanning electron microscopy (FESEM) confirming FESEM a dense sintered akermanite. The diametral tensile strength (DTS) value of akermanite was found to be 5.71 MPa, which was in the range of human cancellous bone i.e. 1.5-38 MPa. Finally, akermanite induced obvious apatite formation on the surface after 7 days of soaking. All the results postulated that the akermanite ceramics might be potentially used as bone repair biomaterials for non-load bearing applications.

1. Introduction

Bone damages mostly due to disease such as osteoporosis or accident have been a significant challenge in the field of biomedical science. Bone graft substitutes including autograft and allograft possess some drawbacks such as bone availability for transplantation, potential disease transmission, and immunogenic response, respectively [1]. Scientists have been making effort to develop bioceramic and bone filler or substitute material, particularly with bioactive fixation capability to the bone tissue through hydroxyapatite layer formation [2].

Calcium (Ca), magnesium (Mg) and silicone (Si) are also known as important elements in the human body which plays a vital role in the bone formation process and their deficiency can result in severe disorders. Calcium is abundantly found in the bone and tooth tissues [3]. Magnesium is important in the regulation of bone tissue growth and repair while silicon plays a vital role in skeletal development [4]. Calcium silicate-based biomaterials ($CaSiO_3$, CS) have shown excellent *in vitro* bioactivity mainly due to the release of Ca and Si ions into the surrounding environment [5]. However, their major drawback is the high dissolution rate which leads to high pH value [6] which is detrimental to cells [7].

One way to control the dissolution rate of these biomaterials is to incorporate another trace element such as magnesium (Mg), zinc (Zn), zirconium (Zr), titanium (Ti) etc. [6]. Akermanite is calcium magnesium silicate-based biomaterials ($Ca_2MgSi_2O_7$) which is formed by the incorporation of magnesium into the calcium silicate ceramics. One of the methods to synthesize akermanite is mechanical activation [8] which is known as one of the important methods to synthesize bioceramic nanopowders [8]. The major objective of the present study is to synthesize akermanite ceramics by high-speed planetary milling and evaluate the *in vitro* bioactivity and mechanical properties for bone and dental applications.

2. Experimental study

2.1 Synthesis of akermanite powders

Akermanite nanopowder was synthesized using a wet high-energy and high-speed planetary ball milling. Briefly, calcium oxide (CaO, Merck, 98%), magnesium oxide (MgO, Merck, 98%), silicon dioxide (SiO2, Sigma-Aldrich, 99.9%), were weighed based on the stoichiometric ratio of akermanite and then were ball milled for 3 h in the planetary ball milling (PM 400-Reutch) using zirconia vial and 37 zirconia balls. The ball to powder weight ratio was 10:1, while the powder to deionized water ratio was 3:1. The speed of vial was set at 500 rpm. Wet milling was adopted for higher efficiency and avoiding any dead zone. After milling, the mixture was dried in oven for 24 h at 100°C. Then, the dried powder was ground using agate mortar pestle and sieved through 200 μm. Finally, sieved powders were pressed into pellets using uniaxial hydraulic press (24T Laboratory hydraulic Press, MTI Cooperation) with the dimension of 13 mm (diameter) under the pressure of 200 MPa. The pellets were then sintered 1200°C for 4 h with a heating rate of 5°C/min.

2.2 Characterization of akermanite powders

The phase composition, lattice parameter and crystallite size of synthesized akermanite powders were determined using X-ray diffraction (XRD, Bruker Advanced X-ray Solution D8, XRD) through CuKα radiation (λ=0.154 nm at 20 kV and 30 mA). The diffraction angles (2θ) between 10° and 90° was scanned with a step size of 0.05° and a count time of 1.25 s per step. In order to determine the phase compositions, the patterns were matched to the International Center for Diffraction Data (ICDD) reference files. The crystallite size of akermanite powders was further calculated from XRD patterns using modified Scherrer equation (Eq.1) [9]:

$$Ln\beta = \ln k\lambda/D + 1n1/\cos\theta \quad (1)$$

where β (in Rad) is the diffraction peak width at half maximum intensity (FWHM), θ is the Bragg diffraction angle(°), D crystallite size (nm), λ the wavelength of the radiation (0.15406 nm), and k is Scherer constant which is usually 0.9. The sum of the absolute error values (i.e. $\pm \Delta Ln\beta)^2$ is decreased by using modified Scherrer equation which yields a single line in which, the line intercept is lnkλ/D. To investigate the crystallographic lattice parameters, Rietveld refinement was also employed using X'Pert software.

Functional groups of akermanite powders were evaluated by transmission Fourier transform infrared spectroscopy (FTIR, PerkinElmer, USA). The spectrum was recorded in the region of 400-4000 cm^{-1} with a resolution of 2 cm^{-1}.

In order to evaluate the morphology and elemental composition, the sintered pellets were coated with a layer of gold (Au) by sputtering (EMITECH K450X, England) and the morphology and microstructure of samples were observed on a field emission scanning electron microscopy (FESEM, ZEISS, Germany) equipped with energy dispersive spectroscopy (EDS) with a voltage of 10 kV

2.3 Shrinkage, porosity, and density

The linear shrinkage (length) of akermanite was calculated based on the following equation:
Linear shrinkage (%) = $(L_g-L_s)/L_g \times 100$ (2)

Where L_g is the bulk length before sintering (green sample) and L_s is the bulk length after sintering (sintered samples), respectively. In addition, the density and porosity of akermanite were calculated based on ASTM B962-17 [10] and using the Archimedes method as follows:
D (%) = $W_1/(W_1-W_3) \times 100$ (3)
P (%) = $(W_2-W_1)/(W_2-W_3) \times 100$ (4)
Where D is the bulk density, W_1 is the weight of dry pellet in air, W_2 is the weight of pellet after saturation in water, and W_3 is the weight of samples suspended in water.

2.4 Mechanical properties

For mechanical evaluation of akermanite, the diametral compression test (DTS) was performed on the sintered compacted pellet of 13 mm through universal experimental instrument (INSTRON 3367) with the speed rate of 1 mm/min. The tensile strength of the pellets parallel to the applied force is calculated as follows [11]:
$\sigma_{tensile} = 2P/\pi dh$ (5)

In equation 5, the d is diameter (mm), h is thickness (mm) of sample and P is the applied force (N).

2.5 In vitro bioactivity evaluation of akermanite powders

The *in vitro* bioactivity of akermanite was evaluated by soaking the samples in simulated body fluid (SBF) solution as described by Kokubo [12]. The sintered pellets were soaked in 25 ml of SBF solution (pH=7.4) at 37°C for 1, 7, 14, 21 and 28 days. After soaking, the samples were rinsed with deionized water and dried at 100°C for 24 h and the apatite formation on the surface was evaluated by FESEM analysis.

3. Results and discussion

In the present study, akermanite was successfully synthesized by mechanical milling and mixing for 3 h and subsequent heat treatment at 1200°C. The XRD patterns of powders sintered at 1200°C in Fig 1 (a) revealed characteristic peaks of akermanite (ICCD No#035-0592) with a small amount of diopside [13]. Furthermore, according to the modified Scherrer equation (Eq.1), Rietveld refinement, phase composition, crystallite size (nm), grain size (nm) and lattice parameters a , b and c were estimated (Table 1). The crystal structure of akermanite is tetragonal with the space group of $P\bar{4}2_1m$ in which, the lattice parameters are a=b and c.

The density and porosity of akermanite were obtained 63.73% and 35.59%, respectively (Table.2). Besides, the linear shrinkage was nearly 17.30% after sintering at 1200°C.The FTI-IR spectra of sintered akermanite are shown in Fig 1(b). As observed, the bending vibration of 3439 cm^{-1} indicated the presence of OH group [14] and the vibration at 1637 cm^{-1} is related to bending vibration of H_2O molecules [15]. The bending vibrations at 511 cm^{-1} could be ascribed to O-Ca-O. In addition, the bending vibration at 573 cm^{-1} could be assigned to the O-Mg-O. In addition, vibrations at 641 and 684 cm^{-1} revealed Si-O-Si bonding. Also, the vibration at 1021 cm^{-1} was related to the symmetric stretching of Si-O-Si [8]. Finally, stretching Si-O bonding revealed vibrations at 848, 981 and 923 cm^{-1}, respectively.

The FESEM micrograph of akermanite before and after immersion in SBF solution are depicted in Fig.2 (a, b). It appeared that akermanite possess uniformly distributed grain microstructure. However, akermanite did not clearly reveal the grain boundaries. In addition, after soaking in SBF for 7 days, an obvious apatite layer was formed on the surface indicating excellent *in vitro* bioactivity of akermanite. The pH value of the solution after 28 days of soaking reached to 7.69-7.72 which is in the range of human biological value compared to $CaSiO_3$ which was reported as 8.8 [16] revealing the chemical stability of akermanite (see Fig.2(c)).

The DTS value of akermanite was 5.71 MPa which is higher than conventional ceramics such as hydroxyapatite (HA) and beta-tricalcium phosphate (β-TCP) as summarized in Table.2, reflecting stronger bioceramics has been developed although with a higher porosity and lower bulk density. This suggests that silicate bioceramic has potentially good future for bone grafting materials.

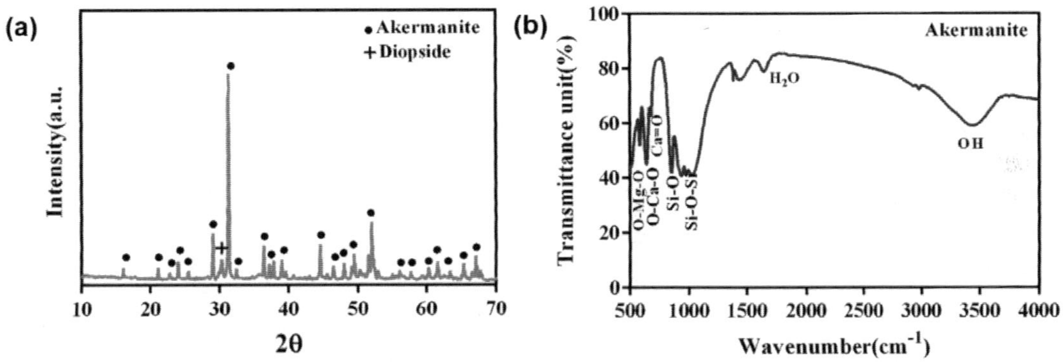

Figure 1. (a) XRD and (b) FTIR of akermanite and sintered at 1200°C

Figure 2. FESEM images of akermanite sintered at 1200°C (a) before and (b) after soaking in SBF solution for 7 days (10x magnification) and (c) The change of pH values in SBF after soaking of akermanite at different period of time

Table 1. The grain size, lattice parameters, and crystallite size of akermanite

Material	Grain size (μm)	*a*(Å)	*b*(Å)	*c*(Å)	*β*(°)	Crystallite size(nm)
Akermanite	0.93±0.25	7.837	7.837	5.011	90	45.69

Table 2. A comparison of mechanical properties of cortical and cancellous bone, hydroxyapatite (HA), akermanite (In this study)

Material	DTS (MPa)	Relative density (%)	Porosity (%)	Reference
Cortical bone	50-150	87-95	5-13	[17]
Cancellous bone	1.5-38	0.05-0.7	30-90	[18]
Calcium silicate cement	2.57	-	-	[19]
Hydroxyapatite	4	65-95	19.4-2.8	[20]
β-TCP	3.82,2.85	85.9,91.2	-	[21]
Akermanite	5.71	63.73	35.59	This study

4. Conclusion

In this study, nano-structured akermanite was successfully synthesized by mechanical milling and mixing and subsequent sintering. The crystallite size of akermanite under 3 h ball milling time and subsequent heat treatment at 1200°C was 45.69 nm. The XRD and FTIR analyses demonstrated the formation of akermanite ceramics. Furthermore, the diametral tensile strength of akermanite was in the range of human cancellous bone suggesting that they might be used for bone tissue repair under non-load bearing application. Finally, akermanite revealed obvious apatite-forming ability after soaking in SBF solution for 7 days of soaking as demonstrated by FESEM analysis.

Acknowledgement

The authors sincerely acknowledge the financial support from Fundamental Research Funds Scheme (6071376).

References

[1] Sarkar S K and Lee B T 2015 *The Korean Journal of Internal Medicine* **30** 279
[2] McEntire B J, Bal B S, Rahaman M N, Chevalier J and Pezzotti G 2015 *Journal of the European Ceramic Society* **35** 4327
[3] Beto J A 2015 *Clinical nutrition research* **4** 1
[4] Castiglioni S, Cazzaniga A, Albisetti W and Maier J A M 2013 *Nutrients* **5** 3022
[5] Li H C, Wang D G and Chen C Z 2015 *Ceramics International* **41** 10160
[6] Mohammadi H, Hafezi M, Nezafati N, Heasarki S, Nadernezhad A, Ghazanfari S and Sepantafar M 2014 *J. Ceram. Sci. Technol* **5** 1
[7] El-Ghannam A, Ducheyne P and Shapiro I 1997 *Biomaterials* **18** 295
[8] Sharafabadi A K, Abdellahi M, Kazemi A, Khandan A and Ozada N 2017 *Materials Science and Engineering: C* **71** 1072
[9] Monshi A, Foroughi M R and Monshi M R 2012 *World Journal of Nano Science and Engineering* **2** 154
[10] ASTM S 2017 *Standard Test Methods for Density of Compacted or Sintered Powder Metallurgy (PM) Products Using Archimedes' Principle*
[11] Bang L T, Ishikawa K and Othman R 2011 *Ceramics International* **37** 3637
[12] Bohner M and Lemaitre J 2009 *Biomaterials* **30** 2175
[13] Li H C, Wang D G, Chen C Z and Shi H 2015 *Materials Letters* **159** 459

[14] Omori K 1971 Analysis of infrared absorption spectrum of diopside *American Mineralogist* **56** 1607
[15] Lin K, Lin C and Zeng Y 2016 *RSC Advances* **6** 13867
[16] Hu S, Ning C, Zhou Y, Chen L, Lin K and Chang J 2011 *Journal of Wuhan University of Technology-Mater. Sci. Ed.* **26** 226
[17] Wang M 2003 *Biomaterials* **24** 2133
[18] Witte F, Hort N, Vogt C, Cohen S, Kainer K U, Willumeit R and Feyerabend F 2008 *Current Opinion in Solid State and Materials Science* **12** 63
[19] Chen Y-W, Hsu T-T, Wang K and Shie M-Y 2016 *Materials Science and Engineering: C* **60** 374
[20] Joon Yoo J, Joong Kim H, Seo S-M and Oh K-S 2014 *Ceramics International* **40** 3079
[21] Tahmasebifar A and Evis Z 2013 *Journal of Ceramic Processing Research* **14** 549-56

IOP Conf. Series: Journal of Physics: Conf. Series **1082** (2018) 012022 doi:10.1088/1742-6596/1082/1/012022

Effect of Co-Dopant on the Phase Structure and Photoluminescence Properties of $Sr_2mgsi_2o_7$: En^{2+} Phosphors

S.Y.Pung*, C.H. Chang, M.N. Ahmad Fauzi and S.R. Kasim

School of Materials and Mineral Resources Engineering, Engineering Campus, Universiti Sains Malaysia, 14300 Nibong Tebal, Penang, Malaysia
E-mail: sypung@usm.my

Abstract. The role of co-dopants, i.e. Dy, Ce, Tb or Y in SMS : Eu, RE^{3+} phosphor sintered using solid state reaction method at 1250°C under forming gas was studied. All SMS phosphors exhibited tetragonal $Sr_2MgSi_2O_7$ as the domain phase and monoclinic $SrSiO_3$ as the minor phase based on XRD analysis. It is noted that only single blue emission (469nm) contributed by Eu^{2+} ions was detected in RTPL measurement regardless types of co-dopants introduced during the sintering process. This result indicates that the co-dopants did not act as luminance centres in the SMS phosphor. The time-decay PL and temperature dependent PL analysis suggest that Ce^{3+} ions acted as sensitizer which improved the luminescence intensity and afterglow property of SMS phosphor. The present of Dy^{3+} ions pro-longed the afterglow property by introducing more traps density in the SMS phosphor structure. The co-doping of Tb or Y into the SMS phosphor generated many non-radiative recombination centres which deteriorated the optical performance of SMS: Eu^{2+}, RE^{3+} phosphor.

1. Introduction

Phosphors have been used in architecture, traffic signs, art and graphics, phosphorescent paint, decoration, watch and clock dials, fishing tackles, toys and other goods for daily use [1-3]. Recently, it has been widely applied in phosphor converted white light emitting diodes (white LEDs). The white LEDs have many excellent properties such as long lifetime (> 100,000 hrs), low power consumption and compact size. However, the conventional yellow emitting phosphor based white LEDs have a poor colour rendering index (<65) due to the lacking of red emission. Thus, a new generation of LEDs i.e. near ultraviolet LEDs combined with multiphase phosphor, with a better colour rendering index is developed. The multiphase phosphor consists of blue emission $Sr_2MgSi_2O_7$(SMS) based phosphor and Eu^{2+}sensitized red-emitting $Ca_3Mg_3(PO_4)_4$ (CMP) phosphor[4-5].

SMS phosphor is an alkaline earth silicate which has attracted attention for the applications in long lasting phosphor too due to its relatively high brightness and chemical stability. Various synthesis methods have been used to prepare this silicate based phosphor material. For examples, sol gel method [6], combustion method [7], and solid state reaction method [8] are the commonly used synthesis techniques for phosphor materials. Nevertheless, these techniques have their own strengths and limitations. The sol gel method requires expensive starting materials and long reaction time whereas the product that synthesized using combustion method has a low brightness and the afterglow lasts shortly. For co-precipitation method, it induces in homogeneity and agglomeration in the powders easily. On the other hand, solid state reaction is the most widely used method even nowadays in the industry due to its simplicity, low cost and easily controlled operation [2, 9].

The blue emission of SMS phosphor could be achieved by doping of Eu^{2+} ions. The Eu^{2+} ion is proposed to substitute into the Sr^{2+} site since it has similar ionic size (1.25 Å) to that of Sr^{2+} ions (1.26 Å) in SMS lattice. The Eu^{2+} ion is unlikely substituted into the Mg^{2+} site (0.72 Å) and Si^{4+} site (0.26 Å) as the ionic radii are small. During the substitution, two Eu^{3+} ions occupy two Sr^{2+} sites and then one Sr^{2+} vacancy is created by charge compensation in vicinity. The Sr^{2+} vacancy acts as electron donor and the two Eu^{3+} ions at the Sr^{2+} sites become acceptors. Thus, two electrons in each vacancy will transfer to two adjacent Eu^{3+} ions and then Eu^{2+} ions are formed [10-11].The blue emission is produced because of the 5d $\rightarrow$4f transition after optical excitation.

As the absorption and emission bands of Eu^{2+} are broad because of the crystal field splitting effect, these ions are well suited for LED characteristics. Nevertheless, the emission peak and its intensity of Eu^{2+} emission can vary as a result of strong dependence on the lattice symmetry, host composition and crystal structure [12-15]. In this work, SMS:Eu^{2+} phosphor co-doped with Dy^{3+}, Ce^{3+}, Tb^{3+}or Y^{2+} were synthesized by solid state reaction. The SMS:Eu^{2+}, Dy^{3+} phosphor was the most studied phosphor [16-17] and thus it was used as control sample. These co-dopants might acts as (i) another luminescent centre, (ii) bridges to transport carriers between trapping and emission centres or (iii) as defect related traps in the host. The influence of co-dopants on their photoluminescence properties, particularly the temperature dependent photoluminescence and time decay photoluminescence were comparatively investigated in detail. The result from these photoluminescent measurements provides a better understanding on the luminescence mechanism and energy transfer in the SMS:Eu^{2+},Re^{3+} phosphor materials.

2. Experimental setup

The SMS: Eu^{2+}, RE phosphors were prepared by solid state reaction technique. The precursors were $SrCO_3$, MgO, SiO_2, Eu_2O_3, CeO_2, Dy_2O_3, Y_2O_3 and Tb_4O_7 powder. The precursors were weighed according to the given stoichiometric amount ($Sr_{1.96}MgSi_2O_7$: $Eu_{0.015}$, $RE_{0.005}$) and mixed thoroughly in an agate mortar. A small amount of H_3BO_3 (10 mol% of silicate) was used as flux. The mixed precursor were sintered at 1250°C for 2 hrs under reductive ambient (95% N_2 and 5% % H_2).The samples were labelled based on their corresponding co-dopants. For example, SMS co-doped with Ce was labelled as SMSECe.

The X-ray diffraction patterns of SMS phosphors were recorded by Philip model PW 1729 X-Ray powder diffractometer (Cu K_{α}, $\lambda = 0.154$ nm) for structural analysis. The weight percentage of phases present in the as-synthesized samples were estimated based on the area under the diffraction peaks of XRD using X'Pert Highscore Plus program (version 2.1). The morphology and composition of the SMS phosphors were characterized using Field Emission Scanning Electron Microscopy (FESEM, Zeiss, Supra 35 VP) and Electron Dispersive X-Ray (EDX). The photoluminescence (PL) properties of SMS phosphors were measured by Spectra Fluorolog-3 from Horiba Jobin Yvon using450 W *Xe* lamp as the excitation source.

3. Results and discussion

Figure 1(a) gives the XRD patterns of the as-sintered phosphor doped with various types of co-dopants. The results show that the phases present in all the samples are mostly composed of tetragonal $Sr_2MgSi_2O_7$ and a small amount of monoclinic $SrSiO_3$. Other than these phases, no other phase was observed in any of the samples. It is clearly seen that the positions and intensity of the diffraction peaks for all the samples are almost the same. This implies that the type of co-dopants has no effect on the host structure. The quantitative phase analysis was estimated for all the samples using X'Pert Highscore Plus program (version 2.1) as shown in Figure 1(b). The weight fraction of $Sr_2MgSi_2O_7$ phase is slightly different by varying the type of co-dopants. SMSECe exhibits the highest weight fraction of $Sr_2MgSi_2O_7$ phase (67.1 wt%), followed by SMSED (66.7 wt%), SMSEY (66.4 wt%) and

SMSETb (65.56 wt%). Although the phosphors contained $SrSiO_3$ phase, it was still named as SMS phosphors throughout the paper for the convenient expression.

The surface morphology of SMS:Eu^{2+} phosphors co-doped with Ce, Dy, Y and Tb are shown in Figure 2 (a) - (d), respectively. Generally, the morphology of all the samples composed of agglomerated irregular grains. The shape and the size of grains are non-uniform, indicating that the type of co-dopants does not have much influence on the surface morphology and grain shape as well as grain size.

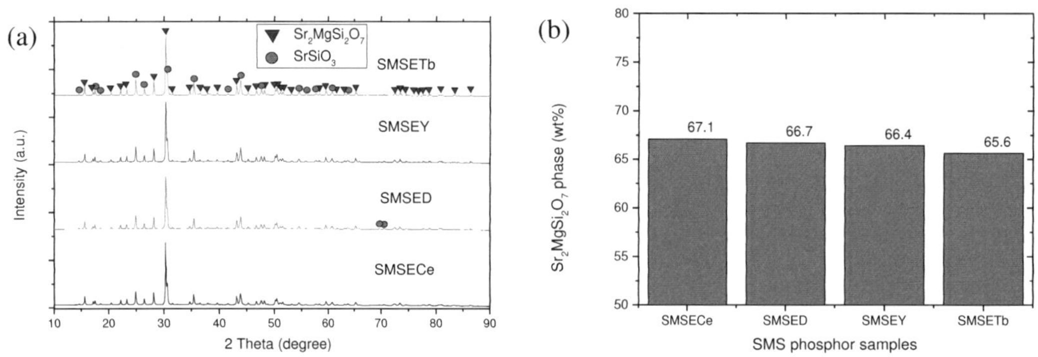

Figure 1. (a) XRD patterns and (b) weight fraction of SMS:Eu^{2+}phosphor co-doped with Ce, Dy, Y and Tb

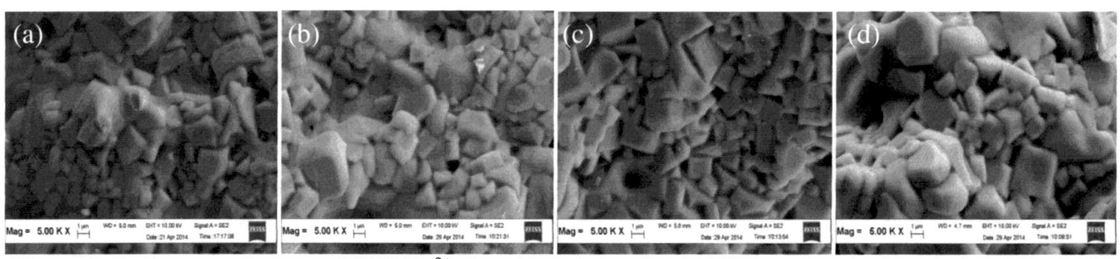

Figure 2. Surface morphology of SMS:Eu^{2+} phosphors sintered at 1250°C co-doped with (a) Ce (b) Dy (c) Y and (d) Tb

The as-sintered SMS:Eu^{2+}co-doped with Dy is yellowish in colour. The blue emission was emitted when it was optically excited using UV light (254 nm) for 5 min. Figure 3(a) shows that the emission spectra of all the samples present a single band peaking at 469 nm when excited at 375 nm. The 469 nm broad band corresponds to the $^8S_{7/2}$ ($4f^7$) ground state of Eu^{2+} ions [9]. However, there is no other emission peak observed in the PL result. The emission peaks of Ce^{3+} (370 nm) [18], Dy^{3+} (480nm) [19], Tb^{3+} (542 nm) [20] and Y^{3+} (601 nm) [21] are absent in the emission spectra. Thus, it indicates that these RE ions did not become the luminescent centres in the SMS host lattice. As no emission in the range of 570 - 750 nm, this suggests that the Eu^{3+} ions have been reduced completely to Eu^{2+} ions.

The excitation spectra at the emission wavelength 469 nm are shown in Figure 3(b). The excitation spectrum has a broad band ranges from ultraviolet to visible light and with the main peak at 423 nm. The broad absorption and emission bands of Eu^{2+} are ascribed to crystal field splitting effect. This phenomenon is suitable for fabrication of white LEDs. In addition, there is no shift of the emission and excitation wavelength in all the samples. This implies that RE co-doping does not cause any significant changes on the emission and excitation spectra.

Time-decay photoluminescence measurement (TDPL) at room temperature was performed to measure the decay of Eu^{2+} emission at 469 nm upon 375 nm excitation. The time decay of 469 nm of the SMS:Eu^{2+}, Ce^{3+} phosphor is shown in Figure 4(a). The decay curves can be fitted into a single exponential function as

$$I(t) = I_o + A_1 \exp(-t/t_1) \quad (1)$$

where A_1 is constant, I_o is the luminescence intensity at time 0, t and t_1 are the decay time for the exponential components, respectively. The above fitting is reasonable as only one Sr^{2+} sites is available in the $Sr_2MgSi_2O_7$ phase [4]. From the fitted function, the measured lifetime of SMS:Eu^{2+}, Ce^{3+}phosphor is 0.5402 ms. The time decay PL of other SMS:Eu^{2+} phosphors co-doped with Dy, Tb and Y are shown in Figure 4(b). The lifetimes determined from the fittings are 0.44603 ms, 0.31092 ms and 0.26921 ms for SMS:Eu^{2+} phosphors co-doped with Dy, Tb and Y, respectively.

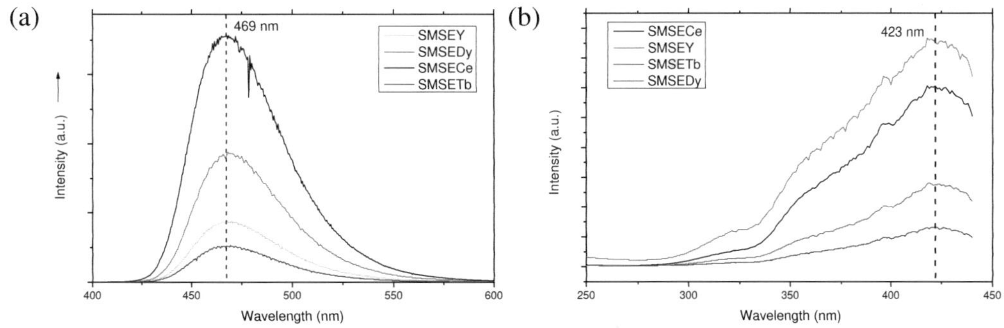

Figure 3. SMS phosphors prepared at different type of co-dopants (a) emission spectra monitored at the excitation wavelength 375 nm and (b) excitation spectra monitored at the emission wavelength 469 nm

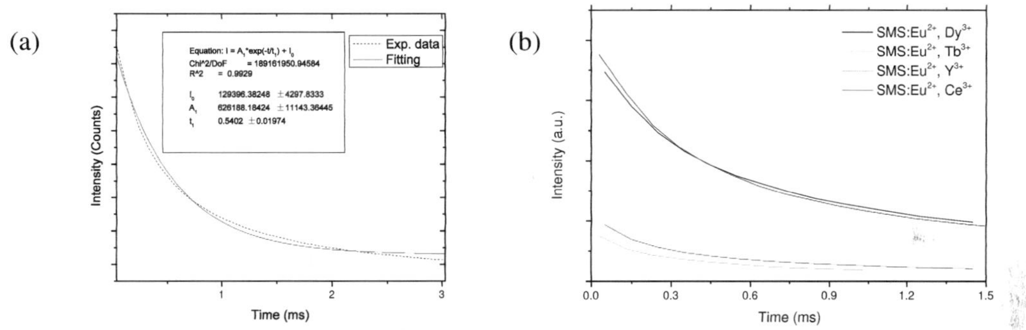

Figure 4. Time decay PL of SMS: Eu^{2+} phosphors co-doped with (a) Ce^{3+} and (b) various co-dopants (excitation wavelength 375 nm)

This result is similar with the works of Brito et al. [22] and Guo et al. [23] whereby Ce and Dy co-doping with Eu contribute to a better luminescence property. This could be due to the co-dopant possess energy levels close to that of the intrinsic trap and therefore increasing the trap density. The higher the trap density is, the better the afterglow property of the phosphor. Other co-dopants (i.e Tb and Y) have a lower trap density and therefore contributing to the poor afterglow property. The poor afterglow property (short lifetime) of SMS:Eu^{2+}, Y^{3+}phosphor could be ascribed to the defect impurities related to Y^{3+} co-dopants that acted as recombination centres rather than traps. It is also suggested that the addition of Tb and Y may destroy some of intrinsic traps in the SMS phosphor. This phenomenon is known as quenching of co-dopants.

Since the afterglow property of SMS phosphor is caused by the thermal detrapping of charge carriers, the intensity of temperature dependent luminescent provides information related to the traps density in the phosphor. Figure 5 (a)-(d) gives the temperature dependent emission spectra of SMS:Eu^{2+} phosphors co-doped with various dopants. It is noted that only one emission peak, i.e 469 nm was observed in all the phosphor samples. Because of the single position of Sr in the SMS host lattice, the 5d states of Eu^{2+} are split into two energy levels, which can be designated by t_{2g} (the lowest) and e_g (the highest). The single blue emission of 469 nm in the phosphors is expected to occur from the lowest t_{2g} level to the $4f^7$ ground state. It is worth mentioning that the relaxed 5d state of Eu^{2+}

is located 0.29 eV below the bottom of the conduction band, otherwise the luminescence emission would be quenched by rapid auto-ionization process [24].

The variation of the relative emission intensity of these phosphors as a function of temperature is plotted in Figure 5(e). It can be clearly seen that SMSECe phosphor has comparatively good temperature quenching resistance as it is still having 38.8% of normalized emission intensity at 300 K. It is commonly known that the emission band of Ce^{3+} ions normally have two maxima located at the emission wavelength of 420 nm and 450 nm, attributed to the crystal field splitting of the 5d band ($^2F_{5/2}$ and $^2F_{7/2}$ states) and the ground state for the $4f^1$ electronic configuration [25]. Since there is no emission peak contributed from Ce^{3+} ions was found in this study, the Ce^{3+} ions were not acted as luminescence centre. Instead, the PL intensity and lifetime of SMS:Eu^{2+} phosphor was enhanced by co-doping with the Ce^{3+} ions as shown in Figure 3(a) and Figure 4(b). Thus, the Ce^{3+} ions was proposed acting as sensitizer. The energy absorbed by the Ce^{3+} ions was transferred to Eu^{2+} ions. This could be verified by the spectral overlap between the PL excitation spectrum of Ce^{3+} and the PL emission of Eu^{2+} in Figure 3(a) and (b). Moreover, the energy transfer rate from Ce^{3+} to Eu^{2+} is generally faster than the emission rate of Ce^{3+} in the dipole-dipole interaction [26], allowing higher emission intensity [Figure 3(a)] and a longer decay lifetime [Figure 4(b)].

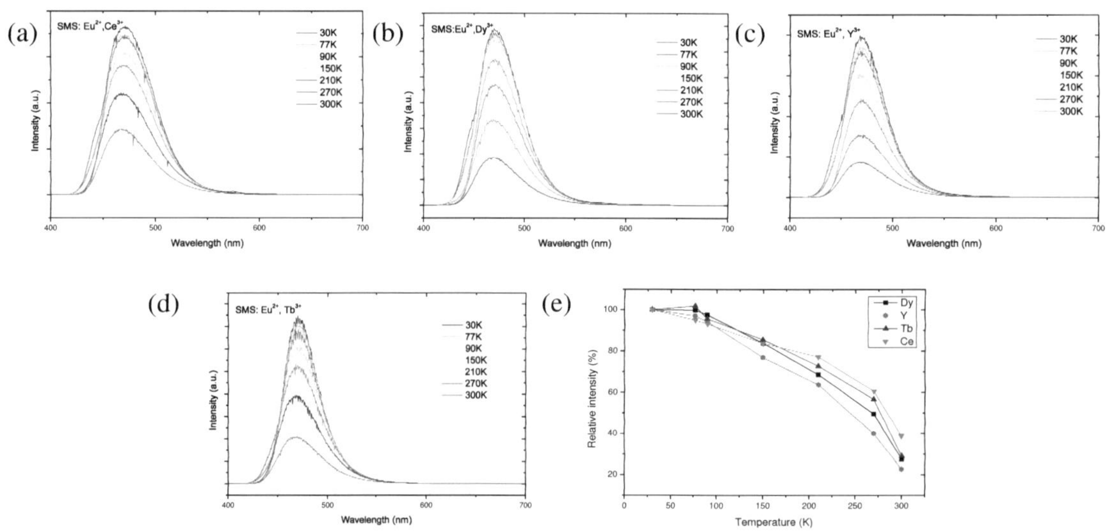

Figure 5. Temperature dependent PL of SMS: Eu^{2+} phosphors co-doped with (a) Ce, (b) Dy, (c) Y, (d) Tb and (e) relative emission intensity of the phosphors as a function of temperature (excitation wavelength 375 nm)

In the case of SMS: Eu^{2+} co-doped with Dy, the unique Dy^{3+} emission peak with the wavelength of 486 nm or 577 nm is also not present [27]. Thus, the co-dopant of Dy^{3+} plays a role as electron-trapping centre rather than a luminescent centre in the host crystal lattice [28]. Exposing SMS: Eu^{2+},Dy^{3+} phosphor to ultraviolet light excites Eu^{2+} to a $4f^6 5d$ state which is located inside the conduction band. A free electron is created by auto-ionization, leaving Eu^{3+} ion behind. The free electron is trapped by Dy^{3+} to form Dy^{2+} (0.66 eV located below the bottom of the conduction band) [24]. The afterglow, i.e. the delay luminescence after excitation, is caused by the thermally activated release of the electron from Dy^{2+} back to the conduction band with subsequent recombination with Eu^{3+} leading to 5d $\rightarrow$ 4f emission at 469 nm. Thus, the persistence time is pro-long. The depth of the trap influences the lifetime of the luminescence of SMS phosphor. In addition to Dy^{3+} ions, other defects such as oxygen vacancies which trap the charge carriers could delay the emission too.

The SMS:Eu^{2+} co-doped with Tb has relatively shorter lifetime, poorer luminescence intensity and thermal quenching resistance in our study. This observation agrees well with the finding of Li et al. that the increase of amount of Tb^{3+} ions reduced the lifetime of Eu^{2+} emission. Under UV irradiation, electrons were optically excited into the lowest excited state ($4f^65d$) of Eu^{2+} from the ground state ($^8S_{7/2}$). Some of these excited electrons relaxed back to the ground state, giving the blue emission at 469 nm. The other electrons of Eu^{2+} transferred to the excited level ($^5D_4$) of Tb^{3+} ions and could lose its energy as phonon. Therefore, the Tb co-dopants acted as non-radiative recombination centres but not traps/luminescence centres in our work. It is worth mentioning that the weak absorption of Tb^{3+} ion at 423 nm due to the $^7F_6 \rightarrow ^5D_4$, transitions and its much narrower FWHM do not favour the co-dopant to act as activator [5]. Lastly, the SMS:Eu^{2+},Y^{3+} phosphor has the poorest temperature quenching resistance as it's relatively emission intensity dropped to 22.6% at 300 K. This is attributed to the effect of Y^{3+} ions which acts as the thermal quenching centre and reduces the luminescence.

4. Conclusions

The SMS: Eu^{2+}, RE^{3+} phosphor were sintered using solid state approachat 1250°C under reducing atmosphere. The effect of co-dopants (RE^{3+}), i.e. Ce, Dy, Y and Tb on the luminescence properties of phosphor was studied. Generally, only one blue emission peak cantered at 469nm was found regardless the types of co-dopants used in the sintering process. This blue emission was ascribed to the Eu^{2+} ion which acts as luminescence centre. The study shows that Ce^{2+} ions were the sensitizer in the SMS:Eu^{2+}, RE^{3+} phosphor by pro-longing the afterglow and by enhancing the luminescence intensity and thermal quenching resistant. The SMS phosphor co-doped with Dy has the comparable emission intensity and long afterglow properties. This is attributed to the Dy ions which introduced more intrinsic traps into the SMS host lattice. The co-doping of Tb or Y was found deteriorated the optical performance of the SMS:Eu^{2+}, RE^{3+} phosphor with poor emission intensity and short decay lifetime.

References

[1] Emily F K 2011 *University of Free State* pp1-27.
[2] Nazarov M 2013 *J .Phys. Sci.* **12** pp102-118.
[3] Wu H *et al* 2009 *J. Alloy. Compd.* **486** pp 549-553.
[4] Xia Z *et al* 2010 *J. Rare Earth.* **28** pp 874-877.
[5] Li B *et al* 2014 *Opt. Mater.* **36** pp 1649–1654.
[6] Ji H *et al* 2007 *J. Sol-Gel Sci. Tech.* **44** pp 133-137.
[7] Xu Y and Chen D 2008 *Ceram. Inter.* **34** pp 2117-2120.
[8] Wu H Y *et al* 2011 *Acta Phys.-Chim. Sin.* **27** pp 1201-1206.
[9] Zheng M *et al* 2013 *ECS Solid State Lett.* **2** pp R19-R22.
[10] Shi Q *et al* 2008 *Mater. Sci. Eng. B* **149** pp 82-86.
[11] Kubo H *et al* 2005 *J. Cryst. Growth* **275** pp 1767-1771.
[12] Kottaisamy M *et al* 2008 *Mater. Res. Bull.* **43** pp 1657–1663.
[13] Pan Y X *et al* 2009 *J. Alloy. Compd.* **488** pp 638–642.
[14] Kim J S *et al* 2006 *Solid. State. Commun.* **137** pp 187–190.
[15] Diaz A and Keszler D A 1996 *Mater. Res. Bull.* **31** pp 147–151.
[16] Shrivastava R *et al* 2014 *Adv. Phy. Lett.* **1** pp 18-20.
[17] Kaya S Y *et al* 2010 *Adv. Sci. Tech.* **62**, pp 88-94.
[18] Li Y *et al* 2009 *J. Electrochem. Soc.* **156** pp J77-J80.
[19] Liu B *et al* 2007 *J. Lumin.* **122-123** pp 121-124.
[20] Wu J *et al* 2013. *J.Chin.Universities* **34** pp 306-312.
[21] Viswanath R *et al* 2014 *J. Nanotechno.* 2014 pp 1-8.
[22] Brito H F *et al* 2011 *J. Therm. Anal Calorim.* **105** pp 657–662.
[23] Guo C *et al* 2007 *J Lumin.* **126** pp 333-338.
[24] Dorenbos P 2005 *Phys. Stat. Sol. (B)* **242** pp R7–R9.
[25] Blasse G and Bril A 1967 *J. Chem. Phys.* **47** pp 5139-5145.
[26] Kwon K H *et al* 2013*J. Nanosci. Nanotechno* **13** pp 4079–4083.
[27] Peng Y *et al* 2004 *J. Chin. Rare Earth Soc.* **22** pp 596–597.
[28] Wu H *et al* 2011 *Opt. Laser Technol.* **43** pp 1104-1110.

Effect of Composition and Temperature to the HA/ß-TCP Composite

N. A. Moslim[1], C. Y. Beh[1], S. R. Kasim[1*] and S. Ramakrishnan[1]

[1]Biomaterials Niche Area, School of Materials and Mineral Resources Engineering, Universiti Sains Malaysia, Penang, Malaysia.

*srshahrizal@usm.my

Abstract. The aim of this study is to fabricate hydroxyapatite (HA) and beta-tricalcium phosphate (β-TCP) composite by varying the HA/β-TCP compositions (100% HA, 80% HA with 20% β-TCP, 60% HA with 40% β-TCP, and 50% HA with 50% β-TCP) and sintering temperatures (1000°C and 1200°C). The powders are mixed with ethanol by using agate mortar, dried in oven (80°C, 24 hours) and uniaxially pressed into 16 mm diameter pellets followed by sintering. Samples are characterized by bulk density testing, phase analysis using X-ray Diffraction (XRD) and Vickers hardness test. Pure HA powder sintered at 1200°C shows the highest density and hardness with 3.090 gcm^{-3} and 421.70 HV respectively. XRD phase analysis show decrease of HA phase at higher sintering temperature due to phase decomposition of HA into β-TCP and alpha-tricalcium phosphate (α-TCP). Both HA/β-TCP composition and sintering temperature are significant factors to the hardness with sintering temperature the more dominant factor.

1. Introduction

Calcium phosphates ceramics are well known in the field of biomaterials due to their occurrence in normal bone and their association with the formation and progression and of enamel and dentin caries. Hydroxyapatite (($Ca_{10}(PO_4)_6(OH)_2$, HA) with Ca/P stoichiometric ratio of 1.67 is one of the common calcium phosphate used in medical applications [1]. It is considered as a high strength and non-resorbable material. Beta-tricalcium phosphate ($Ca_3(PO_4)_2$, β-TCP) with Ca/P ratio of 1.5 is also another type of calcium phosphate that have bioresorbable ability [2]. HA/β-TCP composite; known as biphasic calcium phosphate (BCP) consist of an intimate mixture of HA and β-TCP with varying Ca/P ratios from 1.5 up to 1.67. By varying HA/β-TCP ratios, the properties such as compressive strength, macroporosity, dissolution behavior and the bioresorbable ability of the HA/β-TCP composites can be modified [3]. The lack of bioresorbability of HA in body fluid is unfavorable to the host tissue surrounding the implant. Therefore, β-TCP is being introduced but the presence of β-TCP alone causing too fast resorption in an uncontrollable way. However, by combining HA and β-TCP together, the properties can be combined and their properties can be customized [4]. Despite that, HA/β-TCP composite is still lacking in its mechanical properties such as hardness and fracture toughness which is a major drawback for load bearing applications [5]. Sintering process can enhance the densification of HA/β-TCP composite and thus improving its mechanical properties. In this research, the main objectives are to fabricate and characterize the HA/β-TCP composite with varying the composition of HA/β-TCP and sintering temperature.

2. Materials and method

2.1 Sample preparation

Four different composition of HA/β-TCP were made based on weight percentage (wt%); namely 100% HA, 80% HA with 20% β-TCP, 60% HA with 40% β-TCP, and 50% HA with 50% β-TCP (tabulated in Table 1). The powders were mixed homogeneously using agate mortar with the aid of ethanol as medium. The mixing process was carried out until ethanol solution had evaporated. The mixture was dried for 24 hours at 80°C in oven to completely remove excess ethanol. 1.5g of each powder was pressed into 16 mm diameter pellets using Specac hand press machine. The pellets were sintered at 1000°C and 1200°C for two hours soaking time with heating rate fixed at 5°C/min.

Table 1. Variation of initial composition with subjected sintering temperature

Sample	Sintering temperature (°C)	Initial composition (wt%)	
		HA	β-TCP
R1	1200	100.0	0.0
R2	1200	80.0	20.0
R3	1200	60.0	40.0
R4	1200	50.0	50.0
Q1	1000	100.0	0.0
Q2	1000	80.0	20.0
Q3	1000	60.0	40.0
Q4	1000	50.0	50.0

2.2 Sample characterization

The bulk density of the pellets after sintering was determined using Archimedes principle. The sintered pellets were subjected to XRD phase analysis with 2θ used between 10°-90° (XRD machine: model Bruker AXS D8 Advance). The phase of pellets was analysed with X'pert Highscore Plus software and Rietveld refinement was used to determine the changes in phase composition of pellets. The result obtained was compared with the International Center for Diffraction Data (ICDD). The hardness of pellets was measured via Vickers hardness test model from Future Tech Corporation. Before performing hardness test, pellets were grind with silicon carbide paper and then polished using alumina to obtain a smooth flat surface. All pellets were subjected to 3 kgf indentation load with 10 second dwell time.

3. Results and discussion

3.1 Bulk density analysis of HA/β-TCP composite

It is essential to determine the density of pellets because it can greatly affect the mechanical properties of composite. Bulk density of HA/β-TCP composite are determined and tabulated in Table 2. Density plays an important role as it can affect the mechanical properties of HA/β-TCP composite. Pellet with the composition of 100% HA shown to have the highest density value when sintered at 1200°C with 3.090 gcm^{-3} (sample R1). Conversely, composition 50% HA 50% β-TCP has the lowest density values at 1000°C with 1.975 gcm^{-3} (sample Q4). By comparing in terms of sintering temperature factor, an increasing trend can be observed where the density values increase as sintering temperature increases (refer Table 2). Highest density value obtained from sample R1 while the lowest density value obtained from sample Q4. From this finding, densification process is expected to occur but the stage of densification varies at different sintering temperature. At the early of sintering process, pellets have successfully entered the first stage of densification process where only inter-particle necks form and grow, reaching up to 65% of its theoretical density (~1.63 gcm^{-3}). During intermediate stage, the pores in the pellet will start to shrink but remain open and connected, reach up to around 90% theoretical density (~2.26 gcm^{-3}). Then, in the final stage of densification, the pores become isolated or eliminated, leaving a fully or near fully dense structure (~3.14 gcm^{-3} HA/β-TCP full density) [6]. It can be concluded that pellet sintered at 1200°C (sample R1, R2, R3 and R4) have entered the final stage by having a denser structure compared to samples sintered at 1000°C (sample Q1, Q2, Q3 and Q4). A study reports that approximately 95% of theoretical density was achieved when calcium phosphate – mullite composite was being sintered in the range of 1300°-1350°C [7]. Another study reported that the density of HA increase as the sintering temperature being increased [8]. It was also reported that different holding time during sintering gives different density of HA where longer holding time resulted in higher density of HA. However, the effect of holding time was not being investigated in this study.

Table 2. Variation of bulk density, final composition (obtained via Rietveld refinement) and hardness value of sintered pellets

Sample	Sintering temperature (°C)	Bulk density, gcm^{-3}	Final composition (%)			Hardness value (HV)
			HA	β-TCP	α-TCP	
R1	1200	3.090	56.1	19.0	24.9	421.70
R2	1200	3.039	4.5	91.8	3.7	340.60
R3	1200	2.891	0.0	100.0	0.0	256.37
R4	1200	2.709	0.0	100.0	0.0	231.80
Q1	1000	2.243	73.6	26.2	0.2	70.67
Q2	1000	2.145	2.8	97.2	0.0	68.43
Q3	1000	2.005	0.0	100.0	0.0	57.37
Q4	1000	1.975	0.0	100.0	0.0	55.07

3.2 Phase analysis of HA/ß-TCP composite via XRD

Rietveld refinement is used to determine the percent of phase composition exist in the pellets. These data are presented in Table 2. The diffraction pattern with identified phases for pellets sintered at 1200°C and 1000°C are shown in Fig. 1 and 2 respectively. The presence of α-TCP (ICDD #98-007-8160) in R1 and R2 indicates that HA (ICDD #98-007-7628) had lose its OH functional group through dehydration process and decompose into α-TCP after sintering process [9]. The formation of α-TCP can suppress the densification process thus lowering mechanical strength of HA. The formation of α-TCP probably due to the sintering of β-TCP at 1200°C resulting in the decomposition of HA through dihydroxylation [10]. The complete transformation of HA to β-TCP as in sample R3, R4, Q3 and Q4 is due to the presence of β-TCP (ICDD #98-007-6561) in the initial composition that able to promote thermal decomposition of HA [10]. Thus, higher decomposition of HA into β-TCP is observed with the higher initial amount of β-TCP in the sample. Subsequently, the variation of phase composition is difficult to be controlled, and the large difference in phase composition before and after sintering can be explained via pressureless sintering fabrication route; where the HA/β-TCP mixture is unstable during pressureless sintering, thus altering the final composition of composite [10]. On the other hand,

Q1 which initially has 100% HA have transformed into triphasic calcium phosphate comprises 73.7% HA, 26.2% β-TCP, and 0.2% α-TCP. Q2 pellet which has an initial composition of 2.8% HA and 97.2% β-TCP, remained as a biphasic calcium phosphate with 2.8% HA and 97.2% β-TCP, after sintering. The formation of triphasic calcium phosphate in Q1 when sintered at 1000°C are due to the unstable mixing of HA/β-TCP composite [11]. For the case in pellet R3 and Q3, the complete decomposition of HA also comes to a good agreement with some studies whereby 60% HA can completely transformed to β-TCP [10–12].

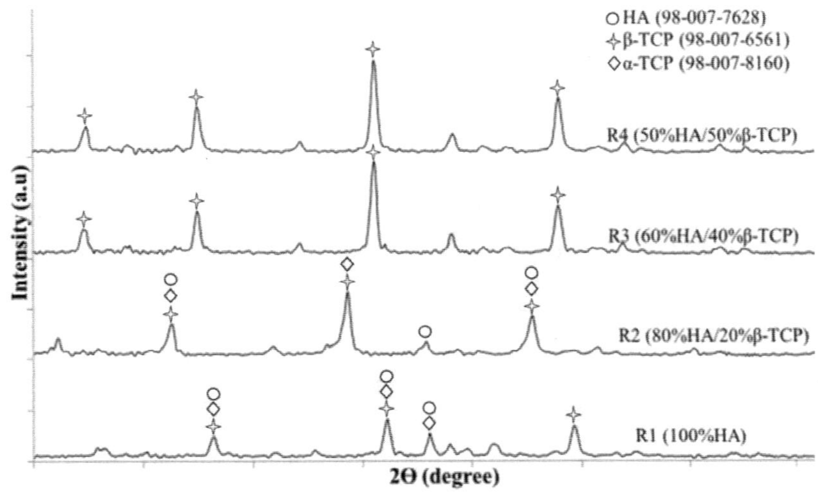

Figure 1. XRD diffraction pattern for pellets sintered at 1200°C

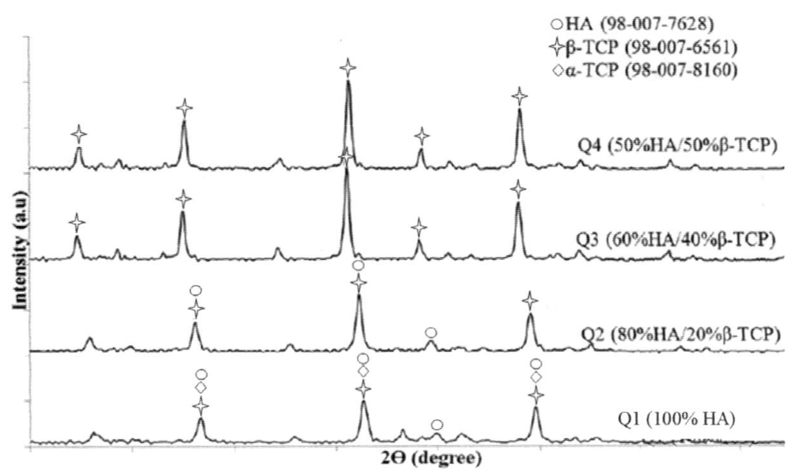

Figure 2. XRD diffraction pattern for pellets sintered 1000°C

3.3 Vickers hardness test

The hardness value obtained is tabulated in Table 2. Higher initial composition of HA in sample R1 resulting in high hardness after sintering. It is due to the better mechanical properties of HA than β-

TCP which relate to the highest bulk density and highest HA present in R1 as mentioned in Table 2 and Fig. 1 and 2 respectively. This had contributed to its high hardness properties. By referring to Rietveld refinement data in Table 2, it is noticeable that both R1 and Q1 has HA phase while R4 and Q4 has only β-TCP phase. These explain the high hardness in R1 with 421.74 HV and low hardness in Q4 with 55.07 HV. However, Q1 (70.67 HV) exhibit lower hardness when compared to R4 (231.80 HV) although Q1 have higher HA initial composition than R4. Such findings are unable to be explained only by the initial HA composition factor; therefore, this finding will be evaluated based on the sintering temperature factor. An increasing trend is seen in Fig. 3; where hardness value increases with increasing sintering temperature. Highest hardness value is obtained from sample R1 with the highest sintering temperature of 1200°C, highest initial HA composition of 100% and highest density, while the lowest hardness value is obtained from sample Q4 with lowest sintering temperature of 1000°C, lowest HA initial composition of 50% and lowest bulk density of 1.975 gcm^{-3}. It shows that temperature plays a significant role in affecting the density which also influences the hardness of pellets. Higher sintering temperature would result in better densification; increasing density of pellets, hence increasing pellets' hardness. It can be claimed that sintering temperature is the more influential factor in producing high hardness HA/β-TCP composite compared to the initial HA composition in pellets.

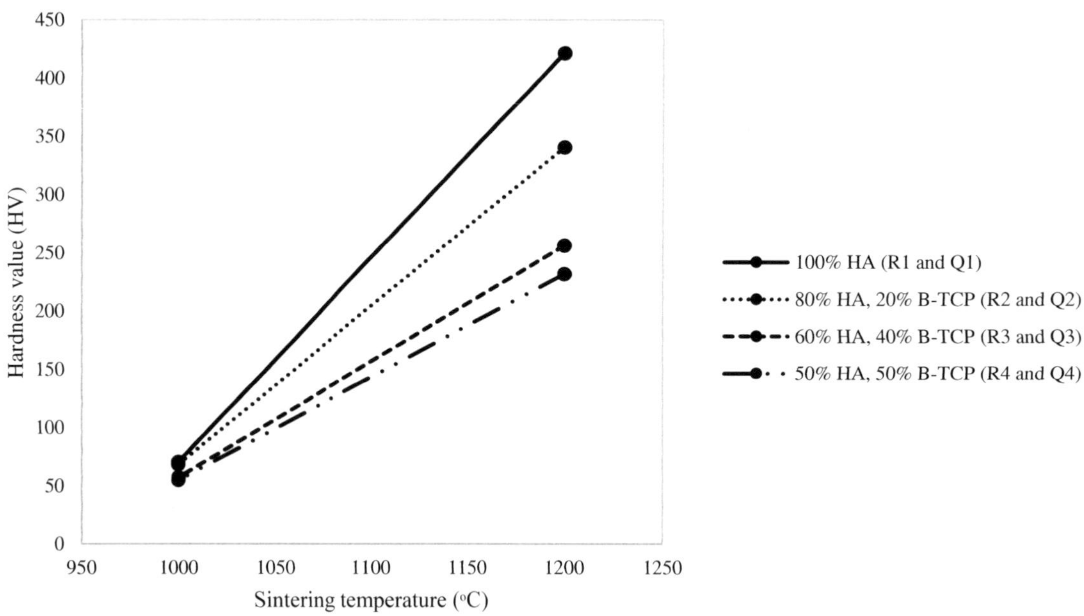

Figure 3. Graph of hardness value versus sintering temperature of pellets with different composition

4. Conclusion

HA/β-TCP composite has been successfully fabricated. Based on characterization results, hardness value increases with higher HA initial composition and sintering temperature. From the XRD phase analysis, most HA will convert into β-TCP after sintering at 1000°C. The increase of hardness of the composite is due to the decomposition of HA to β-TCP observed in samples sintered at 1000°C and 1200°C. Sample with high HA composition after sintering provides higher hardness. However, it should be noted that the effect of sintering at higher temperature is more significant towards increasing the composite hardness compared to the effect of initial HA composition.

References

[1] Kokubo, *Bioceramics and their clinical applications* (Woodhead Publishing, 2008).
[2] Farzadi A, Solati-Hashjin M, Bakhshi F, & Aminian A 2011 *Ceramics International* **37** 65
[3] Lobo S E & Arinzeh T L 2010 *Materials* **3** 815
[4] S. V. Dorozhkin *Advanced Ceramics* 2016 **8** 33
[5] Wade M, Li Y C, & Wahl G M 2013 *Nature Reviews Cancer* **13** 83
[6] L. F. Francis 2015 *Material Processing: A Unified Approach to Processing of Metals, Ceramics and polymers*
[7] Nath S, Biswas K, Wang K, Bordia K, & Basu B, 2010 *Journal of the American Ceramic Society*, **93** 1639
[8] Tolouei R, Ramesh S, Tan C Y, Amiriyan M, & Teng W D 2011 *International Journal of Automotive and Mechanical Engineering* **3** 249
[9] Muralithran G & Ramesh S, T 2000 **26** 221
[10] Nilen R W N & Richter P W 2008 *Journal of Materials Science: Materials in Medicine* **19** 1693
[11] Chaki T K & Wang P E1993 J*ournal Of Materials Science Materials In Medicine* **4** 150
[12] Yetmez M 2014 *Advances in Materials Science and Engineering* **2014**

IOP Conf. Series: Journal of Physics: Conf. Series **1082** (2018) 012024 doi:10.1088/1742-6596/1082/1/012024

Conversion of YBCO Pellet into Single Layer Thin Film using Pulse Laser Deposition

Fasih Ud Din[1,2]*, Abdul Halim Shaari[1], Chen Soo Kien[1], Zainal Abidin Talib[1], Amad Ud Din[3], and Lim Kean Pah[1]

[1]Department of Physics, Universiti Putra Malaysia, 43400 Serdang, Selangor, Malaysia.

[2]Department of Physics, Government College University, Faisalabad, Sahiwal Campus, 57000 Sahiwal, Pakistan.

[3]Department of Electronic Engineering, Advanced Electronics System (AES) Lab, Fatima Jinnah Women University (FJWU), The Mall, Rawalpindi-Pakistan.

Email: uddin.fasih@gmail.com

Abstract. The Pulse laser deposition (PLD) is used to convert bulk YBCO Pellet to high-quality YBCO epitaxial film and superconducting transport properties of YBCO single layer has been investigated using surface techniques and four point probe method. The present study includes thin film fabrication via PLD, X-ray diffraction (XRD), Scanning electron microscopy and four point probe method. This study reveals the relationship between plume dynamics and quality of the single layer film based on the important parameter, such as background pressure, substrate temperature, plume size and film thickness. The plume dynamics role in fabricating high-quality layer and improving the superconducting properties of the single layer has been analysed.

1. Introduction

The discovery of High-Temperature Superconductors (HTS) motivated the practitioner and vendors to use them for specific practical purposes such as better low cost, usage efficiency and the performance of material that was not achievable in the bulk form of material. Due to this reason, nowadays for practical applications most of the HTS materials prepared in the form of thin films. To progress the quality of HTS Layer's, widespread efforts have been made since the discovery of high-T_c oxide superconductors. The Low cost HTS materials were prioritized over low-temperature superconductors as they showed critical temperature when cooled down in liquid nitrogen [1-4].

Thin film deposition techniques engage atoms and molecules of the respective compounds, their transportation and deposition. Thin solid films provide the foundation for numerous devices and have important applications in fields of materials sciences and emerging technologies.

In the PLD of layer's, a transparent Ultra Violet (UV) lens focuses the laser light to solid bulk YBCO pure target sample, melt it and some part of the material is detached in the form of a plume as shown in figure 2. The plume comes in contact with the help of oxygen molecules to close fitted

IOP Conf. Series: Journal of Physics: Conf. Series **1082** (2018) 012024 doi:10.1088/1742-6596/1082/1/012024

substrate a few centimetres away at an angle of 45° [5]. The plume generated by the interaction of laser and pure target YBCO can be used to assist the growth of fine quality layers at relatively low processing temperatures.

For the formation of high-T_c phases, it has been found that the thermal treatment parameters like heating up rate, annealing time and cooling rate play avital role. The researcher believed that YBCO has regained interest as it retains superconductivity under large magnetic fields as compared to the other class of superconductors [6-8].

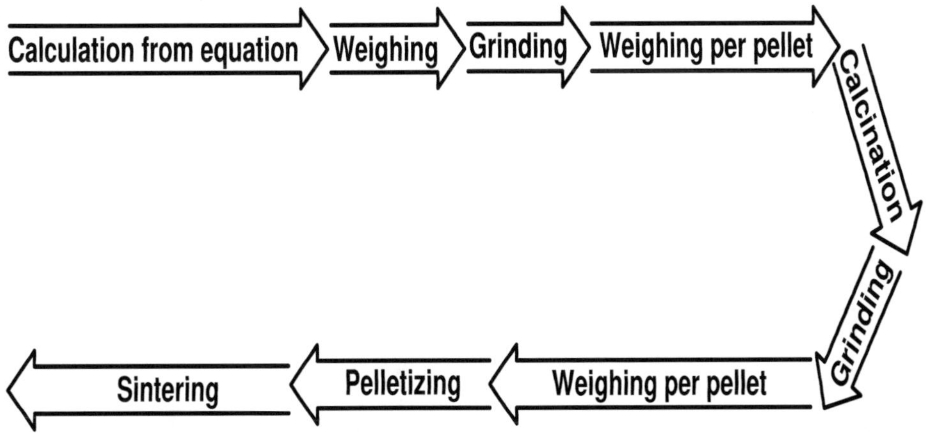

Figure 1. *YBCO* sample preparation using solid state reaction method.

2. Material and Method

The YBCO bulk sample pellet is prepared by Solid-State Reaction Method as shown in figure 1. PLD a novel film fabrication method is used to make the vastly stoichiometric epitaxial films as shown in figure 2. The ultra-high vacuum is created with turbo molecular pump which assists the deposition cavity to remove oxides and impurities. The oxygen gas has the primary role in all this creation process and the oxygen dependency of YBCO layers. The PLD technique is an economical technique for making the epitaxial superconducting YBCO film layers as shown in figure 2.

During this PLD process, the MgO substrate is placed 3-5 cm away from the pure bulk target prepared with solid state reaction method. The substrate object which is attached to the heater can be heated according to its own range from 0 to 600^0C depending upon the system. In case of low range substrate heating system compel the vendor to prepare the samples ex-situ by use of a tubular furnace to alleviate the prepared YBCO thin films. The most important role of oxygen gas is used to progress the blasted particle of to the substrate fitted at some angle from the YBCO target. The YBCO target is hit by converged intense laser light beam. The size and shape of the glowing plume also depend upon the quality of the film. The whole system can properly manage to improve the quality of epitaxial YBCO films [9, 10].

IOP Conf. Series: Journal of Physics: Conf. Series **1082** (2018) 012024 doi:10.1088/1742-6596/1082/1/012024

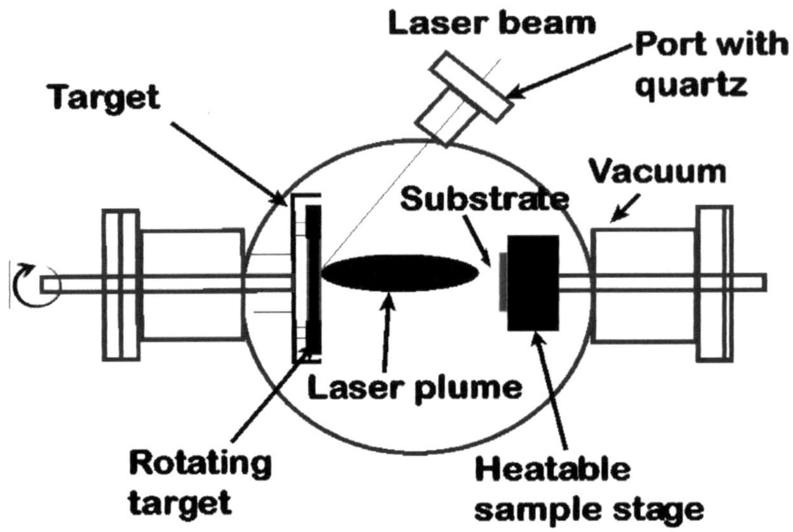

Figure 2. *PLD* setup for converting YBCO pellet into thin film form.

3. Results and discussion

The pellets were prepared by simple solid-state reaction method the calcination temperature was 940°C and sintering temperature was also 940°C, it was then used as a target and was placed at 4.5 cm distance from the substrate MgO in the vacuumed chamber in presence of oxygen as background gas at pressure 200 m Torr. The deposition time was 40 minutes. The deposition was done at 790°C and a post-annealing was applied in-situ at 550°C for 30 min and flow of oxygen pressure of 300 Torr. Later the system was cooled down to room temperature. In order to obtain the best annealing sequence, the annealing temperature was varied. The heating rate was 3°C/min while the cooling rate was 2°C/min in order to improve the quality of the film [11, 12].

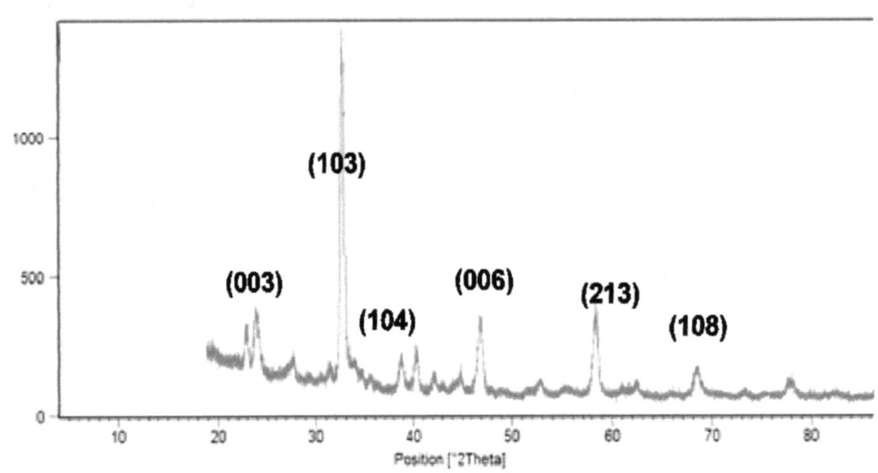

Figure 3. *XRD* pattern of pure YBCO pellet.

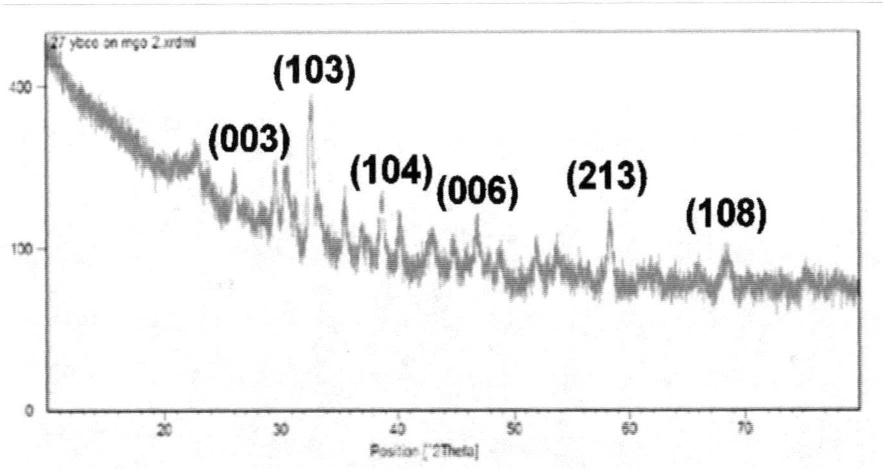

Figure 4. *XRD* pattern of YBCO thin film.

YBCO single crystal material was confirmed by XRD patterns with pointed peaks that portrayed different crystal lattice planes as shown in figure 4 and 5. Those are represented by (hkl) values, which phase appeared and was analyzed and considered appropriately. The XRD analysis was accomplished as per X'pert High score matching with a standard database (ICCD Data).

The Surface topology with ripples, ridges and cones is clearly shown for the bulk YBCO pellet by SEM micrographs for a smooth bulk pellet surface under different magnifications are shown in figure 5. The magnification was 20,000 times. The voids averaging ≈1 μm in width with a non-homogeneous and non-uniform microstructure were observed.

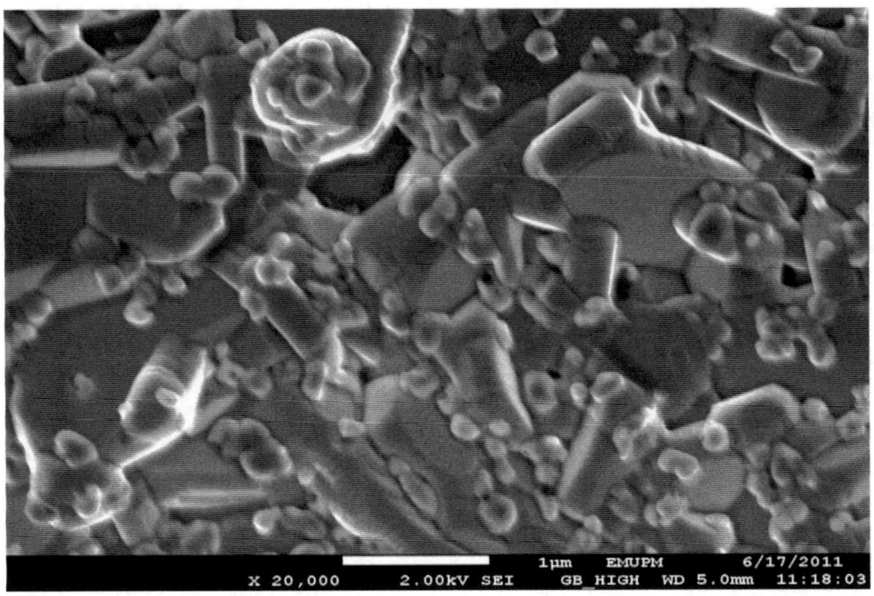

Figure 5. *FESEM* image of bulk pellet with 20,000 times magnifications.

The SEM micrographs of the YBCO smooth surface on MgO under different magnifications showing surface topology of the thin film sample with voids of ≈1μmin width with a non-

IOP Conf. Series: Journal of Physics: Conf. Series **1082** (2018) 012024 doi:10.1088/1742-6596/1082/1/012024

homogeneous, non-uniform microstructure. The magnification was 20,000 times as shown in figure 6. These microstructure properties of bulk pellets and the thin film sample stand in good agreement [13].

Figure 6. *FESEM* image of the thin film with 20,000 times magnification.

The resistance vs temperature curve of the pure bulk was also measured by four point probe method showing the quality of pure YBCO bulk achieving onset resistivity Tc ~ 80° (K) as shown in figure 7(a).

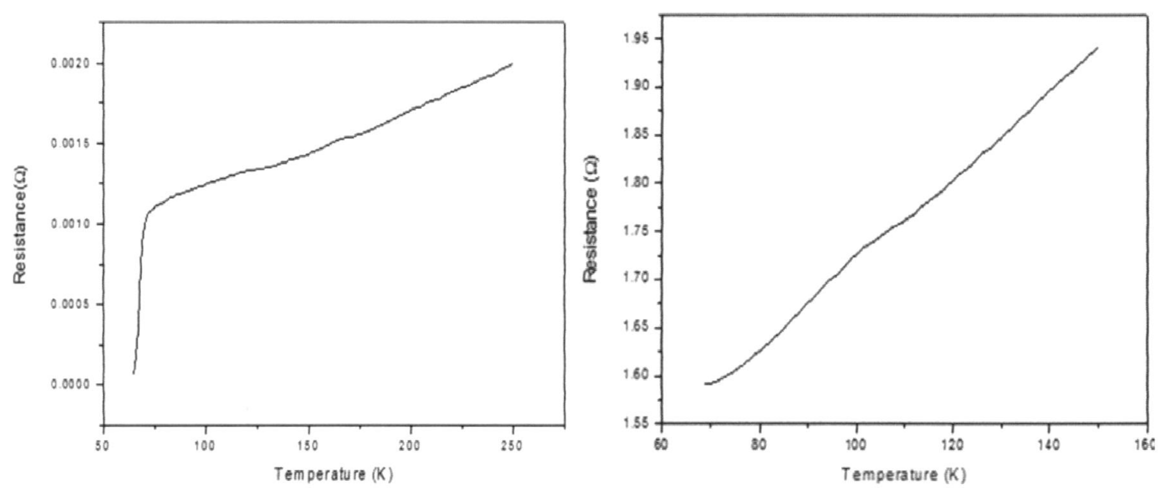

a b

Figure 7. (a) R - T curve showing resistance of pure YBCO pellet (b) R - T curve showing the resistance of pure YBCO thin film.

A typical R-T curve of a pure YBCO thin film deposited on (100) MgO annealed inside the PLD vacuum chamber as shown in figure 7(b). Similarly in order to improve the quality of the deposited film which may be due to the deficiency of low oxygen absorbed during the deposition that effects the superconducting behaviour to some extent. It has been subjected to post annealing under oxygen flow for 2h at 850° (C) inside the tubular furnace which has the 3° (C/min) increasing rate while 2° (C/min) cooling rate. It plays a critical role during the synthesis of YBCO thin films. It has been observed that the electrical properties of high-T_c are also dependent upon the charge carriers in the CuO_2 planes. The doping under oxygen ambient affects the T_c and pushes it toward the highest value. Similarly, in our present case 1.6 holes per CuO_2 planes push the improved T_c value near to 94° (K).

4. Conclusion

This study leads to the successful conversion of pure YBCO bulk into the thin film form under normal pressure. During this study, the method of preparation YBCO bulk and their conversion into a thin film and superconducting electrical properties has been discussed. The laser sintered pellets target interaction includes the role of laser wavelength and pulse rate. The effect of background pressure during the deposition is not superconducting and subjected to high-temperature post-annealing treatment experiments under O_2 for the recovery of superconducting properties during the epitaxial growth. Stream velocity of the hot plasma species scrambled up exponentially with a change in temperature. This conversion from the bulk YBCO to thin films at normal pressure is a very successful effort leading the normal pressure synthesis of pure YBCO layer films possible. The normal pressure synthesis of this compound subjected the limitations on the versatility of this compound for commercial device fabrication.

References

[1] Jannah A N, Halim S A, and Abdullah H 2009*European Journal of Scientific Research 29* 438
[2] Jannah A N, Halim S A,Younus W M M, Zulkifli A, Pah L K,Masrianis A, and Walter C P 2004 *Journal of Solid State Science and Technology Letter* 11 215
[3] Roas B,Schultz L, and Endres G 1988 *Journal of Applied Physics Letters* 53 15 57
[4] Viret M,Lawler J F, and Lunney J G 1993 *Journal of Superconductor Science and Technology* 6 490
[5] Shahid M, Bidin N, DaudY M, and Ashiq G B 2011 *Journal of Intense Pulsed Lasers and Applications in Advanced Physics* 1 35
[6] Shahid M, Bidin N,Daud Y M, and Ullah M I 2011 *Journal of Intense Pulsed Lasers and Applications in Advanced Physics* 1 65
[7] Din F U, Shari A H B,Kamalianfar A,Pah L K, Kien C S, Talib Z A B,Shahid M, and Ashiq G B 2012 *Journal of Ovonic Research* 8 9
[8] Liu L, Zhao Z, Liu H, and Li Y 2010 *Journal of IEEE Transaction on Applied Superconductivity* 20 1553
[9] Augieri A, Galluzzi V, Celentano G, Angrisani A A, Rufoloni A M A, Vannozzi A, Silva E, Pompeo N, Petrisor T, Ciontea L, Gambardella U, and Rubanov S 2009 *IEEE Transaction on Applied Superconductivity* 19 3399
[10] Huhtinen H, Laiho R, Lahderanta E, Paturi P, Raittila J, and Stepano Yu 2000 Ph*ysica C* 341 2377
[11] Khan N A,Din F U, and Khurram A A 2005*Physica C* 415 119
[12] Al-Hada N M, Saion E, Talib Z A and Halim S A 2016 *Polymers* 8 113
[13] Dihom M M, Shaari A H, Baqiah H, Al-Hada N M, Talib Z A, Kien C S, and Abd-Shukor R 2017 *Ceramics International* 7 707

IOP Conf. Series: Journal of Physics: Conf. Series **1082** (2018) 012025 doi:10.1088/1742-6596/1082/1/012025

XRD Analysis of Calcined Magnesium Substituted Biphasic Calcium Phosphate (Mg-BCP)

N. A. Moslim[1], N. Ahmad[1] and S. R. Kasim[1*]

[1]Biomaterials Niche Area, School of Materials and Mineral Resources Engineering, Universiti Sains Malaysia, Penang, Malaysia.

*srshahrizal@usm.my

Abstract. The incorporation of magnesium (Mg) in biphasic calcium phosphate (BCP) was prepared through precipitation method. The concentration of added Mg was varied with respect to calcium (Ca) precursor molarity as such Mg/Ca molar ratio were 0.05 and 0.1, while the (Ca+Mg)/P ratio was maintained at 1.59 throughout the experiment to ensure that calcium deficient apatite was formed. Precipitate obtained was aged for 24 h in ambient condition followed by decantation, centrifugation and drying at 90°C overnight. The as-synthesized powders were calcined at 600°C and characterized by X-ray diffraction (XRD) via Rietveld refinement (X'Pert HighScore Plus software). Both as-synthesized and calcined Mg-BCP powder contains hydroxyapatite (HA) and beta-tricalcium phosphate (β-TCP) phase with Mg in its lattice. The substitution of Mg in calcium-deficient apatite resulted in the formation of biphasic mixtures of β-TCP/HA after calcinations with shifted peaks β-TCP when compared to reference β-TCP peaks. High deficiency of calcium in apatite enables more substitution of magnesium thus stabilizing β-TCP phase when subjected to high temperature.

1. Introduction

For the past few decades, calcium phosphate-based bioceramics are materials of choice for bone tissue engineering applications because of their composition similarity with bone mineral; which are excellent bioactivity and osteoconductivity [1]. One of the most common calcium phosphate used in biological applications is biphasic calcium phosphate (BCP) which contain the hydroxyapatite ($Ca_{10}(PO_4)_6(OH)_2$, HA) and beta-tricalcium phosphate ($Ca_3(PO_4)_2$, β-TCP) phases. Mixture of these two phases ensures that BCP would have ideal biodegradability and biocompatibility compared to the applications of HA or β-TCP alone in bone healing process. Lack of magnesium (Mg) in bone may cause patient to suffer from osteoporosis disease [2] . Doping of calcium phosphate with required amount of Mg gives improvement on mechanical and biological properties of BCP. Due to the smaller size of Mg compared to calcium (Ca), it easily enter Ca lattice and controls the crystallization of HA, affecting the stability of apatite [3]. The synthesis of Mg doped calcium apatites can be achieved through different preparation route as such sol-gel, hydrothermal or precipitation method [4–5].

Current trend of studies focusing on the structural analysis of Mg doped BCP have been discussed via X-ray powder diffraction (XRD) in parallel with Rietveld refinement. So, in this study, we focused on the preparation of Mg substituted BCP through precipitation method as well as to investigate the effect of Mg substitution on XRD peaks. The amount of Mg required for BCP doping usually in between 1-3 wt% of total BCP wt%. Only a few papers have reported and studied magnesium doped BCP obtained via this method.

2. Experimental

2.1 Powder Preparation

The sources for calcium (Ca), phosphorus (P) and, magnesium (Mg) precursors are calcium nitrate tetrahydrate ($Ca(NO_3)_2.4H_2O$, Merck), diammonium hydrogen phosphate ($(NH_4)_2HPO_4$, Merck) and magnesium nitrate hexahydrate ($Mg(NO_3)_2.6H_2O$, Merck) respectively. Pure biphasic calcium phosphate (pure BCP) and Mg substituted BCP (Mg-BCP) samples were synthesized through precipitation method. To prepare pure BCP powder, an appropriate amount of $Ca(NO_3)_2.4H_2O$ was dissolved in water and stirred at the rate of 1000 rpm. $(NH_4)_2HPO_4$ solution was slowly added into the mixture through titration process in a fixed molar ratio (Ca+Mg)/P of 1.59 [6]. The pH of the mixture was controlled between 10-11 with the addition of 25% ammonia solution (NH_4OH). The precipitated suspension was allowed to age to maturity for 24 h at room conditions. After aging, precipitates were decanted and undergo centrifugation followed by drying at 90°C overnight. The dried cakes were ground to fine powders using agate mortar and calcined at kept at 600°C in air for 2 h with heating rate of 5°C/min.

To prepare Mg-BCP powders, same preparing approach of pure BCP was implemented with the subsequent addition of $Mg(NO_3)_2.6H_2O$ whilst stirring. The Mg/Ca molar ratio was varied between 0.05 and 0.1 (coded M-MgBCP1.59 and H-MgBCP1.59 respectively). $(NH_4)_2HPO_4$ solution was added into mixture by titration with the same molar ratio of (Ca+Mg)/P as mentioned above. 25% ammonia solution (NH_4OH) was added to control pH at 10-11. Precipitated product was aged for 24 h at room conditions, decanted, centrifugated and dried at 90°C overnight. The dried cakes were ground to fine powders and kept for calcinations. All of the as-synthesized and calcined samples were submitted for characterizations.

2.2 Characterization

The phase formation of the as-synthesized and calcined sample was characterized using X-ray diffraction (XRD) with Rietveld refinement. X-ray diffraction spectra of the Mg doped BCP powders were analysed over 2θ range of 10-70°C, copper Kα radiation (λ=1.5418Å), with a step scan of 2°/min. The accelerating voltage and current were 40kV and 30mA respectively. Crystallographic identification of the synthesized phases was accomplished by comparing the experimental XRD patterns to International Centre for Diffraction Data (ICDD); #98-010-2187 for hydroxyapatite (HA), #98-000-5558 for monetite, and #98-000-0800, beta-tricalcium phosphate (β-TCP).

3. Results and discussion

3.1 As-synthesized powders

The XRD patterns for the as-synthesized powders are presented in Fig. 1 and Fig. 2. Both pure biphasic calcium phosphate (BCP) and magnesium (Mg) substituted BCP powders indicated the presence of HA except for the difference in phases and peak intensities. As can be seen in Fig.1, the phase with highest intensity in pure BCP powder were monetite. This phase normally exists predominantly before HA formation due to its high phase stability under acidic condition compared to HA [7]. This condition being monitored through pH value (pH less than 5). Mg substitution is characterized by the broad diffraction patterns which indicate the formation of calcium deficient HA; which can be observed in Fig. 2. A little amount β-TCP phase can be found in the as-synthesized Mg substituted BCP powders. Various study report that the diffraction pattern usually resembles that of HA although the Ca/P molar ratio was not fixed at 1.67 [8].

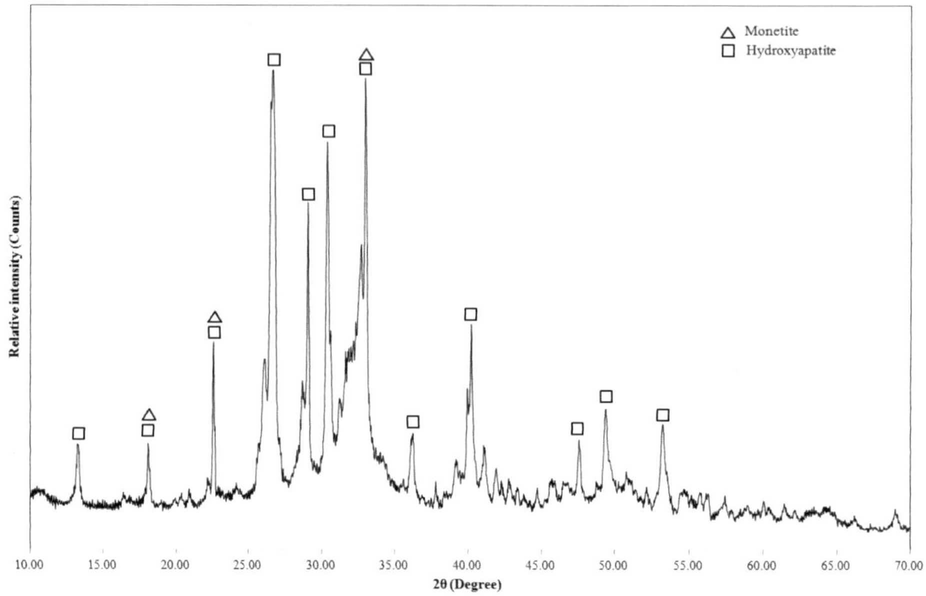

Figure 1. XRD patterns of as-synthesized pure BCP powders

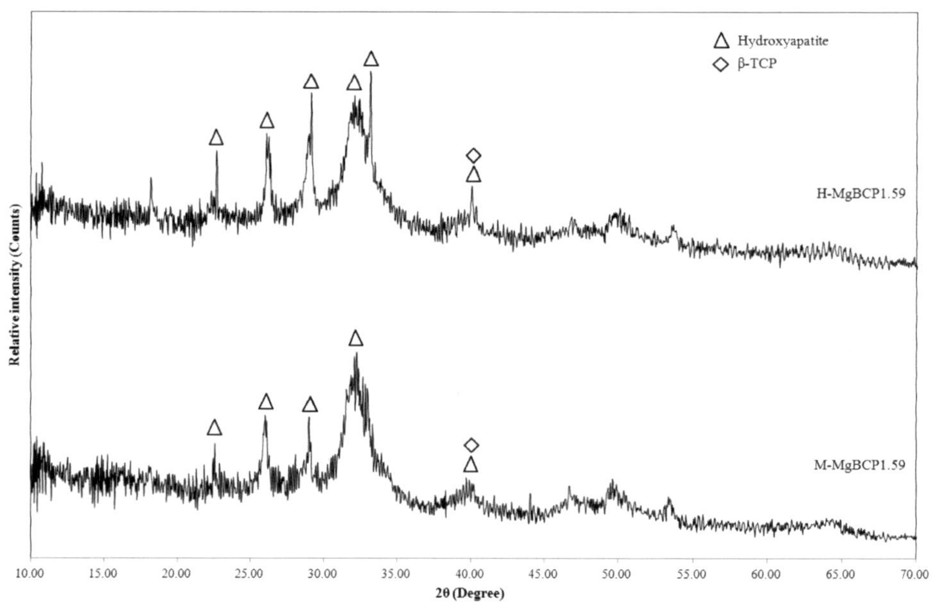

Figure 2. XRD patterns of as-synthesized BCP powders with different Mg concentrations

3.2 Calcined powders and formation of biphasic phases

Calcinations of the powders at 600°C as shown in Fig. 3 indicated the improvement in crystallinity by the resolution of peaks when compared to the as-synthesized powders in Fig. 2. These peaks assigned to the β-TCP phase are viewed by the difference in intensities which related to the level of calcium

IOP Conf. Series: Journal of Physics: Conf. Series **1082** (2018) 012025 doi:10.1088/1742-6596/1082/1/012025

deficiencies. Higher amount of β-TCP phase formed denotes the higher deficiencies of calcium in powders [9]. Considering the substitution of Mg inside BCP lattice, there is slight shift in 2θ and d-spacing in (0 2 1 0) plane of the highest intensity (refer Table 1). The 2θ for both Mg doped BCP samples were higher than pure BCP sample with 31.2573° and 31.2591° for H-MgBCP and M-MgBCP samples respectively; and 30.6063° for pure BCP sample. This shift is mainly caused by stabilization effect of Mg when this ion (ionic radius ~0.65Å) being introduced into Ca ion (ionic radius ~0.99Å) sites; thus resulted in peak displacement toward higher 2θ angles of β-TCP peaks. However, there is no exact match found for the powders peaks with those β-TCP peaks. [5]. As for the d-spacing of these three samples, pure BCP sample happen to have bigger d-spacing with 2.9186Å compared to Mg doped BCP samples with 2.8591Å and 2.8593Å. It suppose that Mg substitute the Ca sites in BCP lattice. While Mg ion have slightly smaller size than Ca ion, the substitution would trigger lattice stresses which later causes an increase in d-spacing of Mg doped samples [10].

$$Ca_{10-x}(Mg_2P_2O_7)_{x/2}(PO_2)_{6-x}(OH)_{2-x} + x/2H_2O \rightarrow (1-x)Ca_{10}(PO_4)_6(OH)_2 + 3x(CaMg_{2x/9})_3(PO_4)_2 + xH_2O \quad \textbf{(Eqn. 1)}$$

The above Eqn. 1 explains that the higher substitution of Mg in Ca sites is in parallel with the transition of HA to β-TCP phase. Higher content of Mg causes HA phase to convert into β-TCP phase. These cause high amount of β-TCP produce in heat treated Mg substituted BCP powder when compared to heat treated pure BCP. It is correlated to the high amount of β-TCP formed in Mg substituted powder compared to pure BCP powder as observed in Table 1. Small difference in the values tabulated in Table 1 probably effect of different Mg concentrations substituted into β-TCP phase. Quantitative determinations of the β-TCP/HA phase ratios of calcined pure BCP, M-MgBCP1.59 and H-MgBCP1.59 powders were found to be 90.5/9.5, 99.2/0.8 and 98.7/1.3 respectively in phase compositional percentage obtained via Rietveld refinement.

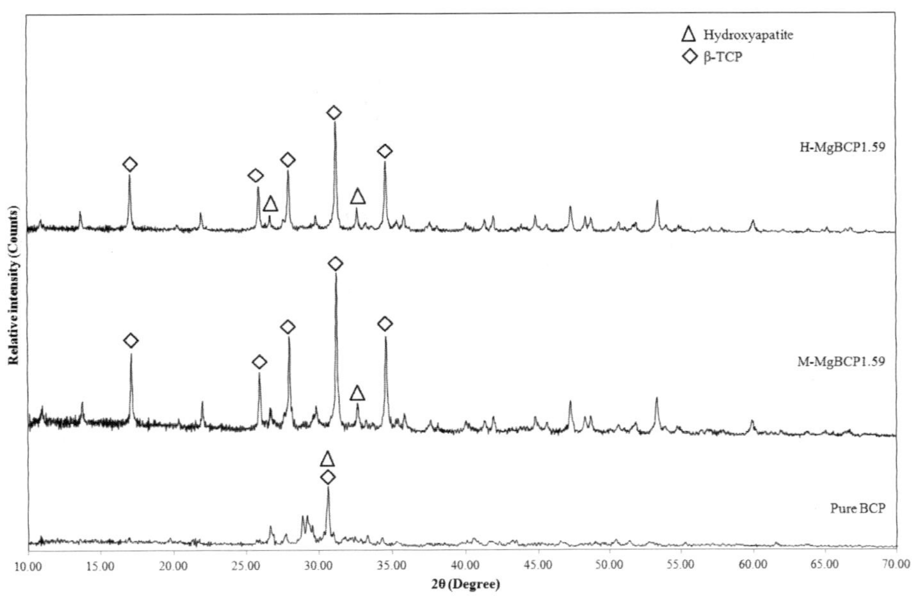

Figure 3. XRD patterns of pure BCP and magnesium substituted BCP powders after calcined at 600°C

Table 1 Comparison of diffraction angle (2θ) and d-spacing values in XRD patterns of calcined samples

Sample	0 2 1 0 plane	
	2θ (°)	d-spacing (Å)
Pure BCP	30.6063	2.9186
M-MgBCP1.59	31.2591	2.8591
H-MgBCP1.59	31.2573	2.8593

4. Conclusion

Mg substituted BCP with varying Mg content has been successfully synthesized through wet precipitation method. After calcinations at 600°C, most of HA phase has been converted into ß-TCP denotes the formation of biphasic mixtures. The addition of Mg in BCP structure causes lattice contraction which can be observed from the larger 2θ of Mg doped BCP samples with 31.2591° and 31.2573° compared to the 2θ of pure BCP sample with 30.6063°. The effect of lattice contraction which causes stresses in lattice was proved from the smaller d-spacing of Mg doped samples with 2.8591Å and 2.8593Å than pure BCP sample with 2.8592Å. The stresses is a result of the substitution of Mg with smaller ionic size into the Ca sites with larger ionic size.

Acknowledgements

The authors are grateful to the Fundamental Research Grant Scheme (FRGS) from Ministry of Education (MOE) for the financial support from the grant FRGS/203.PBAHAN.6071354.

References

[1] Seely Z, Bandyopadhyay A, & Bose S 2007 *Journal of Biomedical Materials Research* **33** 97
[2] Castiglioni S, Cazzaniga A, Albisetti W, & Maier J 2013 *Nutrients* **5** 3022
[3] Habraken W, Habibovic P, Epple M, & Bohner M 2016 *Materials Today* **19** 69
[4] Toibah A R, Sopyan I, Hamdi M, & Ramesh S 2008 *IFMBE Proceedings*, **21 IFMBE** 314
[5] Kannan S, Lemos I A F, Rocha J H G, & Ferreira J M F 2005 *Journal of Solid State Chemistry*, **178** 3190
[6] Stipniece L, Riga Technical University, 2016.
[7] Sunarso, Mohd Noor A F, Kasim S R, Otman R, Ana I D & Ishikawa K 2013 *Journal of Biomaterials and Nanobiotechnology* **4** 273
[8] Kim S R, Lee J H, Kim Y T, Riu D H, Jung S J, Lee Y J, Chung S C, & Kim Y H *Biomaterials* **24** 1389
[9] Ebrahimi M, Botelho M G, & Dorozhkin S V 2017 *Materials Science and Engineering C*, **71** 1293
[10] Farzadi A, Bakhshi F, Solati-Hashjin M, Asadi-Eydivand M, & Osman N A A 2014*Ceramics International* **40** 6021

Effect of Magnesium Oxide on Physical and Biological Properties in β-tricalcium Phosphate Ceramic

M. Chuthathip[1,3], M. N. Ahmad-Fauzi [1], B. I. Yanny-Marliana[1], S. Khairul-Anuar[1], K. Masakazu[2], L. Banhan[3]

[1]Biomaterials Niche Group, School of Materials and Mineral Resources Engineering, Universiti Sains Malaysia, Penang 14300, Malaysia

[2]Graduate School of Biomedical Engineering, Tohoku University, Sendai 980-8579, Japan

[3]Department of Industrial Engineering, Faculty of Engineering, Burapha University, Chonburi 20131, Thailand

Email: srafauzi@usm.my

Abstract. β-TCP is bioresorbable ceramic which has excellent osteoinduction and osteoconduction properties for using as bone cement and bone implant material. However, their natural brittleness and insufficient compaction after sintering under the transformation phase from β- to α-TCP limit their utility for clinical performance. The purpose of this work is to investigate the effect of magnesium ion as the doping element on the β-TCP. The experiment was conducted by introduced 1.0wt% of MgO into the β-TCP structure via simultaneous balls milling and mixing method. Afterward, the powders were pressed into pellets and went through conventional sintering method. The sintered samples were examined by powder X-ray diffraction (XRD), density and biological test for investigation.

1. Introduction

Tricalcium phosphate (TCP) has three polymorphs: β-TCP is stable below 1180°C, α-TCP from 1180°C to 1400°C, and ά-TCP above 1470°C. β-TCP is the preferred phase as a bioceramic substituting graft material due to its good chemical stability and proper bioresorption rate [1]. The low mechanical strength of β-TCP ceramic is, however, an essential concern to use these ceramics as surgical implants. Hence, it is important to improve the mechanical properties by densifying β-TCP ceramics fully since the sintering temperature should be lower than the phase transition of β-to α-TCP [2]. Magnesium is an essential element in the human body, with about 65% of total body magnesium is contained in bone and teeth. An adverse depletion of magnesium affects all stage of skeletal metabolism consequently decreasing bone growth, osteoblastic and osteoclastic activities [3]. Doping of calcium phosphate materials with Mg could improve osteoblastic cellular attachment, proliferation and ALP production. It has been reported that Mg can be incorporated into TCP structure by substitution of Ca and improves the thermal stability of TCP which prevents phase transformation at high temperature [4]. Mg has been used clinically in magnesium phosphate bone cement and in several different bioglass compositions. In this work, the influences of MgO on physical and biological properties of bulk commercial TCP were investigated.

2. Experimental

2.1 Sample preparation

The samples were prepared by mixing 50 g β-TCP commercial powder (Sigma-Aldrich) with an appropriate amount of 1.0 wt% MgO (Sigma-Aldrich) [5] in a 500 ml polypropylene bottle containing 200 ml deionized water and 500 g of 13 mm diameter cylindrical zirconia milling media. The mixtures were milled for 24 hr at 150 rpm and dried in an oven at 100 °C for 24 hr, followed by grinding and sieving to obtain fine powders. Then, 0.8 g of the powders were pressed into a pellet in 14 mm diameter stainless steel mold using a uniaxial press (Specac) by 150 MPa. Finally, the green pellets were sintered at different temperatures (1100°C, 1150°C, 1200°C, 1250°C and 1300°C) by 20°C/min of heating rate and 2 hr of soaking time then cooled down by air atmosphere in the muffle furnace (Lenton UAF 15/15).

2.2 Phase analysis

Phase analysis of sintered doped and undoped β-TCP samples was carried out by powder X-ray diffraction (XRD) (Bruker D2 Phaser) with a copper anode (CuKα, λ=1.5406 Å) target. Each run was performed with 2θ values of 10-90° with a scan rate of 0.5 °/min.

2.3 Linear shrinkage, apparent porosity and relative density measurements

The diameter and thickness of pellets were measured before and after sintering to calculate the linear shrinkage. While the relative density and total porosity were evaluated using Archimedes method. The experiment procedure of density and porosity measurement was carried out according to the ASTM-C830 [6].

2.4 Biological test and microstructure analysis

The bioactivity of β-TCP has is normally evaluated by the ability of apatite formation in simulated body fluid (SBF) which relies on their ability to induce hydroxyapatite (HA) formation [7]. In this work, the SBF solution was prepared according to ISO 23317:2012 (E) [8] and the immerse duration was 3 weeks. Field emission scanning electron microscope (FESEM, ZEISS SUPRA 35VP) was employed to observe the surface morphologies of bone-like apatite layer of sintered TCP and MgO-TCP samples.

3. Results and discussion

3.1 Phase analysis

The XRD patterns of TCP and MgO-TCP samples are shown in Fig. 1 and 2. The spectrums of all samples (Fig. 1) matched well with the Inorganic Crystal Structure Database (ICSD) no. 97500 of β-TCP. Likewise, the spectrums of MgO-TCP (Fig. 2) also exhibited the single phase of β-TCP without the presence of α-TCP phase. It had been documented the β-TCP phase undergoes allotropic conversion to α-TCP beyond 1180°C and the presence of Mg^{2+} in its crystal lattice delay the allotropic conversion till 1400°C [9]. In this study, from the XRD analysis of sintered TCP and Mg-TCP sample presented the single pages of β-TCP even at high sintering temperature at 1300 °C. This suggests that due to the starting material was commercial β-TCP so the sintered samples have a stable phase although sintered at high temperatures. In addition, a shifting in the 2θ and the d-spacing value for doped TCP corroborates the substitution of Ca^{2+} with Mg^{2+} in the β-TCP lattice [5]. The ionic radius of Mg^{2+} (0.69 Å) is smaller than that of Ca^{2+} (0.99 Å), leads to a decrease in the unit cell parameters of the β-TCP lattice, and therefore the shift in the peak to higher 2θ value occurred [10]. Hence, the shifting of the highest intensity diffraction peak at (0 2 10) to higher 2θ of MgO-TCP compared with TCP was observed as shown in Fig. 3. In addition, the shifting of peak to lower 2θ was noticed for both TCP and MgO-TCP samples. Particularly in MgO-TCP samples (Fig. 4), the diffraction peak at (0 2 10) shifted to lower 2θ as sintering temperature increased. This can be accounted as in the higher sintering temperature, an atomic mobility would increase and caused grain growth, which results in the better crystallinity and densification of samples [11, 12].

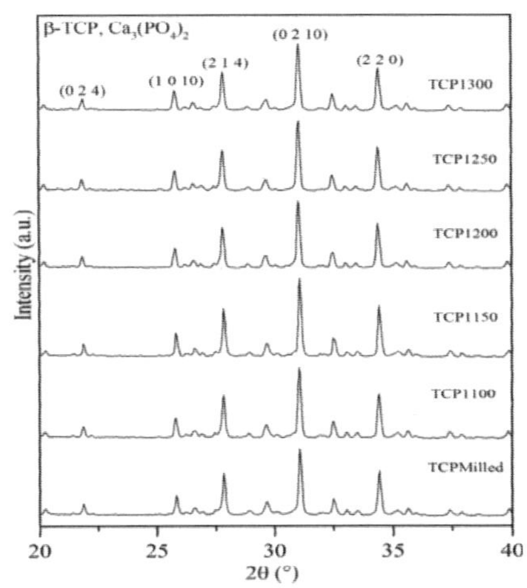

Figure 1. XRD pattern of milled powder and sintered TCP at different temperatures.

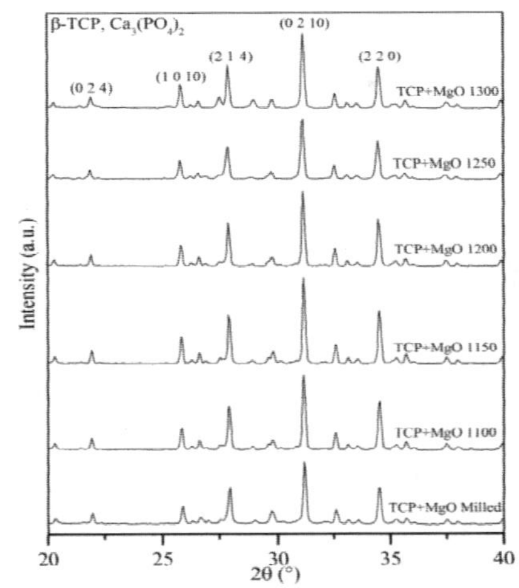

Figure 2. XRD pattern of milled powder and sintered MgO-TCP at different temperatures.

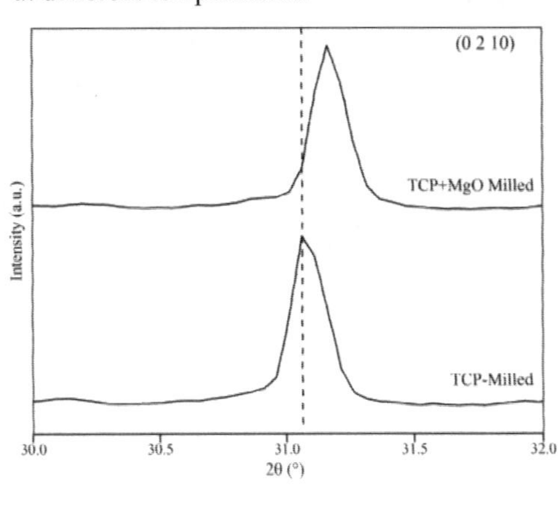

Figure 3. Peak shifted observation of milled TCP and MgO-TCP by XRD in the length of 2θ =30° to 32°.

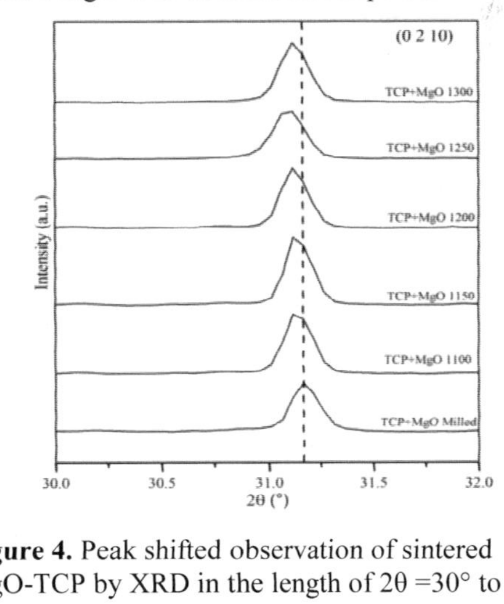

Figure 4. Peak shifted observation of sintered MgO-TCP by XRD in the length of 2θ =30° to 32°.

3.2 Linear shrinkage measurement

The linear shrinkage in diameter and thickness of sintered samples are shown in the Fig. 5 and 6, respectively. Generally, the shrinkage in both dimensions was rapidly increased as sintering temperature increased. However, at the temperature from 1250°C to 1300°C of TCP samples, the diameter shrinkage slightly decreased (Fig. 5), while there is a small increase in thickness shrinkage at the same temperatures (Fig. 6). Conversely, the MgO-TCP samples showed slightly increased in diameter shrinkage (Fig. 5) while slightly decreased in thickness shrinkage (Fig. 6) at the temperature from 1250°C to 1300°C.

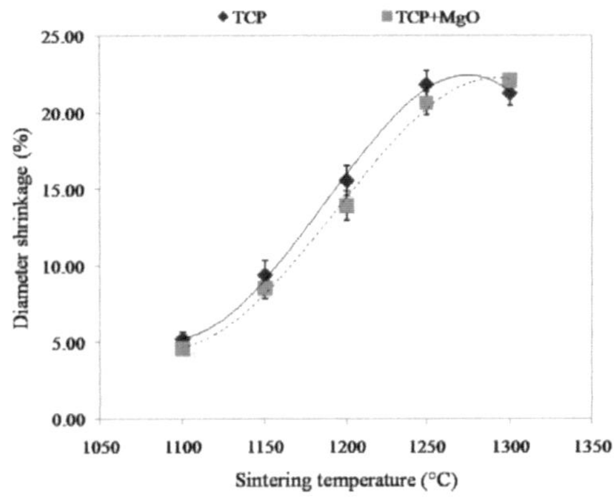

Figure 5. Diameter shrinkage of sintered TCP and MgO-TCP at different temperatures.

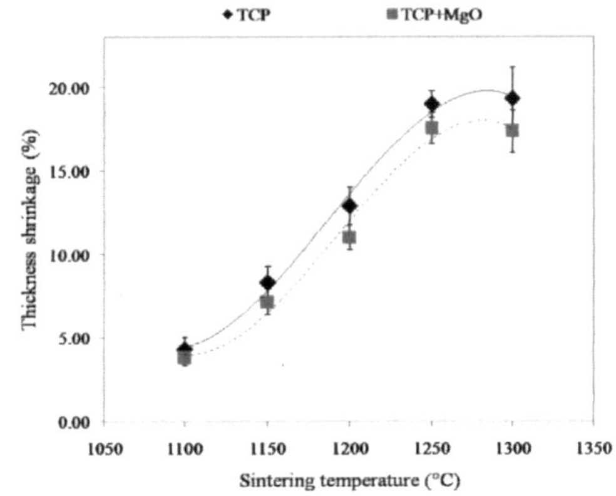

Figure 6. Thickness shrinkage of sintered TCP and MgO-TCP at different temperatures.

3.3 Apparent porosity and relative density measurement

Generally, the total porosity of both sintered TCP and MgO-TCP samples decreased in higher sintering temperature (Fig. 7) which well corresponded with the increase of relative density in the Fig.8. However, the sintered TCP samples showed better porosity and density than the sintered MgO-TCP samples except at the sintering temperature at 1300°C. Interestingly, although the XRD analysis of sintered TCP samples could not detect the α-TCP phase at a high sintering temperature of 1300 °C it clearly presented lower properties of linear shrinkage and density and porosity test compared with sintered MgO-TCP samples. These can be suggested that the α-TCP phase has started to form in the sintered TCP sample at 1300°C but could not detect by XRD analysis due to a very small quantity of α-TCP phase. Moreover, the expansion of sintered TCP during high sintering temperature causes micro-crack which consequently reduces the density property [13]. Since the Mg has the ability to stabilize β-TCP under high sintering temperature by delay the β-to α-phase transformation, the sintered MgO-TCP samples presented the better density and porosity result at the temperature of 1300°C.

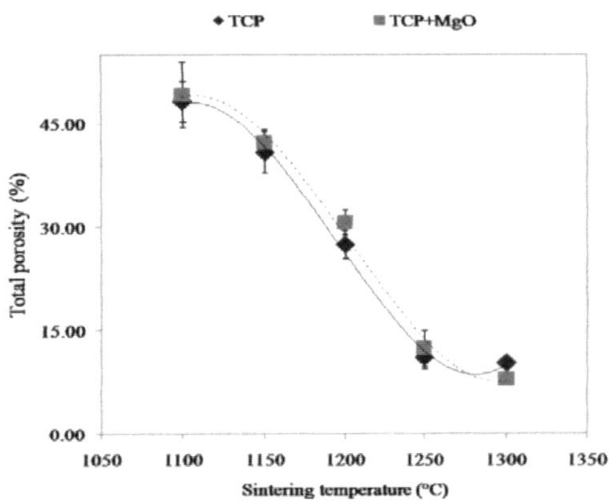

Figure 7. The total porosity of sintered TCP and MgO-TCP samples at different temperatures.

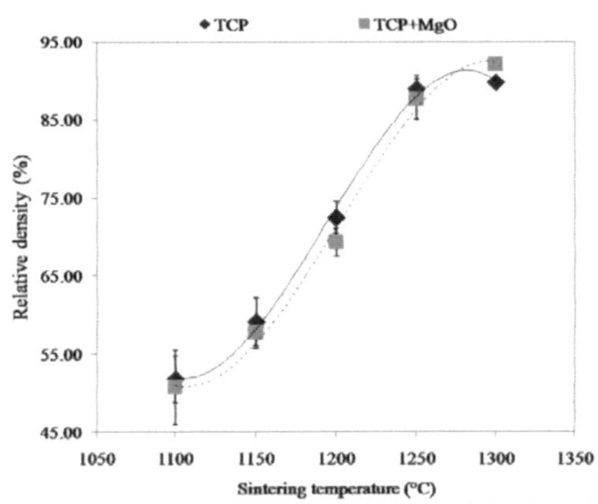

Figure 8. The relative density of sintered TCP and MgO-TCP samples at different temperatures.

3.4 Biological test and microstructure analysis

The investigation of the apatite formation ability of bioceramics in simulated body fluid (SBF) is an original protocol developed by Kokubo et al. [7]. The apatite formations were not observed on the sample surface of sintering temperature at 1100°C, 1150°C and 1200°C (Fig. 9 and 10: a, b, c) but can be seen on the sample surface of 1250 °C and 1300°C in both TCP (Fig. 9d and e) and MgO-TCP samples (Fig. 10d and e). The absent and negligible apatite formation of MgO-TCP samples due to the addition of magnesium into the β-TCP structure. It has reported that magnesium inhibited in vitro apatite formation which effects on the nucleation and crystal growth of calcium phosphates. Since the magnesium substituted into calcium site of β-TCP structure hence they inhibit nucleation of Ca-P phases and disrupt adsorption at active growth site. This kinetically consequently hinder nucleation of hydroxyapatite which results to formation of the apatite layer on the sintered MgO-TCP samples [14-18]. In addition, on the surface of sintered TCP at 1250°C exhibited predominant dense apatite formation which suggests it was the later stage for each granule grows bigger and becomes predominantly plate-like crystals (Fig. 9d) [13]. The apatite formation with similar morphologies has been reported in previous studies [19, 20].

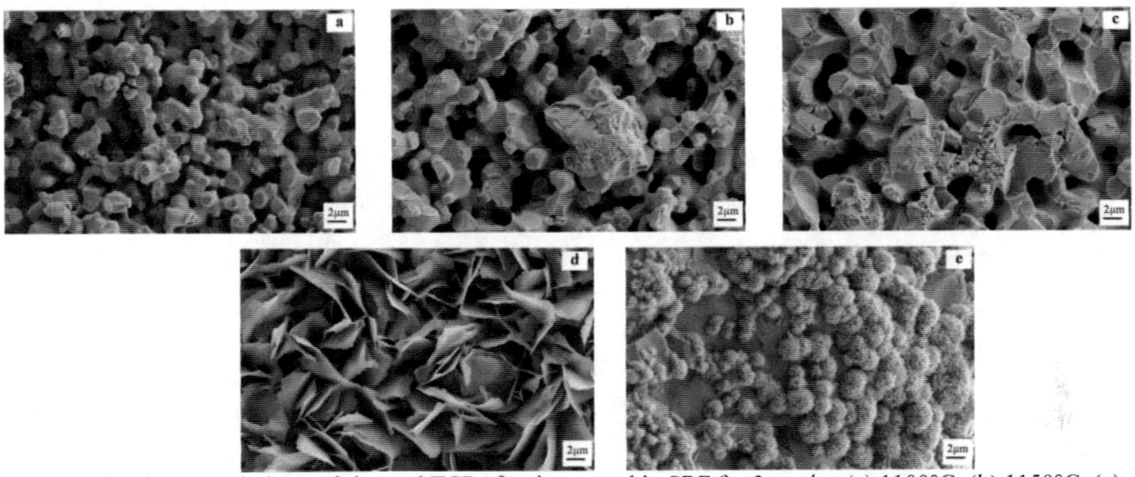

Figure 9. Surface morphology of sintered TCP after immersed in SBF for 3 weeks: (a) 1100°C; (b) 1150°C; (c) 1200°C, (d) 1250°C; and (e) 1300°C.

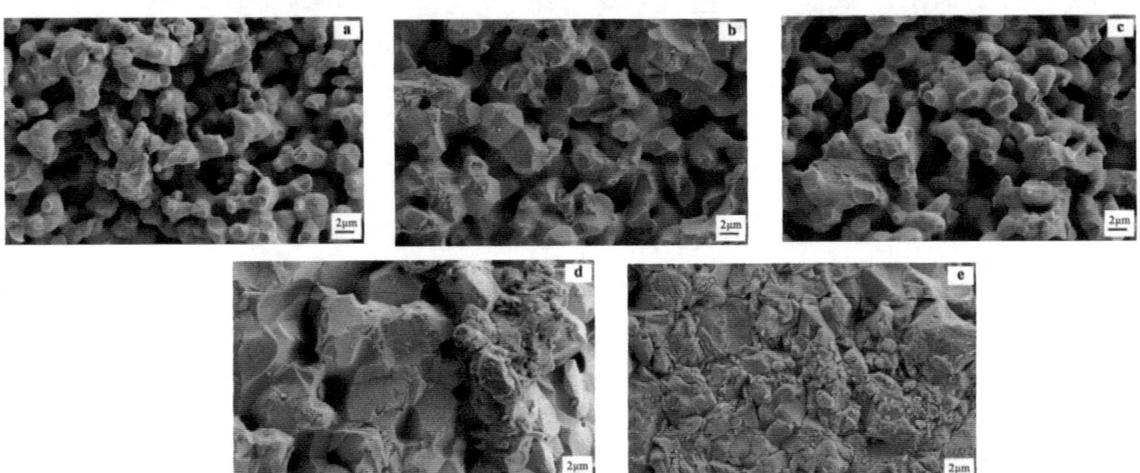

Figure 10. Surface morphology of sintered MgO-TCP after immersed in SBF for 3 weeks: (a) 1100°C; (b) 1150°C; (c) 1200°C, (d) 1250°C; and (e) 1300°C.

IOP Conf. Series: Journal of Physics: Conf. Series **1082** (2018) 012026 doi:10.1088/1742-6596/1082/1/012026

4. Conclusion

The effect of MgO dope β-TCP was investigated. The substituted of Mg^{2+} in β-TCP was confirmed by the peak shift in XRD result and the α- TCP phase was not presented in both TCP and MgO-TCP sintered sample for all sintering temperatures. The linear shrinkage and density and porosity showed better properties as increasing the temperature up to 1250°C meanwhile at the sintering temperature of 1300°C the MgO-TCP samples presented better properties. These suggest that due to the expansion of sintered TCP during high sintering temperature which causes micro-crack in the samples. The biological test revealed that both TCP and MgO-TCP can perform apatite layer on its surface for some sintering temperature which initially confirmed that these materials are biocompatible to use in the clinical application.

Acknowledgment

The authors gratefully acknowledge the financial support from ASEAN University Network/Southeast Asia Engineering Education Development Network (AUN/SEED-Net) of grant number 6050313.

References

[1] Famery M, Richard N and Boch P 1994 Ceramics International 20 327
[2] Ryu H S, Youn, H J, Hong K S, Chang B S, Lee C K and Chung S S 2002 Biomaterials 23 909
[3] Percival M 1999 Applied Nutritional Science Report 5 1
[4] Xue W, Dahlquist K, Banerjee A, Bandyopadhyay A and Bose S 2008 Journal of Materials Science: Materials in Medicine 19 2669
[5] Banerjee S S, Tarafder S, Davies N M, Bandyopadhyay A and Bose S 2010 Acta Biomaterialia 6 4167
[6] Standard test methods for apparent porosity, liquid absorption, apparent specific gravity, and bulk density of refractory shapes by vacuum pressure 2000 ASTM standard-C830
[7] Kokubo T and Takadama H 2006 Biomaterials 27 2907
[8] Implants for surgery-In vitro evaluation for apatite-forming ability of implant materials 2012 ISO 23317:2012(E)
[9] Kumar P N, Mishra S K and Kannan S 2015 Journal of Solid State Chemistry 231 13
[10] Kannan S, Neunhoeffer F G, Neubaure J, Pina S, Torres P M C, Ferreira J F 2010 Acta Biomaterialia 6 517
[11] Ahmad S I, Rao P K and Syed I A 2016 Journal of Taibah University for Science 10 381
[12] Syamimi N F, Matori K A, Lim W F, Aziz S A and Mohd-Zaid M H 2014 Journal of Spectroscopy 2014 1
[13] Ryu H S, Youn H J, Hong K S, Chang B S, Lee C K and Chung S S 2002 Biomaterials 23 909
[14] Qi G, Zhang S, Khor K A, Lye S W, Zeng X, Weng W, Liu C, Venkatraman S S and Ma L L 2008 Applied Surface Science 255 304
[15] Berg C 2017 UPTEC K 17021 Examensarbete 30 hp 3
[16] Kanchana P and Sekar C 2012 Journal of Minerals and Materials Characterization and Engineering 11 982
[17] Cheng P T 1989 Urolithiasis 225
[18] Bachra B N and Fischer H R A 1969 Calcified Tissue Research 3 248
[19] Xin R, Leng Y, Chen J and Zhang Q 2005 Biomaterials 26 6477
[20] Bachra B N and Fischer H R A 1969 Calcified Tissue Research 3 248

Optimization of Preparation Conditions for Acid Red 88 Dye Removal using Rubber Seed Coat Based Activated Carbon

Mohamad Azwan Mad Naser[1], Mohd Nazri Idris[1], Mohd Azmier Ahmad[2]

[1]School of Materials and Mineral Resources Engineering, Universiti Sains Malaysia, Engineering Campus, 14300 NibongTebal, Penang, Malaysia

[2]School of Chemical Engineering, Engineering Campus, Universiti Sains Malaysia, 14300 Nibong Tebal, Penang, Malaysia

Email: srmnazri@usm.my

Abstract. This study investigates the optimal conditions for preparation of activated carbon from rubber seed coat (RSC) for removal of Acid Red 88 (AR88) dye from aqueous solution. The RSC activated carbon was prepared using physiochemical activation method which consisted of sodium hydroxide (NaOH) treatment and carbon dioxide (CO_2) gasification. Central composite design (CCD) was used to determine the effects of the three preparation variables; CO_2 activation temperature, CO_2 activation time and NaOH impregnation ratio (IR) on AR88 percentage removal and RSC activated carbon yield. Based on the CCD, a quadratic model and a linear model were respectively developed for AR88 percentage removal and RSC activated carbon yield. The optimum conditions for activated carbon preparation were obtained by using activation temperature of 775°C, activation time of 2 h and IR of 2.25, which resulted in 38.29 % of AR88 removal and 33.26 % of activated carbon yield.

1. Introduction

Synthetic dyestuffs are extensively used in textile, paper, printing industries and dye houses due to their ease of production, fastness and variety of colors compared to natural dyes [1]. More than 100,000 commercially available dyes are known and approximately one million tons of these dyes are produced annually worldwide [2]. Azo dyes, one of the major classes of synthetic dyes used in textile industries are characterized by presence of one or more azo bonds (-N=N-) in association with one or more aromatic systems. They are known to be toxic and carcinogenic [3]. As one of the azo dyes group, therefore, the removal of Acid Red 88 (AR88) is necessary and very important.

There are several methods are available for removing dyes from wastewater such as membrane separation, aerobic and anaerobic degradation, chemical oxidation, coagulation and flocculation, adsorption and reverse osmosis. Among them, adsorption process is widely used and most efficient method especially if the adsorbent can be produced from low cost or waste materials [4]. Activated carbon is the most employed adsorbent for dye removal from aqueous solution because of its excellent adsorption properties [5]. Based on the literature, many researchers have tried to produce more

IOP Conf. Series: Journal of Physics: Conf. Series **1082** (2018) 012027 doi:10.1088/1742-6596/1082/1/012027

economical activated carbons for removal of various pollutants using alternative cheap and renewable materials such as rambutan peel [6], corn cob [7], orange peel [8], bamboo [9], ginger waste [10], oil palm empty fruit bunch [11], and banana pseudostem fibers [12]. In this work, rubber seed coat based activated carbon (RSCAC) was prepared as an adsorbent to remove AR88 from aqueous solution. The aim of this study was to optimize the physicochemical activation conditions of RSCAC using central composite design (CCD) that involves sodium hydroxide (NaOH) and carbon dioxide (CO_2) as activation agents throughout this study.

2. Experimental study

Acid Red 88 (AR88), 4-(2-Hydroxy-1-naphthylazo)-1-naphthalenesulfonic acid sodium salt supplied by Sigma Aldrich (M) Sdn. Bhd. was used as an adsorbate. AR88 has an empirical formula of $HOC_{10}H_6N{=}NC_{10}H_6SO_3Na$ with molecular weight of 400.38 g/mol. Rubber seed coat (RSC) used for preparation of activated carbon was obtained locally in the area of Selama, Perak, Malaysia. The RSC was firstly washed thoroughly to remove dirt and inorganic matters prior to dried in oven at 100°C for 24 h. The dried precursor was ground into small pieces and was sieved to the size of 1-2 mm. Carbonization step was carried out at 500°C and held for 2 h under nitrogen gas using the furnace. The flow rate of nitrogen gas and the heating rate were held constant at 150 cm^3/min and 10°C/min, respectively. The char produced was mixed with NaOH pellets with different impregnation ratio (IR).

The activation step was done using similar reactor as in carbonization step but at final temperature of 775°C. Once the desired activation temperature was reached, the gas flow was switched to carbon dioxide (CO_2) at the same flow rate for different period of time. The activated carbon was cooled to room temperature under nitrogen flow and washed with hot deionized water to recover the remaining NaOH. The washing process was continued using 0.1M HCl, then washed several times with hot deionized water until the pH of the washing solutions reached 6-7.

Adsorption studies were performed using Erlenmeyer flasks where 300 mg of adsorbent were mixed with 200 mL aqueous dye solutions of 100 mg/L initial concentration in 20 sets of these flasks. The mixture was agitated at 120 rpm at 30 °C until equilibrium was attained. The pH of the solution was natural without any pH adjustment. The concentration of adsorbate was determined using a double UV-Vis spectrophotometer (UV1800 Shimadzu, Japan) at maximum wavelength of 535.5 nm. The percentage removal at equilibrium was calculated by Equation (1).

$$\text{Removal (\%)} = 100 \times (C_o - C_e)/C_o \quad (1)$$

where C_o and C_e are the liquid-phase dye concentrations at initial state and at equilibrium (mg/L), respectively. The RSCAC yield was calculated as the dry weight of final activated carbons, w_c to the dry weight of precursors, w_o as expected by Equation (2):

$$\text{Yield (\%)} = 100 \times (w_c / w_o) \quad (2)$$

In this work, a standard response surface methodology (RSM) design called central composite design (CCD) was used to study the parameter of preparing RSCAC [13]. There are three independent variables studied for the preparation of RSCAC which were activation temperature (x_1), activation time (x_2) and NaOH:char IR (x_3). These three variables together with their respective ranges were chosen based on the literature and preliminary studies [6]. The two responses; AR88 removal (Y_1) and RSCAC yield (Y_2) were used to develop an empirical model which correlated the response to the three activated carbon preparation variables. Scanning electron microscopy was used to study the surface morphology of the precursors, chars and the activated carbon prepared including the pore structure, surface structure and pore arrangement. The analysis was carried out using a scanning electron microscope (Model LEO SUPRA 55VP, Germany).

3. Results and discussion

3.1 Development of regression model equation

The correlation analysis between RSCAC preparation variables and response, AR88 removal and RSCAC yield, were developed using CCD. According to the sequential model sum of squares, the models were selected based on the highest order polynomials where the additional terms were significant and the models were not aliased. The complete design matrixes together with both response values obtained from the experimental work is shown in Table 1.

Run 15-20 at the center point were conducted to determine the experimental error and the reproducibility of the data. AR88 removal and RSCAC yield were found to range from 25.98 to 42.18% and 27.02 to 38.46%, respectively. The test carried out by Fit Summary suggested that quadratic model was statistically significant for AR88 removal (Y_1) and RSCAC yield (Y_2), where the models were not aliased. The final empirical formula models for both responses in terms of coded factors after excluding the insignificant terms are represented as follows:

$$Y_1 = 38.34 + 4.55\,x_1 - 0.30\,x_2 + 1.32\,x_3 - 1.68\,x_1^2 - 1.08\,x_2^2 + 0.16\,x_3^2 - 0.26\,x_1x_2 - 0.73\,x_1x_3 - 0.071\,x_2x_3 \quad (3)$$

$$Y_2 = 33.16 - 3.58\,x_1 - 1.10\,x_2 - 1.34\,x_3 - 0.79\,x_1^2 + 0.097\,x_2^2 + 0.15\,x_3^2 + 0.055\,x_1x_2 - 0.10\,x_1x_3 - 0.13\,x_2x_3 \quad (4)$$

The effect of the particular factor represented by the coefficient with one factor, while interaction between two factors and quadratic effect represented by the coefficient with two factors and those with second order terms, repectively. The positive and negative sign in front of the coefficient from empirical formula models for both responses shows the synergistic and antagonistic effect. The correlation coefficients, R^2 value evaluates the quality of the developed models. In fact, high value for R^2 and low standard deviation was the best indicator in order to develop a good model which is closer to unity as it will give predicted value closer to the actual value for the responses [11].

Table 1. Experimental design matrix for preparation of RSC activated carbon.

Run	Level			RSCAC preparation variable			AR88 removal, Y_1 (%)	RSCAC yield, Y_2 (%)
				Activation temp, x_1 (°C)	Activation time, x_2 (h)	IR, x_3		
1	-1	-1	-1	700	1.00	1.00	28.53	37.31
2	+1	-1	-1	850	1.00	1.00	39.03	28.43
3	-1	+1	-1	700	3.00	1.00	28.75	36.93
4	+1	+1	-1	850	3.00	1.00	39.75	28.62
5	-1	-1	+1	700	1.00	3.50	33.08	36.98
6	+1	-1	+1	850	1.00	3.50	42.18	28.05
7	-1	+1	+1	700	3.00	3.50	34.54	36.42
8	+1	+1	+1	850	3.00	3.50	41.09	27.36
9	- 1.682	0	0	649	2.00	2.25	25.98	35.15
10	+1.682	0	0	901	2.00	2.25	40.85	27.02
11	0	- 1.682	0	775	0.32	2.25	36.73	37.64
12	0	+1.682	0	775	3.68	2.25	33.49	29.53
13	0	0	- 1.682	775	2.00	0.15	37.64	38.46
14	0	0	+1.682	775	2.00	4.35	39.57	29.03
15	0	0	0	775	2.00	2.25	38.06	33.04
16	0	0	0	775	2.00	2.25	37.93	32.04
17	0	0	0	775	2.00	2.25	38.54	33.94
18	0	0	0	775	2.00	2.25	38.48	33.42
19	0	0	0	775	2.00	2.25	38.29	33.26
20	0	0	0	775	2.00	2.25	38.80	33.19

From the experimental data, the R^2 values for Equation (3) and (4) were 0.9689 and 0.8122, respectively. The standard deviations for the two models were 1.09 and 2.29 for Equation (3) and (4), respectively. The R^2 values for AR88 removal was considered relatively high, indicating that the predicted values for AR88 removal of the RSCAC would be more accurate and closer to its actual value. The R^2 for RSCAC yield was considered as moderate to validate the fit, which might lead to larger variation in the RSCAC yield predicted from the model. Although the value for standard deviation slightly high, the predicted values for this model is still considered as suitable to correlate the experimental data. AR88 removal model shows much more accurate prediction and closer to its actual values, compared to RSCAC yield models.

3.2 AR88 removal

Based on analysis of variant (data not shown), both the activation temperature and IR variables were found to have significant effects, meanwhile activation time showed the least significant effect on this response. Figure 1a shows the three-dimensional response surfaces that illustrates the interaction effects of the activation temperature and IR on the AR88 removal. For this plot, the activation time was fixed at zero level (t = 2 h).

Figure 1(a) shows AR88 removal generally increased with increase in activation temperature and IR. The result obtained were in agreement with effect of activation temperature an IR for production of activated carbon from cassava peel and apricot stones [2, 13]. Activation time was not as significant as activation temperature and IR in preparation of activated carbon. Increasing activation temperature and activation time would increase reaction rates, promote enlargement opening of pores, thus increase adsorption uptake of dye onto the activated carbon [14]. However, prolong the activation time might lead to undesirable effect depending on requirement of specific activated carbon application, in term of pores enlargement [15]. IR ratio played critical role in formation of pores and increasing surface area of activated carbon. At high IR, pores development was due to intercalation of sodium metal in the carbon structure. New pores created was due to catalytic oxidation of carbon surface by sodium metallic salt [16, 17].

3.3 RSCAC yield

Figure 1b shows three-dimensional response surface which illustrates the effects of RSCAC preparation variables on the RSCAC yield by plot the effect of activation temperature and activation time on the RSCAC yield, with IR fixed at zero level (IR = 2.25). The RSCAC yield was found decreased with increasing activation temperature and activation time. However, activation time is not really significant for this response.

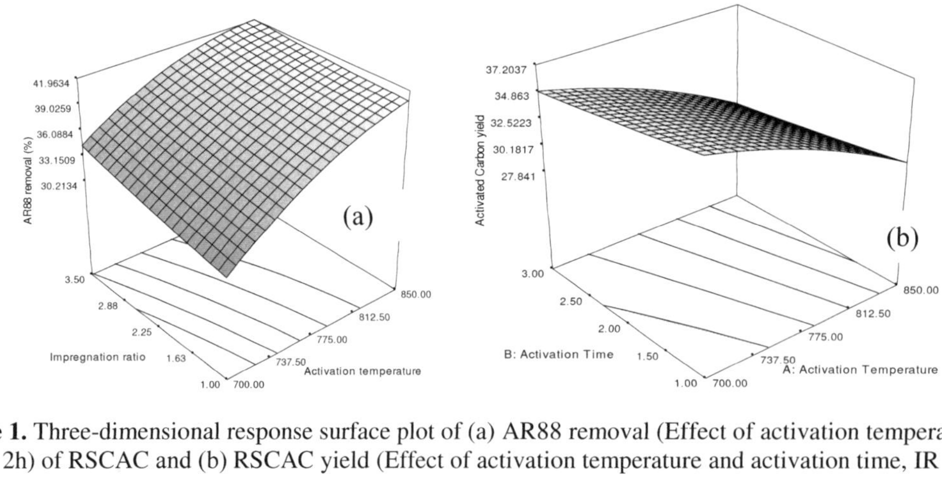

Figure 1. Three-dimensional response surface plot of (a) AR88 removal (Effect of activation temperature and IR, t = 2h) of RSCAC and (b) RSCAC yield (Effect of activation temperature and activation time, IR = 2.25).

This result shown were in agreement with the experimental data by Ahmad et al. [18] which found that activation temperature plays an important role on the yield of activated carbon prepared by rattan sawdust whereas activation time did not show much effect on the carbon yield. In addition, the overall weight loss was found to increase with IR, whereas in decreasing yield of activated carbon at high temperature. The weight loss was due to devolatilization of the precursors, primary to increase of the pore development and create new pores, as a result of intensifying dehydration and elimination reactions. When temperature increasing, C-NaOH and C-CO_2 reaction rate is increased as well, leading towards decreasing of carbon yield [19]. Thus, increasing of IR cause decreasing of RSCAC yield due to oxidation process on the surface of carbon atoms.

3.4 Process optimization

Optimization of process parameters to obtain high carbon yield and high AR88 removal was one of the main objectives of this study. However, it is difficult to optimize both these responses under the same condition because the interest region of factors is different. Since high adsorption capacity always comes with low activated carbon yield, the function of desirability was applied using Design-Expert software version 6.0.6 (STAT-EASE Inc., Minneapolis, USA) in order to compromise between these two responses. In optimization analysis, target criteria was set as maximum values for two response, AR88 removal and RSCAC yield, while the three particular variable (activation temperature, activation time and IR) were set in range being studied. The experimental conditions with the highest desirability were selected to be verified. The predicted and experimental results of AR88 removal and RSCAC yield obtained at optimum conditions are listed in Table 2. The optimum RSCAC was obtained by using activation temperature, activation time and IR of 775°C, 2h and 2.25, respectively. The optimum RSCAC showed AR88 removal of 38.29% and RSCAC yield of 33.26%. It was observed that the experimental values obtained were in good agreement with the values predicted from the models, with relatively small errors between the predicted and the actual values, which was only 0.13% and 0.30%, for AR88 removal and RSCAC yield, respectively.

Table 2. Model validation.

Model desirability	Activation temperature, x_1 (°C)	Activation time, x_2 (h)	IR, x_3	AR88 removal (%)			RSCAC yield (%)		
				Pred.	Exp.	Error (%)	Pred.	Exp.	Error (%)
0.886	775	2.0	2.25	38.34	38.3	0.13	33.16	33.3	0.30

3.5 Surface morphology

The surface morphology of samples was examined using scanning electron microscopy (SEM). Figure 2 shows the SEM micrographs of the samples. RSC surface textures were observed as rough, uneven, undulating and very little pores were presence as observed in other biomass [20]. After carbonization and activation processes, irregular holes and pores were developed and found on the surfaces of the sample. This is probably due to effect of the diffusion of NaOH and CO_2 molecules onto the sample, which create more pores on the sample's surface [21, 22].

IOP Conf. Series: Journal of Physics: Conf. Series **1082** (2018) 012027 doi:10.1088/1742-6596/1082/1/012027

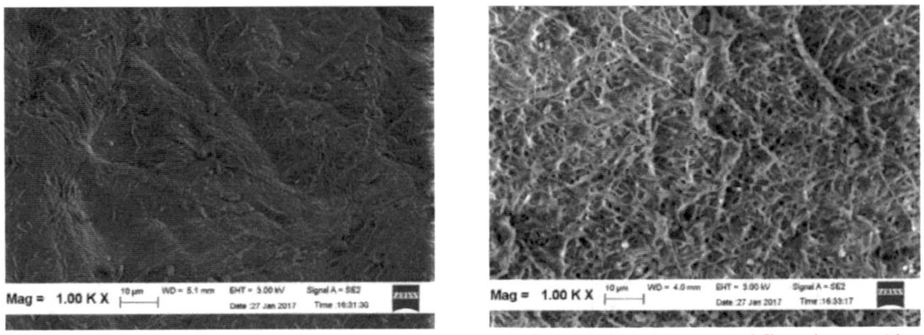

Figure 2: SEM micrographs of (a) RSC and (b) RSCAC (magnifications: 1k x).

4. Conclusion

The response surface methodology (RSM) technique was successfully utilized in optimizing the process condition for the preparation of activated carbon from rubber seed coat. The optimum conditions obtained for preparing the activated carbon were 775°C, 2h and 2.25 for activation temperature, activation time and impregnation ratio, respectively, and resulting in maximum AR88 removal of 38.29% and 33.26% carbon yield. The result of study indicates the potential of rubber seed coat to be a potential feedstock for production of activated carbon by using physiochemical activation method. Thus, indicates the advantages of RSM method by reducing trial numbers, time and cost of formulation development.

References

[1] Claus M M, Mirna J S, Sudaryanto Y and Ismadji S 2002 *Journal of the Taiwan Institute of Chemical Engineers* 40 457
[2] Selvam T, Ucar S and Karagöz S 2003 *Journal of Hazardous Materials* 165 481
[3] Modirshahla N, Behnajady M A, Shamel A and Eskandari H 2010 *Journal of Physical and Theoretical Chemistry of Islamic Azad University of Iran* 7 77
[4] Chandra T C, Mirna M M, Sudaryanto Y and Ismadji S 2007 *Chemical Engineering Journal* 127 121
[5] Ahmad M A 2006 *Adsorption* 62 145
[6] Ahmad M A and Alrozi R 2011 *Chemical Engineering Journal* 168 280
[7] Ahmad M A, Yun E T C, Abustan I, Ahmad N and Sulaiman S K 2011 *International Journal of Engineering & Technology* 11 283
[8] Khaled A, Nemr A E, Sikaily A E and Abdelwahab O 2009 *Journal of Hazardous Materials* 165 100
[9] Hameed B H and El-Khaiary M I 2008 *Journal of Hazardous Materials* 157 344
[10] Ahmad R and Kumar R 2010 *Journal of Environmental Management* 91 1032
[11] Hameed B H, Tan I A W and Ahmad A L 2009 *Journal of Hazardous Materials* 164 1316
[12] Haris M R H M and Sathasivam K 2009 *American Journal of Applied Sciences* 6 1690
[13] Sudaryanto Y, Hartono S B, Irawaty W, Hindarso H and Ismadji S 2006 *Bioresources Technology* 97 734
[14] Tan I A W, Ahmad A L and Hameed B H 2008 *Chemical Engineering Journal* 137 462
[15] Gratuito M K B, Panyathanmaporn T, Chumnanklang R A, Sirinuntawittaya N and Dutta A 2008 *Bioresource Technology* 99 4887
[16] Liu Y, Yupeng G, Yanchau Z, Dongmin A, Wei G Z W, Yuejia M and Zichen W 2011 *Journal of Hazardous Materials* 186 1314
[17] Adinata D, Daud W M A W and Aroua M K 2007 Bioresources Technology 98 145
[18] Ahmad A, Rafatullah M, Sulaiman O, Ibrahim M H and Hashim R 2009 *Journal of Hazardous Materials* 12 678
[19] Ada K, Ergene A, Tan S and Yalc E 2009 *Journal of Hazardous Materials* 54 568
[20] Ahmad M A and Alrozi R 2010 *Chemical Engineering Journal* 165 883
[21] Tan J S and Ani F N 2003 *Carbon* 41 840
[22] Lu C, Xu S and Liu C 2010 *Journal of Analytical Applied Pyrolysis* 87 282

Electrophoretic Deposited LSCF-SDCC-Ag Cathode Coating on Ferritic Stainless Steel Interconnect for SOFC

N.I.Omar[1], K.H.Tan[1], H.Taib[1] and H.A.Rahman[1*]

[1] Faculty of Mechanical and Manufacturing Engineering, Universiti Tun Hussein Onn Malaysia, 86400 Parit Raja, Batu Pahat, Johor, Malaysia

*hamimah@uthm.edu.my

Abstract. The application of electrophoretic deposition (EPD) technique in development of composite cathode as a coating layer on ferritic stainless steel (FSS) interconnect for solid oxide fuel cell (SOFC) has acquired a great interest. The aim of this study is to determine the capability of EPD technique in producing composite cathode coating. The lanthanum strontium cobalt ferrite (LSCF)-samarium doped ceria carbonate (SDCC) composite cathode powder with silver (Ag) addition was fabricated adopting the EPD technique. LSCF-SDCC-Ag suspension was prepared by using organic-aqueous solvent for the deposition. The coated FSS substrate was first heat treated at 600°C for 90 minutes in air. The deposition was carried out at different applied voltages (5, 7 and 10 volt) and deposition durations (5-20 minutes). The increment of applied voltage and deposition duration was found to contribute to the increment of LSCF-SDCC-Ag deposited weight. The highest deposited weight and thickness of coating was 2.47 mg/cm^2 and 62.73 μm respectively attained by deposition of 10 volt for 20 minutes. According to the obtained results, deposition at 10 volt within 5 to 20 minutes has shown a better deposition. It is thus clear that EPD is indeed a feasible technique applicable for the development of composite cathode for SOFC interconnects and yielded better coating thicknesses of LSCF-SDCC-Ag on FSS interconnect for SOFC.

1. Introduction

Solid Oxide Fuel Cell (SOFC) is a highly effective fuel cell which oxidizing fuel from chemical energy into electrical energy [1]. Generally, a single SOFC comprises three components which are anode, cathode and electrolyte. Interconnects are the component to vividly separate each single fuel cell and prevent them from electrically contact and delivering electricity from single fuel cell to external circuit [2]. SOFC traditionally operates in high temperature (approximately 1000°C) while ceramic (for example, $LaCrO_3$) is the common interconnect material [3]. Various efforts have been attempted recently to lower down operating temperatures of solid oxide fuel cell which is ranged between 400°C to 800°C. This has enabled metallic interconnect to be applied instead of ceramic as it is low cost, good manufacturability, excellent electrical conductivity, matched thermal expansion coefficient (TEC) and better compatibility with adjacent components [4]. Ferritic stainless steel is considerably applied as interconnect. This is due to its TEC is very compatible with less than 10 % different from most anode, cathode and electrolyte materials at around 10-13°C^{-1} [5, 6]. There are variety of ferritic stainless steel suitable to be applied as interconnect such as ferritic stainless steels and have been reported of having considerable chromium poisoning issue [7]. Long term working in elevated temperature would diffuse out chromium from interconnect to the air and formed chromia scale depositing on cathode. Consequently, it declines catalytic activity of electrodes and hence deteriorates whole cells conductivity and performance.

In order to impede this undesirable phenomenon, the surface of stainless steel has to be modified by depositing protective coating. Compounds such as LaSrMn (LSM), LaSrCoFe (LSCF) and LaSrCo (LSC) have been deposited on interconnects [8-10]. From new generation mixed ionic and electronic conductor (MIEC) cathode materials, LSCF exhibits excellent conductivity and low polarization resistance [11-13]. LSCF has been successfully deposited on stainless steel with desired coating quality, deposited thickness and area specific resistance (ASR) [14, 15]. However, LSCF is reported of having significant high number of TEC (~16 $^{o}C^{-1}$) which is not beneficial to stainless steel interconnect [16]. LSCF composite has been developed by incorporating electrolyte powder such as samarium doped ceria carbonate (SDCC) and metallic catalytic such as silver (Ag) into LSCF cathode. The TEC of the composite LSCF-SDCC-Ag is reported to be 13 $^{o}C^{-1}$ which is very matched with TEC of ferritic stainless steel in less than 10 % different [17].

In this study, LSCF-SDCC-Ag was coated on SUS 430 ferritic stainless steel by electrophoretic deposition method (EPD). EPD is a method of coating a conductive substrate with particles suspended in a fluid dispersion under the influence of an electric field applied between the work part and the counter electrode[18].

2. Materials and Method

2.1 Sample preparation

SDCC was mixed using a low speed wet ball-milling method containing 80wt% of SDC nanopowders (Sigma Aldrich, Malaysia) and 20wt% of binary carbonates (67 mol % Li_2CO_3 : 33 mol % Na_2CO_3) (Sigma Aldrich, Malaysia). SDCC was dried and later calcined at 680°C for 2 hours. Then, 50 wt% SDCC powder was mixed with 50 wt% perovskite LSCF6428 (Kceracell, Taiwan) by high speed wet ball-milling method. The LSCF-SDCC powder was dried and calcined at 750°C for 2 hours. Low speed dry milling method was carried out to mix 5 wt% of Ag (Sigma Aldrich, Malaysia) with LSCF-SDCC.

2.2 EPD set up

A glass beaker was used as electrolytic cell. SUS430 ferritic stainless steel was selected as substrate and counter electrode with the area of deposition of 1 cm^2. The distance between two parallel electrodes was fixed at 10 mm. 10g/l LSCF-SDCC-Ag suspension was prepared for EPD process by using aqueous-organic suspension with ratio of 1:1 ethanol and deionised water. Polyethylene glycol (PEG) as dispersing agent was added to suspension by 1wt % from the dry weight of the powder. Suspension with pH 8 has been identified by zeta potential analysis (Malvern, UK) to be the most suitable suspension for EPD process. The deposition was carried out at room temperature with different applied voltages (5, 7 and 10) and deposition durations (5-20 minutes). The deposited coatings were dried at room temperature and then heat treated at 600°C in air atmosphere.

2.3 Characterisation

X-ray diffraction (XRD, Bruker D8 Advance, Germany) and Field Emission Scanning Microscopy (FESEM) (JEOL JSM 7600F, Japan) were used to identify the crystalline phase structure, morphology of LSCF-SDCC-Ag composite powder. After EPD process, the LSCF-SDCC-Ag coatings were analysed for its weight deposition and coating thickness. Meanwhile, the coating surface morphology was observed via FESEM (Hitachi, SU1510, Japan).

3. Result and Discussion

3.1 Phase Identification of composite powder

Figure 1 presents the XRD diffraction patterns of the raw cathode composite powders namely LSCF, SDC and Ag and LSCF-SDCC-Ag after going through milling and calcination process. It could be seen clearly that the addition of 5wt % Ag in LSCF-SDCC composite cathode had not caused any changes of LSCF-SDCC (LSCF JCPDS No. 01-089-1268 and SDC JCPDS No. 01-075-0158) lattice structure. After calcination, carbonate was observed to be in amorphous state and thus no distinct phase was detected in the XRD pattern. The crystal phase of Ag was in face-centered cubic with JCPDS pattern No. 00-004-0783. No distinct secondary phase was appeared and it can be concluded that high speed milling method and calcination process at 750°C did not affect the composite cathode structure and calcined composite has distinct crystalline perovskite intensity.

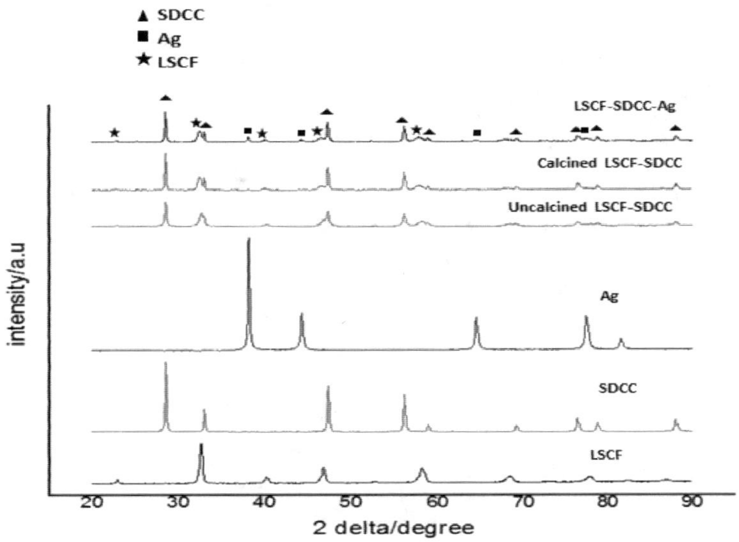

Figure 1. XRD patterns of composite cathode LSCF-SDCC-Ag before and after milling.

3.2 Morphology of composite powder

FESEM micrographs of the uncalcined LSCF-SDCC, uncalcined LSCF, calcined SDCC, calcined LSCF-SDCC and LSCF-SDCC-Ag powders are as shown in Figure 2. Presence of agglomeration and irregular shaped particles were presence in uncalcined powders as shown in figure 2(a) 2 (c). Calcination process removes secondary residue remaining in the ash and convert it to well-defined crystalline perovskites LSCF and SDCC phases as shown in figure 2(b) and 2 (d). During calcination process, the agglomerated particles would strengthened, leading to formation of strong bond between the particles agglomerates and increment in particle size [19]. The growth of particles and changes in shapes can also be clearly seen.

3.3 Weight and thickness of LSCF-SDCC-Ag coating

Figure 3(a) illustrates the highest weight deposition of LSCF-SDCC-Ag (3.70 mg/cm^2) obtained at applied voltage 10 volt with deposition duration of 20 minutes. While, the highest deposition thickness, 62.73 μm was also obtained by deposition at 10 volt for 20 minutes (Figure 3 (b)). It could

IOP Conf. Series: Journal of Physics: Conf. Series **1082** (2018) 012028 doi:10.1088/1742-6596/1082/1/012028

be observed from Figure 4 that the deposited film was very homogeneous, uniform and dense. There were no pores appeared on the film. This may be owing to the fine LSCF-SDCC-Ag powder used and that the particles were well suspended with no agglomeration issues.

3.4 LSCF-SDCC-Ag Coating Morphology

Micrograph of the LSCF-SDCC-Ag coating deposited at 5 volt (Figure 5 (a), (b), (c) and (d) clearly produce non-uniform deposition. This phenomena may occur due to the electrolysis of water and poor particle distribution [20]. Uniform coatings were obtained at higher deposition duration and higher applied voltage. The micrographs of LSCF-SDCC coating at 10 volt indicated lesser crack with denser microstructure, which may be contributed to the increment of deposition duration. Higher deposition can increase the homogeneity and uniform surface structure which contributed to less crack deposition. Hence, a more packed and dense coating layer may be produced.

Figure 2. Morphology of (a) uncalcined LSCF, (b) uncalcined SDCC (c) uncalcined LSCF-SDCC, (d) calcined LSCF-SDCC and (e) LSCF-SDCC-Ag powders.

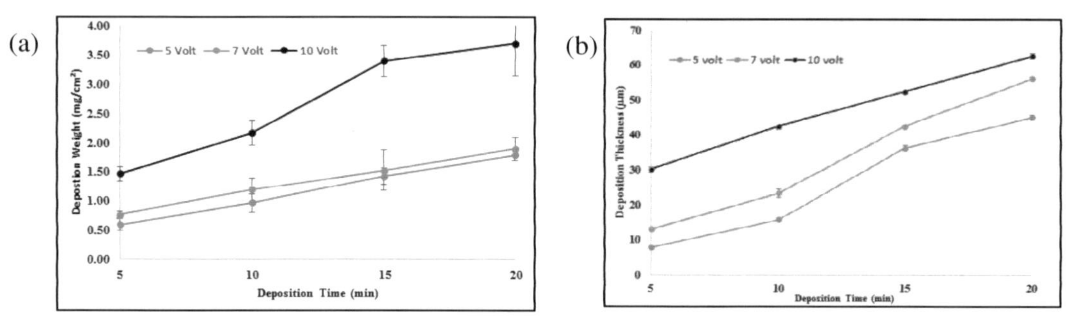

Figure 3. (a) Deposition weight and (b) Deposition thickness as a function of deposition time for LSCF-SDCC-Ag coating on SUS430

Figure 4. Micrograph of coating deposited at applied voltage 10 volt for 20 minutes.

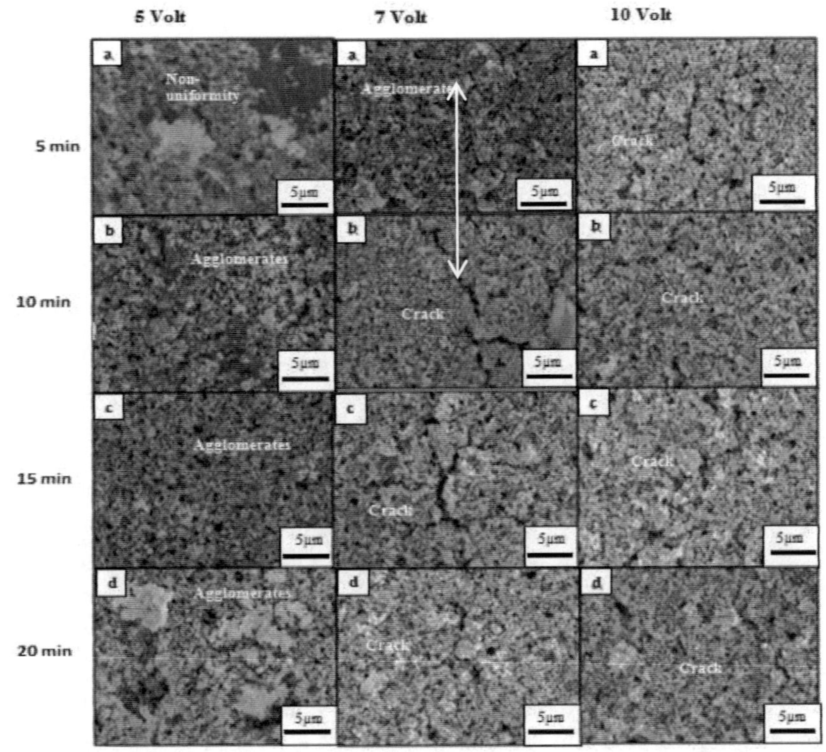

Figure 5. Micrographs of heat-treated LSCF-SDCC-Ag coating deposited at 5 volt, 7 volt, and 10 volt with deposition time (a) 5 min, (b) 10 min, (c) 15 min, and (d) 20 min.

4. Conclusion

Ag exhibits good thermal compatibility with the composite LSCF-SDCC-Ag. Crystalline phase and morphology of composite powder indicated good compatibility after milling process and calcination. LSCF-SDCC-Ag was successfully deposited on the surface of SUS430 ferritic stainless steel with better deposition weight and coating thickness attained at deposition of 10 volt for 20 minutes. Thus the study confirmed the possibility of obtaining LSCF-SDCC-Ag coatings by EPD from mixed aqueous-organic suspensions.

Acknowledgement
The author gratefully acknowledges Universiti Tun Hussein Onn Malaysia (UTHM) and the Malaysia Government for research sponsorship under FRGS Vot 1585.

IOP Conf. Series: Journal of Physics: Conf. Series **1082** (2018) 012028 doi:10.1088/1742-6596/1082/1/012028

References

[1] Mahato, N., et al., *2015* Progress in Materials Science, 2015. **72**: p. 141-337.

[2] Fergus, J., et al., *Solid oxide fuel cells: materials properties and performance*. 2016: CRC press.

[3] Chen, G., et al., *Mn 1.4 Co 1.4 Cu 0.2 O 4 spinel protective coating on ferritic stainless steels for solid oxide fuel cell interconnect applications.* Journal of Power Sources, 2015. **278**: p. 230-234.

[4] Brouzgou, A., A. Demin, and P. Tsiakaras, *Interconnects for Solid Oxide Fuel Cells*, in *Advances in Medium and High Temperature Solid Oxide Fuel Cell Technology*. 2017, Springer. p. 119-153.

[5] Mehran, M.T., et al., *Nano-CeO 2 and-LaCrO 3 dispersed ferritic stainless steels as potential interconnect materials for solid oxide fuel cells.* Journal of Alloys and Compounds, 2017. **709**: p. 453-463.

[6] Bian, L.-z., et al., *Oxidation resistance, thermal expansion and area specific resistance of Fe-Cr alloy interconnector for solid oxide fuel cell.* Journal of Iron and Steel Research, International, 2017. **24**(1): p. 77-83.

[7] Talic, B., et al., *Effect of coating density on oxidation resistance and Cr vaporization from solid oxide fuel cell interconnects.* Journal of Power Sources, 2017. **354**: p. 57-67.

[8] Hwang, H. and G.M. Choi, *The effects of LSM coating on 444 stainless steel as SOFC interconnect.* Journal of electroceramics, 2009. **22**(1-3): p. 67-72.

[9] Przybylski, K., et al., *Oxidation properties of the Crofer 22 APU steel coated with La0.6Sr0.4Co0.2Fe0.8O3 for IT-SOFC interconnect applications.* Journal of Thermal Analysis and Calorimetry, 2014. **116**(2): p. 825-834.

[10] da Silva, A.L.A., G.G. Castro, and M.M. Souza, *Synthesis of Sr-doped LaCrO3 powders by combustion method.* Journal of thermal analysis and calorimetry, 2012. **109**(1): p. 33-38.

[11] Mani, R., et al., *A Study on La0. 6Sr0. 4Co0. 3Fe0. 8O3 (LSCF) Cathode Material Prepared by Gel Combustion Method for IT-SOFCs: Spectroscopic, Electrochemical and Microstructural Analysis.* Research in Chemistry, 2015. **8**(6): p. 389-393.

[12] Agun, L., et al. *Durability and stability of LSCF composite cathode for intermediate-low temperature of solid oxide fuel cell (IT-LT SOFC): Short Review.* in *Advanced Materials Research*. 2014. Trans Tech Publ.

[13] Cheng, Y., et al., *An Investigation of LSF-YSZ Conductive Scaffolds for Infiltrated SOFC Cathodes.* Journal of The Electrochemical Society, 2017. **164**(6): p. F525-F529.

[14] Tsai, M.-J., C.-L. Chu, and S. Lee, *La0.6Sr0.4Co0.2Fe0.8O3 protective coatings for solid oxide fuel cell interconnect deposited by screen printing.* Journal of Alloys and Compounds, 2010. **489**(2): p. 576-581.

[15] Lee, S., et al., *High temperature oxidation behavior of interconnect coated with LSCF and LSM for solid oxide fuel cell by screen printing.* Applied Surface Science, 2010. **256**(6): p. 1817-1824.

[16] Shen, F. and K. Lu, *Perovskite-type La0.6Sr0.4Co0.2Fe0.8O3, Ba0.5Sr0.5Co0.2Fe0.8O3, and Sm0.5Sr0.5Co0.2Fe0.8O3 cathode materials and their chromium poisoning for solid oxide fuel cells.* Electrochimica Acta, 2016. **211**: p. 445-452.

[17] Agun, L., et al., *Influence of Ag on the chemical and thermal compatibility of LSCF-SDCC for LT-SOFC.* 2014.

[18] Besra, L. and M. Liu, *A review on fundamentals and applications of electrophoretic deposition (EPD).* Progress in Materials Science, 2007. **52**(1): p. 1-61.

[19] Hanifi, A.R., et al., *Effects of calcination and milling on surface properties, rheological behaviour and microstructure of 8mol% yttria-stabilised zirconia (8 YSZ).* Powder technology, 2012. **231**: p. 35-43.

[20] Rahman, H.A., et al. *Electrophoretic Deposition of La0. 6Sr0. 4Co0. 2Fe0. 8O3-δ Cathode Film on Stainless Steel Substrates*. in *Advanced Materials Research*. 2010. Trans Tech Publ.

IOP Conf. Series: Journal of Physics: Conf. Series **1082** (2018) 012029 doi:10.1088/1742-6596/1082/1/012029

Properties of BSCF-SDC Composite Cathode Powder for Low Temperature Solid Oxide Fuel Cell

U.A. Yusof[1] and H.A. Rahman[1*]

[1]Faculty of Mechanical and Manufacturing Engineering, Universiti Tun Hussein Onn Malaysia, 86400 Parit Raja, Batu Pahat, Johor, Malaysia

*hamimah@uthm.edu.my

Abstract. Barium strontium cobalt ferrite (BSCF) demonstrates properties, such as ionic conductivity, superconductivity, ferroelectricity, and magnetic resistance. Thus, BSCF has been reported as an impressive and effective cathode material in terms of high conductivity, excellent oxygen transport, catalytic activity and a potential candidate for a low-temperature solid oxide fuel cell (LT-SOFC). In this study, the nanocomposite cathode BSCF–samarium-doped ceria (SDC) was developed for LT-SOFC, and the behavior of the nanoceramic powders was examined. The BSCF–SDC composite cathode powders were mixed through the high-energy ball-milling process at 550 rpm for 2 h. The powders were then calcined at 900 °C and 950 °C for 2 h. The composite cathode powders were characterized based on morphology and grain size by using field emission scanning electron microscope and ImageJ software. The element analysis, chemical bonding, and thermal properties of each sample were studied using X-ray diffraction and energy dispersive spectroscopy respectively. Based on the result obtain, secondary phase has triggered after through milling process and gradually increase when the calcination temperature has increase. It also found that the size of particles become easily agglomerated and bigger at increased of calcination temperature. For particles distribution all major element in composite cathode has uniformly and homogeneous. Results of this study confirm that selecting a suitable calcination temperature is important in eliminating secondary phases in the BSCF–SDC composite cathode and enhancing its properties.

1. Introduction

Fuel cell is a device that efficiently converts chemical energy to electricity in a silent and environmentally friendly way [1]. Among the various kinds of fuel cells, solid oxide fuel cells (SOFCs) as alternatives for energy conversion have shown certain advantages, including wide fuel flexibility, high energy efficiency, and exhaust heat [2]. SOFCs are one of the most important energy conversion devices with high oxygen vacancies [3]. Among the most promising oxides are those of the perovskite (ABO_3) type, which is a structure that combines considerable flexibility and chemical stability, thus permitting custom tailoring of the material properties by doping the A or B sites with different elements and stoichiometries [4].

Among the various possible compositions, $Br_{0.5}Sr_{0.5}Co_{0.8}Fe_{0.2}O_{3-\delta}$/barium strontium cobalt ferrite (BSCF) has emerged as one of the best aspirants for both applications, in particular, considering its high oxygen permeation and oxygen-ion diffusion [5]. Thus, the composition of BSCF was found to exhibit outstanding oxygen permeation fluxes. Under application of relevant temperature, BSCF is not stable when operated in atmospheres containing carbon dioxide due to the formation of carbonates, because of this fact the applicability of BSCF as a cathode for SOFC has been question. When BSCF membrane was being operated at temperature below 800°C, it will make oxygen permeation rate movement slower with time because of the phase transition. By this situation, cubic perovskite phase become not stable in air when below these temperatures [4].

BSCF is a cathode material that has been considered a favorable and highly potential candidate for low-temperature SOFC development in the last two decades. The use of high operating temperature (800 °C–1000 °C) will lead to many disadvantages that include chemical and mechanical instability and high cost during implementation. The conversion of high operating temperature to low operating temperature (400 °C–600 °C) is a possible solution for SOFC commercialization [6]. However, the structural instability of cubic BSCF phase at low temperature is a concern. Therefore, the investigations concerning the phase stability of BSCF and the corresponding properties are expected. In addition, variable in calcination temperatures were carried out to investigate the influence of it on the phase of BSCF-SDC composite cathode.

2. Materials and Method

BSCF–samarium-doped ceria (SDC) composite cathode powders consist of commercial BSCF (Kceracell, Korea) and SDC (Kceracell, Korea) composites that are mixed through high-energy ball-milling (HEBM) process (Fritsch Pulverisette 6, Germany). The composite cathode powders consist of 50 wt% BSCF and 50 wt% SDC powder. The mixing process was performed by using wet-milling method with ethanol as a mixing medium. This method was performed with speed of 550 rpm within 2 h at room temperature. The mixed powder was dried in an oven at a temperature of 80 °C for 12 h after the milling process. The dried mixed powder was placed in an agate mortar and pulverized. Then, the powder underwent calcination process using an electric furnace at a temperature of 900 °C and 950 °C for 2 h. Selection of calcination temperature are support by the literature, where 900 °C are set as a minimum temperature for BSCF phase recovery and elimination of the secondary phases [8]. Then, the calcined composite cathode powder was ground again using the agate mortar to produce fine powders. Table 1 summarizes the sample identification.

Table 1. BSCF–SDC sample identification with different calcination temperatures.

Sample	Milling speed/rpm	Calcination Temperature/°C
BSCF	-	-
BSCF 900	-	900
BSCF 950	-	950
BSCF–SDC	550	-
BSCF–SDC 900	550	900
BSCF–SDC 950	550	950

The composite cathode powder samples were characterized based on their phase identification and crystalline structure by using X-ray diffraction (XRD) (Bruker D8 Advance, Germany). The test was conducted at room temperature and metered by Cu Kα radiation, λ = 0.15418 Å. This diffraction pattern was scanned in the range of 2θ, starting from 20° to 90°, with a step scanning of 0.02°. The analysis data were then defined using Eva Diffraction Plus software.

Element analysis was performed by using energy dispersive spectroscopy (EDS) through a scanning electron microscope (JEOL USA Scanning Electron Microscopes). EDS aims to obtain the area and spot elemental compositions. Images at low magnification in secondary electron imaging mode with an accelerating voltage of 15 kV can be collected. Prior to the process, the sample was in a conductive phase and then was coated with platinum through platinum plating for 60 s with a current of 20 A.

Morphology grain analysis was conducted by using field emission scanning electron microscopy (FESEM) (JSM 6700F-Jeol, Japan). FESEM was used to visualize extremely small topographic details on the surface or the entire or fractioned objects [7]. In this study, samples were made from ceramic materials and were coated with platinum because samples should be in a conductive phase to conduct electron radiation and read the desired image using FESEM. Then, ImageJ software (Version 4.18) was used to measure the average grain size of the sample.

3. Results and Discussion

Figure 1 illustrates the XRD diffractogram of BSCF and BSCF–SDC samples before and after through the calcination process with two temperatures. The XRD result for BSCF powder sample after being calcined at 900 °C and 950 °C denotes that the BSCF spectrum generates the same spectrum pattern with the pure commercial BSCF powder (JCPDS No: 55-0563). The powders exhibit an excellent purity with no secondary peak that is observed during the analysis.

The composite powder of the BSCF–SDC sample indicates that the secondary peaks of $BaCO_3$ (JCPDS No: 71-2394) and Sr_2FeO_4 (JCPDS No: 42-0480) were observed in both milled powders. The formation of the secondary phase might occur because of the heating state, which was generated during the HEBM process. The formation of the secondary phase peak was clearly observed and generated many peaks when the composite powder was calcined at 950 °C. These phases were also known as surface carbonate that is formed from the reaction between alkaline-earth oxide (Ba and Sr) with CO_2 during the calcination process. The formation of secondary phase should be eliminated because this phenomenon will affect the electrochemical performance of the sample. The sample that was calcined at 900 °C and 950 °C exhibits a BSCF peak at JCPDS No: 55-0563 and SDC peak at JCPDS No: 75-0157. This peak finding agreed with the result of the literature, where the minimum calcination temperature is at 900 °C for the BSCF phase recovery and elimination of the secondary phase [8].

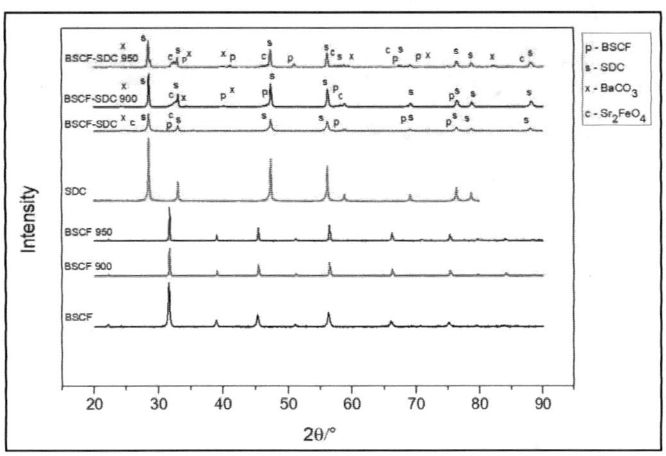

Figure 1. XRD diffractogram of the uncalcined BSCF, SDC, BSCF-SDC and calcined BSCF and BSCF–SDC powders at 900°C and 950°C temperatures.

IOP Conf. Series: Journal of Physics: Conf. Series **1082** (2018) 012029 doi:10.1088/1742-6596/1082/1/012029

Figure 2 depicts the FESEM micrographs of the BSCF and BSCF–SDC samples that were calcined at 900 °C and 950 °C. The above-mentioned results denote that the particle becomes agglomerate as the calcination temperature increases. This phenomenon occurs because the particle size increases with the increase in calcination temperature, which then reduces the surface area. Different states of the particle size can be observed from the sample before and after through the calcination process. The calcination process strengthens the particle bond, but the small particle size will simplify agglomeration.

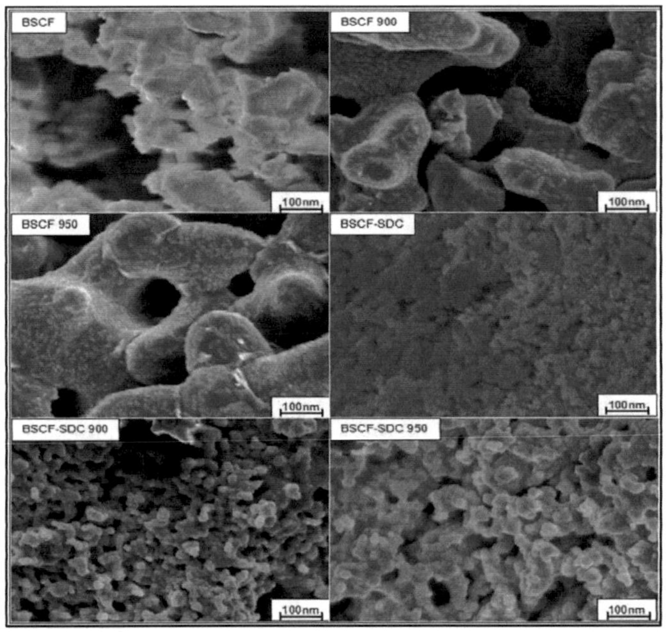

Figure 2. FESEM micrographs of BSCF and BSCF-SDC sample at different calcination temperatures.

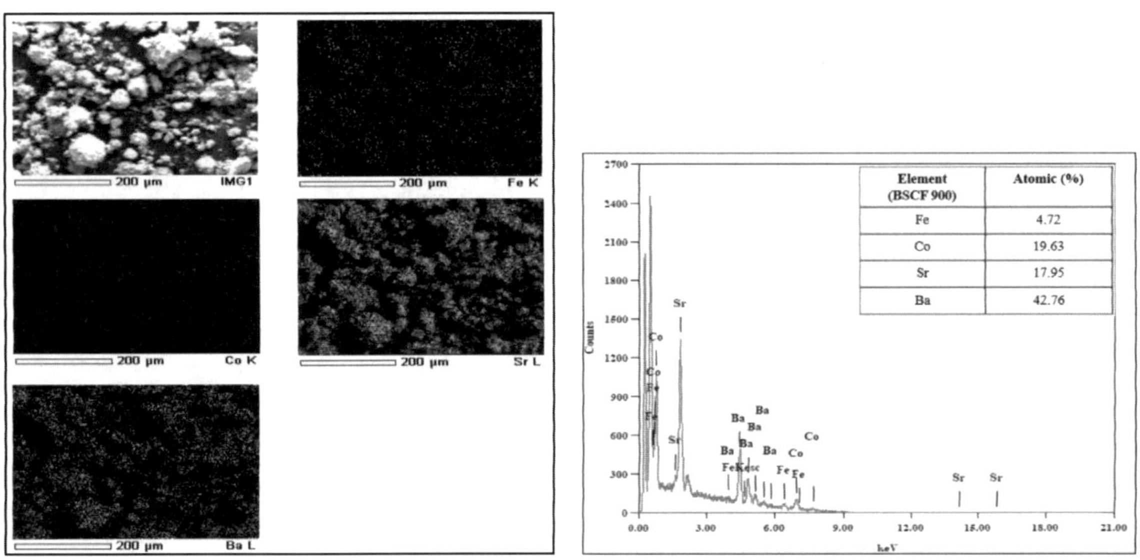

Element (BSCF 900)	Atomic (%)
Fe	4.72
Co	19.63
Sr	17.95
Ba	42.76

Figure 3. EDS mapping of the BSCF calcined at 900 °C.

Figure 3 shows EDS mapping for BSCF powder that calcined at 900 °C. From the result, all element in BSCF which is barium, strontium, cobalt, ferrous have distributed uniformly at that temperature. This is also shows the same result for temperature 950 °C. The quantitative result shows that Sm has displays the higher atomic percent by 42.76 and follow by Sr, Co and Fe.

Figure 4 displays the EDS mapping of the BSCF–SDC calcined at 900 °C and 950 °C. The EDS spectrum reflects the intensity of the elements contained in the BSCF–SDC composite cathode powder. The results indicate that all major elements, such as barium, strontium, cobalt, ferrous, samarium, and ceria, have distributed uniformly and homogeneous. The result also shows that all elements for BSCF-SDC that related to the phases in XRD results has been observed through EDS analysis and it's still exist after milling process at speed of 550rpm and calcination process at 900 °C and 950 °C. The quantitative result of EDS depicts the amount of each element in percentages of the atom. The BSCF–SDC calcined at 900 °C and 950 °C displays the highest atomic percentage for Sm, followed by Co, Fe, Ce, Ba, and Sr. From the analysis, it can say that all element is able to withstand their structure towards the milling and heat treatment process.

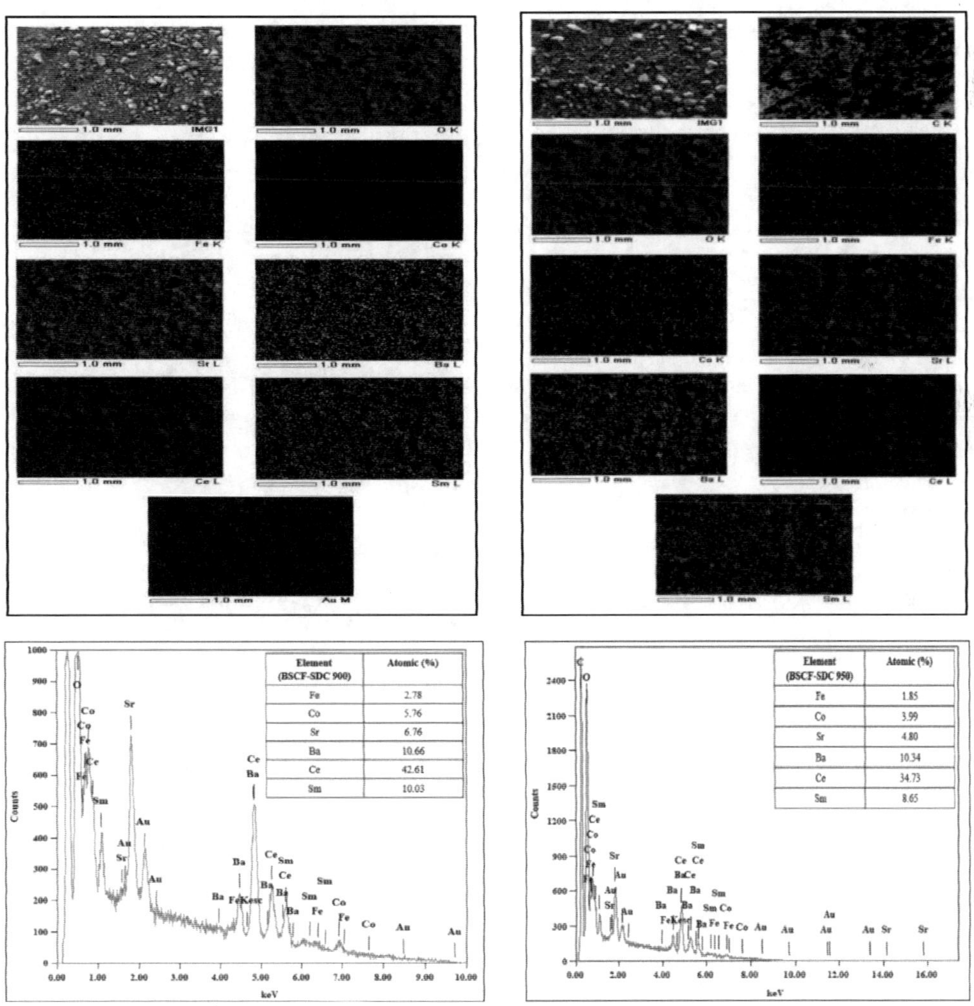

Element (BSCF-SDC 900)	Atomic (%)
Fe	2.78
Co	5.76
Sr	6.76
Ba	10.66
Ce	42.61
Sm	10.03

Element (BSCF-SDC 950)	Atomic (%)
Fe	1.85
Co	3.99
Sr	4.80
Ba	10.34
Ce	34.73
Sm	8.65

(a) (b)

Figure 4. EDS mapping of the BSCF–SDC calcined at (a) 900 °C and (b) 950 °C.

4. Conclusion

The phase stability of the BSCF and BSCF–SDC cathode powders was observed based on the effect of calcination temperature. The XRD result of BSCF-SDC cathode powder has shown that the BSCF phase peak was observed at a temperature of 900 °C. Meanwhile, the secondary phase can be clearly identified at 950 °C of BSCF-SDC cathode powder. The increasing of calcination temperature has led to easy agglomeration for the particle which will limit the surface area of the sample. Based on the EDS result, all element related to the phases in XRD result has been observed trough EDS analysis and its prove that all element still exists and can withstand with the milling speed and calcination temperature. Meanwhile, additional strategic research, such as a method for preparing composite cathode powder and studies on the LT-SOFC itself, should be conducted to overcome and evaluate the stability of the BSCF and BSCF–SDC sample.

Acknowledgements

The author gratefully acknowledges Universiti Tun Hussein Onn Malaysia (UTHM) and the Malaysia Government for research sponsorship under FRGS Vot 1585. This paper was partly sponsored by the Centre for Graduate Studies UTHM.

References

[1] Lee S, Lim Y, Lee E A, Hwang H J, and Moon J W 2006 J. *Power Sources* 157 848
[2] Ding X, Zhu W, Gao X, Hua G, and Li J 2015 *Fuel Process. Technol.* 135 14
[3] Sharma D R K, Rao P V B, Reddy M N, and Srikanth D 2015 *Mater. Today Proc* 2 4442
[4] Niedrig C, Taufall S, Burriel M, Menesklou W, Wagner S F, Baumann S, and Ivers-Tiffée E 2011 *Solid State Ionics* 197 25
[5] Araki W and Malzbender J 2013 *Scr. Mater.* 69 278
[6] Agun L, Ahmad Shah M H, Ahmad S, and Abd Rahman H 2016 *Mater. Sci. Forum* 840 247
[7] Janssen G J 2015 *Radboud Univ. Nijmegen* 1
[8] Hoa K NG, Abd Rahman H and Afandi S, *Mater. Sci. Forum.* 888 47

IOP Conf. Series: Journal of Physics: Conf. Series **1082** (2018) 012030 doi:10.1088/1742-6596/1082/1/012030

Effect of Sintering Temperature and Holding Time to the Final Characteristics of Fecral Powder Compacts Formed at Elevated Temperature Through Die Compaction Method

M.M. Rahman, R.M. Shahmi and H.Y. Rahman

Department of Mechanical Engineering, Universiti Tenaga Nasional, Putrajaya Campus, Jalan IKRAM-UNITEN, 43000 Kajang, Selangor, Malaysia.

mujibur@uniten.edu.my

Abstract. This paper presents an experimental study on the effect of sintering temperature and holding time to the final characteristics of FeCrAl powder compacts formed at 200°C through uniaxial die compaction process. The powder mass was prepared by mechanically mixing iron powder ASC 100.29 with chromium (22 wt%) and aluminum (11 wt%) for 30 min at room temperature. The powder mass was formed at 200°C by applying an axial load of 425 MPa. The as-pressed compacts termed as green products were then cooled to room temperature and subsequently sintered in argon gas fired furnace at a rate of 5°C/min for three different holding times, i.e., 30, 60, and 90 min at three different sintering temperatures, i.e., 800, 900, and 1000°C. The sintered samples were characterized for their density, bending strength, and microstructure. The results revealed that the sample sintered at 800°C for 30 min got higher relative density but lower bending strength. Sample sintered at 1000°C for 30 min achieved the higher bending strength and better microstructure.

1. Introduction

A solid solution of two or more materials is termed as alloy provided that at least one of the materials is a metal. Matrix is the most dominant element of an alloy whereas the other elements are dissolved in it [1-3]. Since every single element has its own weakness which might not fulfil the design requirements, alloy is formed to overcome the weaknesses of individual materials. The main intention of making alloy is to improve the physical, chemical, mechanical, thermal, and electrical properties of materials, which can fulfill the application requirements [4-6]. Alloys are able to withstand structural integrity, surface, and property stability at elevated temperature under high mechanical stress and severe corrosive environment.

The most dominant element or matrix of FeCrAl alloy or usually termed as Fecralloy is the iron (Fe) which has a lot of disadvantages where the most critical ones are the strength degradation and corrosion at high temperature. When iron is mixed with chromium (Cr), a protective layer is formed which resists the corrosion. Since aluminum (Al) is soft and malleable, it enhances the formability [7].

The forming of mechanical components by compacting a single powder or a blend of more than one types of powder in a rigid container to a desired shape is defined as die compaction process [8]. The as-pressed powder mass or green products must have sufficient high density and strength to be handled for further treatment [9]. The advantages of forming of powder mass a above ambient temperature have

IOP Conf. Series: Journal of Physics: Conf. Series **1082** (2018) 012030 doi:10.1088/1742-6596/1082/1/012030

been identified by previous researchers [10-11], however the compaction must be conducted below the recrystallization temperature of the main powder constituent [12].

Sintering is an essential step of a full cycle of powder compaction process, which must be conducted at controlled environment to avoid contamination of the component by other materials [13]. The basic principle of sintering is the achieving rate of the desired degree of bonding among the particles in powder compacts [14]. The heating and cooling rate, sintering temperature, and holding time play important roles in controlling the microstructure and porosity that determine the degree of particles bonding. In solid state sintering, the bonding among particles requires material transport by volume diffusion (migration of vacancies), grain-boundary diffusion, surface diffusion, viscous or plastic flow, and evaporation/condensation of atoms on the surfaces. The densification and the strengthening of particles inside the metal powder compacts reflect the properties of the sintered component. The sintered density relates directly to the dimensional and mass changes, which give perfect understanding of the densification process [15].

It is clear from the explanation above that sintering schedule plays an important role in determining the quality of a final product where the green products are formed through powder compaction route. Current practice of FeCrAl alloy forming is through foundry process which is time consuming and energy inefficient hence expensive. There is also less thorough study found on the forming of FeCrAl alloy through powder compaction route. Therefore, the objective of this paper is to present the effect of sintering parameters, i.e., temperature and holding time to the final characteristics of FeCrAl powder compacts formed at above ambient temperature through die compaction route.

2. Materials and Methods

Iron powder ASC 100.29 manufactured by Höganäs AB with the composition of 1.5 wt% Cu, 0.5 wt% Mo, and 4 wt% Ni balanced with Fe [16] was used as main powder constituent. Chromium (Cr) powder (22 wt%) and aluminum (Al) powder (11 wt%) were used as alloying elements where the balance was Fe powder. The main powder constituent together with the other alloying elements were mixed mechanically for 30 min at room temperature at a rotation of 30 rpm.

The blended powder mass was formed through uniaxial die compaction process at a temperature of 200°C by applying simultaneous downward and upward loading of 425 MPa. The as-pressed powder compact termed as green compact (Figure 1) was subsequently ejected from the die cavity by pushing the bottom punch to upward direction. All the green samples were inspected visually for any defects or cracks. The defect-free samples were then sintered in a custom made argon gas fired furnace (Model: HT3-1400-SIC, S/N: LT007) at a rate of 5°C/min for three different holding times, i.e., 30, 60, and 90 min at three different sintering temperatures, i.e., 800, 900, and 1000°C.

The green as well as the sintered densities of the samples were calculated from the mass and volume of each sample. The relative density was calculated by dividing the density of the individual product by the solid density. The flexure stress was measured through three point bending test machine (Brand: Instron, Model: Instron 3365, S/N: SAA61569) following the standard (ASTM E290-09). Each sample was hold tightly at both ends, force was applied at the middle of the sample until it was broken down (Figure 2). The image of the fractured surface of each sample was captured through scanning electron microscopy (Brand: JEOL, Model: JSM- 6010PLUS/LA).

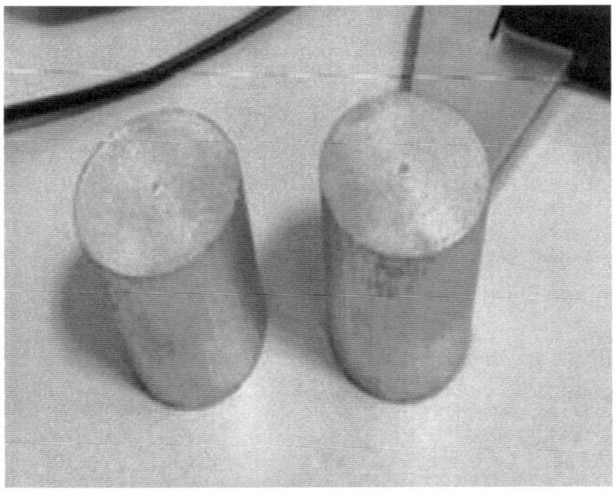

Figure 1. FeCrAl green samples formed at 200°C

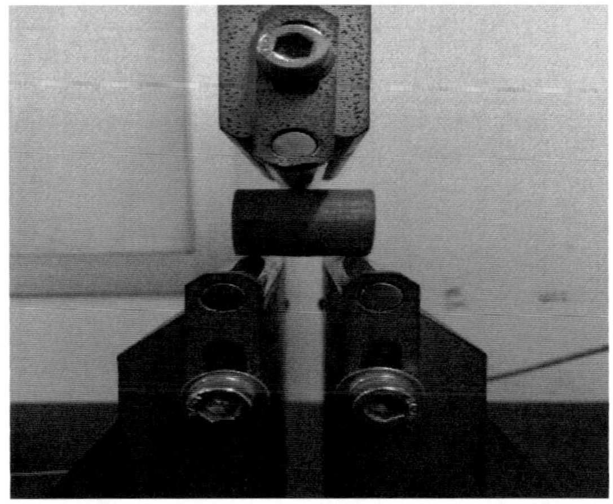

Figure 2. Bending strength measurement (ASTM E290-09)

3. Results and Discussion

The relative densities of products sintered at different temperature for different times are depicted in Figures 3-4. It can clearly be observed that the relative density decreases as the sintering temperature increases irrespective of holding time. Longer holding time also caused the relative density to decrease. This finding is also in-line with volumetric expansion (Figure 5) where sintering at high temperature for a longer period of time caused the samples to swell. Since density was calculated from the mass of sintered product divided by its volume, therefore swelling of sintered product ended up with lower density. The swelling is believed to be due to the competition between classical sintering phenomena induces densification due to necking among powder particles (Figure 6) and material creeping under stresses caused by internal gas pressure [17]. Neck formation is driven by the surface diffusion which is usually the dominant mass transport mechanism during the early stage of sintering. The internal gas includes air trapped during powder forming and oxygen released from particle surface inside the pores that closed during compaction or sintering. Material creeping is believed to be the dominant mechanism at this current sintering temperature and holding time.

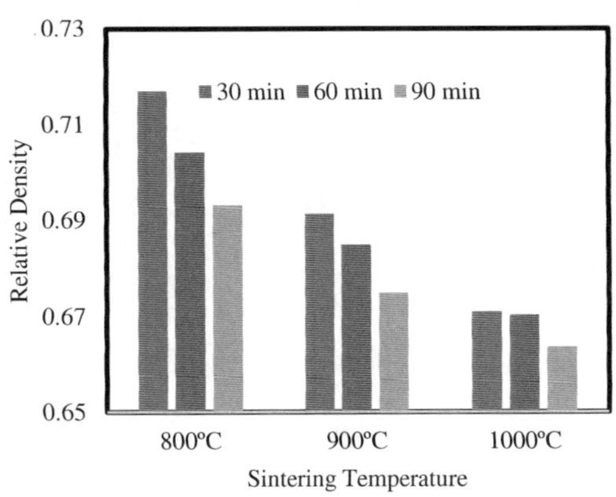

Figure 3. Relative density of the products sintered for different time

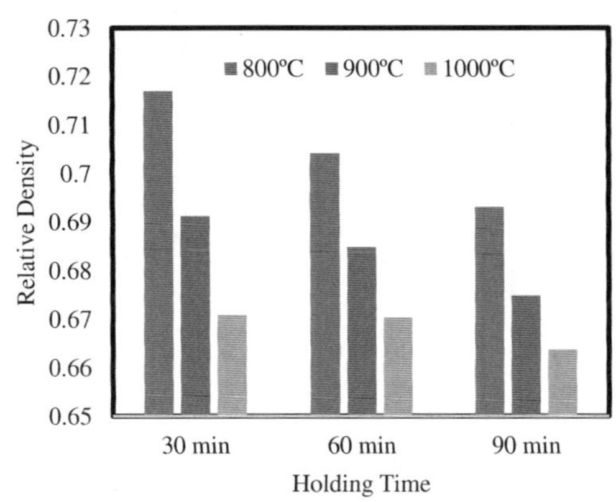

Figure 4. Relative density of the products sintered at different temperature

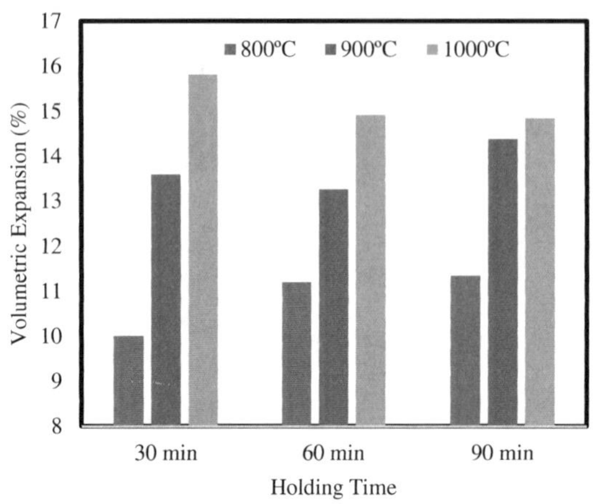

Figure 5. Volumetric expansion of the products sintered at different temperature

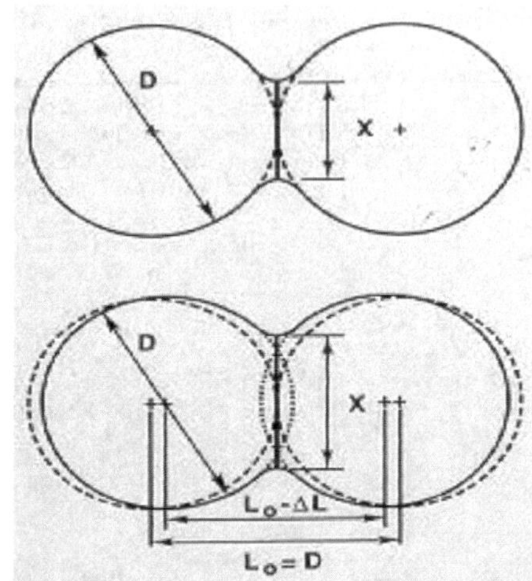

Figure 6. Two sphere model [19]

Flexure stress or bending strength of sintered samples is shown in Figure 7. Sample sintered at 1000°C for 30 min is found to obtain the highest strength. On the other hand, sample sintered at 800°C for 90 min obtained the lowest strength. The strength difference is also significant, i.e., strength of the products sintered at 1000°C is found to be almost double of the strength of the products sintered at 800°C and 900°C. It is believed that the samples were not fully sintered at temperatures 800°C and 900°C. Since the main powder constituent was the iron powder ASC100.29 which has the melting temperature of 1536°C. Therefore, sintering at a temperature lower 60% of the melting temperature of the main powder constituent does not allow the formation of necking which strengthen the sintered products [18].

Figures 8-10 show the microstructures of the products sintered at 800°C, 900°C and 1000°C for 30 min at a rate of 5°C/min. Bigger size and irregular shape pores are observed at the sample sintered at 800°C (Figure 8) as the pore size became smaller with increasing sintering temperature to 900°C (Figure 9) and 1000°C (Figure 10). Spherical shape pores are observed at the samples sintered at 900°C and 1000°C, however interconnected and irregular shape pores are found at sample sintered at 800°C. Interconnected pores weakened the sample (Figure 7) hence less force was required to break down the sample when transverse load was applied at the sample, i.e., lower bending strength. At higher sintering temperature, the diffusion rate increased and the atom diffused to adjacent grain size, the pores shrank and isolated resulting more spherical shape pores.

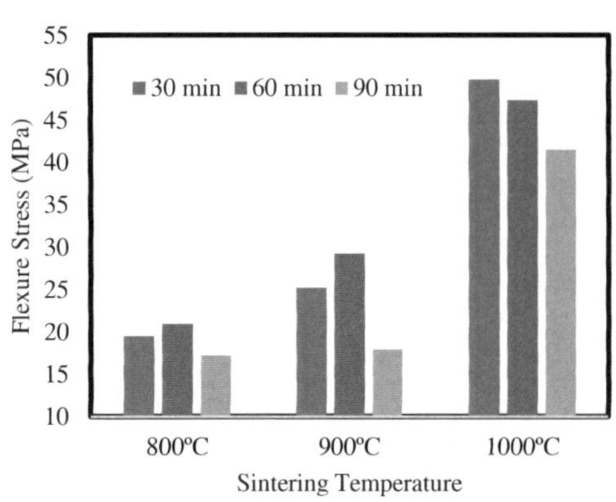

Figure 7. Bending strength of products sintered at different temperature.

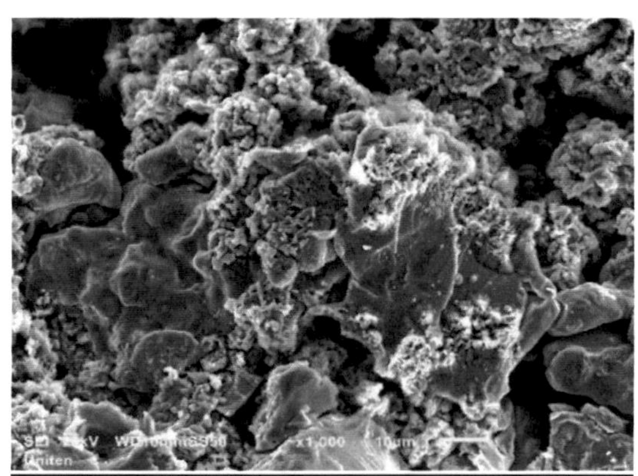

Figure 8. SEM image of the fractured surface of sample sintered at a rate of 5°C/min for 30 min at 800°C.

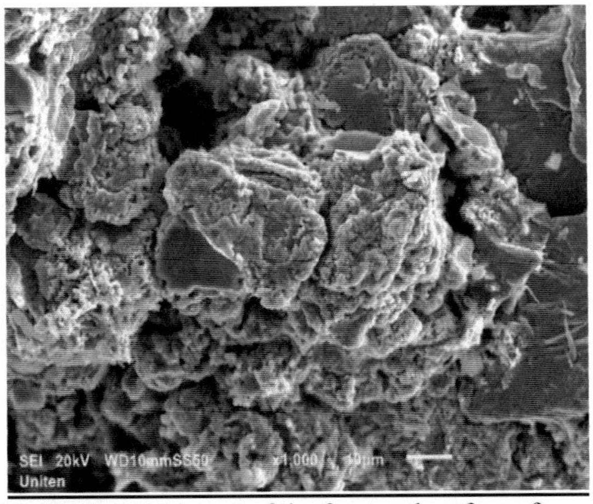

Figure 9. SEM image of the fractured surface of sample sintered at a rate of 5ºC/min for 30 min at 900ºC

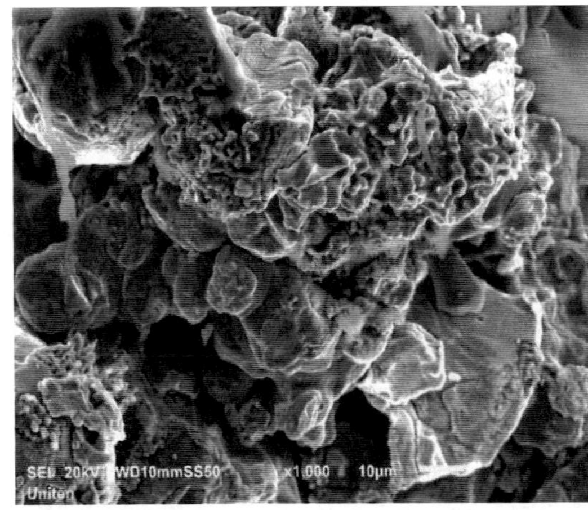

Figure 10. SEM image of the fractured surface of sample sintered at a rate of 5ºC/min for 30 min at 1000ºC

4. Conclusion

The effects of temperature and holding time were studied during sintering of FeCrAl powder compacts formed through uniaxial die compaction process at 200ºC. The results revealed that higher relative density could be obtained by sintering the powder compacts at lower temperature, i.e., 800ºC for 30 min. However, sintering at higher temperature, i.e., 1000ºC allowed the formation of round shape isolated pores hence the bending strength of the compacts sintering at higher temperature was found to be higher. The micrographs also revealed that interconnected pores are visible at the sample sintered at lower temperature, i.e., 800ºC.

Acknowledgement
This research is financially supported by Universiti Tenaga Nasional through the BOLD research grant 10289176/B/9/2017/35.

References

[1] Thakur A and Gangopadhyay S 2016 *International Journal of Machine Tools & Manufacture* 100 25
[2] Pervaiz S, Rashid A, Deiab I and Nicolescu M 2014 *Materials and Manufacturing Processes* 29 219
[3] Callister WD 2012 *Materials Science and Engineering*, 8th Edition, John Wiley & Sons
[4] Rahman MM and Talib AAA 2015 *Materials Today Communications* 4 42
[5] Rahman MM and Yogaswerarow SR 2016 *Advanced Materials Research* 1133 264
[6] Akca E and Gursel A 2015 *Periodicals of Engineering and Natural Sciences* 3(1) 15
[7] Li D and Ghosh A 2003 *Materials Science and Engineering A* 352(1-2) 279
[8] Baccino R and Moret F 2000 *Materials and Design* 21 359
[9] Rahman MM, Ariffin AK, Nor SSM and Rahman HY 2011 *Materials and Design* 32 264
[10] Yahya Z, Rahman MM and Daud M 2017 *Solid State Phenomena* 264 202
[11] Rahman MM, Nor SSM and Rahman HY 2011 *Materials and Design* 32 447
[12] Rahman MM, Ismail, MA and Rahman HY 2017 *Solid State Phenomena* 264 103
[13] Rahman MM, Namasivayam D and Rahman HY 2015 *Malaysian Journal of Microscopy* 11 74

[14] German RM 2013 *Powder Metallurgy* 56(2) 117
[15] Wang YU 2006 Acta Materialia 54(4) 953-961
[16] Hoganas AB 2013 *Material and Powder Properties- Hoganas Handbook for Sintered Components* Hoganas AB Sweden pp. 95-100
[17] Biguereau E, Bouvard D, Chaix JM and Roure S 2016 *Powder Metallurgy* 59(5) 394
[18] Demirskyi D, Agrawal D and Ragulya A 2010 *Materials Science and Engineering A* 527 2142
[19] Kuczynsky GC 1949 *Transactions of the AIME* 85 169.

Effect of Sonication Time on the Production of Graphene by Electrochemical Exfoliation Method

Y.Z.N. Htwe[1], M. Mariatti[1], W.S. Chow[1], Y. Suda[2] and A.A. Thant[3]

[1]School of Materials and Mineral Resources Engineering, Engineering Campus, Universiti Sains Malaysia, 14300 Nibong Tebal, Pulau Pinang, Malaysia

[2]Department of Electrical and Electronic Engineering, Toyohashi University of Technology, Aichi Toyohashi 441-8580, Japan

[3]Department of Physics, University of Yangon, 11041 Kamayut, Yangon, Myanmar

Email: mariatti@usm.my

Abstract. In this study, Graphene has been synthesized by using electrochemical exfoliation method. Effect of sonication times (15, 30 and 45 min) at room temperature on the production of graphene was investigated. The synthesized graphene were characterized by X-ray diffraction (XRD), Fourier transform infrared spectroscopy (FTIR) and Atomic Force Microscopy (AFM). The crystal structure of the exfoliated graphene were analyzed using XRD. The graphene flakes exhibit a sharp peak centered at 26.5°corresponding to the (002) plane of graphene crystal. It is found that, the crystallite sizes is gradually decreased with increased in sonication time. The FTIR spectrum reveals the prominent appearance of C-O-H group which indicate that OH^- radicals exist in conjunction with exfoliated graphene sheets. Graphene flakes with average thickness of 0.2 nm to 0.8 nm were observed using AFM. The electrical conductivity results showed that 45 min sonication time produced the highest electrical conductivity ($1.38\times10^2 sm^{-1}$).

1. Introduction

A single atomic layer of sp^2 bonded carbon material, graphene which is a 2- dimensional hexagonal honeycomb like structure, has attracted significant attention [1, 2]. The planer orbitals are energetically stable and localized sigma bonds with the three nearest neighbour atoms arranged in a honeybomb lattice [3]. This material possess various intriguing properties of interest such as intrinsically super electrical conductivity ($10^6\Omega^{-1}cm^{-1}$), nearly transparent in visible light (97.7 %), high intrinsic carrier mobility (2.5×10^5 $cm^2v^{-1}s^{-1}$), high specific surface area (2630 m^2g^{-1}), excellent mechanical properties (Young's modulus > 1TPa), and high thermal conductivity (above 3000 Wmk^{-1}) [4, 5]. These excellent properties make graphene a promising material for various applications such as nano electronics, composite materials, transparent conductive films, super capacitor, printable electronics and several other fields

IOP Conf. Series: Journal of Physics: Conf. Series **1082** (2018) 012031 doi:10.1088/1742-6596/1082/1/012031

[6]. Graphene has an electrical conductivity greater than any other known material, and its thermal conductivity is higher than diamond [7]. Presently, the most widely used transparent conducting materials are doped metal oxide. Indium tin oxide (ITO) is nature, high surface roughness and growing cost due to shortage of indium. Many other materials have been tried earlier but without much success. Previous studies suggest that in comparison to ITO, graphene is a better material for making transparent and highly conducting electrode.

To date, several methods have been demonstrated for graphene synthesis using both top-down and bottom-up approaches. In recent years, mechanical exfoliation, chemical exfoliation, chemical synthesis and thermal chemical vapour deposition (CVD) synthesis are the most commonly used methods. The electrochemical exfoliation is another technique for exfoliating graphene [8]. Recently, electrochemical exfoliation of graphite has attracted attention due to its easy, fast and environmental friendly nature to produce high-quality graphene. The electrochemical cell is setup with a graphite anode and a metal cathode, such as platinum. Both probes are dipped into an electrolyte solution made of water and ionic liquid or a surfactant [9, 10]. Other chemicals can be added to enhance the conductivity of the electrolyte as well. The surfactant or ionic liquid intercalate the graphite anode, penetrating the layers of graphite. A voltage is applied across the robes, facilitating cleavage of graphene flakes at the anode. The electrolyte then stabilizes the graphene sheets in solution, preventing agglomeration after exfoliation [11]. In this study the synthesis of graphene by electrochemical exfoliation of method is considered. Further the effect of sonication times (15 min, 30 min and 45 min) at room temperature on the production of graphene was investigated.

2. Experimental Method

Graphite rod with diameter of 6.35 mm was supplied by Sigma-Aldrich. Sulphuric acid (H_2SO_4) supplied by Merck assay was used as electrolytes. Ethanol supplied by J.T. Baker was used as a cleaning agent. Distilled water was used in the synthesis process. In this study, a typical electrochemical synthesis of graphene, which a conventional two-electrode system was used. Exfoliation was conducted with a graphite rod which is used as a working electrode and a platinum (Pt) rod was treated as the counter electrode. The platinum electrode was placed parallel to the graphite rod at a distance of about 2 cm. The positive voltage was supplied to the graphite electrode using DC power supply. After the graphite exfoliation was completed, the graphene solution were sonicated for 15, 30, 45 min at room temperature and washed several times by distilled water using vacuum filtration. The powder was obtained after dried in a vacuum oven at 80°C.

2.1 Characterization Techniques

X-ray diffraction (using Bruker D8 X-ray scattering systems) with Ni-filtered Cu-Kα (λ = 1.54021Å) radiation source was used to identify the crystallographic structure of graphene. Fourier transform infrared spectroscopy (FTIR) was obtained by using a Perkin Elmer FTIR spectrometer (spectrum one) at room temperature with a typical wave number of 4000- 500 cm^{-1} in open air at transmittance mode. Atomic force microscopy (AFM) observation was performed using (AFM, Nano Navi SII). Direct current (DC) electrical conductivities of graphene pellets were tested at room temperature according to the ASTM D257-99 standard test method by using a Gwinstek LCR 1817 meter at a voltage of 1V and frequency of 1KHz. The resistivity and conductivity were calculated using Equations (1) and (2), respectively.

$$\sigma = \frac{1}{\rho} \qquad (1)$$

$$D = \frac{0.9\lambda}{B \cos\theta} \qquad (2)$$

Where σ = electrical conductivity, ρ = resistivity, D = crystallite sizes, λ = wavelength.

3. Results and discussion

The crystal structure of the graphite and exfoliated graphene were analysed by XRD. The XRD pattern of graphite powder is shown in Figure 1 (A). The purification of graphite powder has been confirmed from XRD pattern. The XRD diffraction peak at about 26° was observed and confirmed the purification of graphite flake [14]. Figure 1 (B) shows all graphene flakes exhibit a broad peak centered at 26.5° corresponding to the (002) of a graphene crystal [15]. In the present work, the average crystallite sizes are tabulated in Table 1. It was found that the crystallite sizes is gradually decreased with increased sonication time. The acoustic cavitation's energy that generated during the sonication process decrease in crystallite size of synthesized graphene [16]. The reduction of crystallite size is with increasing sonication time.

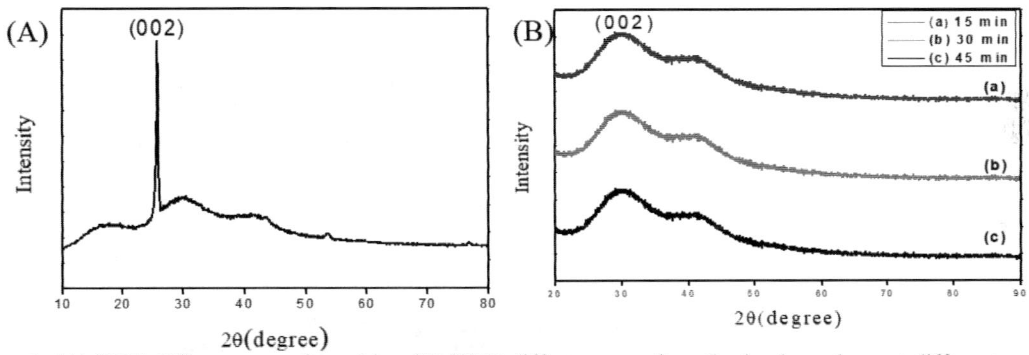

Figure 1. (A) XRD diffratogram of graphite, (B) XRD diffratogram of synthesized graphene at different sonication times (a) 15 min, (b) 30 min and (c) 45 min.

Table 1. The average crystallite size of exfoliated graphene.

No.	Sonication time (min)	Crystallite sizes (nm)
1	15	8.7
2	30	8.5
3	45	8.3

FTIR spectroscopy was carried out to investigate the surface chemical nature of graphite and exfoliated graphene. FTIR spectra of graphite and exfoliated graphene are shown in Figure 2 (a) and (b). The spectrum of graphite had broad absorption peaks at 3440 cm^{-1}, 1623 cm^{-1} and 990 cm^{-1} corresponding to O-H, C-C stretching vibration and CH deformation. Few new peaks appears after exfoliation of graphite, the peak at 1654 cm^{-1} can be attributed to C-O stretching vibration, the small peak at 1579 cm^{-1} and 1582 cm^{-1} represent C-C stretching and 1040 cm^{-1}, 1042 cm^{-1} and 1124 cm^{-1} represent C-O stretching vibration. The broad peaks at 3441 cm^{-1} and 3422 cm^{-1} show the prominent appearance of O-H group clearly shows that OH^{-1} radicals did exist in conjunction with exfoliated graphene sheets [17]. Therefore, the FTIR peaks of the exfoliated graphene support that oxygen functional group was introduced into graphene during the exfoliation process.

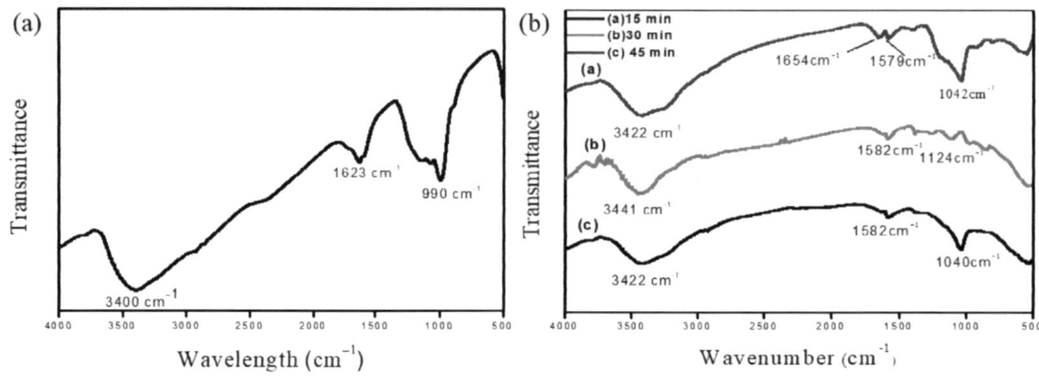

Figure 2. FTIR spectrum of (a) graphite (b) synthesized graphene at different sonication times (a) 15min, (b) 30 min and (c)

Atomic Force Microscopy (AFM) was employed to identify the thickness of the exfoliated graphene. The AFM analysis of the exfoliated graphene at different sonication times (15 min, 30 min, 45 min) are shown in Figure 3, 45 min sonication time produced graphene which is thinner than those 15 min and 30 min sonication times. Table 2 shows the thickness of graphene sheets measured by using AFM analysis.

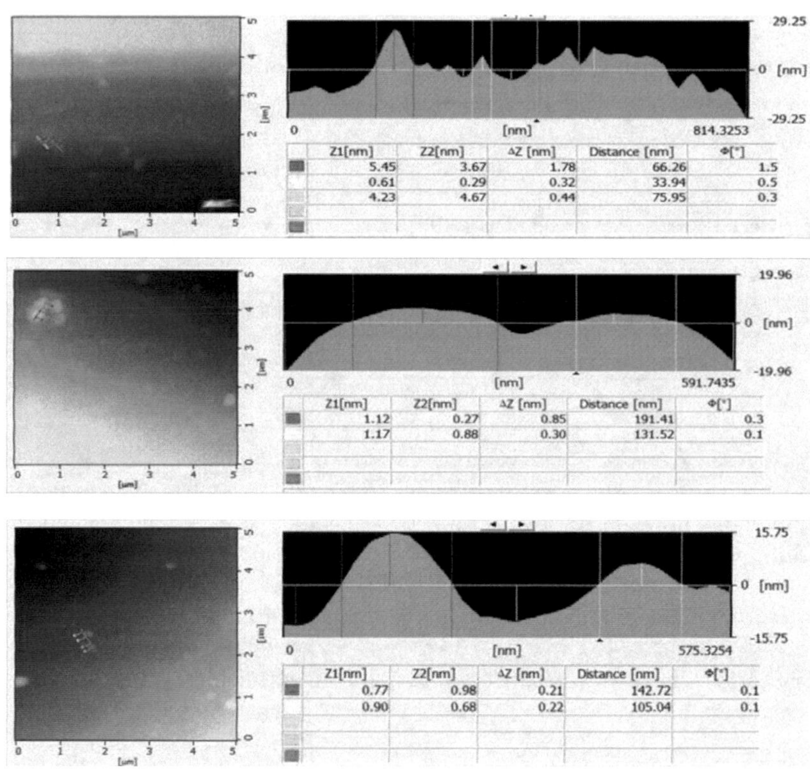

Figure 3. AFM analysis of synthesized graphene at different sonication times (a) 15 min, (b) 30 min and (c) 45 min).

Table 2. The thickness of exfoliated graphene sheets and electrical conductivity values of graphene produced by different sonication times

No.	Sonication time (min)	Thickness (ΔZ), nm	Electrical Conductivity (Sm^{-1})
1	15	0.85	1.29×10^2
2	30	0.60	1.31×10^2
3	45	0.22	1.38×10^2

The electrical conductivity of the synthesized graphene at different sonication times (15 min, 30 min, 45 min) are presented in Table 2. The synthesized graphene at 45 min sonication time showed the highest electrical conductivity. High conductivity of synthesized graphene is due to sufficient time to exfoliate the graphite.

4. Conclusions

In conclusion, characterization results of XRD, FTIR and AFM showed that sonication time of 45 min is suitable to be used to exfoliate graphene from graphite rods. As the sonication time increases, the average crystallite sizes and average thickness values are decreased. The electrical conductivity results showed that 45 min sonication time produced the highest electrical conductivity. Therefore, in this study, 45 min is considered as the maximum sonication time on the production of graphene.

Acknowledgements

The authors gratefully acknowledge ASEAN University Network for Science and Engineering Education Development Network (AUN/SEED-Net) Project and Japan International Corrperation Agency (JICA) for their financial support. And also would like to thank the Universiti Sains Malaysia (USM) for the use of their facilities.

References

[1] Liu, J., et al., 2013 Nano energy 2 377-386
[2] Tripathi, P., et al.,2013 *arXiv preprint arXiv* 1310.7371
[3] Zhou, M., et al., 2013 *Chemical Physics Letters* 572: p. 61-65
[4] Chen, K. and D. Xue, 2014 *Journal of colloid and interface science* 436 41
[5] Papageorgiou, D.G., I.A. Kinloch, and R.J. Young 2017 *Progress in Materials Science* **90** 75
[6] Noroozi, M., et al., 2016 *PloS one* **11** e0152699
[7] Su, C.-Y., et al., 2011 *ACS nano* **5** 2332
[8] Li-Shang, L., A. Westwood, and R. Brydson 2016 *Circles* **11** 12
[9] Lu, J. 2009 *ACS Nano* 3 2367
[10] Ossonon, B.D. and D. Bélanger 2017 *Carbon* **111** 83
[11] Alanyaliog Lu, M. 2012 *Carbon* 50 142
[12] Parvez, K., et al., 2014 *J. Am. Chem. Soc* 136 6083
[13] Durge, R., R. Kshirsagar, and P. Tambe 2014 *Procedia Engineering* **97** 1457
[14] Liu, W.-W., et al., 2014 *Journal of Industrial and Engineering Chemistry* 20 1171
[15] Siyar, M., A. Maqsood, and S. Khan 2014 *Materials Science-Poland* 32 292
[16] Sumari, Roesyadi, Achmad 2013 *Scientific study and Research. Chemistry and chemical engineering , Biotechnology, food industry* 14 229
[17] Siyar, M., A. Maqsood, and S. Khan 2014 *Materials Science-Poland* 32 292

Chlorination of Titanium Carbonitride ($TiC_{0.7}N_{0.3}$) for Production of $TiCl_4$

Najwa Ibrahim, Law Hon Kin, Eltefat Ahmadi, Norlia Baharun, Sivakumar Ramakrishnan and Sheikh Abdul Rezan*

[1] School of Materials and Mineral Resources Engineering, USM Engineering Campus, Pulau Pinang, Malaysia

Email: srsheikh@usm.my

Abstract. The chlorination of titanium carbonitride ($TiC_{0.7}N_{0.3}$) was conducted on the effect of temperature, reaction time and nitrogen gas flow rate for low-temperature production of titanium tetrachloride ($TiCl_4$). Statistical design of experiments (DOE) was employed to analyze the interaction of the parameters. The chlorinated samples were characterized by X-Ray diffraction (XRD), scanning electron microscopy (SEM) and energy-dispersive X-ray spectroscopy (EDX) analysis. The DOE results showed that the flow rate of nitrogen (N_2) gas was the most influential parameter followed by temperature and chlorination time. The highest weight loss for $TiC_{0.7}N_{0.3}$ sample was obtained to be about 79.52% at 500 °C for 60 minutes under N_2 gas with the flow rate of 200 $mL.min^{-1}$. The XRD patterns showed a strong decrease in intensity of $TiC_{0.7}N_{0.3}$ peaks. The XRD and SEM/EDX analyses also showed some free carbon in the residues after the chlorination process that could be formed from the chlorination of TiC in solid solution with TiN.

1. Introduction

Titanium carbonitride (TiCN) is a synthetic material commonly employed for coating of cutting tools. It has high cutting performance and can be used at high cutting speeds [1]. In synthesis of TiCN coatings, TiN and TiC have been used as starting material and prepared by sintering techniques [2, 3]. In addition, Mostert et al. [4], have reported that titanium carbonitride produced by reduction and nitridation of titanium dioxide, ilmenite and/or titanium slag could be chlorinated successfully at 200-500°C. In the low-temperature chlorination process, impurities normally do not react with chlorine gas to any significant extent or may become chlorinated at very slow rate [5-7]. The reason that chlorine gas was used is because of its high reactivity as well as the more environmentally friendly waste. The wastes generated from chlorine metallurgy routes, compared to those obtained from other processes, the chlorine and the chlorides in the effluents can be easily dissolved in water or precipitated and subsequently recycled in the same unit or otherwise [8, 9]. The overall chlorination reaction for Ti(C,N) can be written as below as Eq.(1) . Preliminary chlorination studies on nitrided ilmenite and Ti(O,C,N) has been reported elsewhere [10]

$$TiCN\ (s) + 2Cl_2\ (g) \rightarrow TiCl_4\ (g) + C\ (s) + \tfrac{1}{2}\ N_2\ (g),\quad \Delta G° = -\ 413.67 + 3.5 \times 10^{-3}\ T\ (kJ/mole) \quad (1)$$

2. Materials and Method

Titanium carbonitride powder was used as raw material in this study. The mean particle size of the $TiC_{0.3}N_{0.7}$ was about 40 µm. Aluminum powder (Sigma-Aldrich, CAS 7429-90-5) was used as a catalyst to speed up the reaction between hydrochloric acid (Sigma-Aldrich, CAS 7647-01-0) and potassium permanganate (Sigma-Aldrich, CAS 7722-64-7) to generate significant amounts of chlorine gas. Details of the thermodynamics calculation and experimental use of aluminum metal powder is given elsewhere [7]. Hydrochloric acid was used as scrubber to trap the $TiCl_4$ gas produced from the chlorination process. The solubility limit of $TiCl_4$ in 37 wt.% HCl at 20°C was [11] .The studied parameters were chlorination temperature, chlorination time and nitrogen (N_2) gas flow rate. The pellets were prepared using Specac Hand Press Machine and chlorinated at different chlorination temperatures (400-500 °C), chlorination time of 1-3 hours, and the N_2 gas flow rate of 200-300 ml.min^{-1}). A fully randomized factorial design with different parameters including temperature, chlorination time and flow rate of nitrogen gas (N_2) was applied to evaluate and optimize the chlorination process parameters. The design matrix for this study is presented in Table 1.

Table 1. Design matrix of chlorination of TICN.

Factor	**Unit**	**Level**		
		Low	**Medium**	**High**
Temperature	°C	400	450	500
Soaking Time	Minutes	30	45	60
Flowrate of N_2	cc/min	200	250	300

3. Result and Discussion

The highest weight loss was obtained for the sample chlorinated at 500 °C for 60 minutes when the flow rate of N_2 gas was 200 ml/min. The weight lost for this sample was about 79.52 wt.% obtained for experimental std #4 designated in table 1. Similarly, highest extent of chlorination of titanium (X_{Ti}) obtained was about 62.35% for experimental std #4 based on the ICP-OES data. Table 2 shows the weight reduction (X_R) and extent of chlorination (X_{Ti}) of the chlorinated samples at different conditions, As increase in temperature causes the gaseous reactant to possess more thermokinetic energy, hence higher extent of chlorination of $TiC_{0.7}N_{0.3}$.

Table 2. The weight reduction and extent of chlorination of $TiC_{0.7}N_{0.3}$

STD	**1**	**2**	**3**	**4**	**5**	**6**	**7**	**8**	**9**	**10**	**11**
Weight lost*, (%)	63.90	69.33	74.88	79.52	9.25	21.59	28.11	45.02	47.82	44.65	52.02
Extent of chlorination, X_{Ti}**(%)	51.79	59.53	58.23	62.35	14.17	12.85	21.20	24.95	47.24	46.43	51.13

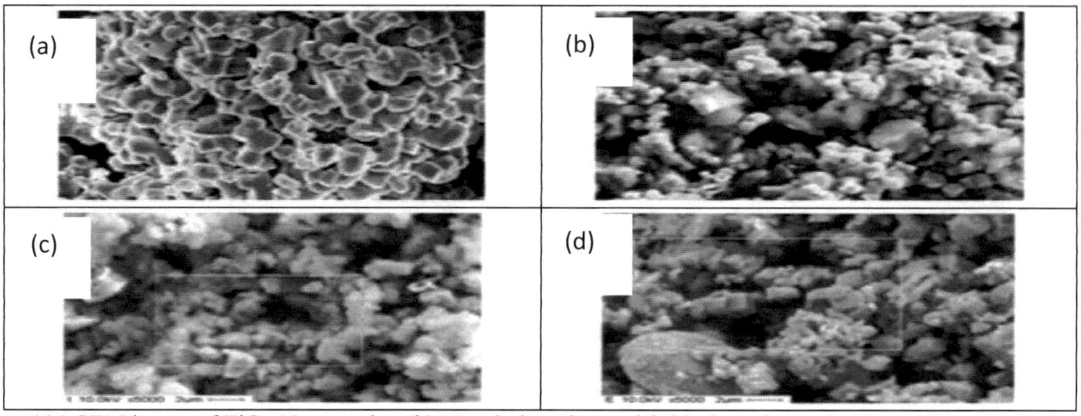

Figure 1(a) SEM image of $TiC_{0.7}N_{0.3}$ powder, (b) Morphology image of chlorinated sample (STD #5) at 400 °C for 30 minutes with flow rate 30 ml/min, (c) Morphology image of chlorinated sample (STD #9) at 450 °C for 45 minutes with flow rate 25 ml/min, (d) Morphology image of chlorinated sample (STD#4) at 500 °C for 60 minutes with flow rate 20 ml/min

Table 3. The elemental analysis by Energy Dispersive X-ray (EDX)

Experimental Std Run	**Elements**					**Weight lost, (%)**
	Ti (wt. %)	Cl (wt. %)	O (wt. %)	C (wt. %)	N (wt. %)	
STD#5	73.15	03.21	00.00	15.40	08.24	9.26
STD#9	12.16	18.17	10.42	39.84	19.41	47.83
STD#4	65.89	02.78	07.15	13.95	10.23	79.52

SEM / EDX analysis provides more detailed insights into the morphological changes in the chlorinated samples. The samples with lowest (STD #5), middle (STD #9) and highest (STD #4) weight lost were analyzed using SEM/EDX. The elemental analysis obtained is given in Table 3 above. It can be seen that with increasing concentration of detected surface chlorine atoms via EDX, the concentration of detected surface oxygen was also increasing. This implied that the $TiCl_4$ residue was trapped in the mesoporous product after chlorination. As the surface was exposed slowly to ambient conditions, it was being hydrolyzed, forming oxides of titanium by reaction in Eq. (2).

$$TiCl_4 + 2H_2O \rightarrow TiO_2 + 4HCl, \quad \Delta G° = -18.13 - 0.05T \text{ (kJ/mole)} \qquad (2)$$

The surface morphology of the chlorinated samples were shown in Figure 1. From these micrographs, it can be concluded that as the weight loss increased, the size of the pores being generated was also increasing. XRD analysis of the residue was used to determine the phases present in the samples after chlorination. The sample designated experimental std #3 showed highest weight loss, which indicated almost all of titanium being removed. Other samples showed similar trends, that with increasing weight loss, the $TiC_{0.7}N_{0.3}$ (ICSD 98-006-3229) XRD peaks intensity was lower. Selected XRD patterns are shown in Figure 2 at different temperature and N_2 gas flow rate. There were trace amount of $TiO_{0.33}$ (ICSD 98-001-4919) phase and $TiCl_3$ were observed in the residue obtained from the sample with highest extent of chlorination. This results showed that upon chlorination, $TiC_{0.7}N_{0.3}$ is being chlorinated into $TiCl_4$ or $TiCl_{2.3}$ (ICSD 98-000-3839). However, due to the porous nature of the chlorinated residual product, these titanium sub-chlorides were being trapped inside the pores. The $TiC_{0.7}N_{0.3}$ was not fully chlorinated due to this inadequate contact with

Cl_2 gas and with subsequent contact with ambient moisture in air, it will eventually hydrolyzed into oxides of titanium by Eq. (4).

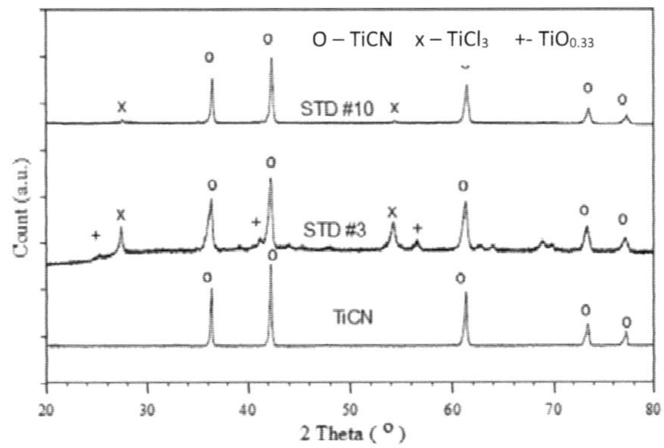

Figure 2. XRD patterns of (a) raw TiCN and the residues obtained from the chlorination (b) at 400 °C for 60 min with 200 ml/min (std #3) and (c) at 450 °C for 45 min with 250 ml/min (std#10)

Statistical design study was carried out using Design Expert v8.0.6 using a factorial model to determine the interaction and the most influential factors in chlorination of $TiC_{0.7}N_{0.3}$. Results obtained are shown Table 2 in the previous section. Then, ANOVA (Analysis of Variance) and model graph analysis was performed in order to interpret the experimental results based on factorial model. The model graphs analysis used in this study included perturbation plots, and 3D cube plot. The results were expressed in the form of percentage of weight lost (wt. %) and X_{Ti} or percentage of titanium being extracted. ANOVA was performed to analyze the results obtained and the respective standard deviation and correlation factor (R^2) for weight loss (%) is 4.4 and 0..9731 respectively, while standard deviation for extent of chlorination was 6.21 and 0.8239 respectively. The predicted value and actual value is shown in Figure 3(a)-(b). It can be seen that in both figures, the experimental data point lies around the predicted linear line and evenly distributed. Hence it can be stated that 95% confidence level can be claimed in experimental data fitting within the range of the parameters investigated.

(a)

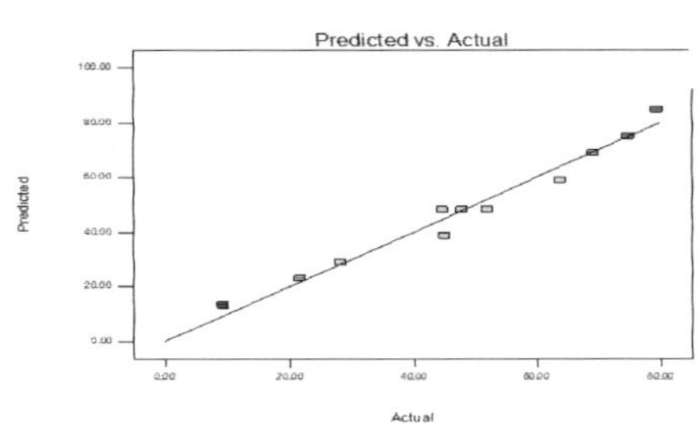

(b)

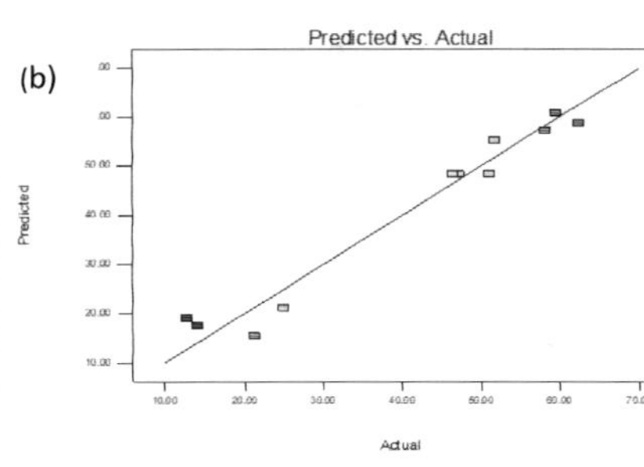

Figure 3. Graph of predicted value against actual value of (a) percentage of weight reduction (wt.%) (b) extent of chlorination, X_{Ti}

The perturbation plots for both responses are shown in Figure 4(a)-(b). From these figures, it can be seen that chlorination temperature and time had a positive impact on the percentage of weight loss, while N_2 flow rate had a negative effect on the weight loss. From the magnitude of the slopes for all 3 factors, N_2 gas flow

IOP Conf. Series: Journal of Physics: Conf. Series **1082** (2018) 012032 doi:10.1088/1742-6596/1082/1/012032

rate showed the highest negative gradient compared to other factors. This implied that it had the most significant effect on the percentage of weight loss and X_{Ti}. It can also be easily explained by the dilution of reactant gas namely Cl_2, which caused the reaction to have slower kinetics. Same trend can be observed for the X_{Ti} data. However, it should be noted that none of the X_{Ti} exceeded 65% which was rather moderate when compared to chlorination of titanium oxycarbonitride [6, 12]. A possible explanation for this was due to the bonding energy of titanium carbonitride ($TiC_{0.7}N_{0.3}$) compared to titanium oxycabonitride ($TiO_xC_yN_z$). In the latter, the electronic bandgap was less than pure TiC and TiN whereas for $TiC_{0.3}N_{0.7}$, the bandgap was about 5.52 eV [13]. Electronic bandgap is an indication of the strength of the chemical bonds between the atoms. The higher bandgap for $TiC_{0.7}N_{0.3}$ made it more unfavorable for the removal for Ti via $TiCl_4$ formation.

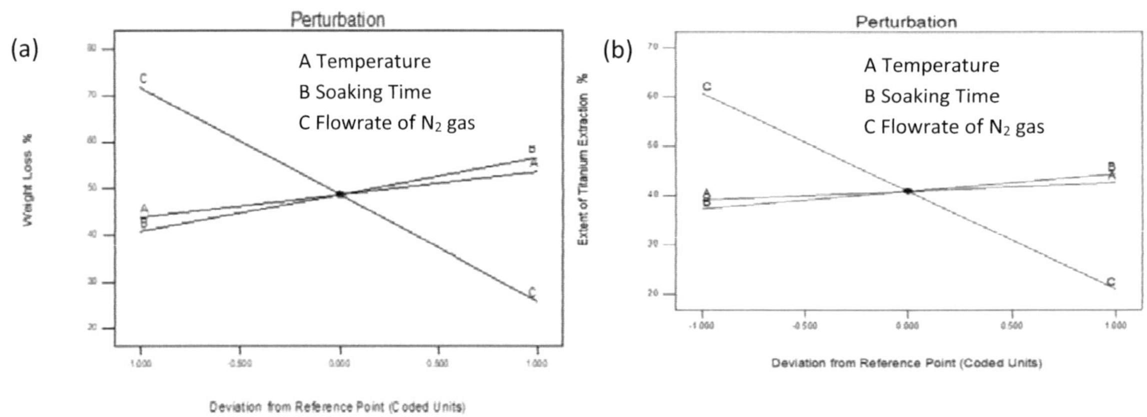

Figure 4. Perturbation plots for (a) weight loss and (b) X_T

4. Conclusion

Titanium Carbonitride ($TiC_{0.7}N_{0.3}$) used in this study was obtained via carbothermal reduction and nitridation of titanium metal with carbon and N_2 gas. It can served as a model for chlorination studies on nitrided ilmenite which contains titanium oxycarbonitride ($TiO_xC_yN_z$). The $TiC_{0.7}N_{0.3}$ was chlorinated to form $TiCl_4$ and C at various reaction temperatures, chlorination times and N_2 gas flow rates. The response of the study was percentage of weight lost and X_{Ti}. Based on the findings of this paper, the most significant parameter was N_2 gas flow rate, followed by temperature and chlorination time. The maximum weight lost obtained was 79.52% whereas the highest X_{Ti} was 62.36%. These results were obtained by at 500 °C at 60 minutes of chlorination time with 20 ml.min^{-1} of N_2 flow rate. From the XRD analysis, samples chlorinated over 45 minutes was contaminated by traces of TiO_2 due to the exposure to ambient moisture. The findings for this work will enable the production of $TiCl_4$ and nanoporous carbon from $TiC_{0.7}N_{0.3}$. It will also serve as good representative for chlorination of titanium oxycarbonitride with low oxygen fraction in nitrided ilmenite.

References

[1] Krar S F and Gill A, *Exploring advanced manufacturing technologies*. 2003: Industrial Press Inc.
[2] Bahl R, Kumar A, Vedawyas M, Patel D 1999 *Applied Physics A: Materials Science & Processing* **69** S643
[3] Jha A and Yoon S 1999 *Journal of materials science* **34** 307
[4] Mostert G J, Bodo R, Rohrmann Roger J, Wedlake Rodney C, Baxter 1993 *Process for the recovery of titanium values*. 1993, Google Patents.
[5] Adipuri A, Li Y, Zhang G, Oleg O 2011 *International Journal of Mineral Processing* **100** 166
[6] Ahmadi, E., Rezan, Sheikh Abdul, Baharun, Norlia, Ramakrishnan, Sivakumar, Fauzi, Ahmad M N Zhang, Guangqing 2017 *Metallurgical and Materials Transactions B* **48** 2354
[7] Najwa I, Ahmadi E, Rahman S A, Fauzi A M N, Rezan S A 2017. *AIP Conference Proceedings* AIP Publishing
[8] Gupta C K and Sathiyamoorthy D 1998 *CRC Press*
[9] Jena P and Brocchi E 1997 *Mineral Processing and Extractive Metullargy Review* **16** 211
[10] Eltefat A, Rezan S A, Hussin H, Baharun N, Ariffin K S, Ramakrishnan S, Fauzi M N Ahmad *Second International Conference on Innovative Advancements in Engineeringand Technology (IAET)* 2016, Jaipur National University: Jaipur National University, Jaipur, Rajasthan, India 5.
[11] Hala J, *Halides, oxyhalides and salts of halogen complexes of titanium, zirconium, hafnium, vanadium, niobium and tantalum*. 1989: Pergamon Press Oxford.
[12] Rezan S A, Adipuri A, Zhang G, and Ostrovski O 2010 *XXV International Mineral Processing Congress (IMPC 2010)* Brisbane, Australia, 2010
[13] Karlsson B, Sundgren J E, and Johansson B O 1982 *Thin Solid Films* **87** 181

IOP Conf. Series: Journal of Physics: Conf. Series **1082** (2018) 012033 doi:10.1088/1742-6596/1082/1/012033

Structural and Electrical Properties of Na_xCOO_2 Thermoelectric Synthesized via Citric-Nitrate Auto-Combustion Reaction

M. S. M. Suan*, M.A. Farhan, K. T. Lau, R. F. Munawar, S. Ismail

Carbon Research Technology Group, Department of Engineering Materials, Faculty of Manufacturing Engineering, Universiti Teknikal Malaysia Melaka, 76100 Durian Tunggal, Melaka, Malaysia

*Email: mohdshahadan@utem.edu.my

Abstract. A series of Na_xCoO_2 thermoelectric has been successfully prepared from nanoparticles powder synthesized via citric-nitrate auto combustion reaction. In this work, the aqueous Co nitrate was mixed with a controlled concentration of aqueous Na nitrate to obtain the mixture solution with Na molar concentration, x from = 0.1 to 0.9. A calculated amount of citric acid and ammonia solutions were added to the mixture as a fuel and ignition agent for the combustion reaction respectively. The drying process at 300 °C on the hotplate with an infrared lamp at the top has turned the mixture solution into precursor gels which yielded fine ashes throughout the auto combustion reaction. The ashes are calcined at 700 °C for 1 h, pelletized by 14.2 MPa of compressive force and sintered at 900 °C for 1 h. The TG/DT analysis of the gel samples show that the decomposition reaction of the Na and Co nitrates occurred in a highly exothermic single-step reaction at around 250 °C to 280 °C and left the desirable mass. The XRD analysis revealed that the calcination process has transformed the mass into pure Na_xCoO_2 phase. The atomic percentage obtained from the EDX indicated that the Na in Na_xCoO_2 yielded with expected x in each sample. The FESEM images of the calcined samples depicted two distinct microstructures appeared in samples with $x \leq 0.3$ which was recognized as Na_xCoO_2 and CoO_2 phases. Whereas for the Na_xCoO_2 samples with $x \geq 0.5$ a single, dense and plate-like microstructure is developed during the calcination process. The pellet sample with $x = 0.7$ exhibits the highest temperature dependence electrical conductivity around 690 $S{\cdot}cm^{-1}$ at 330 K compared to other samples due to effective phonon scattering center provided by Na ions by this concentration.

1. Introduction

In the last 20 years, thermoelectric materials have become significant interest driven by the need for more efficient materials for electronic refrigeration and power generations via direct conversion of waste heat into useful electric energy [1]. Performance of thermoelectric materials indicated by the figure of merit ZT was greatly depended on thermopower or Seebeck effect S, total thermal conductivity κ, and electrical conductivity σ [2]. Active efforts are focusing on increasing the ZT either by lowering the lattice thermal conductivity or most fundamentally by developing new materials with large temperature dependence electrical conductivity. For all known bulk thermoelectric material such as skutterrudites, clathrates, half-Heusler alloys, and complex chalcogenides have considerably high thermopower [3, 4]. However many of these compounds involved un-stable and toxic elements such as tellurium, antimony,

IOP Conf. Series: Journal of Physics: Conf. Series **1082** (2018) 012033 doi:10.1088/1742-6596/1082/1/012033

and germanium. Since the call to develop new thermoelectric materials is strong and will remain so as the technology improves and becomes less expensive, the discovery of ceramic oxide thermoelectric has brought excitement. Ceramic oxide materials have shown promising potential as high temperature thermoelectric materials with better stability, less toxicity and negative environmental effect compared to conventional thermoelectric compounds. The Na_xCoO_2 which consist of two layers; CoO_2 layer and Na layer is one important ceramic oxide thermoelectric material with high electrical conductivity [5, 6]. However, the thermoelectric properties of this oxide are largely varied by its Na concentration because of its roles in adjusting the concentration of electron in CoO_2 layer and acting as phonon scattering centre. It has been reported that the ZT of Na_xCoO_2 with concentration ratio Na/Co = 0.5 and Na/Co = 0.8 at 650 K is 0.02 and 0.4 respectively [7]. Moreover cracking and exfoliation during conventional sintering method are also serious [8]. Therefore the synthesizing method is crucial in order to achieve the accurate concentration of Na in Na_xCoO_2 compound with increased thermoelectric properties. Among the methods, citric-nitrate auto-combustion reaction attracted much attention because of it has successfully been used in synthesizing various ceramic oxides with well-controlled of the concentration includes Al_2O_3, $YBa_2Cu_3O_{7-\delta}$, ZrO_2 and TiO_2 [9-11]. This method is capable to vaporize the impurities hence producing highly pure samples through high exothermic combustion reaction. Thus in this work, Na_xCoO_2 is synthesized using citric-nitrate auto-combustion reaction to yield highly pure samples with controlled Na molar concentrations as formulated, homogeneous and nanostructure powder, and have increased electrical conductivity.

2. Methodology

The aqueous solution of Na nitrate (0.5 M), and Co nitrate (0.5 M) was prepared by dissolving analytical grade and 99.9 % pure of respective reagent powders into distilled water. The determined amount of aqueous $NaNO_3$ was added into aqueous $Co(NO_3)_2$ to obtain the solutions with the mole ratio of Na:Co = x:1 where the x was varied from 0.5 to 0.9. Subsequently, the calculated amount of citric acid was added to the mixture and the pH was adjusted to around 7 by adding ammonia solution. The as-prepared solution was heated on the plate heater at 300 °C with an infrared lamp on top of it to provide a uniform heating. The continuous heating process turned this solution into a gel before combusted and yielded fine ashes. Most of the combustion process involved flame which turned the gel into very fine, spongy and black ashes. The ashes were calcined inside a furnace at 700 °C for 1 h under normal atmosphere and self-cooled to room temperature in the furnace. Calcination process has transformed these highly reactive ashes into a more stable, black and dense powder. Calcined powders were pelletized to the diameter of 10 mm and 2 mm thickness pellet by applying 50 MPa load. The pellets were sintered at 900 °C for 1 h under normal atmosphere. Thermal stability and decomposition characteristic of the precursor gels were determined by thermogravimetric (TG) and differential thermal analysis (DTA). These studies were performed under Argon atmosphere from room temperature to 700 °C by a heating rate of 5 °C min-1 in a NETZSCH thermal analyser unit (409C) and using α-alumina as a reference for DTA. The structure of the samples was characterized using X-ray diffraction (XRD) technique using Ni-filtered Co Kα radiation (Bruker D8-advanced machine). The peak position and intensities were obtained in between 20° and 80° with a velocity of 0.02° s^{-1}. Particles sizes were calculated from the line broadening of the XRD peaks by using the Scherer formula. Lattice structure and parameters of the sample was determined using Ritveld analysis. Field emission scanning electron microscope (FESEM: Zeiss Ultra 40XB) was used to observe microstructure and particle size of the samples. Elemental identification of the samples was determined by EDX technique. The temperature dependence electrical conductivity (σ) of the pellet samples was measured using four-point probe method by cooling the samples from 900 to 300 K.

3. Results and discussion

The TG-DTA curves of the precursor gels are shown in Figure 1. It can be seen that each gel lost its mass with relatively same pattern started with minor lost at the temperature 100 °C - 200 °C, followed by a major drop at temperature 250 – 280 °C and finally constant plateau from 300 °C to 700 °C. As for $x = 0.1$ sample, there is an additional drop of mass occurred at around 400 °C. Figure 1 also showed the DTA curves where the occurrence of an intense exothermic peak at temperatures ranged from 250 °C to 280 °C indicates the main decomposition of precursor gels.

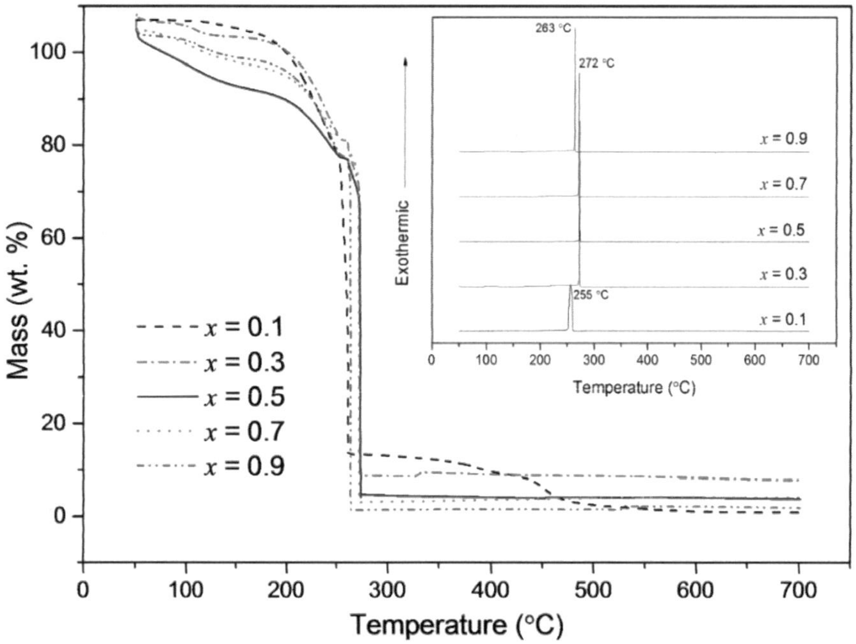

Figure 1. TGA curves of the gel samples with different concentration, x of Na. Inset shows the DTA patterns of the gel samples.

The minor loss of mass at around 100 °C to 200 °C is due to the removal of excesses water and decomposition of citric acid. As the temperature increases, citric acid ($C_6H_8O_7$) began to decompose to aconitic acid ($C_6H_6O_6$) at the temperature 183 °C. Upon heating, aconitic acid is further decomposed to itaconic acid ($C_5H_6O_4$). Polymerization reaction for itaconic acid has occurred when reached its melting temperature of 166 °C which produced itaconic anhydride ($C_5H_4O_3$) with byproduct water and carbon dioxide. Itaconic acid is acted as a fuel once ignited.The combustion reaction of the gel is triggered by the decomposition of ammonia solution and zinc nitrate. Ammonium nitrate, cobalt hydroxide and sodium hydroxide are formed after the combuction reaction. Ammonium nitrate further pyrolysis has ignited the itaconic acid hence triggered the combustion reaction of the gel to yield nascent oxygen, N_2, water, amount of heat energy. The heat energy that released from the reaction is sufficient to remove the water in the sodium hydroxide and cobalt hydroxide and produced Na_xCoO_2.

In the combustion reaction, the citrate ion is acted as the reducing agent while nitrate ion acted as the oxidizing agent. The combustion process is done within a few seconds and emitted a large amount of carbon dioxide and N_2 gases. This kind of combustion mode is known as the rapid deconcentration which occurred for gel samples with $x = 0.3$, 0.5, 0.7 and 0.9. Whereas for the gel with $x = 0.1$, the additional mass loss at 400 °C can be attributed to the deconcentration of itaconic anhydrate at around 390 °C [12] which was formed during the low intensity combustion reaction.

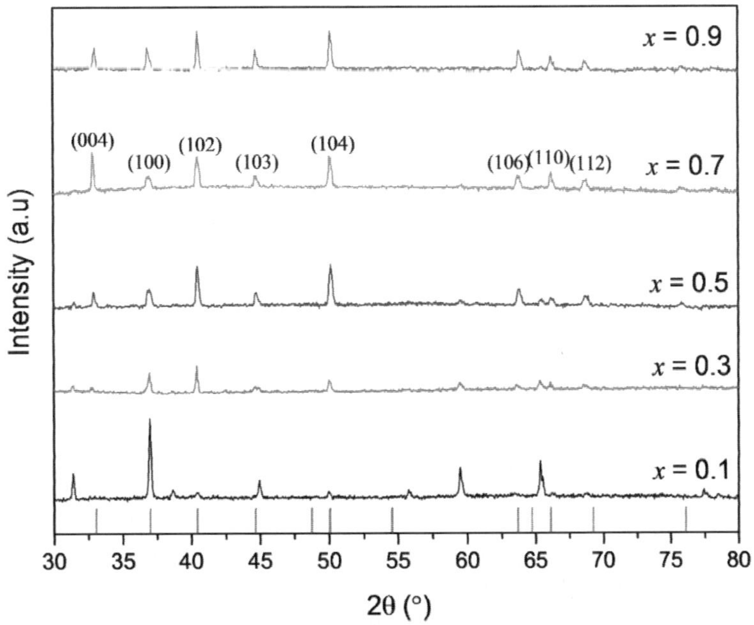

Figure 2. XRD pattern of Na_xCoO_2 calcined powders samples with different concentration, x of Na ions. Bars at bottom indicated Bragg diffraction pattern for $Na_{0.74}CoO_2$.

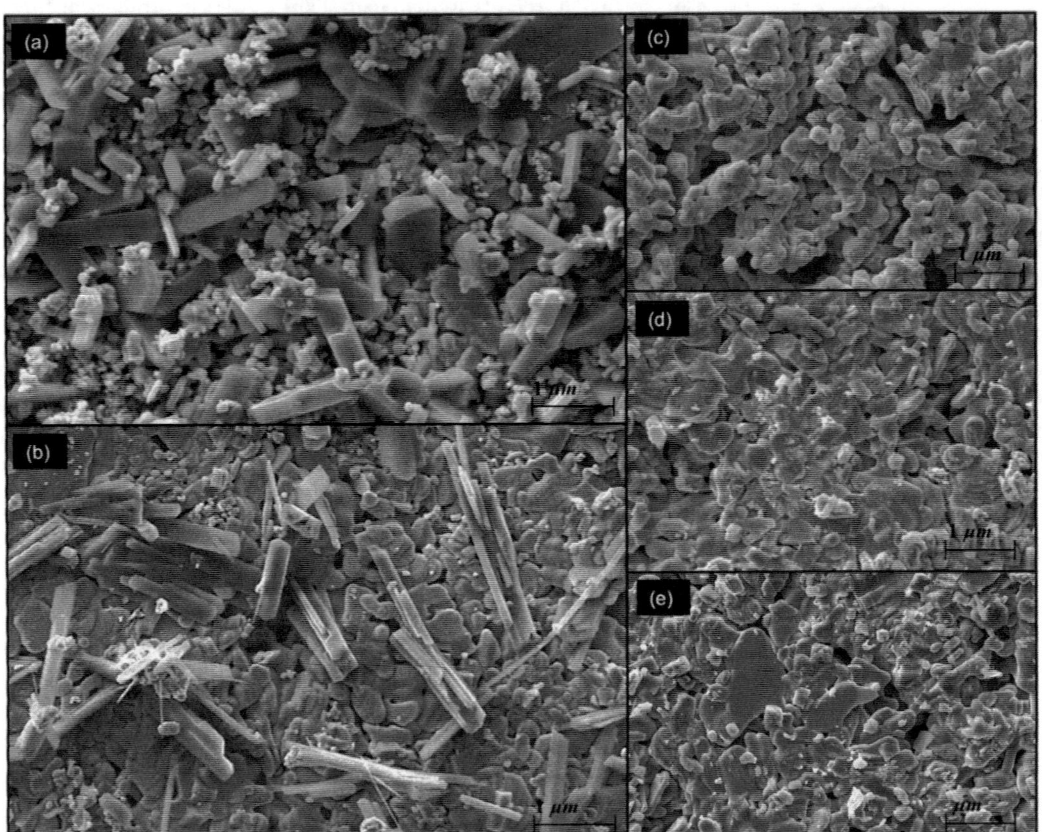

Figure 3. FESEM images of Na_xCoO_2 powders samples with different with different concentration, x of Na ions; (a) $x = 0.1$, (b) $x = 0.3$, (c) $x = 0.5$, (d) $x = 0.7$, and (e) $x = 0.9$.

Figure 2 shows the XRD pattern of Na_xCoO_2 samples with different x after calcination at 700 °C. It can be seen that the XRD peaks of samples with $x > 0.1$ existed at the same Bragg diffraction angles of $Na_{0.74}CoO_2$ peaks indicated by red bars at the bottom of the figure. The most significant peaks for $NaCoO_2$ were attributed from (004), (100), (102), (103) and (104) planes which are in agreement as reported by Ding et al. [13]. Crystallite size of the samples that calculated from the Scherer's equation is in the range from 48.8 nm to 85.2 nm. It can be observed that a strong and narrow peak is due to the excellent crystallinity of Na_xCoO_2. As for the sample with $x = 0.1$, most of the peaks are assigned for CoO_2 phase particularly at 2θ 31.7°, 34.4°, 59.2° and 65.6°. It can be seen from the figure that the significant peaks of CoO_2 and $NaCoO_2$ phases were decreased and increased respectively with the increase in Na concentration. There is no additional peak presence in the patterns other than CoO_2 and Na_xCoO_2 phases indicated that there is high possibility of no impurities existing in the sample. High purity of the samples is attributed to the effectiveness of combustion reaction in burning and vaporizing the un-wanted masses hence the latter was able to yield pure Na_xCoO_2 phase.

Figure 3 shows FESEM images of the calcined Na_xCoO_2 powder samples. It can be observed in Figures 3 (a) and (b) that there are two significant microstructures existed in samples with x = 0.1 and 0.3 respectively. The particle-shaped structure is recognized to be Na_xCoO_2 while the bar structure is representing Co oxide. The increase of Na concentration in the Na_xCoO_2 samples with $x = 0.5$, 0.7 and 0.9 has yielded a single type microstructure as shown in Figures 3 (c), (d) and (e) respectively which is in agreement with the XRD findings. Detail observation on the figure revealed that the Na_xCoO_2 microstructures have been transformed from individual particles into dense, melted, and plate-like microstructure by increased of Na concentrations as described by Baster et al. [14]. The concentrations of Na in the samples is confirmed by the EDX results as listed in Table 1. It is shown that the atomic percent, at. % of Na detected within the samples is almost the same as the expected at. %. The expected atomic mass of Na was calculated based on the concentration of Na in the samples. Thus the resultant atomic mass provides evidence that the Na has merged with CoO_2 structure with suggested formula listed in Table 1.

Table 1. Elemental analysis of powder samples with different concentration, x of Na ions.

Na concentrations	EDX analysis (Atomic %)			Formula	
(x)	Co	Expected Na	Achieved Na	Expected	Achieved
0.1	28.54	2.86	1.78	$Na_{0.1}CoO_2$	$Na_{0.06}CoO_2$
0.3	28.44	8.52	6.58	$Na_{0.3}CoO_2$	$Na_{0.23}CoO_2$
0.5	28.24	14.18	13.94	$Na_{0.5}CoO_2$	$Na_{0.49}CoO_2$
0.7	28.34	19.84	19.26	$Na_{0.7}CoO_2$	$Na_{0.68}CoO_2$
0.9	28.4	25.56	25.06	$Na_{0.9}CoO_2$	$Na_{0.88}CoO_2$

Figure 4 shows the temperature dependence electrical conductivity of Na_xCoO_2 samples. Generally, the electrical conductivity is found to be enhanced with increases of Na molar concentration where the highest value of 690 $S{\cdot}cm^{-1}$ is achieved in the sample with $x = 0.7$ at 330 °C. The electrical conductivity achieved in this work is in the range of the value for the same Na_xCoO_2 concentration reported by Liu et al. [15]. It can be clearly seen that the electrical conductivity in $Na_{0.1}CoO_2$ and $Na_{0.3}CoO_2$ samples are significantly low compared to other samples. This occurrence can be attributed to the small percentage of the yielded Na_xCoO_2 phase in the samples. Once the Na concentration is increased up to $x \geq 0.5$, single phase of Na_xCoO_2 is completely developed hence increase the amount and movement of conducting electron. On top of that, the role of Na as phonon scattering center is the primary mechanism contributed to the increase of electrical conductivity in Na_xCoO_2 samples. The increase of Na ions has lower the lattice vibrations which occurred at higher temperature hence allowed the electrons pass through the lattice structure.

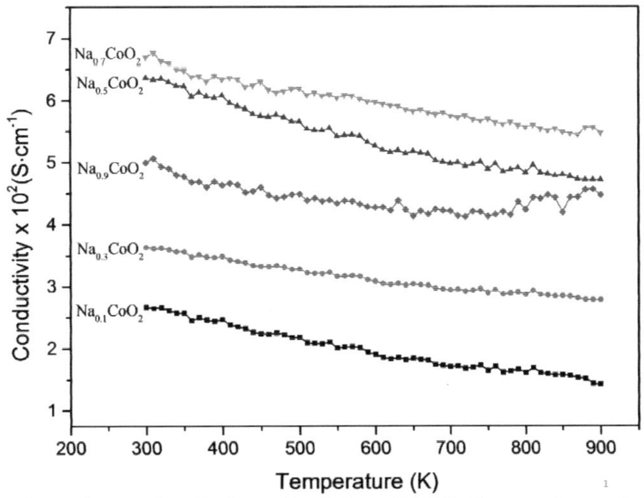

Figure 4. Temperature dependence electrical conductivity of Na_xCoO_2 samples varied by Na concentration, x.

It also can be seen in Figure 4 that the electrical conductivity in the sample with $x = 0.9$ has been reduced. This phenomenon is highly due to the *c* lattice distortion by the Na ions which have increased the travel distance of electron between CoO_2 layers. The increment of phonon scattering center offered by Na ions in this sample resulted in a nearly constant value of the electrical conductivity at increased temperatures.

4. Conclusion

The citric-nitrate auto-combustion reaction is an effective method for synthesizing Na_xCoO_2 compound with well-controlled of Na molar concentration. A single step and high exothermic reaction during auto-combustion have evaporated out the impurities which the latter yielded pure Na_xCoO_2 during calcination. The temperature dependence electrical conductivity of Na_xCoO_2 is maximized in the sample with $x = 0.7$ attributed Na as effective phonon scattering center.

Acknowledgement
Authors are grateful to Universiti Teknikal Malaysia Melaka for the financial support through PJP Research Grant numbered PJP/2016/ FKP-AMC/S01499.

References

[1] Koumoto K, Funahashi R, Guilmeau E, Miyazaki Y, Weidenkaff A, and Wan C 2013 *Journal of the American Ceramic Society*, *96* 1.
[2] Chen Z, Han G, Yang L, and Zou J 2012 *Progress in Natural Science: Materials International*, 22 535.
[3] Tritt T M, and Subramanian M A 2006 *MRS Buletin* 31.
[4] Shiab X, Chenab L and Uherc C *International Materials Reviews* 2016 61 379.
[5] Liu W, Yan X, Chen G, and Ren Z. 2012 *Nano Energy,* 1 42.
[6] Krasutskaya N S, Klyndyuk A I, Evseeva L E, and Tanaeva S *Inorganic Materials* 2016 52 393.
[7] Ito M and Furumoto D 2008 *Journa of Alloys and Compdounds* 450 517.
[8] Schneider C, Schichtel P, Mogwitz B, Beyer A, and Rohnke M 2016 *Physica Stat. Solidi (a)* 213 1284.
[9] Suan M S M and Johan M R 2014 *Materials Research and Innovations* 18 73.
[10] Singh K A, Pathak L C, and Roy S K, 2007 *Ceramics International* 33 1463.
[11] Chandradass J and Kim K H 2009 *Met. Mater. Int.* 15 1039.
[12] Suan M S M and Johan M R 2013 *Physica C* 492 49.
[13] Ding J J, Zhou Y N, Sun Q, and Fu Z W 2013 *Electrochimical Acta*, 87 388.
[14] Baster D, Dybko K, Szot M, Świerczek K, and Molenda J 2014 *Solid State Ionics* 262 206.
[15] Liu P, Chen G, Cui Y, Zhang H, Xiao F, Wang L and Nakano H 2008 *Solid State Ionics* 179 2308.

IOP Conf. Series: Journal of Physics: Conf. Series **1082** (2018) 012034 doi:10.1088/1742-6596/1082/1/012034

Influence of MgO on Sol-Gel Derived SiO_2-CaO-Na_2O-P_2O_5 Bioglass System

S.A.S. Salim[1], H. Mohamad[1], S.N.F.M. Noor[2]

[1]School of Materials and Mineral Resources Engineering, Universiti Sains Malaysia, Nibong Tebal, Penang, Malaysia

[2]Craniofacial and Biomaterials Sciences Cluster, Universiti Sains Malaysia, Advanced Medical and Dental Institute, Penang, Malaysia.

Email: hasmaliza@usm.my

Abstract. Sol-gel derived SiO_2-CaO-Na_2O-P_2O_5 (45S5) bioactive glass (BG) is well known for its enhanced bioactivity properties towards bone regeneration. The synthesis of 45S5 and the addition of MgO (1, 2, 3 and 5 mole percentages) in the 45S5 system was prepared using sol-gel method. The prepared bioactive glass powder was immersed in simulated body fluid (SBF) at room temperature for 7 days. The effect of MgO on the BG structure and bioactivity were characterized through X-ray diffraction (XRD), Fourier transform infrared spectroscopy (FTIR) and Scanning Electron Microscopy (SEM). The results showed that increasing MgO concentration within the 45S5 influenced the glass structural phases formation. In addition, FTIR proved the presence of Si-O bending and Si-O stretching as the main transmission bonds. The addition of MgO affected the apatite layer formation on the glass surface following SBF bioactivity test.

1. Introduction

Bioactive glass (BG) 45S5 was first invented by Professor Larry Hench in 1969 composed of 45% SiO_2, 24.5% CaO, 24.5% Na_2O, 6% P_2O_5 which are able to bond to bone *via* the formation of apatite layer [1]. Bioglass® 45S5 possessed unique properties such as biocompatibility, outstanding bioactive behaviour and regenerative characteristic which greatly facilitate the application in bone regeneration. The interest of 45S5 is geared towards their bioactivity with the surrounding tissues. Bioactivity reflects the material property and it depends on the environment and the solution used for *in vitro* test. The common methods to produce bioactive glass materials are melt-quenching and sol-gel techniques. BG fabricated *via* sol-gel derived have many advantages compared to melt quench including lower fabrication temperature, better controlled homogeneity, larger surface area which increases dissolution rate and hence its bioactivity [2].

The bone-bonding ability of BG occurs through the formation of a biological apatite layer at the glass surface. This layer is observed after immersion in simulated body fluid (SBF), and during *in vivo* experiments. Apatite layer promotes the adhesion of tissues and permits an intimate bone-bonding with the implants. Upon BG contact with SBF, a silica hydrogel layer formation is generated on the

BG surface. The formation of this silica rich layer is related to the dissolution of the glass surface. This layer allows the subsequent crystallization of the apatite-like phase at the glass interface. However, the dissolution kinetics of glass in the biological environment, and consequently the bioactivity of the upper thin layer depend on many variables such as chemical composition, glass structure and interaction with surrounding medium [3]. Bioactivity of glass provides the information regarding the growth of apatite layer, whereas structural and dissolution rate properties appraise the strength of BG. To improve the bioactivity of glass system, a few types of alkaline earth element can be added into the BG composition. Magnesium addition into BG by substituting with calcium may provide excellent bioactivity over time. Hence, magnesium may influence hard tissue mineralization by binding strongly to phosphate through the formation of hydroxyapatite [5], however, contribution of magnesium to the BG properties can act as network modifier and network former [6].

The purpose of this work is to provide information on physical and chemical behavior of sol-gel pure and doped BG 45S5 with MgO (1, 2, 3 and 5 mol.%) and subsequently evaluate their chemical reactivity and bioactivity. Magnesium contribution to the properties of the apatite-like phase requires a broad investigation of doped Mg-based bioactive glass.

2. Experimental study

The 45S5 BG composition consisted of 46.13% SiO_2, 26.91% CaO, 24.35% Na_2O and 2.60% P_2O_5 (in mole percentages, mol.%) and addition of MgO (1, 2, 3 and 5 mol.%) in the BG 45S5 system was prepared using sol-gel technique. Initially, tetraethylorthosilicate (TEOS) was added into a mixture of nitric acid and deionized water, followed by stirring for 60 minutes to allow TEOS hydrolysis. Then, triethylphosphate (TEP) was added into the solution, followed by the addition of calcium nitrate and sodium nitrate at 45 minutes interval. Then, $Mg(NO)_3.6H_2O$ at different compositions (1, 2, 3 and 5 mol.%) was added into the glass mixture using the similar procedure at 45 minutes interval. The obtained solution was stirred overnight to allow completion of hydrolysis. The solution was left for 2 days at 70°C to allow gelation to occur. Gel obtained was then dry for 2 days at 120°C to ensure complete liquid vaporization within the gel. The dried gel was sintered at 700°C for 1 hour with 10°C per minute heating rate. The as-synthesized BG powder was grinded by using agate mortar to avoid agglomeration and stored inside a container in a desiccator prior to test. The BG powder was subjected to FTIR, XRD and SEM analyses. For *in vitro* SBF testing, the SBF solution were prepared by dissolving of NaCl, $NaHCO_3$, KCl, $K_2HPO_4 \cdot 3H_2O$, $MgCl_2 \cdot 6H_2O$, HCl, $CaCl_2$ and Na_2SO_4 following Kokubo's method [7]. In each experiment, the BG samples were removed from SBF after 7 days. Then, the samples were filtered and dried at 70°C and characterized using XRD, FTIR and SEM.

3. Results and discussion

3.1 Fourier transform infrared spectroscopy (FTIR)

Figure 1 shows the BG 45S5 and BG doped MgO (1, 2, 3 and 5 mol.%) infrared spectra in the mid-range of wavelength from 400 cm^{-1} to 2200 cm^{-1} before and after soaking in SBF for 7 days. The transmission spectra of all samples show numerous peaks related to crystalline and amorphous structures. All bioactive glass powders before soaking in SBF solution exhibit vibrational bands around 440 cm^{-1} due to Si-O-Si bending. Whereas for BG 45S5, the main transmission bonds which represent amorphous phase are Si-O-Si bending and Si-O-Si stretching. Characteristic band of sol-gel BG as wide absorption bands occur at 810 cm^{-1}, 850 cm^{-1}, 930 cm^{-1} and a peak at 460 cm^{-1} refers to Si-O-Si bending bond. Besides, there are two small peaks at 580 cm^{-1} and 625 cm^{-1} reflecting a crystalline sodium calcium silicate ($Na_2Ca_2Si_3O_9$). There were also three peaks occurring at 980 cm^{-1}, 1040 cm^{-1} and 1106 cm^{-1} that reflects the Si-O-Si stretching (asymmetric) bands.

IOP Conf. Series: Journal of Physics: Conf. Series **1082** (2018) 012034 doi:10.1088/1742-6596/1082/1/012034

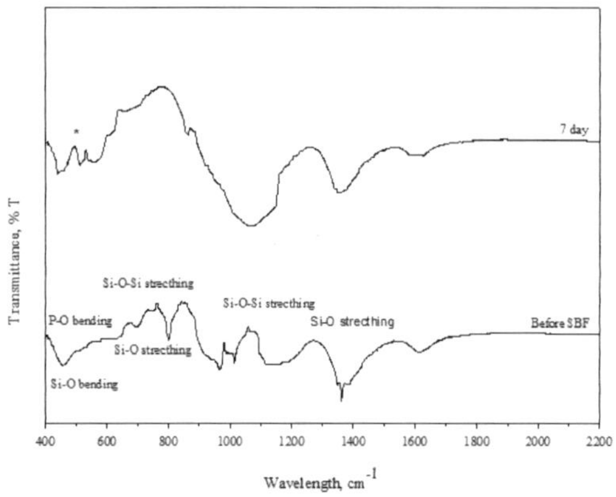

(a) 45S5

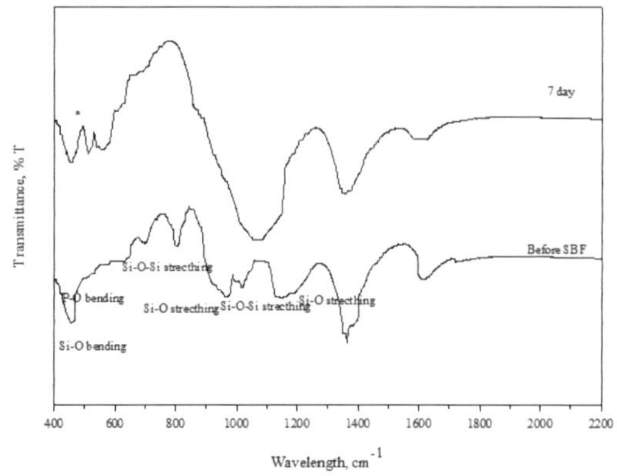

(b) 1 mol. % MgO

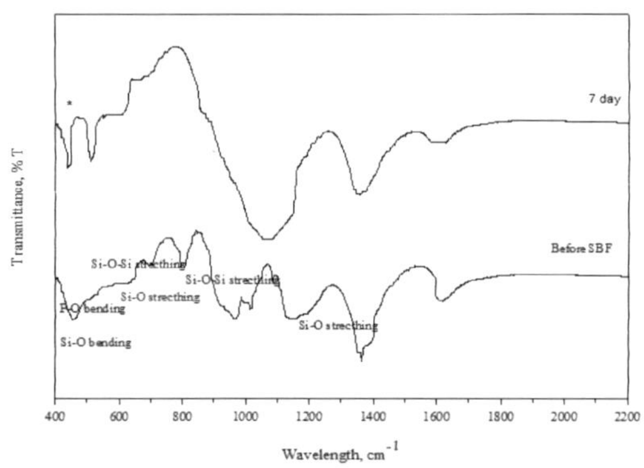

(c) 2 mol. % MgO

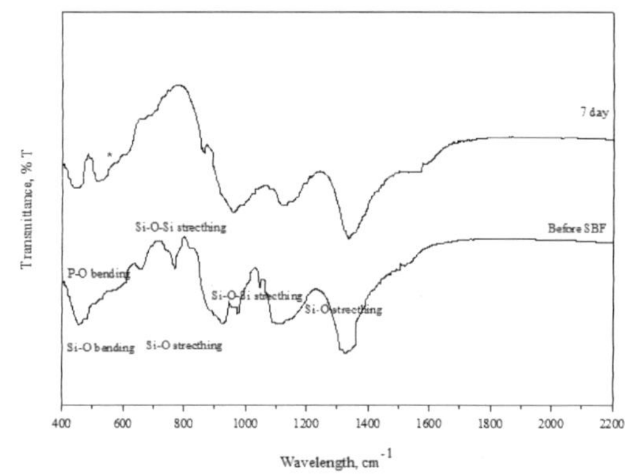

(d) 3 mol. % MgO

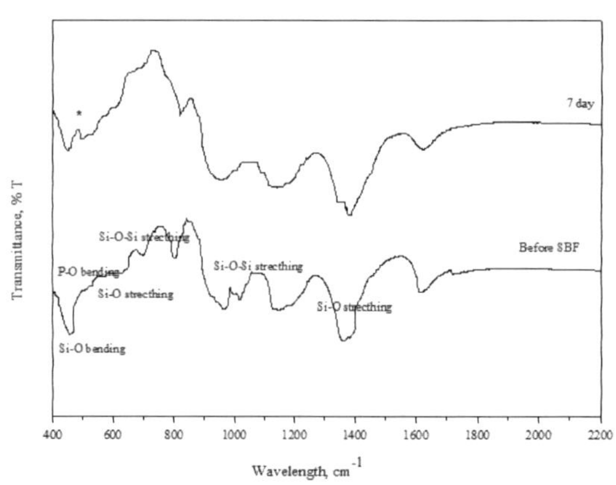

(e) 5 mol. % MgO

Figure 1. FTIR spectra of BG 45S5 and BG doped MgO (1, 2 ,3 and 5 mol.%) before and after soaking in SBF for 7 days.

The FTIR spectra of BG 45S5 after soaking in SBF solution for 7 days reveal Si-O-Si tetrahedral (700-800 cm^{-1}) which indicates the formation of the silica-rich layer and presence of P-O bending peak at 500 cm^{-1} indicating the formation of an amorphous calcium phosphate surface layer.

For BG doped MgO (1, 2, and 3 mol.%) after soaking in SBF solution for 7 days immersion, a band located at 510 cm^{-1} are observable and corresponded with crystalline calcium phosphate. However, this band is not clearly observed in BG doped MgO at 5 mol.% MgO at 7 days. The precipitation of pure hydroxyapatite in SBF is likely less to occur since the surrounding solution is saturated with respect to carbonated apatite where the orthophosphates are substituted with carbonates in the crystal lattice as reported in previous research [8]. Therefore, the phases formed at the surface of the bioactive glass samples were further confirmed by X-ray diffraction.

3.2 X-ray diffraction (XRD)

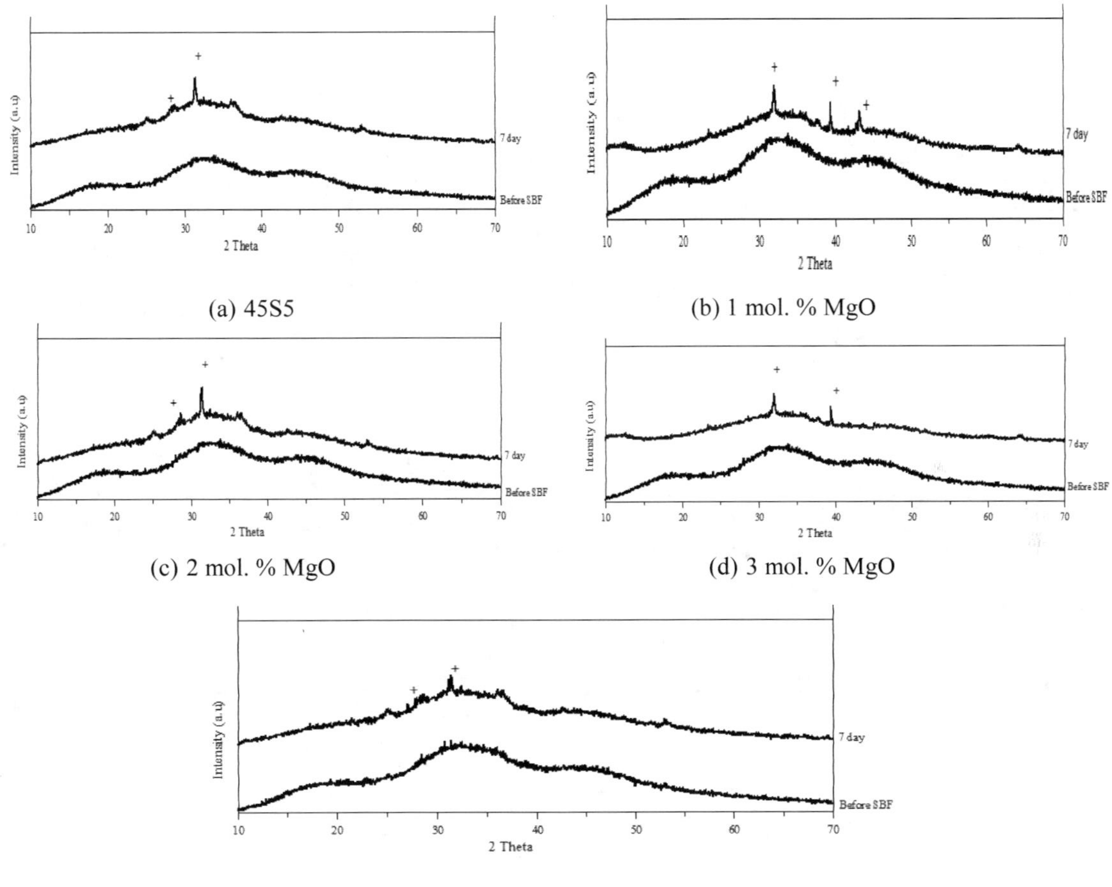

(a) 45S5

(b) 1 mol. % MgO

(c) 2 mol. % MgO

(d) 3 mol. % MgO

(e) 5 mol. % MgO

Figure 2. XRD pattern of BG 45S5 and BG doped MgO (1, 2 ,3 and 5 mol.%) before and after soaking in SBF for 7 days. (+ hydroxy- apatite)

The XRD pattern of the bioactive glass before and after soaking in SBF for 7 days are shown that the pattern are amorphous state indicated the glassy structure. The BG 45S5 apatite phase is detected at 25.7 and 31.9 diffraction peaks following SBF immersion for 7 days. For BG doped MgO (1, 2, and 3 mol.%), the intensity of apatite increases with the increasing addition of MgO since the 1 mol MgO replaces 1 mol CaO, hence, Mg is substituted for Ca in biological apatite. However, the intensity of apatite at 5 mol.% MgO substitution showed less crystalline apatite formation. The retardation of the apatite formation could be attributed by the basis of apatite layer formation mechanism proposed by Hench and co-workers where the initial stage for hydroxyapatite formation on bioactive glass surface was delayed due to the rapid ion exchange between Ca^{2+} and Mg^{2+}as shown by the existence of Si-O-Ca-O-Si and Si-O-Mg-O-Si, non-bridging oxygen bond, H^+ or H_3O^+ ions in SBF solution. Hence, this indicates that the rapid ion exchange corresponds to higher bioactivity of the BG. Consequently, the higher Mg-O bond energy compared to Ca-O resulting in glass containing more MgO content tend to be less reactive in solution and the rate of ions exchange might be slower. Thus, this would result in lower apatite formation on the surface of BG doped with higher MgO content [9].

3.3 Scanning Electron Microscopic (SEM)

Figure 3 shows the SEM analysis for BG 45S5 and BG doped MgO (1, 2, and 3 mol.%) before and after soaking in SBF for 7 days. Based on the figure, before soaking in SBF solution, all BG showed fine particles with a few agglomerations probably due to high water absorption ability of the BG.

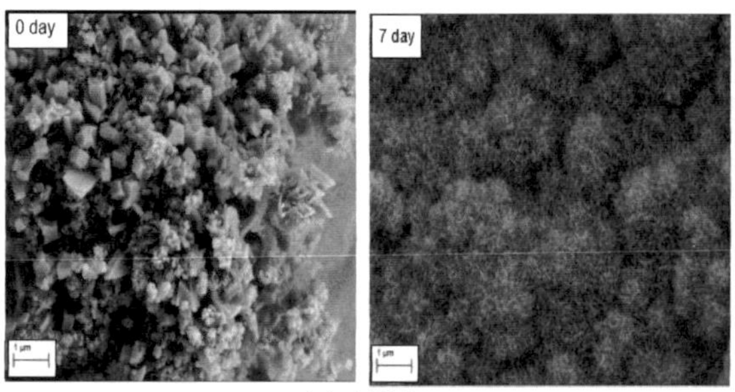

(a) 45S5

(b) 1 mol. % MgO

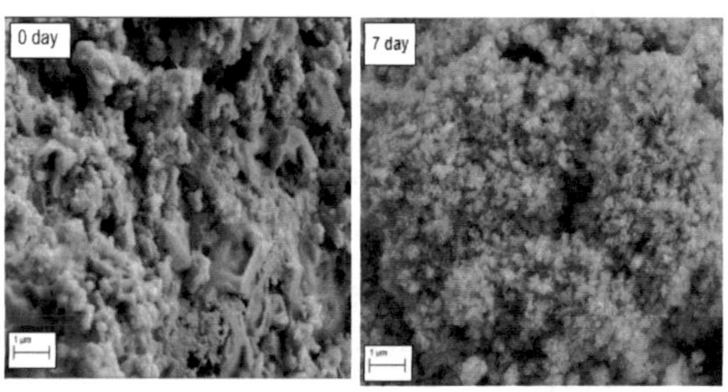

(c) 2 mol. % MgO

(d) 3 mol. % MgO

(e) 5 mol. % MgO

Figure 3. SEM microstructure and EDX for 45S5 and (1,2,3 and 5) mole % MgO before soaking in SBF after soaking in SBF for 7 days.

doi:10.1088/1742-6596/1082/1/012034

After soaking in SBF for 7 days, BG doped MgO showed that the tiny speherical crystalite indicating the homogeneous crystallization behaviours. Such similar behaviour can also be detected on BG particles with MgO addition (1 to 3 mol.%). However, for BG doped with 5 mol.% MgO, the dendritic crystal instead of homogenous spherical crystallite were observed. With even more MgO addition, surface was partially covered with less amount of HA. The growth of HA for BG 45S5 can be explain by ionic dissolution of $Ca^{2+}$and Na^{+}, while for BG doped MgO, it is explained by ionic dissolution of Ca^{2+}, Mg^{2+} and Na^{+}. It is notable that Mg^{2+} has larger ionic field strength (4.73Å) than Ca^{2+} (2.04 Å), indicating that the glass modifying role of Ca is larger than Mg. Therefore, the glass network become tighter when MgO was incorporated into the BG structure [10].

4. Conclusion

45S5 bioactive glass and the addition of MgO (1, 2, 3 and 5 mol.%) in the 45S5 system was synthesized using sol-gel route. The FTIR proved that the main transmission bond was crystalline structure (Si-O-Si bending and Si-O-Si stretching). Whereas the phase found in all glass was calcium phosphate. For SBF *in vitro* study, all glasses exhibited apatite layer growth at their surface after immersion in SBF at 7 days. The incorporation of MgO especially with higher content (5 mol.%) tend to slow down the formation rate of apatite layer formation on glass surface. The influence of MgO on glass properties are explained in terms of internal structure. The obtain result proved that the 45S5 bioglass with addition of MgO in BG showed potential for application in wide range of hard and soft tissue regeneration.

References

[1] Kai Z, and Aldo R B 2017 *Advanced in Colloid and Interface Science* 3018
[2] Hamidreza P, and John N 2013 *Journal of the American Ceramic Society* 96 1643
[3] Vikas A, Singh K J, and Kulwinder K 2014 *Journal of Non-Crystalline Solids* 88 406
[4] Siti N F, Yusuf M M, Abdullah T K and Zuhailawati H 2016 *Procedia Chemistry* 75 82
[5] Devis B, Antonella S, Roberto S, Alenxandre A, Luigi C and Valeria C 2016 *Journal of Materials Science and Engineering C* 72 575
[6] Ma J, Chen C Z, Wang D.G, Jiao Y and Shi J Z 2010 *Journal Colloids and Surface B: Biointerface* 88 81
[7] Anthony L B M, Taek B K, Esther M. V, Kathryn G, Richard K B, Delbert E D, Alexander H, Aldo B,Yong K, Chikara O, Tadashi K, Akiyoshi O, Maria V, Daniel A, Leandro F, Antonio J S, Alexandra V T, Yuliya V, Rui M A, Marta M, Chiara V, Enrica V, Wolfram H, and Julian R J 2015 *Journal Mater Sci: Mater Med* 26 115
[8] Ma J, Chen C Z, Wang D G and Hu J H 2011 *Materials Letters* 65 130
[9] Anthony L B, Taek B, Esther M, Kathryn G, Richard K, Delbert E, Alexander H, Aldo R, Yong K, Chikara O, Tadashi K, Maria V, Yuliya V, Rui M, Marta M, Chiara V, Enrica V, Wolfram H and Julian R *Journal Material Science* 26 115
[10] Ma J, Chen C Z, Wang D.G, Shao X, Wang C Z and Zhang H 2012 *Ceramic International* 38 6677

IOP Conf. Series: Journal of Physics: Conf. Series **1082** (2018) 012035 doi:10.1088/1742-6596/1082/1/012035

Synthesis of LiV_3O_8 Materials using Oxalic Acid as Chelating Agent

N.N.M. Zorkipli[1], N.H.M. Kaus[2], A.A. Mohamad[1]

[1]Universiti Sains Malaysia, Engineering Campus, Penang, Malaysia

[2]Universiti Sains Malaysia, Main Campus, Penang, Malaysia

Email: aam@usm.my

Abstract. LiV_3O_8 was synthesised via the sol-gel method using oxalic acid as the chelating agent. Synthesis was carried out at different calcination temperatures (400 °C, 500 °C and before calcination). The structural analysis determined that the monoclinic structure of pure LiV_3O_8 formed better at a calcination temperature of 500 °C. Morphological analysis demonstrated the existence of the LiV_3O_8 rod structure.

1. Introduction

Lithium vanadate oxide (LiV_3O_8) has drawn considerable attention for use as electrode material, and offers great promise for increasing the energy density of aqueous lithium-ion batteries (LIBs) [1]. The LiV_3O_8 material is beneficial due to its high capacity, easy preparation and excellent stability in air [2, 3]. The synthesising method plays an important role in producing the best quality LiV_3O_8 materials, and can be produced by various synthesising methods for application in aqueous LIBs [4, 5].

Sol-gel is one of the best synthesis methods to produce electrode materials. The sol–gel process is a relatively easy and economical way to produce uniform particle distribution which exhibits homogeneous mixing, better crystallinity, and smaller particle size for high quality stoichiometric targeted products [6, 7]. Because LiV_3O_8 is limited by its poor electronic conductivity, oxalic acid has been added as chelating agent to produce a layer of carbon on the LiV_3O_8 rod [8]. The function of the carbon layer is to increase the electronic conductivity of LiV_3O_8. According to other reports, adjusting the quantity of the oxalic acid in the solution can control the morphology and rod size of the powder by preventing agglomeration [2, 9]. At the same time, the calcination temperature also contributes to the effect on the size [4], distribution [4] and morphology [4, 10] of the LiV_3O_8.

This work aims to study the effect of oxalic acid as chelating agent on pure LiV_3O_8. Several important material characterisations were examined to investigate the quality of the synthesised LiV_3O_8. Structural and morphological analyses are required to determine the synthesis of LiV_3O_8.

IOP Conf. Series: Journal of Physics: Conf. Series **1082** (2018) 012035 doi:10.1088/1742-6596/1082/1/012035

2. Experimental study

Lithium vanadate oxide (LiV_3O_8) was synthesised by the sol-gel method. The molar ratio of lithium hydroxide mono-hydrate ($LiOH.H_2O$) (≥98.0 %, Sigma Aldrich), ammonium vanadate (NH_4VO_3) (99 %, Sigma Aldrich) and saturated oxalic acid ($C_2H_2O_4$) (≥ 99 %, Sigma Aldrich) in the precursor was 1:3:4. The oxalic acid was added as chelating agent, until the colour of mixture changed from buff to brown. After stirring continually at 80 °C, a homogeneous dark blue was obtained. To obtain the LiV_3O_8 sample, the gel was placed in an oven to dry at 120 °C and then calcined in air atmosphere by using a tube furnace at 500 °C for 5 hours. The phase identification and structural analysis were carried out by (XRD, Bruker Advanced X-ray Solutions D8) in the 2θ range of 10° to 80° with monochromatised Cu Kα radiation (λ = 1.5406 Å). A Rietveld refinement was performed on the XRD data by using the Pan Analytical X'pert High Score Plus software to obtain the parameters of the structure. The morphology of the LiV_3O_8 was observed directly by field emission scanning electron microscope (FESEM, Zeiss Supra™, 35VP).

3. Results and discussion

3.1 Structural analysis

The structural analysis was examined by XRD diffraction patterns at different calcination temperatures to determine the phase of the synthesised LiV_3O_8 (Figure 1a). Before calcination, an amorphous phase demonstrated that there was no significant change. At 400 °C calcination temperature, broad peaks were formed and only a few peaks occurred. By comparing this sample with the sample calcined at 500 °C, all reflection peaks of LiV_3O_8 were formed. The main diffraction peaks of (100) and ($\bar{1}11$) were used to calculate the crystallite size using the Scherrer equation. The Bragg reflection peak of LiV_3O_8 was investigated in the crystalline state, and indexed to (ICDD: 98-000-4678). A monoclinic system with P21/m space group was obtained.

This result in the amorphous phase indicated that no LiV_3O_8 existed. The different calcination temperature results at 400 °C and 500 °C demonstrated that different phases occurred between these two calcination temperatures. Several peaks were formed at 400 °C, whereas all reflection peaks were formed at 500 °C. This result proved that lithium vanadate was successfully synthesised at the 500 °C calcination temperature and readily oxidised to LiV_3O_8 in the crystalline phase.

The quantitative XRD analysis (Rietveld refinement method) was carried out to confirm the refined structural parameters and matching between the diffracted and reference patterns of the LiV_3O_8 (Figure 1b, 1c). During the refinement process, goodness of fit (GOF) starts out large when the model is poor, and decreases as the model produces better agreement with the data. After the simultaneous refinement of all the appropriate parameters, a visual fit of the data was obtained [11]. Numerous residual (R) values such as the weighted R profile (R_{wp}) and expected R (R_{exp}) values can be generated from the refinement process. However, the main value always defaults to R_{wp} [12]. At 400 °C, the visual fit of the data was not good, as the peaks did not match each other. The quality indicators were established at: GOF (9.87 %), R_{wp} (19.23 %) and R_p (6.42 %). However, after simultaneous refinement of all appropriate parameters at 500 °C, the visual fit of the data was good, as were the quality indicators: GOF (5.44 %), R_{wp} (14.04 %) and R_p (6.53 %). The value is smaller than the 400 °C sample. The refinement parameters are listed in Table 1.

The structural results indicated that the LiV_3O_8 synthesised at the 500 °C calcination temperature was the purer sample. It achieved an ordered layered monoclinic crystalline structure with high crystallinity and a good degree of refinement. The lattice parameters of the LiV_3O_8 synthesised at 500 °C calcination temperature (a = 6.644 Å, b = 3.599 Å and c = 11.809 Å) were also in good agreement.

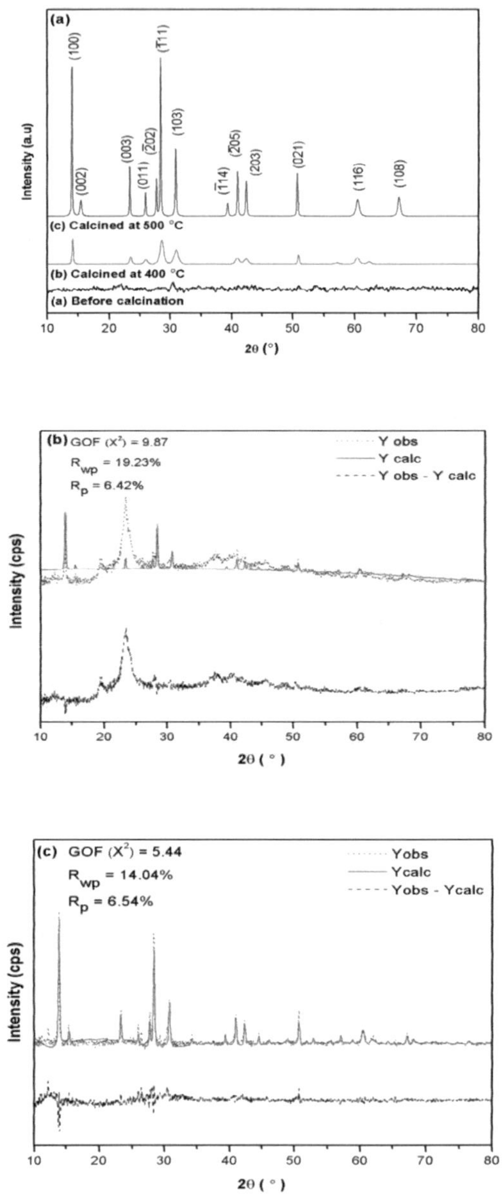

Figure 1. XRD pattern of (a) LiV_3O_8 at different calcination temperatures, (b) observed, calculated and difference profiles from refinement of LiV_3O_8 at 400 °C and (c) 500 °C.

IOP Conf. Series: Journal of Physics: Conf. Series **1082** (2018) 012035 doi:10.1088/1742-6596/1082/1/012035

Table 1. Rietveld refinement parameter of the LiV_3O_8 synthesised at different calcination temperature which possesses a monoclinic crystalline structure and belongs to the P21/m space group .

Parameter	400 °C	500 °C	References (ICDD: 98-000-4678)
a (Å)	6.65308	6.64378	6.6800
b (Å)	3.59539	3.59862	3.600
c (Å)	11.80681	11.80927	11.8377
R_p (%)	6.41784	6.53915	-
R_{wp} (%)	19.23263	14.04188	-
GOF (χ°)	9.86976	5.43557	-
α (°)	90.000	90.000	90.000
β (°)	104.63090	104.49480	104.66300
γ (°)	90.000	90.000	90.000
Crystallite size (nm)	59.61	48.21	-

3.2 Morphological analysis

The morphological analysis was executed by FESEM. The morphologies of the samples before calcination, after calcination at 400 °C, and after calcination at 500 °C were investigated at magnification (10000x) (Figure 2). Undefined shapes and sizes were observed before calcination (Figure 2a). The sample that was calcined at 400 °C demonstrated the beginnings of the rod shape formation with thin rod shapes forming (Figure 2b). The rod shape was clearly observed in the sample that was calcined at 500 °C (Figure 2c).

Based on the morphological result, the sample examined before a calcination process demonstrated undefined shapes due to the lack of heat treatment [13]. Several thin rod structures began to grow at 400 °C after the heat treatment was applied. However, these rods were not clearly defined because the applied calcination temperature was still low. The rod structure was successfully produced at 500 °C, due to the suitable selection of calcination temperature. The distributions of the rods' size demonstrated uniform growth with an approximate diameter of about 590 nm. Hence, 500 °C is chosen as the better calcination temperature.

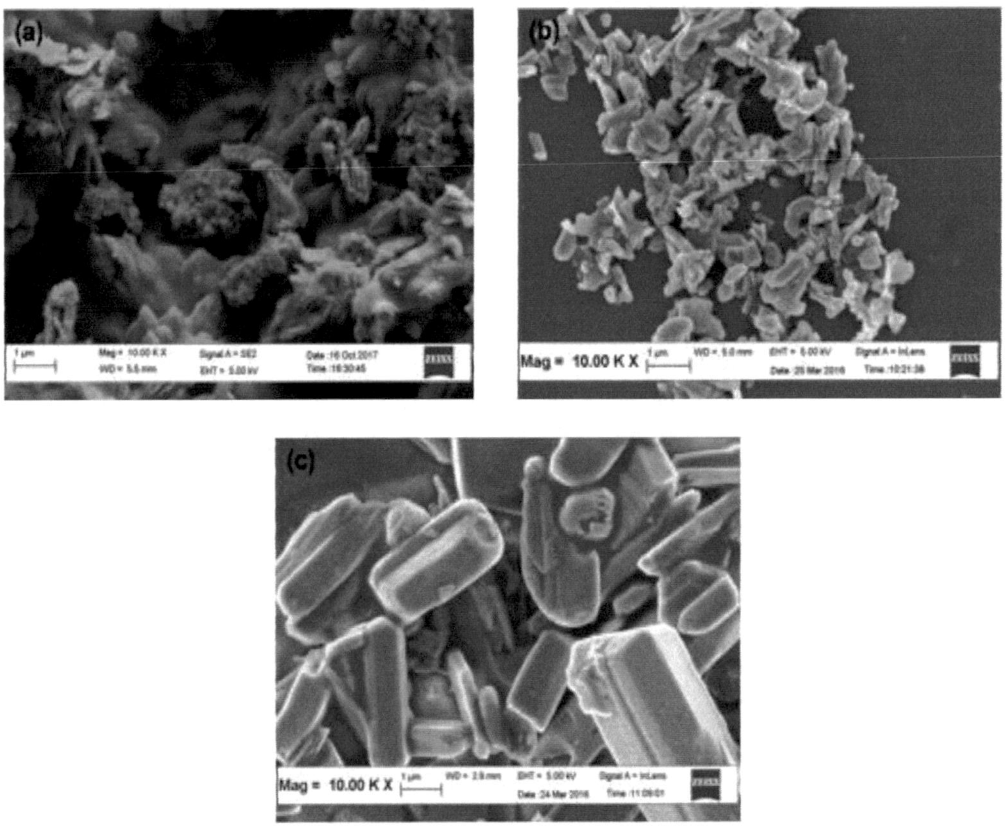

Figure 2. SEM micrograph at different calcination temperature of LiV_3O_8 at (a) before calcination (b) 400 °C and (c) 500 °C

IOP Conf. Series: Journal of Physics: Conf. Series **1082** (2018) 012035 doi:10.1088/1742-6596/1082/1/012035

4. Conclusion

LiV_3O_8 materials were successfully synthesised by the sol-gel method using oxalic acid as chelating agent. Based on a structural analysis, the LiV_3O_8 formed a monoclinic structure due to the use of oxalic acid. The result indicated that pure LiV_3O_8 was obtained at 500 °C calcination temperature. The morphological analysis demonstrated the uniformly grown distribution of rod shapes formed at 500 °C. Hence, 500 °C was chosen as the better calcination temperature compared with 400 °C for LiV_3O_8 materials using oxalic acid as chelating agent.

References

[1] Liu L, Jiao L, Sun J *2009 Journal of Alloys and Compounds* 471 352
[2] Dunqiang Wang LC, Jianfeng H, Jianpeng W 2013 *Ceramic International* 39 3759
[3] Caballero A, Morales J,Vargas O A 2010 *Journal of Power Sources* 195 4318
[4] Liu J, Yi L, Liu L,Peng P 2015 *Materials Chemistry and Physics* 161 211
[5] Kohler J, Makihara H, Uegaito H, Inoue H,Toki M 2000 *Eletrochemica Acta* 46 59
[6] Brinker C J, S G W Sol–Gel Science: The Physics and Chemistry of Sol–Gel Processing. . *Academic Press* 1990.
[7] Fu L, Liu H, Li C *2005 Progress in Materials Science* 50 881
[8] Hao Y J, Lai Q Y, Lu J Z 2006 *Journal of Power Sources* 158 1358
[9] Huang Z, Cao L, Chen L 2016 *The Journal of Physical Chemistry C* 120 3242
[10] Si Y C, Jiao L F, Yuan H T, Li H X,Wang Y M M 2009 *Journal of Alloys and Compounds*486 400
[11] Toby B H R factors in Rietveld analysis: How good is good enough? Powder Diffraction 2012 21 67
[12] Mccusker L B, Dreele R B V, Cox D E, Louer D,Scardi P 1999 *JApplCryst* 1999 36
[13] Almeida MS , Santos M A B, Gonçalves Rd F 2016 *Materials Research* 19 1155

IOP Conf. Series: Journal of Physics: Conf. Series **1082** (2018) 012036 doi:10.1088/1742-6596/1082/1/012036

Modelling of Carbothermal Reduction of Tio2-Fe2o3 Composite with Compressed Natural Gas Using Design of Experiment

Nur Atiqqah Mohd. Moslim[1] , Sivakumar Ramakrishnan[1]*, Sheikh Abdul Rezan, Eltefat Ahmadi[1], Najwa Ibrahim[1], Norlia Baharun[1], M.N. Ahmad Fauzi[1], Abdul Rahman Mohamed [2]

[1] School of Materials & Mineral Resources Engineering, Universiti Sains Malaysia, Engineering Campus, Nibong Tebal, Penang, Malaysia

[2] School of Chemical Engineering, Engineering Campus Universiti Sains Malaysia, Engineering Campus, Nibong Tebal, Penang, Malaysia

*Email: srsivakumar@usm.my

Abstract. An alternative way to process titanium minerals was investigated via carbothermal reduction/nitridation (CTRN) by yielding titanium oxycarbonitride ($TiO_xC_yN_z$). Synthetic anatase (TiO_2) and hematite (Fe_2O_3) were mixed together with graphite/CNT mixture to study the reduction kinetics. The reduction process was carried out at 1150 – 1250 °C for 1 to 4 hours in a gas mixture of hydrogen, nitrogen and compressed natural gas. The carbon to reducible oxygen molar ratio ($C:O_{red}$) was varied in this experiment from 0.5 to 1.0. The extents of reduction and nitridation with the amount of free carbon deposited were studied using 2^3 factorial design to define the optimum parameters in producing $TiO_xC_yN_z$. From Design Of Experiment (DOE) analysis, the most influential factor contributing to the extent of reduction was reduction time, followed by carbon to reducible oxygen molar ratio and reduction temperature subsequently. The highest extent of reduction was 92.68% which was obtained at 1150 °C for 4 hours using sample with $C:O_{red}$ of 1.

1. Introduction

Titania or titanium dioxide (TiO_2) exists in a number of crystalline forms; the most important of which are anatase, rutile and brookite. In this study, anatase, TiO_2 being used together with hematite, Fe_2O_3 to form TiO_2-Fe_2O_3 composite or synthetic ilmenite. The composite was reduced in reducing atmosphere to obtain titanium oxycarbonitride ($TiO_xC_yN_z$) which contains the phase of titanium oxide (TiO), titanium nitride (TiN), titanium carbide (TiC) and solid iron (Fe). At the very end of the carbothermal reduction of the TiO_2-Fe_2O_3 composite, high density of titanium oxycarbonitride ($TiO_xC_yN_z$) would be expected to be formed. Chlorination process is being carried out to form metal chlorides after carbothermal reduction. The decomposition of natural gas used in the experiment provides H_2 and CO gases for the reduction process (Lobo et al., 2013). One of the factors that would influence the reduction kinetics of TiO_2-Fe_2O_3 composite would be the presence of carbon source either in the form of coal, graphite or carbon nanotube. Design of Experiment (DOE) was used in this study by incorporating 2^3 factorial designed experiment including three center points. A kinetic model for cylindrical pellet sample was created using MATLAB software with respect to the physical and thermodynamic properties of the product and reactant. The reaction is expected to show the effects of the reduction time, reduction temperature and also the $C:O_{red}$ molar ratio contributed throughout the process.

2. Experimental study

Sample was prepared via wet mixing, sonication, drying, grinding and pellet pressing. The reduction process was carried out in H_2-N_2 and compressed natural gas (CNG) atmospheres. H_2 and N_2 gases both occupied the volume of 50% each and allowed to flow from the start to the end of experiment with total flow rate of 1.5 L/min. For CNG, it was purged at reduction temperature for 45 minutes. The flow rate for CNG was 0.15 L/min and occupied 10% of gas volume. Both heating and cooling rates for the furnace were adjusted at 10°C/min. Figure 1 shows the experimental setup of reduction process. This setup is the same for oxidation process except for the absence of CNG, H_2 and N_2 gas inlet because the process was being carried out in air only. The reduction process was carried out at 1150°C and 1250°C for 1 to 4 hours by using the samples with $C:O_{red}$ of 1:1, 0.75:1 and 0.5:1 while the oxidation process was carried out in open-end tube furnace at 850°C for 6 hours.

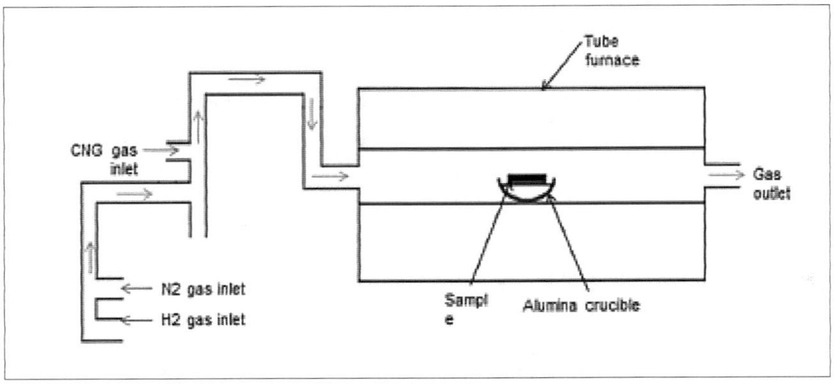

Figure 1. Schematic diagram of reduction process

3. Result and discussion

3.1 Chemical reactions and equilibria

The reactions occured during the reduction process were studied using Outkompu HSC Chemistry software v6.00. The simulation and modelling of the chemical reactions used to predict the reactions that occured at a certain temperature and gas atmosphere. Titanium oxycarbonitride ($TiO_xC_yN_z$) phase existed as a metastable solid solution phase between its boundary counter parts titanium carbide (TiC) and titanium nitride (TiN) [1]. For this reason, only TiC and TiN will take part in the reaction equilibrium of the reduction process. From the overall chemical reaction, natural gas which mainly composed of methane (CH_4) was assumed to be converted into hydrogen and carbon monoxide (CO). Study done by Wullenkord (2010) [6] proved that this argument is valid.

$$2TiO_2 + N_2(g) + 4CH_4(g) = 2TiN + 4CO(g) + 8H_2(g)\ (\Delta G^o = -23.763 \text{ kcal/mole at } 1000°C), \quad (1)$$

$$Fe_2O_3 + 3H_2(g) = 2Fe + 3H_2O(g)\ (\Delta G^o = -1.95\ 7\text{kcal/mole at } 600°C), \quad (2)$$

The simulation proved that the above mentioned reaction occurs below 1000°C (1273 K) and thus agreeing to the actual reaction equation.

3.2 Statistical design of experiment (doe) analysis

Statistical design was performed to find the most and the least effective parameter in the carbothermal reduction of TiO_2-Fe_2O_3 composite powder. The experimental responses were fitted to a factorial model using Design-Expert v8.0.6.1. ANOVA (Analysis of Variance) was also performed to determine the best reduction condition, thus several types of graphs were analyzed. The experimental parameters are shown in Table 1 while the standard deviation and values of R-squared obtained from the responses and ANOVA are given in Table 2. From the value of R-squared in Table 2, it can be concluded that the experimental results are generally agree with the statistical model when the model have high R-squared value with 1 being the highest value.

Table 1. Variable parameter used in Design of Experiment.

	Variable parameter	Unit	Details	
A	Reduction time	Hours	1	5
B	Reduction temperature	°C	1150	1250
C	$C{:}O_{red}$ molar ratio	-	0.5	1.0

Table 2. Standard deviation and R-squared responses from the reduction process.

Response	Standard Deviation	R-Squared
Extent of reduction, X_O	3.87	0.9520
Extent of nitridation, X_N	5.36	0.8060
Carbon deposited, $C_{deposited}$	21.75	0.8310

The relationship between the responses and each of the parameters can be determined using Equations 1, 2 and 3. Negative values in the equation states that the single or a combination of parameters did not contribute much to increase the extent of reduction, extent of nitridation and carbon deposited. For example, the negative values of -5.73C and -1.11AC in Equation 3 indicated that factor C and a combination of factor A and C did not have significant effect on the carbon deposited response.

$$\text{Extent of reduction } X_O = 74.54 + 7.97A - 1.52B + 4.73C - 4.05AB + 1.07AC + 2.30BC + 0.074ABC \quad (1)$$

$$\text{Extent of nitridation } X_N = 43.03 + 2.78A + 2.04B + 3.03C - 3.95AB + 0.70AC - 2.67BC + 0.59ABC \quad (2)$$

$$\text{Carbon deposited} = 45.49 + 22.21A + 8.84B - 5.73C + 16.28AB - 1.11AC - 1.50BC - 0.075ABC \quad (3)$$

Where A is reduction time, B is reduction temperature and C is $C{:}O_{red}$ molar ratio
From Equations 1, 2 and 3 can be observed that the interactions among the three factors can be significant and greater or less than expected from each factor separately [5]. This situation can also be used to explain the influencing factors for carbon deposited where higher carbon to reducible oxygen molar ratio should result in higher amount of carbon deposited

3.3 Effect of reduction temperature and time

Table 3 presents the data for extent of reduction at different reduction times. The results show that the extent of reduction of the samples with 1:1 $C:O_{red}$ increased with reduction time for both temperatures. Extent of reduction was the highest (92.68%) when the reduction process was carried out at 1150°C for 4 hours compared to the other samples reduced in other conditions. At higher temperatures, more carbon was produced from the thermal cracking of the methane gas from CNG which causes high amount of carbon deposition on the surface of the pellets. The excessive amount of carbon deposited would then hinder the reduction process [3].

Table 3. Data for extent of reduction, X_O of TiO_2-Fe_2O_3 composite with 1:1 molar ratio of reducible oxygen.

Temperature, °C	1150		1250	
Reduction time, h	1	4	1	4
Extent of reduction, %	85.84	92.68	69.94	86.29

Longer reduction times allow more reduction of the sample to occur until the reaction completed at some extent. It enables higher amount of hydrogen and carbon to be formed when the reduction time is increased. As carbon forms, reduction extent also increases. The newly formed carbon will aid in the reduction process by acting as reductant [2].

3.4 Effect of carbon to reducible oxygen molar ratio

The effect of carbon to reducible oxygen molar ratio on TiO_2-Fe_2O_3 reduction at 1150°C was carried out in 1 to 4 hours. CNG was purged during the reduction (soaking) time for 45 minutes. Carbon nanotube (CNT) was used as carbon source that function as reductant in the reduction process. Lower extent of reduction obtained when performing the reduction without carbon source. This argument proves to valid from the study by Yin et al. (2016) [7]. From Table 4, it can be seen that higher extent of reduction obtained when using sample of 1:1 carbon to reducible oxygen molar ratio. By adding reductant in the sample, the reduction process can occur faster and more efficiently.

Table 4. Data for extent of reduction with different $C:O_{red}$ molar ratio at 1150 °C.

$C:O_{red}$ molar ratio	0.5:1		1:1	
Reduction time, h	1	4	1	4
Extent of reduction, X_O	63.78	66.64	85.84	92.68

3.5 Phase development in reduction of TiO_2-Fe_2O_3 composite in H_2-N_2-CNG atmosphere

Phase detected by XRD analysis indicates that titanium oxycarbonitride (TiOCN) formed at the end was in solid solution. Figure 2 shows the differences between the XRD patterns of raw material and samples that was reduced at different temperatures. The ICDD number for TiN and solid Fe in the reduced sample are 98-008-9933 and 98-008-7922 respectively. The figure clearly shows a higher amount of solid iron (Fe) was produced. Iron (III) oxide have lower reduction temperature in the range of 220-680°C [4] (Pineau et al., 2006). This temperature is much lower than the reduction temperature of titanium dioxide which is the reason why titanium dioxide did not convert into titanium metal even when reduced at 1250°C. The effects of carbon to oxygen reducible molar ratio on the extent of reduction are presented in Figure 3. From the XRD result, high amount of titanium nitride was formed when reduction of sample with $C:O_{red}$ of 1:1 was executed at high temperature. Reduction increases when temperature was increased. After 1 hour, TiN peak was more affirm at reduction temperature of 1250°C compared to 1150°C.

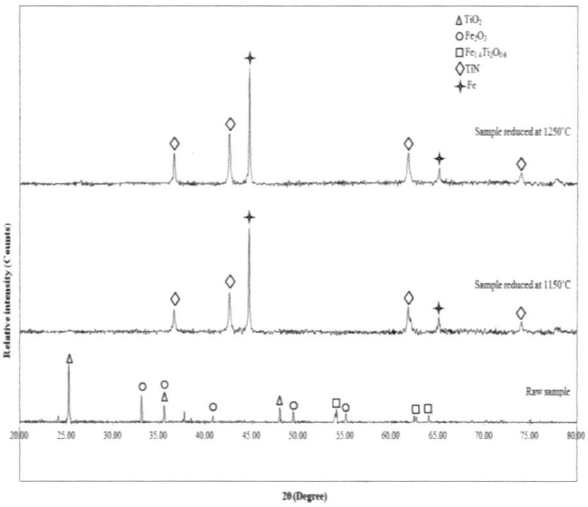

Figure 2. XRD patterns of raw material and reduced samples with 1:1 carbon to reducible oxygen molar ratio for 1 hours in H_2-N_2-CNG atmosphere at different temperature.

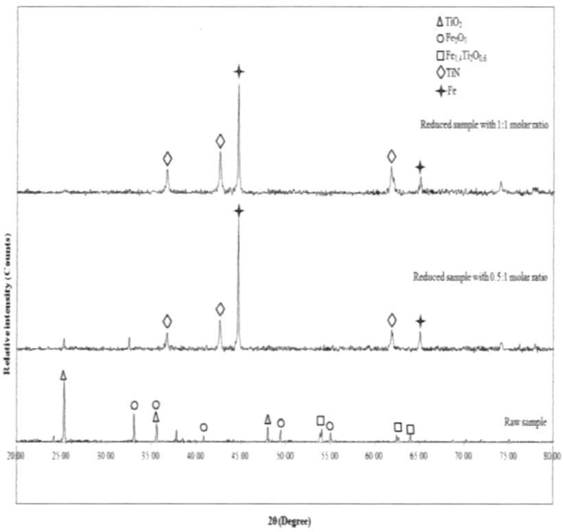

Figure 3. XRD patterns of raw material and reduced samples at 1150°C for 1 hour in H_2-N_2-CNG atmosphere with different molar ratio.

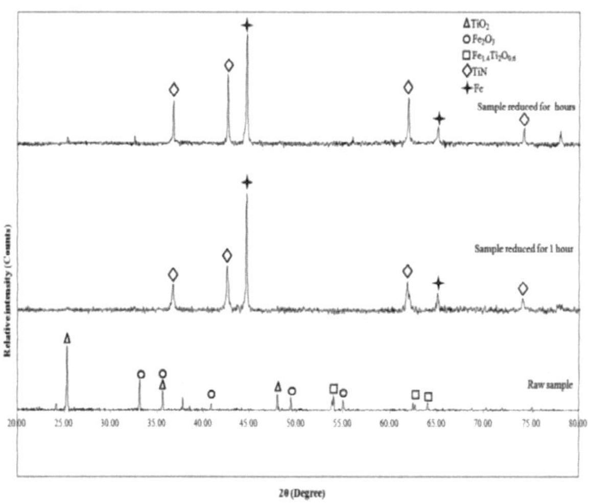

Figure 4. XRD patterns of raw material and reduced sample with 1:1 carbon to reducible oxygen molar ratio a1150°C in H_2-N_2-CNG atmosphere for different duration.

The effects of carbon to reducible oxygen molar ratio shown in Figure 4. Almost like the effect of temperature, larger ratio of carbon results in higher extent of reduction. At 1150°C, TiN peaks shows an increment in the intensity. The presence of active carbon enhanced the reaction kinetics in the sample. This led to high formation of TiN in the compound. Initially, titanium dioxide and hematite formed ilmenite-hematite solution from mixing process. When reduced at high temperature, TiO_2 and Fe_2O_3 convert into another stable phase which is titanium nitride and solid iron.

Figure 4 shows the XRD pattern of raw material, 1 hour reduced sample and 4 hours reduced sample. This pattern however indicates differences in the amount of phases occur when reduced at 1 hour and 4 hours. High amount of TiN formed when reduced for 4 hours compared to the sample that reduced for 1 hour. Longer duration has provided more opportunity for the sample to be reduced at high temperature.

4. Conclusion

In this paper, three parameters were arranged using Design of Experiment. With respect to the response of extent of reduction, reduction time behaves as the most influential factors compared to reduction temperature and carbon to reducible oxygen molar ratio. The highest extent of reduction of 92.68% was obtained at 1150°C with 4 hours reduction time and carbon to reducible oxygen molar ratio of 1:1. It can be concluded that the extent of reduction was increasing with reduction time, carbon to reducible molar ratio and reduction temperature.

Acknowledgement

The authors gratefully acknowledge the financial support from Universiti Sains Malaysia (USM) Fellowship, APEX (1002/JHEA/ATSG4001).This work was also financially supported by USM and Ministry of Higher Education (MOHE) of Malaysia through the following research grants: Fundamental Research Grant Scheme (FRGS) (No. 203/PBAHAN/6071364) and (203/PBAHAN/6071261) from MOHE, USM Research University Individual (RUI) grant (No. 1001/PBAHAN/814273) and Nippon Sheet Glass Research Grant (NSGRG) (No. 304/PBAHAN/650360/N120). Special thanks to USM technicians Mr. Shahrul and Mr. Syafiq for supporting the experimental work.

References

[1] Ramakrishan S, Kwok T C, and Hamid S A R S A 2016 *AIP Conference Proceedings* 175 6050001

[2] Ibrahim N S C, Ramakrishan S, Baharun N, Hamid S A R S A, Alizadeh R, and Roohi P 2015 *Materials Science Forum* 819 6

[3] Zhang G and Ostrovski O 2000 *Metallurgical and Materials Transactions B* 31129

[4] Pineau A, Kanari A and Gaballah I 2006 *Thermochimica Acta* 447 89

[5] Telford J K 2007 *Johns Hopkins Apl Technical Digest* 27 224

[6] Wullenkord M, Funken K H, Sattler C, and Pitz-Paal R 2010 *18th World Hydrogen Energy Conference* 3 201

[7] Yin T W, Ramakrishnan S, Hamid S A R S A, Noor A F M, and Shoparwe N I 2016 *Journal of Physics: Conference Series* 822 012063

IOP Conf. Series: Journal of Physics: Conf. Series **1082** (2018) 012037 doi:10.1088/1742-6596/1082/1/012037

Mathematical Modelling of Titanium Hydride Formation from Titanium Tetrachloride with Magnesium Hydride using Matlab

Sanjith Udayakumar[1], Atif Sadaqi[1], Najwa Ibrahim[1], M. N. Ahmad Fauzi[1], Sivakumar Ramakrishnan[1] and Sheikh Abdul Rezan[1*]

[1]School of Materials & Mineral Resources Engineering, Universiti Sains Malaysia, Engineering Campus, Nibong Tebal, Malaysia

*Email: srsheikh@usm.my

Abstract. Titanium hydride (TiH_2) is an important material for the synthesis of titanium metal by dehydriding reactions. The reaction to synthesize TiH_2 by reacting titanium tetrachloride ($TiCl_4$) with magnesium hydride (MgH_2) was investigated so that the important factors governing the kinetics of the reaction could be determined. Experiments were carried out by varying the reaction conditions to establish the optimum level to achieve highest degree of hydriding (X_h). The combination of reaction parameters that yielded the highest degree of hydriding (38.49%) was a reaction temperature of 300°C at 180 minutes and Argon (Ar) flow rate of 20 cc/min. The reaction products were characterized using scanning electron microscopy (SEM) equipped with energy dispersive X-ray spectroscopy (EDX) and X-ray diffraction (XRD) to confirm the formation of TiH_2. The X_h was measured by means of weight gain of the sample after the reaction with elemental hydrogen content analysis. Based on the experimental results, the reaction was simulated using Shrinking Core model (SCM) with MATLAB and the kinetic parameters were estimated by fitting the model to the experimental data. The simulated isothermal SCM model achieved an X_h of 57% at a reaction time of 180 minutes and Ar flow rate of 20 cc/min. The results of simulation inferred that the isothermal SCM showed good agreement with the experimental outcome providing a good fit to the data. In general, the X_h increased with increasing temperature, time and Ar flow rate within the investigated range of parameters.

1. Introduction

The titanium-hydrogen system has generated much attention in battery technology and energy related applications as titanium has been favoured owing to its high capacity of absorbing hydrogen and its ability to be charged to different hydrogen contents [1]. Besides this, the titanium hydride (TiH_2) is gaining attention for the production of Titanium (Ti) metal powder [2]. Titanium is one of the popular metals that can form hydrides, which are non-stoichiometric compounds with tetrahedral distribution of hydrogen atoms randomly distributed in the lattice sites with respect to the titanium atoms [3]. There were many methods to make Ti metal that are racing to replace the Kroll process [4, 5]. Among them were the ADMA [5], oxycarbonitride [6, 7], OS process [8] and TIRO process [9]. The typical procedure to prepare TiH_2 is a traditional method wherein Ti metal is exposed to H_2 at elevated temperature and/or pressure in order to attain the highest hydrogen content [10]. Alternatively, a low cost starting blends of titanium powder and alloying powders (elemental or master alloy ones) were used for the Blended Elemental Powder Metallurgy (BEPM) method which had relative density lesser than 95% [5]. The benefits of BEPM were minimal material waste and low cost of applications. The development of an innovative and low cost manufacturing non-Kroll process for production TiH_2 powder came as a boon to the Ti industry followed by subsequent improvements [11-13]. This study focus on a new alternative approach of formation of TiH_2 by reacting $TiCl_4$ with MgH_2 through **Eq. (1)**.

$$2MgH_2 + TiCl_{4(g)} \rightarrow TiH_2 + 2MgCl_2 + H_{2(g)} \qquad (1)$$

$$\Delta G° \text{ (kJ/mole)} = -0.0393*T - 488.42$$

The preliminary studies on the feasibility and optimized condition to produce TiH_2 through this method has been published elsewhere [2, 13]. In this study, an attempt was made to study the hydriding kinetics of TiH_2 in isothermal condition by means of modelling the solid-gas reaction using Shrinking Core model (SCM) [14].

2. Methodology

The mathematical model used to compare the kinetics of the hydriding reaction was based on the mass and thermal transport properties of the reactants [15, 16]. Previous application of SCM was limited to Fe_2O_3, TiO_2 and ilmenite systems [17-21]. The kinetic model was developed by using MATLAB software v.2015 with SCM. The transport limited reaction rate of the formation of TiH_2 by using the constructed mathematical models was predicted for isothermal reduction condition. In this study, the reaction started by flowing $TiCl_4$ gas over MgH_2 which was in a cylindrical pellet form and final product was TiH_2. Details of the experimental setup is given elsewhere [2]. The hydriding reaction was simulated by dividing the pellet into equidistant concentric shells, based on the time taken for each shell to react completely. As the reactants comprise of two different phases, a gas-solid system was considered, wherein complete conversion of MgH_2 into TiH_2 is assumed **(Eq. 2).** This assumption was valid due to favourable thermodynamic conditions as given in Eq.1.

$$\text{M (s)} + \text{A (g)} \rightarrow \text{M (s)} + \text{B (g)} \qquad (2)$$

where MgH_2 and TiH_2 are denoted by M, $TiCl_4$ and H_2 gas is denoted as A and B respectively.

2.1 Kinetics of transport model

It is essential to identify the physical and thermodynamic properties of the reactants that affect the kinetics in the system. Among the physical and thermodynamic properties were molecular weight of the species, density, thermal conductivity, emissivity and properties of the gas. The robustness of the model was decided based on the ability of the model to simulate the reaction by involving all the intrinsic and extrinsic parameters that deter the system.

2.2 Determination of core, gas, surface and product temperature

Based on the temperature profiles of the components involved in the reaction in our system as shown in **Figure 1.**, five different temperature zones have been considered for modelling namely, T_g (gas temperature), T_c (core temperature), T_s (surface temperature) and T_p (product temperature). For determining the various assigned temperatures, the system is assumed to be in pseudo-steady state i.e., steady state in a bounded area. It is difficult to predict the variation of temperature in the spherical pellet compared to the gaseous mass transport rate during the reaction. This is due to the fact that heat transfer within the reactant core occurs as a parallel process with the transfer happening through the accumulated solid reaction product and the bulk gas.

(Transfer into region) + (Transfer out of region)- (Production within region) = (Accumulation in region) (3)

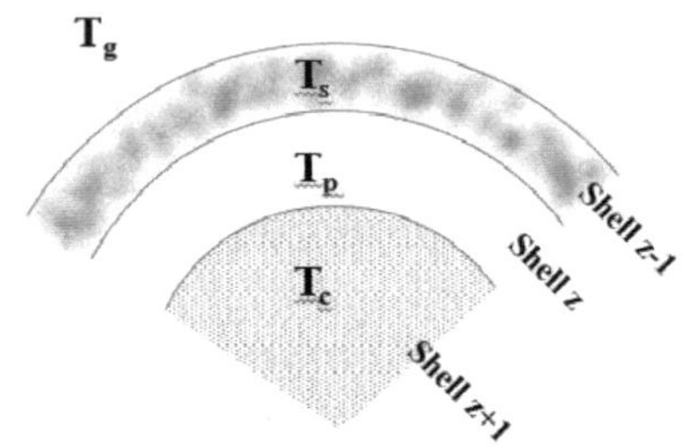

Figure 1. Assignment of temperature zones into concentric shells

In order to simplify expressions of the different temperature zones in MATLAB, a few assumptions were made namely; there was negligible accumulation of heat in the reduction zone, there is only one dimensional heat flow and transfer of heat was conductive, convective and radiative and finally, the average pellet core temperature (T_c) represents the overall heat transfer by the pellet. However, in isothermal condition of the reaction system, the temperature changes are constant in the shells. The overall reaction for **Eq. (1)** of the system was presented by SCM [14] and the degree of the hydriding (X_h) reaction is predicted by simultaneous solution of the derived equations based on **Eq. (3)** in MATLAB. A detailed explanation on the solution was given elsewhere [17, 20]. The magnitude of X_h was equated to the fractional reacted (R') in **Eq. (4)** below.

$$R_o \rho_{MC}(1-(1-R')^{\frac{1}{3}} = K_{eq} t \tag{4}$$

Where **R'**is the fractional reaction, can be represented by several forms, such as instantaneous pellet weight divided by initial pellet weight or remaining volume divided by the initial volume (R' = $(M_o-M_t)/(M_o-M_f)$), $M_o$is the initial weight of the pellet, $M_f$is the final weight of the pellet, M_t is the weight of the pellet at time 't' and ρ_{MC} is the molar density.

3. Results and discussions

Based on the experimental results previous work of Atif Sadaqi [14] and Udayakumar et al. [2], the optimized conditions for achieving the highest X_h was chosen. The model was simulated for a combination of two different reaction temperatures namely, 300°C and 200°C at 180 minutes with Argon (Ar) flow rate of 20 cc/min for isothermal condition. The experimental and simulated outcomes for isothermal conditions are individually interpreted and compared to determine the deviation.

3.1 Isothermal shrinking core model

The major assumptions that are made before simulating Isothermal-SCM were; the temperature is homogenous throughout the reaction and also the temperature of the pellet was uniform and constant throughout the reaction. The X_h determined by simulations using isothermal SCM was shown in **Figure 2**. It can be seen that as temperature increases, the rate of X_h also increases. The reaction was expected to be at 50% completed by 180 minutes at temperature of 200°C and Ar flow rate of 20 cc/min according to this model. Furthermore, it can be noted that an increase in 100°C of reaction temperature allowed the reaction to be completed faster by 8%. There were still a few assumptions made in this model stated by Gower [16] that were hard to be accomplished in the model namely; the sample was assumed to remain intact throughout the reaction and the pellet porosity was also assumed constant. In reality, it was difficult to model the variations in porosity function as the dehydriding reaction was constantly producing pores from the evolution of H_2 gas. Secondly, the intense exothermic reaction during the formation of TiH_2 as well as the thermal expansion of the sample itself caused the pellet to crack and expose multiple new surface areas. As a result of this phenomenon, the reaction rate could be altered. On the other hand, as the reaction between $TiCl_4$ and MgH_2 proceeded, the removal of Mg in the form of $MgCl_2$ occurred which further affected the pellet porosity. This in turn will accelerate the reaction rate which will lead to the deviation of simulation results from experimental.

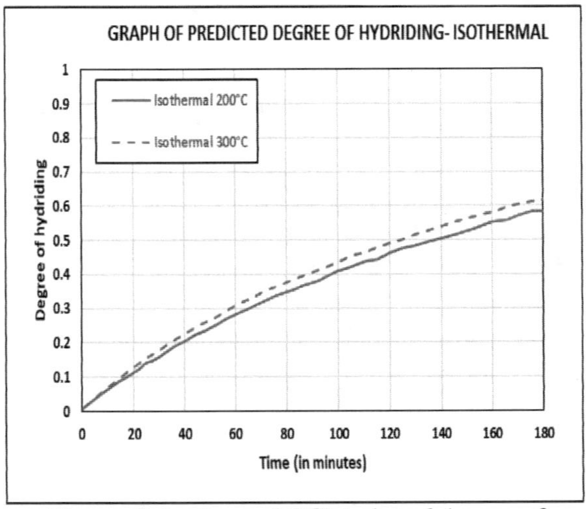

Figure 2. Isothermal SCM plot of degree of hydriding

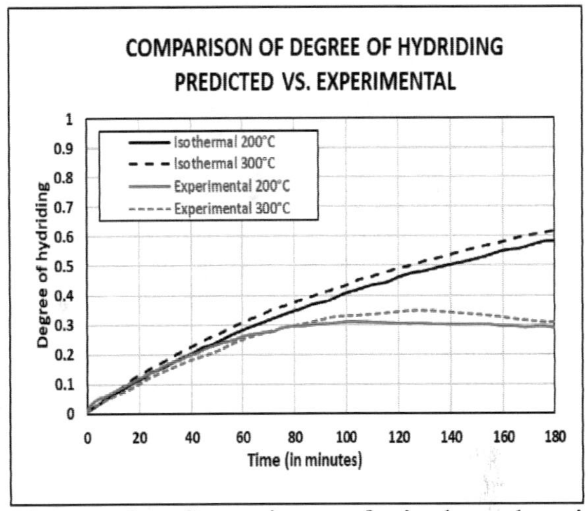

Figure 3. Comparison of isothermal with experimental data by using the model of SCM.

Figure 3 shows the comparison of the experimental and simulated outcomes of the hydriding reaction with Ar flow rate of 20 cc/min, reaction time of 180 minutes and for temperatures 200°C and 300°C. On comparing the curves of simulated (isothermal condition) with experimental data, for both temperatures given in **Figure 3**, the reaction does follow the proposed model initially. However, upon reaching a period of approximately 80 minutes, the deviation started to occur. This showed that the experimental results were much more complex than the present model could predict. The reaction trend slowed down and eventually reached a plateau with final X_h indicating a mass controlled mechanism [22].

3.2 Confirmation test

In an effort to provide additional evidence to support the validity of the SCM results, experimental results were obtained by using Scanning Electron Microscopy equipped with Electron Dispersive X-ray Spectroscopy (SEM/EDX; Leo Supra 35VP). The SEM/EDX analysis were carried out for reactant products. **Figures 4** and **5** showed the SEM micrographs of the hydrided samples at 180 minutes of reaction time, Ar flow rate of 20 cc/min and reaction temperature of 300°C and 200°C respectively. Although the scope of SEM/EDX analysis was restricted to the surface analysis, it helps to understand the progress of the reaction which can be seen by the presence of *Ti* metal peaks below 2 eV. For the stability of TiH_2, electronic charge on the Ti metal should be between 2 -1.50 eV as evident by the numerous stable phases such as TiH_2 –$TiH_{1.50}$. The presence of Ti electron peaks below 2 eV confirms the formation of TiH_2 in the reaction products. Furthermore, hydrogen atoms were not detected by EDX due to the limitation of the Leo Supra 35VP instrument. In **Figure 4,** there were numerous needle like structures which corresponded to the Magnesium Oxychloride ($MgOCl_2$) formed as a result of oxygen contamination due to exposure to ambient moisture. XRD analysis as reported by Udayakumar et al. [2] showed that the bulk product consisted of $MgCl_2$, $TiCl_{2.3}$ and TiH_2. An irregular cluster of mass can be seen in **Figure 5** which is an agglomeration of the TiH_2 and $MgCl_2$ with some unreacted MgH_2. The experimental study was done in Ar atmosphere but a low X_h indicate that the mass transfer in Ar atmosphere was low. Further thermodynamic analysis done by Atif Sadaqi [14] has indicated that H_2 gas atmosphere was more preferred for higher rate X_h. A detailed explanation of the role of H_2 gas was given elsewhere [2].

Element	Wt.%	At.%
O	36.52	49.90
Mg	43.15	38.81
Cl	12.57	07.75
Ti	07.76	03.54

Figure 4. SEM/EDX micrograph of product obtained by hydriding at a reaction temperature of 200°C at 180 minutes with Ar flow rate of 20 cc/min.

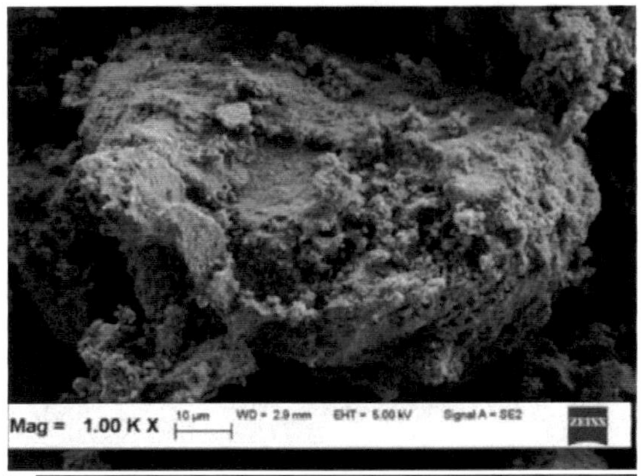

Element	Wt.%	At.%
O	26.27	37.21
Mg	57.80	53.88
Cl	08.23	05.26
Ti	07.70	03.64

Figure 5. SEM/EDX micrograph of product obtained by hydriding at a reaction temperature of 300°C at 180 minutes with Ar flow rate of 20 cc/min

4. Conclusion

Shrinking core model (SCM) was used to study the reaction between MgH_2 and $TiCl_4$ using MATLAB. Results of the simulations were used to compare with the experimental results obtained by previous study on the reaction kinetics [2]. The isothermal SCM gave a X_h of 57% at 180 minutes with temperature of 300°C with Ar flow rate of 20 cc/min. Isothermal SCM showed considerable approximation to the experimental results. SEM/EDX analysis confirmed the occurrence of hydrides in the reacted products. There were some discrepancies between the predicted and experimental results which could be due to the limitation of SCM to predict variations in porosity and temperature in the concentric shells. All in all, both models followed similar trend with the experimental data. Future hydriding experimental work will be carried out in H_2 atmosphere to increase the X_h magnitude. In conclusion, the SCM developed provided an insight into the reaction mechanism and rate controlling steps during hydriding of $TiCl_4$ with MgH_2. These results will pave for innovative ways to synthesize TiH_2.

Acknowledgement
The authors gratefully acknowledge the support from Universiti Sains Malaysia (USM) Fellowship Award. This work was financially supported by USM and Ministry of Higher Education (MOHE) of Malaysia through the following research grants: Fundamental Research Grant Scheme (FRGS) (No. 203/PBAHAN/6071364), USM Research University Individual (RUI) grant (No. 1001/PBAHAN/814273) and Nippon Sheet Glass Research Grant (NSGRG) (No. 304/PBAHAN/650360/N120). Special thanks to USM technician, Mr. Azrul for supporting the experimental work.

References
[1] Zhang H, Kisi EH 1997 Journal of Physics: Condensed Matter 9 L185
[2] Udayakumar S, Sadaqi A, Ibrahim N, Fauzi MNA, Ramakrishnan S, Rezan SA. In: A P K A Razak, editor. Regional Conference on Materials & ASEAN Microscopy Conference 2017 (RCM & AMC 2017); Olive Tree Hotel, Penang, Malaysia: USM 2017

IOP Conf. Series: Journal of Physics: Conf. Series **1082** (2018) 012037 doi:10.1088/1742-6596/1082/1/012037

[3] Mackay KM 1966 Hydrogen compounds of the metallic elements: Spon
[4] Ivasishin O, Moxson 2015 Qian, Ma. In: Froes FH, editor. Titanium Powder Metallurgy. Boston: Butterworth-Heinemann 8 117
[5] Duz V, Matviychuk M, Klevtsov A, Moxson V 2017 *Metal Powder Report* 72 30
[6] Ahmadi E, Rezan SA, Baharun N, Ramakrishnan S, Fauzi A, Zhang G 2017 *Metallurgical and Materials Transactions B* 48 2354
[7] Ahmadi E, Fauzi A, Hussin H, Baharun N, Ariffin KS, Rezan SA 2017 *International Journal of Minerals, Metallurgy and Materials* 24 444
[8] Ahmadi E, Yashima Y, Suzuki RO, Rezan SA 2018 *Metallurgical and Materials Transactions B* 1
[9] Doblin C, Chryss A, Monch A 2012 *Key Engineering Materials* 95
[10] Nagel H, Goretzki H 1975 *Journal of Physics and Chemistry of Solids* 36 431
[11] Kasparov S A, Klevtsov A G, Cheprasov A I, Moxson VS, Duz VA, inventors; Adma Products, Inc., assignee. Semi-continuous magnesium-hydrogen reduction process for manufacturing of hydrogenated, purified titanium powder. United States patent US 8,007,562. 2011 Aug 30.
[12] Klevtsov A, Nikishin A, Shuvalov J, Moxson V, Duz V, inventors; Adma Products, Inc., assignee. Continuous and semi-continuous process of manufacturing titanium hydride using titanium chlorides of different valency. United States patent US 8,388,727. 2013 Mar 5.
[13] Moxson VS, Duz VA, Klevtsov AG, Sukhoplyuyev VD, Sopka MD, Shuvalov YV, Matviychuk M, inventors. Method of manufacturing pure titanium hydride powder and alloyed titanium hydride powders by combined hydrogen-magnesium reduction of metal halides. United States patent US 9,067,264. 2015 Jun 30.
[14] Atif Sadaqi Bin Zainordin. Kinetic Modelling of the Formation of Titanium Hydride from the reaction between Magnesium Hydride and Titanium Tetrachloride using MATLAB. University of Science, Malaysia: University of Science, Malaysia; 2017.
[15] Homma S, Ogata S, Koga J, Matsumoto S 2005 *Chemical Engineering Science* 60 4971
[16] Gower RC. A Determination of the Transport-limited Reaction Rate for a Gas- Solid Reaction Forming a Porous Reaction Product: Lehigh University; 1971.
[17] Giap TW, Ramakrishnan S, Rezan SA, Hamid A, Alizadeh R, Roohi P 2012 *Proceedings of the International Conference on Science, Technology and Social Sciences (ICSTSS)* 441
[18] Chun Chien L, Ramakrishnan S, Abdul Rezan S, Alizadeh R, Roohi P 2018 *Solid State Science and Technology* 25 40
[19] Ibrahim C, Binti NS, Ramakrishnan S, Rezan SA, Baharun NB, Alizadeh R, Roohi P 2015 InMaterials Science Forum 2015 Apr 10 819
[20] Ramakrishnan S, Kwok TC, Hamid SA 2016 *AIP Conference Proceedings* Jul 19 2016 1756 050001
[21] Yin TW, Ramakrishnan S, Rezan SA, Noor AF, Shoparwe NI, Alizadeh R, Roohi P 2017 *Journal of Physics: Conference Series 2017* 822 012063
[22] Gaskell D. An introduction to transport phenomena in materials engineering: Momentum Press; 2012

IOP Conf. Series: Journal of Physics: Conf. Series **1082** (2018) 012038 doi:10.1088/1742-6596/1082/1/012038

ZTA Hardness Enhancement via Hot Isostatic Press Sintering

NA Rejab[1], Z.A. Ahmad[1], M. F. Ab Rahman[1], and N.K. Su[2]

[1]Structural Materials Niche Area, School of Materials and Mineral Resources Engineering, Universiti Sains Malaysia, Engineering Campus, 14300 Nibong Tebal, Penang, Malaysia
[2]Politeknik Tuanku Syed Sirajuddin, pauh Putera, 02600 Arau, Perlis, malaysia

*akmarnik@usm.my

Abstract. The present work mainly deals with the influence of Y_2O_3 stabilize ZrO_2 (YSZ) additions on the phases, microstructure and mechanical properties of Al_2O_3 which is known as zirconia toughened alumina (ZTA) through hot isostatic press (HIP) sintering. Different amounts of 8 mol% of yttria stabilize ZrO_2 (8YSZ) were added into Al_2O_3 and the amount of 8YSZ was varied from 0 wt.% to 25 wt.%. There is significant evidence that microstructural coarsening within ZTA was found to produce an improvement of hardness and fracture toughness. These improvements are the evidence for the toughening mechanism through the transition of tetragonal to monoclinic (t-m) structure provide by 8YSZ which is defined to optimize the materials and mechanical properties of Al_2O_3 ceramics. The 8YSZ grains and Al_2O_3 grains are impartially distributed among each other. The XRD patterns are corresponds to yttria doped zirconia $(Zr_{0.935}Y_{0.065})O_{1.968}$ peaks (ICDD files No. 01-078-1808) and Al_2O_3 (No. 00-010-0173). The existence of $(Zr_{0.935}Y_{0.065})O_{1.968}$ phase relates to the metastable tetragonal form of ZrO_2.

1. Introduction

The excellent properties such as good corrosion resistance, high hot hardness, and high chemical resistance make alumina (Al_2O_3) based ceramic a very promising candidate as material for implant and cutting tools applications. However, their application as cutting tools is restricted due to their brittleness [1–3]. Therefore, many researchers focus on their toughening improvement through secondary phase dispersion approach [4–9], however this approaches sacrificing their hardness for higher quality applications [3,4]. There are several factors that influence the deterioration of the hardness properties. For example, when 5 mol% of Y_2O_3 is added to ZrO_2 as stabilizer, certain amounts of lattice defects, e.g. oxygen vacancies formed in the ZrO_2 lattice [10]. Tekeli et al. [11] state that most promising results are reported from the addition of Al_2O_3 to the cubic YSZ known as zirconia toughened alumina (ZTA), where both enhanced densification behaviour and improvement in electrical conductivity is obtained together with improved mechanical strength. They found that 8 mol% of YSZ (8YSZ) will affect the sintering and grain growth behaviour of the alumina. Furthermore, they found that the alumina with 8YSZ achieved a density of 95% of its theoretical value. However, these findings shows that reduction of hardness value still remain as an issue due to an existence of porosity on the Al_2O_3 crystal lattice. An improvement of porosity related to densification of materials generated during the sintering process.

There are several categories of sintering such as solid state sintering, vitrification, viscous sintering and liquid state sintering [12]. Under some conditions, densification achieved by any of four categories of sintering might be inadequate. A solution to this problem is the application of external pressure during heating, giving the method of pressure sintering (of which hot pressing and hot isostatic pressing are common examples). Pressure sintering has the disadvantage of increased fabrication costs but effective when a microstructure of high density and fine grain size must be guaranteed. To distinguish it from pressure sintering, sintering performed without the application of an external pressure is usually referred as conventional sintering [13,14].

Hot isostatic pressing (HIP) has been used in powder metallurgical applications to densify the materials and for single crystals to reduce the porosity generated during the casting and homogenization processes. HIP can be applied to a wide range of metals, alloys and ceramics [15]. Furthermore, this process has been recognized to be an effective sintering technique to achieve full densification [13]. Porosity reduction by means of HIP has a significant beneficial effect on the certain applications [14]. In this paper, the effectiveness of using HIP towards the microstructure and hardness of alumina-8YSZ (ZTA) ceramic has been investigated. Therefore, the influence of HIP condition is believed to affect the microstructure by reducing the porosities and form densified ceramic.

2. Experimental Method

The composition of 8 mol% of yittria stabilize zirconia (8YSZ) was varied from 0 wt.% to 25 wt.%. The starting materials was mixed continuously for 8 hours and dried in oven at 90 °C for 24 hours [3]. The mixture is then subsequently hydraulically pressed into cylinder mould (Ø 13 mm) at 10 MPa to form green bodies. The pressed samples were sintered in the hot-isostatic pressing (American Isostatic Presses, INC HP630) at 1600 °C for 1 hour soaking time under 150 MPa using Ar gas. The pressureless sintering was carried out to perform a comparison between the two techniques of the same mate- rials. The analysis of structure was carried out by X-ray diffraction (XRD) using the Bruker D8 Advance with CuKa radiation (40 kV, 30 mA) diffracted beam monochromator, using a step scan mode with a step size of 0.1° in the range of 20°–80° of 2ϴ°. Field emission scanning electron microscopy (Carl Zeiss, Germany) was employed to study the microstructure of the grinded and polished samples. Sample grinding process was executed using Metkon 120, 240, 500, 800 and 1200 grit alumina abrasive grade papers. The samples were polished by using Imptech grinder polisher with 1 μm and 0.05 μm Al_2O_3 particle as polishing medium. The density and porosity values were obtained according to the ASTM C 830-00 test procedure. The Vickers hardness test (Shimadzu HSV-20) was carried out to obtain both Vickers hardness values.

3. Results and Discussion

ZTA contains yittria stabilize zirconia particles that increase its fracture by microcracking and by stress induced tetragonal-to-monoclinic phase transformation. Figure 1shows the stacking of XRD analysis for Al_2O_3 ceramics mixed with various amount of 8YSZ (0 wt.% to 20 wt.%). The patterns are correspond to yttria doped zirconia $(Zr_{0.935}Y_{0.065})O_{1.968}$ (ICDD files No. 01-078-1808) and corundum (ICDD files No. 00-010-0173) phases. A quantitative evaluation of phase analysis was performed using the EVA software at 2ϴ° = 17° - 89° and recorded in Table 1. The addition starts at 5 wt.% 8YSZ which shows 98% corundum and 2% of 8YSZ. As the addition increase to 10 wt.% of 8YSZ the value of corundum decrease to 95% while the 8YSZ (wt.%) increase to 5%. In this work, the presence of $(Zr_{0.935}Y_{0.065})O_{1.968}$ phase plays significant roles in determining the properties on Al_2O_3 ceramics. XRD analysis demonstrates the presence of monoclinic ZrO_2 crystal structure, it is an unavoidable phenomena and the small amount of monoclinic ZrO_2 would not influence the composition overall. Additionally, a work directed by Azhar et al. [16] expressed that no less than 15% of monoclinic stage will dependably be available. In this work, $(Zr_{0.935}Y_{0.065})O_{1.968}$ phase proved to preserve fine grains as shown in Figure 2 which is related to the grain boundary inhibition. This finding is in good agreement with previous works [17] which is report that grain boundary inhibition can be due to the pinning effect, which is associated with particles locations at grain boundaries.

Table 1: Phase percentages of Al_2O_3 ceramics with the addition of 8YSZ

8YSZ (wt.%)	Corundum (%)	YSZ (%)
0	100	0
5	98	2
10	95	5
15	93	7
20	92	8
25	90	10

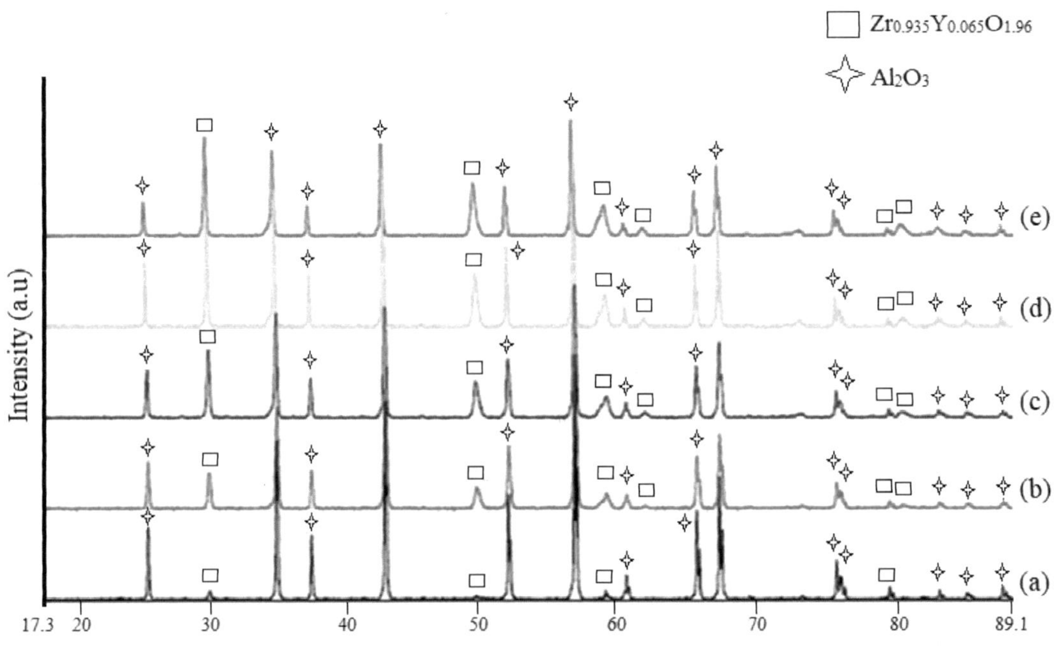

Figure 1. X-ray diffraction patterns of the sintered Al_2O_3 based ceramics added 8YSZ, (a) 5 wt.%, (b) 10 wt.%, (c) 15 wt.%, (d) 20 wt.% and (e) 25 wt.%.

Based on Figure 2, the SEM micrograph for $(Zr_{0.935}Y_{0.065})O_{1.968}$ phase showed the bright grains appeared as small rounded grains and the microstructure of Al_2O_3 with the dark grains and round shapes. There are some large Al_2O_3 grains that are formed due to the coalescence of the grains, which is related to grain boundary migration [18]. The result of grain sizes (Figure 3) were found to be decreased from 14 μm to 5 μm with various 8YSZ additions on Al_2O_3 ceramics, respectively. The significant role of $(Zr_{0.935}Y_{0.065})O_{1.968}$ phase, are more efficient in pinning the grain boundaries towards the grain growth. In the present work, further 8YSZ additions in alumina ceramics show that the distribution of the particles is very uniform and combine with each other to form an overall. Basically, reduce pore and gap exist. Therefore the structure becomes compact, and crystal grains are refined with the adding of 8YSZ, thus mechanical properties of these ceramics are far better than those of pure alumina ceramics. Hardness of ceramics is usually affected by the microstructural features such as multiphases, grain size, porosity and grain boundary constitution [19].

doi:10.1088/1742-6596/1082/1/012038

The effect of various amount 8YSZ additions on the Vickers hardness and porosity is shown in Figure 4. It is observed that the percentage of the porosity dropped, the hardness of the materials is increase accordingly. At the beginning, 5 wt.% 8YSZ has 1650.21 HV and 19.48% porosity. The value of hardness is increase to the highest point which is 1706.74 HV as the percentage of porosity decreased to the lowest point to 1.94% for 15 wt.% 8YSZ addition. At composition higher that 15 wt.% 8YSZ, the value of hardness decreased as the percentage of porosity increased. At 15 wt.% of 8YSZ addition, the grains starts to growth and the 8YSZ start to diffused inside the alumina and gradually filled these pores. Therefore, with the less amount of porosity, the ZTA able to increase the hardness at the optimum composition which is 15 wt.% 8YSZ [20]. However, further addition of 8YSZ will make the Al_2O_3 ceramic become more brittle with the addition of more than 20 wt.% of 8YSZ such as at 25 wt.% . These increase the percentage of porosity up to 15% and reduce the hardness to 1448.9 HV (13% reduced).

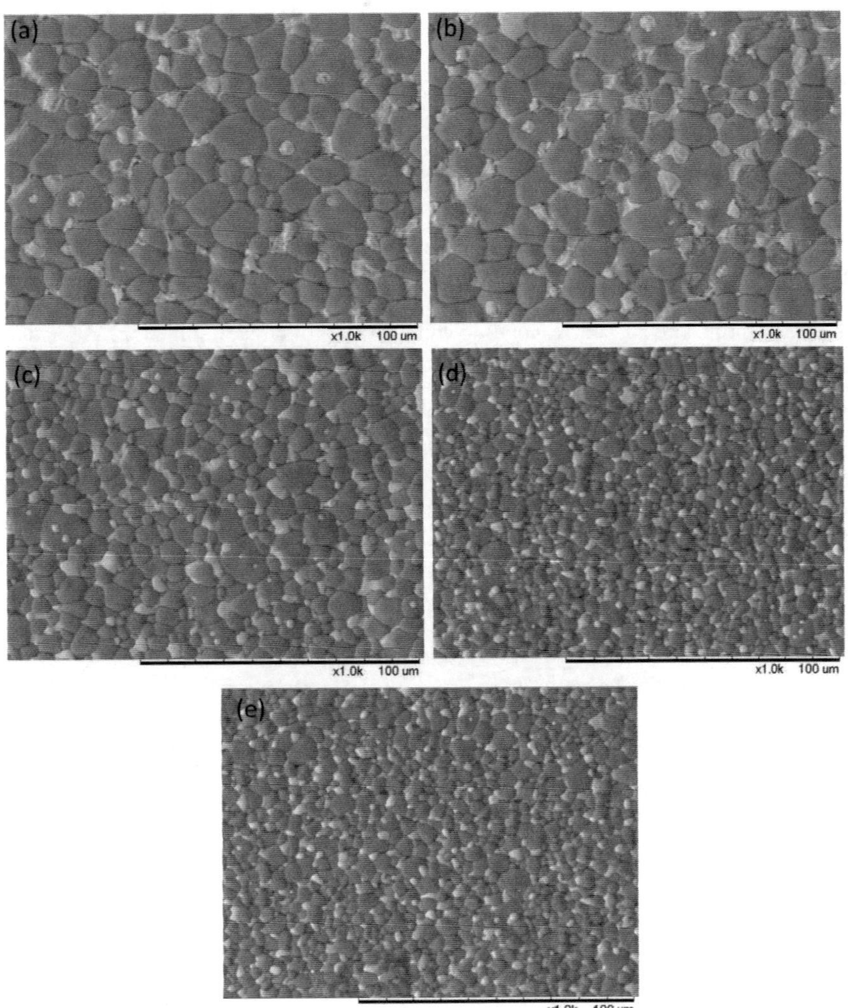

Figure 2. SEM micrograph corresponding to the various additions of 8YSZ on Al_2O_3 ceramics, (a) 5 wt.%, (b) 10 wt.%, (c) 15 wt.%, (d) 20 wt.% and (e) 25 wt.%.

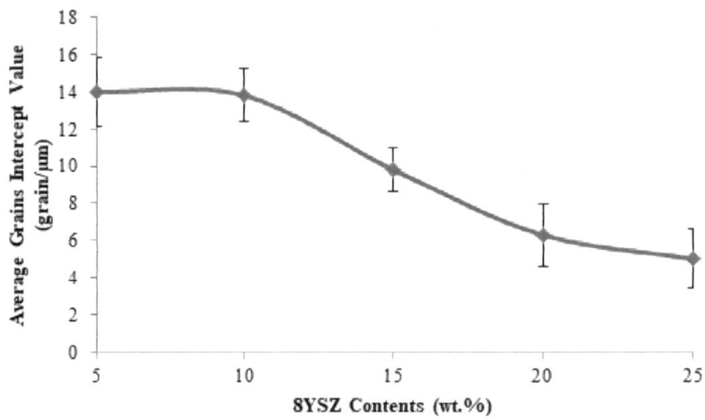

Figure 3. Average grains intercept value for Al_2O_3 ceramics with various 8YSZ contents.

Table 2 shows the optimum values of Vickers hardness reported from previous authors and their similar corresponding compositions of YSZ addition in Al_2O_3 ceramics. Hossen et al. [21] and Kwon et al. [22] found that sintering of 20 wt.% of 8YSZ with conventional method (presureless sinter) at 1600 °C shows maximum hardness are 1175 HV and 1193 HV, respectively. The current work shows that by using HIP the maximum hardness are 1537.26 HV which is significantly improved (~23%). By comparing previous findings and the current work, it's proved that the effectiveness of using HIP towards the Vickers hardness of Al_2O_3 ceramics has been achieved.

Table 2. Authors and their corresponding compositions of YSZ involved in Al_2O_3 ceramic

References	YSZ (wt.%)	Firing Method (1600 °C)	Vickers Hardness
[23]	20	Conventional	1175 HV
[22]	20	Conventional	1193 HV
Current	20	HIP	1537.26 HV

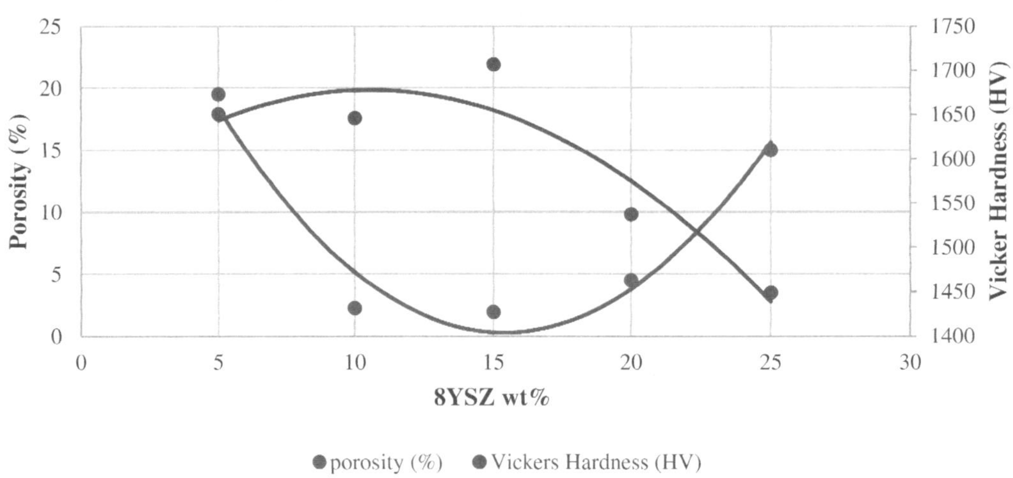

Figure 4. Relationship between hardness and percentage of porosity.

4. Conclusion

The values of the Vickers hardness for the toughened Al_2O_3 ceramics were controlled through addition of different quantities of 8YSZ. Superior properties of ZTA ceramics achieved by adding 15 wt.% of 8YSZ. The ceramics exhibited the highest Vickers hardness (1706.74 HV), and the lowest porosity (1.94%). The presence of 7% of $(Zr_{0.935}Y_{0.065})O_{1.968}$ phase is considered as the major contributor to optimize the mechanical properties of Al_2O_3 ceramics. Therefore, by using hot isostatic press sintering (HIP) technique proved to improve the hardness properties of Al_2O_3 ceramics.

Acknowledgments

This work is sponsored by the Universiti Sains Malaysia (USM) under grant no. 1001/PBAHAN/811212.

References

[1] Oungkulsolmongkol T 2010 *J. Met. Mater. Min.* **20** 71
[2] Azhar A Z A, Choong L C, Mohamed H, Ratnam M M, Ahmad Z A, 2012 *J. Alloys Compd.* **513** 91
[3] Rejab N A, Lee W K, Sktani Z D I, Ahmad Z A, 2017 *Int. J. Refract. Met. Hard Mater.* **69** 60
[4] Marin E, Rondinella A, Zhu W, McEntire B J, Bal B S, Pezzotti G, 2017 *J. Mech. Behav. Biomed. Mater.* **65** 616
[5] Rejab N A, Azhar A Z A, Kian K S, Ratnam M M, Ahmad Z A 2014 *Mater. Sci. Eng. A.* **595** 18
[6] Rejab N A, Azhar A Z A, Ratnam M M, Ahmad Z A, 2013 *Int. J. Refract. Met. Hard Mater.* **36** 162
[7] Rejab N A, Azhar A Z A, Ratnam M M, Ahmad Z A, 2012 *Adv. Mater. Res.* **620** 252
[8] Azhar A Z A, Rejab N A, Hasmaliza M, Ratnam M M, Ahmad Z A 2012 *Adv. Mater. Res.* **620** 35
[9] Rejab N A, Zhwan D I S, Afifah M A, Ahmad Z A, 2016 *Mater. Sci. Forum* **840** 57
[10] Naga S M, Hassan A M, Awaad M, 2015 *Ceram. Int.* **41** 6248–6255.
[11] Tekeli S, Erdogan M, Aktas B 2004 *Ceram. Int.* **30** 2203
[12] De Jonghe L C, Rahaman M N 2003 *Adv. Ceram. Mater. Appl. Process. Prop.* 187
[13] Lee S H, Kupp E R, Stevenson A J, Anderson J M, Messing G L, Li X, Dickey E C, Dumm J Q, VK Simonaitis, Castillo GJ 2009 *J. Am. Ceram. Soc.* **92** 1456
[14] Wang J, Ma J, Zhang J, Liu P, Luo D, Yin D, Tang D, Kong L B 2016 *Opt. Mater. (Amst)* 3
[15] Chang S, Chen S, Huang K, 2012 **53** 1689
[16] Azhar A Z A, Mohamad H, Ratnam M M, Ahmad Z A, J. Alloys Compd. 497 (2010) 316–320.
[17] Chinelatto A, Pallone E, Mechanisms of Microstructure Control in Conventional Sintering, *Cdn.intechopen.com.* (1998)
[18] N A Rejab, A Z A Azhar, M M Ratnam, Z A Ahmad, 2016 *Mater. Sci. Technol.* **2** 160120063414000
[19] S. Tekeli, 2005 *J. Alloys Compd.* **391** 217
[20] Azhar A.Z.A., Ratnam M.M., Ahmad Z.A 2009 *J. Alloys Compd.* **478** 608
[21] Hossen M M,. Chowdhury F U Z, Gafur M A, Abdul Hakim A K M, Hossen M B, *IOSR J. Mech. Civ. Eng.* 2014 **11** 01
[22] Kwon NH, Kim G H, Song H S, Lee H L 2001 *Mater. Sci. Eng. A.* **299** 185
[23] Hossen M M, Phil M, Hakim A K M A 2014 *Eur. Sci. J.* **10** 399

Effect of Glass Addition on the Phase Formation, Microstructures Evolution and Dielectric Properties of CCTO Ceramics for Energy Storage Capacitor

MF Ab Rahman[1], ZA Ahmad[1], JJ Mohamed[2], MF Ain[3], M Ahmadipour[1] and NA Rejab[1*]

[1]Structural Materials Niche Area, School of Materials and Mineral Resources Engineering, Universiti Sains Malaysia, Engineering Campus, 14300 Nibong Tebal, Penang, Malaysia
[2]Faculty of Bioengineering and Technology, Universiti Malaysia Kelantan Kampus Jeli, Karung Berkunci No. 100, 17600 Jeli, Kelantan, Malaysia.
[3]School of Electrical and Electronic Engineering, Universiti Sains Malaysia, Engineering Campus, 14300 Nibong Tebal, Penang, Malaysia

*akmarnik@gmail.com

Abstract. Effect of glass addition on the phase formation and microstructural evolution of CCTO ceramics for energy storage capacitor was studied. CCTO and CCTO-borosilicate glass based ceramics (0-1 wt.%) were prepared by solid state reaction method. The powders were pressed at 250 MPa and were sintered at 1040 °C for 10 hours. XRD analysis illustrates the formation of CCTO phase with minor peaks of CuO phase for CCTO and CCTO-glass composites samples. The glass were observed to be segregated at the grain boundaries of CCTO alongside with CuO compound. The dielectric properties indicate that both ε_r and tan δ were reduced after the addition of glass. Thus, it shows that the borosilicate glass addition into CCTO can affect the physical and dielectric properties of CCTO ceramics.

1. Introduction

Recently, $CaCu_3Ti_4O_{12}$ (CCTO) ceramics had received huge interest because of the attractive feature such as high dielectric constant, ε_r over a wide range of temperature (100 to 600 K) [1]. This special properties had increased the potential of CCTO as a promising material in the microelectronics field. Despite exhibiting high ε_r, CCTO also has high tan δ (tan δ > 0.05) [2, 3]. Usually, high ε_r will result in higher tan δ due to the inherent material properties [4]. Ahmadipour et al. [5] indicated that tan δ of the dielectric material could be from dipolar, distortional, interfacial and conduction loss. Furthermore, tan δ can also be affected by several extrinsic factors, such as microstructural defects, microcracks, porosity and impurities [6, 7]. Normally, high ε_r materials have important applications in high frequency region such as wireless communication devices which are generally operated in the range of few to thousands of MHz frequency. Unfortunately, high tan δ limits the materials applications at high frequencies.

Several approaches have been used to reduce tan δ of CCTO through the application of different processing parameters, doping different elements into CCTO or combining it with another materials to produce composite [8-10]. One of the most interesting approaches is by fabrication of CCTO composites

IOP Conf. Series: Journal of Physics: Conf. Series **1082** (2018) 012039 doi:10.1088/1742-6596/1082/1/012039

via incorporation of glass as additive. This approach is able to improve densification and reduction of tan δ [11].

In current studies, the phase formation and microstructure evolution of CCTO ceramic with borosilicate glass addition and its effect on the electrical properties are still not been explored. Nevertheless, there are lacks of high frequency measurements of CCTO dielectric properties were presented. Hence, in this study, the effects of borosilicate glass addition on the phase formation and microstructures evolution of CCTO ceramics characterized at high frequencies (1 GHz) were investigated.

2. Experimental Method

CCTO was prepared via solid state reaction method. The raw materials of $CaCO_3$ (Sigma Aldrich, >99%), CuO (Sigma Aldrich, >99%) and TiO_2 (MERCK, >99%) powder was wet mixed for 24 hours and the mixture was dried overnight in the oven. The dried mixtures then were calcined at 930 °C for 12 hours. The borosilicate glass (Pyrex™), taken from laboratory beaker was grinded by using planetary ball mill machine to form a finer powder. Subsequently, the glass powder was mixed with CCTO's calcined powder for 24 hours using the following chemical composition: (100-x) % $CaCu_3Ti_4O_{12}$ + x % glass additives (wt.%), where x = 0, 0.01, 0.1, and 1.0 respectively. The mixed powder was pressed into pellets under a pressure of 250 MPa. The sample was labelled as CCTO for pure CCTO sample while BSG0.01, BSG0.1 and BSG1 were labelled for the samples of CCTO with addition of 0.01, 0.1 and 1 wt.% of borosilicate glass, respectively. The pellets then were sintered at temperature 1040 °C for 10 hours. The phase formation of the pure CCTO and CCTO-borosilicate glass composites were analyzed by using X-ray Diffraction (XRD) (model: Bruker Advance D8) while Scanning electron microscopy (SEM) (model: TM 3000) was used to observe the microstructures of the samples. For electrical measurement analysis, the dielectric properties of the samples were measured by using impedance analyzer machine (model: 4291B Hewlett Packard) at 1 MHz.

3. Results and Discussion

Figure 1 shows the XRD patterns of CCTO and various CCTO-borosilicate glass composites (0.01-1 wt.%). Since the maximum amount of each glass added was only 1 wt.%, the major diffraction peaks observed are corresponds to body centered cubic perovskite related structures of CCTO in all composites. Besides, it is observed that minor CuO phase also appeared in XRD patterns for each sample at ~35.5° and ~38.6°. The CCTO and CuO phases are compared with ICDD Data File Card No. 01-075-2188 and 00-045-0937, respectively. The minor CuO phase presented for each sample is from the CuO compound that precipitated at the grain boundary of CCTO. Cu ions were first separated out from CCTO at ~1000 °C where it mostly segregated at the grain boundaries and then were re-oxidized to CuO compound during cooling [12].

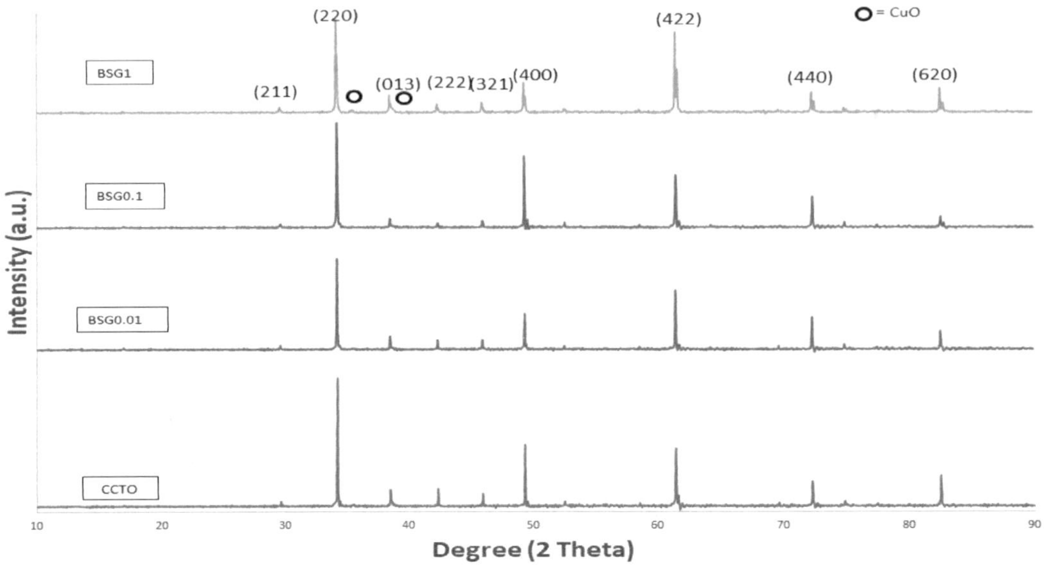

Figure 1. X-ray diffraction patterns of the sintered CCTO and CCTO-borosilicate glass composites .

SEM analysis was carried out on the cross-section area of CCTO and CCTO-borosilicate glass based composite samples and the results are shown in the Figure 2. It was observed that the quantity of the glass increased with the increasing of borosilicate glass compositions for each CCTO-glass based composite. The glass was seen assembled at the grain boundary in the bulk body (in the small circle in each figure). During sintering process, glass additives wet the grains of the matrix material, thus it would facilitate the densification via pore elimination process [13].

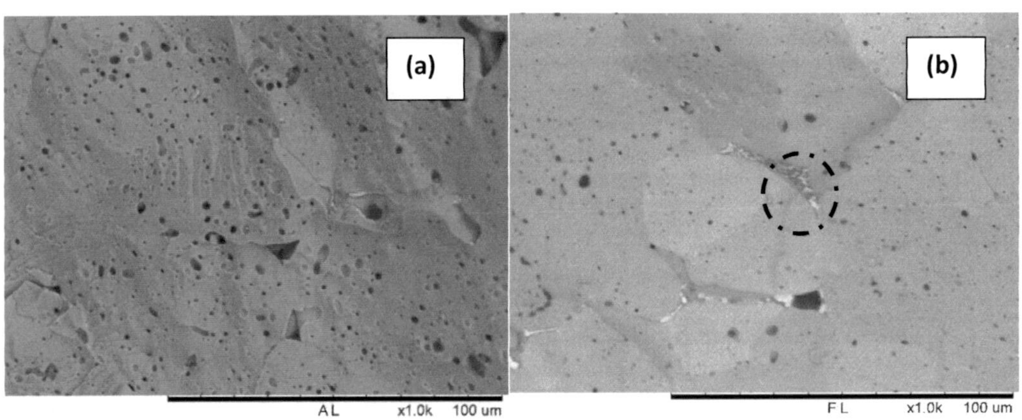

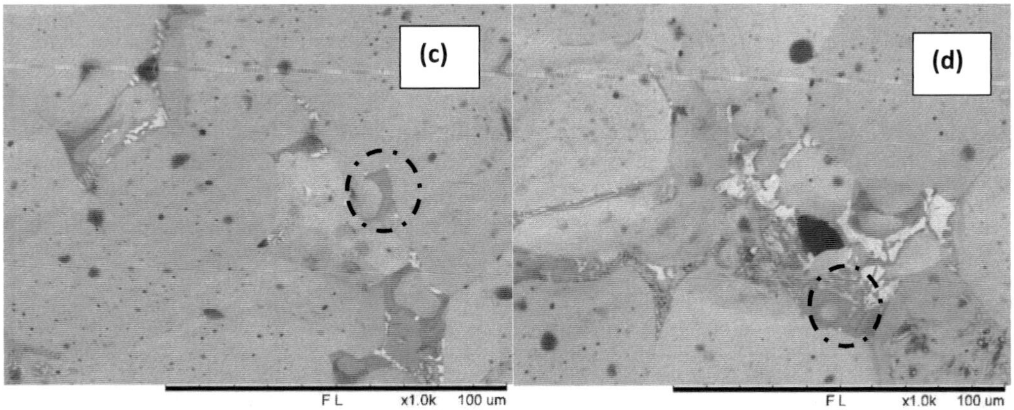

Figure 2. SEM micrograph of cross-section of CCTO and CCTO-borosilicate glass composite samples with different composition of glass (a) 0 wt.%, (b) 0.01 wt.%, (c) 0.1 wt.% and (d) 1.0 wt.%.

The graph of ε_r and tan δ of CCTO and CCTO-borosilicate glass composites is plotted in the Figure 3. It is apparent that the ε_r of CCTO was increased by the addition of borosilicate glass into CCTO where BSG0.01 sample had highest ε_r of 10582 at frequencies 1 MHz compared to CCTO (7309). However, by increasing the glass addition up to 1 wt.%, the ε_r was decreased. BSG1 obtained lowest ε_r of 5214 at the same frequency. Glass addition will act as sintering aid during the sintering process and increase the densification of the ceramic materials, subsequently will increase the ε_r of CCTO. Meanwhile, the addition of low tan δ glass materials into CCTO help to reduce tan δ of CCTO [14]. The tan δ was reduced from 0.44 (CCTO) to 0.3 (BSG1) at 1 MHz.

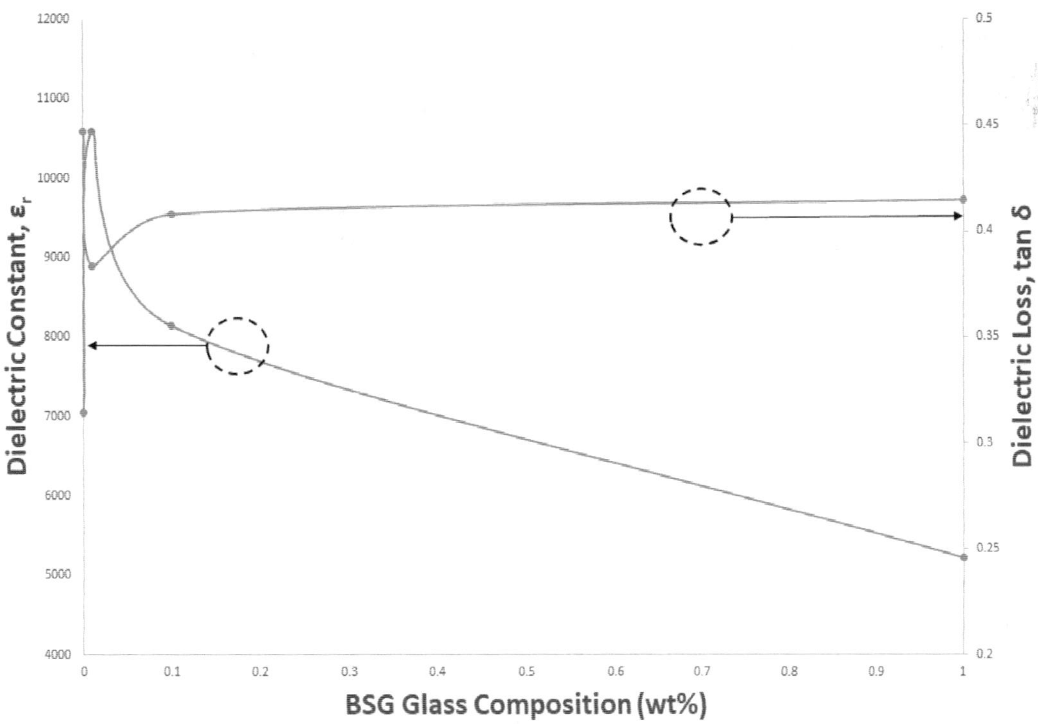

Figure 3. ε_r and tan δ of CCTO and CCTO-borosilicate glass composites at 1 MHz.

4. Conclusion

CCTO and CCTO-borosilicate glass composites were successfully prepared via solid state reaction method. From the XRD analysis, the formation CCTO phase alongside with minor CuO phase were seen for each sample. SEM micrographs show that the quantity of glass was increased with more addition of glass. Both ε_r and tan δ of CCTO was decreased with increasing the addition of borosilicate glass. Thus, it shows that the borosilicate glass addition into CCTO can improve the dielectric properties of CCTO ceramics for energy storage capacitor applications.

Acknowledgments

The authors gratefully acknowledges to the Ministry of Higher Education (MOHE), Malaysia for awarding MyBrain15 and Fundamental Research Grant Scheme (FRGS) under the Grant No. 000284 for supporting this work.

References

[1] Ab Rahman MF, Abu MJ, Karim SA, Zaman RA, Ain MF, Ahmad ZA and Mohamed JJ, 2016 *AIP Conf. Proc.* **1731** 050004.
[2] Almeida-Didry SD, Autret C, Lucas A, Honstettre C, Pacreau F and Gervais F 2014 *J. Eur. Ceram. Soc.* **34** 3649.
[3] Li W, Zhang T, Liu S, Lu Z, and Xiong R 2017 *Ceram. Int.* **43** 4366.
[4] Wan Ali WFF, Othman M, Ain MF, Abdullah NS and Ahmad ZA 2015 *J. Mater. Sci. Mater. Electron.* **25** 560.
[5] Ahmadipour M, Ain MF and Ahmad ZA 2016 *Nano-Micro Lett.* **8** 291.
[6] Alford NM and Penn SJ 1996 *J. App. Phys.* **80** 5895.
[7] Rhouma S, Said S, Autret C, Ameida-Didry SD and Amrani AE 2017 *J. Alloy Compd.* **717** 121.
[8] Ab Rahman MF, Abu MJ, Ain MF, Mohamed JJ and Ahmad ZA 2016 *Procedia Chem.* **19** 910.
[9] Wang Y, Ni L and Chen XM 2010 *J. Mater. Sci. Mater. Electron.* **22** 345.
[10] Thongbai P, Jumpatam J, Yamwong T and Maensiri S 2012 *J. Eur. Ceram. Soc.* **32** 2423.
[11] Mohamed JJ, Ab Rahman MF, Ain MF, Ahmad ZA 2016 *Mater. Sci. Forum* 840 52.
[12] Yuan WX, Luo Z and Wang, C 2013 *Journal of Alloys and Compounds* **562** 1.
[13] Prakash BS and Varma KBR 2007 *J. Solid State Chem.* 180 1918.
[14] Wang B, Pu Y, Wu H, Chen K, Xu N 2012 *J. Mater. Sci. Mater. Electron* 23 612.

Fabrication of Calcium Doped PlSZT Ceramics using High Planetary Mill Method

T.N. I. Tuan Ab Rashid[1], H. Mohamad[2*], Z. A. Ahmad[3] and J. J Mohamed[4]

[1,2,3]School of Materials and Mineral Resources Engineering, Universiti Sains Malaysia, Engineering Campus, 14300 Nibong Tebal, Penang, Malaysia

[4]Materials Technology, Universiti Malaysia Kelantan, Jeli Campus, Beg Berkunci No. 100, 17600 Jeli, Kelantan, Malaysia

Email: hasmaliza@usm.my

Abstract. Ca-doped modified Lead Lanthanum Strontium Zirconium Titanate (PLSZT) ceramic, was synthesized with different concentrations of dopant in order to enhance the properties of PZT. Calcium carbonate, $CaCO_3$ have been used as a source of Ca dopant in the preparation of PCLSZT using high planetary ball mill, palletized and sintered at 1200 °C for 3 hours. The uses of high planetary milling benefit in shorten the time for powder preparation where the calcination was not required. The concentration of $CaCO_3$ was varied from 0.01, 0.03, 0.05 and 0.10 mol% and their effects on phase formation, surface morphology, density and porosity have been studied using X-ray diffraction (XRD) and Scanning electron microscopy (SEM). XRD results showed that all Ca-doped PLSZT samples present a tetragonal phase. Whereas SEM results showed that the grain size of the samples was increased with increasing calcium concentration. In addition, the density was observed to be decreased with increasing amount of Ca concentrations.

1. Introduction

Lead Zirconate Titanate (PZT) is one of piezoelectric material that is well-known in the field of electroceramics which extensively used in the wide range of electronic applications, such as in actuators, sensors and transducers [1]. The PZT materials having perovskite structure, finding great attention in the field of research compared to other electroceramic materials. PZT is most sensitive material to the doping because its properties can be manipulated by additional of foreign ions. In the last few decades, numerous dopants have been added on the PZT systems to modify due to its interesting in dielectric, pyroelectric, piezoelectric and superconducting properties [2].

The substitutions by many oxides in perovskite structure influenced the microstructure, density, piezoelectric and dielectric properties depending on the different synthesis process. By substituting Pb sites with several acceptor or donor cations improved the electrical properties of the electroceramic systems[3]. Among the dopants, the combination of La^{3+} and Sr^{2+} shows good electrical properties for electronic applications [4-6]. However, Devi et al. [5] claimed that the dielectric permittivity was decreased with additions of Sr^{2+}. Therefore, a further study is needed to improve the dielectric properties of PLSZT system.

Kour et al. [7] had been study on the Ca^{2+} doped PZT and profound the dielectric constant increases with an addition of calcium concentration due to the increase in electronic polarization. Thus, an effort has been made through this research focusing on the isovalent dopant (Ca^{2+}) in PLSZT system for piezofan application. Pb^{2+} site was substituted with Ca^{2+} while maintaining the perovskite structure responsible for the strong ferroelectric behavior. Synthesis of Ca^{2+} doped PLSZT was performed by high planetary milling due to it tendency to synthesize compound at room temperature with small size. Besides, the PbO was easy to evaporate and cause a loss during calcination and sintering process. So, this milling process will control the loss of PbO by passes the calcination process.

2. Experimental study

The starting raw materials include lead (II) oxide (PbO), zirconium oxide (ZrO_2), titanium oxide (TiO_2), strontium carbonate ($SrCO_3$), lanthanum oxide (La_2O_3), and calcium carbonate ($CaCO_3$) were weighted based on the stoichiometric of $Pb_{0.93-x}Ca_xLa_{0.02}Sr_{0.05}(Zr_{0.52}Ti_{0.48})O_3$ with x = 0.00, 0.01, 0.03, 0.05 and 0.10. The raw materials were then dry-mixed in a planetary mill for 40 hours followed by pressing into pellets with 13 mm diameter at 200 MPa. Pallet samples were sintered at 1200 °C for 3 hours. The pellet samples then were characterized by X-ray Diffraction (XRD, Bruker AXS D8 Advance Diffractometer) and Scanning Electron Microscopy (SEM, Zeiss SUPRA 35VP) for phase formation and surface morphology analysis, respectively. Meanwhile, the bulk density and porosity of sintered pellets was calculated using Archimedes' Principle and the dielectric properties were measured using LCR meter (RF Impedance Analyzer 4291B Hewlett Packard).

3. Result and Discussion

3.1 Phase formation

The X-ray diffraction patterns of undoped and Ca-doped PLSZT (PCLSZT) samples (different Ca doping mol%) were shown in Figure 1. The XRD patterns indicate that the peaks of all samples were matched with pure perovskite structure. Diffraction pattern of the samples can be assigned to tetragonal related structure accordance with ICSD card No. 98-007-6141. There is no evidence of impurity peak shown in the diffraction patterns.

IOP Conf. Series: Journal of Physics: Conf. Series **1082** (2018) 012040
doi:10.1088/1742-6596/1082/1/012040

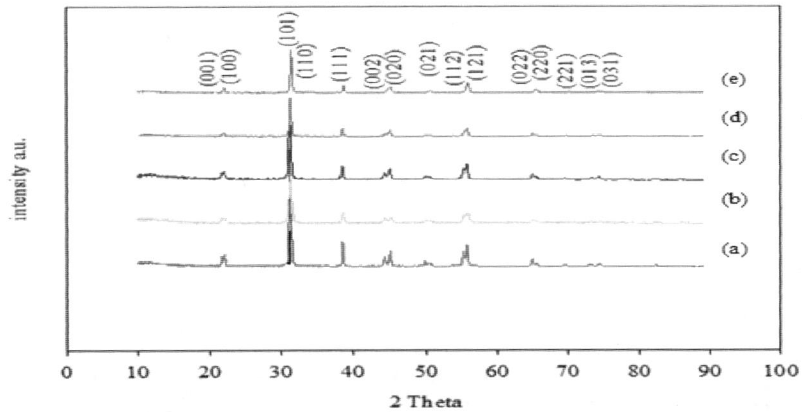

Figure 1. XRD pattern with different Ca doping concentrations (a) undoped PLSZT, (b) 0.01 mol%, (c) 0.03 mol%, (d) 0.05 mol% and € 0.10 mol%.

However, the peaks show the changes in intensity with increasing of Ca concentration where the peaks position was shifted to the right as the calcium concentration increases. The shifted of the peaks was due to the existence of lattice strain in the samples which mainly due to the different in ionic size of Ca^{2+} ion is lighter than Pb^{2+} ion [8]. Furthermore, due to slight variation of atomic radius, strain will be generated in the matrix and cause the reduction in crystallite size in PCLSZT.

3.2 Microstructure Analysis

Figure 2 shows the SEM micrograph of PCLSZT samples. Results show that the average grain size of the sample was slightly decreases from 1.134 µm to 1.056 µm as calcium concentration increases up to 0.03 mol%. The decrease in grain size as calcium concentration increases corresponds to the diffusion of calcium in PLSZT system. According to Kour et al., when the calcium replaced the lead, the decrease in the bond length gives rise to stress in the structure. The stress was increased with the increase in calcium concentration which obstructs the nucleation of the crystallite and reduces the average grain size of the samples [7].

IOP Conf. Series: Journal of Physics: Conf. Series **1082** (2018) 012040 doi:10.1088/1742-6596/1082/1/012040

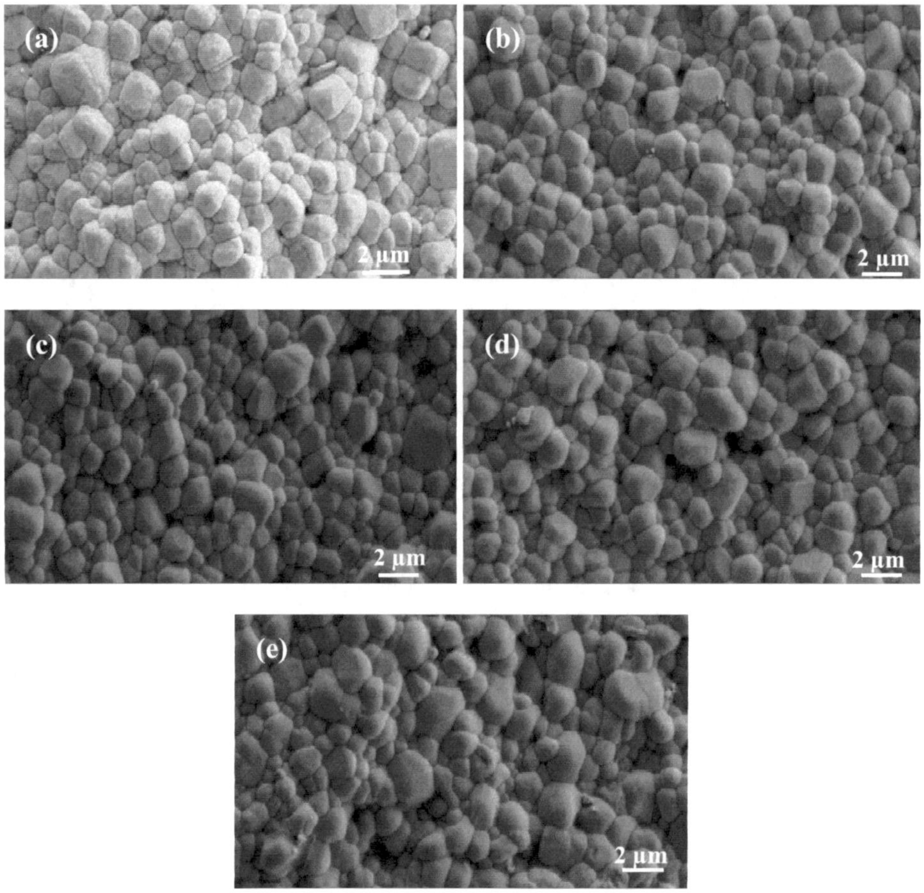

Figure 2. Surface morphology of PCLSZT at different concentrations (a) undoped PLSZT, (b) 0.01 mol%, (c) 0.03 mol%, (d) 0.05 mol% and (e) 0.10 mol%

However, the average grain size were slightly increase at higher concentration of Ca (0.05 and 0.10 mol%). High calcium content seems to promote the grain size enlargement. It is believed the enlargement of grain size was due to the increasing in mass transport at grain boundary and calcium dopant seems to induce that effect. Silva et al. [8] also stated that their study on PZT ceramics doped with calcium shows that the grain size was enhanced with higher amount of calcium doped in PZT.

3.3 Density and Porosity

Figure 3 shows the density and porosity for undoped PLSZT and PCLSZT samples. The density of Ca-doped PLSZT samples was lower than undoped PLSZT sample. This is due to the light weight of Ca^{2+} ion doped into Pb^{2+} site. The formation of cation vacancies resulted the increasing of crystal volume thus decreased in density. This phenomenon was in agreement with Chu et al. [9] who claimed the density of ceramic is reduced as the amount of Ca dopants increased.

IOP Conf. Series: Journal of Physics: Conf. Series **1082** (2018) 012040 doi:10.1088/1742-6596/1082/1/012040

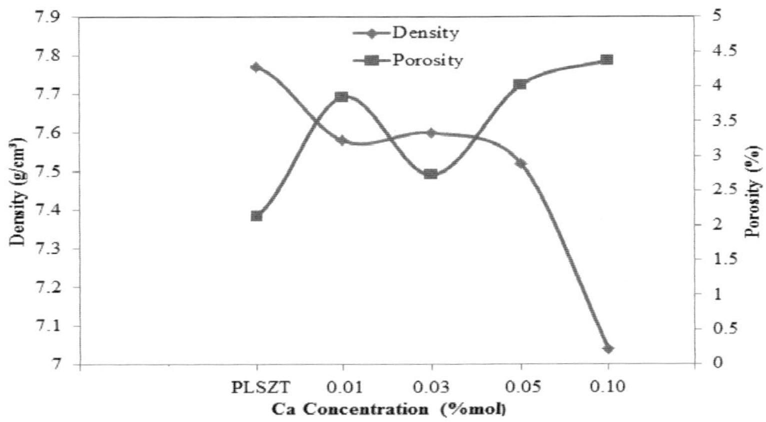

Figure 3. The bulk density and porosity for undoped PLSZT and Ca doped PLSZTs.

3.4 Dielectric properties

The dielectric properties of undoped PLSZT and Ca-doped PLSZT (PCLSZT) were measured at 1 MHz at room temperature and the results obtained was shown in Figure 4.

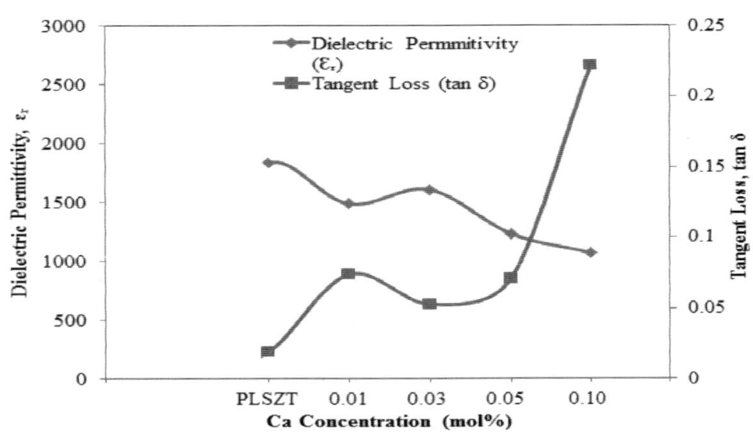

Figure 4. Dielectric permittivity and tangent loss of PZT, PLSZT and PCLSZTs at 1 MHz

Results show that the dielectric permittivity (ε_r) was decreased with increasing of the Ca concentration and vice versa for dielectric loss (tan δ). The decreasing of ε_r was related to the density and porosity present as mentioned by Adel et al. (2013), where the ε_r start to decrease as the density of the sample decrease [10]. Sample doped with 0.10 mol% Ca shows the lowest ε_r due to the lowest density and highest porosity compared to other samples. According to previous research, high porosity in the sample will block the mobility of domain in enhancing the ε_r. Meanwhile, tan δ was increased as increasing of Ca concentration was due to the defect created in Pb site which increases anisotropy in the material. Hence the internal stress was not strong enough to switch the polarization back after the electric field was removed [7].

4. Conclusion

The Ca addition shows promising effects toward microstructure, density led to the improvement of dielectric properties of PLSZT material. XRD results confirmed the formation of perovskite structure. Whereas, the grain size decreases with addition of calcium concentration from 1.134 µm to 1.056 µm but begins to enlarge again at 0.05 mol% calcium added. In addition, the density of the sample decreases with increasing calcium concentration and dielectric permittivity was decreased as increasing in calcium concentration.

Acknowledgement

The authors would like to express their gratitude to Universiti Sains Malaysia and Ministry of Higher Education for funding this project through Fundamental Research Grants Scheme (FRGS 6071300).

References

[1] Setter N and Waser R 2000 *Acta materialia* **48** 151
[2] Voisard C S 1997 *Journal of the European Ceramic Society* **17** 1231
[3] Klissurska R, Brooks K, and Setter N 1999 *Ferroelectrics* **225** 327
[4] Bahanurddin N F K 2015 *Advanced Materials Research* Trans Tech Publ
[5] Devi P 2012 *International J. Edu. Plan. Admin* **2** 97
[6] Kalem V, Çam I, and Timuçin M 2011 *Ceramics International* **37** 1265
[7] Kour P et al. 2014 *Solid State Communications* **190** 33
[8] Silva M S et al. 2015 *Materials Science Forum*. 2015.
[9] Chu S Y and Chen C H 2001 *Sensors and Actuators A: Physical* **89** 210
[10] Adel S and Ahmed B 2013 *Materials Sciences and Application* **4** 478

IOP Conf. Series: Journal of Physics: Conf. Series **1082** (2018) 012041 doi:10.1088/1742-6596/1082/1/012041

Mechanical Properties of Palm Oil Polyol based Polyurethane Foam Reinforced Wollastonite Clay

M.S. Norshahli, L.H. Jun and S.A. Zubir*

School of Materials and Minerals Resources Engineering, Engineering Campus, Universiti Sains Malaysia, Seri Ampangan, 14300 Nibong Tebal, Pulau Pinang, Malaysia.

Email: syazanazubir@usm.my

Abstract. The palm oil polyol based polyurethane (PU) foam reinforced wollastonite clay was prepared to study the effects of wollastonite clay addition towards the mechanical properties of PU foam. The palm oil polyol based PU foam with NCO: OH ratio of 1:1.22 was synthesized by mixing palm oil-based polyol, p-MDI, DMCHA, PMDETA, distilled water and incorporation of different amount of wollastonite clay powder (1, 3, 5 and 7 wt%). The FTIR analysis showed the absence of NCO peak at 2280 cm^{-1} suggesting the complete conversion of isocyanates into urethane during the mixing process. Based on FESEM/EDX, it was found that the wollastonite clay were fitted into the PU matrix, whereby SEM showed an open cell structure of PU foam reinforced wollastonite clay. The density of PU foam reinforced wollastonite clay is increased as the amount of clay reinforcement is increased. This is due to the high viscosity of the mixture which hinder the foam from rising and hence produced a denser foam, thus reduced the pore sizes. Hence, the compressive strength increased as the density and wollastonite weight percentage in the foam is increased.

1. Introduction

Polyurethane (PU) foam is one of the most versatile class of specialty polymer. These foams are classified into three major classes; flexible, rigid and semi-rigid polyurethane foams. Rigid polyurethane foam possess a good adhesion, high durability and strength that make them an indispensable material in the construction industry and useful for insulating materials, polymeric concrete components, sealants and others [1].

Over few decades, research on polyurethane were usually performed using petroleum-based polyol. However, due to the environmental awareness as the arose of issue related to petroleum depletion and the escalation of the petroleum oil's price, had raised concern among researchers to discover the alternative sources for production of polyurethane. Utilizing the natural based polyol mainly from vegetables oil in fabrication of PU foam have been done by researchers mainly using soy-bean oil, rapeseed oil and palm oil. The rigid PU foam derived from soybean polyols were found to have a better thermal and thermo-oxidative stabilities [2]. On the other hand, palm oil polyol has also been reported to exhibit good mechanical properties as it is employed in the fabrication of PU foam [3]. In addition, it is abundant and available in Malaysia. Hence, in this study, POP are used as replacement of petroleum based polyol which is believed improve the properties of the PU foam.

Developing a high class of polymeric materials has become a great interest among researchers in order to produce an improved properties and cost effective polyurethane foam. Besides choosing the types of reactants and processing technique, addition of fillers may potentially improve the properties of the foam. The properties of the polyurethane foam can be well-controlled and improved according to needs by adjusting the filler in terms of the types, shapes, sizes, volume and degree of dispersion [4]. In this present work, we investigate the possibility of using wollastonite clay as reinforcement materials in enhancing the mechanical properties of the rigid polyurethane foam palm oil polyol based for load bearing application. The relationship between pores size, density of the foam and the percentage of fillers addition are critically studied in order to identify the effect of filler addition in the PU foam.

2. Experimental

2.1. Material

Palm oil polyol (POP) and 2,4-aromatic methylene diphenyldiisocyanate (MDI) was obtained from Maskimi Polyol Sdn. Bhd, Kajang Selangor. The wollastonite clay was supplied by Ipoh Ceramic Sdn. Bhd. It was heated at 100°C for 24 hours to remove the residue of water or volatile compounds entrapped in the clay. The calayst, pentamenthyldipropylenatriamine (PMDETA) and dimethyl cetylhexamine (DMCHA) was purchased from Maskimi Polyol Sdn. Bhd.

2.2. Preparation of Palm Oil Polyol Based Polyurethane Reinforced Wollastonite Clay

The PU foam was prepared based on NCO:OH ratio of 1:1.22. Different weight percentage of wollastonite (1, 3, 5 and 7%) were added to reinforce the PU foam. Table 1 shows the composition of each reactant used in synthesizing PU foam. The palm oil polyol (POP), PMDETA and DMCHA were mixed together and distilled water was added as blowing agent. The mixture was mixed using standard propeller with speed 1300 rpm for 2 minutes followed by addition of wollastonite clay at different weight percentage and prolong mixing for one and a half minutes to enhance the homogeneity and obtain greater dispersion of clay in the polyols. MDI was then added into the mixture and stirred for 10 seconds then, the foam was left to rise and set in normal room temperature and atmospheric pressure for 16 hours before demoulded.

Table 1. The composition of reactants used in synthesizing Palm Oil Polyol Based Polyurethane Reinforced Wollastonite Clay.

Sample	Wollastonite Clay Weight Percentage (%)	Wollastonite Clay (g)	p-MDI (g)	POP (g)	PMDETA (g)	DMCHA (g)	Distilled Water (g)
S1	0	0	303.28	306.43	1.21	1.81	0.12
S2	1	6.5	289.58	350.29	1.40	2.10	0.14
S3	3	19.5	283.73	343.21	1.37	2.06	0.14
S4	5	32.5	277.88	336.13	1.34	2.02	0.13
S5	7	45.5	272.03	329.06	1.32	1.97	0.13

2.3. Characterization of Palm Oil Polyol Based Polyurethane Reinforced Wollastonite Clay

2.3.1. Fourier Transform Infrared Spectroscopy (FTIR). The chemical bond characteristic of the foam was determined using FTIR Spectroscopy (PerkinElmer Spectrum One FT-IR Spectrometer). A total of 4 scans were taken for each sample over the wavelength of 4000-600 cm^{-1} at a resolution of 4 cm^{-1}.

2.3.2. Scanning Electron Microscope & Field Emission Scanning Electron Microscope/ Energy-disperse X-ray Spectroscopy (SEM/FESEM). SEM was used to determine the cellular structure differences resulting from the use of different weight percentage of wollastonite clay. The sample of size 10mm x 10mm x 5mm were cut and coated with Au conductive layer were analysed using Hitachi TM3000 with accelerating voltages of 5keV. The wollastonite clay composition in polymeric matrix was captured using FESEM/EDX (SUPRA 35VP, ZEISS).

2.3.3. Free Rise Density (FRD). The paper cup (220 cm^3) are weighed before mixing. The foam was then stabilized for 10 minutes after stop rising. Weight of foam was recorded and the FRD was calculated as shown in equation 1.

$$Free\ Rise\ Density = \frac{[(weight\ of\ cup, g + foam, g) - (weight\ of\ cup, g)]}{volume\ of\ cup, cm^3} \times 1000 \quad (1)$$

2.3.4. Moulded Density (MD). Moulded Density was obtained based on method BS 4270: Part 1:1988 Method 2 by producing foam in the 24.5cm x 24.5cm x 6cm dimensional moulds. The moulded density was calculated based on equation 2.

$$Moulded\ Density = \frac{weight\ of\ foam, g}{volume\ of\ mould, cm^3} \times 1000 \quad (2)$$

2.3.5. Core Density (CD). Core Density was measured using method BS 4370: Part 1: 1988: Method 2. Rigid polyurethane foam was cut into 5 blocks with dimension 50mm x 50mm x 50mm. Each sample was weighted and dimension was recorded. The density of the blocks was calculated using equation 3.

$$Core\ Density = \frac{weight\ of\ block, g}{volume\ of\ block, cm^3} \times 1000 \quad (3)$$

2.3.6. Compressive Testing. The compressive test was carried out using Instron 5982 universal testing machine according to BS4370: Part 1 1988: Method 3. Samples with size of 50 x 50 x 50 mm^3 was compressed to 80% of its original thickness with crosshead speed 5 mm/ min at 23±2°C. The stress-strain curves obtained was used to determine the compressive strength.

3. Result and Discussion

3.1. FTIR analysis

FTIR analysis was done to study the completion reaction and identify the chemical bonding characteristic in PU foam. The foam spectrum of pristine PU foam and PU reinforced wollastonite clay at different percentage are illustrated in Figure 1.

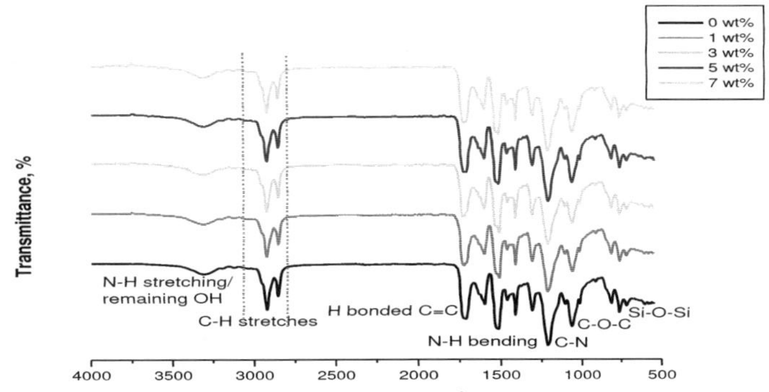

Figure 1. Polyurethane foam spectrum with different weight percentage of wollastonite clay

The presence of hydrogen bonding can be monitored from the group of hydrogen bonded amine (H-NH) and remaining OH groups at 3316 cm^{-1}. The C-H stretches was observed at 2926 cm^{-1} and 2854 cm^{-1} followed by peaks at 1511 cm^{-1}, 1213 cm^{-1} and 1068 cm^{-1} which indicates the urethane linkages of N-H, C-N and C-O which confirmed the reaction between polyol and isocyanates had occured. Hydrogen bonded carbonyl group (H-CO) was observed at 1715 cm^{-1} which indicates the formation of inter-urethane hydrogen bonding between N-H and C-O. It can be seen that there is no trace of NCO peaks at 2280 cm^{-1} suggesting the complete reaction had occurred in formation of PU [5]. Extra peaks were expected to be found in the wavenumber range of 800-1250 cm^{-1} which is assigned to the asymmetric Si–O stretching modes within a SiO4 tetrahedron, the peaks in wavenumber range of 600-800 cm^{-1} are attributed by both Si–O–Si bond bending and symmetric Si–O stretching vibrations, and lastly those in the wavenumber range (400-600 cm^{-1}) are originated from O–Si–O bending and Ca–O stretching modes [6]. FESEM/EDX was used to observe the presence of wollastonite clay particle under the magnification level of 3000x at two different spot. From Figure 2, it was found that the wollastonite clay does fitted into the foam matrix as there is trace of calcium and silicon elements observed for 7% wollastonite reinforced PU foam.

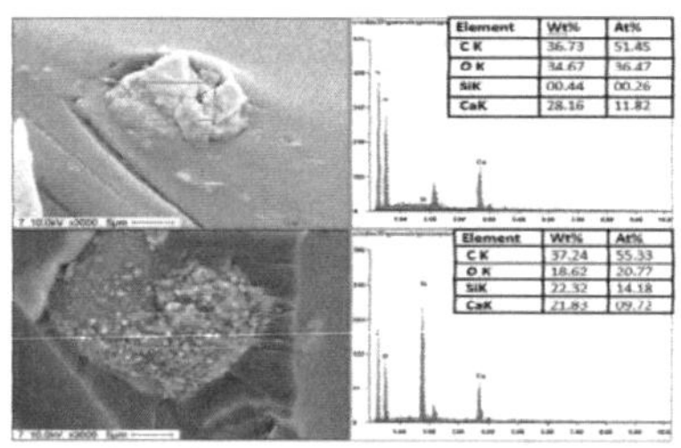

Figure 2. FESEM/EDX results of 7 wt% wollastonite reinforced polyurethane foam

3.2. Density and Pore Size Analysis of Wollastonite Clay reinforced PU foam

The density of the wollastonite clay reinforced polyurethane foam was found to increase with the amount of wollastonite clay. Figure 3 showed the core and mould density calculated using the equation (2) and (3). The wollastonite clay have a relatively higher density value as compared to the polyurethane foam matrix, thus increase the overall density of the samples [7]. Besides, the viscosity of the mixture with wollastonite clay would also affect the density of the foam. The higher the amount of wollastonite clay added into the mixture, will increase the viscosity of the mixture. The high viscosity of the mixture will then retard the foam from rising which therefore produced a denser foam.

Table 2. Pore size of the wollastonite reinforced polyurethane foam for different wt% of wollastonite

Sample No.	Wollastonite Clay Weight Percentages, %	Pore Size, μm Average	Pore Size, μm Standard Deviation
S1	0	799.8	61.34
S2	1	596.2	51.92
S3	3	501.2	35.37
S4	5	484.4	66.10
S5	7	417.4	58.91

The pore size of the wollastonite clay reinforced PU foam is influenced by the amount of wollastonite clay in the mixture. The result shown in Table 2 shows that the pore size of the foam decreased as the amount of wollastonite clay is increased. The cell structure of the wollastonite reinforced polyurethane foam is an open-cell type as shown in Figure 4. In open-cell structure, the cell walls are broken and consist of mainly ribs and struts, thus provide better absorptive capability. It is revealed that S1 have fewer cells and larger cell size as compared to the composites foam having 1, 3, 5 and 7 wt% of wollastonite clay. The average cell size value 800 μm of S1 was decreased to 596, 501, 484 and 417 μm with the incorporation of 1, 3, 5 and 7 wt% clay respectively. S3 showed a uniform cell size distribution as results from uniform distribution of nucleation agents in the foam matrix [8].

The decreasing of cell size due to the existence of wollastonite clay particles that exfoliate in the foaming system acting as a cell opener. During cell formation, the clay particles act as a nucleation site and since in cellular structure, there are more number of cells that start to nucleate at the same time while adequate gas are available for cell growth hence resulting into a more compact size of the cell [9]. Heterogeneous nucleation occurred faster than homogeneous nucleation as the nucleation barrier is much lower at the surface. Therefore, as the amount of wollastonite clay is increased in the formulation of foam, a smaller pore size and denser foam is obtained as compared to lower amount of wollastonite added in the polyurethane foam if the volume of both samples are the same [10].

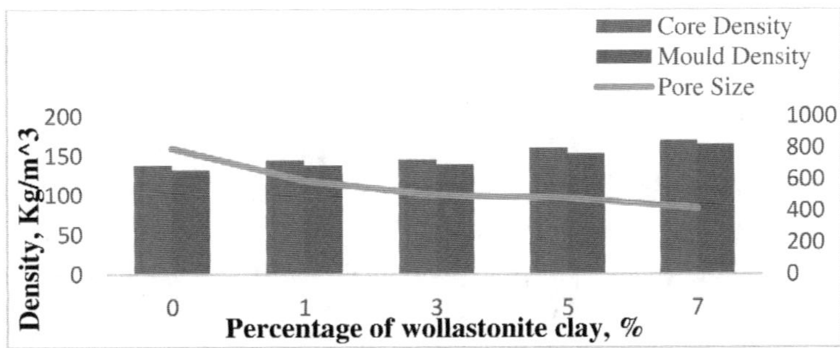

Figure 3. Graph of core density and mould density of the wollastonite reinforced polyurethane foam

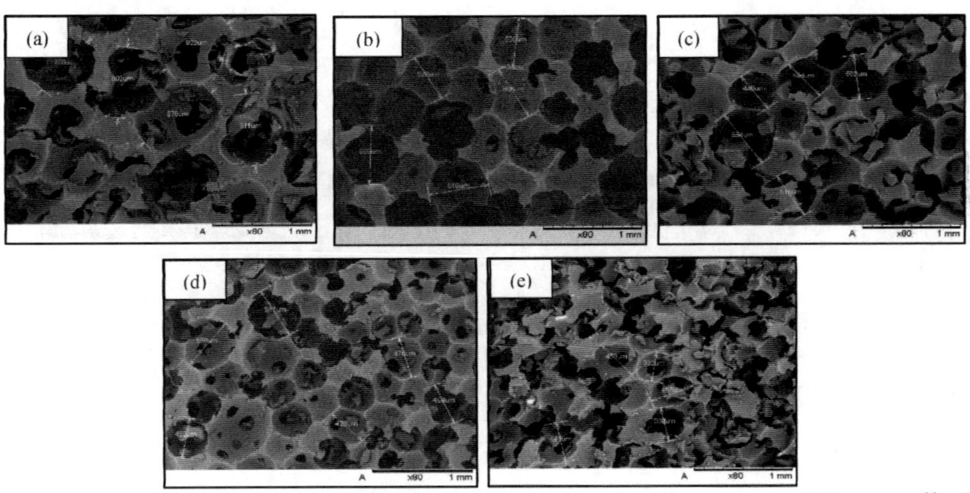

Figure 4. SEM images of wollastonite reinforced polyurethane foam's cell structure at different wollastonite percentage in the foam (a) 0%; (b) 1%; (c) 3%; (d) 5%; (e) 7% at 800X.

3.3. Compressive Strength of Wollastonite reinforced PU foam

Compressive strength is related to the load-bearing capacity that is an important factor in foam application. Thus, the effect of density towards mechanical properties was investigated. The relationship between density, compressive strength and the weight percentage of wollastonite are shown in Figure 5. It can be seen that the compressive strength of the foam linearly increased with the density and wt% of wollastonite in the foam. This is believed to occur due to well distribution of wollastonite clay within PU foam matrix and mixing process. The mixing of the particles in low viscosity monomer using high shear mixer resulting in higher rate of diffusion of composites into the matrix [8]. Thus, improved interlayer interaction between the particles and the PU matrix resulting in denser foam formation with reducing cell size.

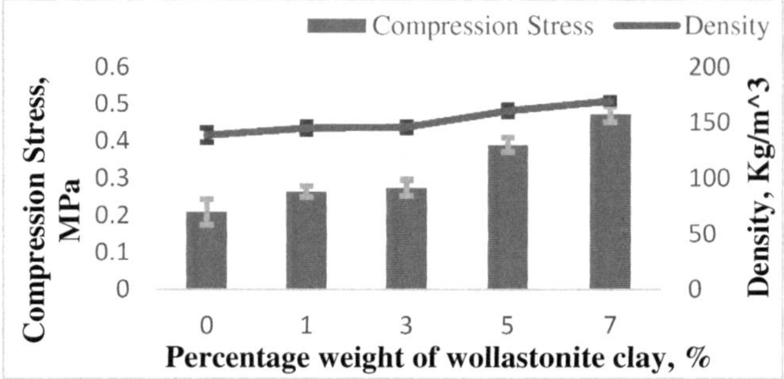

Figure 5. Graph of compressive strength and density against the weight percentage of wollastonite clay

4. Conclusion

The addition of filler have improved the properties of PU foam. The correlation between filler percentage, density, pore size and compressive strength was successfully studied. The increment in filler percentage, density and compressive strength resulted in reducing the pore size, thus, improved the rigidity of palm oil polyol based polyurethane foam.

References

[1] Yang L T, Zhao C S, Dai C L, Fu L Y and Lin S Q 2012 *Journal of Polymers and the Environment* **20** 230-236

[2] Guo A, Javni I and Petrovic Z 2000 Journal of Applied Polymer Science **77**(2), pp.467-473

[3] Dzulkifli M H, Yahya M Y, Akhir F S M and Majid R A 2014 *Jurnal Teknologi* **68**

[4] Madaleno L, Pyrz R, Crosky A, Jensen L R, Rauhe J C M, Dolomanova V, De Barros A M M V, Pinto J J C and Norman J 2013 *Composites Part A: Applied Science and Manufacturing* **44** 1-7

[5] Marques J E 2013 Ignition, Combustion and Detection of Polyurethane Foams with ammonium nitrate and sodium bicabornate additives.

[6] Singh U P & Ray B C 2012 *International Journal of Engineering Research and Technology, ESRSA Publications*

[7] Fan H, Tekeei A, Suppes G J & Hsieh F H 2012 *Inter J Polym Sci* 2012

[8] Lee L J, Zeng C, Cao X, Han X, Shen J and Xu G 2005 *Composites science and technology* **65** 2344-2363

[9] Panda S S, Samal S K, Mohanty S and Nayak S K 2017 *Journal of Composite Materials* 0021998317710707

[10] Ali E S & Ahmad S 2012 *Composites Part B: Engineering* **43** 2813-2816

IOP Conf. Series: Journal of Physics: Conf. Series **1082** (2018) 012042 doi:10.1088/1742-6596/1082/1/012042

Influence of Soda Lime Glass Addition on the Dielectric Properties of CCTO Ceramics

MF Ab Rahman[1], NA Rejab[1], ZA Ahmad[1], JJ Mohamed[2], MF Ain[3], and S Abdullah[1]

[1]Structural Materials Niche Area, School of Materials and Mineral Resources Engineering, Universiti Sains Malaysia, Engineering Campus, 14300 Nibong Tebal, Penang, Malaysia
[2]Faculty of Bioengineering and Technology, Universiti Malaysia Kelantan Kampus Jeli, Karung Berkunci No. 100, 17600 Jeli, Kelantan, Malaysia.
[3]School of Electrical and Electronic Engineering, Universiti Sains Malaysia, Engineering Campus, 14300 Nibong Tebal, Penang, Malaysia

*akmarnik@usm.my

Abstract. The dielectric properties of CCTO ceramics with soda-lime glass addition was investigated. The addition of soda lime glass was varied from 0 - 0.05 wt.% by solid-state reaction. Both XRD patterns of CCTO and CCTO-glass sintered samples showed the formation of CCTO phase with minor peaks of CuO phase, respectively. The phase area analysis of CCTO and CCTO-glass proved that the CCTO phase was found to be decreased while the grain boundaries area of CuO + glass was then increased after the addition of soda-lime glass. The dielectric properties of corresponding samples indicated that both ε_r and tan δ were reduced with glass addition. Thus, it shows that the soda lime glass addition has a significant effect on dielectric properties of CCTO ceramics.

1. Introduction

Nowadays, materials with high dielectric constants and low dielectric loss have attracted a big interest for technological electronic applications such as capacitors, resonators and filters. Hence, $CaCu_3Ti_4O_{12}$ (CCTO) perovskite-like compound has received increasing interest since it has reported to obtain high dielectric constant ($\varepsilon_r \sim 10^4$) at room temperature [1, 2]. However, CCTO also suffering high dielectric loss (> 0.05 at 1 kHz) [3, 4]. Thus, this prevents CCTO to be commercialized. The densification of ceramics body play a vital role to improve the dielectric properties of ceramics. Addition of another material into stoichiometric ratio of CCTO has been shown as an effective method to improve the dielectric properties of CCTO.

There are several researchers reported on the fabrication of CCTO ceramic composites. Previous researchers had put some glass materials into CCTO in order to improve densification of ceramics body thus can reduce dielectric loss [5-7]. Wang et al. (2016) [6] reported that, at 1 MHz, the tan δ of 1wt.% and 3 wt.% of SBS glass added into CCTO were ~0.5 and ~0.7, respectively. Nevertheless, the effect of soda-lime glass addition into CCTO on the dielectric properties of CCTO are still not been explored. To

IOP Conf. Series: Journal of Physics: Conf. Series **1082** (2018) 012042 doi:10.1088/1742-6596/1082/1/012042

the best of our knowledge, the crystallite structure, surface morphology, and dielectric properties of CCTO have been studied for the first time.

Hence, in this study, the effects of soda-lime glass addition on the dielectric properties of CCTO ceramics characterized at high frequencies (1 GHz) were investigated.

2. Experimental Method

CCTO ceramics was prepared via solid state reaction. The raw materials of $CaCO_3$ (Sigma Aldrich, >99%), CuO (Sigma Aldrich, >99%) and TiO_2 (MERCK, >99%) powder was mixed in wet condition for 24 hours and the mixture was dried for overnight in the oven. The dried mixtures were then calcined at 930 °C for 12 hours. The soda-lime glass was milled by using the planetary ball milling machine to form fine powder. The glass powder was mixed with the calcined CCTO powder for 24 hours with the ratio of corresponding powders to the following chemical composition: (100-x) % $CaCu_3Ti_4O_{12}$ + x % glass additives (wt.%), where x = 0, 0.05 and 0.5, respectively. The mixed powder was pressed into pellets shape under a pressure of 250 MPa. The sample was labelled as CCTO for pure CCTO sample while SLG0.05 and SLG0.5 for the samples of CCTO with addition of 0.05 and 0.5 wt.% of soda-lime glass, respectively. The pellets were then sintered at 1040 °C for 10 hours. The XRD (Bruker Advance D8) was used to analyze the phase formation of pure CCTO and CCTO-borosilicate glass composites while Scanning Electron Microscopy (SEM) (TM 3000) was used to observe the microstructures of the samples. For electrical measurement analysis, the dielectric properties of the samples were measured by using impedance analyser machine (4291B Hewlett Packard) at 1 GHz.

3. Results and Discussion

3.1 X-ray diffraction analysis

Figure 1. shows the XRD patterns of CCTO-glass based composites (0.01-0.05 wt.%) compared to pure CCTO. Since the maximum amount of each glass added was only 1 wt.%, the major peaks observed are belong to CCTO in all composites. For each sample, it is observed that minor CuO phase also appeared in XRD patterns. CuO peaks were seen at ~35.5° and ~38.6°. The CCTO phase is compared with ICDD Data File Card No. 01-075-2188 whereas CuO phase is matched with ICDD Data File Card No. 00-045-0937. The CuO phase existing for each sample is from the CuO compound that precipitate or segregated at the grain boundary of CCTO. Similar kind of results were also reported by Ab Rahman et al. (2016) [8] where the presence of CuO compound was observed at the grain boundary of CCTO after being sintered at 1040 °C for 10 hours. According to Yuan et al. (2013) [9], the intergranular CuO phase is due to the instability (or activity) of Cu ions in the CCTO lattice. Cu ions were first separated out from CCTO at ~1000 °C, then were ousted to the surface layer of the pellet and mostly segregated at the grain boundaries and were finally oxidized to CuO compound.

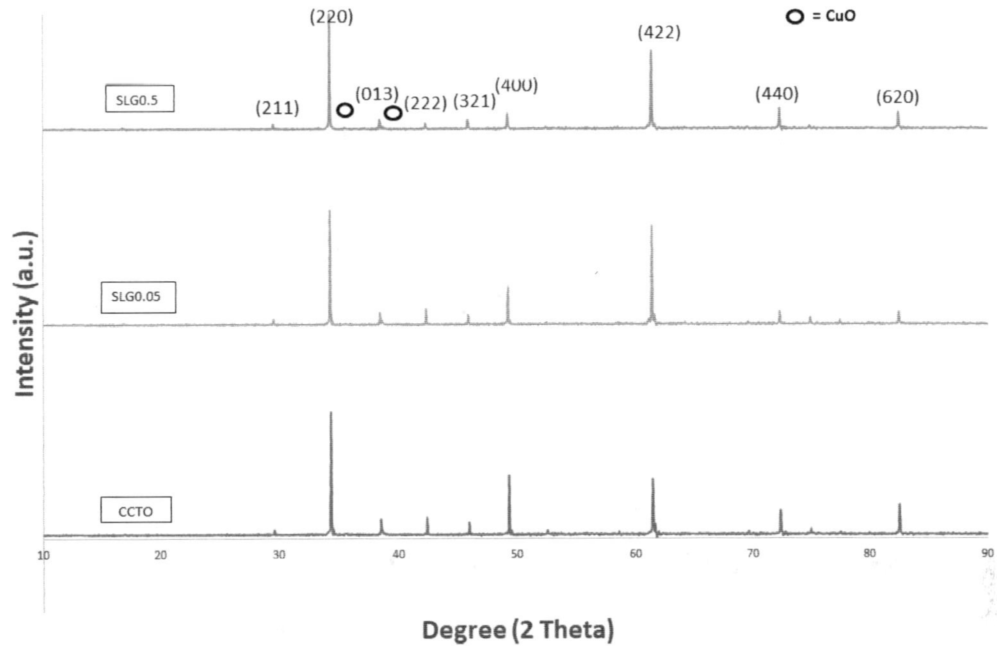

Figure 1. X-ray diffraction patterns of the sintered CCTO and CCTO-soda-lime glass composites

3.2 Phase area analysis

Figure 2. shows the area percentages of the grain and the grain boundary of CCTO and CCTO-soda-lime glass-based composites calculated by using i-Solution DT software where the grain, grain boundaries and the porosity were represented by the orange, green and blue colors, respectively. Based on these figures, it was seen that the grain area was decreased whereas the grain boundaries area was increased by increasing the glass addition into CCTO ceramics. The grain area was decreased from 87.90% (CCTO) to 83.05% (SLG1) while grain boundary area was increased from 9.78% to 15.99% for the same samples. Thus, it shows that the glass addition seems to decrease the grain area and increase the grain boundaries of CCTO. Patel et al. (2015) [10] also have reported where with increasing the glass content, the grain boundary of $BaTiO_3$ ceramics becomes larger and the average grain separation was increased.

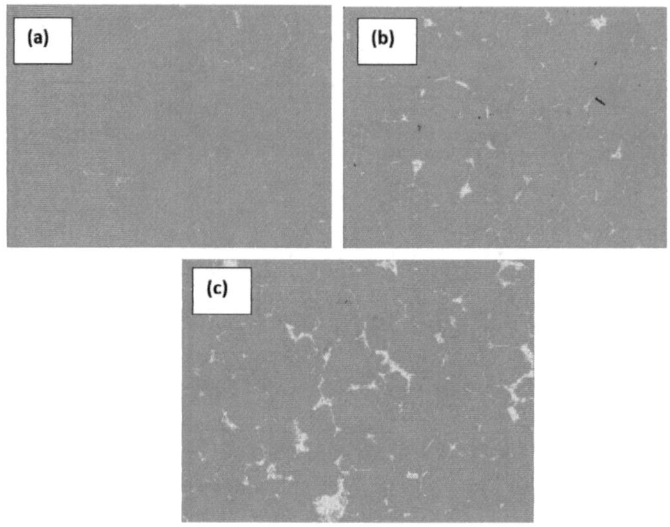

Figure 2. Phase area micrograph of CCTO different soda-lime glass composition (a) 0 wt. %, (b) 0.05 wt. % and (c) 0.5 wt. %.

3.3 Dielectric properties analysis

Figure 3 showed the ε_r and tan δ of CCTO and CCTO-SLG glass composites sample. At frequency 1 GHz, all CCTO-SLG glass composite samples exhibit lower ε_r than CCTO sample where SLG0.5 had lowest ε_r of 59. Meanwhile, the tan δ of CCTO was reduced from 0.76 (CCTO) to 0.54 (SLG0.5). Glass is known to have low tan δ [6]. The addition of low tan δ glass materials into the high tan δ ceramic materials can help to reduce tan δ of ceramic materials.

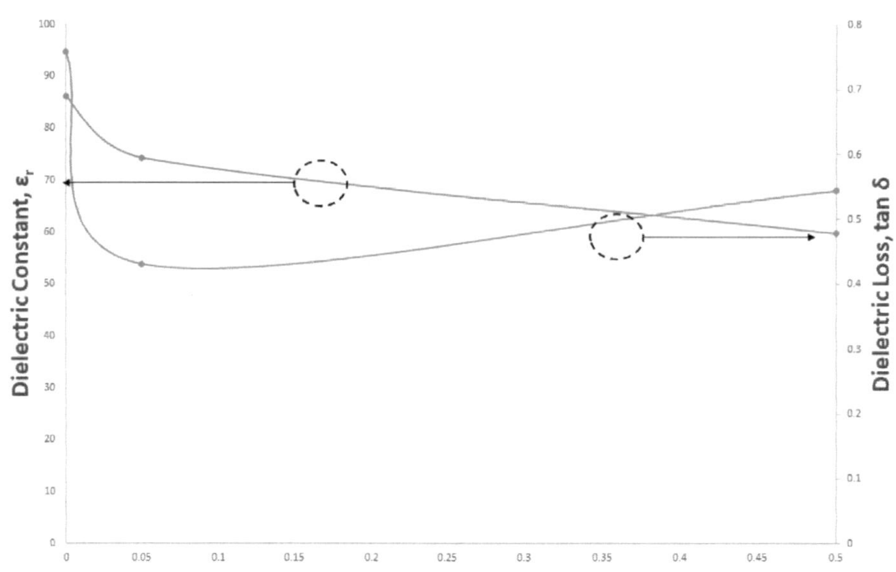

Figure 3. ε_r and tan δ of CCTO and CCTO-soda-lime glass composites at 1 GHz

4. Conclusion

CCTO and CCTO-soda-lime glass composites were successfully prepared via solid state reaction. From the XRD analysis, the formation of CCTO phase alongside with minor CuO phase were found to be seen for each sample which were sintered at 1040 °C for 10 hours. The analysis of area percentage showed that the grain size of CCTO was decreased while the grain boundaries were increased with increasing glass concentration. The ε_r of CCTO was increased while the tan δ of CCTO was decreased with small addition of soda-lime glass. Thus, it shows that the soda-lime glass addition into CCTO can give the greatest effect on the dielectric properties of CCTO ceramics.

Acknowledgments
The authors gratefully acknowledge to the Ministry of Higher Education (MOHE), Malaysia for awarding MyBrain15 and Universiti Sains Malaysia (RUI No. 1001/PELECT/8014009) for supporting this work.

References

[1] Subramanian MA, Li D, Duan N, Reisner BA and Sleight AW 2000 *J Solid State Chem.* **151** 323.
[2] Subramanian MA and Sleight AW 2002 *Solid State Sci.* **4** 347.
[3] Li W, Zhang T, Liu S, Lu Z and Xiong R, 2017 *Ceram. Int.* **43** 4366.
[4] Almeida-Didry SD, Autret C, Lucas A, Honstettre C, Pacreau F and Gervais F 2014 *J. Eur. Ceram. Soc.* **34** 3649.
[5] Prakash BS and Varma KBR 2007 *J. Solid State Chem.* **180** 1918.
[6] Wang B, Pu Y, Wu H, Chen K, Xu N 2012 *J. Mater. Sci. Mater. Electron* **23** 612.
[7] Mohamed JJ, Ab Rahman MF, Ain MF, Ahmad ZA 2016 *Mater. Sci. Forum* **840** 52.
[8] Ab Rahman MF, Abu MJ, Zaman RA, Mohamed JJ, Ain MF, Jamil MKM and Ahmad ZA 2017 *Mater. Sci. Forum* **888** 23.
[9] Yuan WX, Luo Z and Wang, C 2013 *Journal of Alloys and Compounds* **562** 1.
[10] Patel S, Chauhan A, Vaish R and Thomas P 2015 *Journal of Asian Ceramic Societies* **3** 383.

IOP Conf. Series: Journal of Physics: Conf. Series **1082** (2018) 012043 doi:10.1088/1742-6596/1082/1/012043

The Effect of Aging on the Transformation Temperatures and Microstructure of Cualni-Mn Shape Memory Alloys

Ali. A. Shakir[1], T. Abubakar[1]*, Safaa.N.Saud[2]

[1]Faculty of Mechanical Engineering, Universiti Teknologi Malaysia, 81310 UTM Johor Bahru, Johor, Malaysia
[2]Faculty of Information Science and Engineering, Management and Science University, 40100 Shah Alam, Selangor, Malaysia

Email: tuty@utm.my

Abstract. Shape memory alloys (SMAs) are one of the unique classes of smart materials, which have the ability to recover their shape when the temperature is applied. Depending on the alloy composition, shape memory alloys (SMAs) properties are largely related to the thermoelastic transformation, which occurs in the temperature range from -100 to 273 °C. This study presents the effect of aging treatment on the phase transformation temperatures, microstructure and hardness of Cu-Al-Ni-0.4 wt. % Mn SMAs. Various aging times (1-5) hour along with different aging temperatures (300 and 400°C) were applied. The phase transformation temperatures, microstructural changes and hardness were evaluated using differential scanning calorimetry (DSC), optical microscopy and Vickers hardness test. It was found that the phase structure and properties of the alloy are varied in accordance to the aging treatment parameters. It was also observed that during the aging treatment, the grain size of Cu-Al-Ni-Mn SMA are varied in the terms of the size and thickness. Due to this, the values of hardness were observed to decrease at elevated temperatures and prolonged aging times.

1. Introduction

Shape memory alloys (SMAs) are one of the unique classes of smart materials, which have the ability to recover their shape when the temperature is applied [1]. Depending on the alloy composition, shape memory alloys (SMAs) properties are largely related to the thermoelastic transformation, which occurs in the temperature range (-100 to 273 °C) [2]. Their unusual properties are due to reversible changes of crystal lattice under influence of temperature, pressure, electrical field and magnetic, etc. [3]. The transformation of SMAs occurs between two phases' austenite and martensite, which the phases between a high temperature called austenite and low temperature called martensite. The austenite and martensite transformation start and finish temperatures are A_s, A_f, M_s and M_f, respectively [2, 4-7]. The main types of SMAs are Ni-Ti, Cu-based and Fe-based alloys [1, 8]. The main advantages of Cu-based SMAs are their low price, easy fabrication, excellent electrical and thermal conductivity and high transformation temperature compared to other SMAs [8]. The properties of CuAlNi alloys are superior to those of CuZnAl alloys due to their wide range of useful transformation temperature and small hysteresis [9, 10]. Therefore, CuAlNi alloys more suitable to work in high temperature applications. CuAlNi shape memory alloys are less prone to the stability of the obtaining phase due to the high brittleness that may relate to the type and structure of this phase that may obtain a limit by the applications of these alloys [11, 12]. To overcome this problem, there are two ways; the addition of grain-refiners to the melt is one of the most common methods used to refine the grain size of the alloys that are produced in bulk [13-15], and apply aging treatment [16,17], which they are able to adjust the phase diagram and improve the mechanical properties. The present paper is focused on the microstructure and transformation temperature variation of CuAlNi-Mn SMAs alloys after different aging conditions were applied, such as aging time and temperatures.

2. Experimental procedure

The alloy used for this study was prepared by melting the metals of Cu, Al, Ni, and Mn in induction furnace and then pouring into a mild steel mold with a rectangular ingot shape of 270mm x 50mm x 20mm. The chemical composition analysed using an inductively coupled plasma mass spectrometry (ICP-MS) as Cu-11.9 wt. %Al-4wt. %Ni-0.4wt. % Mn SMA. After preparing the samples from the ingot by EDM wire cutting with a final dimension (10mm×10mm×5mm), the samples were homogenized at 900°C for 30 min and then quenching in water bath. Followed by, the aging treatment process performed for two groups of CuAlNi-Mn: these groups consist of five samples treated at different aging temperatures of (300 and 400) °C for various aging time of (1- 5) hr and the aged samples were quenched in water. The microstructure in terms of grain size and morphology was observed by using an optical microscope. The Vickers method with a load of 10 kg and 20s was used to measure the hardness of samples.

3. Results and discussion

<u>3.1 Transformation temperature</u>

In order to determine the phase transformation temperatures of the CuAlNi-Mn shape memory alloy, DSC analysis was carried out. All the transformation temperatures were according to Figure 1. It is well known that SMAs undergo a thermally induced austenite ↔ martensite transformation characterized by the critical temperatures; A_s, A_f, M_s and M_f. Here A_s and A_f refer to austenite transformation start and finish temperatures, respectively, during heating while M_s and M_f refer to martensite transformation start and finish temperatures, respectively, during cooling.

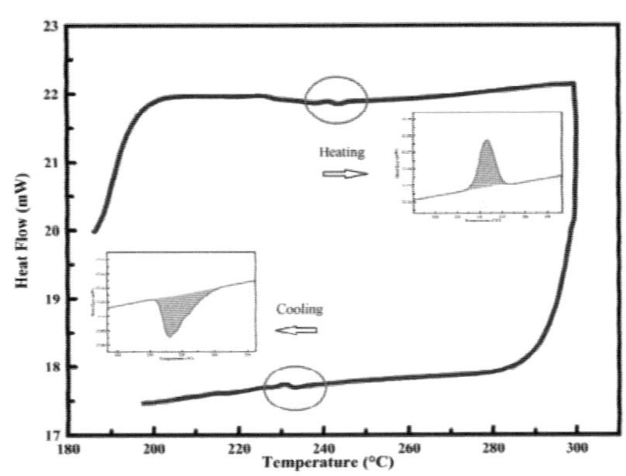

Figure 1. DSC result of unaged and aged samples at 300°C.

Small peaks were identified in the DSC curves as shown in Figure 1. An endothermic peak was observed in the temperature range of 238-245°C, and an exothermic peak was observed in the temperature range of 230-245°C for unaged and aged samples. These peaks are most likely corresponding to the reverse martensitic transformation and forward martensitic transformation. From the obtained result of DSC, the transformation temperatures and the value of the heat flow effected directly on the austenite ↔ martensite transformation curve in the shape and location. The transformation hysteresis (A_f-A_s) and (M_s-M_f), with aging temperatures, remained almost constant. From Figure 2 for aging temperature at 300°C and 400°C noted slight variations in transformation temperatures at increasing aging times at aging temperatures.

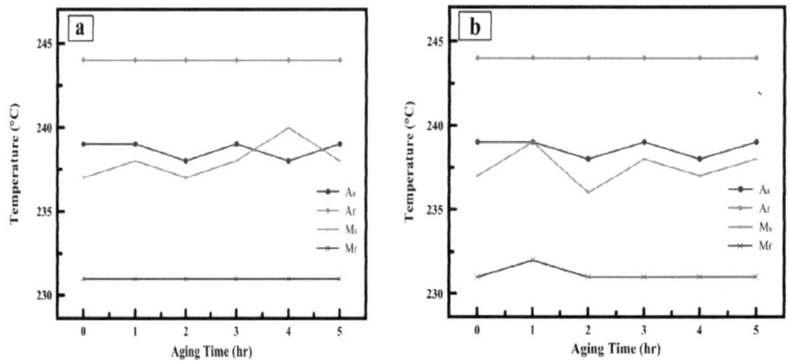

Figure 2. Changes of transformation temperatures with various aging times in CuAlNi-Mn alloy a) at 300°C and b) at 400 °C.

3.2 Grain size and hardness

3.2.1 Grain size. Figures 3 and 4 show the optical micrograph results for unaged and aged treatment at different temperatures; 300 °C and 400 °C and at different times (1, 2, 3, 4 and 5 hrs). The aging parameters were chosen based on the previous research [18]. All samples were etched using solution consist of 2.5g ($FeCl_3.6H_2O$) and 48ml methanol in 10 ml HCl for 20 seconds. Depending on the results shown from previous figures of optical micrographs for CuAlNi-Mn shape memory alloys (SMAs), we can calculate the average value of longitudinal grain size before and after aging [19], for aging we have two aging; temperatures 300 and 400 ° C, and five aging times (1hr, 2hrs, 3hrs, 4hrs and 5hrs). It can be observed that grain size increased with increasing the aged temperature. The grain before aged was (173μm), and at the applying heat-treatment at temperatures 300 and 400 °C, the grain size became at first hour (217.9 and 258.33 μm) and then started to increase with increasing duration of temperatures 300 and 400 °C five hours for each temperatures (302.5 and 290 μm), respectively. From Figure 3, overall it is shown that upon the aging treatment, the grain size increased with increase the aged temperature, the time of aging treatment influence on the grain boundary. The grain size in the optical micrographs increased with increasing the time of aging of each aging temperature compared with unaged sample. The similar optical micrograph changes observed for the aging temperature at 400 °C.

3.2.2. Hardness. The value of Vickers hardness for aged CuAlNi-Mn shape memory alloys for 300°C and 400°C at five different aging times were investigated. The effect of aging time and aging temperature on the hardness of the CuAlNi-Mn shape memory alloys are shown in Figure 4. As overall the aging effect exhibits the decreasing in Vickers hardness value with the increasing of aging time at 300°C and 400°C. The value of hardness is highest for unaged specimen which is 360 MPa, while when undergo the heat-treatment the value of hardness starts to decrease. It proved that the value of hardness not only depend on temperature but also depend on the time of aging treatment. Overall the value of hardness is actually related to the changes of microstructural during aging where the microstructure possesses decreasing in grain size lead to increase the value of hardness as Figure 5.

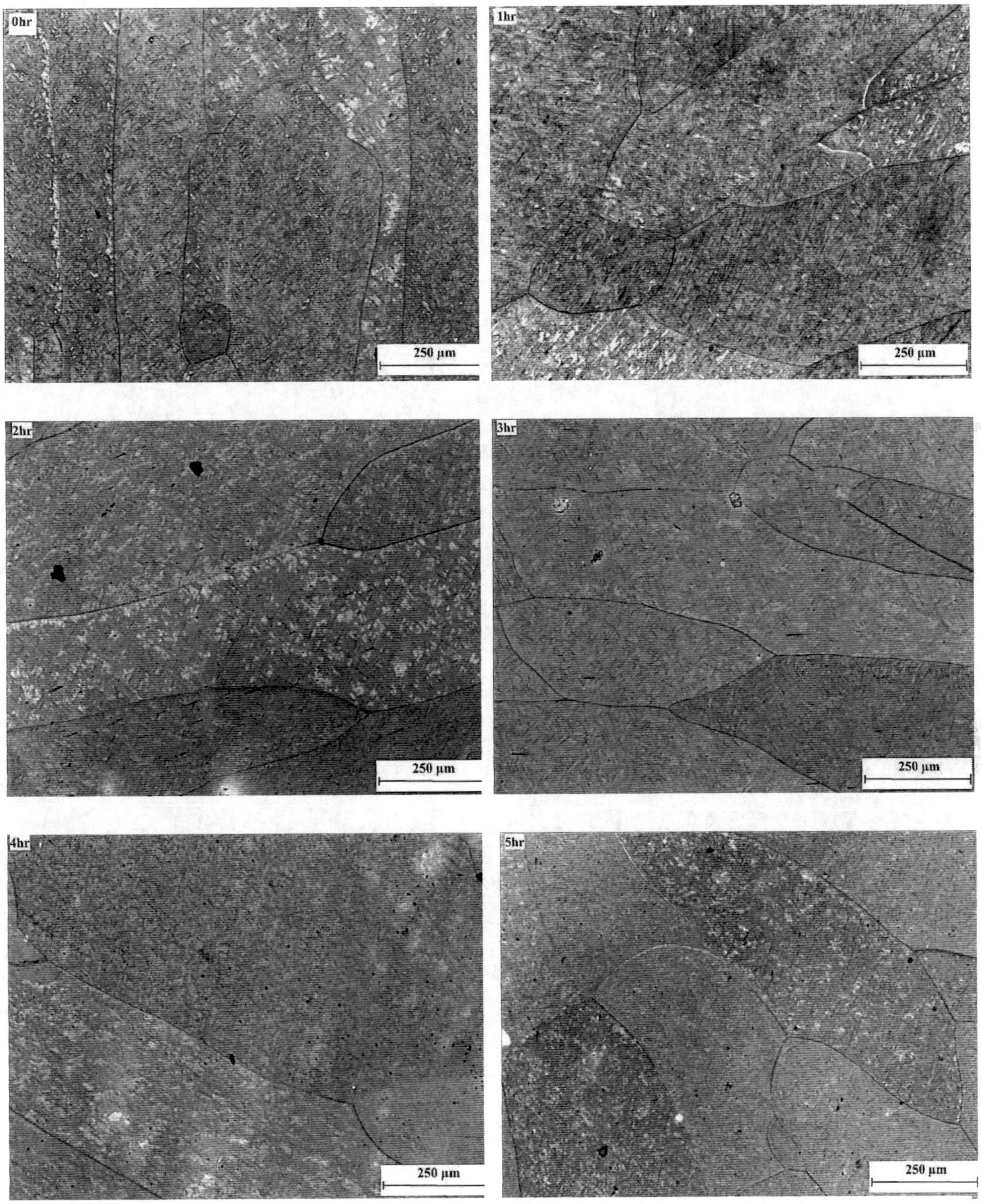

Figure 3. Optical micrographs of unaged and aged samples of CuAlNi-Mn (SMAs) at 300 °C and different aging times

IOP Conf. Series: Journal of Physics: Conf. Series **1082** (2018) 012043
doi:10.1088/1742-6596/1082/1/012043

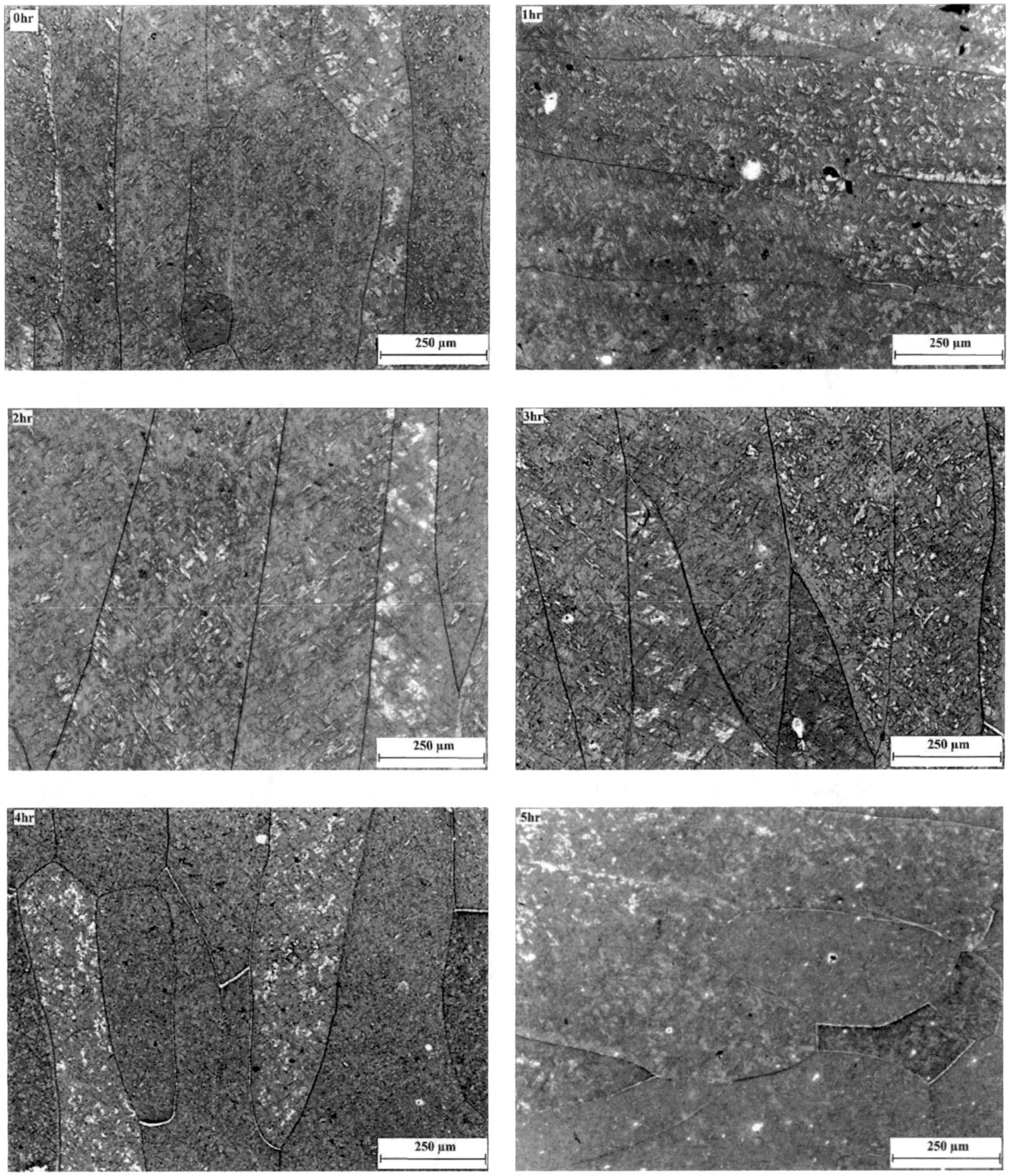

Figure 4. Optical micrographs of unaged and aged samples of CuAlNi-Mn (SMAs) at 400 °C and different aging times

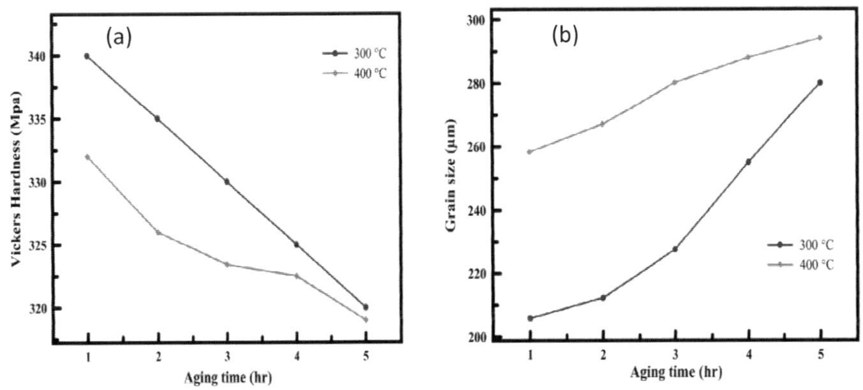

Figure 5. The effect of aging time and aging temperature; (a) on the hardness and (b) on grain size

4. Conclusions

As conclusions, the aging time and aging temperature influence the transformation temperatures, grain size and hardness of the CuAlNi-Mn shape memory alloy. The transformation temperatures (A_s, A_f, M_s and M_f) had a slight variation with increasing aging time of each aging temperature whereas the grain size was observed increasing with the increase of the aging temperatures and aging times. The values of Vickers hardness were observed to decrease at elevated temperatures and longer aging times.

Acknowledgement

The authors gratefully acknowledge the financial support for this research provided by Ministry of Higher Education (MOHE) and Universiti Teknologi Malaysia under Fundamental Research Grant Scheme (FRGS) vote number 4F945.

References

[1] Gandhi V and Thompson BD 1992 *Springer*
[2] Suresh N and Ramamurty U 2008 *Journal of Alloys and Compounds* 449 1
[3] Tatar C and Zengin R 2005 *Thermochimica acta* 433 1
[4] Otsuka K, and Wayman CM 1999 *Cambridge University Press*
[5] Lojen G, Anzel I, Kneissl A, Krizman A, Unterweger E, Kosec B and Bizjak M 2005 *J Mater Process Technol.* 220 9
[6] Izadinia M, and Dehghani K 2011 *Trans Nonferrous Met Soc China* 21 2037
[7] Sobrero CE, Roca PL, Roatta A, Bolmaro RE and Malarria J 2012 *Mat Sci Eng A*. 536 207
[8] Wei ZG and Sandstrom R 1998 *Journal of Materials Science* 33 15
[9] Goji M 2013 *Materiali in tehnologije* 47 2
[10] Wang Z, Liu XF and Xie JX 2012 *Materials Science and Engineering* A 532
[11] de Albuquerque, Victor HC. 2010 *Materials & Design* 317
[12] Xiao Z 2014 *Materials & Design* 58
[13] Sure GN and Brown LC 1985 *Scr. Metall.* 19 401
[14] Sure GN and Brown LC 1984 *Metall. Trans.* A 15 1613
[15] Sampath V 2005 *Smart materials and structures* 14 5
[16] Balo Ş, Nevin and Neslihan S 2012 *Thermochimica* A 536.
[17] Dunne DP Kennon NF 1981 *Met Forum* 4 176
[18] Wei, Z.,Peng, H.,Zou, W and Yang, D. 1997 *Metallurgical and Materials Transaction A* 28(4)
[19] Hansen and Niels 2004 *Scripta Materialia* 51 8

IOP Conf. Series: Journal of Physics: Conf. Series **1082** (2018) 012044 doi:10.1088/1742-6596/1082/1/012044

Hierarchical Porous α-Fe_2O_3 Formation by Thermal Oxidation of Iron as Catalyst for Cr(Vi) Reduction

Subagja Toto Rahmat[1], Ong Yee Chin[1], Wai Kian Tan[2], Go Kawamura[3], Atsunori Matsuda[3], and Zainovia Lockman[1*]

[1] Green Electronic nanoMaterials Group, Nanomaterials Niche Area (School of Materials and Mineral Resources Engineering, Engineering Campus, Universiti Sains Malaysia, 14300 Nibong Tebal, Penang, Malaysia)
[2] Center for International Education (Toyohashi University of Technology, Toyohashi, Aichi, 441-8580, Japan)
[3] Department of Electrical and Electronic Information Engineering (Toyohashi University of Technology, Toyohashi, Aichi, 441-8580, Japan)

Email: zainovia@usm.my

Abstract. α-Fe_2O_3 is a semiconductor photocatalyst that can adsorb and reduce heavy metal ions from contaminated water. Here, hierarchical porous structured α-Fe_2O_3 was synthesised by thermal oxidation of iron wire at 400 °C – 700 °C in the presence of water vapour for 1 hour. The mechanism of formation of the iron oxide nanowires is proposed to follow stress-driven mechanism and when the nanowires merged, nanoblades resulted hierarchical porous structure. Field emission electron microscope (FESEM) images of the oxidized iron had shown the formation of surface oxide comprising of a hierarchical porous structures. High resolution transmission electron microscope (HRTEM) and Raman spectroscopy results confirmed that the iron oxide is consisted of α-Fe_2O_3 in the surface whereas Fe_3O_4 and FeO are in the inner layer. The oxides were immersed in Cr(VI) solution and illuminated under sunlight to produce reducing electrons. The highest reduction precentage of Cr(VI) at pH 2 on the hierarchical porous structure is 80.78% for synthesized sample at 500 °C. It may be due to the higher surface area of the porous hierarchical structure which provide more catalytic reaction sites hence improving the photocatalytic activity.

1. Introduction

Chromium is one example of toxic heavy metal pollutant in industrial effluents. When discharged to water bodies, unlike most organic contaminants, being inorganic, chromium ions do not undergo degradation. This impose severe treats to aquatic life and through bioaccumulation and food chain, chromium ions can find their way to human. The extent of toxicity depends on the specific metal oxidation state with Cr(VI) to be known as highly toxic than Cr(III). At low pH, the most soluble,mobile and toxic forms of Cr(VI) is dichromate ($Cr_2O_7^{2-}$) while at higher pH, Cr(VI) exists as chromate (CrO_4^{2-}) [1]. Both are harmful to human and the environment.

Various methods have been employed to remove Cr(VI) from wastewater such as ion exchange, electrochemical treatment, chemical precipitation, membrane filtration, adsorption, and photocatalytic technology [1 - 2]. Among all these, photocatalytic technology is one of the most efficient remediation strategies. The process utilize oxidation–reduction reactions on a solid photocatalyst surface, converting Cr(VI) to Cr(III). Electron donors are produced from the photocatalysts when illuminated. In this work, we produced photocatalyst in nanoscale as nanoscale material has higher surface area for more sites for sorption and reaction. We choose iron oxide as the catalysts since it is low cost, non-toxic and abundant. More importantly, it can be activated under sun light.

When iron oxide is illuminated under light irradiation with energy higher than its band gap, electron-hole pairs are generated. The photoelectrons must then diffuse to the surface of the oxide and can then be transferred out to adsorbed Cr(VI) ions on the oxide surface. Reduction will take place to produce Cr(III). The use of material in nanoscale can help in improving the charge transfer process as electrons will only need to travel a shorter distance to the surface for reduction to take place. This can reduce recombination, ensuring more reduction process to take part on the surface of the oxide [1, 3]. In this work, iron oxide was fabricated by a simple thermal oxidation process. Iron oxides normally exists in three common forms: hematite (α-Fe_2O_3), maghemite (γ-Fe_2O_3), and magnetite (Fe_3O_4) [4-5]. Hematite (α-Fe_2O_3) is an n-type semiconductor with small energy gap of 2.2 eV and is photoactive. It is extremely stable compared to other phase of iron oxides and hence can be expected to be the main product of thermal oxidation of iron [5]. There have been a number of researches on the application of iron oxides on Cr(VI) removal such as flower-like α-Fe_2O_3 [6], Fe@Fe_2O_3 core-shell nanowires [7], and iron (III)-based metal-organic frameworks [8].

Different approches have been studied to synthesis nanostructure iron oxides such as metal-organic chemical vapour deposition [9], sol-gel deposition [10], hydrothermal [11-12], and thermal oxidation [13-15]. Nevertheless, the assembly of the low dimensional building blocks (particles, wires, tubes, and flakes) into a complex 3-D nanostructures (hierarchical structures) is still considered rather challenging. Hierarchically structured iron oxides can be formed by employing templated methods [6], [16]. But in this work the thermal oxidation process was observed to be successful in producing 3-D hierarchical α-Fe_2O_3 nanostructures. The oxide was then used as a photocatalysts to reduce Cr(VI) under sunlight. To the best of our knowledge no other work has reported on the use of 3-D hierarchical α-Fe_2O_3 nanostructures derived from thermal oxidation of commercially available iron as Cr(VI) reduction catalyst.

2. Methods and materials

1 cm long high purity iron wire (Nilaco Corporation, 99.5%), was used as the starting material. The wire was polished with SiC sand paper #1200 grit to remove impurities and native oxide from the surface. The polished wire was then cleaned in ultrasonic bath with aceton, and ethanol for 3 minutes respectively. After that, it was rinsed by deionised water throughly and dried by air gun.

The wire was then subjected to thermal oxidation process with rapid quenching process. Horizontal tube furnace (Lenton 1200) was heated up to a desired temperature (400 °C – 700 °C) then the wire was pushed into it. After an hour the wire was removed from the furnace and cooled down naturally until room temperature. Another set of experiment was done by varying oxidation times: 15, 30, and 60 minutes. In order to supply water to the furnace, aerosol of water was produced by ultrasonicating the water with nebuliser (Omron NE-U17). The water aerosol was changed to water vapour once it entered the furnace.

For photocatalytic reduction of Cr(VI) experiment, 10 ppm of Cr(VI) 30 mL solution (pH 2) was first prepared. Oxidised iron was then placed in the solution and was then left in the dark for 1 hour to reach equilibrium. Then, it was placed under sunlight in the morning (~ 10 am). 3 mL solution was taken out for 10 minutes interval for an hour and colouration method was used to investigate the success of the reduction process. UV-Vis spectroscopy was used to detect the absorbance at 540 nm using the diphenylcarbazide (DPC) method.

Several characterization techniques were performed to investigate the oxidised samples. Field emission scanning electron microscopy (FESEM – Zeiss Supra 35 VP) and trasmission electron microscopy (TEM - Tecnai G2 20 S-Twin) were used to observe the morphology of the samples. For phase identification, and crystal analysis, X-ray diffractometer (XRD - Bruker Advanced X-Ray Solution D8 with Cu-K_α radiation (λ = 0.154 nm)) and Raman spectrometer (RENISHAW inVia 9P1567) were carried out. UV-visible spectrometer (Varian Cary 50) was used to measure the reduction of Cr(VI).

3. Results and discussion

Figure 1 shows the FESEM images of the oxidised iron wires at 400 °C – 700 °C in dry air. Micrographs of iron oxidised at similar temperature range but in water vapour are shown in Figure 2. As can be seen, oxidation in dry air at 400 °C (Figure 1 (a)) resulted in nanowires formation. However, when water vapour was injected in the furnace, the surface oxide is seen to have hierarchical porous structures (Figure 2 (a)). The structure seems to consist of sheets of oxide, interconnected from one to another. Oxidation at 500 °C resulted in nanoblades (nanosheets) formation (Figure 1 (b)). Upon the addition of water vapour, hierarchical porous structures is also observed but with thicker sheet thickness compared to sample oxidised at 400 °C (Figure 2 (b)). Increase of temperature to 600 °C yield surface oxide with hierarchical porous structure even in dry air condition (Figure 1 (c)). At 700 °C in both conditions, the walls of the porous structure thicken (Figure 1 (d) for air and Figure 2 (d) for water vapour). This observation shows that the surface morphology of oxidised iron is dependent on the oxidising atmosphere and temperature.

It is known that during oxidation process, iron oxide will form when iron lost its electrons and react with oxygen ions from the surrounding [14]. It is obvious that oxidation requires oxidant for oxide formation and in here; oxidant was supplied by the surrounding air or water vapour [16]. It is well documented that oxidation of iron will result in multilayered oxide with the surface oxide comprised of α-Fe_2O_3. Depending on the oxidation temperature, inner oxide scale can consist of either Fe_3O_4, or a mixture of Fe_3O_4 and FeO phases [14], [17]. In order to investigate the phase of the oxide formed, HRTEM imaging was done on the sheet as seen in Figure 3, interplanar distance determined from the sheet is 0.25 nm which is corresponding to the (1 0 0) plane of α-Fe_2O_3.

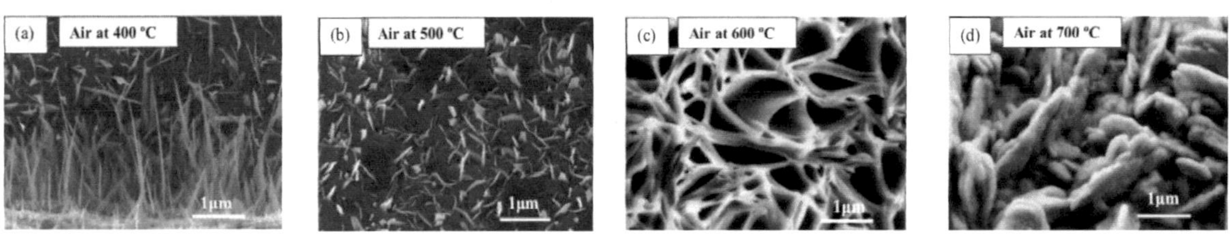

Figure 1. FESEM images of iron wires oxidised at (a) 400 °C, (b) 500 °C, (c) 600 °C, (d) 700 °C in dry air for 1 hour.

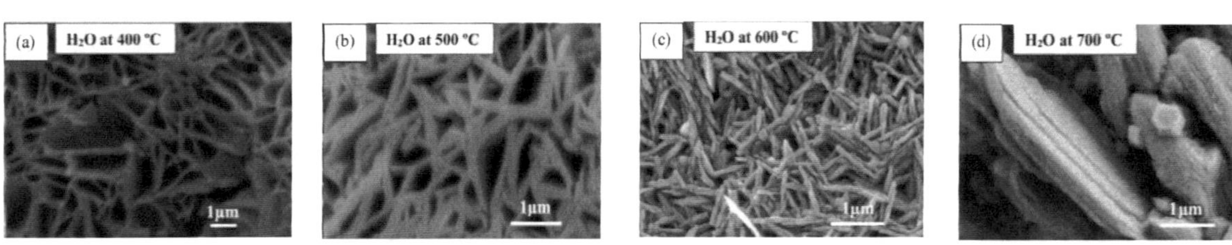

Figure 2. FESEM images of iron wires oxidised at (a) 400 °C, (b) 500 °C, (c) 600 °C, (d) 700 °C in water vapour for 1 hour

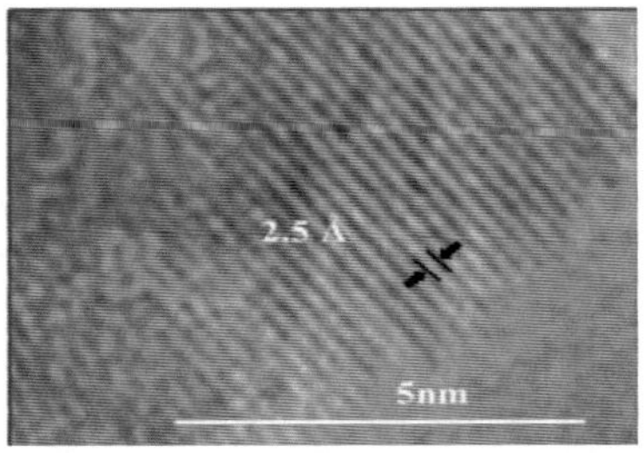

Figure 3. HRTEM image of nanostructured formed on iron oxidised at 600 °C (1 h) in water vapour

Figure 4 is the corresponding Raman shift of iron wires oxidised at different temperature in the presence of water vapour. Only sample oxidised at 500 °C shows rather intense Raman peaks. For this sample, bands at 225, 247, 292, 412, 500, and 612 cm^{-1} are seen which correspond to α-Fe_2O_3. However, for sample oxidised at 400 °C and 600 °C, α-Fe_2O_3 peak at 247 cm^{-1} is rather strong as well. Weak bands at 380 and 717 cm^{-1} can be attributed to γ-Fe_2O_3. A small peak of Fe_3O_4 at 667 cm^{-1} is also detected for sample oxidised at 500 °C. It can be concluded that α-Fe_2O_3 is the main phase formed on the iron wire, followed by traces of γ-Fe_2O_3 and Fe_3O_4.

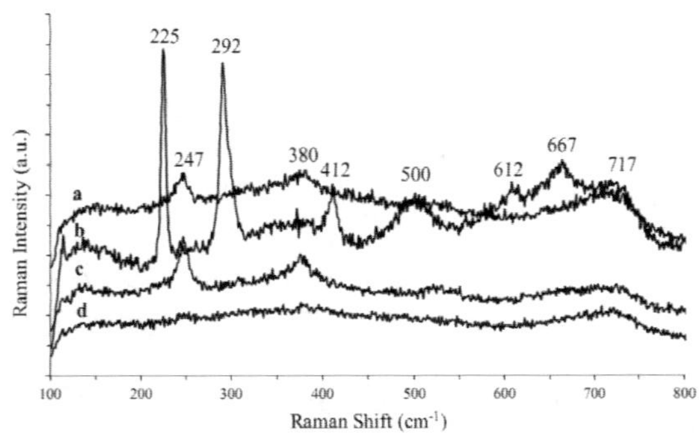

Figure 4. Raman shift of oxidised iron wires at (a) 400 °C, (b) 500 °C, (c) 600 °C, and (d) 700 °C.

The growth mechanism of hierarchical porous structure iron oxide by thermal oxidation with rapid quenching process consists of three main parts: (1) initial oxidation, (2) oxide scale formation, and (3) assembly of nanostructure. In the initial oxidation of iron wire in dry air, chemisorption of oxygen ions (O^{2-}) on iron substrate occurred. It can stimulate iron ion (Fe^{2+}) to diffuse from iron substrate to the adsorbed oxygen ions plane [18]. The growth of iron oxide layers can be described by the outward diffusion of Fe^{2+} through the oxide layer and inward diffusion of O^{2-} across the oxide layer [18][19]. While in water vapour atmosphere, the vapour will be adsorbed at first and dissociated as express in equation 1 and 2 [20]. Water vapour acts as a source of hydrogen which dissolves as hydrogen ion, $H_i^{\bullet}$ (or proton) in oxide of iron. The formation of proton resulting from dissociation of water will change the defect properties of oxide and affect to the increasing of diffusion between hydroxyl ions and iron ions.

$$H_2O_{(ads)} + V_o'' \leftrightarrow OH_o' + H_i^{\bullet} \quad (1)$$
$$2\,H_i^{\bullet} + 2e^- \leftrightarrow H_2 \quad (2)$$

During oxidation process, several iron oxide phases were formed on the substrate. The difference of volume between the oxide phases at high temperature will induce either compressive stress or tensile stress within these layers. The growth of nanostructures on the surface is believed to be the result of the release of this stress through boundary from the interface layer $Fe_2O_3 | Fe_3O_4$ to the outermost surface of Fe_2O_3 which is stress free. This process is called stress-driven mechanism [13].

IOP Conf. Series: Journal of Physics: Conf. Series **1082** (2018) 012044 doi:10.1088/1742-6596/1082/1/012044

The formation of hierarchical porous structures started from the formation of nanowires and nanosheets during oxidation process. It can be seen in figure 5 which only nanowires and few nanosheets occur for 15 minutes of oxidation, and more nanosheets, less nanowires for 30 minutes of oxidation. It proves that nanostructure grows wider as oxidation time goes longer. It can be concluded that when nanowires are close enough to each other, the surface diffusion will lead to lateral growth which cause the nanowires merge together at the bottom part and continue to grow to a sheet-like form. In the other hand, when oxidation time reach 60 minutes, nanosheets evolve to hierarchical porous structures. It is believed that lateral diffusion is predominant resulting the nanosheets merge with the neighbouring nanosheets to form hierarchical porous structures.

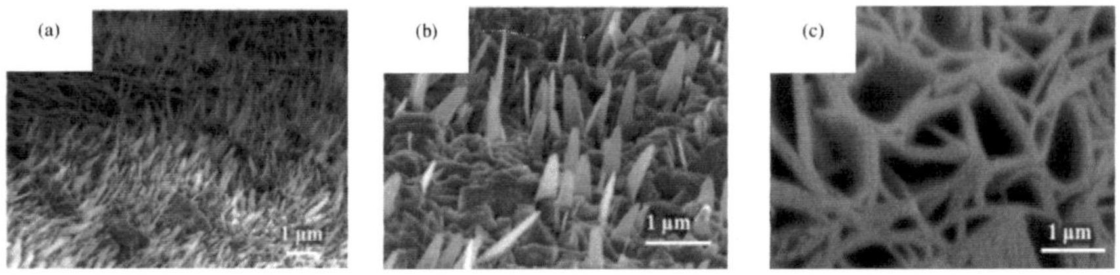

Figure 5. Oxidised iron wire at 500 °C in water vapour for (a) 15 min, (b) 30 min, (c) 60 min

Also, as the oxidation is conducted in water vapour atmosphere, the inward transport of hydroxyl ions (OH^-) originated from water vapour and outward transport of iron ions (Fe^{2+}) will form iron oxides and H^+ ions which create defects in the oxides leading to the faster oxidation process and the formation of hierarchical porous structures. Figure 2 proves that it can be formed at samples oxidised in water vapour atmosphere. For samples oxidised in dry air, the same structure is not observed for samples oxidised at 400°C, and 500°C (figure 1 (a and b)). However, as diffusion is temperature activated process, this can explain why at higher temperature in dry air atmosphere (600°C), the hierarchical porous structure is observed due to the temperature was increased which favoured the growth and merging of nanostructures.

In photocatalytic reduction experiment, oxidised iron wire in the presence of water vapour at 400°C - 700°C were used to assess the performance of hierarchical porous structures iron oxide on Cr(VI) reduction. As seen in figure 6, sample oxidised at 500°C gives the best performance among all samples (80.78%). This result is better than previous works [22][23][24] which the photoreduction performances are below 60%.

The better performance of hierarchical porous structures iron oxide can be attributed to the highly porous structure and high surface area of the iron oxides. This provides more site for Cr(VI) ions to be adsorbed and increase the number of catalytic reaction sites which benefits the improvement of the photocatalytic activity of the iron oxide photocatalyst.

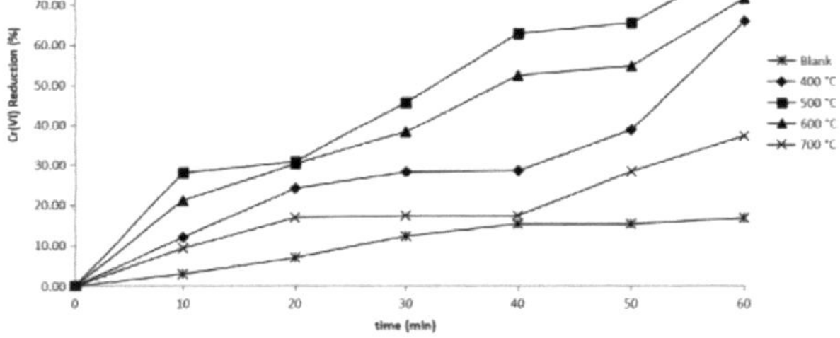

Figure 6. Reduction of Cr(VI) in solution without catalyst and in the presence of synthesised iron oxide at 400°C - 700°C in water vapour atmosphere

4. Conclusion

Hierarchical porous structures iron oxide was successfully fabricated through rapid quenching process by thermal oxidation. It was found that hierarchical structure formed in the presence of water vapour during oxidation at 400°C-500°C. The mechanism of formation of nanostructured iron oxide is suggested to follow the stress-driven mechanism and the fast growing blades would consume the smaller nanowires and resulted in hierarchical porous structure.

The performance of hierarchical porous structures iron oxide was investigated by photocatalytic reduction experiment for Cr(VI). The reduction percentage of Cr(VI) by sample oxidised at 500°C is the highest (80.78%) followed by sample oxidised at 600°C, 400°C, and 700°C, respectively.

Acknowledgements

ASEAN University Network for Science and Engineering Education Development Network (AUN/SEED-Net), Japan International Cooperation Agency (JICA) Grant No. USM CR 1701 and Research University Grant (USM-TUT Collaboration) 1001.PBAHAN.870048.

References

[1] Mohan, D., K. P. Singh and V. K. Singh 2005 *Ind. Eng. Chem. Res.* 44 1027–1042.
[2] Chong M N, Jin B, Chow C W K, Saint C 2010 *Water research* 44 2997-3027.
[3] Ghorab M F, Djellabi, Messadi R 2013 *e3s Web of Conferences.*
[4] Xu P, Zeng G M, Huang D L, Feng C L, Hu S, Zhao M H, Lai C, Wei Z, Huang C, Xie G X, Liu Z F 2012 *Science of the Total Environment* 424 1-10.
[5] Cornell R M, Schwertmann U 2006 *The Iron Oxides: Structure, Properties, Reactions, Occurrences and Uses.*
[6] Zhong L S, Hu J S, Liang H P, Cao A M, Song W G, Wan L J 2006 *Advanced Materials* 18 2426-2431.
[7] Ai Z, Cheng Y, Zhang L, Qiu J 2008 *Environmental science & technology* 42 6955-6960.
[8] Shi L, Wang T, Zhang H, Chang K, Meng X, Liu H, Ye J 2015 *Advanced Science* 2.
[9] Cha H G, Kim C W, Kim Y H, Jung M H, Ji E S, Das B K, Kim J C, Kang Y S 2009 *Thin Solid Films* 517 1853-1856.
[10] Woo K, Lee H J, Ahn J P, Park Y S 2003 *Advanced Materials* 15 1761-1764.
[11] Tang B, Wang G, Zhuo L, Ge J, Cui L 2006 *Inorganic Chemistry* 45 5196-5200.
[12] Almeida T P, Fay M, Zhu Y, Brown P D 2009 *The Journal of Physical Chemistry C* 113 18689-18698.
[13] Hiralal P, Unalan H E, Wijayantha K G U, Kursumovic A, Jefferson D, MacManus-Driscoll J L, Amaratunga G A J 2008 *Nanotechnology* 19 455608.
[14] Yuan L, Wang Y, Cai R, Jiang Q, Wang J, Li B, Sharma A, Zhou G 2012 *Materials Science and Engineering: B* 177 327-336.
[15] Budiman F, Kian T W, Razak K A, Matsuda A, Lockman Z 2016 *Procedia Chemistry* 19 586-593.
[16] Yuan R, Fu X, Wang X, Liu P, Wu L, Xu Y, Wang X, Wang Z 2006 *Chemistry of materials* 18 4700-4705.
[17] Mathiazhagan P, Khanna A S 2009 *Arabian Journal for Science and Engineering* 34 159.
[18] Khanna A S 2002 *Introduction to High Temperature Oxidation and Corrosion.*
[19] McCafferty E 2010 *Introduction to Corrosion Science.*
[20] Nasibulin A G, Rackauskas S, Jiang H, Tian Y, Mudimela P R, Shandakov S D, Nasibulina L I, Jani S, Kauppinen E I 2010 *Nano Research* 2 373-379.
[21] Yuan J, Wang W, Zhu S, Wang F 2013 *Corrosion Science* 75 309-317.
[22] Xiao D, Dai K, Qu Y, Yin Y, Chen H 2015 *Applied Surface Science* 358 181-187.
[23] Du Y, Tao Z, Guan J, Sun Z, Zeng W, Wen P, Ni K, Ye J, Yang S, Du P, Zhu Y 2015 *RSC Advances* 5 81438-81444.
[24] Zhang S, Xu W, Zeng M, Li J, Xu J, Wang X 2013 *Dalton Transactions* 42 13417-13424.

IOP Conf. Series: Journal of Physics: Conf. Series **1082** (2018) 012045 doi:10.1088/1742-6596/1082/1/012045

Assessment of Rhodamine B Dye Removal by ZnO Nanodisks under Visible Light

L.A. Thi[1], S.Y. Pung[1*], S. Sreekantan[1], A. Matsuda[2] and H.D. Phu[3]

[1]School of Materials and Mineral Resources Engineering, Universiti Sains Malaysia Engineering Campus, Nibong Tebal, Pulau Pinang, Malaysia
[2]Department of Electrical and Electronic Information Engineering, Toyohashi University of Technology, Toyohashi, Aichi, Japan
[3]Faculty of Materials Technology, University of Technology, Vietnam National University, HoChiMinh City, Vietnam.

E-mail: sypung@usm.my

Abstract. The organic dye from the textile and dye industries are highly hazardous to aquatic living and human. They cause serious damage to the surrounding environment. Zinc oxide (ZnO) photocatalyst has attracted much attention because of its high efficiency applications, especially in organic dyes removal, for instance Rhodamine B (RhB) dye. The precipitation method was used to synthesize ZnO nanodisks (NDs) by using zinc acetate dehydrate, aluminium sulphate and ammonia hydroxide as precursors. The X-ray powder diffractometry (XRD) shows that the as-synthesized particles consisted of ZnO and zinc aluminium carbonate hydroxide hydrate ($ZnAlOHCO_3$) phases. These diffraction peaks matched with the Joint Committee on Powder Diffraction Standards (JCPDS) No. 98-002-7791 and No. 98-010-5856, respectively. The morphology of ZnO NDs was clearly observed by scanning electron microscope (SEM). An average diameter and thickness of ZnO hexagonal disk were 2.12 μm and 74.9 nm, respectively. The UV-Visible spectroscopy showed that the photodegradation efficiency of ZnO NDs in removal of RhB dye in aqueous solution was 39% under visible light illumination of 5 hours. The possible photodegradation mechanism was studied by the examination of active species e.g. $^{\bullet}HO$, $^{\bullet}O_2^-$ and h^+ through adding their respective scavengers such as methanol, benzoquinone and potassium.

1. Introduction

ZnO is a II-VI group direct semiconductor with wide bandgap (3.3 eV) in the near-UV spectral region and large binding energy (60 meV) at room temperature [1, 2]. It exhibits in a wide range of nanostructures such as nanodisks [3, 4], nanorods [5, 6], nanobelts [7, 8], nanowires [9, 10], nanorings [11]. The variety of nanostructures of ZnO make its unique and possessed remarkable performance in photonics, sensors, photovoltaic devices and photocatalysts applications [12]. Particularly, ZnO photocatalysts have been received much attention for wastewater treatment such as removal of organic – inorganic pollutants [13-19], antibacterial [20, 21] and water splitting [22].

As the preferred growth direction of ZnO is in [0001], ZnO nanorods (NRs) will be formed under normal synthesis condition particularly when no foreign catalyst is used [23]. The application of ZnO NRs as photocatalyst for organic pollutants removal is widely studied because of its simplicity, low cost, nontoxic, high degradation efficiency, and excellent stability in nature [24-25]. Nevertheless, using ZnO NDs as photocatalyst is relatively less known and explored although this nanomaterial was used as building blocks in lasers, sensors and photonic crystal [26-27]. As compared to NRs, ZnO NDs have large (002) polar surface (disk surface), which either terminated by positively charged Zn^{2+} or negatively charge O^{2-}. The present of this polar surface might be able to enhance the photocatalytic performance under visible light illumination. This is crucial for realization of ZnO nanomaterials as photocatalyst because the energy of incoming solar spectrum consists of less than 4% ultraviolet (UV) light but more than 50% visible light [28]. Thus, it is necessary to investigate the photocatalytic activity in organic dye removal of ZnO NDs by using visible light illumination."

In this work, the photocatalytic performance of ZnO NDs in organic dye removal under visible light irradiation was studied. ZnO NDs were synthesized via precipitation method [29]. Aluminium sulphate was used as surface directing agent in order to obtain the disk-like nanostructures. In addition, Rhodamine B (RhB) was used in this study. It represents one of the most important dyes, which has been extensively used in the textile industry due to its high chemical stability. Hence, the decomposition of RhB dye is a challenging task in waste water treatment.

2. Materials and Methods

ZnO NDs were synthesized via precipitation method according to the method described in Ref. [23]. In this process, zinc acetate dehydrate (99.99%, Merck, product number 108802), aluminium sulphate (98%, Systerm, product number Al831-90) and ammonia hydroxide (25%, Merck, product number 105432) were mixed in de-ionised water and stirred at 60 ^{0}C for 2 hours. White precipitate was formed and settled down on the bottom of the beaker. The precipitate then was rinsed with de-ionized water and centrifuged to separate the precipitate and solution. Lastly, the white ZnO powders were dried in an oven at 100 ^{0}C for 8 hours. The X-ray diffraction (XRD) patterns of the ZnO NDs were recorded using X-ray diffractor (Bruker Advanced X-ray Solution D8 Diffractometer. The morphology of particles was clearly observed by using field-emission scanning electron microscope (FESEM, 35VP Supra Carl-Zeiss) and high resolution transmission electron microscope (HRTEM, Tecnai G2 20 S-TWin).

Photocatalytic activity of ZnO particles was studied by monitoring the intrinsic absorption peak of RhB dye. The concentration of RhB solution was fixed to 5 ppm for all experiments. The aqueous suspension (250 ml) of RhB and photocatalyst (ZnO NDs) were placed in a visible chamber. Visible light irradiation was provided by a lamp with the strongest emission at wavelength~555 nm, placed at a distance of about 20 cm above the sample's surface. The photodegradation of RhB solutions using NDs as photocatalysts was recorded by UV-VIS spectrometer (Carry 50). The photodegradation efficiency (PE) was calculated by using following equation:

$$PE\ (\%) = \frac{C_o - C_t}{C_o} . 100 \qquad (1)$$

where C_o and C_t are the concentration of the RhB dye solution at the initial and time t, respectively.

The scavenger study was carried out to find out the main reactive species that responsible for the degradation of dyes. In this study, potassium iodide, methanol, benzoquinone were used as scavengers for h^+, $^{\bullet}OH$, $^{\bullet}O_2^-$, respectively.

3. Results, Discussion

3.1 Structural and morphology of ZnO NDs

The ZnO NDs were synthesized using ammonia hydroxide, aluminium sulphate and zinc nitrate acetate as precursors. The growth mechanism of ZnO was discussed by Thein et al [23]. The XRD pattern of the as-synthesized particles is shown in Figure 1. It is matched with diffraction peaks of JCPDS No. 98-002-7791, indicating that the particles were hexagonal wurtzite ZnO [24]. The XRD at

31.18, 33.91, 35.69, 46.85, 56.05, 62.32 and 67.44 are correspond to (010), (002), (011), (102), (110), (103) and (112) crystal planes of hexagonal ZnO structures, respectively. Furthermore, ZnO NDs also exhibited extra peaks at 11.63, 23.01, 33.91 and 38.73 which were contributed from zinc aluminium carbonate hydroxide hydrate ($ZnAlOHCO_3$, JCPDS No. 98-010-5856). The phase percentage of zinc oxide and zinc aluminium carbonate hydroxide hydrate are 81% and 19%, respectively. The present of aluminium is further verified by EDX analysis, i.e. 7.05 at % aluminium was detected in the ZnO NDs as shown in Fig 2(a). The source of aluminium was come from aluminium sulphate. The complex agent, i.e. $Al_2(SO_4)_3$, was used to suppress the preferential growth of ZnO in [0001], allowing the growth in six symmetric direction $\pm[\bar{1}100], \pm[\bar{1}010], \pm[0\bar{1}10]$to form hexagonal structure [23]. Figure 2(b) shows the morphology of ZnO hexagonal NDs with the diameter of 2.12 μm and thickness of 74.9 nm.

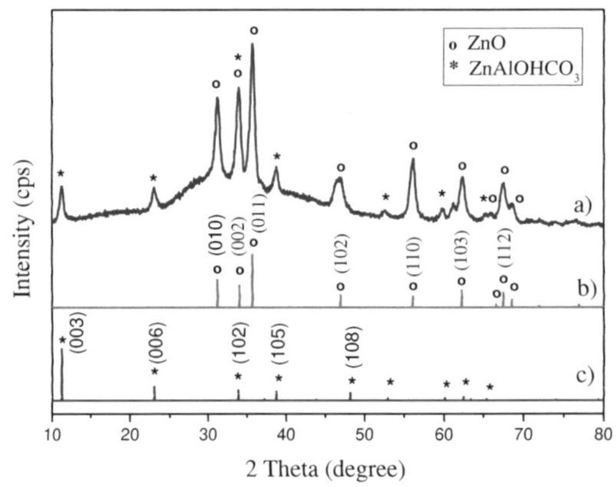

Figure 1. X-ray diffraction pattern of a) ZnO NDs, b) ZnO JCPDS No. 98-002-7791 and c) $ZnAlOHCO_3$ JCPDS No. 98-010-5856.

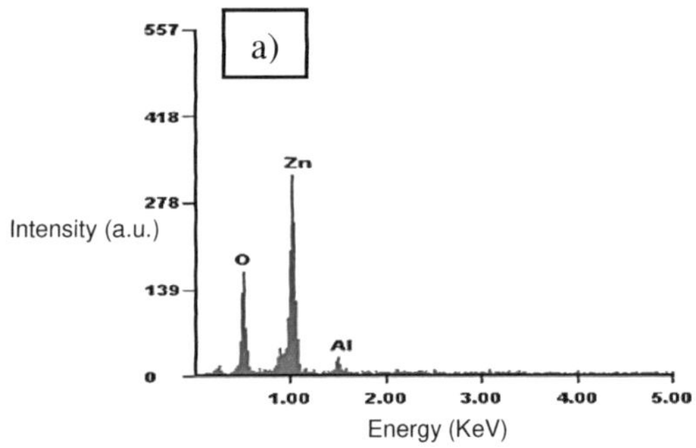

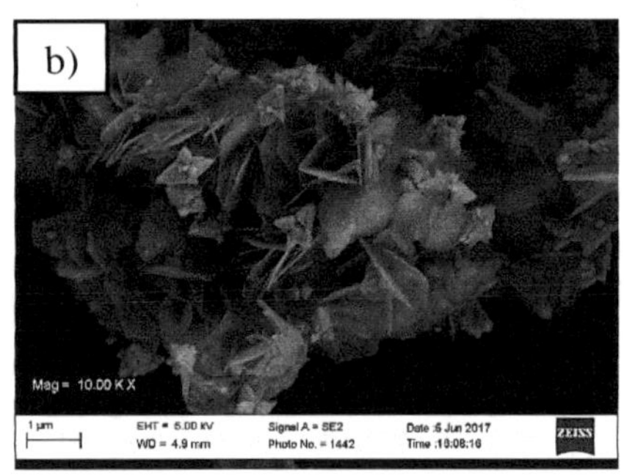

Figure 2. (a) EDX spectrum and (b) SEM image of ZnO NDs

3.2 Photocatalyst study

RhB belongs to the oxygen-containing heterocyclic xanthene dyes family. The RhB contains two key components: chromophores and auxochromes. The chromophores are light absorbing group in the molecule. Figure 3(a) shows the characteristic absorption peak of RhB dye is 554 nm. This is due to the n →π transition of the chromophores, C=N, C=C groups, for instant [25]. To achieve this transition, the RhB molecule need to be excited by high level of energy. Hence, it is neither degraded in the dark in the presence of a photocatalyst nor under illumination of visible light in the absence of a photocatalyst [26]. On the other hand, auxochromes are functional groups of molecule attached to chromophores group such as -NR, -COOH. These functional groups do not absorb radiation itself but has a shifting effect on main chromophore peaks. Figure 3(b) shows the absorption peak was shifted from 554 nm to 542 nm after 5 hours irradiation of visible light. The longer time RhB dye is exposed under visible light, the more blue shift present in the absorption peak; For instant, the peak shift at 542 nm was preferred to N,N-diethyl-N'-ethylrhodamine [27, 28].

The photodecomposition of RhB molecules by ZnO NDs mainly occurred at auxochromes groups under visible light illumination. This phenomenon is called the deethylation path way [29-31]. As a result, the photodegradation efficiency of RhB aqueous solution of ZnO NDs increased gradually and reached 39% after 5 hours (Fig 3(b)) irradiation of visible light. The first-order kinetic is used to describe the absorption rate base on absorption capacity [32]. The linear form of first-order rate was given by equation (2) [33].

$$\ln(C_t) = \ln(C_o) - Kt \qquad (2)$$

where K is the first-order rate constant. Figure 3(c) showed the linear plot of $\ln(C_o/C_t)$ against time preferred the first-order kinetic of discoloration of RhB/ZnO solution with the rate constant of 0.098 min^{-1}.

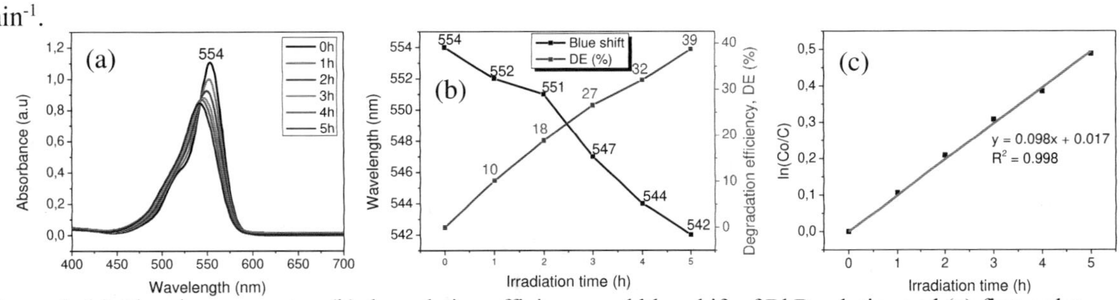

Figure 3. (a) Absorbance spectra, (b) degradation efficiency and blue shift of RhB solution and (c) first-order kinetic plot of ZnO NDs under visible light irradiation

3.3 Scavenger study and photo-degradation mechanism

The scavenger study was carried out to find out the main reactive species responsible for the degradation of RhB dye in aqueous solution of ZnO NDs. Methanol is used as a diagnostic tool of •OH radical mediated oxidation mechanism [34]. The presence of potassium iodide from a solution provide iodide ion as a scavenger which reacts with h^+; that can be used to identify the involvement of hole in the direct oxidation of organic substrate. On the other hand, benzoquinone can be used to detect the superoxide anion due to their ability to trap this ion radical by an electron transfer mechanism [35].

Figure 4(a) shows the scavenger study of RhB with the present of ZnO NDs under visible light irradiation. The photodegradation efficiency of RhB solution in ZnO NDs went down dramatically from 38.65% to 34.19%, 26.55% and 1.88% when methanol, potassium iodide and p-benzoquinone were added into the solution, respectively. When $^{\bullet}OH$ scavenger, methanol (Me) was put into the degradation systems, there was no significant changes occurred on dye degradation except for the hypochromic shift of the absorption band. Hence, $^{\bullet}OH$ radicals were not the major oxidation species in this process. Moreover, it is known that there is limited h^+ arises on the surface of ZnO NDs under irradiation of visible light because of its wide bandgap. Thus, the low degradation rate in the presence of potassium iodide (KI) may be due to electron transfer between KI and two dye molecules [36]. Electron transfers between KI molecules make the dye radical ions return to the ground states and result in the small degradation efficiency, 0.066 min^{-1} of rate constant as shown in Figure 4(b). It means that h^+ was not the main factor causing the discoloration of dye. Besides, p-benzoquinone (BQ) was used to examine the role of $^{\bullet}O_2^-$ radical. The great influence of BQ on degradation efficiency was only 1.88% after 5 hours illumination of visible light. This is due to the ability of BQ to quench $^{\bullet}O_2^-$ radical and inhibited the degradation of RhB dye. Figure 4(b) shows that the $^{\bullet}O_2^-$ radical was the main species scavenger in dye photodegradation oxidation reaction with the slowest rate constant 0.007 min^{-1}.

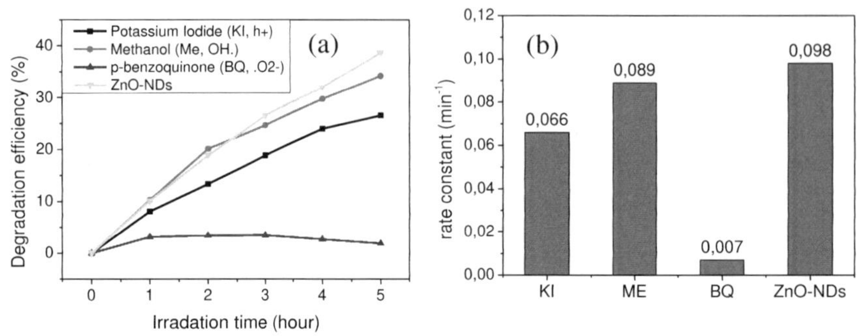

Figure 4. (a) The degradation efficiency with (b) rate constant of RhB by ZnO NDs and scavenger agents under visible light.

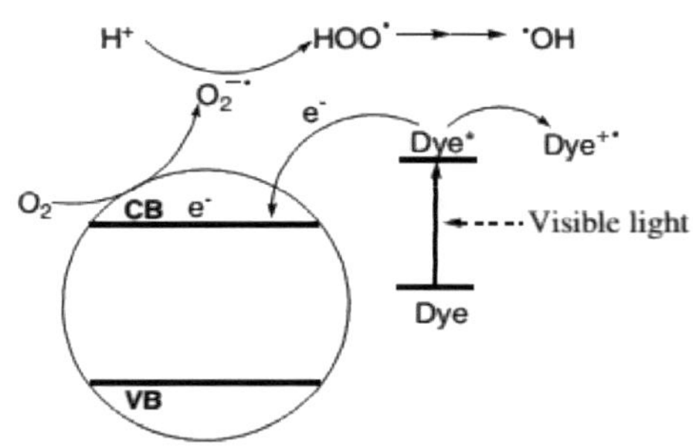

Figure 5. Photocatalyst mechanism of ZnO NDs in degradation of RhB under visible light

Base on the understanding of ROS in photocatalytic process, the mechanism is illustrated in Figure 5. In this strategy, the ZnO NDs were not be able to absorb light directly [29, 30, 37]. Instead, the dye molecules acted as a sensitizer by absorbing the visible light and achieved excited state (Eq. 3). This excited dye (Dye*) injected an electron to the conduction band of ZnO NDs in Eq. 4-5. These injected electron on the conduction band can reduce the molecular oxygen to yield the oxidizing species such as H_2O_2, $^{\bullet}O_2^-$ and $^{\bullet}OH$ radicals (Eq. 6 - 8). These active radicals degrade the RhB dye via N-deethylation process into N,N-diethyl-N'-ethylrhodamine, N,N-diethylrhodamine, N-ethyl-N'-ethylrhodamine, N-ethylrhodamine [27]

$$Dye + h\upsilon\ (visible) \rightarrow Dye^* \quad (3)$$
$$Dye^* \rightarrow Dye^+ + e^- \quad (4)$$
$$ZnO + e^- \rightarrow {}^{\bullet}ZnO^- \quad (5)$$
$${}^{\bullet}ZnO^- + O_2 \rightarrow ZnO + .O_2^- \quad (6)$$
$${}^{\bullet}O_2^- + H_2O + e^- \rightarrow 2H_2O_2 \quad (7)$$
$$H_2O_2 + e^- \rightarrow {}^{\bullet}OH + OH^- \quad (8)$$

4. Conclusion

In summary, ZnO NDs were able to degrade RhB dyes under visible light irradiation. It achieved 39% photodegradation efficiency after 5 hours of light exposure. The result also showed that photodegradation of RhB dye followed the first-order kinetic. Based on the scavengers study, the $^{\bullet}O_2^-$ free radicals were known as the major agent of the discolorations of RhB via utilizing of ZnO NDs. The degradation mechanism of photo-oxidation reaction of RhB dye by ZnO NDs was proposed attributed to the free radicals produced from photoelectrons originated from optically excited RhB dye molecule itself. The scavenger test is an excellent probe to investigate the mechanism of the photocatalytic degradation of organic pollutants, and provides much help on the design of novel visible-active photocatalyst.

Acknowledgements

The authors would like to thank Ministry of Higher Education, Malaysia for providing the research funding under Fundamental Research Grant Scheme (FRGS) (203.PBAHAN.6071327) and AUN/SEEDNet under Collaborative Research Program (CR) (304.PBAHAN.6050354) for providing the research funding to conduct this project.

References

[1] Wang Z L 2004 *Journal of Physics: Condensed Matter* 16 R829.
[2] Janotti A and Van de Walle C G 2009 *Reports on progress in physics* 72 126501.
[3] Qadir M I, Rana S J, Nur O, Willander M, Larsson L, and Holtz P O 2014 *Ceramics International,* 40 2435-2439.
[4] Seo H K and Shin H S 2015 *Materials Letters* 159 265-268.
[5] Zhang X, Qin J, Xue Y, Yu P, Zhang B, Wang L 2014 *Scientific reports* 4.
[6] Yu T Y, Chen Y C, Chiu W T, Luo Y, Wang S S, Chen H 2016 *Journal of New Materials for Electrochemical Systems* 19 181-183.
[7] Kennedy O W, Coke M L, White E R and Shaffer M 2017 *Materials Letters* 12 234
[8] Zou Q, Mo S, Wang Y, Dang M, Qin G, Fu X 2017 *Modern Physics Letters B* 31 1750250.
[9] Roso S, Güell F, Alanis P R M and Llobet E 2016 *Sensors and Actuators B: Chemical* 230 109-114.
[10] Tsai P H and Wang C Y 2016 in *Meeting Abstracts* 1254-1254.
[11] Li C, Lin Y, Li F, Zhu L, Sun D, Shen L 2015 *RSC Advances* 5 80561-80567.
[12] Udom I, Ram M K, Stefanakos E K, Hepp A F and Goswami D Y 2013 *Materials Science in Semiconductor Processing* 16 2070-2083.
[13] Liu J, Zhao Y, Ma J, Dai Y, Li J and Zhang J 2016 *Ceramics International* 42 15968-15974.
[14] Yang G C and Chan S W 2009 *Journal of Nanoparticle Research* 11 221.
[15] Qamar M, Gondal M, and Yamani Z 2011 *Journal of hazardous materials* 187 258-263.
[16] Siboni M S, Samadi M, Yang J and Lee S 2011 *Environmental technology* 32 1573-1579.
[17] Mijin D, Savić M, Snežana P, Smiljanić A, Glavaški O, Jovanović M *Desalination* 249 286-292.
[18] Lee K M, Lai C W, Ngai K S and Juan J C 2016 *Water Research* 88 428-448.
[19] Jiang L, Wang Y and Feng C 2012 *Procedia Engineering* 45 993-997.
[20] Zhou G, Li Y, Xiao W, Zhang L, Zuo Y, Xue J 2008 *Journal of Biomedical Materials Research Part A,* 85 929-937.
[21] Karvani Z E and Chehrazi P 2011 *African Journal of Microbiology Research* 5 1368-1373.
[22] Yuefan W, Lin K, Junhua K, Hong L, Zhihui J, Xuehong L 2012 *Nanotechnology* 23 23540.
[23] Thein M T, Pung S Y, Aziz A and Itoh M 2015 *Journal of Experimental Nanoscience* 10 1068-1081.
[24] Li X, Sun P, Yang T, Zhao J, Wang Z, Wang W 2013 *CrystEngComm* 15 2949-2955.
[25] Chen X, Xue Z, Yao Y, Zhu F and Hong C 2012 *International Journal of Photoenergy* 2012.
[26] Rochkind M, Pasternak S and Paz Y 2015 *Molecules* 20 88.
[27] Wang P, Cheng M and Zhang Z 2014 *Journal of Saudi Chemical Society* 18 308-316.
[28] Chen F, Zhao J and Hidaka H 2003 *International Journal of Photoenergy* 5 209-217.
[29] Chen C, Ma W and Zhao J 2010 *Chemical Society Reviews* 39 4206-4219.
[30] Zhao J, Chen C and Ma W 2005 *Topics in catalysis* 35 269-278.
[31] Boxi S S and Paria S 2015 *RSC Advances* 5 37657-37668.
[32] Soni H, Kumar N and Kumar R N 2016 *Desalination and Water Treatment* 57 19857-19864.
[33] Pung S Y, Lee W P and Aziz A 2012 *International Journal of Inorganic Chemistry* 2012.
[34] Chen Y, Yang S, Wang K and Lou L 2005 *Journal of Photochemistry and Photobiology A: Chemistry* 172 47-54.
[35] Palominos R A, Mondaca M A, Giraldo A, Peñuela G, Pérez-Moya M and Mansilla H D 2009 *Catalysis Today* 144 100-105.
[36] Yin M, Li Z, Kou J and Zou Z 2009 *Environmental science & technology* 43 8361-8366.
[37] Pare B, Jonnalagadda S, Tomar H, Singh P and Bhagwat V 2008 *Desalination* 232 80-90.

IOP Conf. Series: Journal of Physics: Conf. Series **1082** (2018) 012046 doi:10.1088/1742-6596/1082/1/012046

Room Temperature Synthesis and Characterizations of ZIF-8 Formation at Water-Fatty Alcohols Interface

H. Misran[1], N. Mahadi, S. Z. Othman[1], Z. Lockman[2], N. Amin[1] and A. Matsumoto[3]

[1]Universiti Tenaga Nasional, College of Engineering, Selangor, Malaysia

[2]Universiti Sains Malaysia, Engineering Campus, Penang, Malaysia

[3]Toyohashi University of Technology, Toyohashi, Japan

Email: Halina@uniten.edu.my

Abstract. In this study, Zn based ZIF-8 metal-organic framework were successfully synthesized in a green synthesis approach eliminating hazardous solvents and the need to use surfactant. In this method, the syntheses were done at room temperature followed by a simple hydrothermal method using palm oil derived fatty alcohols (PODFA) with 12 carbon chain in various solvents. The effect of several co-solvents on the formation of ZIF-8 materials and on morphology were investigated. The peak positions agreed well with the simulated ZIF-8. However, as the co-solvents became more polar the (011) peak shifted to a higher 2θ value suggesting that the interplanar *d*-spacing were reduced. The addition of fatty alcohols promoted the formation of MOF-5 and ZIF-8 at the water-fatty alcohols interfaces by viscosity lowering effect similar to those when using commercial surfactants. The mechanism of the formation was suggested to be similar to those at water-alcohol insoluble monolayer interfaces at the liquid-gas interaction. As the polarity of solvents increased, the miscibility of Zn^{2+} ions and deprotonated methylimidazole linkers were enhanced causing higher interaction with metal ions which resulted in a smaller unit cell. ZIF-8 exhibited elongated rhombic dodecahedron and truncated dodecahedron morphology at less than 100 nm.

1. Introduction

Metal-organic frameworks (MOFs) materials are a class of porous polymeric hybrid organic-inorganic materials that are rapidly expanding. MOFs consist of metal oxide interconnected by organic ligands to link them together resulting in a geometrically well-defined crystalline structure [1]. MOF materials have great potential as adsorbents and catalysts due to their extremely large surface area, well-ordered nanoporous structures and diverse surface functionalities from the organic-inorganic framework that were available for easy functionalization [2-4].

Furthermore, the metal-organic pore framework resulted in flexible and permanent porosity. Based on these excellent structure-related properties, MOF materials are prominent advanced porous materials for adsorptive removals by hydrogen bonding [5]. Adsorptive removal or separation field is the most important applications of MOF materials due to their ease of surface modifications to increase the hydrogen bonding sites desired for adsorption separation. Compared to other advanced porous materials such as zeolites, activated carbons and mesoporous silica, MOF materials pore framework was versatile and flexible to accommodate several size and chemical reactivity of the adsorptive molecules. The surface structure of MOF materials can be easily modified by changing metal and organic precursors, impregnation or loading with enzymes and drugs, grafting of short functional ligands at metal or organic linkers as well as making composites with other nanostructured materials [6]. Abundance of hydrogen bonding sites in either pristine or modified MOF was necessary requirements for surface structure-related properties for adsorptive molecules to interact with the surface [5-8].

The adsorption of water by porous solids is important for many applications requiring capture and release of water especially in energy applications [9]. Microporous oxides such as zeolites are commercially used to capture water in applications as dehumidifiers. Temperature-triggered capture and release of atmospheric water is expected to be useful in climates where there is a large temperature difference between day and night. In most water capture applications, three criteria are important in the design of a viable porous material [10]. First, the pore filling or condensation of water into the pores of the solid must occur at low relative pressure (relative humidity) and exhibit a steep uptake behavior. The former is important because water is being captured from media where it is present at low concentrations, and the latter is related to the working capacity of the material [10, 11].

On the other hand, zeolitic imidazolate frameworks (ZIF-8) synthesized using Zn or Co metal ions are the extension of metal organic framework MOF-5 exhibited higher hydrostability and chemical stability [12]. ZIF-8 structure was cubic I-43m space group analogous to zeolites due to metal-imidazole-metal (M-I-M) framework that was similar to Si-O-Si chemical bond in zeolite built upon 4 connected nets of tetrahedral units wherein lies the metal ions that are linked through N atoms in ditopic imidazolate anions [7]. Current synthesis method of both MOF-5 and ZIF-8 widely employed solvothermal method with the use of surfactants and corrosive solvents such as chloroform and dimethylformamide (DMF) [5-8].

Thus, in this paper we report for the first time a simple approach combining both room temperature mixing followed by low temperature hydrothermal treatment of MOF-5 and ZIF-8 using palm oil derived fatty alcohols (PODFA) as replacement to surfactants with water as the green solvent. In order to study the effect of co-solvents with different polarity on formation of MOF materials at water-fatty alcohol interfaces, the co-solvent were changed accordingly.

2. Experimental Procedure

ZIF-8 were synthesized using green approach employing PODFA, water, short-chain alcohols while eliminating the use of trimethylamine (TEA). The co-solvents were needed to promote stability in the "pre-Ouzo" region for a surfactantless microemulsion [13].

2.1 Materials and Methods

Zinc (II) nitrate hexahydrate (>99%, $Zn(NO_3)_2.6H_2O$) were purchased from Sigma Aldrich, ethanol (99.8%, C_5H_6O), 2-Methylimidazole (>99%, 2-MIM) were purchased from Merck. Palm oil derived fatty alcohol (PODFA), dodecyl alcohol (dodecanol, C12) were purchased from Emery Oleochemical (M) Sdn Bhd. Methanol, ethanol, n-butanol and acetone (analysis grade) were purchased Sigma Aldrich. All materials were used in as-received condition without further purification. The synthesis approach using PODFA followed the procedure as reported [14]. Generally, the metal organic framework materials were made up of two precursors mixtures. The first solution comprised of 2-MIM in a co-solvent of either methanol (MetOH), ethanol (EtOH), *n*-butanol (ButOH) and acetone (Ace). The second solution consisted of zinc salt in water. Both metal and organic precursor solutions were mixed and stirred thoroughly at room temperature. Then, an amount of PODFA as a modulating agent was added into the solution to promote the formation of "pre-Ouzo" effect in the surfactantless mixture. The ZIF-8 control sample was synthesized without using any PODFA with methanol as the polar co-solvent. Then, the resulting viscous solution was aged overnight at room temperature to promote the growth of ZIF crystals. The stirring power, temperature and time were kept constant during the synthesis. The co-solvents were changed accordingly for subsequent ZIF-8 synthesis. The obtained solids were filtered and washed for several times using distilled water and dried overnight at 358 K to obtain the as-synthesized, dried ZIF-8 materials.

2.2 Characterizations

As-synthesized ZIF-8 samples were subjected to X-ray diffraction (XRD) analyses using Shimadzu XRD-6000 with CuK_α radiation at λ=1.54 nm at 30 kV and 20 mA, scan speed of 2°/min and scan step of 0.02°. The range were at 2θ = 5° - 50°. The scanning electron microscopy (SEM) images were done Zeiss Supra VP 35 at acceleration voltage of 5 kV for ZIF-8 samples. The TGA analyses were done using TGAQ500 at temperature range of 100 °C to 850 °C. UV-Vis analyses were done using UV-vis Perkin-Elmer Lambda 35 using 1 cm quartz cell at ca. 200 nm to 600 nm.

3. Results and discussion

The X-ray diffraction patterns of as-synthesized ZIF-8 samples prepared using various polarity solvents with or without PODFA additions are shown in Figure 1. All XRD patterns exhibited good agreements with the simulated patterns as reported suggesting successful formation of ZIF-8 [12,13]. The first 6 peaks at ca 2θ 7.3°, 10.2°, 12.7° were attributed to the reflections from (011), (002), (112) planes of a rhombic dodecahedron with truncated corners. The subsequent three peaks could be assigned to (022),(013) and (222) planes of a truncated rhombic dodecahedron. Adding the PODFA resulted in the increment of (112) reflection peaks originated from truncated rhombic plane compared to (011) plane reflection peaks of actual rhombic crystal structure.

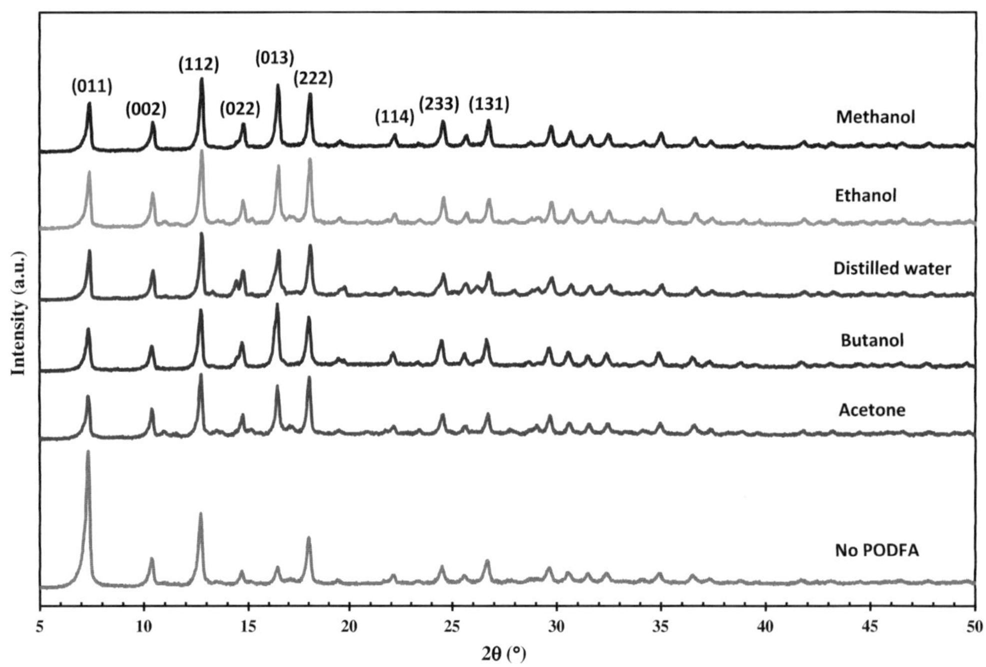

Figure 1. X-ray diffraction patterns of ZIF-8 synthesized with/without PODFA with various co-solvents.

IOP Conf. Series: Journal of Physics: Conf. Series **1082** (2018) 012046 doi:10.1088/1742-6596/1082/1/012046

SEM images of ZIF-8 prepared with or without PODFA in various solvents are shown in Figure 2. ZIF-8 prepared in methanol without PODFA addition exhibited smaller crystal structures at ca. 100 nm. As the polarity of solvents increased the miscibility of organic linker and metals salts with PODFA were increased at the water-fatty alcohols interface. Hence, ZIF-8 crystal evolution rates were changed resulted in various morphology [8]. In high polarity solvent such as methanol, ZIF-8 exhibited highly faceted rhombic dodecahedron. The resulting morphology observed when using ethanol as solvent was faceted rhombic dodecahedron and thinner dodecahedron flakes. However, in less polar solvent such as in *n*-butanol, the morphology changed to thinner dodecahedron to almost pill

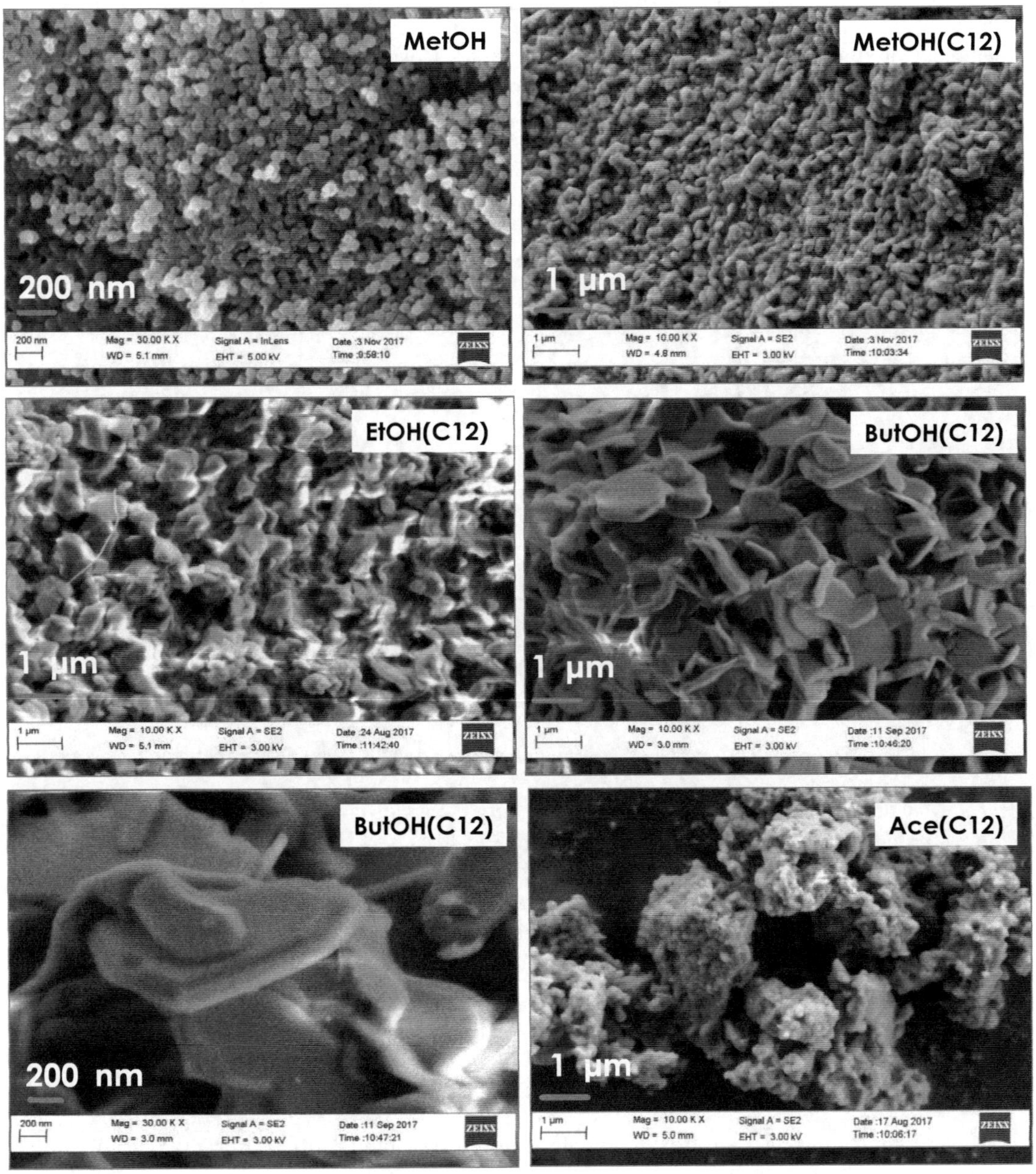

Figure 2. SEM images of as-synthesized ZIF-8 with/without PODFA in various solvents.

-like morphology without distinct facets. In acetone the resulting ZIF-8 exhibited globular aggregates at ca. less than 100 nm. The various morphology observed in this study was due to the hydrogen bond donation ability of solvent molecules that governed the crystallization rates. Higher crystallization rate resulted in less faceted structure. Furthermore, the various solvents used gave different capabilities to dissolve and to solvate Zn^{2+} and NO^{3-} ions as well as the organic ligand because of their different dielectric constants, dipole moment and van der Waals volume [8]. The TGA spectrum of a representative ZIF-8 synthesized using PODFA of C12 in ethanol as solvent is shown in Figure 3. The spectrum showed ca. 2 % weight loss in the temperature range of 25–150 °C corresponding to the removal of solvent molecules which were methanol molecules in this case. Then, there existed a gradual weight loss of ca. 25 % till 350 °C were observed corresponding to the removal of guest molecules from the cavities and some unreacted species from the surface of the materials. Therefore, a guest-free phase Zn $(MIM)_2$ was formed until 350 °C.

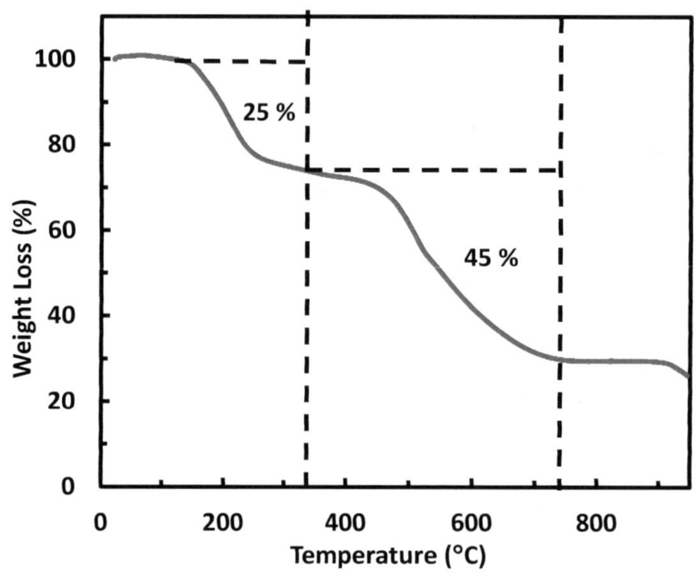

Figure 3. TGA curve of ZIF-8 synthesized using C12 PODFA in methanol.

After 350 °C, the TGA curve exhibited a sharp weight loss of ca. 45 % up to the temperature of 700 °C. It was due to the thermal decomposition of ZIF-8 as 2-MIM decomposed at high temperature leading to the formation of zinc oxide (ZnO) as the final calcination product of ZIF-8 nanocrystals. These results suggested that ZIF-8 prepared in this study exhibited higher thermal stability as reported [7]. The representative UV-visible absorption spectra of ZIF-8 prepared using methanol as solvent with and without the presence of PODFA are shown in Figure 4. Both ZIF-8 samples exhibited a characteristic absorption peak at ca 215 nm [8]. A broad hump centered started at ca. 320 nm to ca. 380 nm suggesting the interactions of solvent molecules with the ZIF-8 framework.

4. Conclusion

ZIF-8 materials were successfully synthesized at room temperature in various solvents with different polarity in the presence of palm oil derived fatty alcohol (PODFA) having C12 carbon chain other than those already reported for the synthesis of this MOF. Monodispersed ZIF-8 with truncated rhombic dodecahedron was obtained in highly polar solvent of methanol without TEA with the aid of

PODFA that increase the immiscibility of precursor mixtures. Thus, the method used in this study was successful in eliminating TEA and surfactant to produce ZIF-8 at room temperature.

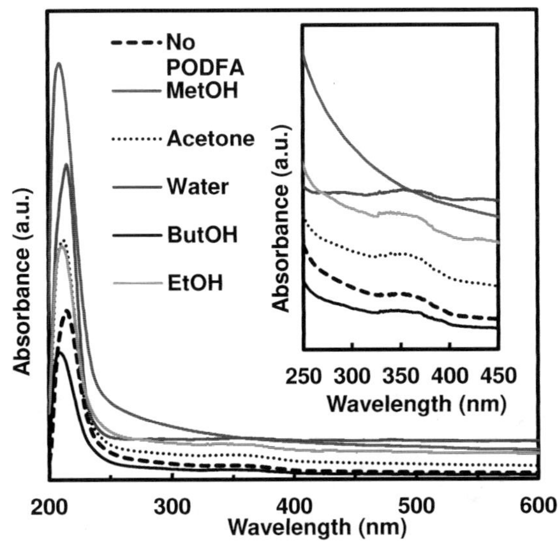

Figure 4. UV-visible spectra of ZIF-8 prepared with various solvents.

Acknowledgement

The authors thank Ministry of Education (MOE) under FRGS grant scheme (Grant number: FRGS20150116) and UNITEN for BOLD grant 10289176/B/9/2017/9.

References

[1] Diercks C, Yaghi O M 2017 *Science* 355, 923.

[2] Kim H, Cho J, Narayanan S, Yang S, Furukawa H, Schi S, Li X, Zhang Y, Jiang J, Yaghi O M, Wang E N 2016 *Science Representative* 1, 19097.

[3] Schoedel A, Yaghi O M 2016, *Macrocyclic and Supramolecular Chemistry*, Ed. R. Izatt, Wiley-VCH Verlag GmbH & Co. KGaA, Weinheim, Germany 200-219.

[4] Li H, Eddaoudi M, O'Keeffe M, Yaghi O M 1999 *Nature* 402, 276–279.

[5] A. Fracaroli, P. Siman, D. Nagib, M. Suzuki, H. Furukawa, F. D. Toste, O. M. Yaghi 2016 *Journal of American Chemical Soc*iety 138, 8352–8355.

[6] P. W. Seo, I. Ahmed, S. H. Jhung 2016 *Science Representative* 6, 34462.

[7] Y. Guan, J. Shi, M. Xia, J. Zhang, Z. Pang, A. Marchetti, X. Wang, J. Cai, X. Kong, 2017 *Applied Surface Science* 423, 349–353.

[8] E. L. Bustamante, J. L. Fernández, J. M. Zamaro 2014 *Journal of Colloid and Interface Science* 424, 37-43.

[9] A. J. Rieth, S. Yang, E. N. Wang, M. Dincă 2017 *ACS Central Science* 3(6) 668-672.

[10] H. Furukawa, F. Gándara, Y.-B. Zhang, J. Jiang, W. L. Queen, M. R. Hudson, O. M. Yaghi 2014 *Journal of American Chemistry Society* 136, 4369-4381.

[11] A. Matsumoto, H. Misran, K. Tsutsumi, 2004 *Langmuir* 20(17), 7139-7145.

[12] H. Zhang, J. James, M. Zhao, Y. Yao, Y. Zhang, B. Zhang, Y.S. Lin 2017 *Journal of Membrane Science* 532, 1–8.

[13] S. Schöttl, D. Horinek, 2016 *Current Opinion in Colloid & Interface Science* 22, 8-13.

[14] H. Misran, S. Z. Othman, A. Manap, N. I. M Pauzi, S. Ramesh 2014 *Science of Advanced Material* 6(7), 1638-1644.

IOP Conf. Series: Journal of Physics: Conf. Series **1082** (2018) 012047 doi:10.1088/1742-6596/1082/1/012047

The Effect of Silica Nanoparticles Stability in Biological Media

A. Ahmad[1], N.D Zakaria[2], Z. Lockman[1] and K.A.Razak[1,2,*]

[1] School of Materials and Mineral Resources Engineering,
Engineering Campus, Universiti Sains Malaysia, 14300 Nibong Tebal, Penang, Malaysia.
[2] NanoBiotechnology Research and Innovation (NanoBRI), Institute for Research in Molecular Medicine,
Universiti Sains Malaysia, 11800 Penang, Malaysia.

Email: khairunisak@usm.my

Abstract. The stability and level of aggregation of nanoparticles (NPs) in physiological conditions or different media are important for biomedical applications. The interaction of NPs in different media could affect the physicochemical properties of NPs. In this study, two different sizes of amorphous silica nanoparticles (SiNPs) encapsulated dye were synthesised using the micelle entrapment method. The SiNPs encapsulated dyes suspension was mixed with different concentration of salt solution, NaCl and mouse serum and incubated at 37°C to mimic human body environment to study the interaction of SiNPs encapsulated dyes in physiological conditions. Particles agglomeration or aggregation of SiNPs encapsulated dyes in NaCl solution and mouse serum were investigated and analysed. The absorbance spectra and the stability efficacy were recorded and calculated using UV-Vis spectrometer, while the particle size was measured using Zetasizer particle analysis and transmission electron microscope (TEM). The results obtained showed that 53 nm of SiNPs was more stable compared to 30 nm both in NaCl solution and mouse serum.

1. Introduction

Amorphous SiNPs encapsulated dyes have been widely explored especially for bio-imaging and bio-labeling applications. Amorphous SiNPs becomes popular due to the desired properties of high hydrophilic property that could give high solubility in water [1]. The solubility of amorphous SiNPs in water at body pH is normally around 130–150 ppm (µg/mL) [2]. Apart from that, SiNPs is optically transparent hence does not affect the fluorescence properties of encapsulated dyes. SiNPs also provides protection for dye molecules from the buffer. For biomedical application, it is important to determine the toxicological properties cause by SiNPs. However, very few and limited initial study on the stability effect of SiNPs in biological media can be found and is still not fully understood.

The stability and level of aggregation of NPs in physiological conditions or different media are important parameters prior to toxicological studies [3]. The interaction between the NPs in biological environment

IOP Conf. Series: Journal of Physics: Conf. Series **1082** (2018) 012047 doi:10.1088/1742-6596/1082/1/012047

could affect the physicochemical properties of NPs such as particle size distribution, aggregation/agglomeration, surface charge and surface chemistry. Hence, the stability of NPs is influenced by the local environment conditions such as temperature, pH and electrolytes concentration [4,5]. In most studies, NPs for biomedical applications are usually suspended in highly salted solutions or serum-containing media, conditions likely to have a large influence on the hydrodynamic diameter hence affect the stability of nanoparticles [6]. Thus, it is important to understand the effect of size of SiNPs in biological media.

In this study, two different sizes of amorphous silica nanoparticles encapsulated 1,1'-Dioctadecyl-3,3,3',3'-tetramethylindocarbocyanine perchlorate (DiI) dye (SiDiI) were synthesised using the micelle formation method. The particle agglomeration and stability of different sizes SiDiI in biological media that mimic the realistic biological conditions were determined and analysed.

2. Experimental

The silica nanoparticles were synthesised using the micelle entrapment method. Tween 80 (Polysorbate 80, Sigma Aldrich) as surfactant was dissolved in distilled water. 200 µl of ammonium hydroxide (Sigma Aldrich) was added as the pH adjuster to basic condition (pH 9-10). Then, different volume (2 and 6 ml) of co-solvent, 2-butanol (Merck) as the synthesis parameter was added into solution and stirred continuously to obtain different size of SiNPs. The solution was transferred into a preheated bioreactor at 50°C with 320 rpm stirring speed for 1 hour. After that, 2 ml solution of DiI fluorescence dye (Sigma Aldrich) was added into the micelle suspension and followed by organosilica precursor, Vinyltrimethoxysilane (VTMS) (98%, Sigma Aldrich). The mixture was continuously stirred overnight. The excess surfactant and solvent were removed by dialysis using 12 kDa dialysis tube (Fisher Scientific) in 4°C water for 4-5 days. The dialysis water was changed twice daily. After dialysis process, the silica encapsulated Dil dye (SiDil) was re-concentrated using 30 kDa cellulose membrane (Fisher Scientific). The concentrated of SiDiI suspension were surface modified with D-glucose (Sigma Aldrich) and trimetaphosphate (TSTMP, Sigma Aldrich). The hydrodynamic particle size of SiDil suspension was analysed using Zetasizer particle analysis. The morphology and size distribution were observed using transmission electron microscope, TEM.

SiDil suspension was further tested in biological media and serum to study its stability. The SiDil suspension (2%) was mixed with 0.1, 0.5 and 1.0 M NaCl (Sigma Aldrich) solution or 5, 10, and 25 % of mouse serum (Sigma Aldrich). The stability of SiDil suspension in biological medium was further analysed by incubation in a water bath at body temperature, 37 ± 0.5 °C for 5 hours. The absorbance value before and after incubation of the filtered samples was analysed using the UV-Vis spectrophotometer. The stability efficacy of SiDil suspension was calculated using the Equation 1:

$$\text{Stability efficacy }\% = \frac{\text{Concentration SiNPs encapsulated fluorescence dye}_{\text{tn (after filtration)}}}{\text{Concentration of SiNPs encapsulated fluorescence dye}_{\text{t0}}} \times 100 \quad (1)$$

3. Results and discussion

Two different sizes of amorphous silica nanoparticles encapsulated DiI (SiDiI) were synthesised using the micelle entrapment method. In this study, micelle entrapment method was chosen because of the ability to produce good monodispersed spherical particles of amorphous SiNPs. Micelles act as a temporary template to define the morphology of silica nanoparticles (SiNPs). Micelle structure formed from the interaction of nonionic surfactant, Tween 80 with aqueous solvent containing water and co-solvent at the critical micelle concentration (CMC). The addition of organosilica precursor, VTMS, produced the silanol groups and deposited around the micelle structure by hydrolysis and condensation reactions. The hydrolysis reaction was enhanced with the addition of ammonia by producing more hydroxyl ions (OH^-) to react with organosilanes. The condensation of alcohol and water caused interaction of silanols with each other and formed crosslinking between Si–O–Si chains until all Si precursors reacted.

In this study, 2ml and 6ml of 2-butanol that acted as co-solvent were used as the synthesis parameter to produce different size of SiDiI. The effect of volume of co-solvent during the synthesis method on the particle formation was determined using Zetasizer particles analyser and TEM. Figure 1 shows the hydrodynamic particle size distribution of SiDiI using 2 and 6 ml co-solvent measured using Zetasizer. SiDiI produced using 2 and 6 ml 2-butanol have hydrodynamic particle size of 32.6 and 59.0 nm with PDI values of 0.27 and 0.09, respectively. Particle size of SiDiI increases with increasing volume of 2-butanol resulted from the reducing hydrolysis rate of silica precursors during the silica formation. In this method, hydrolysis rate was affected by the nature of the solvent involving steric effect and hydrogen bonding between silica precursor and solvent. Hydrolysis rate of silica precursor is faster in water with low amount of alcohol as the co-solvent. Hydrolysis rate reduces with increasing amount of alcohol. In this work, high volome of alcohol such as 2-butanol (6ml), the effective availability of water molecules are limited causes slower hydrolysis rate of silica precursor [6]. As a consequence, less nucleation of particles took place during the hydrolysis which lead to the formation larger particle size [7].

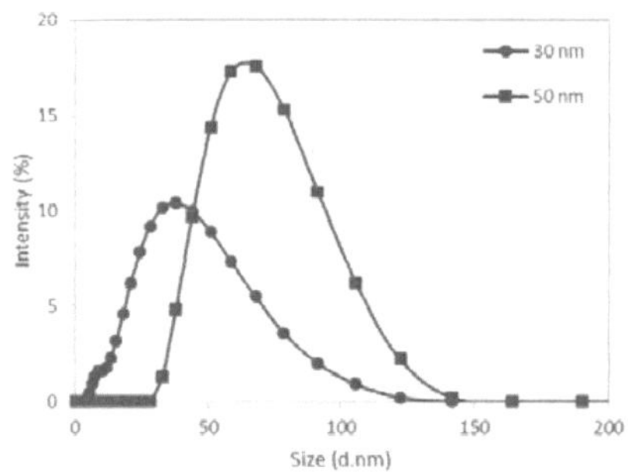

Figure 1. Hydrodynamic size distribution of SiDiI produced using different volumes of 2-butanol (2 and 6 ml).

The morphology of SiDiI produced using different 2-butanol volume is shown in

Figure 2. Spherical shape and well dispersed of SiDiI were observed from TEM images of both samples. 100 particles were measured by using the ImageJ software and the average particle size of SiDiI was calculated and tabulated in Table 1. The average particle size of SiDiI synthesised using 2 and 6 ml of 2-butanol is 30 and 53 nm, respectively. The particle size of SiDiI observed using TEM is smaller than hydrodynamic particle size measured by Zetasizer due to the Brownian motion that influences the hydrodynamic particles in colloidal form [6].

Table 1. Hydrodynamic size, polydispersity index (PDI) and TEM size of SiDiI produced using different volume of 2-butanol (2, 4 and 6 ml)

Volume of 2-butanol (ml)	Hydrodynamic size (nm)	Polydispersity index	TEM size (nm)
2	33	0.27	30
6	59	0.09	53

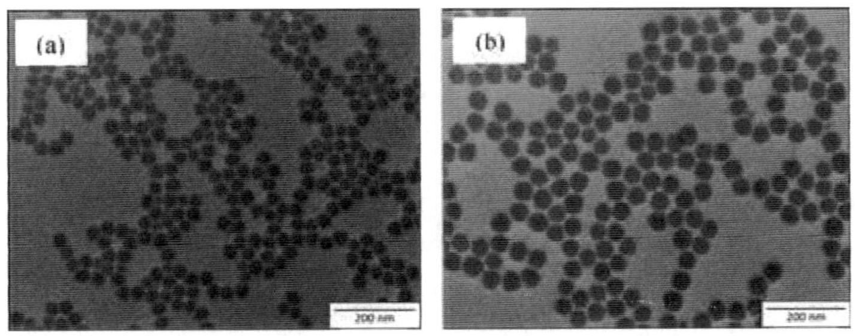

Figure 2. TEM images and histograms of particle size distribution of SiDil produced using different volume of 2-butanol : (a) 2 ml and (c) 6 ml.

Two different sizes of SiDiI suspension of 30 and 53 nm were added into different concentration of sodium chloride salt solution, NaCl. NaCl was chosen since the ionic strength in the cell culture media study mainly arises from this salt with only 0.15 M [8]. In this study, 20 % of SiDiI with sizes 30 and 53 nm were mixed with different concentrations of NaCl medium: 0, 0.1, 0.5 and 1.0 M and incubated in water bath for 24 hours at 37°C to study the stability effect on particle agglomeration. The absorbance value of SiDiI suspension in NaCl solution was measured by using UV-Vis spectroscopy to identify the presence of particle agglomeration. The stability efficacy was calculated from the measured absorbance value by the ratio between the maximum absorbance value of SiDiI after 24 hour incubation after filtration and maximum absorbance value of SiDiI before incubation. After 24 hours of incubation, the stability efficacy of SiDiI at NaCl concentration at 0.0, 0.1, 0.5 and 1.0 M for size 53 nm decreasing from 71, 68, 54 and 51 %, respectively. Similar trend was observed for SiDiI with size 30 nm with stability efficacy of 67, 56, 44 and 38 % with the increasing of NaCl concentration, respectively, as shown in Figure 3(a).

The results revealed that the increase of NaCl concentration caused colloidal suspension became unstable and started to flocculate [8]. In nanoparticle suspensions, particles have high tendency to aggregate more quickly at higher ionic strengths. This behavior can be explained by Derjaguin, Landau, Verwey, and Overbeek, DLVO theory where the agglomeration and stability of particle dispersions are influenced by the attractive and repulsive forces between individual particles [9]. In this case, the increase of NaCl concentration led to increasing of ionic strength in the suspension medium. At high NaCl concentration, ions exist in the dispersion covers the silica nanoparticles and neutralize the charge of their surface. In this condition, the thickness of double layer of SiDiI was reduced and the repulsive force between the nanoparticles was reduced. Hence, the total potential energy of interaction became more attractive and van der Waals forces became dominant resulted to particles agglomeration [10]. In addition, the result shows that SiDiI with size 53 nm has higher stability in NaCl solution than and SiDiI with size 30 nm revealed that SiDiI with size 30 nm has high particle agglomeration (less stable). The particle agglomeration of SiDiI was proven by the effective hydrodynamic particle measured by Zetasizer as shown in Figure 3(b). This is due to the large surface area of smaller particle size, which has high tendency of agglomeration between each particle [5]. On the other hand, small particles have less surface charge per particles leading to weaker electric double layer repulsion forces hence subsequently have greater aggregation tendency [11]. Thus, it can be concluded that at the same ionic strength, smaller particles have faster rate of aggregation than larger particles

IOP Conf. Series: Journal of Physics: Conf. Series **1082** (2018) 012047 doi:10.1088/1742-6596/1082/1/012047

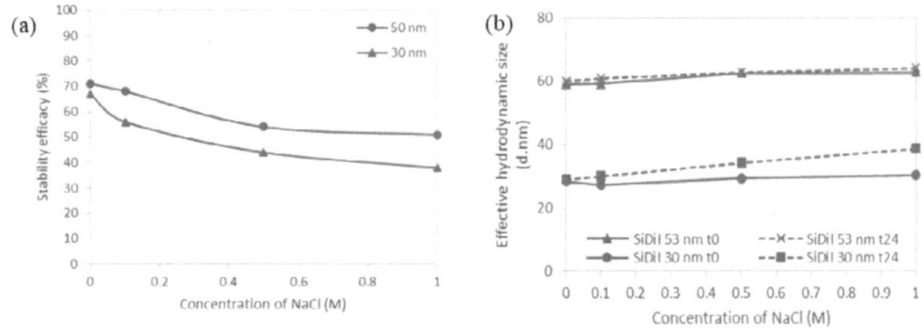

Figure 3. (a) Stability efficacy of SiDiI size 30 and 53 nm in different concentration of NaCl solution and (b) Effective hydrodynamic size of SiDiI size 30 and 53 nm before and after 24 hours of incubation at 37°C

The particles agglomeration of SiDiI was further studied in different concentration of mouse serum. 5, 10 and 25 % of mouse serum were mixed with 30 and 53 nm of SiDiI and incubated for 5 hours at 37°C. The absorbance properties were measured before and after incubation and after filtration to investigate stability of SiDiI. The filtration efficacy of SiDiI was derived from the maximum absorbance value of SiDiI before incubation and after filtration after 5 hours incubation. From the result showed in Figure 6, increasing of mouse serum concentration from 5, 10 and 25 % decreased the SiDiI stability after incubation from 89, 81 and 70 %, respectively, for 53 nm SiDiI. While the stability efficacy of 30 nm SiDiI decreased from 75, 52 and 64 %, respectively. The decrease of stability efficacy with increasing of mouse serum concentration revealed the formation of particles agglomeration in mouse serum at higher concentration. This could be due to the formation of protein corona on the surface of NPs resulted from the interaction of the NPs with the proteins presence in the biological media. Protein corona formed on the surface of NPs lead to modification on properties of NPs, hence, de-stabilised the colloidal systems of NPs favouring the formation of agglomerates [5,12]. Thus, more protein molecules adsorbed on the NPs surface would form more particle agglomeration of NPs [4].

The different agglomeration behaviour of different size of SiDiI was observed from the stability efficacy as shown in Figure 4. SiDiI with size 53 nm had high stability efficacy of 70 % when interacted with high concentration of mouse serum. With the same concentration of mouse serum, the lower filtration efficacy of SiDiI in size 30 nm was observed at 64 %. This result revealed that 53 nm SiDiI formed less agglomeration compared with 30 nm SiDiI. This result is in agreement with Halamoda-Kenzaoui et al., (2015), where smaller NPs has high tendency to agglomerate than larger NPs. In addition, Clemments et al. (2015) also demonstrated that smaller particles adsorbed large amount of protein due to their larger external surface area. Large amount of adsorbed protein lead to the greater accumulation of protein on the surface of NPs, hence, cause agglomeration [13, 14]. Although all the SiDiI samples have same surface chemistry and only differed in size, they behaved differently in cellular media containing the proteins as also observed by other authors [5].

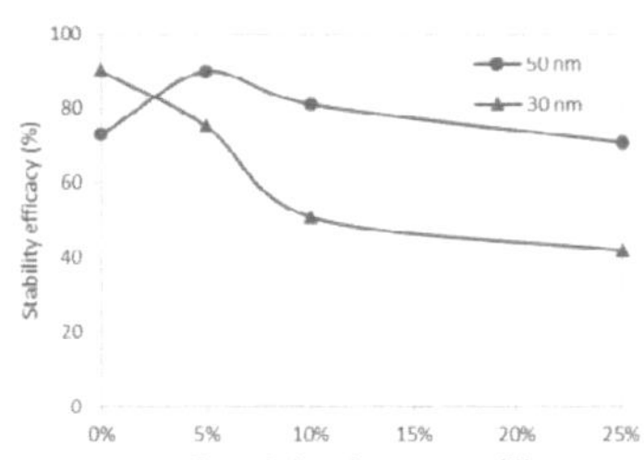

Figure 4. Stability efficacy of SiDiI size 30 and 53 nm in different concentration of mouse serum

4. Conclusion

Spherical shaped and monodispersed SiDiI were successfully produced using the micelle formation method. 30 and 53 nm of SiDiI were produced by varying the volume of co-solvent. The particle size of SiDiI increased with increasing volume of co-solvent. Further study on the stability of 30 and 53 nm SiDiI in NaCl solution and mouse serum showed similar trend. The stability of SiDiI for both sizes decreased as the concentration of NaCl solution and mouse serum increased. Furthermore, 53 nm SiDiI size had higher stability in both NaCl solution and mouse serum compared to 30 nm SiDiI.

Acknowledgment

The authors appreciate technical support from School of Materials & Mineral Resources Engineering, and Institute for Research in Molecular Medicine, USM. This research was supported by Research University (RU) grant 1001/Pbahan/870028. One of the author appreciates scholarship from Ministry of Higher Education Malaysia (MyBrain15).

References

[1] Iler K R 1979 *Colloid and Surface Properties and Biochemistry of Silica.*
[2] Viitala R, Areva S, Jokinen M & Koskinen M 2008 *Sol-Gel Methods for Materials Processing.*
[3] Pavlin M & Bregar V B 2012 *D. J. Nanomaterials & Biostructures (DJNB)*, 7.
[4] Pfeiffer C, Rehbock C, Hühn D, Carrillo-Carrion C, De Aberasturi D J, Merk V, Barcikowski S & Parak W J 2014 *J. of The Royal Society Interface* 11 20130931
[5] Halamoda-Kenzaoui B, Ceridono M, Colpo P, Valsesia A, Urbán P, Ojea-Jiménez I, Gioria S, Gilliland D, Rossi F & Kinsner-Ovaskainen A 2015 *PloS one* 10 e0141593
[6] Dos Santos Neves C S 2014 Development of fluorescent silica nanoparticles encapsulating organic and inorganic fluorophores; synthesis and characterization. PhD, de Ciências e Tecnologia da Universidade Nova de Lisboa.
[7] Song A, Zhang J, Zhang M, Shen T & Tang J A 2000 *Colloids and Surfaces A: Physicochemical &Eng. Aspects*, 167, 253
[8] Orts-Gil G, Natte K, Drescher D, Bresch H, Mantion, Kneipp J & Österle W 2011 *Journal of Nanoparticle Research,* 13, 1593
[9] Jiang J, Oberdörster G & Biswas P 2009 *Journal of Nanoparticle Research* 11 77
[10] Jenkins S, Kirk S R, Persson M, Carlen and J Abbas Z 2008 *The Journal of Chemical Physics* 128 16471
[11] Hotze E M, Phenrat T & Lowry G V 2010 *Journal of environmental quality* 39 1909
[12] Moore T L, Rodriguez-Lorenzo L, Hirsch V, Balog S, Urban D, Jud C, Rothen-Rutishauser B, Lattuada M & Petri-Fink A 2015 *Chemical Society Reviews* 44 6287
[13] Clemments A M, Botella P & Landry C C 2015 *ACS Applied Materials & Interfaces* 7 21682
[14] Albanese A, Walkey C D, Olsen J B, Guo H, Emili A & Chan W C 2014 Secreted biomolecules alter the biological identity and cellular interactions of nanoparticles

IOP Conf. Series: Journal of Physics: Conf. Series **1082** (2018) 012048 doi:10.1088/1742-6596/1082/1/012048

Synthesis of α -Fe_2O_3 Nanowires/MWCNTS Composite for Photocatalyst

F.I. A. Bakar, S. Ismail*, A. Nurulnabilah, and M.S Shahadan

Carbon Research Technology, Faculty of Manufacturing Engineering, Universiti Teknikal Malaysia Melaka, Hang Tuah Jaya, 76100, Durian Tunggal, Melaka, Malaysia.

Email: *Syahriza@utem.edu.my

Abstract. Multi-walled carbon nanotube (MWCNTs) was decorated on nanowires iron oxide (α-Fe_2O_3) thin film (α-Fe_2O_3/MWCNTs) by spin coating method. Nanowires hematite was produced by thermal oxidation of Fe substrate at 400 °C for 120 minutes. The aim of this nanocomposite α-Fe_2O_3/MWCNT is to improve the photocatalytic efficiency of the α-Fe_2O_3 nanowires. The structural characterization of the as-prepared composites were analyzed by x-ray diffraction (XRD), raman spectra, and photoluminescence spectra (PL) to identify the structure of pure α-Fe_2O_3 and decorated of MWCNTs. While, the morphological observation, interaction of MWCNTs and nanowires structure on hematite phase was done by scanning electron microscopy (FESEM). It was found that, the optimization dispersion of MWCNTs in the dimethyl form amide (DMF) solvent through acid treatment with 3 Molar nitric acid (HNO_3), shows the homogenized well-decorated MWCNTs on the surface of the nanowires structure, which can eliminate the limitation of iron oxide and help in electron transfer from α-Fe_2O_3 to MWCNTs.

1. Introduction

Over the past several decades, exponential population, rapid development of industrialization have been accompanied by a sharp modernization and metropolitan growth. The industrial activities leading to deleterious interferences of the vast generation, such as food processing, textiles, leather, paper, synthetic dyes have accounted for causing water pollution.

Therefore, developing advanced methods for purification of water is important for monitoring the environmental and human health. According Jantawasu et al., advanced oxidation processes (AOPs) are one of the promising methods for this issue due to its competency in degrading and mineralizing organic pollutant rather than other methods [1]. α-Fe_2O_3 is one of the challenging materials in this case that has a suitable properties to be one of the attractive semiconductor in such application compared to other semiconductor such as ZnO and TiO_2. This is due to low cost, most stable phases and narrow band gap of 2-2.2 eV [2]. Furthermore, it can absorb in the red light of the visible range, solar light up 620 nm, which is around 43% in the solar spectrum [3]. The photoactivity α-Fe_2O_3 is generally hampered by poor charge transportation, short holes diffusion length of 2-4 nm and its leading to high recombination photogenerated electron-hole pairs [4].

IOP Conf. Series: Journal of Physics: Conf. Series **1082** (2018) 012048 doi:10.1088/1742-6596/1082/1/012048

To overcome the limitation of α-Fe_2O_3 nanostructure, a relatively new method was created to minimize the recombination of charge pairs and improved the charge transportation in nanostructured. Thus, the properties of α-Fe_2O_3 is modified by formation of composite by combining with carbon nanomaterial's such as carbon nanotube (CNT) [5-6]. This composite sample have fascinating properties such as large surface area, inert chemical properties, one-dimension structure and strength that has not been shown by any known materials. Recently, CNT have drawn much attention because of their unique characteristics such as excellent stability, small size, high porosity, higher surface area and layered structure provide enough adsorption sites as well as good support catalytic assistant materials [6]. Furthermore, CNT has excellent electrical, mechanical, and thermal properties. This effect is believed to originate either from the favourable adsorption properties of MWCNTs. Their ability to trap electrons and thus reduce the electron–hole recombination. However, good combination between α-Fe_2O_3 and MWCNTs has become the challenge. Where, problems faced in transferring the MWCNTs to the application. The MWCNT had very strong van der wall interaction, thus it's very easy to agglomeration and entanglements. Therefore, this work is to propose a good solvent for MWCNT dispersion and functionalize the CNT in the carboxyl group to provide extra oxygen so that more hydrophilic surface structure can be generated [7]. This would help to increase the ion-exchange which leads to a well decorated CNT on the α-Fe_2O_3.

Nowadays, many researchers studied on focus on attachment of α-Fe_2O_3 nanoparticle to the MWCNT in various applications [8-10]. However, this study is focusing on nanocomposite of α-Fe_2O_3 nanowires/MWCNTs by convenient method is proposed.

2. Methodology

High-purity iron foil (99.99 %) was used with a thickness of 0.25 cm. The substrate was subjected to an ultrasonic cleaning in ethanol for 10 minutes and rinsed with distilled water. Then, the cleaned Fe substrate is dried in an air stream and placed on a substrate heater in chamber furnace at 400 °C and oxidation for 120 minutes. The heating rate around rate ~ 10°C min^{-1} and cooled down at room temperature to consist of scarlet layer homogeneously. This work was conducted under ambient environment.

MWCNTs were supplied by CNano Technology. MWCNTs had an average length and a diameter of 10 μm and 11 nm with ~ 95 % purity. In order to improve their solubility and remove residual catalyst, the nanotubes were functionalization with –COOH group and carried out using HNO_3 by chemical technique. For the oxidation treatment, 0.3 g MWCNTs were treated into 3 molars HNO_3 with 100 ml solution followed by distilled water. The mixture were sonication using the sonicating bath (100w, 50Hz) for 2 hours at 70 °C. Thus, was promotes MWCNTs disentanglement within acid solution. The purified MWCNTs were obtained after filtering and thoroughly rinsed with distilled water until reaching pH 7. Finally, the slurry is dried in drying oven at 80 °C for 24 hours.

In a composite sample, there have two different solutions for dissolution of the MWCNTs, which are organic solvent such as acetone, DMF and aqueous solvent using surfactant is sodium dodecylbenzenesulfonate (SDBS) with distilled water with amount of 9 mg/ml. The MWCNTs suspension was prepared using UP 100H ultrasonic horn with 0.25 mg/ml for 45 minutes. The deposition of MWCNTs on the α-Fe_2O_3 thin film were carried out by spin coating method. The spinning time and speed was set for 30 sec, 4000 rpm/s for acetone and SDBS. While, DMF solvent was set for 120 sec and 3000 rpm/s. The finally, the sample was contain SDBS solvent was immersed in the Deionized water (DI water) for 1 hours to assist the removal of the surfactant.

Then, to evaluate the incorporation of α-Fe_2O_3/MWCNTs composite, surface morphologies were characterized by FE-SEM S4800, Hitachi. While, the crystalline structure of α-Fe_2O_3/MWCNTs composite was characterized by XRD, Cu Kα radiation, λ= 1.5406Å in the range 20 ° to 80 ° and Raman spectra were recorded at room temperature using the UniRAM-3500 performed at a laser wavelength of 523 nm. Besides, PL spectra were measured on FLS920 Edinburgh instrument, with Xenon lamp 450Watt.

IOP Conf. Series: Journal of Physics: Conf. Series **1082** (2018) 012048 doi:10.1088/1742-6596/1082/1/012048

In this work, the nanocomposite photocatalyst α-Fe_2O_3/MWCNTs was verified by the degradation of 10 ppm methyl orange. The degradation was done under UV-light irradiation. During the experiment, α-Fe_2O_3/MWCNTs photocatalyst film were put in 20 ml of methyl orange containing beaker. The solution of MO with photocatalyst is then kept in the dark for 1 hours for the adsorption and desorption equilibrium. Then, exposed to UV light for 5 hours and the solution were taking out every 2 hours during irradiation for analyzing the absorbance change of MO by UV-VIS spectroscopy (UV-1700).

3. Results and discussion

3.1 Electron microscopy analysis

Nanocomposite sample with the pristine MWCNTs and functionalized MWCNTs with COOH is shown in Figure 1 (a) and (b) respectively. From Figure 1(a) i- 1(a) iii, it is seen that, samples with pristine MWCNTs leads to the agglomeration of MWCNTs on the nanowires structure composite for all solvents. This is due to the strong van der wall interaction of MWCNTs that contributed to the clustering of MWCNTs on the α-Fe_2O_3. While for functionalized MWCNTs (Figure 1(b) i - 1(b) iii), surface reactivity of MWCNTs is significantly enhanced, given the better decorated composite. It showed that, metal catalyst and amorphous carbon were removed and wettability of MWCNTs was enhanced. In Figure 1 b (i), shows the coverage of MWCNTs in acetone are non-uniform and has a small aggregate of MWCNTs on nanowire structure. But the combined effect of surfactant with SDBS in water-soluble (Figure 1 b (iii)) and dispersion in DMF solvent (Figure 1 b (ii)) has given the well-dispersed and decorated of MWCNTs on the nanowires structure. This is because SDBS surfactant has a benzene rings that could contribute to increase the dispersion power of surfactant by interaction of π- π, thus leading MWCNTs prolongs the slurry suspension time [11]. Furthermore, the tail of the surfactant work as hydrophobic properties and it react to attach on the surface of MWCNTs to overcome bundled of MWCNTs, thus stabilized nanotubes. While, dispersion in DMF solvent shows almost a fully covered MWCNTs with a fairly thin film due to the availability free electron pair and high solvatochromic parameter [12]. It has made this solvent as one of the best solvent in the dispersion of MWCNTs. The high polar wettability of DMF offers the fast and all directional solution spread-out during spinning. Thus, the low surface tension of DMF solvent at the liquid/solid interface contributed to the improved uniformity of the layer and boost the coverage of thin film [13].

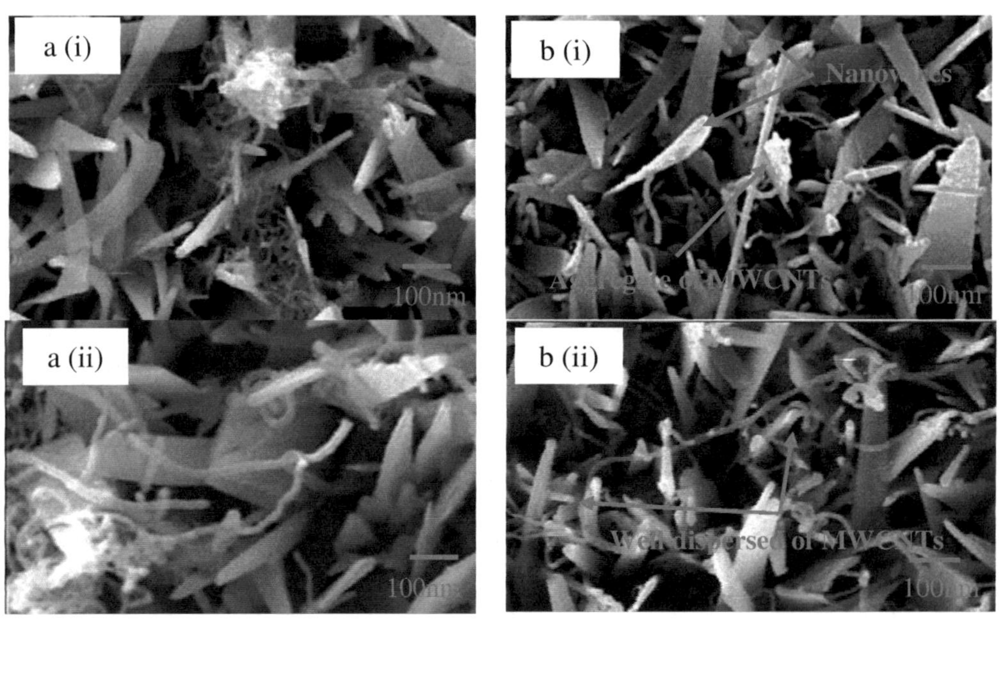

IOP Conf. Series: Journal of Physics: Conf. Series **1082** (2018) 012048 doi:10.1088/1742-6596/1082/1/012048

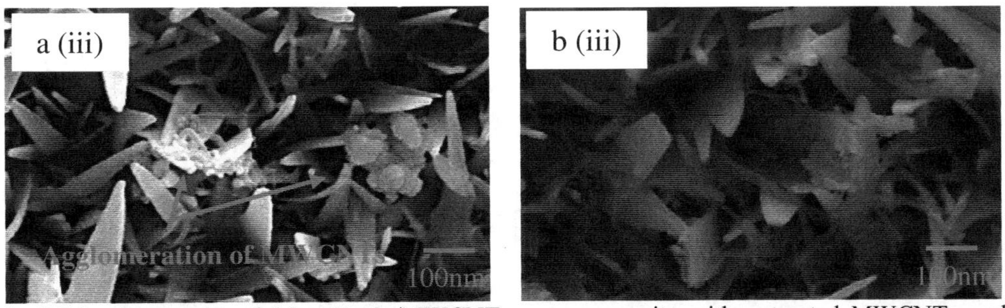

Figure 1. SEM micrographs of (a) α-Fe_2O_3/MWCNTs nanocomposite with untreated MWCNTs and (b) α-Fe_2O_3/MWCNTs nanocomposite with treated MWCNTs. Different solvent used is for WMCNTs dispersion: i) acetone, ii) DMF, iii) aqueous solvent (SDBS)

3.2 XRD analysis

Figure 2 indicates XRD patterns of pristine α-Fe_2O_3 and α-Fe_2O_3/MWCNTs thin film. The pristine α-Fe_2O_3 and nanocomposite of α-Fe_2O_3/MWCNTs in various solvent exhibited dominant peaks of α-Fe_2O_3 phase at 33.3, 35.8, 41.1, 43.8, 49.8, 54.5, 57.8, 62.8, 64.4 which corresponds to (104), (110), (113), (202), (024), (116), (122), (214) and (300) planes. These standard peaks attributed to α-Fe_2O_3 phase for pure rhombohedral, space group: R3c with lattice constant, a = 5.0082Å and c = 13.6467 Å (JCPDS No. 01-072-6226). Therefore, the crystal structure of [110] showing the favorable growth of nanostructured on the hematite phases. In nanocomposite, no diffraction peak of MWCNTs was observed even in functionalized of MWCNTs, due to the low content of MWCNTs and low diffraction pattern intensity [14]. Furthermore, no significant changes occur between the pristine α-Fe_2O_3 and MWCNTs. The phase purity of α-Fe_2O_3 in composite structure is maintained and MWCNTs are not incorporated into the lattice of α-Fe_2O_3.

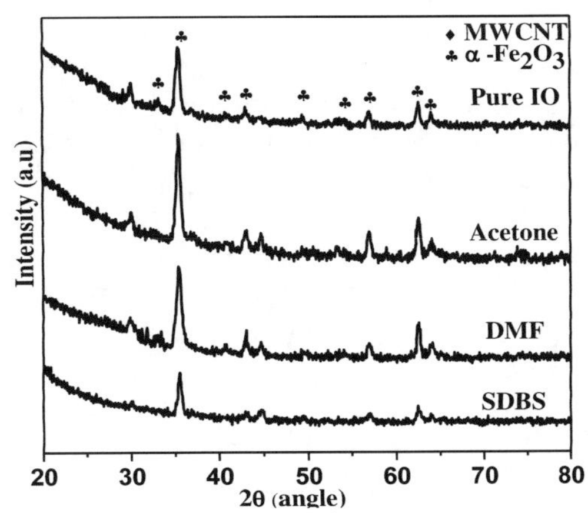

Figure 2. XRD patterns for pristine α-Fe_2O_3 and α-Fe_2O_3 /MWCNTs nanocomposite with, treated MWCNTs in various solvent

3.3 Raman spectra analysis

Raman spectroscopy was carried out to confirm presence of carbon nanotube in the α-Fe_2O_3 /MWCNTs sample. Figure 3 shows the Raman spectra of α-Fe_2O_3 and α-Fe_2O_3/MWCNTs. In the case of a modified sample of α-Fe_2O_3/MWCNTs, two peaks of MWCNTs at 1339 cm^{-1} and 1580 cm^{-1} can be attributed to the D and G band respectively (Figure 3 (a)). This Raman analysis of functionalized MWCNTs shows the intensity of D band become higher compared to G bands, it's due to the presence of more defects on the surface of MWCNTs in functionalization process and good interaction in composite sample. While, another five peaks rise in the composite sample for hematite phase was detected due vibration modes of

Fe-O bonds for α-Fe_2O_3 at lower Raman shift (218, 277, 393, 483 and 592 cm^{-1}). This is classified as "internal mode" ($2A_{1g} + 3E_g$). This results aligned with Wang et al., [15]. In composite, it can also be observed that the average I_G/I_D ratio in DMF solvent are 0.91, which are higher compared to the other solvent. This has caused the dispersion and decorated of MWCNTs on the nanowires structure in DMF solvent is in better condition and quality. Furthermore, in the case of nanocomposite α-Fe_2O_3/MWCNTs as shows in Figure 3 (b), small shift indicating a weak shift compared to pure α-Fe_2O_3. It is reported, that a good contact and direct interaction between α-Fe_2O_3 nanowires structure and MWCNTs is created.

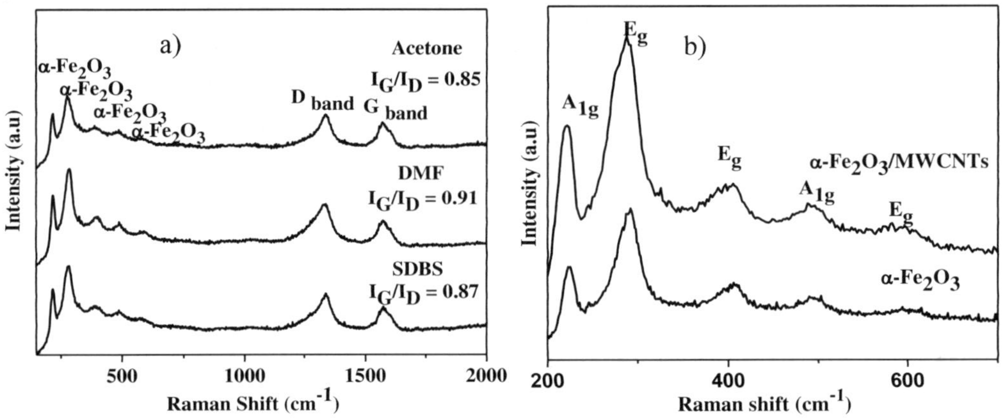

Figure 3. The Raman spectra of α-Fe_2O_3 and α-Fe_2O_3/MWCNTs (a- showing full range in composite and b – showing low range)

3.4 Photoluminescence spectra analysis

Figure 4 shows the PL analysis of pure α-Fe_2O_3 and α-Fe_2O_3/MWCNTs composite using the DMF solvent. PL analysis can be used to evaluate the mobility of the charge carriers to the surface and also its recombination process involving the electron-hole pairs [14]. As seen in Figure 4, the intensity of α-Fe_2O_3/MWCNTs nanocomposites were significantly decreased compared the pure α-Fe_2O_3 indicating that, there is inhibition of electron charge recombination with the presence of MWCNTs. The weak intensity of the composite becomes rise due to the well-decorated MWCNTs on the nanowires structure and amount of MWCNTs was added. It is believed that, MWCNTs could be as electron storage to trap part of the electron generated from the photoexcitation of the α-Fe_2O_3 nanowires due to the defect structure and it contained more oxygen during the functionalized of MWCNTs.

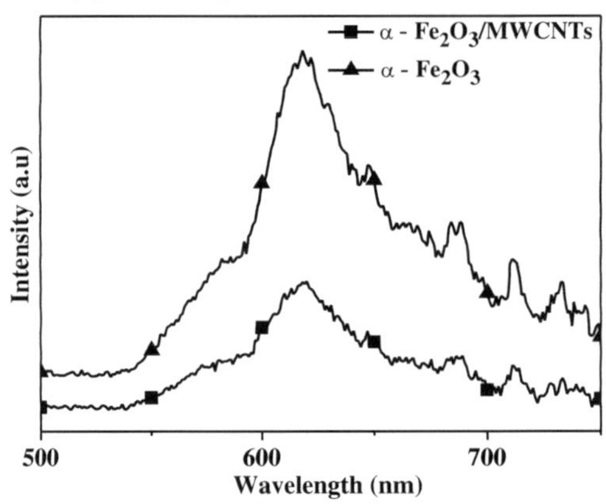

Figure 4. Photoluminescence spectra of α-Fe_2O_3 and α-Fe_2O_3/MWCNTs composite

3.5 Photocatalytic reaction analysis

In the photocatalytic part, degradation of methyl orange is carried out to establish the photocatalytic activity of the pristine α-Fe_2O_3 and α-Fe_2O_3/MWCNTs nanocomposite as shown in Figure 6. The pure α-Fe_2O_3, has a poor photocatalytic indicated only 28 % degradation efficiency through 5 hours of UV irradiation. While after adding MWCNTs, α-Fe_2O_3/MWCNTs composite with DMF solvent shows dramatic enhancement of degradation efficiency compared to pristine α-Fe_2O_3. This is caused, the usage of MWCNTs tends to reduce the recombination process and effective in accepting, and storing of electron.

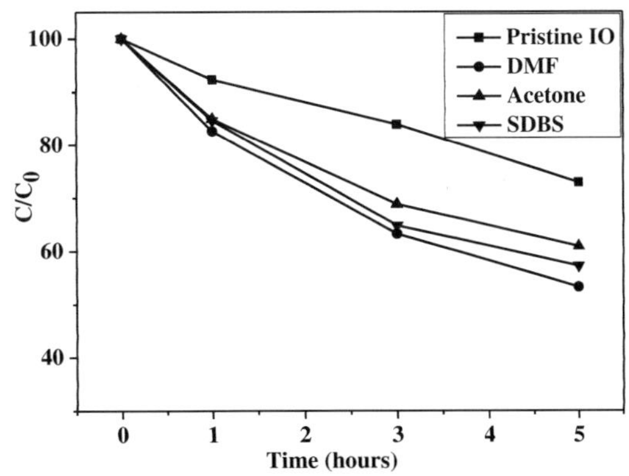

Figure 5. Photocatalytic degradation of methyl orange in 5 hours

4. Conclusions

In summary, the study of α-Fe_2O_3/MWCNTs were successfully prepared by thermal oxidation and spin coating. From this study, photodegradation of the dye and efficiency of composite attribute to a good transfer charge in α-Fe_2O_3 nanowires due to the presence of MWCNTs. Thus, minimized the recombination of electron-hole pairs and MWCNTs were found to be efficient assistant for photocatalyst as it help to create a trap region in α-Fe_2O_3 band gap. The photocatalytic activity of α-Fe_2O_3 can be improved up to 50 % by having a low content of MWCNTs compared pure iron oxide. The higher adsorption capacity by α-Fe_2O_3 /MWCNTs is mainly attributed to adsorption sites and defect from the functionalization of MWCNTs in HNO_3.

Acknowledgement

The authors would like to thank Universiti Teknikal Malaysia Melaka for financial (PJP/2016/FKP-AMC/S01499) and facility support.

References

[1] Jantawasu P, Sreethawong T, and Chavadej S 2009 *Chemical Engineering Journal* 55 223
[2] Rai S, Ikram A, Sahai S, Dass S, Shrivastay R, and Satsangi V R 2015 *Renewable Energy* 83 447
[3] Frites M, Shaban Y A and Khan S U 2010 *International Journal of Hydrogen Energy* 35 4944
[4] Zhao Z, Zhou X W and Liu Z 2015 *In Proceedings Of the 4th International Conference on Mechatronics, Materials, Chemistry And Computer Engineering* 39 2317
[5] Ashkarran, Fakhari M and Mahmoudi M 2013 *RSC Advances* 3 18529
[6] Mohd Adnan M A, Julkapli N M and Abd Hamid S B 2016 *Reviews in Inorganic Chemistry* 36 77
[7] Atieh M A, Bakather O Y, Al-Tawbini B, Bukhari A A, Abuilaiwi F A and Fettouhi M B 2011 *Bioinorganic chemistry and applications* 2010.
[8] Zhang Z J, Lu C Y, Huang W, Guan W S and Peng Y X 2014 In *Advanced Materials Research* 915 933
[9] Asmaly H A, Abusaud B, Saleh T A, Gupta V K, Atieh M A 2015 *Journal of Saudi Chemical Society* 19 511
[10] Liu W, Jiang X, and Che X 2014 *Applied Surface Science* 320 764
[11] Vaisman L, Wagner H D, and Marom G 2006 *Advances in colloid and interface science* 128 37
[12] Sun G, Liu Z and Chen G 2010 *Nano* 5 103
[13] Choi J Y, Alford T L and Honsbery C B 2014 *Langmuir* 30 5732
[14] Pradhan G K, Padhi D K and Parida K M 2013 *ACS applied materials & interfaces* 5 9101
[15] Wang T, Li Y, Wang L, Liu C, Geng S, Jia X, Yang F, Zhang L, Liu L, You B and Ren X 2015 *RSC Advances* 5 60114

IOP Conf. Series: Journal of Physics: Conf. Series **1082** (2018) 012049 doi:10.1088/1742-6596/1082/1/012049

Development and Evaluation of Colloidal Gold Lateral Flow Immunoassays for Detection of *Escherichia Coli* O157 and *Salmonella* Typhi

Y. Mohammad Lukman[1], Z. Nor Dyana[2], N. Rahmah[2] and A. R. Khairunisak[1, 2] *

[1]School of Materials and Mineral Resources Engineering, Universiti Sains Malaysia, 14300 Nibong Tebal, Penang, Malaysia
[2]Nanobiotechnology Research and Innovation (NanoBri), Institute for Research in Molecular Medicine, Universiti Sains Malaysia, 11800 Penang, Malaysia

Email: kayron77@gmail.com

Abstract. Lateral flow immunoassays (LFIAs) are rapid and simple detection methods for detecting bacterial contamination in food and drinking water such as *Escherichia coli* O157 and *Salmonella* Typhi. *Escherichia coli* O157 can cause haemorrhagic colitis, diarrhoea and haemolytic uraemic syndrome, whereas *Salmonella* Typhi is responsible for typhoid fever in human. Colloidal gold nanoparticles (AuNPs) synthesized using citrate reduction method is commonly used as labels for LFIAs. However, in this study, AuNPs synthesized using the seeding-growth method was evaluated as label in LFIAs. The 15 nm AuNPs seed was first synthesized using the citrate reduction method followed by growth to 40 nm AuNPs using mild reduction agent, hydroxylamine. This monodispersional AuNPs was then conjugated with anti-gram-negative endotoxin antibody at 3 μL/mL. The conjugate then immersed to conjugate pad of glass fiber at OD 10. Two sets of nitrocellulose membrane (NC) were used to immobilize anti-*E. coli* O157 antibody as the test dot at 1.7 mg/mL and anti-mouse IgG antibody as the control dot at 1.0 mg/mL. Another set of NC immobilized with 1.0 mg/mL anti-*Salmonella* sp. antibody as the test dot and the same control dot as first NC set. Spiked water and milk sample was prepared by adding 10-fold dilution ATCC stains of *E. coli* O157 and *Salmonella* Typhi to water and milk samples. LFIA to *E. coli* O157 in water and milk sample was able to detect as low as 7.8×10^5 and 3×10^6 CFU/mL, respectively, whereas LFIA to *Salmonella* Typhi was able to detect as low as 3×10^8 and 3×10^7 CFU/mL in water and milk sample, respectively. Specificity for both strips sets was 100%. This method is very useful for monitoring bacterial contamination in food and drinking water towards ensuring food and water safety.

1. Introduction

Current trends in nutrition and food technology have attracted attention of food microbiologist to ensure food supply is safe before entering the market. Bacterial pathogens encountered to human illness in the last decades through consumption of undercooked or minimally processed ready-to-eat meats, dairy products or fruits and vegetables [1-3]. Two of most commonly food borne pathogens are *Escherichia coli* O157 and *Salmonella* Typhi.

Escherichia coli O157:H7 is one of an *Enterohemorrhagic E. coli* (EHEC) that causes diarrhea associated with abdominal cramps, hemolytic uremic syndrome, hemorrhagic colitis, thrombotic thrombocytopenic purpura and presentation of mild fever [4, 5]. *E. coli* O157:H7 infection is hardly found in Malaysia but it is required for precaution by screening all imported foods and beverages. *Salmonella* Typhi on the other hand is responsible for enteric fever causes by typhoid fever [6]. The above bacterial infection if not treated rapidly can increase rate of morbidity and mortality [7]. The conventional and traditional method for detection of these bacterial are culture method, Polymerase Chain Reaction (PCR), enzyme-linked immunosorbent assay (ELISA) and immunological techniques [8]. Unfortunately, these methods require technical expertise, expensive equipment, laborious protocol and impractical for on field detection. Thus, it is a must to develop a rapid, simple and sensitive method for detection of these food borne pathogens.

The lateral flow immunoassay strips (LFIA) is a paper-based biosensor device applying a modern version of immunoassay with advantages of rapid, low limit of detection, high sensitivity, good specificity, low manufacturing cost, less sample operation volume, robustness, user-friendly format and no complicated equipment needed [9]. The LFIA strips used nanomaterials as label which improve their performance for various applications [9, 10]. Colloidal gold nanoparticles (AuNPs) is the most widely used label owing to easy synthesis and modification, stable with time, size-tunable, biocompatible and have an intense red colour that easy to be detected even by naked eye or usually using a colour reader to achieve better detection limit [9].

Citrate reduction method is the most reported method for the synthesis of AuNPs with different size by using trisodium citrate as the reducing agent [5, 10-12]. However, this method hardly produces monodisperse spherical AuNP when the size exceeds 30 nm [13]. Nearly all AuNPs based LFIA use 40 nm in size, thus this method seems to be unsuitable. Moreover, AuNPs synthesized by this method requires more antibody to be stable due to unspecific interaction of the AuNPs surfaces with the sites of the antibody and therefore increasing the production cost of LFIA [14].

To overcome these problems, the seeding-growth method was used to synthesis AuNPs. In this method, smaller particles known as seeds can be grown into larger particles in controlled manner by using hydroxylamine (NH_2OH) as a mild reducing agent that reduces and prevents the formation of new nuclei to allow the growth process [15, 16]. Thus, in this work, 40 nm AuNP was synthesized by seeding-growth method to produce monodisperse spherical AuNP as a label in LFIA. Then, the performance of this LFIA for detection of *Escherichia coli* O157 and *Salmonella* Typhi in spiked water and milk samples was evaluated accordingly.

2. Experimental

2.1 Synthesis and Characterization of Colloidal Gold Nanoparticle (AuNP)

40 nm colloidal AuNPs was prepared by using the seeding-growth method as reported previously [14]. The prepared seeds then were grown to 40 nm AuNPs. About 4 mL seed was added to 50 mL deionized water while stirring followed by the addition of 1 mL hydroxylamine as a mild reducing agent to the solution. Then, 1% (w/v) of chloroauric acid was added dropwise. The synthesized AuNPs was characterized using UV-Vis-NIR spectrophotometer (Model UV-3600, Shimadzu), particle size analyzer (Model ZEN 3600, Nanoseries, Malvern Instruments) and transmission electron microscopy (TEM) (Model FEI CM12, Version 3.2 image analysis system, 120 kV).

IOP Conf. Series: Journal of Physics: Conf. Series **1082** (2018) 012049 doi:10.1088/1742-6596/1082/1/012049

2.2 Conjugation of 40 nm AuNPs with Antibody

The AuNPs was first adjusted to pH 7 with 0.2 M K_2CO_3. Volume of anti-gram-negative endotoxin antibody (GNE) was optimized to stabilize AuNPs. The volume of GNE that retained the original red wine colour of AuNPs after addition of 10% sodium chloride (NaCl) was chosen. The optimized volume was added into AuNPs. After incubated at room temperature for 30 min, 1% (w/v) of bovine serum albumin (BSA) was added to the conjugate and centrifuged at 10000 rpm for 10 min. Supernatant was then removed and this step was repeated twice. In the last washing, pellet was re-suspended in 1% BSA. This conjugate then was characterized by using UV-Vis-NIR spectrophotometer and optical density (OD) was measured.

2.3 Preparation of LFIA Strips

Two sets of LFIA strips were prepared, one set for detection of *E. coli* O157 and another one for detecting *Salmonella* Typhi. An HF90 nitrocellulose membrane was used to immobilize capture antibodies. For *E. coli* O157 LFIA strips, 0.5 µL at 1.7 mg/mL of anti-*E. coli* O157 antibody was pipetted at the lower part of NC membrane as the test dot. Anti-*Salmonella* sp. antibody was pipetted at 1.0 µL of 1.0 mg/mL as the test dot for *Salmonella* Typhi LFIA strips. Both sets were pipetted with 1.0 µL at 1.0 mg/mL of anti-mouse IgG antibody at upper part on the NC membrane as the control dot. The NC membrane was then dried for 2 h at 37 °C. Blocking solution (50% v/v) of western blot solution in 0.01 M TBS pH 7.5 was flowed to NC membrane. This blocking solution was used to block the free site of the membrane. Blocked NC membrane then was re-dried overnight at 37 °C.

The conjugate pad was prepared to hold AuNPs-GNE conjugate. The AuNPs-GNE conjugate was diluted in solution containing 0.2 mM Boric buffer, 1% BSA and 5% trehalose to be at OD 10. Glass fiber as conjugate pad was pipetted with 30 µL of the prepared OD 10 AuNP-GNE conjugate. This conjugate pad then dried overnight at 37 °C. Sample pad from cellulose material was left untreated. Complete LFIA strips were then formed by assembled conjugate pad, NC membrane, sample pad and absorbance pad together.

2.4 Samples Preparation, Visual Limit of Detection (LOD) and Cross-Reactivity Testing of LFIA

American type culture collection (ATCC), ATCC® 43888™ of *E. coli* O157:H7, ATCC® 10536™ of *E. coli*, ATCC® 12022™ of *Shigella flexneri* and ATCC® 14028™ *Salmonella* Typhi were grown in trypticase soy broth and adjusted to McFarland standard 1.0 (approx. 3×10^8 CFU/mL). The ATCC of *E. coli* O157 and *Salmonella* Typhi were diluted in 10-fold from 3×10^8 to 3×10^1 CFU/mL. The dilution then was spiked to sterile water and milk. ATCC of *E. coli* and *Shigella flexneri* at 3×10^8 were also spiked to sterile water and milk for specificity testing.

The LFIA strips were pipetted with the spiked sample on the sample pad. The sample was then flowed to the conjugate pad, NC membrane and finally to the absorbent pad by capillary action. *E. coli* O157 and *Salmonella* Typhi were pipetted to LFIA containing anti-*E. coli* O157 and anti-*Salmonella* sp. antibody respectively. Uninoculated sterile water and milk were used as negative control and flowed to both LFIA sets. After that, a drop of Chase buffer containing 0.035% Tween-20 was used to wash LFIA for removing background stain. Intensity, present or absent of control and test dots were observed for each set.

3. Results and Discussion

3.1 Characterization of Colloidal AuNPs

The colloidal AuNPs was successfully synthesized using the seeding-growth method. The size of 15 nm AuNPs seed and 40 nm AuNPs were confirmed from TEM images and particle size analyzer (Figures 1 and

IOP Conf. Series: Journal of Physics: Conf. Series **1082** (2018) 012049 doi:10.1088/1742-6596/1082/1/012049

2). This result was comparable to previous study where AuNPs synthesized by the seeding-growth method was highly monodispersed with uniform size [14]. Normal distribution histogram with average 17 nm for AuNP seed and 42 nm for 40 nm AuNPs (Figure 2a – b) was obtained from particle size analyzer. The size obtained by particle size analyzer is slightly bigger than TEM because particle size analyzer measures hydrodynamic size of the particle which include thickness of the electrical double layer (solvation shell) as well as the presence of adsorbed substance such as stabilizer [17]. The UV-Vis-NIR spectroscopy analysis showed the maximum peak at absorbance (OD) around 1 for both seed and 40 nm AuNPs (Figure 2c). The maximum peak for AuNPs seed was at wavelength (λ_{max}) 519 nm whereas for 40 nm AuNPs at 527 nm. This also confirmed that AuNPs seed is smaller than 40 nm AuNPs because the maximum peak was red shifted form AuNPs seed to 40 nm AuNPs [14, 17].

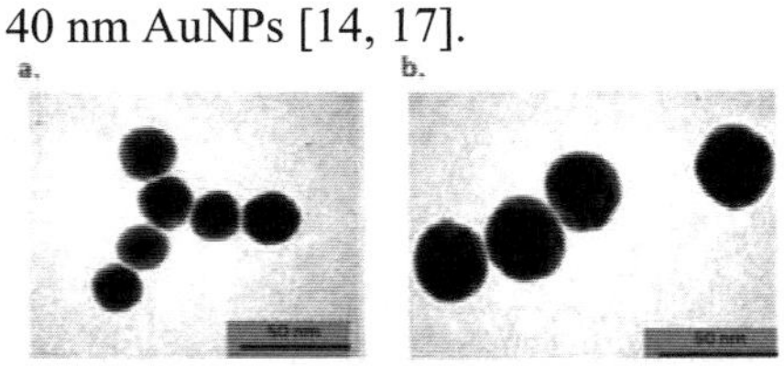

Figure 1. TEM images of (a) 15 nm AuNPs seed, and (b) 40 nm AuNPs

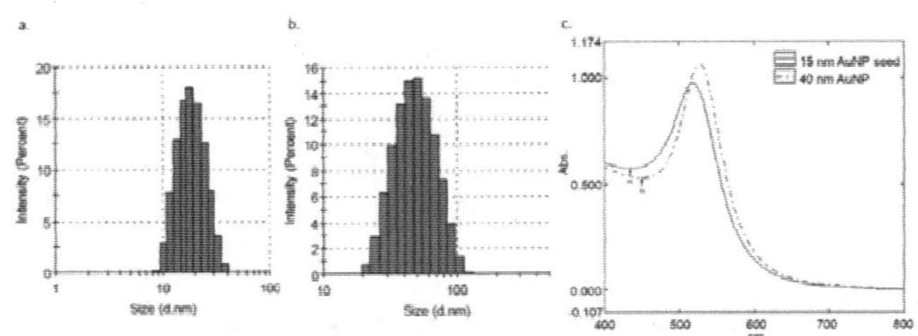

Figure 2. Histograms of size distribution by intensity of particle size analyzer for (a) 15 nm AuNP seed, (b) 40 nm AuNP, and (c) UV-Vis-NIR spectra for 15 nm AuNPs seed and 40 nm AuNPs

3.2 Characterization of Antibody-AuNP Conjugates

Stable conjugate was achieved when the entire surface of AuNPs was covered by sufficient amount of antibody. Unstable conjugate agglomerated when 10% NaCl was added thus change the color of the solution. This happened when negative ionic double layers around AuNPs was broken and aggregated [14, 18, 19]. The presence of enough antibody around AuNPs prevented flocculation. In this work, 1 mL of 40 nm AuNPs at pH 7 needed 3 µL anti-GNE antibody to stabilize.

Successfulness of conjugation was confirmed by comparing the UV-Vis-NIR spectra before and after conjugation. The peak of UV-Vis-NIR spectra after conjugation was red shifted from 527 nm to 532 nm which confirmed the attachment of antibody to the surface of AuNP. The concentration of the conjugates was taken by determine the absorption value as optical density (OD). In this work, the OD of 40 nm AuNP after conjugated was at OD 24.

3.3 LFIA Test

The LFIA test was interpreted by observing the formation of red color dot/s on the NC membrane. If the sample contains *E. coli* O157 and/or *Salmonella* Typhi, the bacteria binds to the gold conjugated anti-GNE antibody in the conjugate pad and formed antibody-antigen complex. The antibody-antigen complex then moved by capillary action to the test dot. The reaction then occurred where the complex was captured by their respective specific anti-*E.* coli O157 or anti-*Salmonella* sp. antibody that was immobilized as the test dot on the NC membrane. Accumulation of many complex at the test dot formed a visible red dot. The visibility and intensity of red dot were proportional to the amount of bacterial in the sample [10, 20]. The excess AuNPs conjugated anti-GNE antibody then kept flowing and reached the control dot. In here, AuNPs conjugated anti-GNE antibody which is a mouse IgG was captured by anti-mouse IgG antibody that was immobilized as the control dot on the NC membrane. The control dot then appeared as the second dot on the upper part of NC membrane which then validated the test. Negative result was interpreted when only the control dot appeared, whereas if no control dot appeared either present or absent of the test dot, the test is considered as invalid. The reaction occurred within 15 min. Chase buffer as wash solution was applied after the reaction to wash away the red background on NC membrane. Tween-20 in the Chase buffer functions as detergent because the background residual contained both soluble and insoluble matrix.

3.4 Visual LOD and Cross-Reactivity Testing of Spiked Samples

In this study, visual LOD and cross-reactivity of spiked samples of two commonly food borne pathogens, *Escherichia coli* O157 and *Salmonella* Typhi in sterile deionized water and packed full cream milk samples (purchased from local mart) was determined. Four sets of LFIA strips each to detect *E. coli* O157 (Figure 3a – b) and *Salmonella* Typhi (Figure 3c – d) in water and milk was used. For detection of *E. coli* O157 spiked in water and milk samples, the lowest detectable of LFIA strip was 7.8 x 10^5 CFU/mL (Figure 3a-8) and 3.0 x 10^6 CFU/mL (Figure 3b-2), respectively. The LFIA strip also was able to detect as low as 3.0 x 10^8 CFU/mL (Figure 3c-1) and 3.0 x 10^7 CFU/mL (Figure 3b-1) of *Salmonella* Typhi spiked in water and milk sample, respectively.

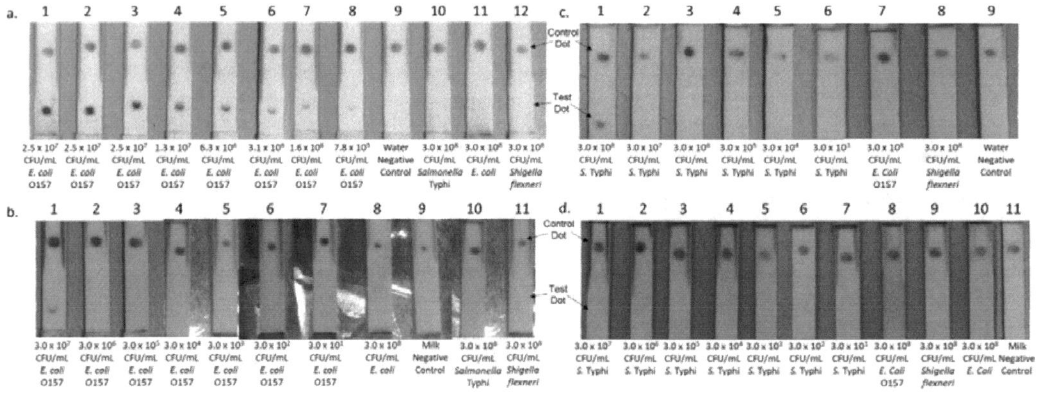

Figure 3. LFIA strips for detection of: (a) *E. coli* O157 in water sample, (b) *E. coli* O157 in milk sample, (c) *Salmonella* Typhi in water sample and (d) *Salmonella* Typhi in milk sample

The visual LOD or sensitivity of LFIA strip to *E. coli* O157 both in spiked water and milk samples was comparable with the previous studies [21-23]. The result also showed that the LOD for spiked milk sample is lower than water sample. Milk sample used in this work was full cream milk which contained high amount of lipid that may interfere the reaction of this test [24, 25]. The milk sample also needed to be diluted with water before spiked with bacterial and pipetted to LFIA strips. The undiluted milk sample was unable to flow in LFIA strip. However, visual LOD of LFIA strip to *Salmonella* Typhi was low. This could be due to low concentration of anti-*Salmonella* sp. antibody used as the test dot compared to the concentration used for anti-*E. coli* O157.

Overall, the visual LOD for both bacterial can be improved by implementing the enrichment method as previous studies [5, 21, 22]. However, adding this method will prolong the turnaround time for the test. This is because enrichment method needs 18 – 24 hour of sample incubation. Furthermore, additional reagents are also needed as enrichment medium. Thus, this will increase cost as well as labor cost and impractical for on field testing application.

In term of specificity, no cross-reactivity was detected for both strips when other bacterial were used as samples (Figure 3a-10,11,12; 3b-8,10,11; 3c-7,8 and 3d-8,9,10). Unfortunately, for anti-*E. coli* O157 strip, very pale red color was detected when *E. coli* non O157 was used in the water sample (Figure 3a-11). This is due to very close relation of genetic material between *E. coli* O157 and *E. coli* non O157 strains [26]. However, this false positive can be reduced by washing step using Chase buffer. Besides, all test strips could produce result within 10 – 15 min which can be considered as rapid test and comparable to the previous studies [5, 12, 21, 23].

4. Conclusions

Monodisperse 40 nm AuNPs was successfully synthesized using the seeding-growth method. GNE antibody was also successfully conjugated with 40 nm AuNPs at pH 7. The LFIA strip was successfully optimized and used for detection of food-borne pathogen (*E. coli* O157 and *Salmonella* Typhi). This LFIA could detect bacterial in water and milk spiked samples as low as 7.8 x 10^5 CFU/mL. Both LFIA strips have higher specificity which only detect *E. coli* O157 and *Salmonella* Typhi but not to other cross related Enterobacteriaceae. The assay time was 10 – 15 minutes and could be considered as rapid test.

Acknowledgments
This work was funded by Transdisciplinary Research Grants Scheme (TRGS) 203/Bahan/6763001.

References

[1] S.F. Altekruse, N. Bauer, A. Chanlongbutra, R. DeSagun, A. Naugle, W. Schlosser, R. Umholtz and P. White 2006 *Emerging Infectious Diseases* 12 1848
[2] M. Lynch, J. Painter, R. Woodruff and C. Braden 2006 *Surveillance for Foodborne-disease Outbreaks: United States, 1998--2002, US Department of Health and Human Services, Centers for Disease Control and Prevention (CD)*
[3] A.K. Bhunia 2008 *Advances In Food and Nutrition Research* 54 1-44
[4] H. Pennington 2010 *Lancet* 376 1428-1435
[5] X. Zhao, X. He, W. Li, Y. Liu, L. Yang and J. Wang 2010 *Afr. J. Microbiol. Res* 4 663-670
[6] Z.A. Bhutta 2006 *BMJ* 333 78-82
[7] I.H. Khan, M.A. Sayeed, N. Sultana, K. Islam, J. Amin, M.O. Faruk, U. Khan, F. Khanam, E.T. Ryan and F. Qadri 2016 *Clinical and Vaccine Immunology* 23 403-409
[8] M.P. Doyle, R.L. Buchanan 2012 Food microbiology: fundamentals and frontiers: *American Society for Microbiology Press*
[9] D. Quesada-González and A. Merkoçi 2015 *Biosensors and Bioelectronics* 73 47-63
[10] X. Huang, Z.P. Aguilar, H. Xu, W. Lai and Y. Xiong 2016 *Biosensors and Bioelectronics* 75 166-180
[11] P. Moongkarndi, E. Rodpai and S. Kanarat 2011 *J Vet Diagn Invest* 23 797-801
[12] C Song, C Liu, S. Wu, H Li, H. Guo, B.Yang, S. Qiu, J. Li, L. Liu, H. Zeng, X. Zhai and . Liu 2016 *Food Control* 59 345
[13] K.R. Brown, D.G. Walter and M.J. Natan 2000 *Chemistry of Materials* 12 306-313
[14] S.R. Makhsin, K.A. Razak, R. Noordin, N.D. Zakaria and T.S. Chun 2012 *Nanotechnology* 23 495719
[15] N.R. Jana, L. Gearheart and C.J. Murphy 2001 *Chemistry of Materials* 13 2313-2322
[16] G. Bauer, J. Hassmann, H. Walter, J. Haglmüller, C. Mayer and T. Schalkhammer 2003 *Nanotechnology* 14 1289
[17] E. Tomaszewska, K. Soliwoda, K. Kadziola, B. Tkacz-Szczesna, G. Celichowski, M. Cichomski, W. Szmaja and J. Grobelny 2013 *Journal of Nanomaterials* 2013 60
[18] I.H. El-Sayed, X. Huang and M.A. El-Sayed 2005 *Nano Lett* 5 829-834
[19] S. Kumar, J. Aaron and K. Sokolov 2008 *Nat Protoc* 3 314-320
[20] J.H. Kim, J.H. Cho, G.S. Cha, C.W. Lee, H.B. Kim and S.H. Paek 2000 *Biosensors and Bioelectronics* 14 907-915
[21] J. Wang, W.N. Chen, K.X. Hu, W. Li and Wei Sheng Yan Jiu 35 439-441
[22] H. Qi, Z. Zhong, H.X. Zhou, C.Y. Deng, H. Zhu, J.F. Li, X.L. Wang and F.R. Li 2011 *Int J Nanomedicine* 6 3033-3039
[23] M. Suria, A. Mohd Afendy, M. Noor Azlina and I. Zamri 2015 *International Food Research Journal* 22
[24] J. Tate and G. Ward 2004 *The Clinical Biochemist Reviews* 25 105
[25] Johan Schiettecatte, Ellen Anckaert and Johan Smitz 2012 *Advances in Immunoassay Technology* Dr. Norman H. L. Chiu
[26] N.T. Perna, G. Plunkett III, V. Burland, B. Mau, J.D. Glasner, D.J. Rose, G.F. Mayhew, P.S. Evans, J. Gregor and H.A. Kirkpatrick 2001 *Nature* 409 529-534

IOP Conf. Series: Journal of Physics: Conf. Series **1082** (2018) 012050 doi:10.1088/1742-6596/1082/1/012050

Fatigue and Mechanical Properties of Graphene Nanoplatelets Reinforced Nr/Epdm Nanocomposites

N. Mohamad[1,3], K. I. Karim[1], M. Mazliah[1], H. E. Ab Maulod[2,3], J. Abd Razak[1,3], M. A. Azam[1,3], M. S. Kasim[1,3], R. Izamshah[1,3]

[1]Fakulti Kejuruteraan Pembuatan, Universiti Teknikal Malaysia Melaka, Hang Tuah Jaya, 7600 Durian Tunggal, Melaka, Malaysia

[2]Fakulti Teknologi Kejuruteraan, Universiti Teknikal Malaysia Melaka, Hang Tuah Jaya, 7600 Durian Tunggal, Melaka, Malaysia

[3]Advanced Manufacturing Centre, Universiti Teknikal Malaysia Melaka, Hang Tuah Jaya, 7600 Durian Tunggal, Melaka, Malaysia

Email: noraiham@utem.edu.my

Abstract. Engine mounting suffers from fluctuating forces which may results in sudden failure due to fatigue. Graphene nanoplatelets have attracted great interest due to their exceptional physical, mechanical and thermal properties. Natural rubber/ethylene propylene diene monomer (NR/EPDM) reinforced with graphene nanoplatelets (GNPs) were investigated to improve the properties of NR/EPDM blend prior to fatigue and mechanical properties for engine mounting service. Firstly, the NR/EPDM blends and nanocomposite were tested for tensile properties and followed by the fatigue properties (tensile mode) using displacement-controlled. Tensile plot shows that the NR/EPDM nanocomposite has higher tensile value of 95% compared to NR/EPDM blend. The S-N curve illustrates that NR/EPDM blend and NR/EPDM nanocomposite break by the same cycle which is 100 000 cycles. The fracture behaviour of loading conditions was supported via morphological analysis using scanning electron microscopy (SEM) and X-ray diffraction (XRD) analysis.

1. Introduction

Nowadays, multiple studies reported the enhanced strength of rubber composite using graphene as reinforcement [1]. Nanomaterials like graphene are well known to have exceptional properties in mechanical, thermal and electrical [2]. NR/EPDM blend have been explored over three decades because NR/EPDM could possess good physical properties with adequately high aging and ozone resistance [3]. Combining NR/EPDM blend with nanofiller (GNPs) is not a new brand material. It has been studied by several researchers on mechanical properties [4], damping properties [5], processing parameters [6], cure characteristics [7] and etc. However, study on fatigue properties (dynamic behaviour) relative to tensile properties (static behaviour) of NR/EPDM filled GNPs are still scarce.

IOP Conf. Series: Journal of Physics: Conf. Series **1082** (2018) 012050 doi:10.1088/1742-6596/1082/1/012050

This paper is focusing on fatigue and mechanical properties of graphene nanoplatelets reinforced NR/EPDM nanocomposites.

Fatigue is one of the major failure mechanisms in engineering structures [8]. Due to Araham et al. [9], mechanical fatigue of elastomers is manifested in a progressive reduction of the physical properties as a result of a crack propagation during continuous dynamic excitation. Though there are levels of stress or strain below each elastomer will suffer fatigue damage, such limits are not well established. Typically, the fatigue process involves a period during which cracks nucleate in regions that were initially free of observed cracks, followed by a period during which nucleated crack grow to the point of failure. Analysis approaches that are currently available for predicting fatigue life in rubber, including both crack nucleation as well as crack growth approaches, are reviewed by Mars and Fatemi [10].

2. Experimental study

Natural rubber (SMR 20) grade and ethylene propylene diene monomer (EPDM Buna® EPT 9650) used in this study were supplied by Rubber Research Institute of Malaysia (RRIM). GNPs was treated using 70/30 of ethanol/water using vibration sonicator for 2 hours before dried in the oven at 60 °C until it reaches the moist bulk form. The compounding process was performed using a Haake internal mixer. 50 grams at one time was mixed with fill factor of 0.7. Both NR and EPDM were first masticated at 30 °C for 10 minutes using two-roll mill prior to rubber blend preparation. The formulations recipe for NR/EPDM blends and nanocomposite were summarized in Table 1. Firstly, formulation recipe in Table 1 were mixed accordingly.

The blend and nanocomposites were filled into mold cavity. NR/EPDM blends and nanocomposites then were compressed using GT7014-A hot press from Gotech. The samples were prepared under compression of 1800 kg/cm^2. From this stock, sheets for both NR/EPDM blend and nanocomposite was vulcanized approximately 2 mm thick with a semi-efficient vulcanization system with a hot press at 150 °C. The samples were cut into 64 mm x 10 mm dog bone sample size. Then the samples were tested for tensile and fatigue properties according to ASTM D1822 using Universal Tensile Machine (Toyoseiki Strograph) and ASTM D7791 using Fatigue Machine (5 kN Shimadzu 4830) testing machine. The sample were subjected under prescribed sinusoidal displacement with stretch ratio of 20. The stretch rate of the material is about 2 s^{-1} with the frequency imposed for all loading conditions was 2 Hz. For each test the number of cycles required for the complete rupture of each specimen. The test was stopped at 2 million cycles if no crack or rupture is observed in the specimen.

Table 1. NR/EPDM Formulation Recipe

Ingredients	Rubber blend (phr)*	Rubber nanocomposite (phr)*
NR (SMR 20)	70	70
EPDM	30	30
GNPs	0	3
ENR-50	10	10
Zinc oxide	5.0	5.0
Stearic acid	2.0	2.0
Sulphur	1.5	1.5
MBTS[a]	1.0	1.0
TMTD[b]	0.3	0.3
6PPD[c]	2.0	2.0

[*] Part per hundred rubber
[a] 2,20-dithiobis (benzothiazole)
[b] Tetramethylthiuram disulphide
[c] N-(1,3-Dimethylbutyl)-N'-phenyl-p phenylenediamine

After that, the fatigue fractured surface morphologies were investigated using a scanning electron microscopy, model EVO-50 from Zeiss. The samples were placed onto aluminium stubs and sputter coated with thin layer of gold, about 20 mm thickness. All samples were examined with secondary electron imaging mode. Meanwhile, XRD analysis was performed to analyse the crystanillity of the rubber using PANalytical *X'Pert PRO diffractometer* and the data was analysed using X'pert PRO software. The x-ray source used was CuKα radiation with λ = 1.5418 nm and all samples were characterized under conditions of 2θ scanning range (5 to 40°), 1 sec of dwell time, and step size of 2° in 2θ min. Finally, the samples with 1mm thickness were analysed using DMTA to measures the stiffness and damping behaviour of sample, which are reported as storage modulus. Dynamic mechanical thermal analysis was performed using a Perkin Elmer DMTA-7e in the temperature scan mode with a parallel plate. The measurement was carried out at a heating rate of 10°C min-1 over a temperature range of -100°C to 100°C.

3. Results and discussion

3.1 Uniaxial test

Figure 1 shows the tensile graph for both blend and nanocomposite. The additional of GNPs filler in NR/EPDM blend had cause an increased in pattern of tensile strength and elongation at break. About 95% and 60% of drastic improvements were observed for tensile strength (TS) and E_B respectively. The increment of elongation at break (E_B) with the additional of GNPs with two-dimensional multilayer structures have a unique frictional interlayer sliding that usually exist because of the weak Van Der Waals forces bonded in the GNPs layers which may assist the elongation of produced nanocomposite blends [11]. Thus, the additional of GNPs may serve as dual functions as and reinforcing agent to the NR/EPDM blend due to the TS increment, as well as an elasticity enhancer due to the increment in E_B [12].

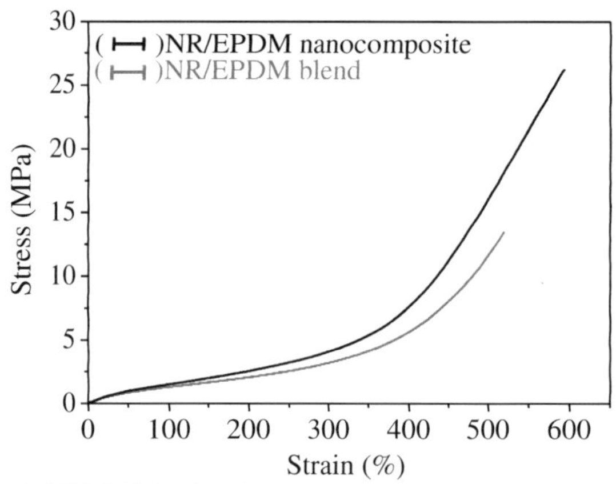

Figure 1. Stress vs Strain for NR/EPDM blend and nanocomposite.

Table 2 shows the comparison of tensile properties and fatigue properties of NR/EPDM blend and nanocomposite. For modulus at 100% (M100) and modulus at 300% (M300), the increasing value of NR/EPDM nanocomposite for both M100 and M300 compared to NR/EPDM blend attributes the stiffness properties. The additional of GNPs loading producing higher M100 and M300. The percentage of improvement in modulus values was about 18% and 32% for M100 and M300, respectively. Increased modulus at M100 and M300 of rubber elongation indicated a good rubber-filler interaction between the NR/EPDM rubber blend and GNPs nanofiller. Furthermore, lamellar structure of the graphene allowed better surface wetting and enhanced polymer-matrix interactions between the nanofiller and rubber matrix, thereby led to better stress transfer that yield better reinforcements effects on the GNPs [13].

NR/EPDM blend and nanocomposite having the same number of fatigue life however NR/EPDM blend sample breaks without having the initial crack compared to NR/EPDM nanocomposites which rapidly propagated after initiation of crack at around 30 000 cycles until it reaches a complete rupture. The additional of GNPs to rubber compounds can have a pronounced strengthening effect, depending on the volume fraction used. It can be seen that, by adding GNPs to NR/EPDM generally crack before it fully ruptured due to pull out of fillers which is in good agreement with the fracture surface morphology shows in Figure 2. NR/EPDM nanocomposite indicate thicker shear yield of matrix. The thicker the shear yield of matrix, the higher the crack resistance. Besides, in displacement control, the utilization of GNPs, maximizing the compound stiffness will maximize the energy release rate, resulting in maximum fatigue life compared to NR/EPDM blend.

Table 2. Uniaxial test results of NR/EPDM blend and nanocomposite.

	Tensile properties				Fatigue properties	
Formulation	**TS (MPa)**	**M100 (MPa)**	**M300 (MPa)**	**E_B**	**Peak Stress (MPa)**	**Fatigue life, N_f (cycles)**
NR/EPDM blend	13.46±4.8	1.29±0.13	3.08±0.42	200±44	1.51±0.56	100, 000
NR/EPDM nanocomposite	26.24±2.6	1.52±0.08	4.07±0.22	320±31	1.60±0.78	100, 000

3.2 Morphological Analysis

Figure 2 shows the fatigue fracture surfaces of NR/EPDM blend and nanocomposite at small-scale with 300x magnification. Figure 2 (a) shows a flaw located on the fatigue crack surface. Crack growth of rubber is assumed to start from very small flaws [14]. Figure 2 (b) and (c) illustrate the fracture surface of NR/EPDM blend and nanocomposite after fatigue failure. Both materials producing two zones of crack which illustrated by the rough and smooth region. The rough region (Zone 1) was a result from the cracks that initiated from mechanism of the pulled-out particle around 10 μm and propagate slowly under cyclic loading [15]. Meanwhile, the smoother fracture surface (Zone 2) was attribute to the catastrophic failure that occurred as a result from the lack of ability of NR/EPDM nanocomposite to resist the growth of cracks as the cyclic loading increases. The incapability to resist crack propagation is due to the filler sheets that are no longer able to transfer the stress more evenly to the neighbouring matrix as the fatigue test progress. The increase in surface roughness indicates that the additional of graphene nanoplatelets resulted in the increase in rubber stiffness which consequently cracks to occur rapidly compared to rubber blend [16]. Figure 2 (d) and (e) show larger number of humps appear on the fracture surface of NR/EPDM nanocomposite compared to NR/EPDM blend which has smaller number of humps. In other words, the NR/EPDM blend has smoother surface than NR/EPDM nanocomposite.

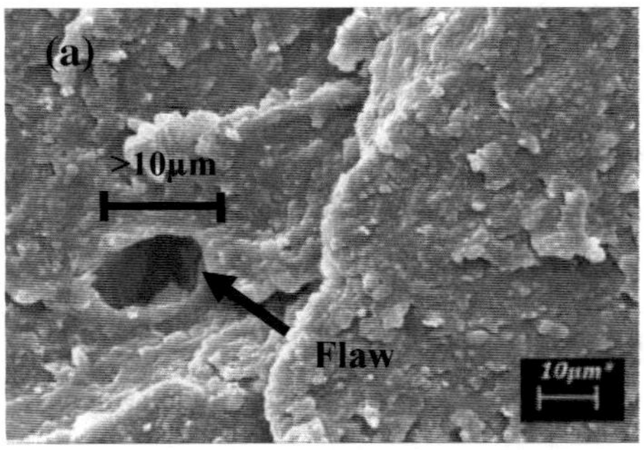

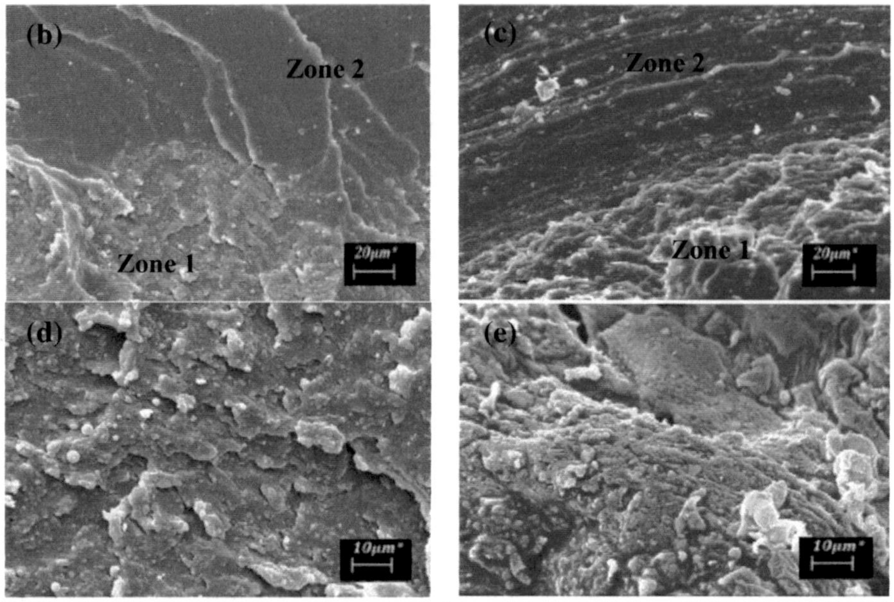

Figure 2. (a) Crack initiation of fatigue fracture. Fatigue failure surfaces of (b) NR/EPDM blend at 300x magnification, (c) NR/EPDM nanocomposite at 300x magnification, (d) NR/EPDM blend in zone 2 at 500x magnification and (e) NR/EPDM nanocomposite in zone 2 at 500x magnification

3.3 X-Ray Diffraction Analysis

XRD patterns for NR/EPDM blend and nanocomposite depicted in Figure 3. The existence of broad diffraction peak in NR/EPDM blend can be referred as amorphous structure which remains unchanged in both pictograms upon blending with GNPs [17]. NR/EPDM nanocomposite exposed the [002] diffraction at 2θ = 26.6° corresponding to an interlayer d-spacing (0.34 nm) of C in graphene platelets [18]. It is attributed to the layered crystalline GNPs weakly with an increment in GNPs loading up to 3 wt%. Crystalline phase is arranged in ordered with high cohesion. Therefore, the higher the crystallinity, the greater the strength of rubber thus reducing the degree of freedom for the molecular chain to move. Peak at 2θ = 34.3° in the spectrum was originated from the (002) plane of ZnO which appeared as the vulcanization agent [19].

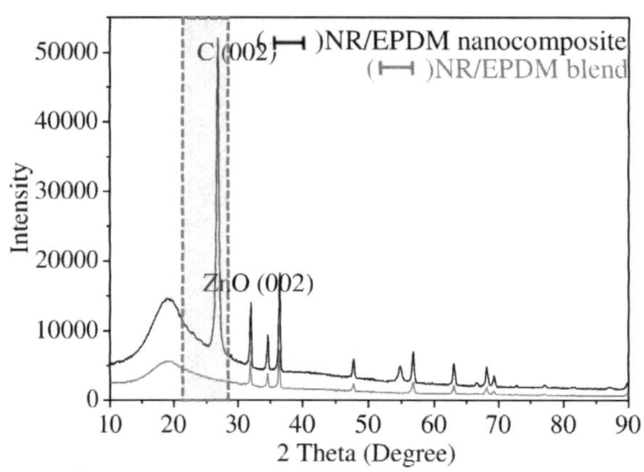

Figure 3. XRD of NR/EPDM blend and nanocomposite

Conclusion

As the conclusions, the NR/EPDM blends and nanocomposite were successfully produced and tested for tensile properties and followed by the fatigue properties (tensile mode) using displacement-controlled. NR/EPDM nanocomposite exhibits superior tensile properties with 95% than its host NR/EPDM blend. Furthermore, it shows compatible dynamic properties where both NR/EPDM nanocomposite has higher peak stress with 6% compared to NR/EPDM blend and had similar fatigue life of about 100000 cycles. The properties were in line with the morphological analysis on its fracture surfaces observed via scanning electron microscopy (SEM) as well as structural, compositional and thermal properties. Therefore, NR/EPDM nanocomposite has advantage on engine mounting industry due to its significant properties of tensile and fatigue.

Acknowledgement

The authors acknowledge UTeM for funding under the project grant of PJP/2016/FKP/H16/S01484 and Carbon Research Technology research group to complete this study.

References

[1] Srinivasaro Y, Hanum Y S R, Chin C H, Nandakumar K, Sabu T 2013 *Advanced Material Research* 812 263-266
[2] Zhu Y, Murali S, Cai W, Li X, Suk J W, Potts J R, Ruoff R S 2010 *Advanced Materials* 22 3906-3924
[3] Sae-oui P, Sirisinha C, Thepswan U, Thapthong P 2007 *Polymer Testing* 26 1062-1067
[4] Razak J A, Ahamad S, Ratnam C, Shueb M I, Mohamood M A, Yaakub J, Shamsuri S R, Mohamad N 2014 *Advanced in Environmantal Biology* 3289-3299
[5] Razak J A, Ahmad S, Ratnam C T, Mahmood M A, Yaakub J, Mohamad N 2015 *Polymer Research Journal* 9 1 1-14
[6] Mohamad N, Yaakub J, Razak J A, Yaakob M Y, Shueb M I, Muchtar A 2014 *Journal of Applied Polymer Science* 131 17
[7] Yaaub J, Mohamad N, Razak J A, Jano Z, Shueb M I 2014 *American-Eurasian Journal of Sustainable Agriculture* 4 148-155
[8] Schütz W 1996 *Engineering Fracture Mechanics* 54 2
[9] Abraham F, Alshuth T, and Jerrams S 2002 *Materials and Design* 263
[10] Mars W and Fatemi A 2002 *International Journal of Fatigue* 249
[11] BLW and Zhong H, 2011 *Journal of Material Science* 46
[12] Galpaya D, Wang M, Liu M, Motta N, Waclawik E, Yan C 2012 *Graphene* 1 30-49
[13] Sridhar V, Lee I, Chun H H and Park H, 2013 *eXPRESS Polymer Letter* 7
[14] Roland C M and Smith C R 1985 *Rubber Chamical Technology* 58
[15] Ismail H, Muniandy K, Othman N 2012 *Bio Resources* 7 1
[16] Ismail H, Mat N S C, Othman N 2017 *Journal of Vinyl & Additive Technology*
[17] Motaung T E, Luyt A S, Thomas S 2011 *Polymer Composites* 32 8 1289-1296
[18] Kumarasinghe, A R, Samaranayake L, Bondino F, Magnano E, Kottegoda N, Carlino E, Ratnayaket U N, de Alwist A A P, Karuunaratne V and Amaratunga G A 2013 *Journl of Physical Chemistry* 117 18 9507-9519
[19] Chang S J, Young A J 2007 *Nanotechnology* 18 225-603

IOP Conf. Series: Journal of Physics: Conf. Series **1082** (2018) 012051 doi:10.1088/1742-6596/1082/1/012051

Comparison of KOH and TMAH Etching on Sinw Arrays Fabricated via AFM Lithography

Nurain Najihah Alias[1], Khatijah Aisha Yaacob[1*], Siti Noorhaniah Yusoh[1] and Ahmad Makarimi Abdullah[1]

[1] School of Materials & Mineral Resources Engineering (SMMRE), Engineering Campus, Universiti Sains Malaysia, Seri Ampangan, 14300 Nibong Tebal, Penang, Malaysia.

Email: mraisha@usm.my

Abstract. Silicon nanowire (SiNW) arrays were fabricated by using top-down fabrication approach via atomic force microcopy (AFM) lithography on silicon-on-insulator (SOI) (100) substrate. Local anodic oxidation (LAO) technique carried out by AFM lithography to draw oxide pattern on top of the SOI substrate. The next steps to fabricate silicon nanowires involve chemical etching of the unmasked silicon by wet etching with the addition of organic additive. In this research, isopropyl alcohol (IPA) was used to improve the surface smoothness of silicon nanowires. Two types of etchant, potassium hydroxide (KOH) and tetramethylammonium hydroxide (TMAH) with and without admixture of IPA at different etching time (10s, 20s, and 30s) were studied. After etching, the local oxide pattern was removed using dilute hydrofluoric acid (HF). The relationships between etching time with wall angle and surface roughness of the produced SiNWs were characterized in detail by using AFM and SEM. The result revealed that the wall angle for SiNW arrays that etched using TMAH was higher compare to KOH. Increasing the etching time produced large wall angle (θ) for both etchants. It was also found that the surface roughness for SiNW arrays was decreased with increasing time. SiNW arrays etched using KOH have higher surface roughness compared to TMAH. The obtained results also show that the etching depths increased with increasing etching time.

1. Introduction

Nanotechnology enclosed many processes that are vital in the fabrication of integrated circuits, memory devices, display units, biochips and biosensor. One of the key processes in nanofabrication is the creation and construction of functional unit in the size regime of less than 100 nm. Nanowire is one-dimensional (1D) nanostructure that has thickness, diameter or length in range of 0.1 nanometers to 100 nanometers. Silicon nanowire (SiNW) represents a class of 1D nanostructure of nanoscience and technology. There are two major processes involve in fabricating SiNWs; bottom-up and top-down processes. AFM lithography is one of the top-down processes that have been used during fabrication of SiNW arrays. The AFM lithography works based on the principle of interaction between the probe and substrate separation in close contact condition < 1 nm. So, the AFM lithography can be operate in the contact mode [1-3] and noncontact mode [4]. Etching process is a must and plays a critical role when it is related to lithography technique.

The two major types of etching processes are wet etching [5] and dry etching [6]. Wet etching process involves the using of liquid chemicals or etchants in order to take off the substrate material. Dry etching using gaseous these gaseous diffused and reacted into the substrate and the byproduct expelled through the vacuum system. There are three types of dry etching; chemical reactions (by using reactive plasma or gases), physical removal (generally by momentum transfer), and a combination of chemical reactions and physical removal. On the other hand, wet etching is only involved a chemical process. Depending on the silicon wafer orientation and the type of etchant being used, wet etching can be either isotropic or anisotropic [7, 8]. Isotropic etching used wet etchants that removes the material uniformly in all directions, whereas anisotropic etching used wet etchants that are orientation dependent and these etchants will etch in different direction at different rates. Anisotropic wet etching is commonly used to fabricate simple microstructures and nanostructures on a single crystal SOI wafer [9]. The most frequently used anisotropic etchant include tetramethyl ammonium hydroxide (TMAH), potassium hydroxide (KOH), sodium hydroxide (NaOH), thylenediamine pyrocatechol water (EDP) and hydrazine/water [9, 10]. Nevertheless, hydrazine/water and EDP are not ideal because of their toxicity, instability and difficulty of handling. The KOH solution becomes a popular anisotropic etchant because of its good etching performance and non-toxic compared with NaOH. However, there is potassium ionic contamination for KOH etchant. TMAH also has attracted much attention because it was found that TMAH produced an excellent etching characteristis, non-toxic and without metal ions [11].

In the present work, a simple, low cost wet chemical etching method was used for the fabrication of SiNW arrays fabricated via AFM lithography on SOI wafer. SOI wafer was used because it has buried oxide that act as a barrier to prevent bulk leakage during electron movement from source to drain. The solution of KOH and TMAH with and without admixture of isopropyl alcohol (IPA) at different etching time (10s, 20s, and 30s) was studied. The IPA is the most commonly used additives and has been known as a very effective additive to increase the smoothness of silicon etched in anisotropic etchant. Its addition has proven useful in the case of silicon etching with TMAH and KOH solutions for obtaining smoother surfaces [11, 12]. The structural characterizations of SiNWs such as wall angle, surface roughness and etching depth were examined by atomic force microscope (AFM) and scanning electron microscope (SEM). The excellent geometrical features and surface of SiNW arrays are crucial for obtaining a semiconductor device with outstanding performance depending on the device application.

2. Experimental procedure

2.1 Fabrication of SiNW arrays

By using AFM lithography, the SiO_2 nanowire were patterned on the clean surface of SOI (100) wafer. The SiO_2 nanowire were designed by using a Nanonavi program vector scan under the scanning probe microscope (SPM) machine (SPI3800N/4000) at a temperature of 24–26 °C with relative humidity of 55–65%. Chromium/platinum (Cr/Pt) cantilever tip (Budgetsensors) with 13 kHz of resonant frequency and 0.2 N/m of force constant under contact mode was used at 9 V bias voltage with 0.4 μm/s writing speed. After the AFM lithography process, thin SiO_2 nanowire pattern were formed on the SOI wafer surface and functioning as the masking layer for the following etching processes.

IOP Conf. Series: Journal of Physics: Conf. Series **1082** (2018) 012051 doi:10.1088/1742-6596/1082/1/012051

2.2 Etching process of silicon and oxide tracks

There were two major steps of etching in order to create the SiNW arrays; etching of silicon layer and etching of oxide pattern. For this work, 25 wt% of KOH and 25 wt% of TMAH with and without admixture of 10 vol % of IPA were used to etch the silicon layer. This etching process was done in three different etching times which were 10 seconds, 20 seconds and 30 seconds. Heating temperature and stirring can be affected towards the formation of SiNW arrays during etching process. For this work, the mixture of KOH and TMAH with IPA was heat up to ~ 65°C - 70 °C and stirred at 600 rpm. The second etching process was done to remove the oxide pattern by using dilute HF. The SOI wafer with SiNW pattern was immersed into the solution of 1: 30, HF: DIW for five seconds. Figure 1 shows the illustration of the steps during fabrication of SiNW arrays.

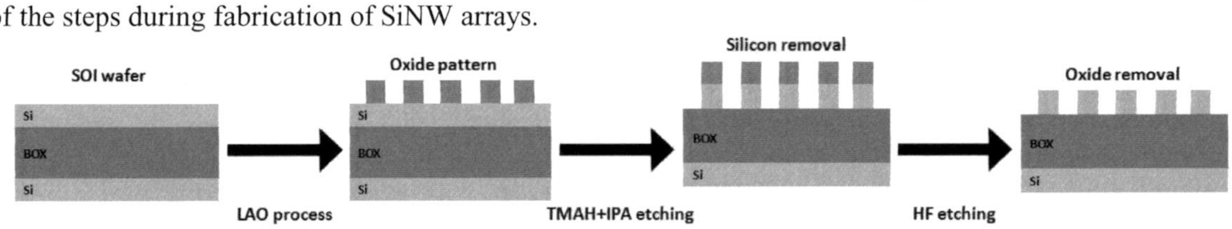

Figure 1.The illustration of the steps during fabrication of SiNW arrays.

3. Experimental result and discussion

3.1 Effect of KOH and TMAH etching on SiNW arrays

Throughout this experiment we compared the effect of 25 wt% KOH and 25 wt% TMAH etching on SiNW arrays at different etching time (10 sec, 20 sec and 30 sec). Figure 2 shows the results of wall angle and surface roughness of SiNWs after etching process. The value of wall angle was varies based on the etching time. The wall angle for standard anisotropic etching of the silicon (100) wafer produced a wall angle with value of 54.7°. In this study, we obtained different value of wall angles at different etching time. It was clearly shown that wall angle of TMAH with value 40.08°, 46.57° and 47.96° was higher compare to KOH with value 13.67°, 25.83° and 34.12° at 10 sec, 20 sec and 30 sec of etching time respectively. The value of the wall angle decrease when the height and width of SiNWs are small and large, respectively. Different values of wall angle were obtained because different etching times produced different width and etching depth which affected wall angle value. Therefore, this situation can influence the wall angle measurement. Equation 1 shows the method to calculate the wall angle for every sample and the calculation of wall angle for SiNWs was illustrated in figure 3, where *E*d is etching depth, and *a* is the adjacent segment that can be measured by using AFM.

$$\theta = \tan^{-1}\frac{Ed}{a} \qquad \text{Equation 1}$$

IOP Conf. Series: Journal of Physics: Conf. Series **1082** (2018) 012051 doi:10.1088/1742-6596/1082/1/012051

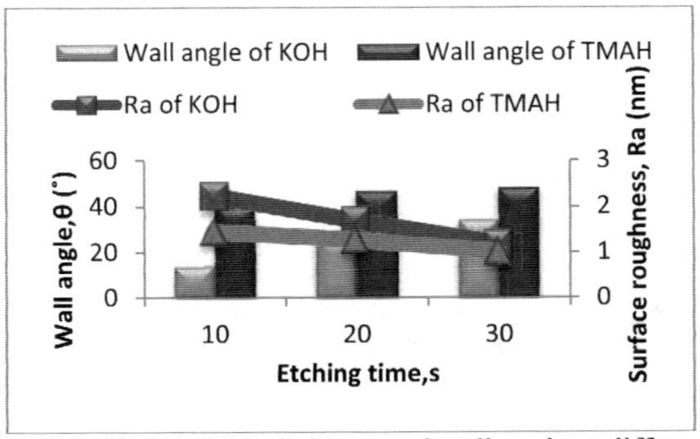

Figure 2. Surface roughness and wall angle at different etching time for KOH and TMAH.

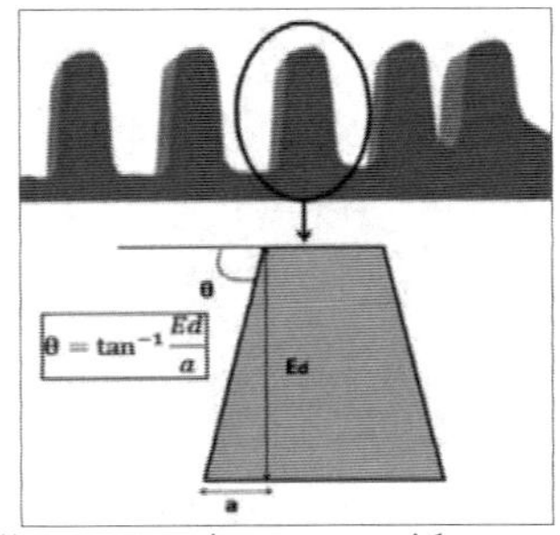

Figure 3.Silicon nanowire arrays with trapezoidal cross-section.

Table 1 shows the AFM topography of SiNW arrays on SOI substrates after etched with KOH and TMAH and the results were depicted in the form of graphs in figure 2 to facilitate understanding. From figure 2, it shows that surface roughness for SiNWs etched by KOH and TMAH decreased by increasing etching time. However, surface roughness of SiNWs that etched using KOH is higher than TMAH. The value of surface roughness for KOH was 2.23 nm, 1.69 nm and 1.19 nm and surface roughness for TMAH was 1.45 nm, 1.27 nm and 1.02 nm for 10 sec, 20 sec and 30 sec of etching respectively. Compared with figure 3 (a) and (b), it can be observed that the TMAH etching produced a very clean surface. Nevertheless, there were a lot of insoluble floccules were residual on the surface for the KOH etching, which may contaminate the silicon surface and affect the surface roughness.

Table 1. AFM topography of SiNW arrays on SOI substrates after etched using KOH and TMAH.

	10 second	20 second	30 second
KOH	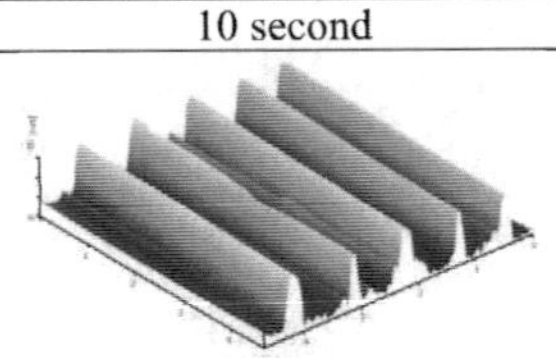 Ra=2.23nm	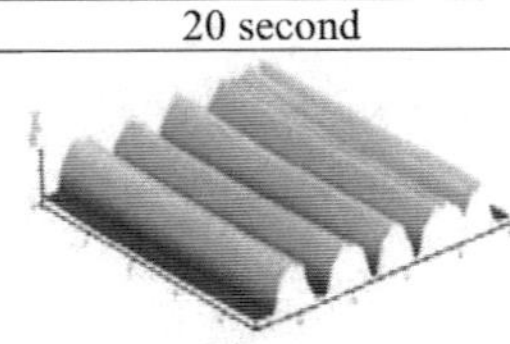 Ra=1.69nm	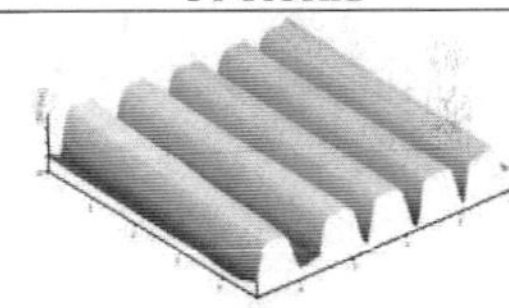 Ra=1.19nm
TMAH	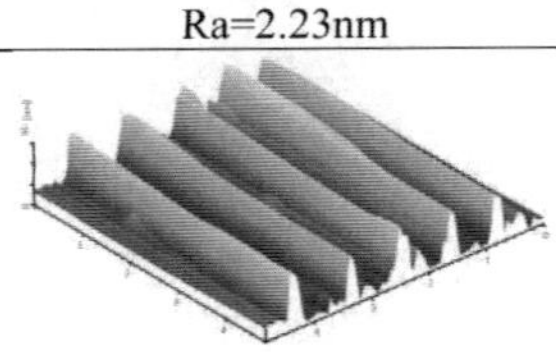 Ra=1.45nm	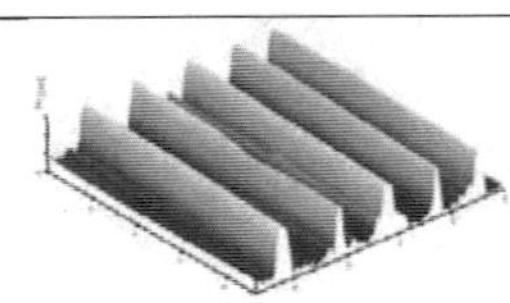 Ra=1.27nm	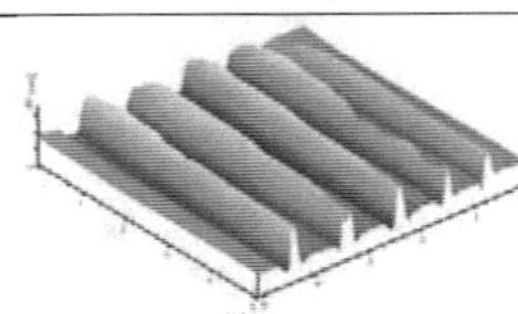 Ra=1.02nm

IOP Conf. Series: Journal of Physics: Conf. Series **1082** (2018) 012051 doi:10.1088/1742-6596/1082/1/012051

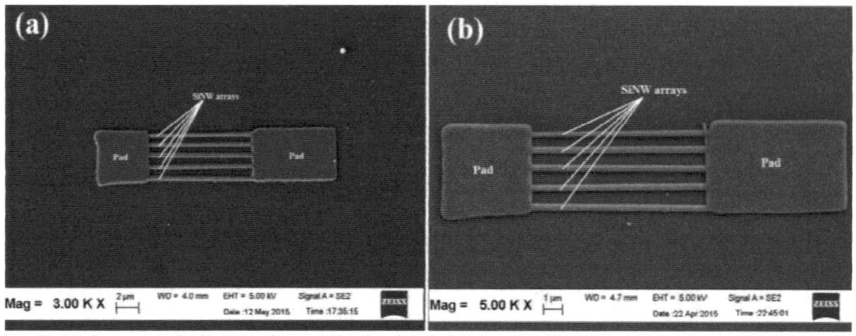

Figure 4.SEM image of SiNW after etched with (a) 25 wt% KOH and (b) 25 wt% TMAH at 30s.

3.2 Effect of KOH and TMAH etching with addition of IPA on SiNW arrays.

Table 2 shows AFM topography of SiNW arrays on SOI substrates after etched with KOH and TMAH with the addition of IPA at different etching time. Figure 5 shows the variation of surface roughness for SiNWs at different etching time with different etchant, KOH and TMAH with and without addition of IPA which was measured using AFM by taking the average profiles of five SiNWs. From figure 5, it shows that surface roughness was decreased respectively when the etching time increased. It is also found that the addition of IPA for both of etching solution drastically reduced the roughness of SiNWs. This clearly indicates that the addition of IPA was very efficient in providing smoother etched surfaces as well as making the etching process more controllable. This phenomenon occurs because the 10 vol % of IPA in KOH and TMAH solution is concentrated, giving rise to the aggregation of IPA molecules and formation of a monolayer on the Si surface as obtained in the previous work that carried out by Yusoh and Yaacob. The formation of this monolayer promotes the adsorption of IPA molecules on the Si surface with the hydrocarbon chains binding to the hydrogen-terminated Si surface [9]. In addition, the hydroxyl groups (OH^-) of IPA are oriented toward the solution. This adsorption would prevent the OH^- from etching the Si surface and leads to the reduction in the etching rate.

Table 2. AFM topography of SiNW arrays on SOI substrates after etched using KOH and TMAH with addition of IPA.

	10 second	20 second	30 second
KOH+IPA	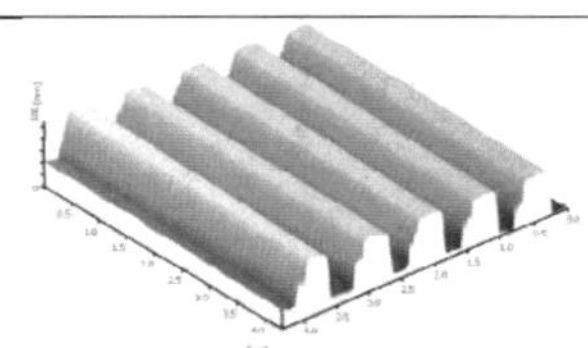 Ra=0.75nm	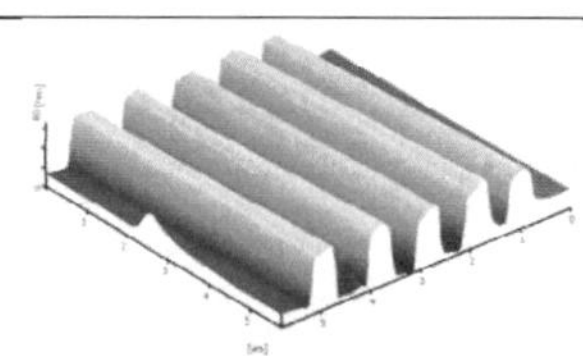 Ra=0.48nm	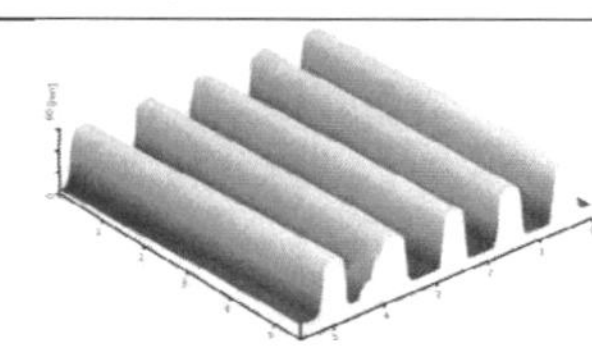 Ra=0.32nm
TMAH+IPA	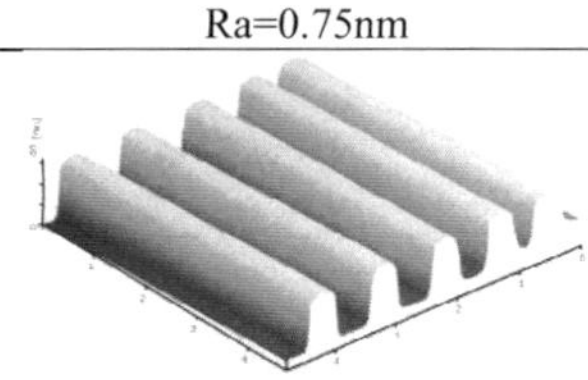 Ra=0.61nm	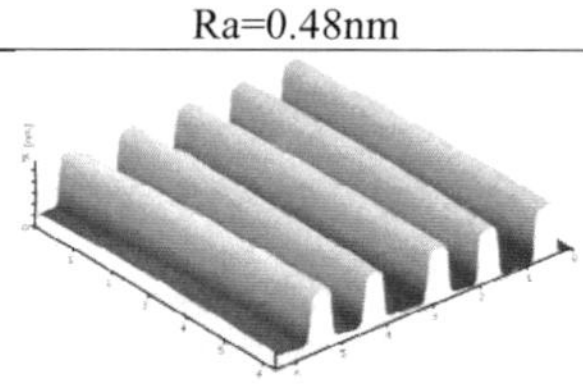 Ra=0.43nm	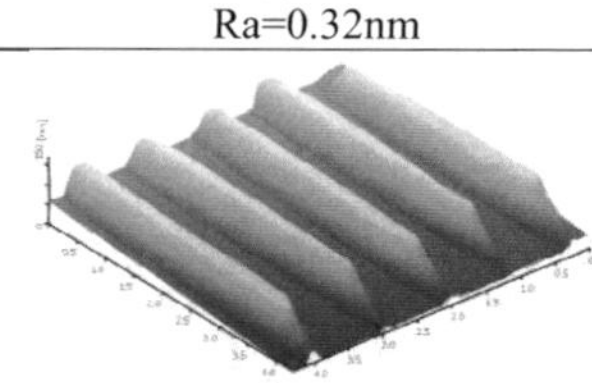 Ra=0.27nm

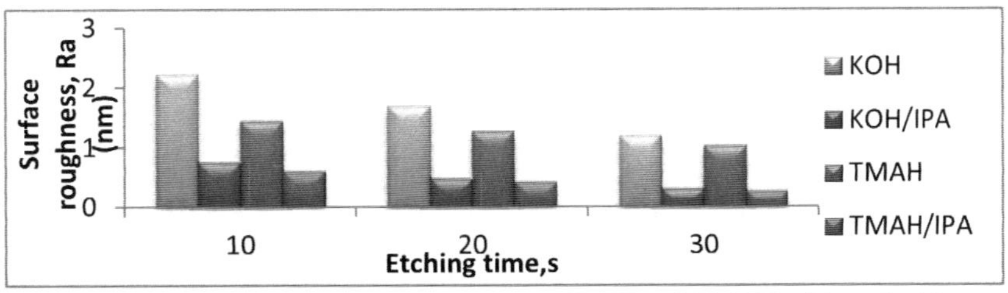

Figure 5. Surface roughness of SiNW arrays after etched using KOH and TMAH with and without addition of IPA at different etching time.

Hydrogen ions were produced during the wet etching process. This presence of hydrogen can be observed through the generated bubbles coming off near to the SOI surface and influence the surface roughness. The density and size of hillocks was affected by hydrogen bubble that produced during etching process [9]. The existence of hillocks after etching is the main contribution to the rougher surface. The addition of IPA produced a smooth surface (Figure 5) because it promotes the wettability of the etchants and decreases the formation of the hydrogen bubbles. As can be seen, the addition of IPA into KOH solution gave high surface roughness of SiNWs compare to TMAH solution. The lowest value of surface roughness obtained by SiNWs etched using TMAH with addition of IPA with 30 second of etching time, 0.27 nm.

4. Conclusion

In this work, KOH and TMAH comparative etching experiments on a SiNW arrays have been carried out at different etching time (5 sec, 10 sec and 15 sec) with and without addition of IPA. The thickness and height of SiNW arrays was controlled by the etching time. Nevertheless, the surface roughness of the SiNW arrays was influenced by the presence of IPA solution in the etchant (KOH and TMAH) and changed considerably with the changes in etching time. It has been found that TMAH solution has higher value of wall angle after etching process compare to KOH solution. More importantly, the surface roughness analysis shows that TMAH etching can obtain smooth surface compared with KOH etching. Unlike the clean surface for TMAH etching, KOH etching always results in the contamination of residual insoluble floccules on the silicon surface. The addition of IPA improved the surface roughness of SiNW arrays for both KOH and TMAH. However, the lowest surface roughness was obtained from SiNWs that etched using TMAH with addition of IPA with 30 sec of etching time.

References

[1] A. Dehzangi, F. Larki, S. D. Hutagalung, M. G. Naseri, B. Y. Majlis, M. Navasery *2013 PLOS ONE* 8 e65409
[2] M. T. Asmah, S. D. Hutagalung, and O. Sidek 2013 *Journal of Physics* 431 012005
[3] S. N. Yusoh and K. A. Yaacob 2015 *International Journal of Electroactive Mater* 3 6
[4] S. D. Hutagalung, C. L. Kam, and T. Darsono 2014 *Sains Malaysiana* 43 267
[5] C. Pakpum and N. Pussade 2015 *Applied Mechanics and Materials* 804 12
[6] I. Hotovy, S. Hascik, M. Gregor, V. Rehacek, M. Predanocy, and A. Plecenik 2014 *Vacuum* 107 20
[7] J. Rouhi, S. Mahmud, S. D. Hutagalung, N. Naderi, S. Kakooei, and M. J. Abdullah 2012 *Semiconductor Science and Technology* 27 11
[8] T. S. Monteiro, P. Kastytis, L. M. Gonçalves, G. Minas, and S. Cardoso 2015 *Micromachines* 6 1534
[9] S. N. Yusoh and K. A. Yaacob 2016 *Beilstein J. Nanotechnol* 7 1461
[10] D. Shankar, M. Imran, P. Kumar, R. Pal, P. Datta, and R. Chatterjee 2011 *Microsystem Technology* 17 1621
[11] K. B. Sundaram, A. Vijayakumar, and G. Subramanian 2005 *Microelectronic Engineering* 77230
[12] M. Kramkowska and I. Zubel 2009 *Procedia Chemistry* 1774

Tailoring Parameters to Produce Nanowires on Metal Surface via Surface Oxidation Process

Subagja Toto Rahmat[1], Chan Yi Hong[1], Faisal Budiman[1], Wai Kian Tan[2], Go Kawamura[3], Atsunori Matsuda[3], and Zainovia Lockman[1*]

[1] Green Electronic nanoMaterials Group, Nanomaterials Niche Area (School of Materials and Mineral Resources Engineering, Engineering Campus, Universiti Sains Malaysia, 14300 Nibong Tebal, Penang, Malaysia)
[2] Center for International Education (Toyohashi University of Technology, Toyohashi, Aichi, 441-8580, Japan)
[3]Department of Electrical and Electronic Information Engineering (Toyohashi University of Technology, Toyohashi, Aichi, 441-8580, Japan)

Email: zainovia@usm.my

Abstract. Thermal oxidation of iron, tungsten, titanium and copper was done in dry air to produce Fe_2O_3, WO_3, TiO_2 and CuO nanowires, respectively, at 500 – 600 °C. Whilst nanowires were observed on iron and copper, no nanowires were seen on titanium and tungsten. Keeping the temperature constant, oxidation of titanium and tungsten was continued but in the presence of water vapour. Through this method nanowires were successfully formed on titanium but not on tungsten. Potassium halide vapour was then introduced to the oxidation furnace for tungsten and after an hour of oxidation, nanowires were produced on the tungsten surface. Nevertheless, the growth mechanism of the nanowires on tungsten is thought to be very much different than iron, copper and titanium whereby on tungsten catalytic induced reaction is likely. Oxide nanowires on iron and titanium were further studied as oxide films on copper and tungsten were not well adhered to the substrate. Under sunlight radiation, ~ 90% of Cr(VI) was successfully removed on Fe_2O_3 nanowires compared to on TiO_2 nanowires (~ 40 % removal) after an hour process. This is due to the smaller energy band gap of Fe_2O_3 compares to TiO_2.

1. Introduction

Metal oxidation can be considered as a degradation process of metal occurring when the metal is subjected to oxygen and/or other oxidising environment. Transition metals like copper, iron, tungsten and titanium in ambient conditions are thermodynamically unstable and hence they will oxidize [1]. A thin layer of native oxide (transition metal oxide (TMO)) will be formed on their surfaces. When exposed to high temperature the growth of the TMO increases rapidly, forming a thick scale of oxide. Growth mechanisms of TMO scale follow several well-documented laws with Wagner's model for high-temperature oxidation of metals as one of the most accepted. Apart from growth mechanisms the phases of oxides that formed i.e. the oxidation products and perhaps their position within the TMO

IOP Conf. Series: Journal of Physics: Conf. Series **1082** (2018) 012052 doi:10.1088/1742-6596/1082/1/012052

scale have also been studied [1]. In this work, the surface oxide is off importance since the surface oxide was exploited for its absorbance and photocatalytic properties. For these properties, the morphologics of the surface oxide are also important. Here the surface oxide is tailored to comprise of nanowires. During growth of TMO at high temperature, the rate limiting step is the lattice diffusion of cations through point defects in the scale. Therefore, at high temperature the scale would be expected to exhibit a flat oxide layer at the scale|gas interface. The oxide is also anticipated to compose of columnar grains. At lower oxidation temperature, grain boundary diffusion dominates. The resulting oxide grains are often observed to be shallow-dish shaped, with slight ridges at their boundaries. At this temperature regime, prolonging oxidation times may induce the growth of oxide resulting in elongated, whiskers structure. On the other hand at an intermediate temperature, growth of whiskers (nanowires) is more obvious [2]. Temperature is therefore an important parameter to determine the morphology of surface oxide. Classification of temperature regime is dependent on the melting points of the oxidized metal. In here, oxidation was conducted at 500 or 600 °C on metal of different melting point: copper (1085 °C), iron (1538 °C), titanium (1,668 °C) and tungsten (3,422 °C). Apart from temperature, oxidation condition for instance the presence of water can also influence nanowires formation especially on metal with rather high melting point [3 – 6]. An extremely high melting point metal like tungsten on the other hand appears to be much more difficult to be oxidized to form nanowires with or without the presence of water vapour. Catalysts are required for the growth of tungsten oxide nanowires on tungsten.

As already mentioned, thermal oxidation of iron, tungsten, titanium and copper was done to produce Fe_2O_3, WO_3, TiO_2 and CuO, respectively. These oxides are typical binary oxide semiconductors with applications ranging from photocatalysts to energy devices. In this work, the four metal oxides were made in a form of nanowires and the photoactivity of the oxide was studied by assessing their ability to reduce Cr(VI) to Cr(III) [7]. To the best of our knowledge, no other works have reported on the use of these oxides in a form of nanowires derived from oxidation of their parent metal as photocatalysts for Cr(VI) ions reduction. Cr(VI) is a typical example of heavy metal ion and has long known to be toxic to humans and animals. The toxicity of chromium depends on the oxidation state with Cr(VI) being more toxic than Cr(III). Moreover, Cr(VI) is more readily absorbed when inhaled or consumed leading to various harmful effect to human. Some of the Cr(VI) compounds are also corrosive and due to this, high dosage may results in systemic toxicity. Therefore method to remove them is necessary. In here, as mentioned, TMO nanowires formed were used to remove Cr(VI) by adsorbing and reducing the ions to Cr(III).

2. Methods and materials

Iron, copper, tungsten and titanium foils or wires were degreased, cleaned and rinsed in deionised water. They were then placed in an alumina crucible and positioned at the hot zone in an horizontal tube furnace (Lenton 1200). They were subjected to oxidation process at either 500 °C or 600 °C for < 2 hours. The oxidised samples were quenched after dwelling time was reached and cooled down naturally until room temperature. A set of oxidation was also done with water mist injected to the furnace. A nebulizer was used to produce the water mist (Omron NE-U17). After reaching the desired dwelling time, the sample was quenched from the furnace. Imaging was then done on the surface of the oxidised samples using field emission scanning electron microscopy (FESEM – Zeiss Supra 35 VP). The morphology desired would be in a form of nanowires. For phase identification, and crystal analysis, X-ray diffractometer (XRD - Bruker Advanced X-Ray Solution D8 with Cu-K_{α} radiation (λ = 0.154 nm)) was carried out on the samples. As no nanowires were formed on tungsten, a catalytic route was selected whereby potassium halide solution like (potassium fluoride, chloride or halide) was nebulized to the furnace during oxidation. To assess the photoactivity of the oxidised metals, 10 ppm of Cr(VI) solution was produced by dissolving 0.8481 mg of $K_2Cr_2O_7$ in 30 mL deionized water. HCl was added in to reduce the pH of the solution to 2. Of all four metals oxidised, only oxidised Ti and Fe samples were tested as the surface oxide was adhered to the substrate. The sample was then placed in the

IOP Conf. Series: Journal of Physics: Conf. Series **1082** (2018) 012052 doi:10.1088/1742-6596/1082/1/012052

solution and was left in the dark for 1 hour to reach equilibrium. Only, oxidised titanium, copper and iron were subjected to this process as surface oxide on tungsten peeled off at some parts of the foil. Cr(VI) solution was then placed under sunlight in the morning (10 am) and left for reduction. 3 mL Cr(VI) solution was taken out from the tube at 10 minutes interval until 1 hour. The reduction of the solution can be seen from colour changes from yellowish to clear liquid. UV-VIS spectrometer was used to assess on the reduction of the Cr(VI).

3. Results and discussions

Fig. 1 shows oxide surfaces on (a) tungsten, (b) iron, (c) titanium and (d) copper after oxidation at either 500 or 600 °C in dry air. Morphologies of oxidised titanium and tungsten are consisted of rather flat oxides whereas on copper and iron whiskers are obvious. The whiskers grew perpendicular to the substrate with tips in nanoscale whereas based in submicron scale. This is more obvious for the case of iron as compared to copper. The whiskers termed nanowires in this work are seen to be covering the surface of iron rather evenly but not so on copper. It is known that the driving force for oxide scale formation is the free energy change associated with the formation of the oxide [1]. On titanium, copper, iron and tungsten, the surface oxides are TiO_2, CuO, Fe_2O_3 and WO_3. Surface oxide is often the stable oxide phase which will not dissociate in ambient oxygen pressure. The standard free energy of oxide formations is a function of temperature and during the process of oxidation the stable oxide will grow and crystallize. The corresponding dissociation pressures of oxides can be referred to Ellingham/Richardson diagram and based on the diagram, stable oxide formed on the four metals can be predicted. On Fe and Cu, the nanowires are likely to be and α-Fe_2O_3 [4, 5, 6] and CuO [3].

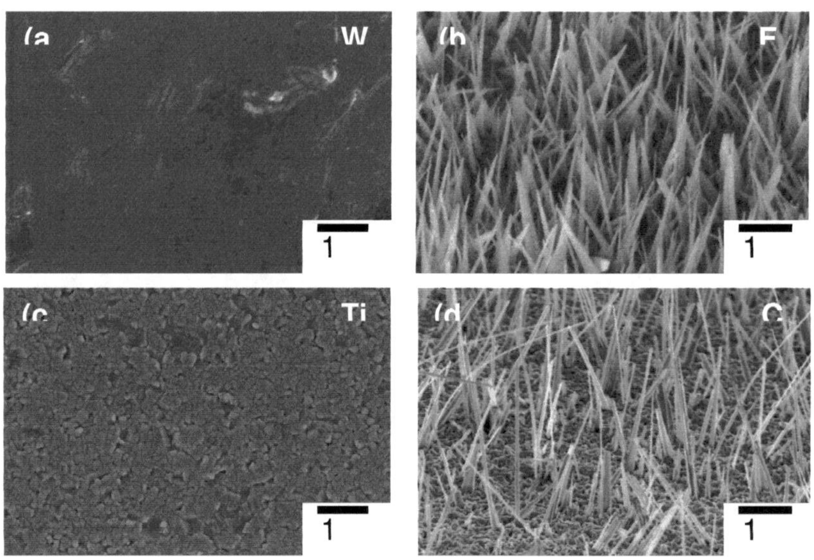

Figure 1. FESEM morphologies of oxidised (a) tungsten (600 °C), (b) iron (500 °C), (c) titanium (600 °C) and (d) copper (600 °C). Oxidation was done in dry air.

As mentioned, melting points of copper, iron, titanium and tungsten are 1085 °C, 1538 °C, 1,668 °C and 3,422 °C respectively. If intermediate oxidation temperature can be defined as $0.5T_m \leq T \leq 0.75T$ whereas low temperature is $< 0.5T_m$, then 500-600 °C can be regarded as intermediate oxidation temperature regime for copper and low temperature regime for iron, titanium and tungsten. At low or intermediate temperature regime the rate of metal oxidation is likely to be controlled by transport along defects in the oxide especially dislocations. Dislocations can provide transport paths for elongated

IOP Conf. Series: Journal of Physics: Conf. Series **1082** (2018) 012052 doi:10.1088/1742-6596/1082/1/012052

surface structure formation and hence explaining the nanowires development. The primary mechanism is via surface diffusion along a tunnel (dislocation core) centered on one dislocation or a bundle of parallel dislocations. Dislocations could either be inherited from the metal substrate, lattice mismatch or volume expansion, or generated by oxidation growth stresses. Growth stresses can occur due to the layered nature of the growing scale. Oxidation of copper results in scale with Cu_2O as the first layer (inner oxide) on Cu follows by CuO as the outer oxide [3]. Iron on the other hand is consisted of magnetite, Fe_3O_4 as the inner oxide and hematite, α-Fe_2O_3 as the outer oxide [4-7]. Due to the different molar volumes between the oxide phases, stress is generated at the interface region between the oxide phases forming dislocation which promote the nanowires growth. Growth occurs from the base to the tip with flux of cations to the surface provided by surface diffusion along the dislocation pipe. To compare the physical appearance of oxidized Cu to Fe, it was observed that the oxide on Cu was easily peeled off and hence limiting its applicability as a photocatalysts.

It appears that 600 °C was not adequate for nanowires growth on titanium despite the melting point of iron and titanium is not very much different. Morphologic feature is therefore not dependent solely on reaction temperature but also upon the gaseous environment; choice of gas, pressure and flow rate. Oxidation condition was then altered by injecting water vapour to the furnace whilst keeping the oxidation temperature at 600 °C. Similarly we performed oxidation of tungsten at 600 °C in water vapour as well. Fig. 2 shows the surface of (a) titanium and (b) tungsten after oxidation in the presence of water vapour. As can be seen, on titanium, nanowires are observed with diameter of ~ 15 nm. It is evident here that water vapour has helped in the formation of TiO_2 nanowires by direct heating in water vapour at 600 °C.

However, no nanowires can be observed on tungsten. Formation of TiO_2 nanowires at low temperature oxidation in water vapour can then be explained by the more rapid dissociation of adsorbed H_2O compared to O_2 molecules [2]. If the oxidant dissociation step is accelerated, then the arriving cations from the titanium are incorporated into lattice growth locally at the tip, resulting in the increase of nanowires length. Therefore, even at rather lower temperature regime, oxidation of titanium can be resulted in nanowires formation provided water vapour is used.

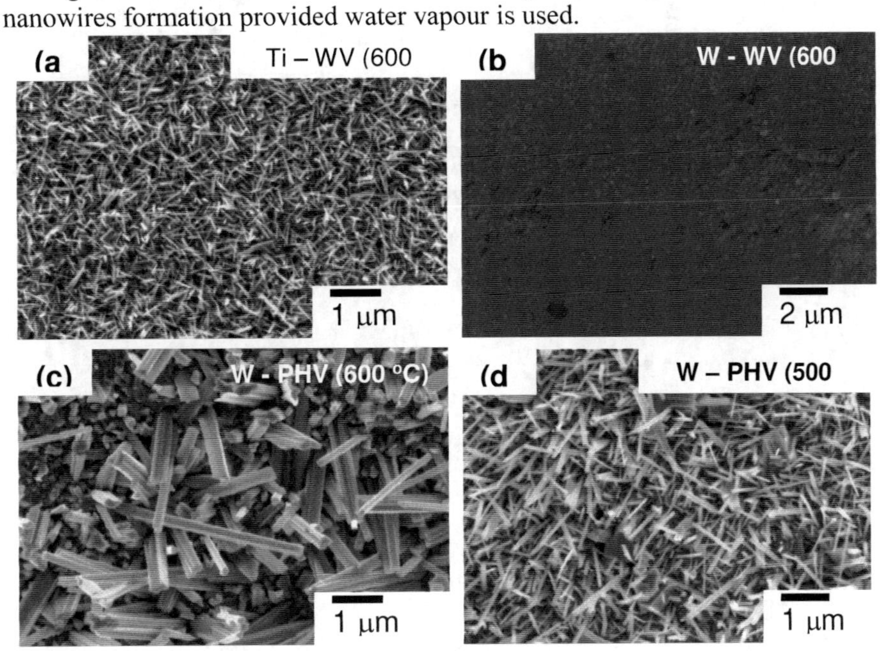

Figure 2. Surface oxide on (a) titanium and (b) tungsten in water vapour (WV) at 600 °C, (c) tungsten in potassium halide vapour (PHV) at 600 °C and (d) tungsten in potassium halide vapour (PHV) at 500 °C.

IOP Conf. Series: Journal of Physics: Conf. Series **1082** (2018) 012052 doi:10.1088/1742-6596/1082/1/012052

Oxidation of tungsten was then done in the presence of potassium halide solution vapour at 600 °C. Fig. 2 (c) is the surface morphology of the oxide. Formation of elongated structure is obvious; indicating the success in the use of potassium halide vapour to aid in nanowires formation but the diameter is rather big. Oxidation temperature was then reduced to 500 °C and as evidence in Fig. 2 (d) the surface oxide is consisted of much smaller whiskers with diameters ranging from 20 – 60 nm. This indicates the success in the formation of nanowires on tungsten at low temperature provided vapour of potassium halide was injected during oxidation process. Note that all oxidation processes were done in one step (i.e. direct heating process) at < 2 h and hence the rate of formation of these nanowires can be considered rather high with less complex process compares to reported works in literature [8]. The formation mechanism of nanowires on tungsten is different than that of nanowires on copper, iron or titanium. It is speculated here that the vapour-solid-liquid process is responsible on the formation of the nanowires with potassium halide acting like crystal droplets at 500 – 600 °C [8]. Nonetheless, the use of potassium halide complicates the formation process. Moreover, the adherence of the oxide to the foil was found to be rather poor making it not suitable to be used as photocatalysts. XRD patterns for oxidised Ti and Fe are shown in Fig. 3 (a) and (b) respectively. The XRD patterns reveal the presence of mostly rutile TiO_2 on Ti agreeing with previous works on oxidation of titanium [9] whereas on iron peaks from hematite (H), α-Fe_2O_3 are detected along with magnetite (M), Fe_3O_4 peaks. Similar findings have been reported by [4-6].

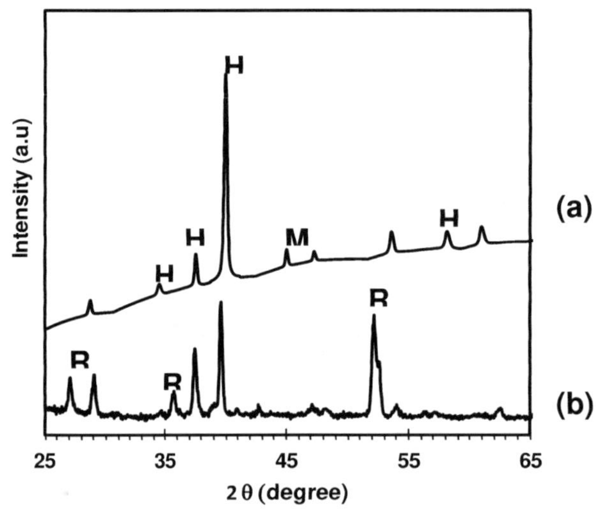

Figure 3. XRD analysis of (a) α-Fe_2O_3 (hematite (H)) and Fe_3O_4 (maghemite (M)) nanowires and rutile (R) TiO_2 nanowires formed by thermal oxidation. Unmarked peaks are from the parent substrate.

Cr(VI) removal experiments were then done on these two samples: rutile TiO_2 and hematite α-Fe_2O_3 nanowires under morning sunlight as mentioned. Photocatalytic reduction is proposed to be one of the main mechanisms in Cr(VI) removal whereby excited electrons can be transferred to the adsorbed Cr(VI) ions on the nanowires. It is apparent from the plot in Fig. 4 that removal on (b) hematite is much more (~ 90 % removal) compared to on (a) rutile nanowires (~ 40 % removal) after an hour of exposure to sunlight. This is obvious since hematite has narrower band gap (2.5 eV) than rutile and hence photoelectrons generation are thought to be much more compared to rutile. TiO_2 is a wide gap semiconductor with energy gap of ~ 3.1 eV and can only be excited under the UV part of the solar spectrum.

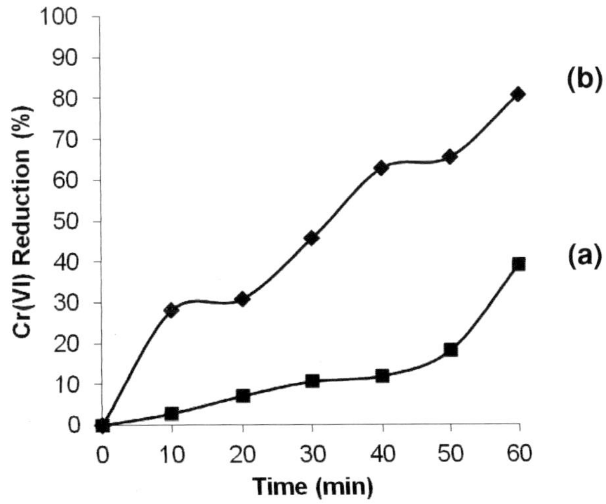

Figure 4. Cr(VI) reduction (%) on (a) TiO_2 nanowires and (b) α-Fe_2O_3 nanowires under sunlight

4. Conclusion

Growth of nanowires was observed on the surfaces of iron and copper in dry air but not on titanium and tungsten at 500 or 600 °C. Nanowires can however be formed on titanium when the furnace was injected with water vapour and on tungsten, potassium halide vapour was needed for nanowires formation. Nevertheless, the growth mechanism of the nanowires on tungsten is thought to be very much different than iron, copper and titanium whereby on tungsten catalytic induced reaction is likely. On iron, copper and titanium, stressed induced mechanism and surface diffusion via dislocation core are the main mechanism of formation. Majority of oxide nanowires on iron are α-Fe_2O_3 whereas on titanium, rutile TiO_2 was the dominant phase. As both of the oxides are semiconductors, electron-hole pairs are expected to be formed when the oxides were irradiated under sunlight. Photoelectrons were used to reduce Cr(VI) with 90 % Cr(VI) reduction after an hour of exposure on α-Fe_2O_3 compared to on 40 % removal on TiO_2 nanowires.

Acknowledgements
Japan International Cooperation Agency (JICA) Grant No. USM CR 1701 and Research University Grant (USM-TUT Collaboration) 1001.PBAHAN.870048 and ASEAN University Network for Science and Engineering Education Development Network (AUN/SEED-Net) are acknowledged.

References

[1] Khanna A S, ASM International, 2002 *Introduction to High Temperature Oxidation and Corrosion*.
[2] Raynaud G M, & Rapp R A, (1984) *Oxidation of Metal* 21: 89
[3] Yuan L and Zhouz G, (2012) *Journal of The Electrochemical Society*, 159 (4) C205-C209
[4] Budiman F, Kian T W, Razak K A, Matsuda A, Lockman Z 2016 *Procedia Chemistry* 19 586-593.
[5] Budiman F, Bashirom N, Tan W K, Abdul Razak K, Matsuda A, Lockman Z, *Applied Surface Science*, 380, 172-177 (2016)
[6] Hiralal P, Unalan H E, Wijayantha K G U, Kursumovic A, Jefferson D, MacManus-Driscoll J L, Amaratunga G A J (2008) *Nanotechnology* 19 455608.
[7] Ku Y, I.-L. Jung, *Water Research*, 35 (1) (2001), 135-142
[8] Qi H, Wang C, Liu J, (2003) *Advanced Materials* 15 5 411-414

IOP Conf. Series: Journal of Physics: Conf. Series **1082** (2018) 012053 doi:10.1088/1742-6596/1082/1/012053

Optimization of Sodium Potassium Niobate Thin Films with Different Deposition Layers

U.A.A. Azlan[1], N.A. Azmi[2], M.A.M. Hatta[2], M.W.A. Rashid[2], M.A.A.M. Abid[2]

[1]Faculty of Manufacturing and Mechanical Engineering Technology, Universiti Teknikal Malaysia Melaka, Hang Tuah Jaya, 76100, Durian Tunggal, Melaka, Malaysia

[2]Faculty of Manufacturing Engineering, Universiti Teknikal Malaysia Melaka, Hang Tuah Jaya, 76100, Durian Tunggal, Melaka, Malaysia

Email: umar@utem.edu.my

Abstract. Sodium Potassium Niobate, KNN with different deposition layers on Si substrate was successfully prepared by sol-gel spin coating and subsequently annealed at 650°C for 10 min. Several deposition layers of KNN were systematically studied with different characterizations. The KNN thin films with five deposition layers showed a good result in terms of its homogeneity and uniformity of thickness that was measured to be around 280 nm. Further study on the surface roughness was obtained by AFM and its value of around 1.698 nm was recorded from the scanned area of 1 μm × 1 μm of thin films. It was also reported that the NbO_6 octahedral vibrational modes of υ6, υ5, υ4, υ2, υ1, υ3, υ1+υ5 was detected in a wide range of 200 cm^{-1} to 900 cm^{-1}, shown by Raman analysis. The elemental composition was also studied and the results were confirmed with deconvolution of XRF spectrum.

1. Introduction

For the past 10 years, ferroelectric ceramics with a perovskite structure, a tungsten bronze structure and bismuth layer-structured (BLSF) have been reported to replace the lead-based materials. Recently, there is more findings with better results has been reported from various perovskite-structured ferroelectrics such as barium titanate, bismuth titanate, bismuth sodium titanate, potassium niobate, sodium potassium niobate [1]. Among these, sodium potassium niobate ($K_{0.5}Na_{0.5}NbO_3$) or KNN has been nominated as one of the most promising candidate to replace the wide usage of lead-based materials especially the lead zirconate titanate [2]. This compound is a combination of solid solution ferroelectric $KNbO_3$ and anti-ferroelectric $NaNbO_3$ that has been the most studied lead-free piezoceramics materials due to its relatively high piezoelectric constant (416 pC/N) and high Curie temperature (420°C) [3-5].

Thin film is a layer of material that resembles a coating with a thickness ranging from nanometres into micrometres. In general, the precursor can be made by various wet chemistry techniques. One of them is a sol-gel process, whereby it is typically accomplished by two primary reactions of hydrolysis and condensation. The most common starting materials, or precursors, used in the preparation of the sol are water-sensitive metal alkoxide complexes (M(OR)x, whereby R = alkyl group (e.g., CH_3, C_2H_5, CF_3,

IOP Conf. Series: Journal of Physics: Conf. Series **1082** (2018) 012053 doi:10.1088/1742-6596/1082/1/012053

etc.). Metal alkoxides are good precursors because they readily undergo hydrolysis; that is, the hydrolysis step replaces an alkoxide with a hydroxide group from water and a free alcohol is formed [6]. There are a few methods that have been used for deposition of the thin films. Evaporation, sputtering, spin coating, pulsed laser deposition (PLD), molecular beam epitaxy (MBE) and aerosol deposition are the physical deposition techniques. Additional deposition techniques that have been used are chemical-based deposition, chemical vapour deposition (CVD), metal chemical vapour deposition (MCVD), platting, spin coating, plasma enhanced chemical vapour deposition (PECVD), atomic layer deposition (ALD) and chemical solution deposition (CSD). Among these techniques, spin coating is chosen to fabricate a good physical of thin film with uniform coating layers. In this work, a study on KNN thin films with different deposition layers was reported in this paper.

2. Experimental study

The precursor was started by dissolving CH_3COOK and CH_3COONa into 2-ME in a beaker with constant stirring at room temperature. $Nb_2(OC_2H_5)_{10}$ was also prepared with a constant stirring at room temperature in another beaker containing 2-ME with the addition of acetylacetone. Both of the precursors were mixed together for another constant stirring at 60°C for an hour. After that, the deposition process was done by depositing the precursor onto the Si substrate with different layers using the spin coating machine. One minute of pyrolysis was taking place between each of the layers. To end the deposition process, the wet films were annealed at 650°C for 10 minutes by using the rapid thermal processing (RTP) furnace. The deposition process can be clearly seen as shown in Figure 1.

Raman spectroscopy has been used to investigate the molecular vibration modes for KNN thin films to find the octahedral bonding between the Nb and O ions. The samples were measured by *UniRAM-II Micro Raman Mapping System* in single mode laser with 532 nm DPSS laser. A FESEM (Model: *Hitachi 51400*) was used to investigate particle morphology, thickness and grain microstructure of the annealed thin films. Electron microscopic images were taken at various magnifications using secondary electron (SE) mode and back-scattered electron (BSE) mode with the high electron high tension (EHT) was set at 10 kV. *Hitachi AFM5100N* was employed to measure the roughness of KNN thin films. This instrument uses a cantilever with a sharp probe tip that scans the surface of the specimens. The results of this analysis will be presented in 3D image form. The scan was conducted on 1 μm x 1 μm area of thin films. Elemental analysis for the thin films samples was conducted using XRF microscope and EDX. This analysis was conducted by Micro-XRF on SEM (Bruker) with the x-ray source of 50 kV, a current of 600 μA and rhodium (Rh) as the X-ray source of the target.

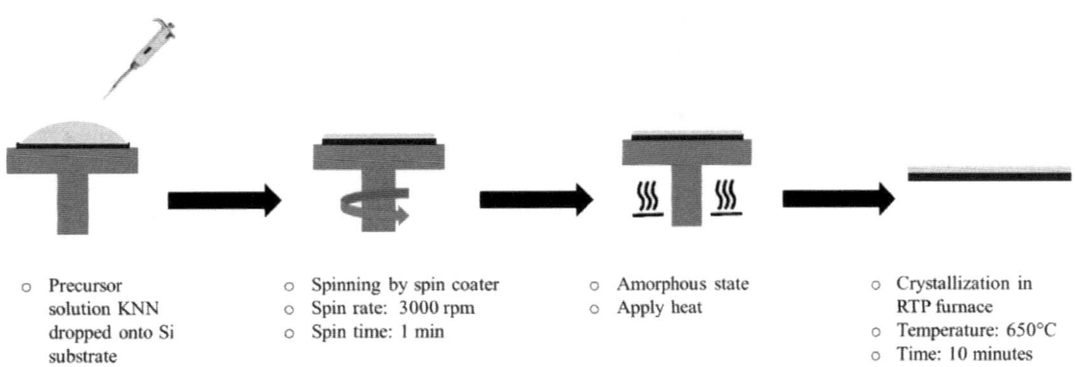

Figure 1. Deposition process of KNN using spin coating technique.

IOP Conf. Series: Journal of Physics: Conf. Series **1082** (2018) 012053 doi:10.1088/1742-6596/1082/1/012053

3. Results and discussion

The Raman spectra for KNN with different deposition layers are shown in Figure 2. The profile in Figure 2 is consistent with the previous report, claimed that the scattering mode of secondary phase are not detected in spectrum [7]. It was also reported that the vibrations of the NbO_6 consisted of 1A1g (υ1) + 1Eg(υ2) + 2F1u(υ3, υ4) + F2g(υ5) + F2u(υ6), whereby 1A1g (υ1) + 1Eg(υ2) + 1F1u (υ3) are the stretching modes and the rest are the bending modes [8]. The NbO_6 octahedral vibrational modes of υ6, υ5, υ4, υ2, υ1, υ3, υ1+υ5 appear in a wide range of 200 to 900 cm^{-1}. Notably, υ1 (617 cm^{-1}) and υ5 (301 cm^{-1}) that were detected in this analysis are also having the relatively strong scattering in the single layer deposition of KNN. At the signal of 520 cm^{-1} it was found that a high intensity peak for each deposition layer is attributed by Si peak. As reported, the broad peak at both modes υ1 and υ5 is a clear signal disorder in symmetry perovskite as detected in Raman spectrum [9]. Besides that, these modes are also common in the perovskite structure of KNN, showing that the distortion of perovskite structure through Nb-O bonding.

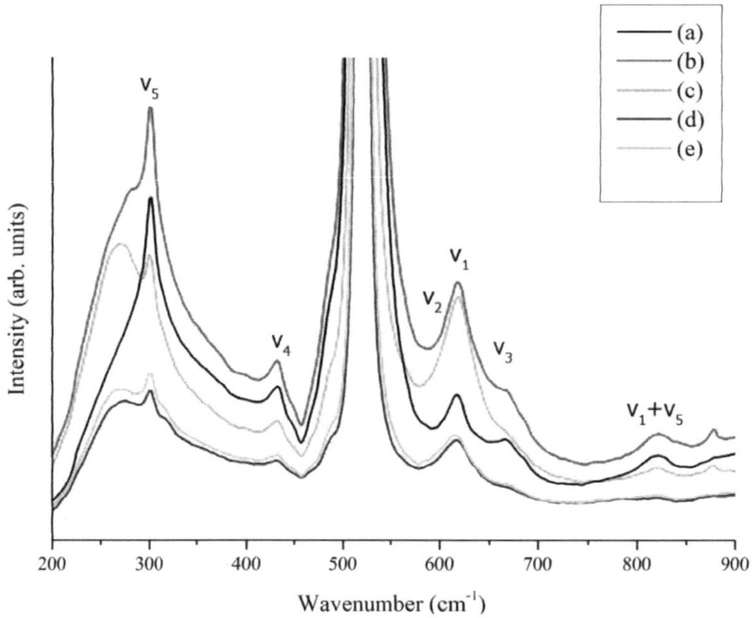

Figure 2. Raman spectra analysis for KNN with different deposition layers: (a) 1, (b) 2, (c) 3, (d) 4, and (e) 5.

The FESEM images of annealed KNN thin films with different deposition layers are shown in Figure 3. As seen, the diverse sizes of grain in each deposition layers can be seen clearly without any crack and with a smooth surface. The inhomogeneity of KNN particles can be seen on the lowest deposition layer of KNN thin films in Figure 3(a) due to the rapid volatilization of K and Na elements thus, reducing the formation of the uniform particle at a single layer of KNN films. However, the films deposited with three deposition layers in Figure 3(b) showing some agglomeration of particles due to the high temperature annealing process. The thin films with five deposition layers in Figure 3(c) showed a better improvement in homogeneity and uniformity as compared with one and three deposition layers. In addition, the thin films with five deposition layer also pleasantly denser from the other films. The uniformity of KNN thin films is also in agreement with a report of PZT thin films for the fifth deposition layer with dense uniform grains and compact crack-free structure [10]. In Figure 3(d), the thickness of the films with five deposition layers was measured to be around 284 nm. Further analysis was also carried out on the films with six deposition layer. Nevertheless, the formation of cracks on the sixth deposition layer was found clearly in the FESEM image as shown in Figure 3(e). This can be explained that the repetitive process of coating and pyrolysis for each deposition layers also contributed to the thermal expansion of thin film that exceeded the substrate and leads to change in crystal structures volume which resulted in residual stresses induced in thin films propagating the severe cracks [11]. Figure 3(f) shows the AFM images of KNN thin films with five deposition layers. From the AFM analysis of the surface morphology and topography, the root mean square roughness (Rq) was determined. The value of surface roughness was obtained from the scanned areas of 1 μm × 1 μm of thin films which indicated that all the deposited films have nano-crystalline growth patterns of 1.698 nm.

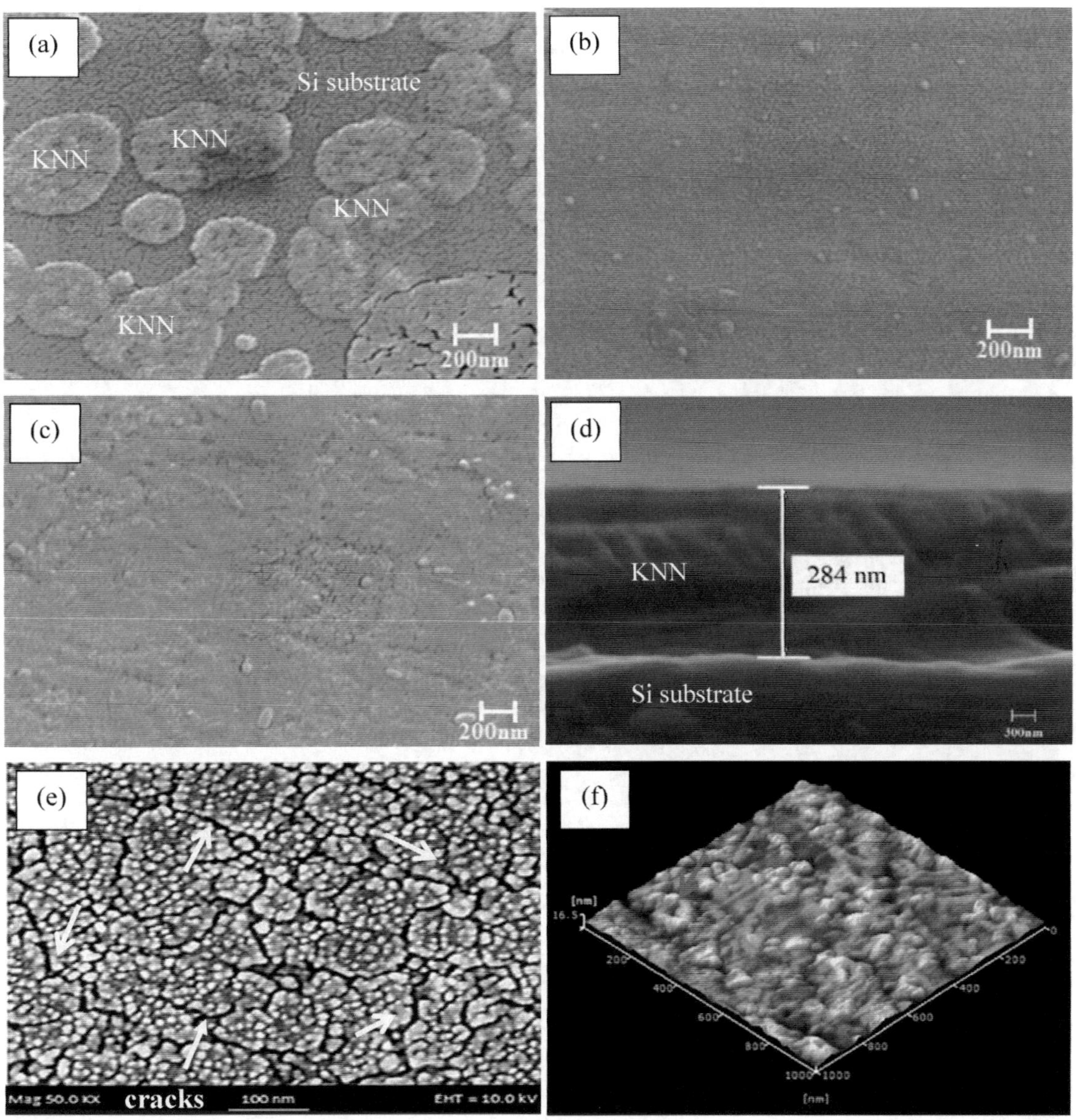

Figure 3. FESEM images of KNN thin films with different deposition layers: (a) 1, (b) 3, (c) 5, (d) thickness of KNN with 5 layers deposited on Si substrate, (e) formation of cracks at six deposition layers and (f) AFM images of 5 deposition layers.

The elemental composition of KNN with five deposition layers was confirmed with de-convolution of XRF spectrum as shown in Figure 4. Elements which are K, Na, and Nb are present in the KNN compound. Note that Rh is the X-ray source, which is used as Rh target. XRF penetrated deep into the layers and does not detect the gaseous type of elements, thus the oxygen gas was not presented in the analysis graph. Distribution in KNN grains was consistent with its XRF spectrum which portrays the lower excitation energy that shows the stability of alkaline elements (K and Na). The result in Figure 4 was found to be consistent with the EDX spectrum in Figure 5, which also portrays the lower excitation energy that shows the stability of alkaline elements of both K and Na due to the volatility of these materials which highly reacted with moisture. It shows that the composition of all elements is slightly changed from the actual numbers because of thermal effect due to alkaline ions character that is easy to volatile on high temperature.

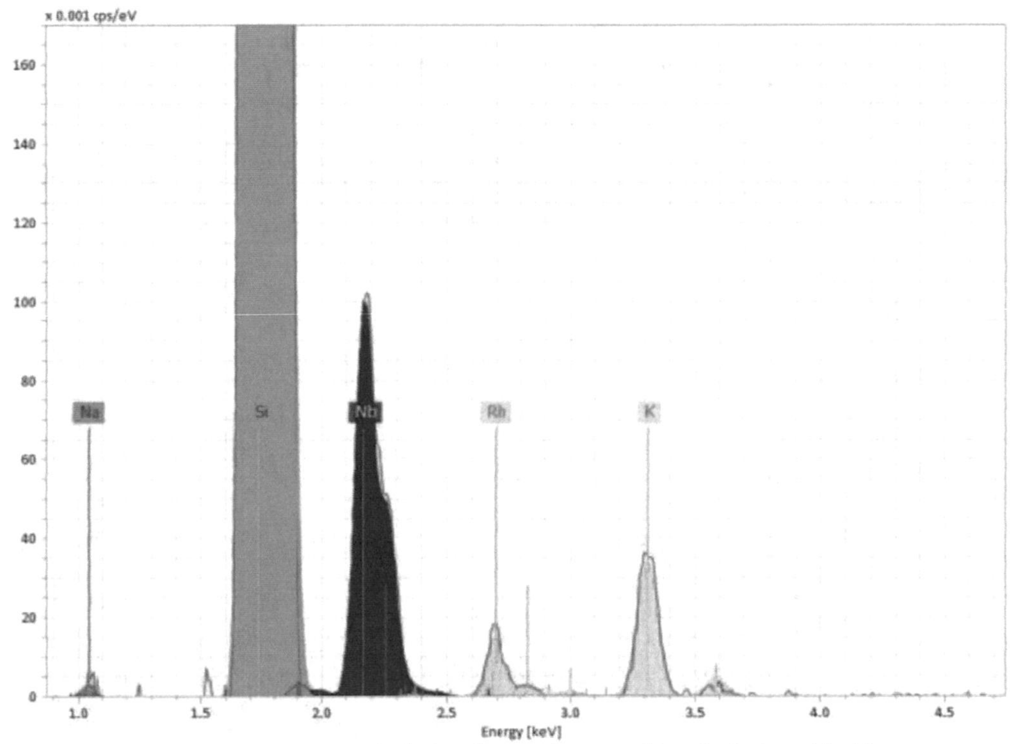

Figure 4. XRF spectra analysis for elemental compositional of KNN thin films.

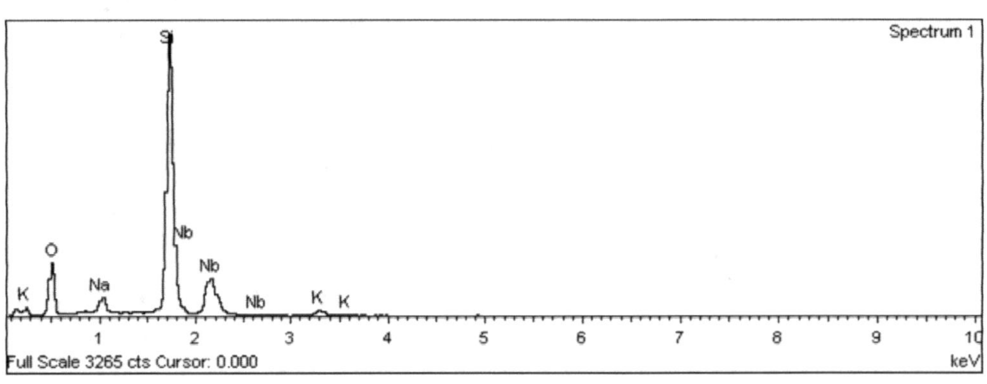

Figure 5. EDX spectrum analysis of K, Na, Nb and O elements.

4. Conclusion

The KNN thin films with different deposition layers was successfully prepared by sol-gel spin coating. It can be concluded that the KNN with five deposition layers are well distributed in terms of homogeneity and uniformity of its morphology. The thickness of the same films was around 284 nm. The surface roughness was measured to be around 1.698 nm with nano-crystalline growth patterns. The presence of the following elements, K, Na, and Nb was confirmed with de-convolution of XRF spectrum.

References

[1] Takenaka T 2013 *Handb. Adv. Ceram.* (2nd Ed.). Japan: Elsevier. Inc.
[2] Maziati Akmal MH, Warikh ARM, Azlan UAA, Azam MA and Ismail S 2016 *Materials Letters* 170
[3] Saito Y, Takao H, Tani T, Nonoyama T, Takatori K, Homma T, Nagaya T and Nakamura M 2004 *Nature* 432 7013
[4] Rödel J, Jo W, Seifert KTP, Anton EM, Granzow T and Damjanovic D 2009 *Journal of the American Ceramic Society* 92 6
[5] Shrout TR and Zhang SJ 2007 *Journal of Electroceramics* 19 1
[6] Azlan UAA, Warapong K, Ahmad Fauzi MN, Khairunisak AR and Srimala S 2012 *In Tech*
[7] Nakashima Y, Sakamoto W, Shimura T and Yogo T 2007 *Japanese Journal of Applied Physics* 46 10B
[8] Kakimoto K, Akao K, Guo YP and Ohsato H 2005 *Japanese Journal of Applied Physics* 44 9B
[9] Mahesh P and Pamu D 2015 IOP *Conference Series: Materials Science and Engineering* 73
[10] He G, Zhang Y, Peng C and Li X 2013 *Applied Surface Science* 283
[11] Prosser JH, Brugarolas T, Lee S, Nolte AJ and Lee D 2012 *Nano Letters* 12 10

IOP Conf. Series: Journal of Physics: Conf. Series **1082** (2018) 012054 doi:10.1088/1742-6596/1082/1/012054

Effects of Oxalic Acid on UV-C Sensing Property of Tomato Thin-Film Based Photodetector

Myo Myo Thu[1], Kuan Yew Cheong[*1] and Atsunori Mastuda [2]

[1,*1] Electronic Materials Research Group, School of Materials and Mineral Resources Engineering, Engineering Campus, Universiti Sains Malaysia, 14300 Nibong Tebal, Penang, Malaysia.

[2] Department of Electrical and Electronic Information Engineering, Graduate School of Engineering, Toyohashi University of Technology, Japan

Email: srcheong@usm.my

Abstract. New concept of UV photodetector with self-powered and biodegradable features are needed to solve energy and environment. In this work, tomato thin-film was used as an active organic layer in UV photodetector. The effect of oxalic acid in the extraction of tomato juice on the formation of a solid thin film that dried at 120°C for 30 min has been investigated. The purpose of adding oxalic acid is to extract ascrobic acid from tomato. A comparison between tomato juice extracted with and without oxalic acid on the structural, chemical, optical, electrical and UV sensing properties of tomato thin-film have been investigated. The UV photodetector was consisted of a sandwich structure of glass substrate/tomato thin-film/interdigitated aluminium electrode. The photodetector operated at 5 V with tomato thin-film extracted from tomato juice with oxalic acid revealed the highest responsivity percentage difference (89.60%) in UV-C with quantum efficiency (η) of 1.0335x10-5 %, responsivity (R) of 2.1178x10-6 A/W, and detectivity (D*) of 1.1649x1021 Jones. While without oxalic acid, the highest responsivity percentage difference (70.6%) in UV-C was operated at 0 V.

1. Introduction

UV region of the electromagnetic spectrum covers the wavelength range between λ~10 nm and λ~ 400 nm. It is divided into the three spectral bands: UV-A (400–320 nm), UV-B (320–280 nm) and UV-C (below 280 nm). UV-C is very energetic and harmful to life but it has been filtered out by ozone layer. This type of UV is technological useful whereby it can be engineered out to kill bacteria and gems, fused for flame detection (fire alarm, missile warning or combustion engine control), forensic analysis, protein analysis and DNA sequencing, space observation (VUV/EUV ranges), and advanced lithography [1-4]. Currently, the most widely used semiconductor-based UV photodetector materials is either from inorganic based such as silicon (Si), wide and direct bandgap semiconductors such as SiC, III-nitrides compounds (AlN, GaN, InN), and selected II-V compounds or organic based synthetic polymer [5-7]. However, these materials are difficult to decompose and degrade when it is being disposed. Therefore, bio-organic materials have been considered as an active or passive material for this application [8,9].

In this work, tomato as a natural source of organic material has been chosen to be used as an active material and converted to a solid thin film for the use of UV sensing properties. The tomato (Lycopersicon esculentum) is a member of the Solanaceae family and is one of the most widely consumed vegetables. Tomato is a major source of antioxidants and various vitamins, which are activated by UV radiation, transferring of electrons that involves either releasing or trapping of electrons may occur [10]. If an external voltage apply, the current will produce with respect to the UV radiation due to the electron transportation phenomenon would serve as a basis in the photodetector of this work. In order to use tomato as a solid-state UV photodetector, tomato juice must be extracted out from the fruit and the juice mixed with oxalic acid must then be formulated before it is deposited on a treated glass substrate and processed it into a solid and dried thin film. In this work, a systematic investigation has been performed in order to extract, formulate, and process the extracted tomato juice with and without oxalic acid into a functional solid thin film with the objective of using it as a sensing layer of UV-C radiation with no (self-powered) or minimum operating power that is able to address issue of electronic waste.

2. Materials and Methods

Firstly, the extraction process of tomato juice consisted of four stages, namely washing, homogenizing, centrifuging and vacuum filtering. A typical fresh tomato was between 38 and 56 g and it was washed with distilled water in order to remove dirt. Then, it was cut into half for seeds removal before homogenized in a food blender (Philips HR2061). After that, to prepare the acid, 2.1 g of oxalic acid (solid) were dissolved in 350 ml of deionized (DI) water in a glass beaker and stirred vigorously for 15 min. The oxalic acid were dissolved in deionized water and appeared as a transparent solution. After that, 50 g of tomato juice was mixed into the oxalic acid. And then, the homogenized solution were transferred into a centrifuged (Hettich Rotina 38) operated at 3000 rpm for 30 min in order to remove any fibres from the juice by passing through a filter paper (Whatman No. 1) using a vacuum filter. The extracted juice was collected in a plastic bottles and stored in refrigerate condition at 5°C for further usage. The extracted tomato juice was then spin coated (3,000 rpm, 30 s) on a pre-cleaned glass substrate and dried at 120°C for 30 min in an oven (air ambient). Then, an array of Al interdigital electrode was thermally evaporated via a shadow mask. Detail of this has been reported in Ref.[10]. An atomic force microscope (AFM) (Nanonavi SPA400-DFM, SII Nanotechnology) was used to investigate surface topography of the dried tomato thin film. Chemical functional groups and absorbance characteristics of dried tomato thin-film was obtained from Fourier transform infrared spectroscopy (FTIR) (Perkin Elmer Spectrum 1) and UV-Vis spectrophotometer (Cary 50-Varian), respectively. MProbe 20, based on diffusive spectral reflectance technique was employed to obtain the thickness and refractive index of the sample. UV sensing property was characterized from the metal-insulator-metal test structure using UV-B = 302 nm and UV-C = 254 nm radiation sources with intensity of 0.0018 and 0.0013 A/cm2, respectively. The current-voltage (I-V) and current-time (I-t) characteristics of the photodetector was measured from an Agilent HP4156C Precision semiconductor parameter analyzer (SPA).

3. Result and Discussion

Table 1 compares the root-mean square surface roughness, physical thickness and refractive index (n) value of tomato thin-film with and without oxalic acid. Using the same processing condition, except with addition of oxalic acid, the thickness of tomato thin-film is approximately 3 times thinner than the thin-film without addition of oxalic acid. The thinner film is slightly smoother on its surface and higher n value than its counterpart. The higher n value indicates that the thin-film tomato is relatively denser when oxalic acid is added during the extraction process.

Table 1. A Comparison of Thickness, Root-Mean Square (RMS) Surface Roughness, and Refractive Index (N) Value of Tomato Thin-Film With and Without Oxalic Acid Dried At 120°C.

Sample	Thickness (nm)	RMS Surface Roughness (nm)	Refractive Index (n)
with oxalic acid	6.30	10.3500	0.7052
without oxalic acid	19.21	11.5300	0.6912

Chemical functional groups available in tomato thin-film with and without oxalic acid were identified by Fourier Transform Infrared (FTIR) Spectroscopy. Figure 1 (a) shows the absorbance spectra of tomato thin-film with and without oxalic acid. Tomato thin-film without oxalic acid shows chemical function groups associated to the three main vitamin groups, namely Carotenoids (lycopens), Phenolic, and Ascorbic Acid (Vitamin C) as well as lipids. This has been reported in Ref:[10]. The similar absorbance spectra at approximately same wave number except for the variation in intensities of the absorbance has been revealed in tomato thin-film with oxalic acid. In general, the intensity of absorbance spectra of tomato thin-film with oxalic acid is lower than tomato thin-film without oxalic acid. This may due to difference in the amount being tested. As shown in Table 2, sample with oxalic acid is thinner hence; the intensity is lower in the FTIR measurement. In addition, there are a few absorbance spectra located at wavenumber below approximately 1380 cm-1 are not detectable in tomato thin-film with oxalic acid if compared with its counterpart. These spectra are assigned to ascorbic acid [11]. Chakraborty et al. (2013) proved that by adding oxalic acid in tomato juice, concentration of ascorbic acid can be reduced [11]. Figure 1 (b) shows the absorbance spectra as a function of wavelength (λ) measured by UV-Visible spectrophotometry. According to the result, a distinct absorbance spectrum is recorded in the range of λ = 200 to 340 nm for the two samples, except with a lower intensity being recorded by the sample with oxalic acid. The lowering in intensity may be attributed to difference in the thickness of the thin-film as oxalic acid sample shows a thinner layer. As reported by Yao et al. (2004), phenolic compounds (quercetin and kaempferol) in the tomato is governing the absorption of UV in this particular wavelength [12].

Dark and UV respond on test structure consists of tomato thin-film with and without oxalic acid was measured and their current density (J) – voltage (V) responds was recorded (Fig. 2). The dark current density with respect to the applied voltage (5 to -5 V) is compared with the current density when illuminated with two different wavelengths (254 and 302 nm) within the UV range. In general, the trend of respond between dark and UV illumination is the same for tomato thin-film with and without oxalic acid with lower J being recorded at shorter wavelength and higher J at longer wavelength with respect to dark condition. Comparing the J-V results obtained from these two samples, there is variation in the differences between dark and UV responds of the current density values. These variation is translated into R and D*, which will be reported in the subsequent paragraphs.

Table 2. Assignment of Chemical Functional Groups of Respective Absorbance Spectra.

Chemical Compound in tomato	Functional	120°C (without oxalic acid)	120°C (with oxalic acid)	References
Phenol (Vitamin group)	3420-3250cm-1 (-OH)	3399.93	3392	[13]
Ascrobic acid (Vitamin group)	1400-1440cm-1 Carboxylic aicd, (OH bending and 1063-1015cm-1 (primary alcohols)	1063.93 1412.92	1068.89	[14]
Lycopene (Vitamin group)	900-650cm-1, CH- out of plane(lycopence),	779.91	-	[10]
Lipid	2990-2850cm-1 (C-CH3, -CH2-)	2919.5	2929.6	[10]
Carbonyl compound	1870-1620cm-1 (C=O),	1630.93	1634.69	[10]

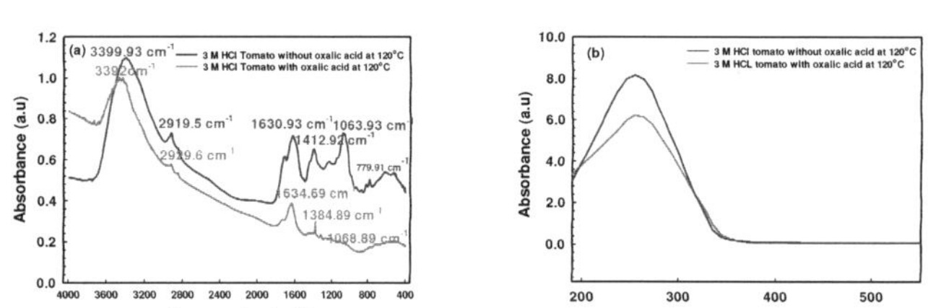

Figure 1. (a) FTIR absorbance spectrum and (b) UV-Visible absorbance spectrum of tomato with and without oxalic acid at 120°C drying temperature. The result of sample without oxalic acid has been presented in Ref:[10].

IOP Conf. Series: Journal of Physics: Conf. Series **1082** (2018) 012054 doi:10.1088/1742-6596/1082/1/012054

The performance of UV sensing properties for the tomato thin-film with and without oxalic acid based UV photodetector was determined from the J-V results (Fig. 2). Figures 3 compares the R and D* values obtained from photodetector with tomato thin-film with and without oxalic acid and measured at -5 V (Fig. 3 (a)), 5 V (Fig. 3 (b)) and 0 V (Fig. 3 (c)) that had been illuminated by UV light of 254 and 302 nm, respectively. According to the result, the R and D* values are higher for tomato thin-film with oxalic acid if compared with its counterpart when measured under illumination of both wavelength at 5 V (Fig. 3 (a)), -5V (Fig. 3 (b)) and 0 V (Fig.3 (c)). However, when the photodetector measured by longer wavelength (302 nm), a reversed trend is being observed only at 5 V. In order to identify a tomato thin-film based UV photodetector that can be selectively and sensitively detecting a specific wavelength for a particular application, the results presented in Fig. 4 can be used as a guide. In the figure, the percentage differences between R values for both samples showed only the UV-C region (254 nm), is calculated based on the results presented in Figs. 3. (a,b,c). The percentage differences between R values of λ = 254 nm and 302 nm (ΔR %) are calculated based on the following equation:

$$\Delta R\% = \frac{\Delta R_{254nm} - \Delta R_{302nm}}{\Delta R_{254nm}} \times 100 \qquad \text{(Eq. 1)}$$

Where, R_{254}nm and R_{302}nm are the responsivity obtained from λ = 254 nm and 302 nm, respectively. The photodetector prepared with oxalic acid is selectively and sensitive to UV-C (254 nm), which is similar to the photodetector prepared without oxalic acid. The only difference is the operating voltage of the photodetector with the former one operating at 5 V and the latter one working at 0 V. This indicates that the chemical functional groups that responsible for the sensing property are different with tomato thin-film prepared by additional of oxalic acid is having purer amount of phenolic compounds while without oxalic acid the thin film is having additional of ascorbic acid. With these differences, the operating voltage is affected. The external quantum efficiency of photodetector with tomato thin-film with and without oxalic acid dried at 120°C illuminated with 254 and 302 nm is compared and shown in Fig. 5. In general, the efficiency of the photodetector prepared with oxalic acid is much higher if compared with those without oxalic acid when the photodetector is illuminated with shorter wavelength (254 nm). This is totally different when the photodetector is illuminating with wavelength of 302 nm. In order to investigate the respond time of the photodetector, tomato thin-film with oxalic acid dried at 120°C and measured at 5 V had been performed. The photodetector was illuminated with a 302-nm UV light for a 5 periodic on and off cycle with an interval of 15 seconds each. The current density as a function of time (J-t) responding to the cycles is presented in Fig. 6 (a). When the photodetector prepared with oxalic acid is exposed to UV light, the J value increases and reaches a relatively stable value before starting to decay when the light is switched off. When the light is switched off the current density decreases substantially until it is stabilized for photodetector prepared with oxalic acid but its counterpart reveals a rapid increment in current density until it is plateau off gradually. By consecutively switching on and off the UV light, the on-off ratio of current density are decreasing with no significant changes of the current density when the UV light is switched off. The extracted respond times for raising (Fig. 6 (b)) and falling (Fig. 6 (c)) for the five cycles are presented in Table 3. The average times for raising and falling (with oxalic acid) are 0.048 s and 0.043 s, respectively if compared with its counterpart, the respective average times are relatively fast.

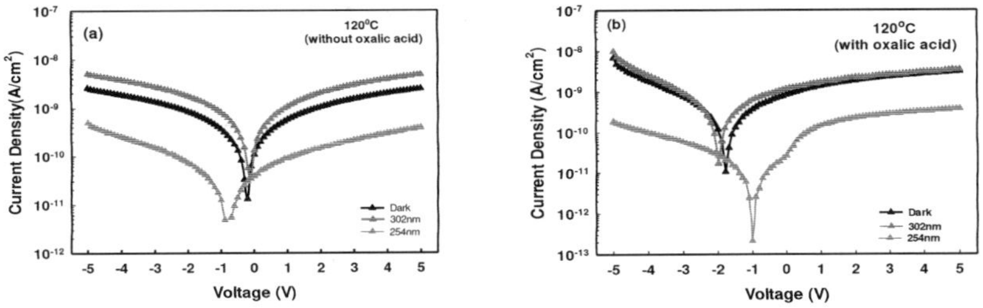

Figure 2. J-V characteristic of thin-film tomato (a) without oxalic acid and (b) with oxalic acid based UV photodetector. The result of sample without oxalic acid has been presented in Ref:[10].

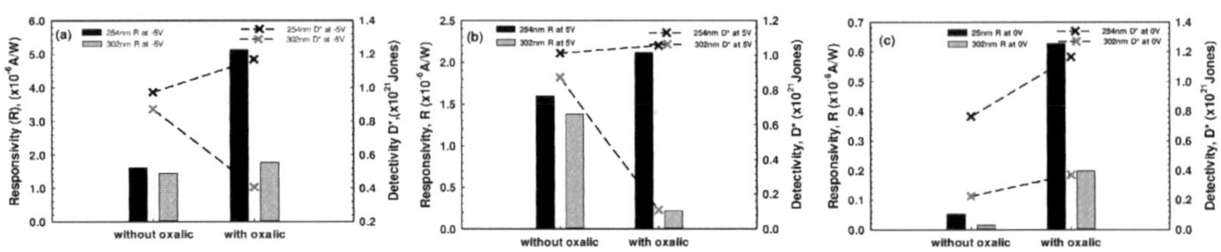

Figure 3. The R and D* values obtained from photodetector with thin-film tomato with and without oxalic acid with (a) -5 V and (b) 5 V and (c) 0 V under UV light of 254 and 302 nm. The result of sample without oxalic acid has been presented in Ref:[10].

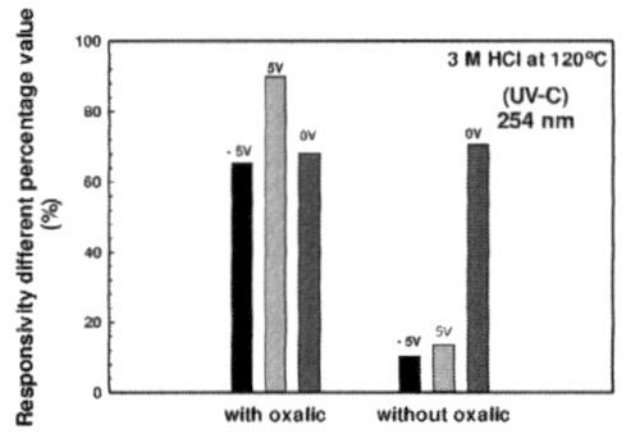

Figure 4. The percentage in positive and negative region shows a higher responsivity of the photodetector with respect to the shorter (254 nm) wavelength. The result of sample without oxalic acid has been presented in Ref:[10].

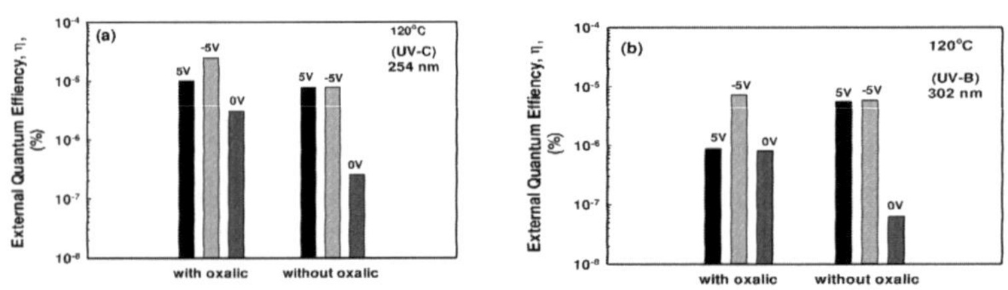

Figure 5. External quantum efficiency value of thin-film tomato with the applied voltage 5V, -5V and 0V under UV light of (a) 254 and (b) 302 nm. The result of sample without oxalic acid has been presented in Ref:[38].

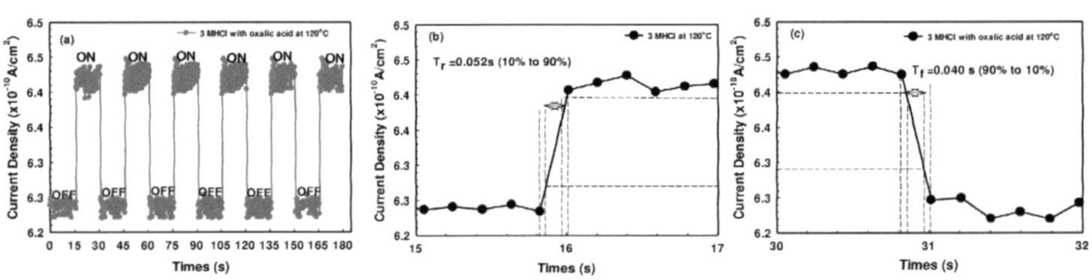

Figure 6. (a) Responses of current density of tomato thin-film (with oxalic acid) as a function of time for 5 switching cycles of UV (wavelength of 302 nm) on and off and measured at 5 V, and respond times for (b) raising and (c) falling of tomato thin-film with oxalic acid.

Table 3: The Extracted Respond Times For Raising and Falling (Fig. 9) For The Five Cycles of Thin-Film Tomto Based UV Photodetector.

Sample	Cycle	Raise time (t_r)	Decay time (t_f)
With oxalic acid	1st	0.052 s	0.040 s
	2nd	0.036 s	0.032 s
	3rd	0.058 s	0.051 s
	4th	0.049 s	0.049 s
	5th	0.043 s	0.041 s
	Average	0.048 s	0.043 s

4. Conclusion

In this work, tomato juice has been extracted, formulated, and processed into a solid thin film deposited on a glass substrate for the use of detecting UV radiation, in particular UV-C. The effects of oxalic acid to form a solid thin film on the structural, optical, chemical, and UV detecting properties have been investigated. The device extracted by adding with and without oxalic acid at 120°C, the highest responsivity percentage difference (89.60%) in positive region (UV-C) is recorded in photodetector with tomato thin-film with oxalic acid dried at 120oC and measured at 5 V with quantum efficiency (η) of 1.0335x10-5 % , R of 2.1178x10-6 A/W, and D* of 1.1649x1021 Jones. While a self-powered (V = 0 V) UV photodetector that was able to detect UV-C can be produced by drying the tomato thin-film without oxalic acid at 120°C with external quantum efficiency of 2.53x10-7%. Because the total concentration of phenolic compounds in the extracted tomato juice will be much higher. If the tomato is extracted without using oxalic acid, treatment of glass substrate is capable to produce a UV sensing properties that can sense UVC. When the oxalic acid is added during the extraction, the sensing property is able to detect UVC. The change of chemical functional groups with respect to the effect of oxalic acid have been evaluated with the identification of three groups of vitamins and lipids. The electrical property of the dried tomato thin-film with respect to the chemical functional groups and antioxidant activity when illuminating under UV-C wavelengths had been proposed. The response time for raising and falling for all testing less than 0.3 s observed in this work enables the tomato thin-film been used as a UV photodetector at this specific wavelength. Hence, the quick response and stability enable the obtained tomato thin-film device to be used as UV sensing properties of the photodetector.

References

[1] W. Kowalski 2009 *UVGI disinfection theory, in: Ultraviolet germicidal irradiation handbook Springer* 17-50.
[2] L. Shi, S. Nihtianov 2012 *Comparative study of silicon-based ultraviolet photodetectors IEEE Sensors Journal* 2453-2459.
[3] J. Hwang, S. Wang, S. Hwang 2016 *Journal of Alloys and Compounds* 618-621.
[4] H.-S. Kim, M.D. Kumar, M. Patel, J. Kim 2016 *Sensors and Actuators A: Physical* 35-41.
[5] J. Hwang, D. Wu, S. Hwang 2015 *Materials Science in Semiconductor Processing* 132-135.
[6] V. Consonni, G. Feuillet 2014 John Wiley & Sons,.
[7] B. Parida, S. Kim, M. Oh, S. Jung, M. Baek, J.-H. Ryou, H. Kim 2017 *Materials Science in Semiconductor Processing* 29-34.
[8] Z.X. Lim, S. Sreenivasan, Y.H. Wong, F. Zhao, K.Y. Cheong 2016 *MRS Advances* 2513-2518.
[9] Z.X. Lim, K.Y. Cheong 2015 *Physical Chemistry Chemical Physics* 26833-26853.
[10] M.M. Thu, K.Y. Cheong, and A. Matsuda 2017 submitted to *Applied Surface Science*.
[11] N. Chakraborty, R. Ghosh, S. Ghosh, K. Narula, R. Tayal, A. Datta, S. Chakraborty 2013 *Plant Physiology* 364-378,.
[12] L.H. Yao, Y. Jiang, J. SHI, F. Tomas-Barberan, N. Datta, R. Singanusong, S. Chen 2004 *Plant Foods for Human Nutrition (Formerly Qualitas Plantarum)* 113-122,.
[13] N. Sinelli, A. Spinardi, V. Di Egidio, I. Mignani, E. Casiraghi 2008 *Postharvest Biology and Technology* 31-36.
[14] F. Van de Voort, "Fourier transform infrared spectroscopy applied to food analysis", Food ResearchInternational, vol. 25, No. 5, pp. 397-403, 1992.
[15] I. Bunghez, M. Raduly, S. Doncea, I. Aksahin, R. Ion, "Lycopene determination in tomatoes by different spectral techniques (UV-VIS, FTIR and HPLC)", Digest Journal of Nanomaterials & Biostructures (DJNB), vol. 6, No.3, pp. 1349-1356, 2011.
[16] A. Rohman & Y. B. C. Man. "The use of Fourier transform mid infrared (FT-MIR) spectroscopy for detection and quantification of adulteration in virgin coconut oil", Food Chemistry, vol. 129, N0. 2, pp. 583-588, 2011.

IOP Conf. Series: Journal of Physics: Conf. Series **1082** (2018) 012055 doi:10.1088/1742-6596/1082/1/012055

Wetting characterization of Flip chip's Lead-Free Solder Interconnect Using Surface Evolver.

S.F. Lim[1], M.S. Nurulakmal[1*]

[1]School of Materials and Mineral Resources Engineering, Universiti Sains Malaysia, Engineering Campus, Penang, Malaysia

[1]Email: limshawfa@gmail.com, *srnurul@usm.my

Abstract. Smaller pitch of first interconnect has posed various challenges to the electronic packaging. The driver to smaller pitch was driven by package and die size miniaturization. In order to enable smaller pitch of first interconnection, copper pillar was introduced. In this paper, surface evolver software was used to predict the solder geometry of solder cap copper pillar onto copper pad process. The study has shown that copper pillar solder joint geometry can be successfully simulated and agreed with experiment. The relationship between various solder joint influencing factors such as die bump diameters, copper pad geometry, solder height and solder volume were established. The optimum die to copper pad width ratio can be obtained through this simulation work. The information can be used to estimate critical volume of solder needed for new, smaller pitch die bump and copper pad design which help to save money and time by avoiding a large number of experiments prior to mass production.

1. Introduction

1st interconnect between die bump and substrate bump in flip chip packages has evolved from solder bump- pre-solder bump substrate joint to copper pillar- pre-solder bump substrate and further evolved into solder cap copper pillar on copper trace. Small die size requirements as a result of transistor size reductions have driven the need to adopt finer die interconnect pitches. Copper pillar with solder cap is able to provide interconnect with smaller solder joint diameter. The adoption of copper pillar was mainly for smaller bump pitch especially below 150 micron. Copper pillar has shown improvement in electro-migration performance over SnAg and SnPb bump for the same current/temperature condition and similar bump/UBM geometry [1-2]. Additionally, copper pillar has superior characteristic which allow high density input/output (> 200 IO) and current capacity afforded by copper pillar bumping. It is easier to control the pillar bump diameter and stand-off height, which are critical for fine-pitch packages.

Solder cap copper pillar on copper trace provided desirable advantage over other solder joint interconnects as it enables finer pitch interconnect joining for miniature device with minimum solder bridging yield losses. Under solder cap copper pillar process, a solder cap copper pillar directly bond onto copper trace and formed a interconnect joint under thermal compression bonding reflow. The surface of the pads on the substrate is Cu-OSP (Organic Solder ability Preservative) and a pre-solder is not required. Thermal compression bonding was used as it is able to control solder reflow spread and the standoff height versus mass reflow. Solder die bump- pre-solder bump substrate joint have been

IOP Conf. Series: Journal of Physics: Conf. Series **1082** (2018) 012055 doi:10.1088/1742-6596/1082/1/012055

reported to result in defects during manufacturing process such as solder bridging when die bump pitch goes below 140 micron [3]. In this study, the open source Surface Evolver software will be used to simulate the solder joint wetting characteristic under different solder height position, solder volume and different copper pad width. The solder wetting characteristic such as contact angle, stand-off height and its solder surface angular values were measured. The simulation result able to model the solder shape seen in manufacturing process.

2. Experimental study

Solder shape geometry was predicted for solder cap copper pillar direct bonding on the copper pad using surface evolver [4]. Surface Evolver is a computer program that able to predict the solder shape formation. The shape will evolve through minimizing the energy of an initial surface subject to constraints that the user defined. Different pad geometry, solder volume, and solder height were part of the factors used in this simulation work. The solder joint height in this simulation was user defined, varied from 8 micron to 45 micron to simulate the solder joint height formation which is controlled by thermo-compression process. The simulation work in this study was compared with published solder joint geometry in experiment [5] to validate the model. In this simulation, copper pillar height 55 micron with diameter 7 micron with different solder cap height was bonded onto rectangle shape copper pad 125 micron length x 85 micron pad width. As this is bump on trace process, the length was constraint at 125 micron which was longer than is needed to cause any impact to the solder shape. Only copper pad width will have influence to the shape formation. From the model established, various solder volume, joint height and pad width were simulated to predict the solder shape. Copper pillar geometry was fixed as it does not reflow under the lead free reflow profile peak at 260 ^{0}C. Solder cap was melted and spread on the copper pad. The spread of the solder depends on the wetting angle between solder and copper pad, volume and pad geometry.

Solder shape geometry in the experiment includes solder fillet, solder spread and wetting angle. Solder shape from simulation was characterized in a similar way and compared to the published solder shape geometry in the experiment.

R_r= Right solder joint fillet radius, R_l = Left solder joint fillet radius
S_r = Right solder spread from the die bump edge, S_l = Left solder spread from the die bump edge
θ_r = Right wetting angle, θ_l = Left wetting angle

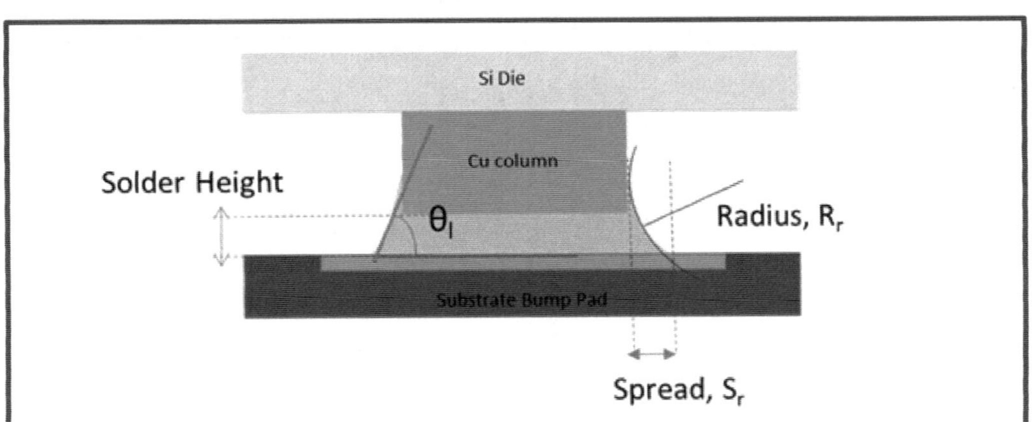

Figure 1. Schematics of solder shape geometry characterization.

Table 1. Solder joint geometry with variable solder cap volume and copper pad width

Copper pillar diameter (micron)	Copper pillar height (micron)	substrate Pad with (micron)	Solder height (micron)	Die bump offset (micron)	Solder cap volume (Kilo-micron^3)	Copper pad width (micron)
70	55	85	8	0	89.8 , 56.3, 30.8	45, 55, 65, 75, 85

3. Results and discussion

Figure 2 was obtained from the simulation with solder height controlled to 8 micron, solder volume 89.8 kilo-micron^3 on copper length 125 micron and 85 micron pad width. The solder joint shape geometry was characterized with solder spread on the copper pad (S_r). The solder shape geometry characterization result was recorded in Table 2. Figure 2 shows the slight concave shape which are resulted from large contact area of solder on copper pad 125 micron length where the solder able to spread. From the side view, Figure 3 is for the pad with 85 micron pad width, and solder spread is constraint by solder resist near to the pad. From the result, the simulation data collected was found to agree with solder shape geometry from the experiment. Solder spread and solder fillet were well correlated as shown in Table 2.

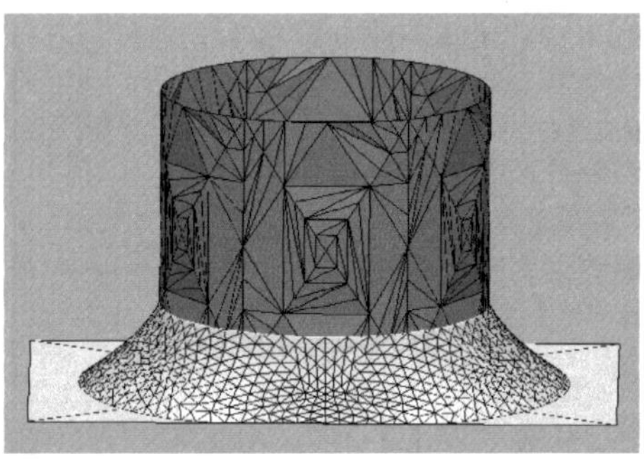

Figure 2. Solder joint shape of 125 length x 85 micron width with 70 micron copper pillar diameter- front view

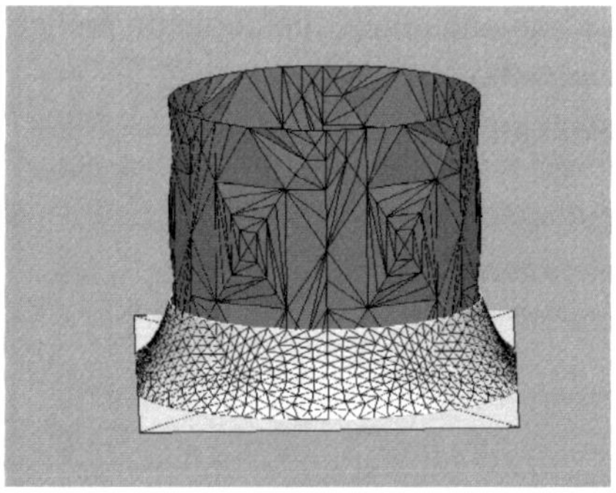

Figure 3. Solder joint shape of 125 micron length x 85 micron width with 70 micron copper pillar diameter-side view

Table 2. Comparison solder joint shape geometry between simulation and experiment.

Sample	Right Angle (Degree)	Left Angle (Degree)	Right Radius Solder Fillet (micron)	Left Radius Solder Fillet (micron)	Right Spread (micron)	Left Spread (micron)
Experiment Data	40	40	38.5	38	18.3	18.3
Simulation data	40	40	38.8	38.8	18	18

Different solder joint height effect to solder shape geometry was shown in Figure 4. It can be observed that the solder joint height shall be < 15 micron before hourglass shape started to appear for a given solder volume. The hour glass shape has reduced the solder joint strength mechanically under temperature cycle for copper pillar-solder joint [6] as the crack initiation started at the necking area of tin solder. Since right and left side shape is symmetry, the left side of the solder shape characteristic was plotted against solder height as shown in Figure 5. Solder spread is getting smaller when the solder joint height was increased. While solder radius is on decreasing trend until solder joint at 45 micron, the radius started to increase with small solder spread and slimmer solder joint shape like "necking". The relationship is important for package engineer to determine the optimized solder volume for a given die bump/pad width ratio.

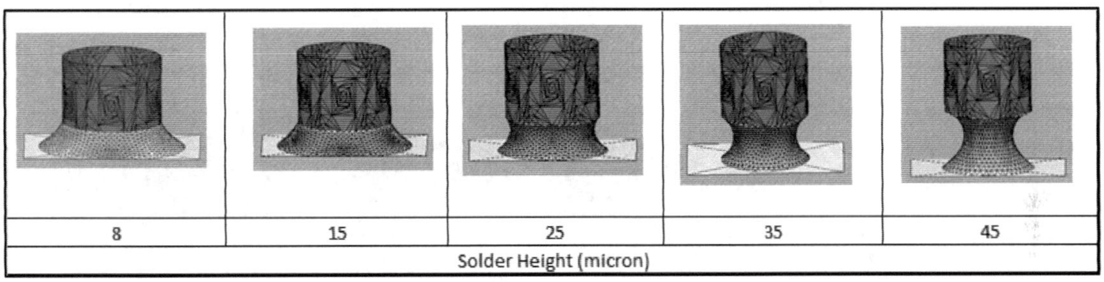

Figure 4. Solder shape geometry at different solder joint height at constant solder volume.

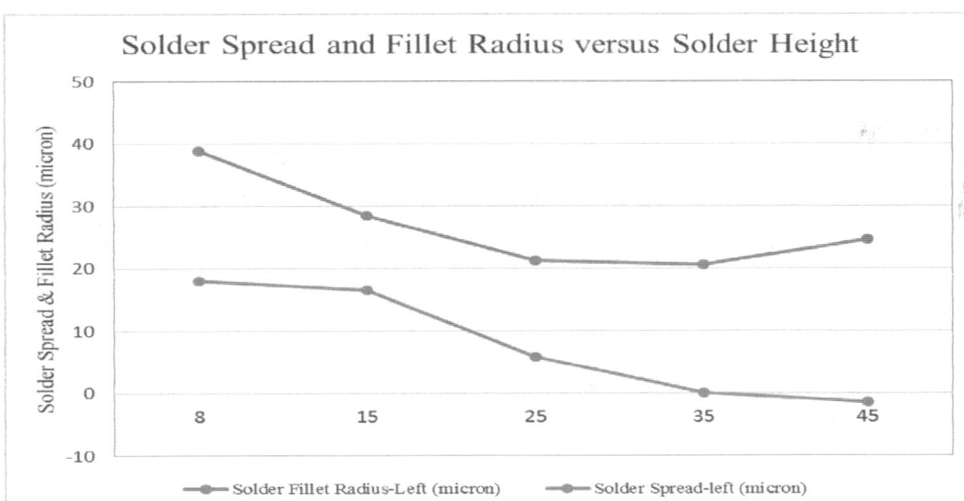

Figure 5. Solder spread and fillet radius at different solder joint height with constant solder volume.

Figure 6 shows the effect of die bump/copper pad width ratio and different solder volume towards solder joint geometry. When the die bump/copper pad width ratio increased, the solder shape changed from inwards solder fillet to outwards fillet like barrel shape. When the volume of solder decreased, necking started to appear. Critical volume for a cylindrical shape from the calculation was 3.08×10^3 mm^3 for 8 micron solder joint height and 0.8 ratio of die bump to copper pad width. However, there is still minor necking on the shape as it is due to surface tension effect on copper pad where some of solder volume was spread out. From the modelling, critical volume was found at 3.10×10^3 mm^3 about 0.65% delta. It has shown that the model established was accurate.

Die bump to copper pad width ratio	Solder Height (micron)	Vol ($8.98x10^{-5}$ mm^3)	Vol ($5.63x10^{-5}$ mm^3)	Vol ($3.08x10^{-5}$ mm^3)
2.0	8	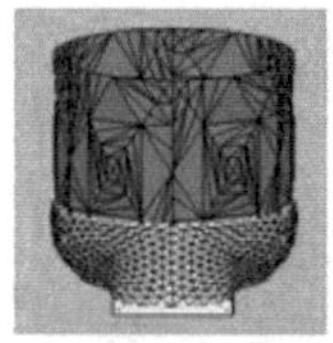	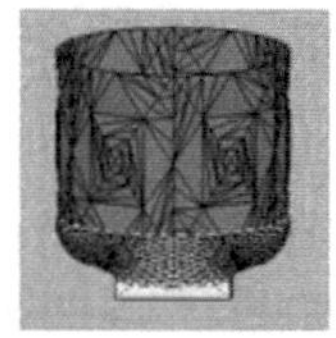	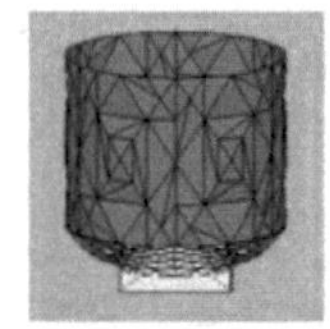
1.3	8	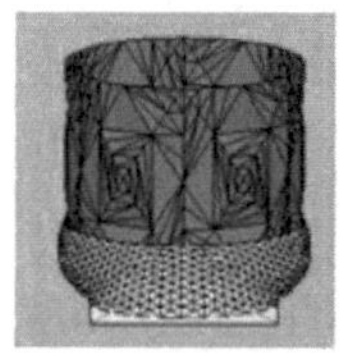	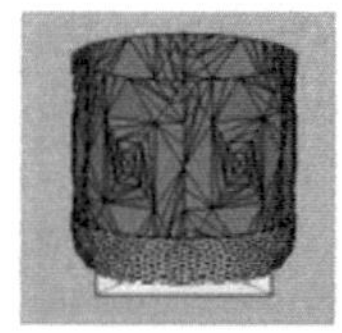	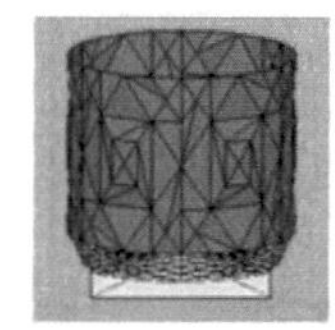
0.8	8	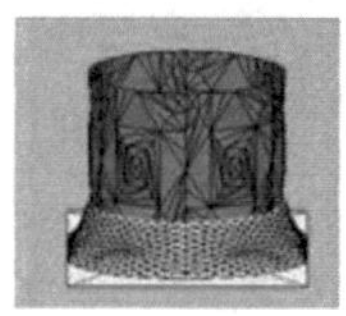		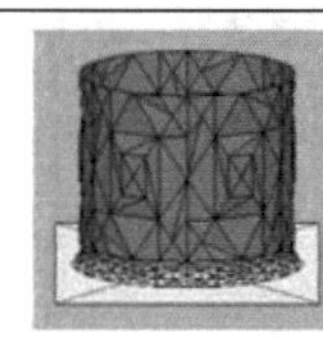

Figure 6. Solder shape geometry at different copper pad with and solder volume

It was reported that 35 micron solder cap ($8.98x10^{-5}$ mm^3) and 85 micron pad width showed better stability from 1100h High Temperature Storage (HTS) test experiment at 175 ^{0}C compared to 25 micron ($5.63x10^{-5}$ mm^3) [5]. This was due to larger solder thickness to allow better creep and thermomechanical stress. The experimental data [5] has shown that smaller die bump to copper pad width ratio with solder cap 35 micron is preferred. To cater for smaller pitch of first-level interconnect, reducing copper pillar diameter to UBM ratio is preferred as it helps to reduce the stress on low-k electric layer. As a result, smaller space between 2 adjacent solder joints. It is preferred to have smaller die bump to copper pad width ratio from assembly process perspective since higher die bump to copper pad width ratio has increased the chances of solder bridging between 2 adjacent solder joint interconnect. It was due to the solder geometry that tends to be barrel shape when die bump to copper pad width increased which create an easier path for solder bridging to adjacent trace or bump. Surface evolver can be used to estimate the bump shape given a known die bump to copper pad width ratio and solder volume which will help to decide the critical distance required for trace to trace spacing.

4. Conclusion

From the simulation work, the solder joint shape of the copper pillar solder cap onto copper pad was successfully predicted and agreed to publish experimental data. The validated model was extrapolated to predict the relationship of die bump/pad width ratio with different solder volume. Higher solder volume is required for lower die bump/pad width ratio. It is found that critical volume from this simulation work was 0.65% delta from the calculation and extrapolation model is deemed reliable.

References

[1] Yeoh A, Chang M, Pelto C, Huang TL, Balakrishnan S, Leatherman G 2006 *Electronic Components and Technology Conference* 1613
[2] Syed A, Dhandapani K, Nicolls L, Moody R, Berry CJ and Darveaux R 2010 *IMAPS Device Packaging Conference* 169
[3] Gerber M, Beddingfield C, and O'Connor S 2011 *Electronic Components and Technology Conference* 612
[4] Brakke K A 2013 *Surface Evolver Manual Version 2.70*
[5] Ebersberger B, Lee C 2008 *Electronic Components and Technology Conference* 62
[6] Huang YW, Zhan CJ and etc 2014 *Electronic Components and Technology Conference* 1911
[7] Wang YW, Lu KH, Jay Im and Ho Paul S 2010 *Electronic Components and Technology Conference*
[8] Eslampour H, Joshi M, Kang KT, Bae H, Kim YC 2012 *IEEE* 905

IOP Conf. Series: Journal of Physics: Conf. Series **1082** (2018) 012056 doi:10.1088/1742-6596/1082/1/012056

WORM Memory Effects in Gold Nanoparticles Loaded Aloe Vera Gel Films

Z.X. Lim[1], K.Y. Cheong[1]

[1]Electronic Materials Research Group, School of Materials & Mineral Resources Engineering, Universiti Sains Malaysia, Engineering Campus, 14300 Nibong Tebal, Penang, Malaysia

Email: srcheong@usm.my

Abstract. A bio-organic memory device based on a thin layer of gold nanoparticles (AuNPs) loaded *Aloe vera* gel is demonstrated. The device functions in analogous to an electric fuse, in which the irreversible dielectric breakdown in the *A. vera* gel film causes permanent switching from a high-resistance state to a low-resistance state. The concentrations of AuNPs loaded in the *A. vera* gel film can alter its optical bandgap, thereby affecting the device operational properties. The demonstration further expands the electronic application of naturally occurring *A. vera* gel.

1. Introduction

Rapid changing consumer behaviors and growing environmental awareness have driven a paradigm shift in the way natural resources are being utilized. The immediate impact would be the ongoing dominance of inorganic materials, notably Si, Ge, and GaAs, for a wide variety of electronic applications. Technological evolution over the past decades has enriched mankind with various electronic devices that are being massively produced at rock bottom costs. Consequently, the useful lifespan electronic devices are getting shorter, as they are considered obsolete when new devices with more advanced functionality start to make inroads to the market [1-3]. This planned obsolescence of electronic devices will eventually lead to the sustainability issues when massive amounts of electronic waste are being piled up at landfill sites around the world [4-6]. The impending Internet of Things that encompasses almost every aspects of daily lives requires new functional materials which are biocompatible, biodegradable, renewable, energy-efficient, and low-cost to ensure a sustainable ecosystem within the electronics manufacturing industry.

The paradigm shift toward sustainable electronics has opened up new opportunities for bio-organic materials [7-11], which are generally derived from living or once-living organisms that are found in abundance in the nature. These materials are not only renewable, biodegradable, biocompatible, and environmentally friendly for addressing the environmental issues concerning with electronic waste hazards, but also exotic for next-generation electronics realization. One such natural bio-organic materials is the gel extracted from *Aloe barbadensis* Miller (*Aloe vera*), a succulent plant that has been widely cultivated throughout the world. While *A. vera* gel is commonly used for various medical, cosmetic, and nutritional purposes [12-15], its solid-state electronic application remains largely unexplored. *A. vera* plant has been demonstrated to show responses toward various electrical signals *in vivo* [16-18], making it a promising candidate for utilization in solid-state

IOP Conf. Series: Journal of Physics: Conf. Series **1082** (2018) 012056 doi:10.1088/1742-6596/1082/1/012056

electronics. Our research group has attempted to develop solid-state electronics on the basis of *A. vera* gel thin films as both passive and active layers [19-26].

Here, we demonstrate write-once-read-many (WORM) memory effects in a simple metal-insulator-metal (MIM) device with *A. vera* gel film functioning as the middle insulating layer. The device works in analogous to an electric fuse. Owing to the dielectric properties of the *A. vera* gel film, the pristine device is in its high-resistance state (HRS). Upon undergoing dielectric breakdown, the device is electrically switched to a low-resistance state (LRS). Because dielectric breakdown is an irreversible process, a recovery back to the initial HRS is impossible. The effect of gold nanoparticles (AuNPs) loading in the *A. vera* gel film on the device operational properties is systematically studied.

2. Experimental Procedure

A. vera gel was extracted from *A. vera* leaves measuring about 30–50 cm in length cut from a 3-year-old plant that is cultivated in the garden of School of Materials & Mineral Resources Engineering, Universiti Sains Malaysia Engineering Campus, Penang, Malaysia (Figure 1). Freshly harvested leaves were washed thoroughly with flowing distilled water to remove surface dirt and sap. After removing the spikes along the margins, the epidermis of the leaves was separated from the inner parenchyma using a knife and spoon. The epidermis was discarded, while the parenchyma were washed with distilled water to remove exudate before being homogenized in a food processor (Philips, The Netherlands) at low speed for about 5 min. Then, the homogenized *A. vera* slurry was centrifuged (Rotina 38, Hettich, Germany) at 3,000 rpm for 30 min to separate heavy fibers from the supernatant, which contains the *A. vera* gel. The *A. vera* gel was recovered and vacuum filtered through a filter paper in a sintered glass funnel. The centrifugation and vacuum filtration steps were repeated for at least 10 times to obtain the fiber-free *A. vera* gel for use in subsequent device fabrication.

The device has a simple metal-insulator-metal (MIM) structure, which were fabricated bottom-up according to the following procedure. First, a glass slides was cut into a dimension of 1.5 cm × 1.5 cm for use as the substrates, which were cleaned sequentially in an ultrasonic bath containing acetone (CMOS grade, J.T. Baker), ethanol (95 v/v%, Merck Millipore), and deionized water for 30 min each. Then, 10 mL *A. vera* gel was added with different volumes of (0.2, 0.5, 0.7, and 1 mL) gold nanoparticles (AuNPs) that was purchased from Sigma-Aldrich (Cat. No. 741949-100ML, 5 nm, OD = 1). The mixture was homogenized ultrasonically and was spin-coated onto the glass substrates pre-coated with a layer of 100-nm Al. The *A. vera* gel-AuNPs film was dried subsequently in an oven (UNB100, Memmert, Germany) at 90°C for 5 min before an array of 100-nm Al as the top electrodes was thermally evaporated (K950X, Quorum Emitech, UK) at a base pressure of ~10^{-4} Torr via a metal mask (feature size 1 mm in diameter).

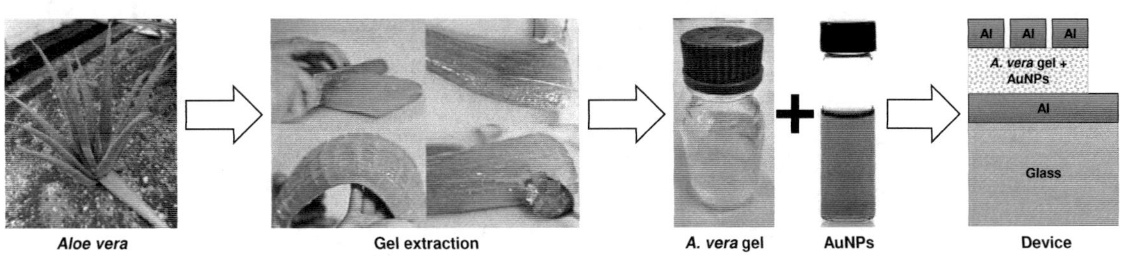

Figure 1. The workflow from *A. vera* gel extraction to device fabrication.

The surface morphology and thickness of the *A. vera* gel-AuNPs film were characterized using scanning electron microscopy (SEM; Supra 35VP, Zeiss, Germany). The optical properties of the A. vera gel-AuNPs film were studied using UV-visible spectroscopy (Cary 50, Varian, USA). The current–voltage (*I–V*) characteristics of the device was measured using a semiconductor parameter analyzer (4156C, Agilent Technologies, USA).

3. Results and discussion

Figure 2 shows the surface morphogy of the *A. vera* gel film loaded with different concentrations of AuNPs. The *A. vera* gel film forms a flat and smooth surface when being deposited on the glass substrate (Figure 2a). However, the surface morphology becomes bumpy with the addition of different volumes (0.2–1.0 mL) of AuNPs (Figure 2b–e). The uneven surface structures mimick the shape of nanoparticles with diameters ranging 3.5–18.5 nm. Obviously, the AuNPs are well-dispersed throughout the film as there is no indication of agglomeration. It has been reported that *A. vera* gel film loaded with nanoparticles demonstrated better insulating properties that render its application as the dielectric layer in electronic devices [19-21,26]. The addition of nanoparticles can result in the creation of defects in the form of grain boundaries, local strains, impurities, broken or cross-linked biopolymer chains. Consequently, more localized states serving as electronic charge traps present in the film leading to better insulating properties.

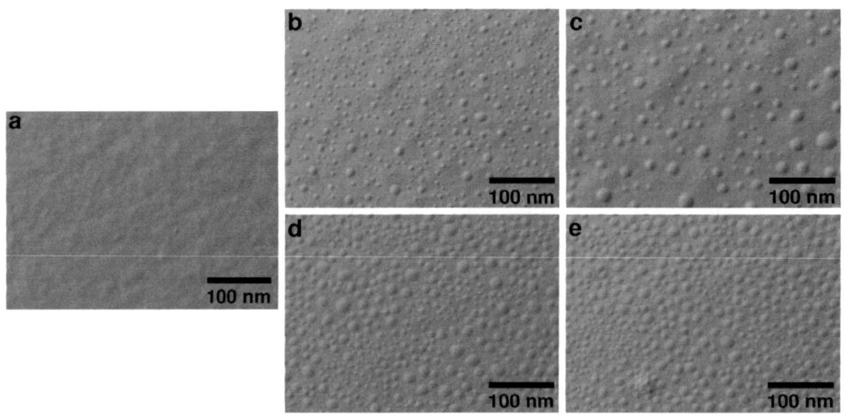

Figure 2. Surface morphology of the *A. vera* gel film loaded with (a) 0 mL, (b) 0.2 mL, (c) 0.5 mL, (d) 0.7 mL, and (e) 1.0 mL of AuNPs.

Figure 3 shows the cross-sectional SEM images of the *A. vera* gel film loaded with different amount of AuNPs on a glass substrate. The thickness of each film is measured from the SEM images and is summarized in Table 1. The addition of AuNPs appears to have no effect on the film thickness, as there is only a difference of 12.3 nm between the thickest (*A. vera* gel + 0.5 mL AuNPs) and thinnest (*A. vera* gel + 1.0 mL AuNPs) films.

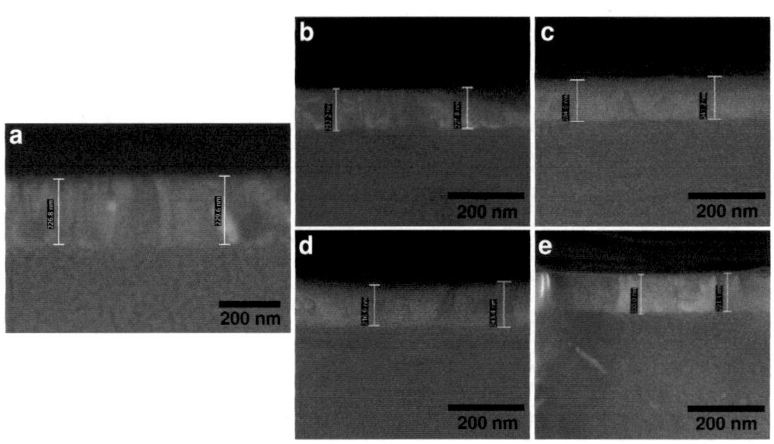

Figure 3. Cross-sectional SEM images showing thickness of the *A. vera* gel film loaded with (a) 0 mL, (b) 0.2 mL, (c) 0.5 mL, (d) 0.7 mL, and (e) 1.0 mL of AuNPs.

IOP Conf. Series: Journal of Physics: Conf. Series **1082** (2018) 012056 doi:10.1088/1742-6596/1082/1/012056

Table 1. Thickness of the *A. vera* gel film loaded with different amount of AuNPs.

Sample	Thickness 1 (nm)	Thickness 2 (nm)	Average Thickness (nm)
A. vera gel	226.8	229.6	228.2
A. vera gel + 0.2 mL AuNPs	232.2	227.8	230.0
A. vera gel + 0.5 mL AuNPs	234.5	241.2	237.9
A. vera gel + 0.7 mL AuNPs	216.6	243.4	230.0
A. vera gel + 1.0 mL AuNPs	230.0	221.1	225.6

Figure 4a shows the current–voltage (*I–V*) characteristics of the device measured using a single DC voltage sweep from 0 to 15 V at 0.01 V/step under ambient conditions. Notice the occurrence of dielectric breakdown in the *A. vera* gel film without AuNPs loading at 9.36 V, corresponding to an electric field strength of (9.36 V)/(228.2 nm) = 0.41 MV/cm. This breakdown voltage shifts to higher values, *i.e.* 9.93 V and 10.34 V, when the AuNPs loading is increased to 0.2 mL and 0.5 mL respectively. However, as the AuNPs loading in the film is further increased to 0.7 mL and 1.0 mL, the breakdown voltage is shifted toward lower values, *i.e.* 8.76 V and 7.83 V, respectively. Figure 4b shows the UV-visible spectra of the *A. vera* films with different AuNPs loading. Apparently, the *A. vera* gel films display UV opacity attributes, due to their high absorbance of wavelengths that fall within the range of 200–400 nm. Their optical energy bandgap can be extracted from the UV-visible spectra (Figure 4b) using Tauc's relation: $(\alpha hc/\lambda)^{1/2} \propto (hc/\lambda)$, where α, h, c, and λ refer to the absorbance, Planck's constant, speed of light, and wavelength respectively. Figure 4c presents the optical bandgap of the *A. vera* gel films as a function of AuNPs loadings. The maximum optical bandgap is achieved by the *A. vera* gel film loaded with 0.5 mL AuNPs. As the optical bandgap increases, the breakdown voltage is shifted to a higher value.

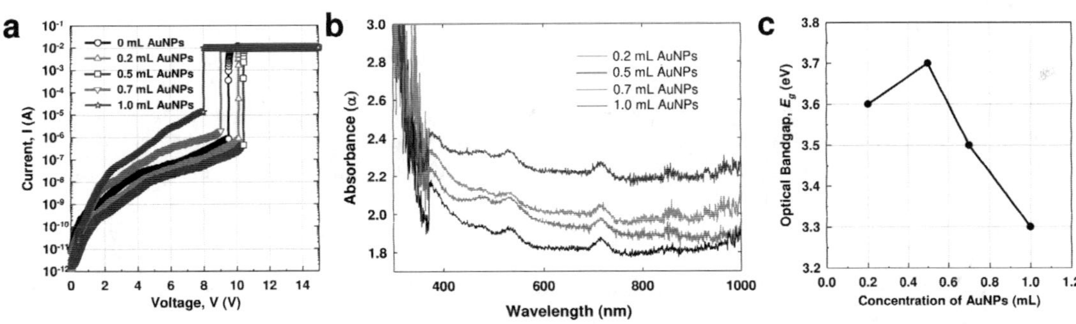

Figure 4. (a) *I–V* characteristics of the device. (b) UV-visible spectra of *A. vera* gel films loaded with AuNPs. (c) Optical bandgap of the film extracted from the UV-visible spectra.

4. Conclusion

WORM memory effects in AuNPs loaded *A. vera* gel films are demonstrated using a simple MIM device structure. The AuNPs loading modifies the surface morphology of the *A. vera* gel film so that it becomes rough and uneven, but it does not affect the film thickness. The breakdown voltage increases when 0.2–0.5 mL of AuNPs is loaded into the *A. vera* gel film. Conversely, the breakdown voltage decreases when 0.7–1.0 mL of AuNPs is loaded. The AuNPs loading can alter the optical bandgap of the *A. vera* gel film causing the changes in breakdown voltage.

Acknowledgements

This work was financially supported by the Malaysian Ministry of Higher Education (MOHE) and Universiti Sains Malaysia (USM) under the Fundamental Research Grant Scheme (203/PBAHAN/6071301), Research University–Individual Grant (1001/PBAHAN/814216), and Postgraduate Research Grant Scheme (1001/PBAHAN/8036018). ZX Lim gratefully acknowledges MOHE for supporting his graduate study in USM under MyBrain15–MyPhD scholarships program. The authors acknowledge the financial aid provided by the School of Materials & Mineral Resources Engineering, USM to attend RCM & AMC 2017.

References

[1] Bauer S and Kaltenbrunner M 2014 *ACS Nano* 8 5380
[2] Geyer R and Blass VD 2010 *International Journal of Advanced Manufacturing Technology* 47 515
[3] Robinson BH 2009 *Science of The Total Environment* 408 183
[4] Irimia-Vladu M, Glowacki ED, Voss G, Bauer S and Sariciftci NS 2012 *Materials Today* 15 340
[5] Savage N 2015 *IEEE Spectrum* 52 18
[6] Cucchiella F, D'Adamo I, Koh SCL and Rosa P 2015 *Renewable and Sustainable Energy Reviews* 51 263
[7] Mühl S and Beyer B 2014 *Electronics* 3 444
[8] Irimia-Vladu, Troshin PA, Reisinger M, Shmygleva L, Kanbur Y, Schwabegger G, Bodea M, Schwödiauer R, Mumyatov A, Fergus JW, Razumov VF, Sitter H, Sariciftci NS and Bauer S 2010 *Advanced Functional Materials* 20 4069
[9] Irimia-Vladu M, Sariciftci NS and Bauer S 2011 *Journal of Materials Chemistry* 21 1350
[10] Raeis-Hosseini N and Lee JS 2017 *Journal of Electroceramics* 39 223
[11] Jung YH, Chang TH, Zhang H, Yao C, Zheng Q, Yang VW, Mi H, Kim M, Cho SJ, Park DW, Jiang H, Lee J, Qiu Y, Zhou W, Cai Z, Gong S and Ma Z 2015 *Nature Communications* 6 7170
[12] Eshun K and He Q 2004 *Critical Reviews in Food Science and Nutrition* 44 91
[13] Rodríguez RE, Martín JD and Romero CD 2010 *Critical Reviews in Food Science and Nutrition* 50 305
[14] Hamman J 2008 *Molecules* 13 1599
[15] Capasso F, Borrelli F, Capasso R, Di Carlo G, Izzo AA, Pinto L, Mascolo N, Castaldo S and Longo R 1998 *Phytotherapy Research* 12 S124
[16] Volkov AG, Reedus J, Mitchell CM, Tucket C, Forde-Tuckett V, Volkova, MI, Markin VS and Chua L 2014 *Plant Signaling and Behavior* 9 e29056
[17] Volkov AG, Nyasani EK, Tuckett C, Scott JM, Jackson MMZ, Greeman EA, Greenidge AS, Cohen DO, Volkova MI and Shtessel YB 2017 *Bioelectrochemistry* 113 60
[18] O'Neal L, Ebere LC, McIntyre R, Volkova-Gugeshashvili MI, Markin VS and Volkov AG 2013 *ECS Transactions* 50 7
[19] Khor LQ and Cheong KY 2013 *Journal of Materials Science: Materials in Electronics* 24 2646
[20] Khor LQ and Cheong KY 2013 *Advanced Materials Research* 858 74
[21] Khor LQ and Cheong KY 2013 *ECS Journal of Solid State Science and Technology* 2 P440
[22] Lim WF, Quah HJ, Sreenivasan S and Cheong KY 2015 *Materials Technology: Advanced Performance Materials* 30 A29
[23] Lim ZX and Cheong KY 2015 *Physical Chemistry Chemical Physics* 17 26833
[24] Lim ZX, Sreenivasan S, Wong YH, Zhao F and Cheong KY 2016 *MRS Advances* 1 2513
[25] Lim ZX, Sreenivasan S, Wong YH, Zhao F and Cheong KY 2017 *Procedia Engineering* 655
[26] Lim ZX, Sreenivasan S, Wong YH and Cheong KY 2017 *AIP Conference Proceedings* 1865 050006

IOP Conf. Series: Journal of Physics: Conf. Series **1082** (2018) 012057 doi:10.1088/1742-6596/1082/1/012057

Numerical Investigation on the Effect of Squeegee Angle during Stencil Printing Process

M.S. Rusdi[1], M.Z. Abdullah[2], M.S. Abdul Aziz[1], M.K. Abdullah[3], A.A. Bakar[3], M.H.S. Abd Samad[4], P. Rethinasamy[4], Sivakumar Veerasamy[4], C.Y. Khor[5]

[1]School of Mechanical Engineering, Universiti Sains Malaysia, Engineering Campus, 14300 NibongTebal, Penang, Malaysia.

[2]School of Aerospace Engineering, Universiti Sains Malaysia, Engineering Campus, 14300 NibongTebal, Penang, Malaysia.

[3]School of Materials and Mineral Resources Engineering, Universiti Sains Malaysia, Engineering Campus, 14300 NibongTebal, Penang, Malaysia.

[4]Celestica Malaysia Sdn. Bhd., Plot 15, Jalan Hi-Tech 2/3 Phase I, Kulim Hi-Tech Park,09000 Kulim, Malaysia.

[5]Faculty of Engineering Technology-Mechanical Department, Universiti Malaysia Perlis (UniMAP), Aras 1, Blok S2, Kampus UniCITI Alam, Sungai Chuchuh, Padang Besar, 02100 Perlis, Malaysia

Email: msharizal@usm.my

Abstract. The high requirement of smaller size, lighter weight, and high performance Printed Circuit Board (PCB) in electronic packaging has contributed to the wide application of stencil printing for soldering process. However, during stencil printing stage contributes major concern compared to other stages in Surface Mount Technology (SMT). Unsuitable process parameters can cause the soldering defects that can lead to product failure in further processes in the production line. An investigation has been conducted to predict the real-time observation of solder paste Sn96.5Ag3.0Cu0.5 (SAC305) filling process into stencil apertures as well as print quality in stencil printing by using Computational Fluid Dynamics (CFD) approach. A 3-Dimensional stencil printing model was developed and simulated in ANSYS Fluent 17 of different angles. It is found that squeegee angle 60° to 80° has potential to obtain the good print quality of solder paste.

IOP Conf. Series: Journal of Physics: Conf. Series **1082** (2018) 012057 doi:10.1088/1742-6596/1082/1/012057

1. Introduction

This Stencil printing is a process of depositing a viscous fluid which is solder paste onto a bare substrate such as PCB through the aperture openings of a stencil. It is originally adapted from the screen printing process. This process is one of the widely used methods of soldering in SMT today as a solution for higher pin count and fine pitch size. This is due to the high requirement of smaller size, lighter weight, and high-performance PCB in electronic packaging. The use of stencil printing for the deposition of conductive interconnects was developed during the late 1960s as companies looked to increase the densities of the products, improve assembly time and drive down production costs. This process was termed surface mount assembly (SMA) and is currently used at least 70% of electronic packaging nowadays. However, stencil printing is identified as the largest contributor to soldering defect by 60% averagely [1].

Numerical studies have been performed to simulate stencil printing process by using CFD approach at macroscopic and microscopic scales by using a 2-D model that involves Finite Volume Method (FVM), Lattice-Boltzmann Method (LBM), and Discrete Element Method (EDM). Unfortunately, the volume fraction of solder paste is far lower than normal due to the unrealistic representation of spherical particles by a 2-D cylinder that causes artificial blockage of fluid passage at greater volume fraction [2].

Neural network approach model has been proposed to solve the fine pitch printing quality issue of the non-linear behavior stencil printing through the volume prediction of solder paste deposits. The proposed approach is found to be an effective way to predict and control the printing quality in SMA such as paste volume deposited on PCB pad with small prediction errors which are less than 7% [3]. In stencil printing, the stencil experiences pressure exerted by the squeegee on its surface that might affect the behavior of the structure. The deformation behavior of the stencil which can affect printing quality has been investigated by using Finite Element Method (FEM) associated with the uneven surface of Printed Wiring Circuit (PWB). Higher surface gap differences between the stencil and PWB affect the stencil bending behavior that can lead to excessive solder paste deposits and solder bridges issue [4].

Moreover, the effect of squeegee speed variation and solder paste with different density on stencil printing process has been studied by using CFD approach through the flow characteristic of solder paste simulation. Based on the study, the increment of squeegee speed can cause the shear stress increase proportionally to the increment of solder paste's shear strain [5]. Therefore, in this study a 3-Dimensional stencil printing model has been simulated numerically and analyze in order to solve the complex and unpredictable printing problems in SMT.

2. Methodology

The A stencil printing model is created in 3-Dimensional by using ANSYS Design Modeler. It consists of two significant components for the simulation which are aperture bodies and printing body as shown in Figure 1. The thickness of the aperture bodies is 0.1 mm as well as the thickness of the stencil. The angles tested in the simulation are 50°, 60°, 70°, and 80°. The angles is varied to create different pressure distribution along the stencil [6]. The specification of the selected apertures design is 225 I/O; 1.5 mm pitch.

IOP Conf. Series: Journal of Physics: Conf. Series **1082** (2018) 012057 doi:10.1088/1742-6596/1082/1/012057

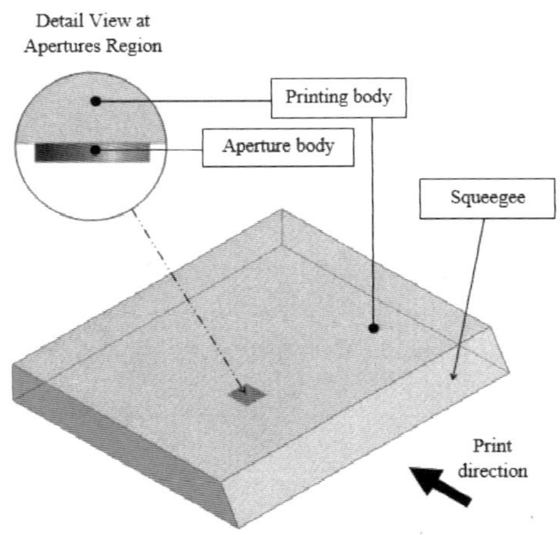

Figure 1. 3-Dimensional stencil printing model

ANSYS Workbench meshing is used to mesh the model and name their boundary conditions accordingly (Figure 2). The filling of solder paste inside the solder deposit bodies due to motion of the angled squeegee is simulated in ANSYS Fluent 17. The element size of 2.8 mm is selected from several of grid dependency test. The model is run with transient calculation in the simulation. The Volume of Fluid (VOF) multiphase model is selected with the implicit formulation due to presence of air and solder paste, which is Sn96.5Ag3.0Cu0.5 (SAC305).

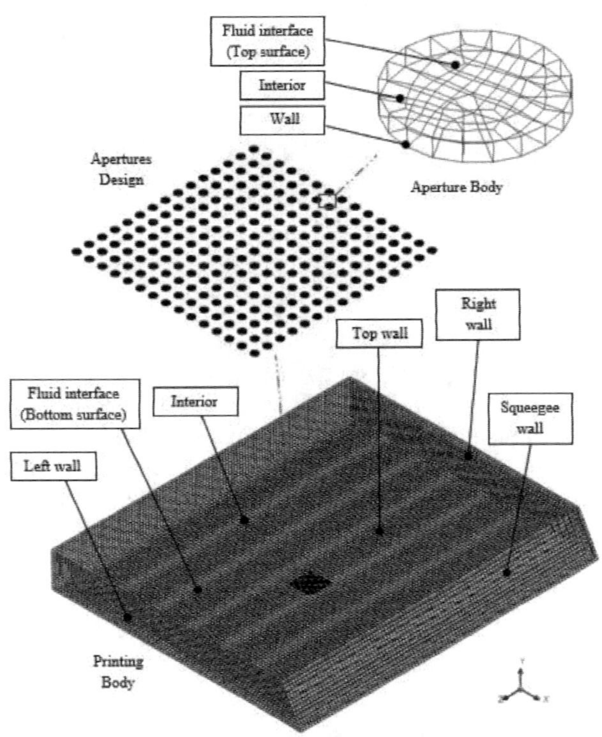

Figure 2. FLUENT structured meshed model and boundary conditions

3. Results and discussions

The flow front and filling of the solder paste into the apertures is observed with respect to the different angle of squeegee at constant print speed, 35 mm/s. The average volume fraction of solder paste inside the apertures through filling stages is shown in Figure 3. At 25 % filling stage of apertures, the filling pattern among the squeegee angles was shown significantly on volume fraction. 50° squeegee angle has the highest volume fraction, while 80° squeegee angle has the lowest average volume fraction. The trend of the filling continues to 50% filling stage but started to have approximately the same average volume fraction at 75%. All squeegee angles achieved 100% filling stage except for 50° squeegee. However, a slight drop of average volume fraction has been observed due to the presence of unfilling aperture holes.

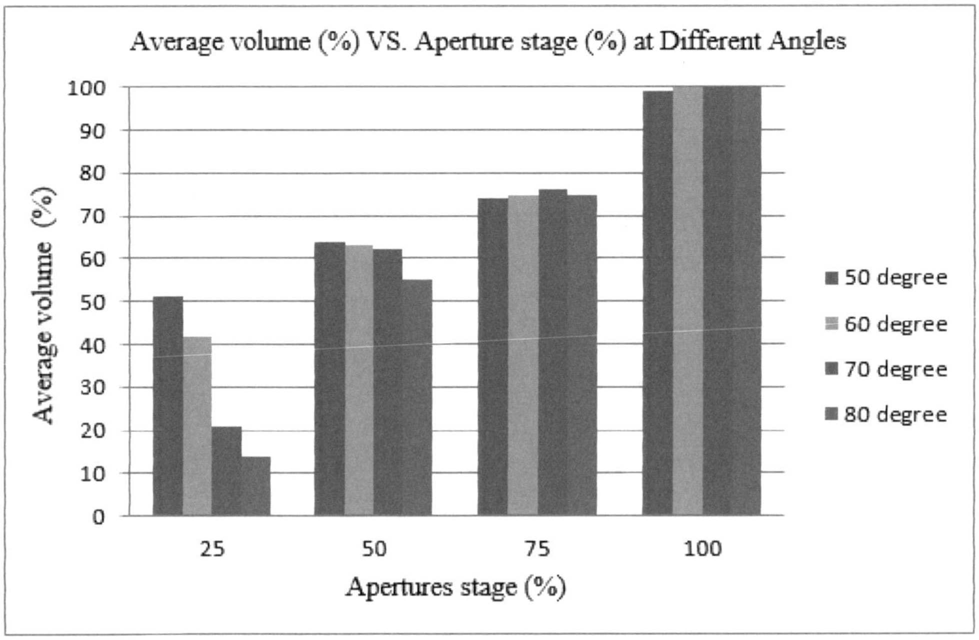

Figure 3. Average volume (%) over apertures stages (%) at different squeegee angles, θ

Figure 4 shows the SAC305 flow front at different squeegee angles and stages. Different squeegee angle has different solder paste pressure distribution along the stencil. Lower squeegee angle have higher solder paste pressure distribution [6]. The solder paste pressure is important for the filling of the aperture. With given same volume of solder paste, lower squeegee angle have bigger contact area between the solder and the stencil, hence it is found that the solder paste fill the aperture faster. Higher pressure distribution and higher contact area of the solder cause the 50° angle to fill faster than 80°.

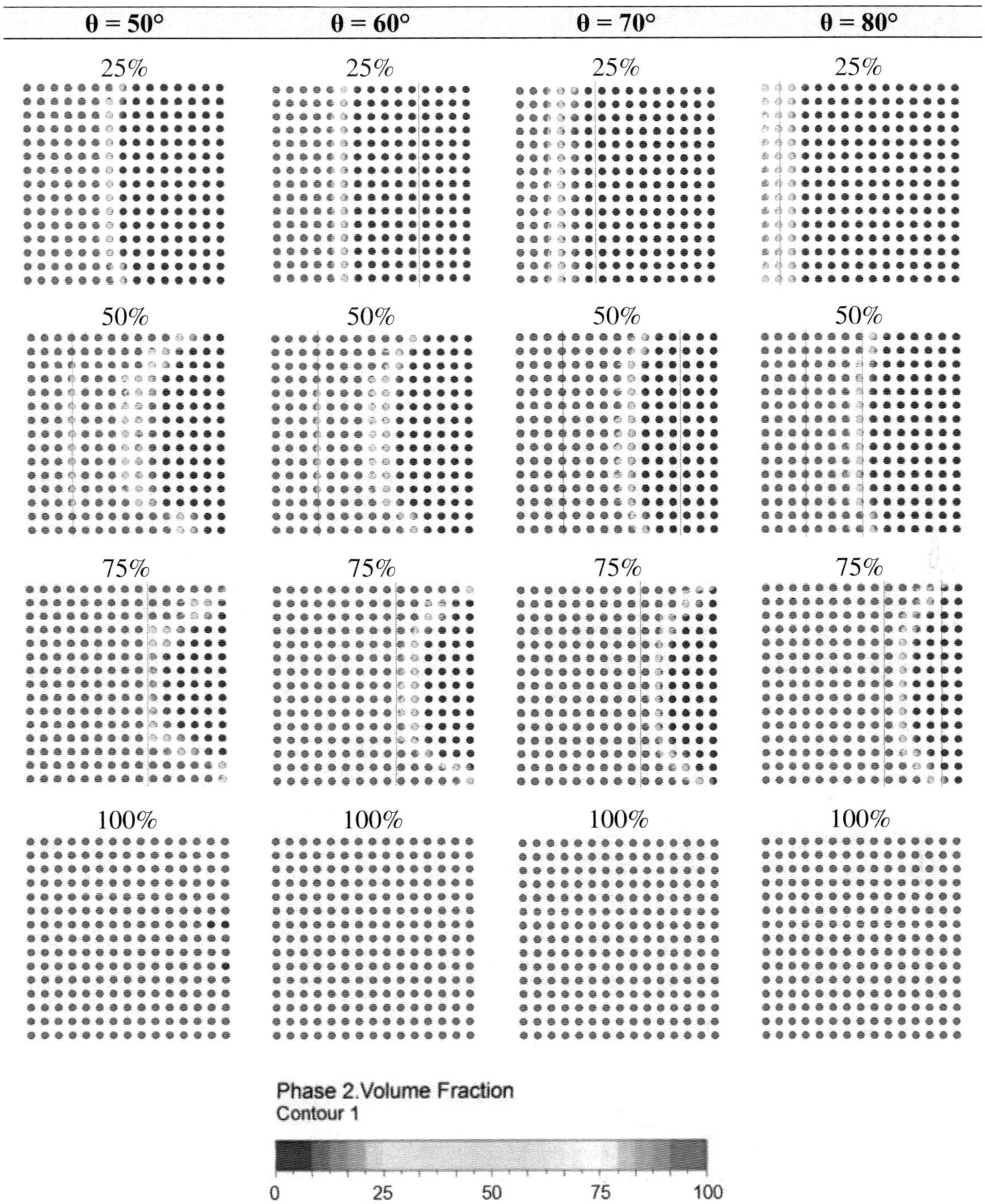

Figure 4. SAC305 flow front by stages at different squeegee angles, **θ** in apertures

IOP Conf. Series: Journal of Physics: Conf. Series **1082** (2018) 012057 doi:10.1088/1742-6596/1082/1/012057

4. Conclusion

The 3-Dimensional stencil printing modelling by using CFD approach shows good prediction on the stencil printing performance. All squeegee angles were successfully filled all the apertures with solder paste except for 50° squeegee angle. Thus, operating stencil printing process by using squeegee angle at 60° to 80° is possible to achieve good solder paste print quality.

References

[1] T. N. Tsai 2008 *Comput. Ind. Eng* 54 374
[2] G. P. Glinski, C. Bailey, and K. Pericleous 2000 *International Symposium on Electronic Materials and Packaging (EMAP2000)* 364
[3] T. Yang, T. N. Tsai, and J. Yeh 2005 *Eng. Appl. Artif. Intell* 18 335
[4] O. Krammer, L. M. Molnár, L. Jakab, and A. Szabó 2012 *Microelectron. Reliab* 52 235
[5] V. Thakur, S. Mallik, and V. Vuppala 2015 *Int. J. Recent Adv. Mech. Eng* 4 1
[6] O. Krammer 2015 *38th International Spring Seminar on Electronics Technology (ISSE)* 343

IOP Conf. Series: Journal of Physics: Conf. Series **1082** (2018) 012058 doi:10.1088/1742-6596/1082/1/012058

Influence of Grain Refinement via ECAP to the Corrosion Behaviour of SAC 305 Solder

Nur Adriana Nazifa A.B. [1,a], **Tuti Katrina A.**[1,b], **Nurulakmal M.S.** [1,c*]

[1]Universiti Sains Malaysia, Engineering Campus, Penang, Malaysia

Email: a) adriana.nazifa11@gmail.com, b) tutikatrina@usm.my,
c) *srnurul@usm.my

Abstract. Equal Channel Angular Pressing or ECAP has been generating interest due to the ability to form bulk ultra-fine grains (UFG) or nanostructured materials. This formation of ultra-fine grains leads to a favorable combination of strength and ductility. Solder materials, which function as the mechanical bridge joining chip to substrate, or package to mother board, could very well benefit from this combination. High strength would offer better joint, and enhanced ductility would enable solder to have better fracture toughness and drop-impact resistance as solder joint regularly being subjected to vibration and impact. This paper reports about corrosion behaviour of ECAPed SAC 305 alloy formed via route Bc up to four passes. The focus is on the effect of grain refinement on corrosion behaviour of bulk solder in 3.5 wt.% NaCl solution. The resulting microstructure was observed by Scanning Electron Microscope (SEM). The grains formed after different passes of ECAP were observed, along with the distribution of intermetallic compound or IMC within the bulk solder. Finer grains were formed with increasing number of ECAP passes leading to reduction of corrosion rate and increased strength of SAC305. The high fraction of grain boundaries in the microstructure with increased of ECAP passes reduced the corrosion rates by accelerating the passivation process.

1. Introduction

Solder is widely used to connect chips to their packaging substrates in flip chip technology as well as in surface mount technology [1]. Among the alternative lead-free solders, Sn-Ag-Cu (SAC) alloy is seen as the most likely candidate due to their low melting temperature (~217°C), better solderability, good mechanical properties and slower growth of the intermetallic at the interface compared with Sn-Pb and Sn-Ag solder alloys [2]. As solder joint serves as mechanical and electrical connection, reliability of the joint is critical for the performance of the device. Exposure to corrosive environment is one of the risk to reliability especially if the device or electrical component is used in aggressive environment such as high humidity or in coastal atmosphere [3]. Mohanty and Lin (2008) studied corrosion behaviour of Pb-free Sn-8.5Zn-0.05Al, Sn-8.5Zn-0.05Al-0.02Ga, Sn-8.5Zn-0.05Al-0.05Ga, Sn-8.5Zn-0.05Al-0.2Ga and Sn-3Ag-0.5Cu alloys in 3.5 wt.% NaCl solution. From this research, the corrosion potential (E_{corr}) value for Sn-8.5Zn-0.05Al-xGa alloy is shifted toward more negative values with increasing of Ga content. While, the E_{corr} value for Sn-3Ag-0.5Cu alloy is shift towards more noble values. The corrosion rate increase with increasing in Ga content from 0.02 to 0.2 wt.%. But, a significant increase in the corrosion rate is found for Sn-3Ag-0.5Cu alloy [4].

Equal-Channel Angular Pressing (ECAP) is one of the modern techniques utilizing severe plastic deformation (SPD) to fabricate bulk ultrafine-grained (UFG) and nanostructured materials that will enhance their physical, mechanical and corrosion properties [5]. This process will also alter the microstructure of Sn-Ag-Cu alloy by reducing the large dendrites into fine and equiaxed grains. Although the refined grains provide improvement in mechanical properties, the effect to corrosion behaviour of SAC may not be as positive. This is added with the possible influence of increased dislocation density and fragmentation of intermetallic particles due to ECAP process. For Sn-Ag-Cu lead-free solder, it has been found that Ag_3Sn precipitates are also present within the Sn matrix. Gong et al. (2010) have reported that during ECAP process, Ag_3Sn could be broken into small fragments and more uniformly distributed within Sn matrix which resulted in better corrosion resistance in 3.5 wt.% NaCl solution compared to SAC solder with large Ag_3Sn structure [6]. However, there was limited explanation of Sn-Ag-Cu lead-free solder on why ECAP process was able to improve the corrosion resistance. Sadawy and Ghanem, (2016) have reported the increase of ECAP number of passes increases the corrosion resistance of the bronze alloy [7]. Therefore, the present work focused on the effect of ECAP as a grain refinement technique on the microstructure evolution and corrosion behaviour of the Sn-Ag-Cu alloy attempting to study the correlation between grain refinement, fragmentation of IMC particles, dislocation density and the corrosion resistance.

2. Experimental study

This experiment was carried out using commercial SAC305 solder bar. The solder bar was cut by using horizontal band saw into smaller pieces and melted at 350°C using lab crucible furnace. The molten solder alloy was then casted in steel mold with shape and size following the ECAP mold cavity. An ECAP die with a channel intersection angle (φ) and an angle at the arc of curvature (ψ) where the two parts of the channel intersect 120° and 20° respectively was used in the ECAP process, which was estimated to yield a shear strain of approximately 1.83 for each pass. ECAP process was carried out up to 4 passes at room temperature by route Bc. During the ECAP process a layer of lubricant was applied to reduce the friction between the die and the sample. The solder alloys were cut to smaller pieces using horizontal band saw after ECAP were ground with SiC abrasive paper grit 100 until 2000 before polished using 1 μm alumina powder. Once mirror-like surface was obtained, the samples were chemically etched with 5 % HNO_3 -2 % HCl-93 % CH_4O etching solution to reveal the microstructure before observed using Scanning Electron Microscope (SEM) equipped with Energy Dispersive X-Ray(EDX).

For the electrochemical behavior, the as-cast and ECAPed SAC305 samples were tested by using potentiostat (AUTOLAB PGSTAT). A three-electrode cell composed of the sample as working electrode, Pt as counter electrode, and saturated calomel electrode (SCE) as reference electrode was used for the tests. The as-cast and ECAPed SAC305 discs with exposed surface area of 1.0 cm^2 were used as the working electrodes. The solder materials were soldered to copper wires with a plastic insulating layer. Then the samples were mounted with epoxy. Surface of samples were prepared via grinding using SiC papers from 600 to 2000 grit. Potentiodynamic polarization tests were carried out at a scan rate of 1.0 mV/min at room temperature (25 ± 0.1°C). Potentiodynamic polarization curves were recorded in the potential range -1000 to +1000 mV vs. SCE. The corrosion experiment was carried out in 3.5 wt.% NaCl solution as electrolyte. The surface morphology and phase analysis of samples after polarization was observed by SEM and XRD respectively.

3. Results and discussion

3.1 Microstructural observation

The microstructure of the as-cast and ECAPed SAC305 of 2 and 4 passes are shown in Figure 1. In Figure 1(a), the as-cast microstructure consists of light areas identified as large β-Sn dendrites and surrounded by thin, gray network of the eutectic region consisting of Cu_6Sn_5 and Ag_3Sn IMCs within Sn matrix. In the microstructure of as-cast SAC305, the needle or rod like Ag_3Sn and irregular but slightly equiaxed shaped Cu_6Sn_5 can also be found in the β-Sn dendrites. The average grain size of as-cast is 16.8 μm. While the average grain size of β-Sn structure in Figure 1(b) and Figure 1(c) were 6.61 μm and 5.51 μm respectively. The β-Sn structure has been reduced to smaller grains with thinner and more evenly distributed eutectic network following ECAP process. Besides that, Ag_3Sn needles were observed to be broken into small fragments and distributed more uniformly within the Sn matrix. Ag_3Sn refinement along with finer Sn grains are expected to lead to higher strength of ECAPed solder than the as-cast SAC305 alloy.

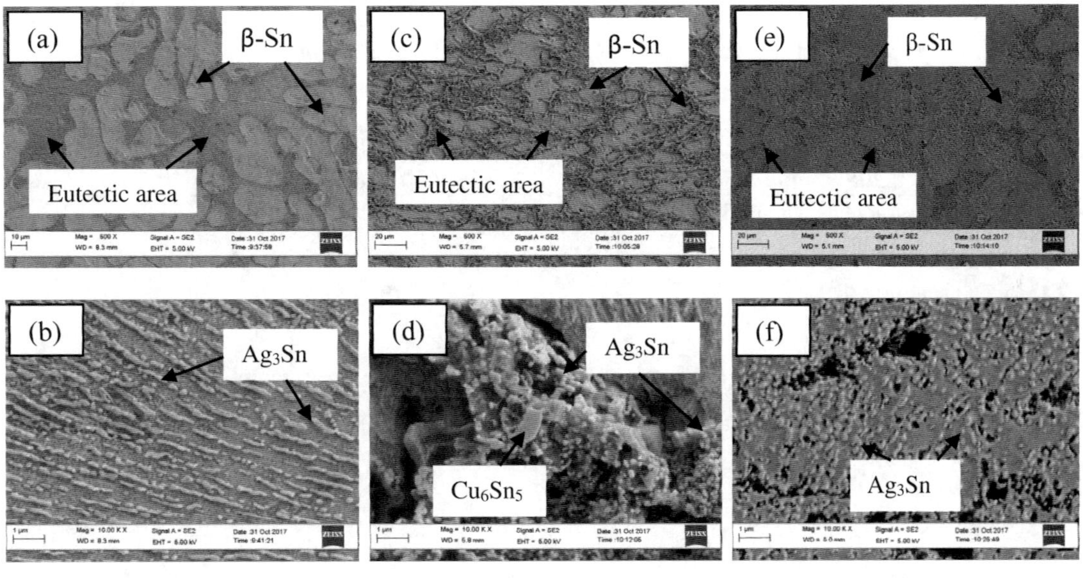

Figure 1. Bulk microstructure of SAC305 solder alloy (a)(b) as-cast, (c)(d) 2 passes ECAP and (e)(f) 4 passes ECAP.

3.2 Potentiodynamic polarization

Potentiodynamic polarization curves of the as-cast and ECAP processed samples were shown in Figure 2. There are no significant differences in polarization curves. For as-cast condition SAC305 there was a passive process beginning at -0.749 V/SCE and extended to -0.415 V/SCE. The passivation process is due to the formation of Sn oxides on the surface [8,9]. The passive behaviour after ECAP condition process of 2 passes occurred at -0.757 V/SCE until -0.397 V/SCE. Further ECAP up to 4 passes seems to enhance the passivation process of SAC305 alloy as the passive behaviour occurs at lower value of -0.743 V/SCE and extended to -0.438 V/SCE. Then, the passivation disappears when a sharp increase of anodic current density occurs due to the breakdown of passive film. As can be observe in Figure 2, the pseudopassivation range value of as-cast SAC 305 solder extends from -0.188 to +0.062 V vs. SCE. In contrast, 2 and 4 passes SAC 305 solder exhibits a pseudopassivation range extending from +0.149 to +0.199 V vs. SCE and +0.095 to +0.241 V vs. SCE respectively when surface film begins to break down in the solution and the anodic current density rises again. Due to a larger pseudopassivation range, the corrosion resistance of 4 passes SAC 305 solder is better than that 2 passes and as-cast condition. The improved passivation behaviour of ECAP processed SAC305 alloy could be due to the small grain size which promotes uniform distribution of the elements and facilitating rapid formation of the passive film [10,11]. The anodic polarization curves show that the corrosion potential (E_{corr}) shifts to more noble potential and the corrosion rate decreased from 0.047 mpy to 0.026 mpy with increasing ECAP number of passes up to 4 passes, and this is shown in Table 1. It can be observed that further grain refinement due to the increased of ECAP number of passes resulted in increased amount of grain boundaries. Therefore, the difference between the rate of anodic and cathodic reactions are expected to alter significantly which leads to more uniform attack and lower corrosion rate [12].

Table 1. Results obtained from potentiodynamic polarization measurements for the as-cast and ECAPed SAC305 solder alloy in 3.5 wt.% NaCl solution

Samples	Tafel slope anodic process (Ba) (V/dec)	Tafel slope cathodic process (Bc) (V/dec)	Ecorr (V/SCE)	Icorr (A)	Corrosion rate (mpy)	Polarization resistance (Ω)	jcorr (A/cm)
As cast	0.189	0.244	-0.749	5.71E-06	0.047	8124.9	5.71E-06
2 passes	0.206	0.254	-0.757	4.16E-06	0.033	9649.2	4.16E-06
4 passes	0.109	0.352	-0.743	3.32E-06	0.026	10867	3.32E-06

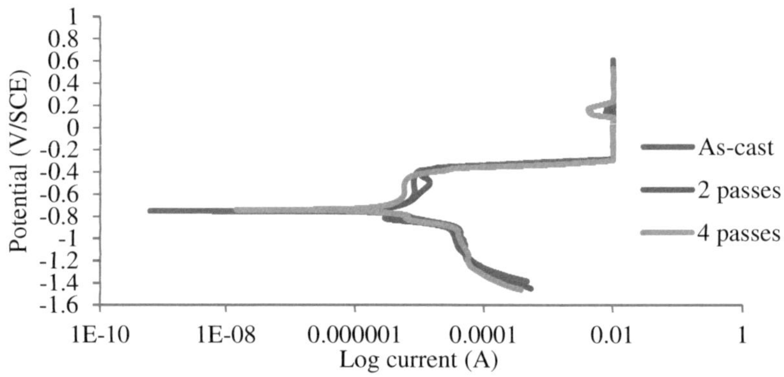

Figure 2. Tafel polarization curves of the as-cast SAC305 solder alloy and deformed by ECAP in 3.5 wt.% NaCl solution

Since all corrosion tests were performed in aerated NaCl solution, the cathodic branch of polarization curves maybe ascribed to the oxygen reduction reaction.

$$O_2 + 2H_2O + 4e^- \rightarrow 4OH^-$$

From the microstructure shown in Figure 3, it can be observed that the corrosion product on surface of SAC305 samples has needle-like shape. Pits were found on as-cast SAC305 surface as shown in Figure 3(a)(b). EDX analysis at the pit confirmed the presence of corrosion product constituted by tin oxychloride. The corrosion product film for sample with ECAP 2 passes contains many agglomerates and the ones on surface of sample with 4 ECAP passes is uniform and relatively smoother. The corrosion product on the ECAPed 4 passes SAC305 surface was observed to be more compact and better adhered thus providing better corrosion resistance. Both samples of 2 and 4 passes shown in Figure 3(c)(d) and 3(e)(f) respectively seems to have uniform corrosion product covering the surface compared to pitting on surface of SAC305 sample. This could be due to the grain refinement via ECAP process and better distribution of IMC particles that encouraged more uniform attack on the surface.

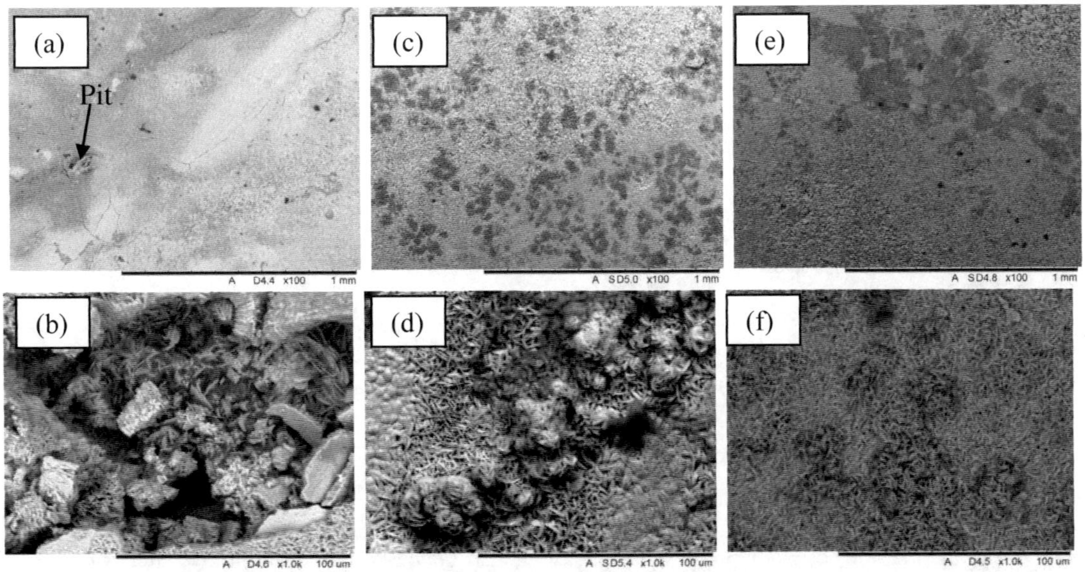

Figure 3: Microstructure of the corrosion product and EDX spectrum of solder alloy (a)(b) as-cast, (c)(d) 2 passes and (e)(f) 4passes after the completion of polarization

Figure 4 shows the XRD pattern for as-cast and ECAPed SAC305 solder alloy after electrochemical corrosion tests. Several phases are detected at the exposed surface which are $Sn_3O(OH)_2Cl_2$, SnO_2, $SnCl_2$, Cu_6Sn_5, Ag_3Sn and β-Sn. The surface is fully covered with corrosion products of which $Sn_3O(OH)_2Cl_2$ could be the dominant ones. The formation of oxides of tin may take place through the following reaction [13].

$$Sn + 2OH^- \rightarrow Sn(OH)_2 + 2e^-$$

$$Sn(OH)_2 \rightarrow SnO + H_2O$$

From the XRD pattern, the corroded product on SAC305 is confirmed to be $Sn_3O(OH)_2Cl_2$ that formed according to following reaction [14].

$$3Sn + 4OH^- + 2Cl^- + 6e^- \rightarrow Sn_3O(OH)_2Cl_2 + H_2O$$

The formation of oxides at the sample surface enhanced passivation and thereby lead to reduction of the corrosion rate. Electrochemically, Ag_3Sn is nobler than Sn. Therefore, if there is an electrochemical coupling between the large Ag_3Sn and Sn phases, large Ag_3Sn plates could acts as a cathode to drive corrosion on Sn phase. For as-cast SAC305, enhanced localized corrosion could occur because of the low area ratio of anode (Sn phase) to cathode (Ag_3Sn phase) in localized regions [15].

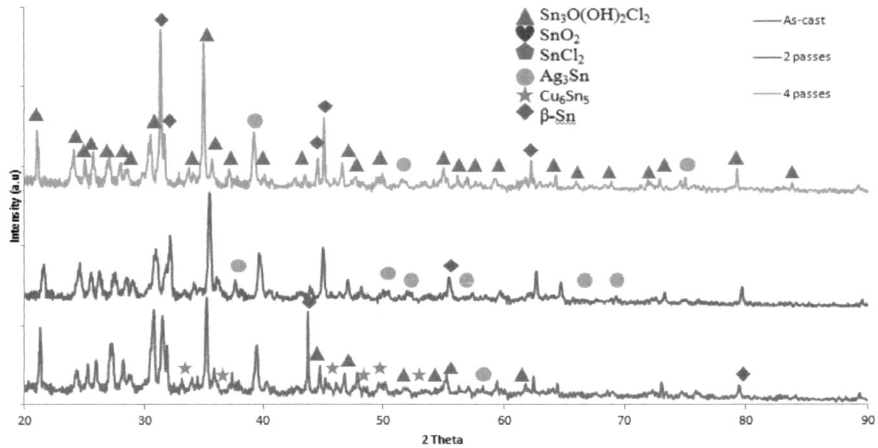

Figure 4. XRD pattern of the corrosion products on as-cast and ECAPed SAC305 solder alloys after polarization test in 3.5 wt.% NaCl solution

4. Conclusion

The ECAP process resulted in grain refinement in the SAC305 solder. After anodic polarization in 3.5 wt.% NaCl solution, the ECAPed SAC305 solder had a more adherent and compact surface film of corrosion products than the as-cast solder. 4 passes ECAPed SAC305 solder with Ag_3Sn needles break into small fragments exhibited better corrosion resistance than as cast commercial SAC305. This could be due to the more compact and continuous oxide film containing $Sn_3O(OH)_2Cl_2$ and SnO_2 formed on the surface of solder preventing further oxidation. The increase of ECAP number of passes increases the corrosion resistance of the SAC305 alloy. The corrosion potential of SAC305 alloy shifts to more noble position with increasing ECAP number of passes from 2 to 4 passes.

Acknowledgement

The authors would like to acknowledge the financial support for this study from AUNSeed-Net JICA (Grant number: 6050355).

References

[1] Zeng, K. & Tu, K.N. *Materials Science and Engineering R* 38 55
[2] Kanlayasiri, K. & Ariga, T. 2012 *J. Alloys Compound* 504 15
[3] Li, D.D., Conway, P.P. & Liu, C.Q. 2008 *Corrosion Science* 50 995
[4] Mohanty, U.S. & Lin, K.L. 2008 *Corrosion Science* 50 2437
[5] Zhao, Y., Guo, H., Ning, Y. & Yao, Z. 2013 *Mater Des* 46 889
[6] Gong, J.C., Liu, C.Q., Conway, P.P. & Silberscmidt, V.V. 2010 *Materials Science and Engineering A* 527 2588
[7] Sadawy, M.M & Ghanem, M. 2016 *Defence Technology* 12 316
[8] Rosalbino, F., Angeline, E., Zanicchi, G., Carlini, R. & Marazza, R. 2009 *Electrochmic. Acta* 5 7231
[9] Wang, M.N., Wang, J.Q., Feng, H. & Ke. W. 2012 *J.Mater Sci. Mater Electron* 23 148
[10] Miyamoto, H.H., Mimaki, T.V. & Hashimato, S. 2008 *Corrosion Science* 50 1215
[11] Zhang, L., Ma, A., Jiang, J., Yang, D., Song, D. & Chen, J. 2013 *Corrosion Science* 75 434
[12] Mohamed Ibrahim, A.E.A. & Sadawy, M.M. 2015 *Trans Nonferrous Met. Soc. China* 253865
[13] Mohran, Hossnias, Abdel-Rahman, E.S. & Hany, M.A.E. 2009 *Journal of Solid State Electrochemistry* 13 1279
[14] Fayeka, M., Haseeb, A.S.M.A. & Fazal, M.A. 2017 *Sains Malaysiana* 46 295
[15] Wang, M.N., Wang, J.Q., Feng, H. & Ke. W. 2012 *Corrosion Science* 63 20

IOP Conf. Series: Journal of Physics: Conf. Series **1082** (2018) 012059 doi:10.1088/1742-6596/1082/1/012059

The Effect of Increasing Silicon on Mechanical Properties of Ductile Iron

W. Arshad[1*], A. Mehmood[2], M. F. Hashmi[1], Obaid ur Rauf[1]

[1]Failure Analysis Center, Institute of Space Technology, Islamabad-44000, Pakistan

[2]Faculty of Mechanical Engineering, University of Engineering and Technology Lahore (FSD Campus)-38000, Pakistan

Email: *13msdmewarshad@smme.edu.pk;

Abstract. Ferritic-pearlitic ductile iron has variation in hardness while the ferritic ductile iron has homogenous hardness. Ferritic-Pearlitic ductile iron has a wide range of hardness that varies between 170-230 HB. Due to variation in hardness, it is difficult to machine ferritic-pearlitic ductile iron as compared to ferritic ductile iron. To overcome this issue, it is desirable to get ferritic microstructure without compromising the mechanical properties of ductile iron. Silicon is a ferritic stabilizer and it strengthens the ferrite phase through solution hardening. Current work is done to see the effect of silicon content on the mechanical properties of ductile iron. The silicon content is increased from 2.1% to 3.6% in different experiments to examine, and then evaluated its effect on ductile iron. It is observed that the increase in silicon content increases the ferrite phase and decreases pearlite phase, resulting in reduced tensile strength and an increase in elongation. However, further increase in the silicon content resulted in increased tensile strength as well as in percentage elongation. Effect of silicon on percentage elongation, hardness, and impact strength are observed in these experiments. Moreover, the effect of silicon on microstructure in all experiments is also studied which has a direct impact on mechanical properties of ductile iron.

1. Introduction

Ductile iron is an alloy that contains spheroidal graphite nodules which gives ductility to it. Ductile iron is also known as SG iron or Spheroidal Graphite Iron. Many of the automotive parts are now manufactured using ductile iron. K. D. Mills added magnesium in grey cast iron in the form of Cu-Mg alloy at International Nickel Company in 1943. Keith Mills observed that there were no flakes rather graphite nodules taken their place in the microstructure [1]. To achieve good quality of ductile iron, sulphur and phosphorus content must be minimized, and their level is to be kept less than 0.03 wt.% sulphur and 0.1% phosphorus. Other trace elements should be kept low to make perfectly round or spherical nodules [2]. Carbon, Carbon equivalent, Silicon, Manganese, Aluminum and phosphorous are the significant components in the production of ductile iron and all these elements should be controlled carefully [3].

According to ISO 1083 the ductile iron grade EN-GJS-400-15 has ferritic microstructure (A single phase matrix) [4]. This category of ductile iron has uniform hardness than ferritic-pearlitic ductile iron (Two phase ductile iron). Casting made from ductile iron grade EN-GJS-500-7 has a range of hardness which varies between 170-230 HB [5]. This variation is because pearlite content is affected by cooling rate [6]. Cooling rate would be different in different section of the casting by varying section thickness [7]. Castings with varying hardness leads to difficulties in machining. One way of reducing these variations is to choose a material that gives uniform hardness [8]. Ferritic ductile iron gives significantly more uniform hardness as compared to ferritic-pearlitic grades. Different components manufactured from ferritic-pearlitic grade often covers the entire of the standard hardness range. For example, ductile iron grade EN-GJS-600-3 has a hardness range between 190 – 270 HB [9] and are thus difficult to machine. So, a material with uniform hardness will have nearly the same hardness at different sections having different thickness [10].

Spheroidal cast iron (SG Iron) or ductile iron contains graphite nodules which are linked with thermally induced stress/strain [11]. Thermal cycling yields stress/ strain cycles in the material through differential expansion and contraction causing due to temperature gradients. The amount of such thermal stress increases with increased temperature, larger range on which the temperature is changed/cycled and high rates of cooling and heating [12]. The effect of these thermal stresses on mechanical properties of ductile iron greatly depends upon size, shape and morphology of the graphite nodules [13].

Cracks due to thermal cyclic heating start from the surface [14]. Under cyclic heating or cooling, structural stability and oxidation resistance are very important parameters [15]. Therefore, SG iron with high silicon content is more suitable for high temperature applications. This is because silicon raises the eutectoid temperature, which leads to high temperature stability [16]. Austenite and ferrite stabilizers widen the respective phase fields. The effect of alloying elements on the Fe-Fe3C phase diagram is strongly reflected in the eutectoid temperature, which is raised or lowered by the alloying addition. The austenite stabilizers lower the eutectoid temperature, thereby widening the temperature range over which austenite is stable. Similarly, the ferrite formers raise the eutectoid temperature, thereby restricting the γ–phase field [17]. As silicon is a ferrtie stabilizer, it raises the eutectoid temperature and improves the performance at elevated temperature [18]. In contrast, increasing the silicon content creates more thermal stresses on casting surface and eutectic cell boundaries under cyclic heating/cooling process [19]. Increasing the silicon content is suitable for high temperature applications, but too much silicon (above 6%) causes brittleness in the material which can result surface cracks on the surface of the casting [10]. This is due to the fact that too much silicon causes thermal stresses on casting surface during thermal cycling. It is confirmed from many reports that increasing the silicon content causes thermal cracking in the early stages of thermal cycling due to thermal stresses [20].

In ductile iron each graphite nucleus gives rise to a single graphite nodule, thus nucleation establishes the final nodule count. Accordingly, by increasing the nodule count in cast iron leads to increasing strength and ductility, improved microstructural homogeneity, and reduction in the chilling tendency [10]. Accordingly, it can be affirmed that graphite eutectic cells or nodules play a significant role in the exhibited properties becoming important factors in the foundry practice [21]. The cycling heating creates more thermal stresses on eutectic cell boundaries.

Ductile iron grade EN-GJS-500-7 is difficult to machine as compared to EN-GJS-400-15 due to pearlite content. Therefore, it is desirable to achieve the properties of EN-GJS-500-7 without compromising the difficulty of machining. In this research work, silicon content is increased in different experiments to analyze its impact on ductile iron, so that a single phase material can be obtained which has the properties of EN-GJS-500-7 and is relatively easy to machine.

2. Experimentation

To analyze the effects of silicon, its content was increased gradually in different experiments. Moulds were prepared using green sand moulding. As sand plays an important role in the sand casting process, different parameters were kept in control for a sound casting. Different sand test results are shown in Table 1. The sand was tested using methods of physical test for foundry sand IS 1918.

During experimentation, samples were prepared through the casting process using standard DIN EN 1563. Total eleven experiments were prepared in which silicon content was increased gradually and each experiment was repeated three times. The drawing of the casting prepared for the samples is shown in Figure 1, and the dimensions of the casting are given in Table 2.

Ductile iron was formed by treating molten metal having low sulfur content with FeSiMg additive. The resulting product was inoculated with ferro-silicon just before pouring it into moulds. Then experiments were carried out to investigate the effect of silicon addition in ductile iron. Composition of ductile iron is given in Table 3.

Table 1. Molding sand parameters

No	Sand Test	Range
1	Compactability	30-40%
2	Specimen weight	137-144 (g)
3	Compression strength	12-26 N/cm^2
4	Splitting strength.	6-8 N/cm^2
5	Permeability	90-95
6	Moisture	3.0%-4.5%

Table 2. Separately cast sample drawing dimensions

Dimensions	Millimeter (mm)
U	10
V	100
X	50
Y	150
z	200

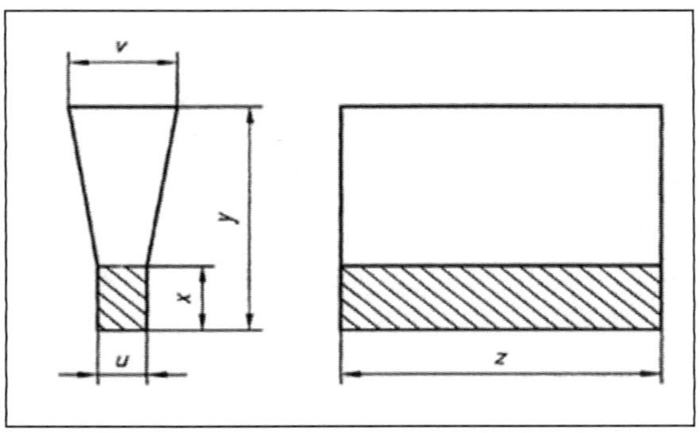

Figure 1. Separately cast sample drawing

Table 3. Ductile Iron composition

Element	Percentage Composition
C	3.7-4%
Mn	0.2-0.3%
Mg	0.03-0.05%
S	<0.02%

C.E meter was used to measure C and Si percentage before pouring of ductile iron melt into moulds. C.E meter was accurate, up to ±0.2%. After Mg treatment ductile iron melt was inoculated. Composition of ductile iron, which includes carbon, silicon, magnesium, manganese and sulfur content was checked using q6 Columbus spark emission spectrometer after inoculation. This instrument gives improved stability and reproducibility. The accuracy of this instrument is ±0.1%. So silicon amount was measured at two stages. Before inoculation using C.E meter, and after inoculation using spark emission spectrometer. Only spectrometer results are given in Table 3. This is because silicon at the final stage is relevant. The composition of silicon before inoculation is not relevant at this stage.

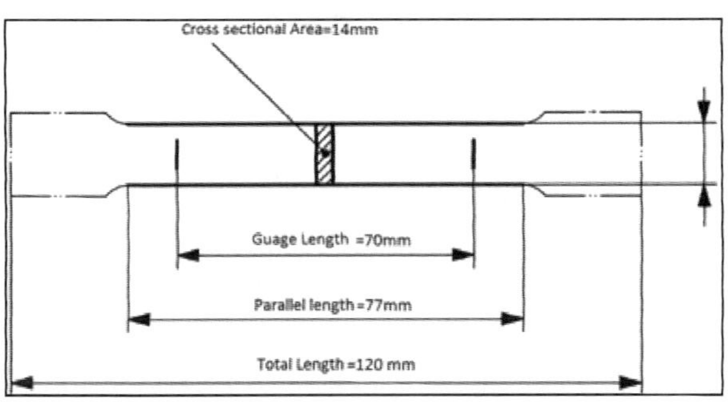

Figure 2. Tensile specimen drawing

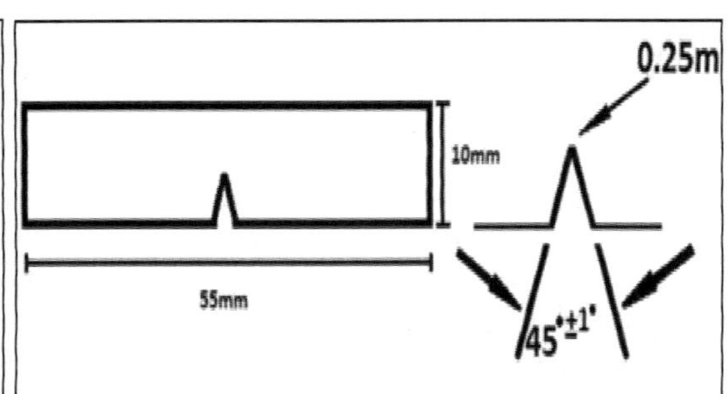

Figure 3. Impact testing sample drawing

The silicon content was increased in different experiments while keeping all other elements constant (Table 3) to study the effect of silicon on mechanical properties of ductile iron. Samples for tensile strength and microstructure were taken from the casting to check the mechanical properties and microstructure of each experiment. Tensile specimens were prepared using ISO 6892-1 standard. The drawing of the specimen is given in Figure 2. The stain rate was set to be 0.002s-1. The impact testing samples were prepared according to standard test "Method for impact testing of Metallic Materials E23". The drawing of the sample for the impact test is shown in Figure 3.

3. Results and discussion

Silicon is added to the ductile iron in different percentages to observe its impact on the tensile strength, % elongation and hardness. The silicon was added in these percentages; 2.17%, 2.18 %, 2.2%, 2.21%, 2.27%, 2.3%, 2.45%, 2.8%, 3.17%, 3.45% and 3.63% for examination and evaluation. Carbon content in ductile iron was kept between 3.7 and 3.9 while the silicon content was increased from 2.1% to 3.6% in different experiments. As silicon content was increased, the ultimate tensile strength decreased (Figure 4) while elongation increased up to a certain point (Figure 5). As silicon content was increased from 2.1 % to 2.3 %, the tensile strength shows some variations, but after 2.3 %, the tensile strength decreases up to 2.8 % silicon content and then it started to increase again. Similarly, as the silicon content was increased from 2.1% to 2.3%, the elongation decreased up to 2.3% silicon content, but after that it started to increase as shown in Figure 5. This trend of tensile strength and elongation can be easily understood from the microstructures. It is clear from the Figure 8 that as silicon content was increased, the pearlite content in the microstructure shows a decreasing trend and an increase in the ferrite phase can be observed. This is the fact because of which a decrease in tensile strength is observed until 2.8% of the silicone content. After this silicon helps to strengthen the ferrite phase due to which tensile strength increased.

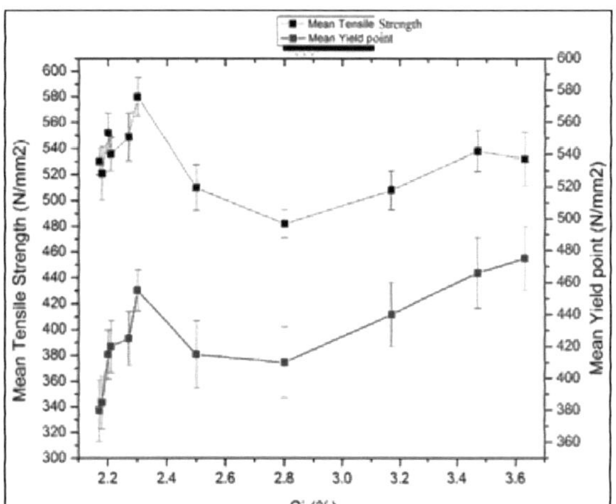

Figure 4. Silicon effect on tensile strength of ductile iron

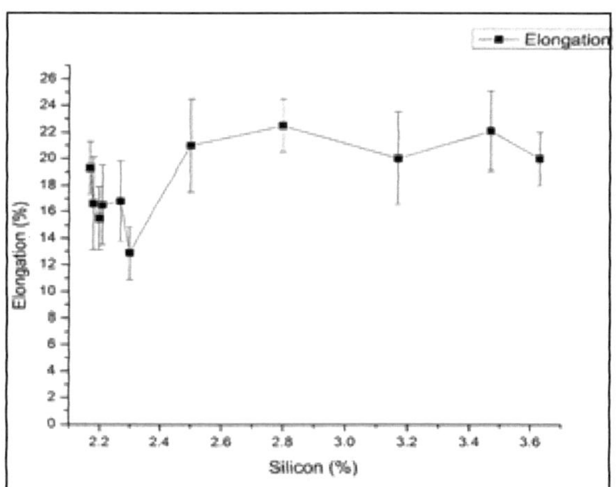

Figure 5. Silicon effect on % elongation on ductile iron

There are various parameters like melt temperature, composition, cooling rate, surrounding temperature, etc. which determines the mechanical properties of the material. All experiments are designed to maintain these parameters at a constant level in every experiment except silicon, but there is a possibility of small variations in these parameters that can affect the mechanical properties. To overcome this issue each experiment is repeated. To make the data easily understandable, an average value as well as standard deviation error bars is given in Figure 4 and Figure 5. Average of yield strength, tensile strength and elongation in Figure 4 and Figure 5 clearly show a significant effect of silicon on these properties which can be correlated with the microstructure in Figure 8.

As the silicon content was increased, hardness decreased as shown in Figure 6. This is because the ferrite was increasing, and pearlite content was decreasing. It was also observed that as the ferrite content was increased, and fully ferritic ductile iron was obtained, there was less variation in hardness as compared to when there was pearlite at lower silicon content. Five different points from the casting were selected on which hardness was measured to observe variation in hardness.

As the silicon content was increased, impact strength was increasing. This is because as silicon increases the pearlite content decreased and ferrite content increased which gives rise to impact strength. The pearlite consist of lamellar structure of ferrite and cementite and it is obvious that cementite (Fe3C) is more brittle as compared to ferrite. So, a material whose microstructure has higher pearlite content leads to lower the impact strength.

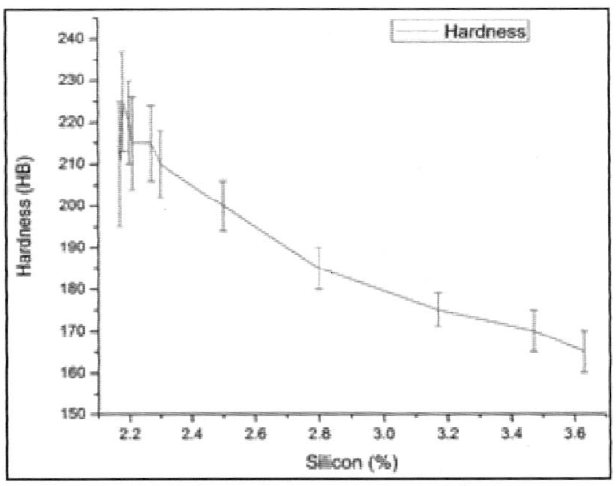

Figure 6. Hardness results

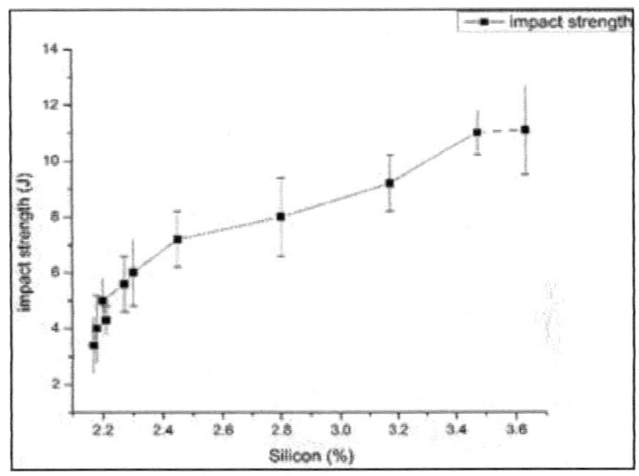

Figure 7. Impact test results at room temperature

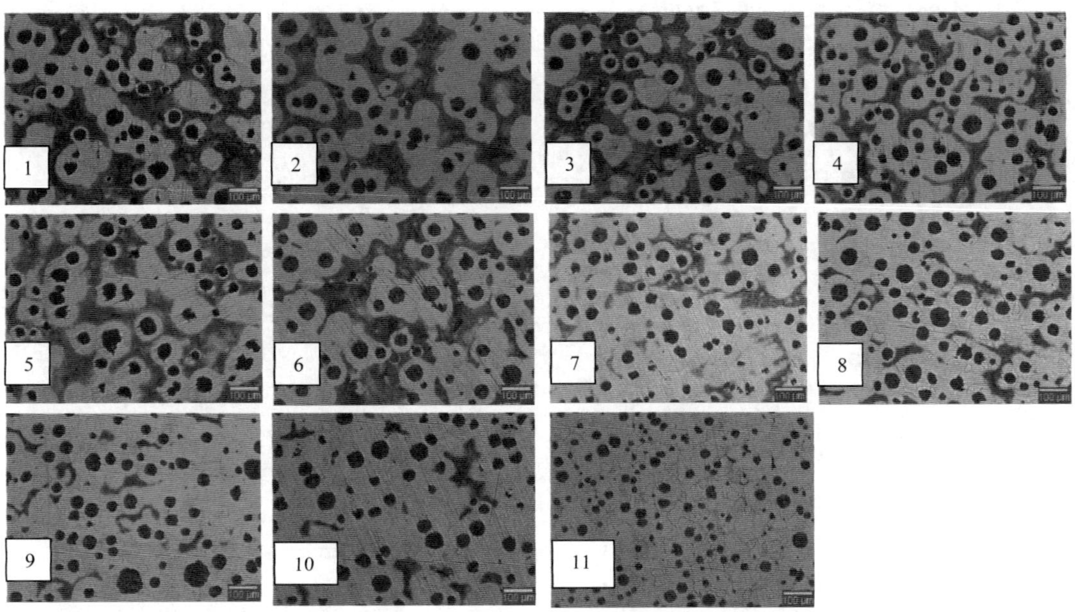

Figure 8. Etched with 2% Nital. 1) Pearlite 45% ferrite 55% nodularity 80% 2) Pearlite 43% ferrite 57% nodularity 85% 3) Pearlite 40% ferrite 60% nodularity 90% 4) Pearlite 38% ferrite 62% nodularity 90% 5) Pearlite 35% ferrite 65% nodule 80% 6) Pearlite 30% ferrite 70% nodularity 90% 7) Pearlite 25% ferrite 75% nodularity 85% 8) Pearlite 20% ferrite 90% nodularity 80% 9) Pearlite 5% ferrite 90% nodularity 90% 10) Pearlite 1% ferrite 90% nodularity 90% 11) ferrite 100% nodularity 90%

IOP Conf. Series: Journal of Physics: Conf. Series **1082** (2018) 012059 doi:10.1088/1742-6596/1082/1/012059

4. Conclusion

The silicon content was increased from 2.1% to 3.6% in different experiments. The tensile strength decreased when silicon content was increased up to 2.8%. Further increase in silicon content led to increase in tensile strength and yield strength. Hardness decreased gradually as silicon content was increased, and variation in hardness at different points was decreased as ferritic ductile iron was achieved. As the silicon percentage of ductile iron was increased, the pearlite content in the matrix become decreased and the ferrite portion increased led to improved impact value. By further increasing the silicon content, the matrix of ferritic pearlitic ductile iron becomes fully ferritic in which variation in hardness was less as compared to ferritic pearlitic ductile iron. It was observed that tensile strength and yield strength decreased up to a certain point and then start increasing. So by increasing the silicon content a fully ferritic ductile iron was achieved, which has a tensile strength of ferritic pearlitic ductile iron. Hardness showed a decreasing trend as pearlite was decreasing in the microstructure. So it will be easy to machine, as this material hardness and variation in hardness at different sections was low as compared to ferritic-pearlitic ductile iron. This high silicon ductile iron has better impact strength at room temperature as compared to ferritic pearlitic ductile iron, which was due to the low pearlite content in the microstructure.

5. References

[1] S. K. Swain 2008
[2] W. F. Smith and J. Hashemi 2011 *Foundations of materials science and engineering: McGraw-Hill*
[3] D. Ednmnds, "GRAPHITE PHASE FORMATION IN Al-ALLOYED DUCTILE IRONS."
[4] "ISO 1083 Spheroidal graphite cast irons - Classification," 2004.
[5] Y. Li and X. Chen 2001 *Materials Science and Engineering: A* 308 277
[6] Y. Sakuma, O. Matsumura, and H. Takechi 1991 *Metallurgical Transactions A* 22 489
[7] L.-E. Björkegren 1994 *Sweedan Patent*
[8] K. H. L-E Björkegren, B. Johannesson *AFS Transaction* 96 139
[9] J.-c. Liu 2015 *China Foundry* 12
[10] H.-M. Lin, T.-S. Lui, and L.-H. Chen 2004 *Materials Transactions* 45 569
[11] L. Chuzhoy, R. DeVor, S. Kapoor, and D. Bammann 2002 *Journal of manufacturing science and engineering* 124 162
[12] D. Holmgren, R. Källbom, and I. L. Svensson 2007 *Metallurgical and Materials Transactions A* 38 268
[13] S. F. Al-Saffar, K. I. Khalaf, and A. N. A. Raghif 2008 *Iraqi Journal of Applied Physics*
[14] S. M. C. C. P. Cheng, T. S. Lui and L. H. Chen 1997 *metallurgical transaction*
[15] J. Mallia and M. Grech 1997 *Materials Science and Technology* 13 408
[16] K. Roehrig 1978 *AFS Trans* 75
[17] M. Maalekian 2007 *Technische Universität Graz: Christian Doppler Laboratory for Early Stages of Precipitation*
[18] R. Okunnu 2015
[19] M. C. R. a. G. P. Reddy 1989 *AFS Trans.* 97 351
[20] P. Lan and J. Zhang 2014 *Materials & Design* 54 112
[21] E. Fraś and M. Górny 2008 *Archives of Foundry Engineering* 8 53

IOP Conf. Series: Journal of Physics: Conf. Series **1082** (2018) 012060 doi:10.1088/1742-6596/1082/1/012060

Effect of indium in Cu-Zn-Al shape memory alloys

C Chanmuang[1,2*], S Niyomsoan[1*] and N Chomsaeng[1]

[1] Integrated and Innovative Jewelry Materials Research Unit, Faculty of Gems, Burapha University, Chanthaburi Campus, Chanthaburi 22170, Thailand

[2]Institut für Mineralogie und Kristallographie, Universität Wien, Althanstaße 14 (UZA 2), A-1090 Wien, Austria

*E-mail : chutimun.chanmuang@univie.ac.at; sniyomsoan@gmail.com

Abstract. Shape memory alloys (SMAs) are among the "smart" materials that can virtually remember their original shape; after being deformed, they may return to their initial shape through heating. The shape memory effect derives from a martensitic transformation induced by stress and/or temperature change. Copper-based SMAs are a more economical alternative compared to Ni-Ti alloys. They have low cost and various interesting mechanical and thermal properties and high practical value for the applications. In this present work, Cu-Zn-Al shape memory alloys with 0.1–0.5 wt.% In were studied. Investment casting (lost-wax casting) was performed at 1100 °C melting temperature and 650 °C mold temperature. Shape memory effect of the alloys was investigated with bending test and then heating for shape recovery. Scanning and transmission electron microscopes were used for microstructural characterization along with Energy X-ray diffraction and selected area diffraction spectroscopy. XRD measurement was also performed to understand effect of indium on microstructure. The results clearly presented that the addition of In in Cu-Zn-Al SMAs caused a precipitation of the second phase and significantly alternated the shape memory effect. The microstructure and orientation relationship of the typical and twinned martensite were discussed.

1. Introduction

Shape memory alloys (SMAs) are a group of "smart" materials that can "remember" their original shape, i.e. they can return to their pre-deformed shape after being heated. The shape-memory effect derives from a martensitic transformation due to the reversible thermo-mechanical behavior [1]. It is a diffusionless phase transformation between high-temperature austenitic phase and low-temperature martensitic phase. Copper-based SMAs are an economical alternative among SMAs because of their low cost and ease of production [2]. The Cu-based SMAs also obtain various interesting mechanical and thermal properties [3, 4] and practical value in modern applications [5-7]. Cu-Al-Ni, Cu-Zn-Al, and Cu-Al-Mn alloys are potential candidates in the Cu-based SMAs [8-10]. Their shape memory properties are largely influenced by microstructure according to the metal preparation process [11, 12]. An investment casting has been introduced to cast the Cu-based SMAs to the applicable shape which successfully retains their shape memory effects [2, 5].

In this present work, the Cu-Zn-Al shape-memory alloy with various indium additions was shaped with the investment casting process. The study was to investigate the microstructural influence using optical microscope, scanning electron microscope (SEM) along with the chemical composition analysis

IOP Conf. Series: Journal of Physics: Conf. Series **1082** (2018) 012060 doi:10.1088/1742-6596/1082/1/012060

by an energy-dispersive X-ray spectroscopy (EDS). In addition, XRD measurement and transmission electron microscope (TEM) were used for phase characterization. Bending test with subsequent heating was performed to follow the shape-memory effect of the alloys with the indium addition.

2. Experimental study

The Cu-Zn-Al-In shape-memory alloys were prepared from a commercial grade metals following the nominal compositions in table 1. The casting process was done by the so-called lost-wax casting technique. The alloys were melted at 1100 °C in a graphite crucible in an INDUTHERM VC400 induction furnace under nitrogen gas atmosphere. A molten alloy was then cast into a hot gypsum mold heated at 650 °C. The mold was subsequently quenched in water at room temperature. So the metal sample was suddenly cooled down. The as-cast samples were in a rectangular shape with 0.5 mm thickness.

Shape memory behavior of the alloys was investigated by bending test as schematically shown in figure 1. The sample was bent to some degrees of deformation angle (ϕ) and then heated with a lighter (roughly 700 °C) for recovering its shape. The recovery angle (θ) was measured afterward. The percentage of strain recovery, η, expressing a degree of shape memory effect (SME) was calculated following equation (1) [13].The thermomechanical test of each alloy was repeated at least 10 times to investigate a repeatability of SME.

$$\eta = \left(\frac{\phi-\theta}{\phi}\right) \times 100\% \quad (1)$$

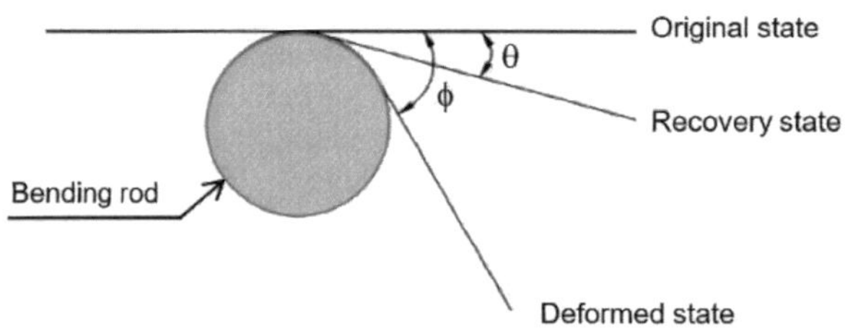

Figure 1. Scheme of bending test for shape memory effect investigation.

The samples were etched with nitric acid solution (1:1 water to nitric acid volume ratio) to reveal their microstructure prior to an observation with an optical microscope. A JEOL JSM-6335F scanning electron microscope operated at 15 kV was used to obtain microstructure images along with chemical investigation using energy dispersive X-ray spectra (EDS). X-ray diffraction (XRD) measurement was performed in a range of 20 – 60° with 0.02° step size using a Rigaku X-Ray diffractometer with CuK_a radiation (40 KV and 30 mA). TEM specimens were prepared using FEI Quanta 3D FEG focused ion beam. Transmission electron images and selected area diffraction were examined with a Philips CM200 operated at 200 eV and equipped with 2D Phaser X-ray diffractometer.

Table 1. Nominal composition and the first stain recovery of Cu-based SMAs.

Alloy	Nominal composition (wt.%)				e/a	Measured 1st stain
	Cu	Zn	Al	In	ratio	recovery, η (%)
A	*bal.*	20.8	5.8	0.0	1.443	67 %
B	*bal.*	20.8	5.8	0.1	1.444	67 %
C	*bal.*	20.8	5.8	0.3	1.447	56 %
D	*bal.*	20.8	5.8	0.5	1.449	67 %

3. Results and discussion

The shape memory effect of the Cu-Zn-Al SMAs with indium addition was expressed in term of the strain recovery (η), in table 1, from the first bending test with subsequent heating. The investigation showed that the addition of indium maintained the recovery level of the Cu-Zn-Al SMAs at 67%. Unfortunately, with 0.3 wt.% In, SME of Alloy C was reduced to 56% strain recovery. After ten repetitions of the SME test as plotted in figure 2(a), the strain recovery decreased gradually with a number of SME tests. A proportion of the sample undergoing thermal transformation decreased due to an increase of a number of dislocations generated by the martensitic transformation and a decrease of total transformation energy [13].

The indium addition also improved the shape recovery level. In comparison with the alloy without indium (Alloy A), a reduction of η according a number of martensitic transformations was delayed when adding 0.1 wt.%In and 0.5 wt.%In in the alloys (Alloy B and Alloy D, respectively). Moreover, both Alloy B and Alloy D could survive twenty SME tests while only ten tests could be performed in Alloy A before breakage. The introduction of indium in the Cu-Al-Zn SMAs could enhance the shape memory effect and prolong their ductility. However, among the studied alloys, Alloy C (0.3 wt.%In) not only gained the lowest recovery levels at each SME test and the fastest reduction of the recovery, but could also survive only ten SME tests.

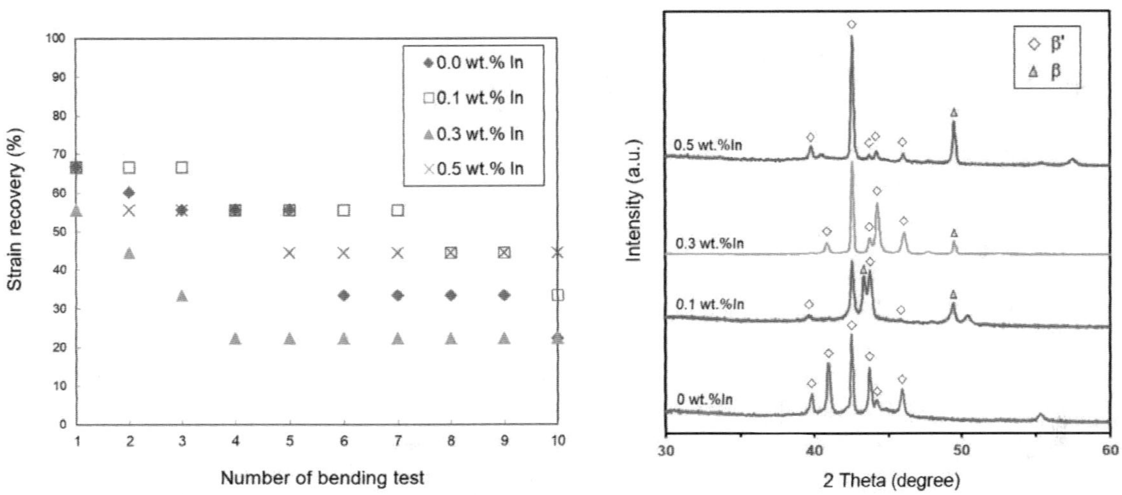

(a) (b)

Figure 2. (a) Repeatability of SME and (b) XRD patterns of the as-cast Cu-Zn-Al-In SMAs.

The X-ray diffraction pattern of the as-cast Cu-Zn-Al SMAs with In contents in figure 2(b) indicated the β′ martensitic phase present in every alloy. The β′ peaks belonged to 18R type martensite which inherits from β_1 or β_3 ordered bcc structure [4, 14-16]. The as-cast Alloy A was in the β′(18R) martensite phase at room temperature. As listed in table 1, the electron to atom (e/a) ratios of every alloy was about 1.443 – 1.449, which indicated the coexistence of the α' and β' martensite morphologies [17]. They are both rhombohedral type structure with different stacking sequence [14].

With the addition of indium, the bcc β parent phase was observed in the XRD pattern which was typically found in quenched Cu-based SMAs [18]. The β peaks shifted gradually rightward with the indium contents increasing from 0.1wt.% to 0.5 wt.%. The shift indicated a more compaction of the cubic structure with the addition of indium. Moreover, the β peaks were relatively higher in comparison with that of the β′ martensite peaks, with an increase of the indium contents. β phase possesses bcc structure and can be diffusionlessly transformed to rhombohedral or hexagonal type martensite [14].

Microstructures of the samples was examined using optical microscope. Figure 3 shows the optical micrographs of the as-cast Cu-Zn-Al SMAs with different indium contents. Martensite plates in V-shaped pattern in some grains and needle-like structure in others were observed in every sample. Two morphologies of the martensite structure indicated the presence of two kinds of martensite phase. 18R and 2H structures have been found in the V-shape martensite with stacking fault and in the needle-like martensite with several twin boundaries [15]. 2H martensite is inherited from β_3 parent phase [14] which could be transformed under quenching the alloys. However, XRD peaks of the H2 (hexagonal) and 18R (rhombohedral) structure superimpose so they could not be differentiated in the XRD pattern.

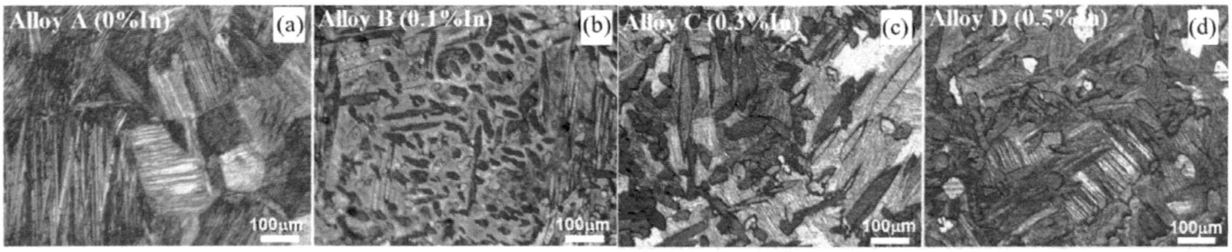

Figure 3. Optical micrographs of the as-cast alloys.

Without indium, Alloy A obtained larger grain size (more than 100 μm) which contained both forms of martensite orientated in different directions. The addition of indium induced a formation of elongated grains which corresponded with the prior β parent phase distributed among martensite matrix. Size of the parent phase increased with indium contents. The addition of indium provided more nucleation sites for the β phase, which obstructed grain growth of the prior parent phase. Grain size of the SMA alloys was then reduced significantly as shown in figure 3.

Higher magnification of SEM images were illustrated in figure 4. The matrix of all alloys showed needle-like structure of the β' martensitic phase. An amount of the prior β parent phase increased with indium content. Inside the parent phase, there was plate-like martensite with less than one micron width. The plate-like structure refers to the 2H martensite structure [15]. The presence of the β phase, as indicated by the XRD analysis, indicated that the martensitic transformation was partially occurred from quenching process during the sample preparation.

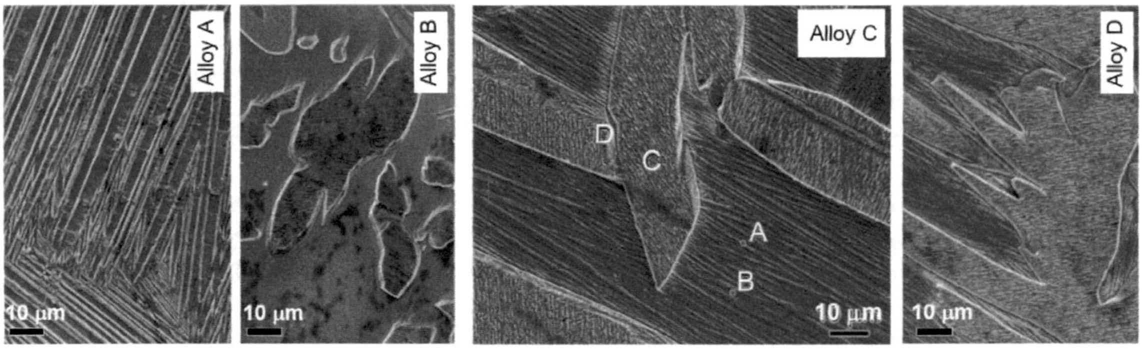

Figure 4. SEM-SE micrographs of the as-cast alloys.

To identify an elemental composition of the two-phase areas as indicated with alphabet A – D in figure 4(c), EDS analysis was performed and tabulated in table 2. The bright and dark domains in the matensite phase were marked as A and B, respectively. Likewise, the bright and dark domains of the prior β parent phase were marked as C and D, respectively. The EDS analysis revealed that Cu and Zn appeared in all domains. Copper was preferentially observed in the dark phase. Concentration of aluminum was higher in the bright area. Indium was only found in the martensitic phase and the concentration was higher in the bright area. There was no indium observed in the parent phase both in bright and dark domains.

Table 2. EDS analysis of Alloy C at the points marked as A, B, C and D in figure 4(c).

Element	Martensite phase composition (wt.%)		Parent phase composition (wt.%)	
	Point A (Bright)	Point B (Dark)	Point C (Bright)	Point D (Dark)
Cu	70.76	79.95	77.00	78.58
Zn	17.62	14.61	17.38	16.56
Al	8.75	4.20	5.62	4.86
In	2.87	1.24	-	-

Figure 5(a) shows a bright field TEM image of the as-cast Alloy D which gained pronounced SME. Shape of the martensite phase was plate-like with a smooth and sharp phase boundary. Moreover, a ten nanometer-size substructure of martensite was observed inside a hundred nanometer-wide martensite plate. Electron diffraction patterns in figure 5(b) and figure 5(c) were obtained from the typical- and twinned-martensite at the areas marked as A and B, respectively, in figure 5(a).

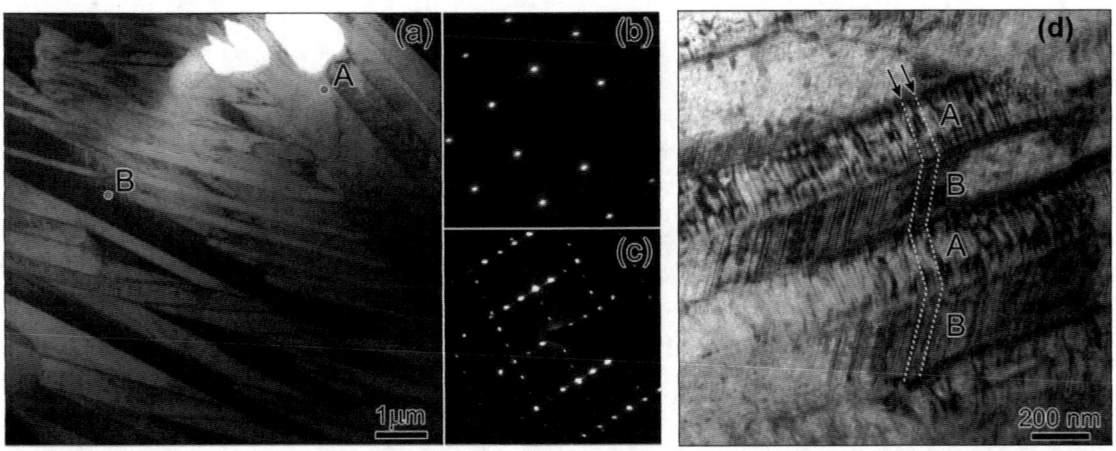

Figure 5. TEM bright-field micrograph and the selected-area diffraction patterns of the Alloy D; (a) typical martensite plate configuration, (b) and (c) SADP of martensite and twinned-martensite taken at an area marked as A and B, respectively, and (d) the twinned-martensite plate.

TEM bright-field micrograph of the twinned-martensite plate in the Alloy D was shown in figure 5(d). Dashed lines show twin planes in this alloy which confirmed the presence of 2H type martensite. The twin planes of plate A and B oriented in the directions which were perpendicular to the twin plane at 90° and 55°, respectively. The presence of extensive twinning in the alloy was one of the requirements for the excellent shape memory effect. The introduction of indium in the Cu-Al-Zn SMAs could enhance the shape memory effect and prolong their ductility because of the abundant moveable twinning.

4. Conclusion

The indium addition in Cu-Zn-Al SMAs with an investment casting process significantly influenced the shape memory effect;

(a) The typical microstructure of Cu-Zn-Al-In alloys was plate-like martensite. The substructure of martensite consisted of twins which was important for the shape memory effect.
(b) The investment casting process using generally for jewelry casting could retain martensite in the Cu-Zn-Al SMAs.
(c) The amount of indium addition (0.1 – 0.5 wt.%) in the Cu-Zn-Al shape memory alloy resulted in a significant increase of shape memory recovery due to an abundant substructure twining that accommodates the shape memory effect.

Acknowledgements

An appreciation for financial support goes to National Research Council of Thailand (NRCT) under grant number 150/2558. Prof. Dr. Thomas Waitz at University of Vienna is gratefully acknowledged for TEM investigation. Support on sample preparation from B. Kumnerdphonpittaya, O. Juntakool and N. Ponsantia are greatly appreciated.

References

[1] Otsuka K and Wayman C M 1998. *Shape Memory Alloys*
[2] Niyomsoan S, Vongchaisri K and Chanmuang C 2017 *The 7th National and International Graduated Study Conference* 20
[3] Guilemany J M, Fernandez J and Zhang X M 2006 *Mater. Sci. Eng.,* A 438 726
[4] Natali S, Volpe V, Yortea L, Di Cocco V and Iacoviello F 2013 *Convegno Nazionale IGF XXII* 132
[5] Niyomsoan S, Mattipong R, Chanmuang C and Chomsaeng N 2014 *Proceedings of Burapha University International Conference 2014* 580
[6] Zang J, Wang Q Z, Yin F X, Cui C X, Ji P G and Li B 2016 *Mater. Sci. Eng.,* A 664 215
[7] Chanmuang C, Juntakool O, Kumnerdponpittaya B, Niyomsoan S and Chomsaeng N 2014 *Proceedings of The 4th International Gem and Jewelry Conference 2014* 317
[8] Recarte V, Pérez-Landazábal J I, Ibarra A, Nó M L and San Juan J 2004 *Mater. Sci. Eng.,* A 378 238
[9] Graczykowski B, Mielcarek S, Breczewski T, No M L, San-Juan J and Mroz B 2013 *Smart Mater. Struct.* 22
[10] Karagoz Z and Aksu Canbay C 2013 *J. Therm. Anal. Calorim.* 114 1069
[11] Pourkhorshidi S, Parvin N, Kenevisi M S, Naeimi M and Ebrahimnia Khaniki H 2012 *Mater. Sci. Eng.,* A 556 658
[12] Xiao Z, Li Z, Fang M, Xiong S, Sheng X and Zhou M 2008 *Mater. Sci. Eng.,* A 488 266
[13] Muesing W E 1982 Thermal martensitic transformation cycling in Cu-Zn-Al shape memory alloys *Theses and Dissertations*
[14] Ahlers M 1986 *Prog. Mater. Sci.* 30 135
[15] Wu S K, Chan W-J and Chang S-H 2017 *Metals* 7 397
[16] Araujo C and Gonzalez C 2005 *Proceedings of COBEM 2005: 18th international congress of mechanical engineering*
[17] Gil F J and Guilemany J M 1992 *Mater.s Res. Bull.* 27 117
[18] Xiaomin C, Feng H, Na L and Xingwen W 2008 *J. Wuhan Univ. Technol. Mat. Sci. Edit.* 23 717

IOP Conf. Series: Journal of Physics: Conf. Series **1082** (2018) 012061 doi:10.1088/1742-6596/1082/1/012061

Chlorination of Iron-Containing Nitrided Ilmenite Prepared by Carbothermal Process with Polystyrene and Coal

E. Ahmadi, F. Amali Alis, M.N. Ahmad Fauzi, N. Baharun, H. Hussin, K.S. Ariffin, S. Ramakrishnan and S.A. Rezan*

School of Materials and Mineral Resources Engineering, Engineering Campus, Universiti Sains Malaysia, 14300 Nibong Tebal, Pulau Pinang, Malaysia

Email: srsheikh@usm.my

Abstract. This work sheds light on the effect of carbothermal reduction and nitridation of ilmenite concentrate with the mixtures of Mukah coal and polystyrene as reducing agents under H_2/N_2 gas atmosphere on the chlorination process. Pre-reduced samples of ilmenite were chlorinated at 400-500°C for 1-3 hours without removing metallic iron. The characterization of the synthesized materials was made by X-ray diffraction (XRD), scanning electron microscopy (SEM), and inductively coupled plasma-optical emission spectroscopy (ICP-OES) analyses. Design of Experiment (DOE) analysis was also applied to evaluate the significant factors and their interactions in low-temperature chlorination process. Statistical analysis indicated that Mukah-Balingian coal-to-polystyrene (C/PS) ratio was the most significant factor influencing the chlorination results. In addition, the highest extent of chlorination ~ 95.01% was achieved with the highest carburized and nitrided samples at 500 °C for 1 hour. The present study can be useful for the chlorination of different types of pre-reduced ilmenite concentrates with high efficiency of titanium tetrachloride production.

1. Introduction

Ilmenite ($FeTiO_3$) composed of iron (Fe) and titanium oxide (TiO_2) is an important world reserve (~ 90%) for production of TiO_2 pigment and titanium metal [1, 2]. A number of procedures have been developed for extracting of titanium metal from $FeTiO_3$ and Ti ores [2-4]. Commercialized processes for upgrading ilmenite to synthetic rutile and separation of iron include but not limited to: (a) pyrometallurgical method in which a mixture of ilmenite concentrate and anthracite as reductant is charged into an electric arc furnace (EAF) to produce a slag containing about 86 %TiO_2 and pig iron. The slag is processed further for removing iron by acid leaching process using 18–20 % HCl solution at 125 °C for production of high-grade TiO_2 feedstock (~ 95 %TiO_2) by upgraded slag (UGS) process, and (b) Becher process that applies both pyro-, and hydrometallurgical routes for oxidation/reduction followed by the aeration acid leaching method using a solution containing ammonium chloride (NH_4Cl) ions as a catalyst to accelerate the metallic iron removal, respectively[2, 3, 5-8].

The study of extracting titanium metal from ilmenites of different grades and selective chlorination process of unreduced and/or reduced and nitrided ilmenite concentrates for developing a low-cost Ti metal technology has been a constant challenge for last decade due to a complex mineralogical composition of ilmenite ores[1, 3, 5, 6, 8-10]. Previously, high-grade TiO_2 feedstocks were chlorinated at high temperatures of about 900-1100 °C to produce titanium tetrachloride ($TiCl_4$) in which majority of the impurities such as iron and other undesired metal oxides in feedstock were converted to metal chlorides [5]. However, the present study focused on the low-temperature chlorination of a mixture of Fe and titanium oxycarbonitride ($TiO_xC_yN_z$) synthesized by carbothermal reduction and nitridation (CTRN) process. In this work, Malaysian ilmenite concentrate was reduced and nitrided by using a mixture of Muka coal and polystyrene (PS) as an alternative reductant. The feasibility of the chlorination process of different mixtures of $TiO_xC_yN_z$ and metallic iron synthesized by CTRN process was investigated in production of $TiCl_4$ at low temperatures.

2. Experimental study

The raw Malaysian ilmenite concentrate was supported from Chee Ng Minerals Sdn. Bhd. located in Kampar, Perak. It was composed of Ilmenite, $FeTiO_3$ (ICSD# 98-001-7094), Iron oxide, Fe_2O_3 (ICSD# 98-009-8279) and small amounts of Protoenstatite, $Mg.SiO_2$ (ICSD# 98-001-2489). The ilmenite samples were mixed with 85% Mukah Coal + 15% Polystyrene (C85P15), 75% Mukah Coal + 25% Polystyrene (C75P25) and 65% Mukah Coal + 35% Polystyrene (C65P35) as reductants. In this work, titanium oxycarbonitride ($TiO_xC_yN_z$) and iron (Fe) mixtures with mean particle size of about 150 μm were synthesized by CTRN process from Malaysian ilmenite at 1250 °C under an H_2-N_2 atmosphere for 3 hours. The reduced and nitrided samples were uniaxially pressed under a pressure of 100 kPa into cylindrical pellets weighing about 2.0g. Pellets with different compositions were chlorinated under flowing nitrogen as carrier gas with a constant flow rate of about 20 ml/min at temperatures in the range of 400 to 500 °C for 1, 2 and 3 hours. The Cl_2 gas for this study was generated by the reaction between potassium permanganate ($KMnO_4$) and hydrochloric acid (HCl) by adding 0.5 g aluminium powder as catalyst to enhance the chlorine gas generation. The moist Cl_2 gas was also passed through concentrated sulfuric acid (1.0 M), and phosphorus pentoxide (P_2O_5) powder to remove moisture prior to the chlorination process. The produced $TiCl_4$ was captured by 37.0% hydrochloric acid, HCl Sigma-Aldrich (CAS# 7647-01-0) in the gas scrubbers and the liquid samples taken from the scrubbers were analyzed by Inductively Coupled Plasma-Optical Emission Spectroscopy (ICP-OES, Perkin Elmer Optima 4300 DV ICP System). Phase characteristics of the unchlorinated and the residues obtained from the chlorination of the samples were carried out by using X-ray diffraction (Bruker D8-advance) and field emission scanning electron microscope (FESEM, Zeiss SUPRATM 35VP) analyses. In addition, a fully randomized factorial design with various combinations of temperature, time and Muka coal to Polystyrene ratios was applied to evaluate and optimize the chlorination process. The design matrix as well as factors levels are presented in Table 1.

Table 1. The design matrix as well as factors levels

Run (#)	Temperature (°C)	Time (hr)	Coal to Polystyrene ratio
1	-	-	-
2	+	-	-
3	-	+	-
4	+	+	-
5	-	-	+
6	+	-	+
7	-	+	+
8	+	+	+
9	0	0	0
10	0	0	0
11	0	0	0

Factors	Code	Units	Levels		
			-1	0	+1
Temperature	A	°C	400	450	500
Time	B	hr	1	2	3
Coal to Polystyrene	C	ratio	C85P15	C75P25	C65P35

IOP Conf. Series: Journal of Physics: Conf. Series **1082** (2018) 012061 doi:10.1088/1742-6596/1082/1/012061

3. Results and discussion

The reduced and nitrided samples were characterized by XRD method. The rietveld refinement method was also applied for the quantitative phase analysis of XRD patterns of the samples. The results for the extents of reduction, nitridation and carburization were determined based on the elemental analysis of the samples by CHN and LECO analyses, and identified phases by XRD analysis are summarized in Table 2.

Table 2. The extents of reduction (X_o), nitridation (X_N) and carburization (X_c) for the reduced samples.

Sample	X_O, %	X_N, %	X_c, %	Identified phases from XRD	R_{wp} (%) -GofF
C85P15	88.74	76.74	2.74	80.8 $TiO_{0.23}C_{0.03}N_{0.75}$ + 19.2 Fe	8.89-3.05
C75P25	86.66	74.50	0.83	79.3 $TiO_{0.26}C_{0.01}N_{0.73}$ + 20.7 Fe	8.82-2.86
C65P35	86.09	73.87	0.30	81.1 $TiO_{0.28}C_{0.003}N_{0.72}$ + 18.9 Fe	7.49-2.03

The XRD patterns revealed that titanium oxycarbonitride, $TiO_xC_yN_z$ (ICSD# 98-008-9933) and iron, Fe (ICSD# 98-002-2044) were the main phases for the reduced and nitrided samples using the mixtures of Muka coal and polystyrene as reductants. The results were also in agreement with the previous findings which reported the main phases after carbothermal reduction and nitridation of titania and ilmenite were $TiO_xC_yN_z$ and Fe phases [8, 11].

The reduced samples were chlorinated in a tube furnace at 400-500 °C for 1-3 hours in order to produce $TiCl_4$ solutions. Table 3 summarizes the quantitative analyses of the solid residues collected from alumina crucibles after the chlorination of the pellets containing metallic Fe and $TiO_xC_yN_z$ at various operating conditions based on the XRD patterns for the highest and lowest chlorination times and temperatures. The phases determined in the solid residues were $TiCl_3$ (ICSD# 98-004-8808), SiO_2 (ICSD# 98-002-8673), TiO_2 (ICSD# 98-007-6757), Ti_3O_5 (ICSD# 98-007-1966) and Fe_2O_3 (ICSD# 98-001-2634). As is observed in Table 3, small amounts of titanium sub-oxides were also formed that can probably be due to the aluminum catalyst used for Cl_2 gas generation producing oxygen gas as explained by Eqs.(1-3) [5, 12].

$$KMnO_2 + 4HCl \rightarrow MnO_2 + KCl + 2H_2O + 1.5Cl_2 (g);$$
$$\Delta G^0 = 13.437 - 0.1178T\ kJ/mol \quad \text{(Eq.1)}$$

$$KMnO_4 + 8HCl (g) + Al \rightarrow MnCl_2 + KCl + Cl_2 (g) + 4H_2O + AlCl_3;$$
$$\Delta G^0 = 140.38 - 0.1746T\ kJ/mol \quad \text{(Eq.2)}$$

$$2KMnO_4 + 2Al + 2H_2O \rightarrow 2MnO_2 + K_2O + O_2 (g) + Al_2O_3 + 2H_2 (g);$$
$$\Delta G^0 = 210.58 - 0.0386T\ kJ/mol \quad \text{(Eq.3)}$$

Table 3: Quantitative analysis of the residues resulted after chlorination of reduced samples at different temperatures and times and extent of chlorination calculated based on ICP analyses of the solutions prepared by chlorination process.

Run (#)	Identified phases from XRD	R_{wp} (%) -GofF	Extent of chlorination, (%)
C3	0.9 $TiCl_3$+32.6 SiO_2+66.4TiO_2	7.40-1.08	94.98
C4	0.8 $TiCl_3$+5.2SiO_2+94.0Ti_3O_5	8.83-1.92	94.68
C5	59.6 $TiCl_3$+30.8SiO_2+9.5Ti_3O_5	10.68-2.68	89.50
C8	4.0 $TiCl_3$+7.4 SiO_2+88.7Fe_2O_3	6.86-1.24	88.92

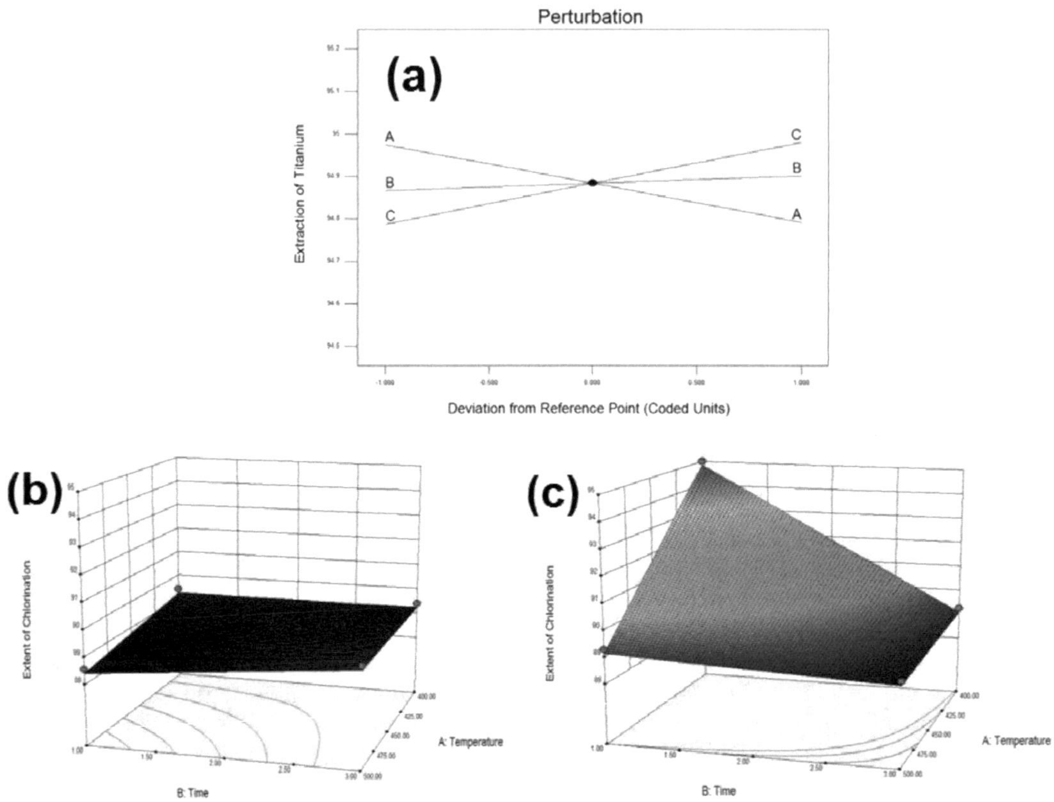

Figure 1. DOE results for (a) Perturbation plots, (b) and (c) 3D surface maps of time and temperature in response to the extents of chlorination of the iron-containing samples reduced with lowest and highest ratios of C/PS in reductant, respectively.

Figure 1 shows the perturbation plots and 3D surface maps of time and temperature in response to the extents of chlorination for the reduced and nitrided samples with lowest and highest ratios of Muka coal-to-polystyrene (C/PS) in required reductants, respectively. The C/PS ratio, denoted as C in the perturbation plots, had the highest gradient and indicated the most significant factor effect on the process, as shown in Figure 1(a). Chlorination time (factor B) showed the lowest gradient among the investigated factors. However, decreasing the chlorination time to 1 hour at 400 °C resulted in $TiCl_3$ formation which was one of the remarkable findings of the present study as compared with the chlorination of the iron-free titanium oxycarbonitride samples[8, 12]. It is observed from Figures 1 (b)-(c) that the extents of chlorination increased with increasing C/PS ratios. The chlorination of $TiO_xC_yN_z$ proceeds by the chlorination of TiC, TiN and TiO phases, and the standard Gibbs free energy change for TiC chlorination is more negative than TiN and TiO phases. Thus, the samples with high carbon content could be easily chlorinated at this range of temperature although it may have significant impact on the mass transfer of gaseous reactants and products. At the chlorination condition, Fe contents of the samples were converted into ferric chloride ($FeCl_3$). The NaCl trap inside an alumina fiber was utilized to reduce the $FeCl_3$ accumulation at the outlet of the chlorination reactor to minimize iron contamination in $TiCl_4$ product. A liquid phase between NaCl and $FeCl_3$ can occur at temperatures of above 157°C [13], and this technique was used to prevent the contamination of $TiCl_4$ product from undesired $FeCl_3$ (g) formed from the iron in chlorination of $TiO_xC_yN_z$.

The results of the present work obviously indicated that the carbothermal reduction and nitridation of the samples with a mixture of coal-to-polystyrene in high ratios were more feasible. Moreover, the presence of iron in the reduced samples had remarkable impact on the composition of the chlorination resultant. More importantly, the $TiCl_3$ phase was determined to be formed at lowest chlorination temperature and time of about 400 °C and 1 hour; respectively, rather than $TiCl_4$ formation based on the rietveld refinements of XRD patterns presented in Table 3. From the DOE analysis, it can be concluded that there was a remarkable agreement between the estimated values and actual responses which can validate the statistical analyzing method. The SEM images of as-received ilmenite, nitrided ilmenite and the chlorinated samples are presented in Fig. 2. The morphology of nitrided ilmenite and chlorinated samples observed to be porous in structures.

Figure 2. SEM images of (a) as-received ilmenite, (b) nitrided ilmenite and (c) chlorinated sample.

4. Conclusion

In summary, chlorination of titanium oxycarbonitride samples obtained from carbothermal reduction and nitridation were performed at temperature of about 400 -500 °C for 1-3 hours. The results showed that approximately 95% of titanium compounds including titanium carbide and titanium nitride and oxides were chlorinated and transformed into $TiCl_4$ and $TiCl_3$. Metallic iron was transformed into $FeCl_3$ which was trapped by using NaCl at the end of chlorination reactor. The extents of chlorination increased with increasing temperature and increasing the C/PS ratio in reductant. Based on the experimental results, it was observed that C/PS ratio had the most significant effect towards the chlorination process. The highest extent of chlorination of 95.01% was obtained at 500 °C for 1 hour with a C/PS ratio of 1.90.

IOP Conf. Series: Journal of Physics: Conf. Series **1082** (2018) 012061 doi:10.1088/1742-6596/1082/1/012061

Acknowledgements

The authors gratefully acknowledge the financial support from Universiti Sains Malaysia (USM) Fellowship, APEX (1002/JHEA/ATSG4001). This work was also financially supported by USM and Ministry of Higher Education (MOHE) of Malaysia through the following research grants: Fundamental Research Grant Scheme (FRGS) (No. 203/PBAHAN/6071364), 203/PBAHAN/6071261 from MOHE, USM Research University Individual (RUI) grant (No. 1001/PBAHAN/814273) and Nippon Sheet Glass Research Grant (NSGRG) (No. 304/PBAHAN/650360/N120). Special thanks to USM technician Mr. Kemuridan Bin Md. Desa for supporting the experimental work.

References

[1] Habashi F, Kamaleddine F and Bourricaudy E 2014 *Conference of Metallurgists Proceedings (COM 2014), Montreal: Canadian Institute of Mining, Metallurgy, and Petroleum*
[2] Zhang W, Zhu Z and Cheng CY 2011 *Hydrometallurgy* 108 177
[3] Ahmadi E, Fauzi A, Hussin H, Baharun N, Ariffin KS and Rezan SA 2017 *International Journal of Minerals, Metallurgy, and Materials* 24 444
[4] Rezan SA, Zhang G and Ostrovski O 2012 *ISIJ International* 52 363
[5] Ahmadi E, Rezan SA, Baharun N, Ramakrishnan S, Fauzi A and Zhang G 2017 *Metallurgical and Materials Transactions B* 48 2354
[6] Rezan SA, Zhang G and Ostrovski O 2012 *High Temperature Material Processing* 31 381
[7] Rezan SA, Zhang G and Ostrovski O 2011 *Journal of the American Ceramic Society* 94 3804
[8] Adipuri A, Li Y, Zhang G and Ostrovski O 2011 *International Journal of Mineral Processing* 100 166
[9] Ahmadi E, Rezan SA, Hussin H, Baharun N, Ramakrishnan S, Ariffin KS and Fauzi MNA 2016 *INROADS-An International Journal of Jaipur National University* 5 11
[10] Ostrovski O, Zhang G and Anacleto N 2003 *10th World Conference on Titanium; Ti-2003 Science and Technology* 121
[11] Ibrahim N, Ahmadi E, Rahman SA, Fauzi MA and Rezan SA 2017 *AIP Conference Proceedings* 1805 1
[12] Ahmadi E, Sereiratana EM, Rezan SA, Yeoh FY, Ahmad Fauzi MN and Zhang G 2016 *Materials Science Forum* 860 111
[13] Charles M Cook Jr and Wendell E Dunn Jr 1961 *The Journal of Physical Chemistry* 65 1505

IOP Conf. Series: Journal of Physics: Conf. Series **1082** (2018) 012062 doi:10.1088/1742-6596/1082/1/012062

Performance Evaluation and Microstructural Characteristics Of Improved Tool Steel Alloy XW42 By WEDM

M.A. Rahim[1], M. Minhat, N.I.S. Hussein, M.N. Rahman

[1] Faculty of Manufacturing Engineering, Universiti Teknikal Malaysia Melaka, Hang Tuah Jaya, 76100 Durian Tunggal, Melaka, Malaysia

Email: kbmesin@gmail.com

Abstract. M3 is a newly improved cold die steel, an upgrade from the familiar cold die steel XW42. Only little information of these steel is available in literature for its machining characteristics since it is a newly improved die steel. This paper attempts to show the effects of machining factors on the responses consist of material removal rate (MRR), surface roughness (SR) and cutting speed (CS) of WEDM-ed M3 die steel. The machining factors that had been investigated were pulse on time (T_{on}), Voltage (V) and wire tension (WT). Other wire electrical discharge machine (WEDM) performance characteristics such as white layer thickness formation, microstructural and metallurgical alterations were also investigated. The technique of analysis of variance (ANOVA) was used to confirm the variables affecting the responses. The experimental results revealed that T_{on}, V and WT greatly influenced the MRR and SR. With high toughness of M3, no micro cracks were observed on the machined surface. Debris and craters were significantly reduced at low T_{on}, medium V and high WT. SEM analysis of white layer thickness exposed a deterioration about 53% from the base material (XW42) while maintaining the surface roughness at 192μm.

1. Introduction

In a down- stream industries including die making industry, which will usually be affected by any technological changes, it is novel to produce various kinds of dies. In order to produce quality die, an improved die material and die-making processes are continuously being introduced by the die-making industry. M3 is a newly improved cold die steel from Universiti Technical Malaysia (UTeM). It is an upgrade over the known cold die tool steel XW42. M3 removes the disadvantages of insufficient toughness, resulting from the lack of material composition and XW42 high temperature tempering. Very little information is available in literature for its machining characteristics since M3 is a new developed die steel.

[1] labelled that the selection of optimal process parameters for obtaining higher cutting efficiency or accuracy, material removal rate (MRR), lower white layer thickness (WLT) and surface roughness (SR) is still not fully solved, even with the most up to date Computer Numerical Control (CNC) WEDM. This is mainly due to the large number of process variables and complicated stochastic process mechanisms in WEDM as captured by [2]. Despite this, the correlation of process parameters and machining performance are difficult to be established precisely.

[3] reviewed the literature from the period and found that lower WLT, good SR or MRR can be gained through the selection of optimum cutting parameters. Improper selected parameter may also

result in short circuiting of wire, wire rupture, high SR, lower CS and low MRR. The issue has recently grown in importance where more investigations have been carried out to improve these outcomes. Conversely, the problem in WEDM process is still unambiguous. The aim of this study is to clarify several aspects of the machining factors on pulse on time (T_{on}), voltage (V) and wire tension (WT) of WEDM-ed M3 die steel. Machining response includes material removal rate (MRR), surface roughness (SR) and cutting speed (CS). Analysis of variance (ANOVA) is used to study the effects of these machining factors. Another objective of this paper is to acquire the best machining parameters to obtain the lowest white layer thickness of M3 in WEDM. This indicates the need to understand the various perceptions of these improved materials within the WEDM parameter since information is either unavailable or scanty.

2. Experimental study

Figure 1 shows a typical newly cast M3 steel alloy through double austenitisation heat treatment process including tempering to relief internal stresses. All samples were examined for microstructural examinations and mechanical testing after grind and machined to the dimension of 10 mm x 10 mm x 20 mm for hardness and white layer thickness through WEDM and 10mm x 10 mm x 55 mm for Charpy impact test.

Figure 1. A newly cast M3 material consists of cerium, lanthanum and niobium

SF, MRR, and CS would be the dependable variables while independent variables would be the T_{on}, V and wire tension [28]. The compositions of the improved cold work tool steel based on XW42 as shown in Table 1 was in range of AISI D2 or SKD 11. The weight of the material would be measured to identify the MRR. The M3 had a bulk hardness of 52.3HRC with the toughness of 2.94J while the base of XW42 was 51.8HRC and 6.84J in toughness.

Table 1. Chemical composition of improved XW42-M3

Element	Mat.	C	Si	Mn	P	S	Cr	Mo	Ni
Cu	W	V	Nb	Ti	Ce	La	Fe		
Weight, Wt (%)	M3	1.95	0.34	0.32	0.01	0.02	12.7	0.95	
0.35	0.12	0.1	0.85	0.01	0.01	0.01	82.1		

A numerically controlled WEDM machine, Sodick AG600L would be used. Electrode wire was winded at constant speed where current was being supplied into the wire. The new features of the WEDM automatically adjusted the wire tension and speed to avoid machine breakdown due to wire rupture. Brass wire of 0.20 mm diameter was mainly used for electrode. Working fluid (de-ionized water) was supplied from both top and bottom. These machining conditions were chosen based on machine maker manual and recommendation. Three responses were considered for the experiment including material removal rate (MRR), surface roughness (SR) and cutting speed (CS). The samples mass were measured before and after the machining process. The machining time could be obtained from the machine monitor. From these parameter, one could gain the MRR.

The SR could be achieved through the aid of Portable Mitutoyo Surface Roughness Measurement SJ 301. Five readings were taken and the final reading would be the average value of the SR. The CS value could be obtained from the monitor system of the WEDM machine.

IOP Conf. Series: Journal of Physics: Conf. Series **1082** (2018) 012062 doi:10.1088/1742-6596/1082/1/012062

The specimens for microstructural observations were prepared using standard metallographic polishing techniques with 0.05 μm Nano-polish Alumina and cleaning with ethanol (after ultrasonic bath to remove foreign embedded particles). This was followed by etching with 2% Nital (5 ml Nitric acid + 100 ml Ethanol) and then with Murakami reagent for 10 s each [4] were done with grinder and polisher double platen and ultrasonic cleaning. Scanning electron microscope (SEM) equipped with energy dispersive x-ray spectroscopy (EDS) was employed for morphology, microstructure and elemental analysis.

Preliminary work on WEDM based on full factorial design (FFD) was undertaken by [5]. For this current research, three factors including T_{on}, V, and WT would act as the independent variables [6]. FFD (2k) with no replication, where k was the factor, would be implemented in this experiment. There were three factors as stated above. Low and high levels of each factors were chosen based on machine manual and machine maker recommendations. No replication with number of eight experiments and 4 center points, therefore, the total number of experiment for each sample was 12.

3. Results and discussion

A complete values of factors and responses are shown in Table 2. MRR ranges from 0.19 to 0.23 kg/s, SR ranges from 1.92 up to 3.73 μm while CS ranges from 235 to 398 mm/min. The run order is randomized to offset any lurking variables, such as machine warm-up to satisfy the statistical requirement of independence of observations. Randomization acts as insurance against the effects of lurking time related variables. By randomizing the order of experimentation, it can reduce the chances in mistakenly attribute the happenstance effect to the non-randomized factor.

Table 2. A complete matrix table with factors value

Run	A	B	C	Material Removal Rate (MRR) (kg/s)	Surface Roughness (SR) (μm)	Cutting Speed (CS) (mm/min)
1	6	8	100	0.23	3.47	3.98
2	6	8	100	0.21	3.54	3.96
3	10	6	120	0.19	3.62	3.42
4	2	6	80	0.2	2.03	2.35
5	2	10	80	0.21	2.08	2.38
6	2	6	120	0.21	1.92	2.41
7	10	10	120	0.21	3.62	3.28
8	10	6	80	0.25	3.39	2.38
9	10	10	80	0.23	3.73	3.28
10	6	8	100	0.21	3.33	3.90
11	2	10	120	0.2	3.16	2.41
12	6	8	100	0.23	3.55	3.92

Table 3 shows data in ANOVA including the sum of squares for model and residual on the first column. It is found that the pulse on time (T_{on}) contribute the highest with 31.53% while the interaction between voltage (V) and wire tension (WT) contribute about 24.36% followed by T_{on} and V interaction 20.89%. The next column shows the degree of freedom corresponding to the sum of squares. Each effect is based on two averages, high versus low, so it contributes 1 degree of freedom (df) for the sum of squares. There are also 3 df for the three effects in the model pool and 8 df for the eight effects in the residual pool. The ratio of sum of square and degree of freedom results in the mean square and the F values are obtained through the ration between MS Model / MS Residual which forms the value of 5.03 for the model. The F value of 5.03 is bracketed by the critical values for 5% and 0.1% risk which end up with more than 95% confident that MRR is significantly affected by one or more of the effects chosen for the model.

The F-tests for each effect which are based on 1 df as numerators and 8 as denominators, the critical F at 5% for these df (1 and 8) is 5.318. The actual F-values for all three individual effects exceed the critical value F and they are all significantly supporting the half normal plot.

Table 3. ANOVA for material removal rate

Source	**Sum of Squares**	**df**	**Mean Square**	**F Value**	**Prob>F**	**Contibution (%)**
Model	8.009 E-5	3	2.670 E-5	5.03	0.0302 sig.	
A	6.311 E-5	1	6.311 E-5	11.89	0.0087	31.53
AB	4.581 E-5	1	4.581 E-5	8.63	0.0188	20.89
BC	4.875 E-5	1	4.875 E-5	9.18	0.0163	24.36

The model tested in the ANOVA can provide a detail in mathematical equation which can be used to predict a given response in coded form with reference to Table 2:

$$MRR = 0.01917 + 0.01\ A - 0.025\ AB - 0.005\ BC \quad \text{(Equation 1)}$$

The average value of all actual responses is represented by the value for the intercept ($\beta 0$) of 0.01917. The factor A (coefficient 0.01) causes a bigger effect than factors AB (coefficient -0.025) and BC (coefficient -0.005) (See Equation 1).

Table 4. ANOVA for surface roughness and cutting speed

ANOVA for Surface Roughness						
Source	Sum of Squares	df	Mean Square	F Value	Prob>F	Contibution (%)
Model	3.82	3	1.27	7.48	0.0104 sig.	
A	3.34	1	3.34	19.62	0.0022	64.45
ANOVA for Surface Roughness						
Model	1.22	3	0.41	0.76	0.5478 sig.	
A	0.99	1	0.99	1.84	0.2125	17.86

Since there are very small contribution of voltage and wire tension towards material removal rate, surface roughness and cutting speed, there is no need such clarification in this paper. Table 5 shows that only pulse on time which affect highest contribution towards these responses such as 64.45% for surface roughness and 17.86% for cutting speed. There is no significant changes in all responses due to the changes in all factors in this experiment with respect to cutting speed (CS). The model tested in the ANOVA can provide a detail in mathematical equation which can be used to predict a given response in coded form with reference to Table 2:

$$SR = 0.29 + 0.26\ A - 0.18\ B - 0.31\ C \quad \text{(Equation 2)}$$

The average value of all actual responses is represented by the value for the intercept (β0) of 0.29 with factor C causes a bigger effect than factor B (See Equation 2).

Basically, the machine parameters are adjusted based on the machine maker manual with the justification that the cutting process will undergo without the interruption from wire rupture with 3 sets of value which are: $T_{on} = 6\mu s$, V: 8V, WT: 100N, T_{on}: 10μs, V: 6V, WT: 120N and T_{on}: 2μs, V: 6V, WT: 80N.

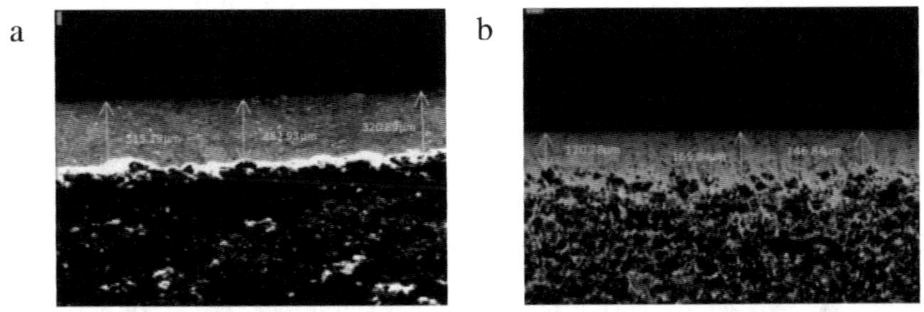

Figure 2: Comparison of white layer thickness between XW 42 and improved XW 42-M3 with niobium, cerium and lanthanum (Sample 6)

Figure 2 introduces micrographs of white layer for sample XW42 and sample 6 (improved XW42-M3) respectively. Figure 2 (a) and (b) shows that the WLT with the lowest value from both (XW42) and sample 6 (Improved XW42-M3) can be found when the optimal machine set up consists of T_{on}: 2μs, V: 10V and WT: 80N and T_{on}: 2μs, V: 6V and WT: 120N. On the average, the white layer thickness value for both material is 306.3μm for XW42 and the value for improved XW42 is 144.32μm. These micrographs verify that the maximum width of the white layer is at the lowest levels T_{on} and WT with highest level of V for XW42.

The improved XW42-M3 micrographs proves that the lowest width of the white layer can be achieved with the highest level of WT, mid -level of V and lowest level of T_{on}. This is because heat energy increases with T_{on} and V. Therefore, greater heat is produced on the machined surface and leads to greater white layer thickness (WLT) on the machined surface. Based on this, using low T_{on} with medium V and high WT yields the best condition for better cutting and low WLT.

From Figure 3, it can be concluded that larger shallow craters are formed at high voltage which resulted in decreased SR due to expansion of the plasma channel in the discharge gap. It can also be found that some of the SR values are increased with high voltage. This phenomena happens because of the debris presence in the working gap which increases the discharging and melts the particle on the workpiece surface due to low flushing efficiency.

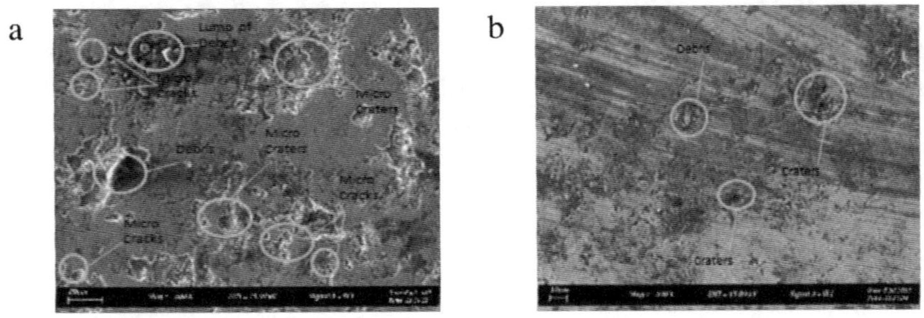

Figure 3: Surface microstructure of sample machined WEDM of XW42 and improved XW42-M3 with lowest surface roughness of 1.92μm and 1.92μm at T_{on}: 2μs, V: 10V, WT: 80N, and WLT_{avg}: 306.3μm; T_{on}: 2μs, V: 6V, WT: 120N, and WLT_{avg}: 144.32μm

The WED machined microstructure surface shows the formation of debris, craters and cracks but no cracks are found because of higher toughness of improved XW42. As shown in Figure 3 (a), due to higher V and low WT, it can be detected that there are more formations of craters, debris and cracks on the machined surface. Large volumes of molten metal at high discharge energy are flushed away due to pressurized waves and form a large crater on the machined surface. Figure 3 (b) shows that lower craters and debris are formed through the use of lower V and higher WT. At low discharge energy, a small volume of metal is melting very small craters and debris and thus improves the surface quality. Debris are formed during re-solidification where some air get trapped within the melted region and when these bubbles collapse, they form globules or debris.

4. Conclusion

Returning to the question posed at the beginning of this study, it is now possible to state that in order to have an appropriate or thin white layer thickness on the improved XW 42, high carbon high chromium (HCHCr) steel alloy, a low level of pulse on time at 2μs, medium voltage at 6V and high level on wire tension at 120N are required with a very fine surface roughness of 192μm. One of the most significant findings to emerge from this study is that through the addition of rare earth elements consisting of cerium and lanthanum with niobium at weight percentages of 0.01%, the toughness is being increased at 18% even though the bulk hardness deteriorates at 4%. The second major finding to emerge from this study is that the white layer thickness decreases at about 53% in width. Multiple ANOVA analyses reveal that T_{on}, V, WT and some interactions between T_{on}, V and WT play an important role in affecting the responses of MRR and SR.

The following conclusion can be drawn from the present study that higher bulk hardness is a promise to gain a thin white layer. It is proven that with bulk hardness value at 52.3HRC of improved XW 42 as compared to the XW42 (51.8HRC), this alloy can have a nice thin white layer of 144.32μm. The current findings add to a growing body of literature on improved cold work tool steel alloy with HCHCr can withstand certain impact toughness and produce good surface finish products if certain improvements are made towards the chemical composition and heat treatment process.

Acknowledgement

This work is partially supported by Universiti Teknikal Malaysia Melaka (UTeM), under the project grant reference number PJP/2017/FKP-AMC/S01560. The authors would like to thank UTeM for its funding support.

References

[1] G. Rajyalakshmi and P. Venkata Ramaiah 2013 *Int. J. Adv. Manuf. Technol.* 69 1249
[2] G. Gautier, P. C. Priarone, S. Rizzuti, L. Settineri, and V. Tebaldo 2015 *Procedia CIRP* 31 203
[3] P. Sharma, D. Chakradhar, and S. Narendranath 2016 *J. Mater. Eng. Perform.* 25 3672
[4] K. B. Small and D. A. Englehart 2008
[5] K. Kanlayasiri and S. Boonmung 2007 *J. Mater. Process. Technol.* 192 459
[6] S. Sarkar, M. Sekh, S. Mitra, and B. Bhattacharyya 2008 *J. Mater. Process. Technol.* 205 376

Properties of Porous Magnesium Using Polymethyl Methacrylate (PMMA) as a Space Holder

P.P.Tan, H.Mohamad and A.S.Anasyida

Structural Materials Niche Area Group, School of Materials and Mineral Resources Engineering, Engineering Campus, Universiti Sains Malaysia, 14300 Nibong Tebal, Penang, Malaysia

Email: anasyida@usm.my

Abstract. Porous magnesium has been recognized as a promising biodegradation metal for bone substitute application because of its excellent biocompatibility, low density, ability to biodegrade in vivo and excellent mechanical properties. In the present work, porous magnesium was fabricated by powder metallurgy using spherical polymethyl methacrylate (PMMA) as a space holder. To determine the optimum sintering temperature for porous magnesium fabrication, the porous magnesium was fabricated using double step sintering process at various sintering temperature (550°C, 585°C and 620°C) in first stage of research. The porous magnesium fabricated was then characterized for morphology, porosity, density and compressive strength. Density of porous magnesium increases and the porosity decreases with increasing sintering temperature. The mechanical characterization indicated that porous magnesium sintered at 620°C exhibited the highest compressive strength and density with optimum range of porosity of 39.37%. For second stage of research, porous magnesium with porosities of 39.38 – 40.82% was produced with different sizes of PMMA particles (38-63 µm, 63-90 µm and 90-125 µm). The compressive yield strength ranges between 19.95 MPa to 23.28 MPa and increases with decreasing PMMA particles size and porosity. These results proven that the PMMA has a potential to be used as space holder in porous magnesium for biomedical application.

1. Introduction

Magnesium has been recognized as a promising biodegradation metal especially for orthopedic applications. Comparing to other biomedical materials, they possess several advantages such as light weight, ability to degrade in vivo without any harmful effects and have elastic modulus which is closer to that of cortical bone. Currently, fabrication of magnesium with interconnected pores and suitable porosity is possible by using power metallurgy (PM) technique based on space holder particles. For this reason, space holder plays a very importance role in producing porous magnesium with desired porous architecture, mechanical properties and has no adverse effect on the resultant structure. Previously, Čapek and Vojtěch [1] synthesized porous magnesium using ammonium bicarbonate as space holder. They found that the total porosity of materials produced by powder metallurgy using the spacer particles is higher than the sum of the added ammonium bicarbonate. They also reported that the expansion of gases (NH_3, H_2O and CO_2) generated in the decomposition of ammonium bicarbonate could contribute to the increased in porosity. In addition, MgO was formed by the reaction of Mg and decomposed space-holding particles that containing oxygen [2]. It also have been reported that the residue of the decomposed space-holding particles was found in the sintered porous Mg using ammonium bicarbonate [3] and carbamide [4] as space holders.

Any reaction between decomposed space-holding particles and magnesium framework may deteriorate the mechanical properties of the resulting scaffolds [5]. Hence, an alternative space holder materials with very worst affinity with Mg such as poly (methyl methacrylate) (PMMA) was utilised by Bi et. al [2]. Their studied are focusing on the different percentage of PMMA but the effect of processing parameters such as sintering temperature and space holder sizes on porous structure and mechanical strength of porous were not detailes studied. Sintering temperature are important in determining the mechanical properties of porous magnesium as it governed the bonding between metal matrix particles in porous magnesium [5]. Moreover, compressibility of porous magnesium are mainly depending on the porosity and pore size [6].

In this work, PMMA space holder was used in fabrication of porous magnesium for biomedical purpose with the aim to reduce the undesired reaction between decomposed space-holder particles and magnesium matrix. At the same time, sizes of space holder and sintering parameter such temperature were studied to obtain an optimum parameter for fabrication of porous Mg. The pore morphology, pore size and distribution, total porosity and compressive strength of sintered porous magnesium were systematically discussed in this research.

2. Methodology

Porous magnesium was prepared by a powder metallurgical method using space holder technique. Pure Mg powder and PMMA particles were used as starting and space holder material respectively. The experiment was divided into two stages. In the first stage, the Mg powders were mixed with PMMA particles and with an Mg/PMMA weight ratio of 80:20. Then the powder mixtures were then compacted and heat treated using a two-step sintering process (550°C, 585°C and 620°C) under high purity argon atmosphere and tested for microstructure, density, porosity and compressive strength based . After obtaining optimum sintering temperature for fabrication of porous Mg, the effect of pore size on the compressibility of porous magnesium were studied. Mg powders were mixed with three different sizes (38-63 μm, 63-90 μm and 90-125 μm) of PMMA particles and sintered with optimum parameter acquiring from the first stage. The sintered porous Mg were then characterized for density and compression test.

3. Results and Discussion

3.1 Microstructure Characterization of Porous Magnesium

The microstructure of porous Mg sintered at three different temperature is shown in Figure 1. The sintered porous Mg had an open-cell structure with uniform pore structure and were distributed homogeneously. This indicated that PMMA space holder and Mg powder produced a homogeneous mixture. The pores shape formed is not perfectly spherical shape. However, there is no significant differences in pore morphology noticed for all the sintered samples. There has two types of pores as shown in Figure 2. The first type of pore (indicated by yellow arrow) has irregular shape. This pore may formed due to expansion of trapped gas during sintering process and incomplete compaction during pressing [3]. The first type of pore also known as micro-pores. This type of pore will reduce the compressive strength of porous Mg due to decrement of load-bearing cross-section area of cell wall [7]. The second type pore (indicated by red arrow) are isolated from each other and have size larger than 100 μm, which is suitable for bone cell ingrowth [8]. There are most likely originated from the decomposition of PMMA particles. Čapek and Vojtěch [3] and Wen et al. [6] using ammonium bicarbonate and carbamide as space holder material have also claimed that two types of pores were formed during porous Mg fabrication. However, compared with other organic space holders such as carbamide and ammonium bicarbonate, PMMA is a polymer with very poor affinity with Mg [2]. As a result, a more uniform pore in the sintered Mg can be easily obtained using PMMA with almost perfectly spherical shape.

IOP Conf. Series: Journal of Physics: Conf. Series **1082** (2018) 012063 doi:10.1088/1742-6596/1082/1/012063

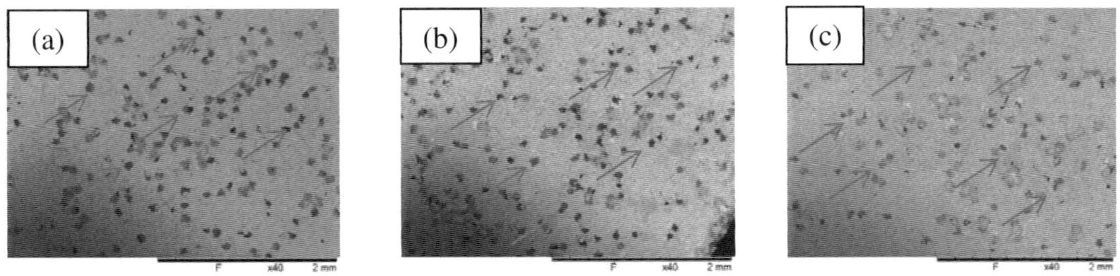

Figure 1. SEM micrograph of porous magnesium sintered at (a) 550°C (b) 585°C and (c) 620°C (⟶ pores)

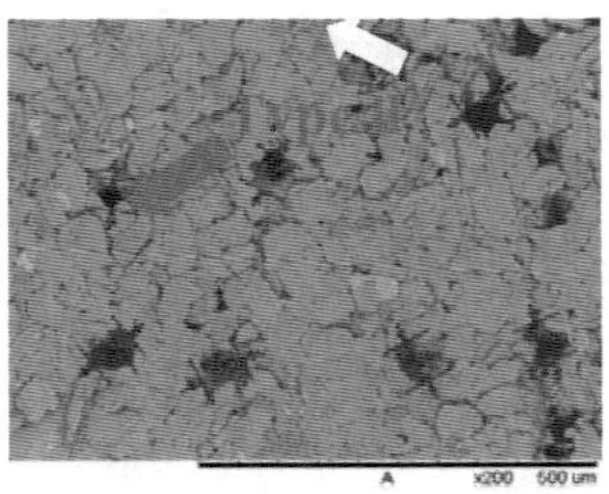

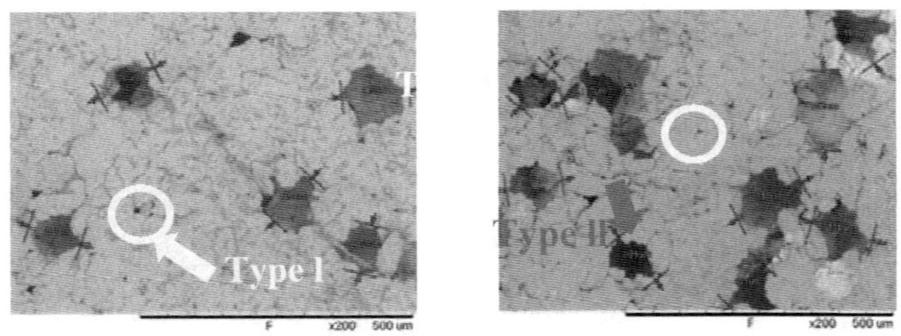

Figure 2. SEM micrograph of porous magnesium sintered at (a) 550°C (b) 585°C and (c) 620°C (type I and II pores are indicated with arrow)

3.2 Sintered Density and Porosity of the Porous Magnesium

Figure 3 shows density and porosity versus sintering temperature. It can be seen that the porosity of porous Mg linearly decreases with increasing density and sintering temperatures. The figure also indicated that samples sintered at 620°C more compacted as compared to samples sintered at 550°C and 585°C. As the sintering temperature increases, the number of inter-particle contacts of Mg increases. This help to improve the growth of sintering necks that ultimately contribute to densification and the decrement of micro-pore sizes [7]. Since the porosity is inversely proportional to the measurement of bulk density, the porosity of porous Mg decreases as the density and sintering temperature increases. However, partial melting of matrix powder and evaporation of some alloying element may occur if the sintering temperature is too high as reported by Dewidar et al. [9]. The density and porosity achieved for the samples are closed to natural bones (density :1.8-2.1 g/cm^3 and porosity :20-70%) [3, 10].

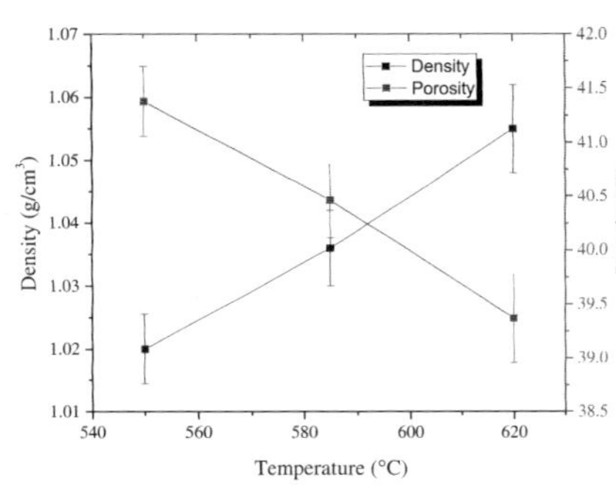

Figure 3. Density and porosity of porous magnesium versus sintering temperature

3.3 Compressive Behavior of Porous Magnesium

Figure 4a exhibits uniaxial compressive stress–strain curves of the porous magnesium sintered at different temperature. The result shows that all compressive stress–strain curves of porous Mg showed ductile behavior under compression and consist of three regions: linear elastic region, plateau region and the densification stage. When the load raised, the stress–strain curve switch from the linear elastic regime to the plateau regime. At plateau stage, the flow stress with some serrations become relatively consistent with the increasing strain. According to Hao et al. [11], present of serrated plateau was mainly due to repeated break and squeeze of the cell wall as the cell wall contain voids and interstices. In fact, the serrated plateau reveal the brittle deformation mode occurs at the cell walls after the yield. As the compression get going, collapse of the cell wall happened caused by rapid propagation of cracks. However, it was observed that porous Mg sintered at 620 °C exhibits different pattern of plateau curve as compared to samples sintered at 550 °C and 585 °C. It showed an increase in compressive strain with the increasing compressive stress, rather than a constant stress plateau. Plateau region are ended after all the voids and pores are virtually eliminated. At last stage, densification stage, it happened beyond 0.45 compressive strain for all sintered samples. The stress increases steeply with the strain increasing slightly due to rapid rise of the resistance of porous Mg as the Mg matrix start to be compressed. In general, the curves showed that the porous Mg that sintered at 620 °C yielded at a higher stress state than those sintered at 550 °C and 585 °C.

Compressive yield strength of the porous magnesium versus sintering temperature are display in Figure 4b. The result indicated that the compressive yield strength of the porous Mg increased with an increase in the sintering temperature. The average yield strength increase from 26.42 MPa to 28.93 MPa with the increasing of sintering temperature from 550 °C to 620 °C. This result can be explained by increase in the number of Mg inter-particle contact during solid state sintering as the sintering temperature increases [12]. In addition, reduction of micro-pores and densification could occur because of enhancement of growth of sintering necks. The yield strength was matched with the average density obtained in the density and porosity test. According to Gibson-Ashby theory, the most influential structural characteristic of porous metal affecting the mechanical properties is its relative density [12]. For this reason, the yield strength of porous Mg increases with the decreasing of total porosity. However, Arifvianto and Zhou [5] have revealed that sintering temperature must not be too high as it could lead to partial melting that influenced resulting pore shape and pore sizes. Overall, the yield strength values of porous Mg are fall within the range of cancellous bone [3]. However, the porosity the porosity range should be take into account since a suitable range of porosity is crucial for bone substitute application.

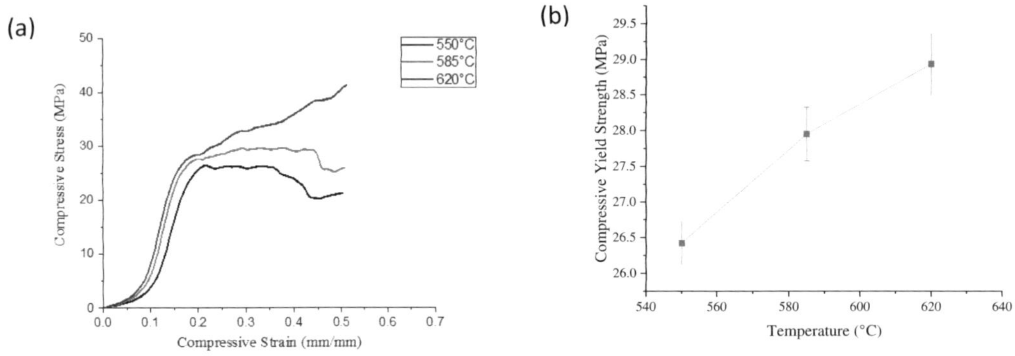

Figure 4 (a) Uniaxial compressive stress–strain curves of the porous magnesium sintered at different temperature, (b) Compressive yield strength of the porous magnesium specimens versus sintering temperature

Among the porous magnesium samples, the sample sintered at 620°C exhibited the highest compressive yield strength and density with optimum range of porosity at around 39.37%. Hence, sintering temperature of 620°C was used on the second stage of experiment, to study the effect of different sizes of space holder on density, porosity and compressive properties.

3.4 Sintered Density and Porosity of the Porous Magnesium with Various PMMA particle sizes

The result of density and porosity of porous magnesium prepared with various size of PMMA particles is shown in Table 1. The average density of sintered porous Mg prepared with 38-63 µm, 63-90 µm and 90-125 µm of PMMA are 1.058 g/cm^3, 1.043 g/cm^3 and 1.030 g/cm^3 respectively. The porosity slightly increased from 39.38 % to 40.82 % with increasing particle size of space holder particle, although identical weight ratio of Mg/PMMA was used for all samples. Also, it is noted that the total porosity of all porous Mg samples was higher than the sum of the added PMMA. The main reason of deviation of total porosity from designed values was due to present of micro-pores. Since the total porosity of the porous Mg is calculated from the sum of macro-porosity and micro-porosity, therefore the total porosity greater than designated value.

Table 1. Density and porosity of porous magnesium prepared with various size of PMMA particles

Particle Size (µm)	Density (g/cm^3)	Average Density (g/cm^3)	Porosity (%)	Average Porosity
38-63	1.058	1.058	39.20	39.38
	1.062		38.97	
	1.055		39.37	
63-90	1.044	1.043	40.00	40.06
	1.040		40.23	
	1.045		39.94	
90-125	1.017	1.030	41.55	40.82
	1.037		40.40	
	1.035		40.52	

3.5 Compressive Behavior of Porous Magnesium with various size of PMMA particles

The stress-strain curve of porous magnesium prepared with different size of space holder is shown in Figure 5 and Table 2 summarized the values of compressive strength for all the samples. In all cases, the stresses strain curves consist of three distinct zones, i) elastic deformation zone (0-0.2 of strain) where stress and strain increases linearly until yield point ii) plateau zone (0.2-0.8 of strain) where serration of flow stress occurs and iii) densification zone (≥0.8 of strain) where the limit stress rapidly increases. It was observed that porous Mg prepared with 38-63 µm yield at a higher stress than those with 63-90 µm and 90-125 µm. The decreasing trend in yield strength for porous Mg containing 63-90 µm and 90-125 µm PMMA particles are correlated with the increasing of porosity. The sizes of space holder have a great influenced on the cell wall. Larger particles size of space holder formed thinner cell wall thus not enough to support the load during compressive test. As a result, the failure of porous Mg prepared with 90-125 µm occurs at relative lower stress compared to others. Moreover, interconnected pores that act as crack initiation might present in porous Mg with higher porosity and causing the stress easily propagate through the framework of samples [13]. Gibson and Ashby revealed that the collapse stress of porous metal is not affected by the pore size [6]. However, the results in the present analysis showed that compressive strength increased with decrease in pore size. Also, it is necessary to determine the upper limit in pore size as bigger pores size could compromise the mechanical properties of porous Mg through increases in void volume. In general, the mechanical properties of the porous Mg fall within the range of those of cancellous bone (2-180 MPa). For that reason, it is proven porous magnesium with PMMA as a space holder is one of the potential candidates for biomedical application especially for hard tissue generation.

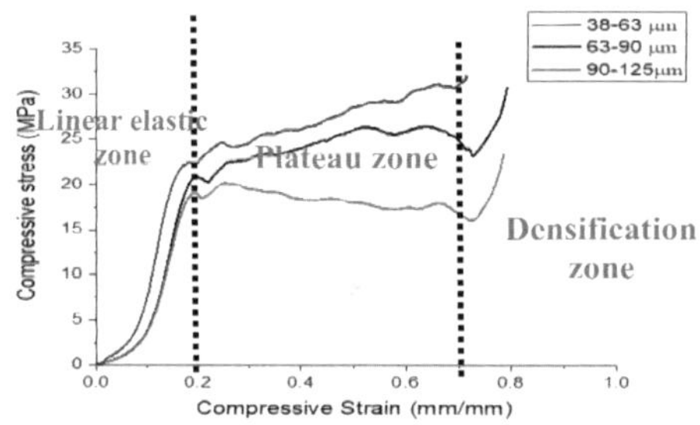

Figure 5. Stress–strain curves of porous magnesium prepared with (a) 38-63μm (b) 60-125 μm and (c) 90-125 μm of PMMA particle

Table 2. Compressive yield strength of porous magnesium prepared with various size of PMMA

Particle Size of PMMA (μm)	Compressive Yield Strength (MPa)	Average (MPa)
36-63	24.45	23.28
	22.89	
	22.50	
63-90	21.38	21.62
	22.54	
	20.95	
90-125	18.14	19.95
	19.07	
	22.63	

4. Conclusion

Porous magnesium with PMMA space holder was successfully prepared by powder metallurgy method using space holder technique.The compressive yield strength ranges between 19.95 MPa and 23.28 MPa and increases with decreasing PMMA particles size and porosity. The mechanical properties of porous Mg produced is in the range of compressive strength of natural bone

Aknowledgements
The authors would like to acknowledge the grant provided by the Ministry of Education through the FRGS 201/PBahan/6071303.

References

[1] Čapek J and Vojtěch D 2013 *Materials Science and Engineering:C* 33 564.
[2] Bi Y, Zheng Y and Li Y, 2015 *Materials Letters* 161 583.
[3] Čapek J and Vojtěch D 2014 *Materials Science and Engineering:C* 35 21.
[4] Zhuang H, Han Y and Feng A 2008 *Materials Science and Engineering:C* 281462.
[5] Arifvianto B and Zhou J 2014 *Materials* 7 3588.
[6] Wen C E, Yamada Y, Shimojima K, Chino Y, Hosokawa H and Mabuchi M 2004 M*aterials Letters* 58 357.
[7] Dizlek M E, Guden M, Turkan U and Tasdemirci A 2009 *Journal of Materials Science* 44 1512.
[8] Rouwkema J, Rivron N C and Van Blitterswijk C A 2008 *Trends in Biotechnology* 26 434.
[9] Dewidar M, Mohamed, H F and Lim J K 2008 *J. Mater. Sci. Technology* 24 931.
[10] Staiger M P, Pietak A M, Huadmai J and Dias G 2006 *Biomaterials,* 27 1728.
[11]Hao G L, Han F S and Li W D 2009 *Journal of Porous Materials* 16 251.
[12]Yilong L, Guibao Q, Yang Y, Xuewei L and Chenguang B 2016 R*are Metal Materials and Engineering* 45 2498.
[13] Aida S F, Zuhailawati H and Anasyida A S 2017 *Procedia Engineering* 184 290.

Effect of SiC concentration on electrochemical and mechanical behavior of Zn-SiC composite coating

T M Al-Dhire, H Zuhailawati[1] and A S Anasyida

[1]School of Materials & Mineral Resources Engineering, Universiti Sains Malaysia Engineering Campus, Nibong Tebal, Penang, Malaysia

E-mail: zuhaila@usm.my
Tel: +6045996101
FAX: +604 594 1011

Abstract. Electrodeposition of zinc with SiC particles is a promising process that could effectively replace chromate process. Increasing of SiC concentration in the bath solution could increase the incorporation of SiC particles into zinc matrix and help in improvement of corrosion resistance and hardness of composite coating. However, it is difficult to obtain composite coating with uniform, dense and high SiC content. The aim of this study to investigate the effect of SiC concentration on electrochemical and mechanical behavior of Zn-SiC composite coating. Zn-SiC composite coatings were successfully deposited on mild steel plate using electrodeposition technique. Zinc sulfate bath containing 2 micron-sized of SiC particles were prepared to deposit the composite coating. The deposition was carried out under operating conditions of current density 40 mA/cm^2 and 600 rpm of agitation rate. Various concentrations of SiC (0, 5, 10, 15 and 20g/l) were investigated. Electrochemical behavior of Zn-SiC coating was performed using linear polarization method. The hardness of the coating was carried out using Vickers hardness tester and the measurements were taken through cross section of the coating. Coating thickness and morphology of the coating were examined using SEM. The experimental results observed an improvement in corrosion resistance and hardness of the coating with increasing SiC content. The microhardness value of 166 (Hv) was obtained at optimum concentration of 20 g/l. The corrosion results showed that the corrosion rate of 6.05 x 10^{-1} mm/yr and 3.0 x 10^{-1} mm/yr were obtained for Zn (0 g/l of SiC) and Zn-SiC (20g/l of SiC) coatings respectively. SEM images of the surface and cross section of the coating showed an embedding of SiC into zinc coating and its particles were distributed along cross section of the coating. Finally average thickness of the coating was recorded at 19 microns.

1. Introduction

The deterioration of metal structures due to corrosion is one of the most serious problem in the world. Corrosion causes durability reduction of steel structures which leads to a loss of hundreds of billions every year. Recently sacrificial metal coating especially zinc coating is shown the most effective method to protect steel against corrosion. Several methods have been developed to deposit zinc coatings, one of which is electrodeposition. The purpose of electrodeposition process is generally to enhance the characteristics of the metal surface so as provide a coating that its ability to withstand corrosive agents, resistance to abrasion and other desired properties or combination of them. However the service life of electrodeposited zinc coating maybe reduced due to formation of white-rust corrosion on the surface of zinc coating [1]. To prevent white rust corrosion, post plating treatment such as chromate is applied [2]. Due to pollution problems associated with using chromate process and environmental impact issues, this process was avoided and no longer in use [3]. The soft zinc coating on the steel surface is susceptible to abrasion damaging [2] and this could lead to initiate the corrosion at steel surface. The strategy available to improve the corrosion behavior of zinc coating and its hardness is the deposition of zinc with ceramics and metal oxides particles. The composite coating is then obtained from electrodeposition bath containing zinc salts and dispersed composite materials. Various materials such as SiC, TiC, Al_2O_3, TiO_2, WC, ZrO_2 and SiO_2 are used to produce composite coating with Cu, Ni, Cr, Fe and Zn. These composite coatings were produced for various industrial applications where desired coating properties are required such as excellent wearing properties, good corrosion resistance and high hardness [4-10]. In recent years the interest in producing composite coatings have been increased rapidly and more attention has been focused on developing of zinc based composite coatings with dispersion particles such as carbides, silicides, nitrides and oxides. Among these particles SiC offers good hardness and wearing properties as well as improvement in corrosion behavior.

The composite coating of Zn-Al_2O_3 with good hardness and corrosion properties was achieved by Sancaglu et al. [11]. Incorporation of CeO_2 with zinc was studied by Ranganathan et al. [12] and reported that the composite coating of Zn-CeO_2 was exhibited an improvement in corrosion resistance. Popola et al. [13] used SiO_2 particles to produce Zn-SiO_2 composite coating on steel surface and concluded that the Zn-SiO_2 composite coating has good corrosion behavior and hardness property.

Few studies were conducted on Zn-SiC composite coating. The corrosion behavior of Zn-SiC composite coating was investigated by Sajjadnejad et al. [2]. They studied the corrosion and hardness behavior of Zn-SiC composite coating at concentration of 5 g/l of SiC and concluded that incorporation of SiC into zinc matrix enhanced hardness and corrosion resistance. The goal of this study is to investigate the influence of SiC concentrations (0, 5, 10, 15 and 20g/l) on morphology, hardness and corrosion behavior of Zn-SiC composite coating.

2. Materials and methods

A glass cell was used for electrodeposition of Zn-SiC composite coatings. The plating solution was zinc sulfate solution. The composition of plating bath is presented in table 1. Zinc sulfate is used as a source of zinc ions in the solution. Adding sodium chloride could improve the electrical conductivity of the solution. Boric acid is used as buffers and fixes pH during electrodeposition, SDS (Sodium Dodecyl Sulfate) is effective surfactant which could reduce hydrophobicity and help in charge stabilization of the suspension particles. The electrodeposition cell consists of anode and cathode. Zinc with (99%) purity and dimension of 30 mm x 20mm x 3mm was used as anode while the cathode was mild steel (20mm x 20mm x 3mm). One side of the steel specimen was subjected to deposition while other sides were covered with epoxy resin. After preparing bath solution, SiC particles with average diameter 2 micron-sized obtained from Alfa Aesar were added to the bath solution and stirred using magnetic stirrer for 18 hrs to allow SiC particles to be suspended in the bath of solution.

The surface preparation of steel specimen involved abrasion of the steel surface with various SiC papers from # 200 to # 2000 grits and polished to 0.05 microns using alumina powder and then ultrasounded in acetone for 5 minutes and finally washed with distilled water. Zinc anode was immersed in HCl (10%) for 30 s and washed with water for activation purpose. The anode and the cathode were immersed in the plating bath and connected to DC power supply. The electrodeposition process started when DC power supply was switched on. All experiments were performed under operating conditions of 600 rpm of stirring rate, current density of 40 mA/cm^2, deposition time of 25 min and various SiC concentrations (0, 5, 10, 15 and 20) were used. pH of solution was kept constant at value of 4.5 during the deposition process. After the deposition process completed, the coated samples were rinsed with distilled water and dried.

Table 1. Chemical composition of zinc sulfate bath

Chemicals	Concentration (g/l)
Zinc sulfate ($ZnSO_4$)	200
Boric acid (H_3BO_3)	16
Sodium Dodecyl Sulfate (SDS)	0.4
Sodium chloride (NaCl)	40

The microstructure and surface morphology were characterized using scanning electron microscopy SEM (Zeiss Supra Model 35VP) and the composition of the coating was identified by energy dispersive system (EDX). The microhardness of Zn-SiC coatings were evaluated using microhardness tester (Shimazdu Co.Ltd., Japan) where 10 gr load and 10 seconds duration time were used. Electrochemical behavior of composite coatings were investigated using linear polarization technique (Auto lab PGSTAT-30). The coating weight and coating thickness were measured and recorded.

3. Results and discussion

3.1. Coating weight measurements

The coating weight was estimated by measuring the sample weight before and after the coating and the difference in the coating weight per unit area was expressed as the coating weight. The coating weight of the Zn-SiC composite coating at various concentrations of SiC was estimated. It was observed that the concentration of 20 g/l of SiC exhibited higher coating weight (0.01685 gr/cm^2) compared with other concentrations. This was due to higher incorporation of SiC particles in the coating.

3.2. Microhardness measurements of Zn-SiC composite coating

Microhardness values of the composite coating with various of SiC concentrations (0 to 20 g/l) were measured (figure 1). The results demonstrated that the microhardness raised with increasing in concentration of SiC in the plating bath. This attributed to the higher hardness of SiC particles. The hardness of Zn-SiC composite coatings were depended on the hardness of SiC particles and as SiC concentration increased, the higher hardness was achieved where it was recorded at 166 and 80 Hv for 20 and 0 g/l of SiC content respectively. Incorporation of SiC into zinc matrix leads to increase in microhardness of Zn-SiC composite coating by refining the coating structure. The coating has many SiC particles which can inhibit further propagation of dislocations and resulted in increase in hardness of Zn-SiC composite coating [14,15].

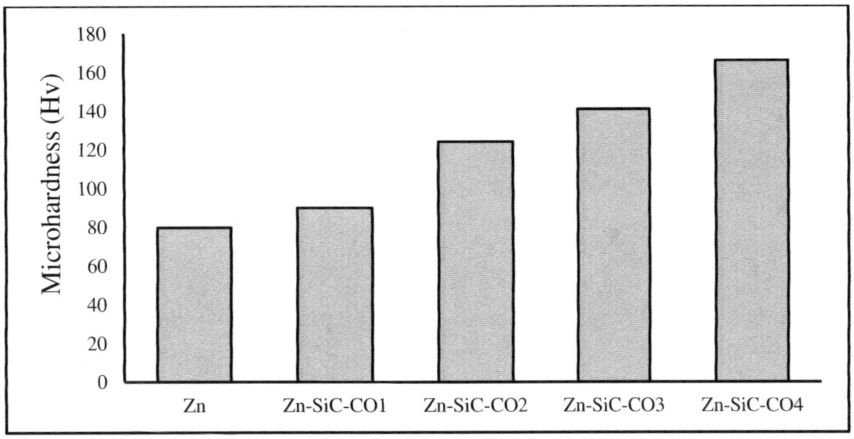

Figure 1. Microhardness of Zn and Zn-SiC composite coatings.

3.3. Morphology and composition of Zn-SiC composite coating

SEM micrographs of Zn-SiC composite coating obtained at various concentrations of SiC are shown in figure 2. It was observed that SiC concentration increased from 5 to 20 g/l in the bath, the incorporation rate of SiC was increased and the steel surface was fully covered with zinc matrix contained SiC particles. Zn-SiC composite coatings have microstructure consisted of large and small grains. These structures were due to the embedment of SiC particles in zinc matrix that perturbs crystalline growth of the metal deposit in some regions [16,17]. Majority of Zn-SiC composite coatings present hexagonal grains parallel to each other. The chemical composition of Zn-SiC composite coatings were obtained using EDX analysis as shown in figure 2(c) and (f). Zn-SiC composite coatings were mainly contained zinc and silicon. At lower concentration of SiC (5 g/l) indicted that zinc matrix contained less amount of SiC particles while at higher concentration of SiC (20 g/l) show higher incorporation of SiC into zinc matrix with less porosity. Incorporation of SiC into zinc matrix is clearly shown in figure 2(b) and (e).

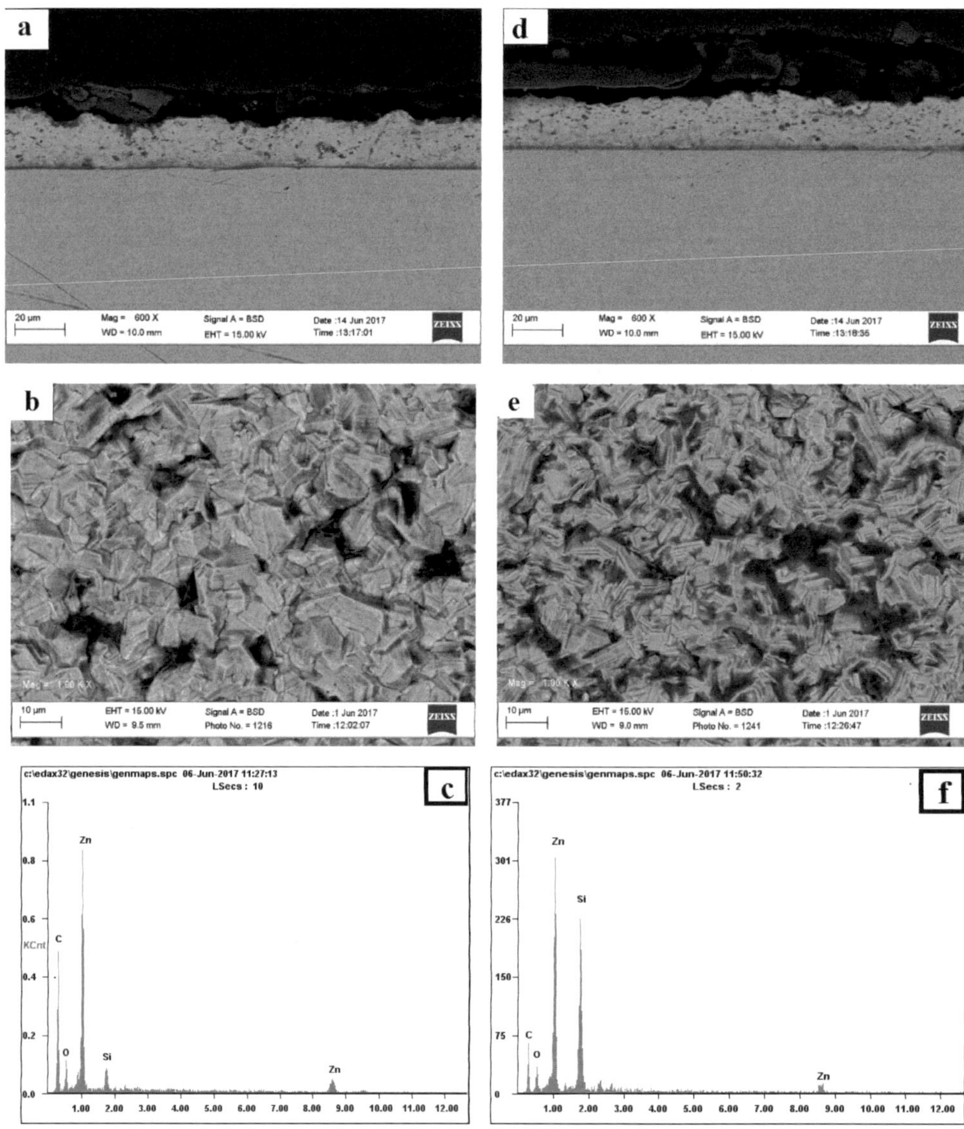

Figure 2. SEM and EDX of Zn-SiC composite coatings respectively (a,b and c) for concentration of 5g/l and (d,e and f) for concentration of 20g/l of SiC.

IOP Conf. Series: Journal of Physics: Conf. Series **1082** (2018) 012064 doi:10.1088/1742-6596/1082/1/012064

3.4. Electrochemical studies

Corrosion tests were conducted using testing solution of 3.5% NaCl. Corrosion characteristics of Zn and Zn-SiC composite coating such as corrosion potential (E_{corr}), corrosion current density (i_{corr}) and corrosion rate were determined from polarization curve using Tefal extrapolation method (figure 3 and 4). Table 2 presents the values of EIS studies. As shown in figure 3 and 4, increasing in SiC concentration has positive effect on corrosion behavior of Zn-SiC composite coatings and Zn-SiC coatings were shifted towards the positive direction compared to zinc coating. Almost all polarization curves of Zn-SiC composite coatings show transition to passivation where SiC particles lead to positive shift in potential from -1160 mV to -1100 mV for zinc and Zn-SiC-CO4 respectively. This demonstrates that the presence of SiC in zinc matrix improved the corrosion behavior. This attributed to formation of protective layer of Zn-SiC where the composite coating surface exhibited dense coating and SiC particles filled the gaps and defects between the grains of zinc matrix.

Table 2. Corrosion behavior of Zn and Zn-SiC composite coatings.

Coatings	i_{corr} (A/cm^2)	E_{corr}(mV)	Corrosion Rate(mm/yr)
Zn-SiC-CO4 (20 g/l of SiC)	2.09×10^{-5}	-1100	3.0×10^{-1}
Zn-SiC-CO3 (15 g/l of SiC)	2.31×10^{-5}	-1117	3.47×10^{-1}
Zn-SiC-CO2 (10 g/l of SiC)	4.79×10^{-5}	-1140	4.18×10^{-1}
Zn-SiC-CO1 (5 g/l of SiC)	2.91×10^{-5}	-1147	4.36×10^{-1}
Zn (0 g/l of SiC)	4.0×10^{-5}	-1160	6.05×10^{-1}

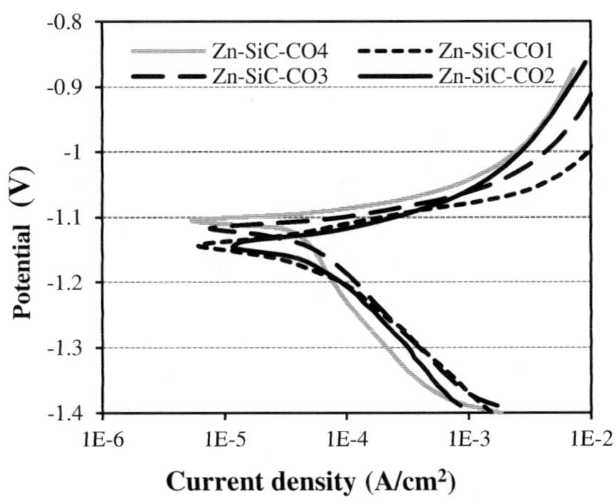

Figure 3. Polarization behavior of Zn-SiC coatings at various concentrations.

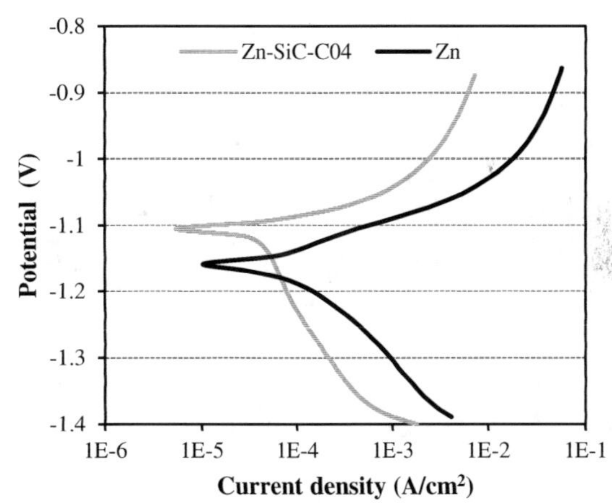

Figure 4. Polarization behavior of Zn (0 g/l of SiC) and Zn-SiC-CO4 (20 g/l of SiC) coatings.

It has been reported that the presence of SiC particles in zinc matrix could lead to formation of corrosion microcells in which SiC particles act as cathode and zinc matrix acts as anode. Because the standard potential of SiC more positive than zinc so that the formation of these microcells facilities anodic polarization, as result of this the localized corrosion was inhibited and homogenous corrosion could occur only [18].

4. Conclusion

Zn-SiC composite coatings were successfully deposited on steel surface using zinc sulfate bath and under direct current condition. SEM images of Zn-SiC composite coating prove the incorporation of SiC into zinc matrix. The surface morphology of Zn-SiC composite coating was smooth and the gaps in the surface of the coating were filled with SiC particles. These conditions resulted in improvement of corrosion behavior of Zn-SiC composite coating where SiC particles act as inert barriers to the initiation of corrosion defects and modify the microstructure of Zn-SiC composite coating. Zn-SiC-CO4 with concentration of (20g/l SiC) was recorded the lowest corrosion rate and was positively shifted in potential. Microhardness of Zn-SiC composite coatings were increased compared to pure zinc coating. This was due to incorporation of SiC particles in coating structure and inhibiting dislocation propagation. All SiC concentrations used in this study show an improvement in the electrochemical and mechanical properties of Zn-SiC composite coating with the best results obtained at 20g/l of SiC.

References

[1] Vathsala K and Venkatesha T V 2011 *Applied Surface Science* 257 8929
[2] Sajjadnejad M, Mozafari A, Omidvar H and Javanbakht M 2014 *Applied Surface Science* 300 1
[3] Langard S and Norseth T 1979 *Hand Book of the Toxicity of Metals*
[4] Zheng H Y, An M Z and Lu J F 2008 *Appl. Surf. Sci.* 254 1644
[5] Benea L, Bonora P L, Borello A and Martelli S 2001 *Wear* 249 995
[6] Sathiyanarayanan S, Azim S S and Vekatachari G 2007 *Electrochem. Acta* 52 2068
[7] Rezrazi M, Dochi M L, Bercot P and Hihn J Y 2005 *Surf. Coat. Technol.* 192 124
[8] Praveen B M, Venkatesha T V, Naik Y A and Prashantha K 2007 *Surf. Coat. Technol.* 201 5836
[9] Praveen B M and Venkatesha T V 2008 *Appl. Surf. Sci.* 254 2418
[10] Aruna S T, Bindu C N, Selvi V E, Grips V K W and Rajan K S 2006 *Surf. Coat. Technol.* 200 6871
[11] Sancakoglu O, Culha O, Toparli M, Agaday B and Celik E 2011 *Materials and Design* 32 4054
[12] Ranganatha S, Venkatesha T V, Vathsala K and Punith kumar M K 2012 *Surface, Coatings Technology* 208 6472
[13] Popoola A P I, Aigbodion V S and Fayomi O S I 2016 *Journal of Alloys and Compounds* 654 561
[14] Vaezi M R, Sadrnezhaad S K and Nikzad L 2008 *Physicochemical and Engineering Aspects* 315 176
[15] Rostami M, Fahami A, Tabrizi B N and Kahrizsangi R E 2013 *Applied Surface Science* 265 369
[16] Zhang Y, Peng X and Wang F 2004 *Materials Letters* 58 1134
[17] Li J, Sun Y, Sun X and Qiao J 2005 *Surface and Coatings Technology* 192 331
[18] Farzaneh, Mohammadi M, Ehteshamzadeh M and Mohammadi F 2013 *Applied Surface Science* 276 697

IOP Conf. Series: Journal of Physics: Conf. Series **1082** (2018) 012065 doi:10.1088/1742-6596/1082/1/012065

Microstructure and Wear Properties of T6 Heat Treated Semisolid A356-TiB_2 composite

N.A.M.Zulkamal, L.M.M.Nasir, A.S.Anasyida

Structural Materials Niche Area Group, School of Materials and Mineral Resources Engineering, Engineering Campus, Universiti Sains Malaysia, 14300 Nibong Tebal, Penang, Malaysia

Email: anasyida@usm.my

Abstract. The aim of this study is to investigate the microstructure and wear properties of T6 heat treated semisolid A356 composite containing 1.5 wt.% and 2.63 wt.% TiB_2. The semisolid A356-TiB_2 composite was prepared using a cooling slope casting technique. The sample, then was solution treated at different temperatures (500°C, 520°C and 540°C) for 5 hours soaking time. Then, age hardening was conducted at different temperatures (120°C, 150°C and 180°C) for 8 hours. Semisolid A356 composite was characterized for microstructure, microhardness and wear properties. Microstructure of semisolid A356 composite was refined with a higher amount of TiB_2. High solution treated temperature (540°C) resulted in the finer size of Si particles in the semisolid A356 alloy. Microhardness of solution treated semisolid A356 alloy achieved higher hardness at 540°C with a hardness value of 75.16 HV. Solution treated of semisolid A356 composite containing 1.5 wt.% TiB_2 has the smallest Si particles and the highest hardness of 119.7 HV. Among the samples, A356-1.5 wt.% composite after aging at 180°C showed the highest hardness (129.7 HV). Wear resistance of semisolid A356 alloys was improved by the addition of TiB_2. Wear surface morphology revealed that the adhesive and abrasive wear mechanism present.

1. Introduction

Aluminium–silicon casting alloys, especially A356 alloy is widely used in the automotive industry because of their good specific mechanical properties [1]. From these alloys we can produce components by semisolid metal processing (SSM), with good surface finish and practically free of porosity. Further enhancement of this alloy can be done by incorporating hard particles in the alloy. Among the reinforcement, TiB_2 has shown an excellent property in A356 alloy using conventional casting. Thus, by incorporating TiB_2 particles in semisolid A356 alloy would be expected to give an improvement by hard particles reinforcement and globular structure.

One of the advantages of SSM processes is that laminar flow during the die-filling which avoids the problems of oxide and gas entrapment and also reduces the shrinkage problems during solidification [2]. Thus, blistering during heat treatment SSM components can be prevented. Therefore, the SSM components are subjected to aging heat treatments to increase their tensile strength and yield strength while preserving a high ductility. The main heat treatment procedures are homogenization, annealing and precipitation hardening (T6 heat treatment). The T6 heat treatment involves solution heat treatment, soaking and aging. The heat treatment principles of conventional as cast aluminium alloys are well understood and their properties have been well studied. However, due to the different microstructure and the solidification of SSM components, the heat treatment conditions which have been optimized for conventionally cast

IOP Conf. Series: Journal of Physics: Conf. Series **1082** (2018) 012065 doi:10.1088/1742-6596/1082/1/012065

materials might not apply to SSM component and the properties produced might not be the same. Therefore, this study was initiated to evaluate the effect of T6 heat treatment on semisolid A356-TiB_2 composite in term of microstructure and wear properties.

2. Methodology

Semisolid A356 composite containing TiB_2 was melted in furnace at 800°C for 1 hour with a heating rate of 25°C/minutes. The molten alloy was stirred at 800°C using stirrer to homogenize the molten mixture and then the temperature was reduced and maintained at 680°C. At the same time, pure magnesium was added into the molten metal and the time was held about 10 minutes. The layer slag formed on top of molten Al was skimmed and removed. Then, the molten metal was poured onto a steel slope and flow down into a pre-heated steel mould. The parameters for slope casting are; slope angle of 60°, cooling slope distance of 200 mm and pouring temperature of 680°C. Then, the sample was left to be cooled down to room temperature for 30 minutes. The dimension of the sample obtained was 14 mm in diameter and 80 mm in length. The samples then were solution treated at various temperature (500˚C, 520˚C and 540˚C) with 5 hours soaking time followed by quenching in water at room temperature. Then, the samples were age hardened at different temperatures (120˚C, 150˚C and 180˚C) for 8 hours. The composites were characterized for microstructure, microhardness and wear properties using optical microscope, Vickers microhardness and pin-on-disc tribometer.

3.Results and Discussion

3.1 Microstructure

Figure 1 shows the microstructure of semisolid A356-TiB_2 composite after solution treated at 500˚C, 520˚C and 540 ºC for 5 hours soaking time. There are four phases; Al, Si, $TiAl_3$ and TiB_2 present in the semisolid A356 alloy containing TiB_2 as confirmed by XRD analysis (Figure 2). During solution treatment, the Si eutectic was sphreroidized and formed rounded shape at all solution treated temperature. Furthermore, the effect of TiB_2 particles is clearly seen which decreased the size of Si needles to finer size as solution treated temperature increases. This observation was similar with Deepak et al. [3] who studied age hardening behavior of thixoformed A356–5TiB_2 in-situ composite. The grain diameter after solution treated at 540°C showed a reduction from 63.04 µm to 35.53 µm for 1.5 wt.% TiB_2 and 57.85 µm to 43.80 µm for 2.63 wt.% TiB_2. Microstructure showed a decreasing in the size of Si particles due to the increment of TiB_2. Semisolid A356 composite containing TiB_2 underwent ageing treatment at 120˚C, 150˚C and 180˚C for 8 hours as indicated in Figure 3. From the figures, the microstructures are relatively look similar therefore aging treatment did not have a significant effect on the microstructure. The microstructure change was mainly caused by solution treatment. The solution process separated TiB_2 and silicon at eutectic region, however, it did not redistribute the particles in the aluminium matrix.

3.2 Microhardness

Figure 4 shows the microhardness of semisolid A356 and semisolid A356 composite containing 1.5 wt.% TiB_2 and 2.63 wt.% TiB_2 after aged treatment. Based on the result, it showed an increment occurred in microhardness as the aging temperature increases. The high hardness achieved at aging temperatures 180ºC for semisolid A356 and semisolid A356-TiB_2 composite. This might be the peak aging temperature, which produced semi coherent precipitate with the matrix [4]. As the result, the hardness for both semisolid A356 alloy and semisolid A356 composite increases.

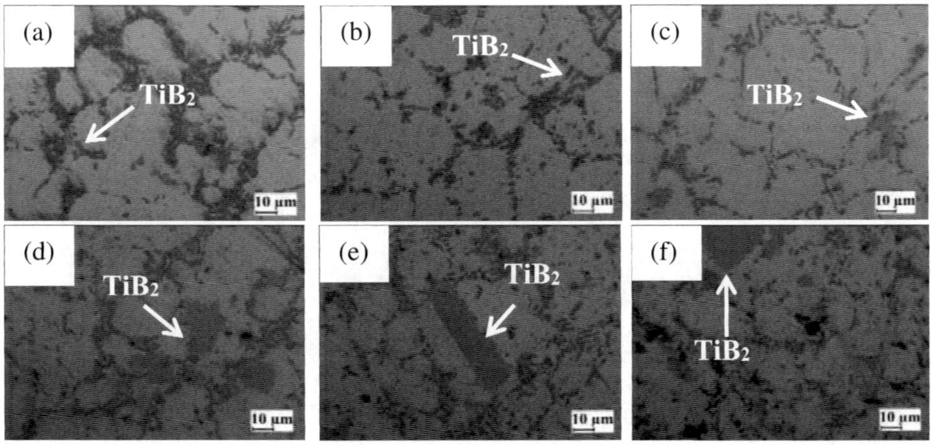

Figure 1. Semisolid A356 alloy composite containing 1.5 wt.% TiB_2 solution treated at (a) 500°C, (b) 520°C, (c) 540°C and 2.63 wt.% TiB_2 at (d) 500°C, (e) 520°C and (f) 540°C for 5 hours soaking time

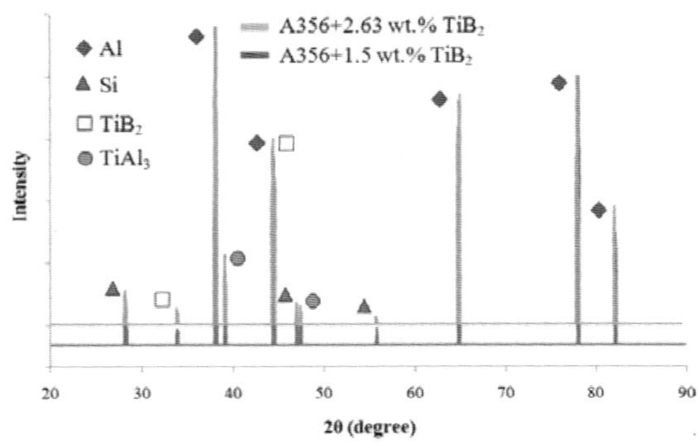

Figure 2. XRD analysis of semisolid A356 composite containing 1.5 wt.% TiB_2 and 2.63 wt.% TiB_2

Figure 3. Semisolid A356 composite containing TiB_2 was aged at different temperature (a-c) 1.5 wt.% TiB_2 at 120°C, 150°C and 180°C, (d-f) 2.63 wt.% TiB_2 at 120°C, 150°C and 180°C

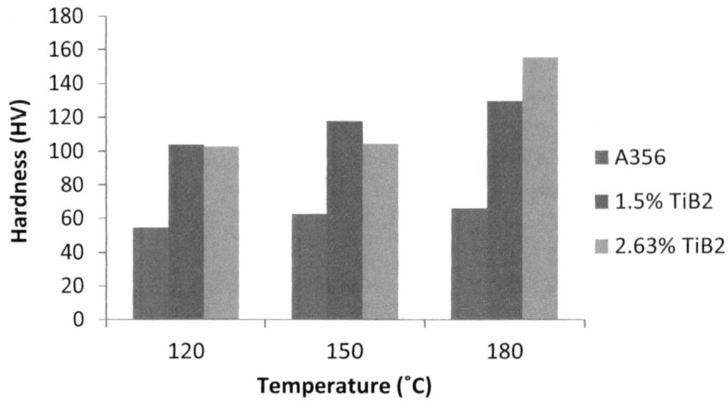

Figure 4. Microhardness of semisolid A356 alloy and semisolid A356 composite with addition of 1.5 wt.% TiB_2 and 2.63 wt.% TiB_2 after aged treatment

3.3 Wear Properties

3.3.1. Volume loss. Figure 5 shows the volume loss for A356 semisolid and semisolid A356-TiB_2 composite. Among them, semisolid A356 alloy has the highest volume loss followed by semisolid A356 containing 2.63 wt.% and 1.5 wt.%. Thus incorporating of TiB_2 particles have lowered the volume loss of A356 alloy due to present of Si-phase and intermetallic compound of $TiAl_3$ and TiB_2. These constituents are very hard and brittle. The work done by Nieto et al. [5] have reported the improved wear rate of aluminium by addition of boron carbide. The performance of Al-Si alloy was improved with the addition of TiB_2 and heat treatment which refined the eutectic Si. The uniform distribution of reinforcement help to improve the mechanical properties and enhanced the machinability [6]. In addition, with the cooling slope casting, globular Al phase produced surrounding by the eutectic silicon matrix also contributes to the better wear properties than as-cast A356 alloy. Similarly, Yasmin et al. [7] have studied the tribological properties of aluminium–silicon eutectic base alloy under dry sliding conditions and have reported that the full heat treatment has shown a great influence on the wear properties of the aluminium–silicon piston alloy. The volume loss of semisolid A356-TiB_2 composite with 2.63 wt.% was high compared to composite with 1.5 wt.% TiB_2 due to high amount of intermetallic compound present in Al matrix which weaken the structure of Al matrix.

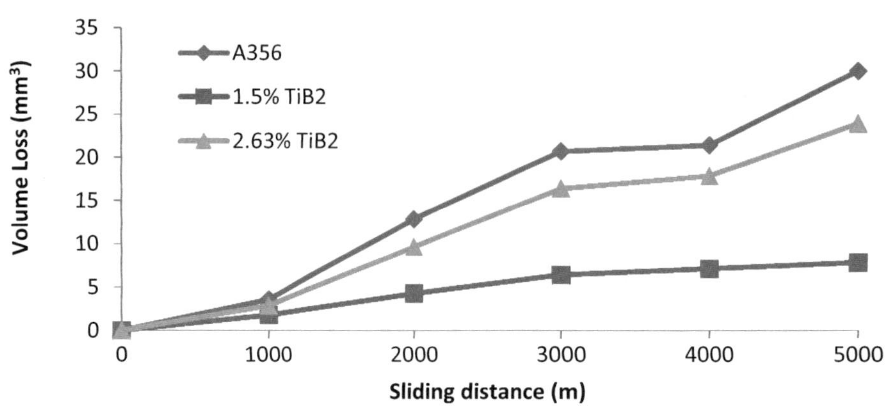

Figure 5. Volume loss of semisolid A356 alloy and semisolid A356-TiB_2 composite

3.3.2. Specific wear rate, K′. Figure 6 shows the specific wear rate of semisolid A356 alloy and semisolid A356-TiB_2 composite at applied load of 70 N. The semisolid A356 alloy has the highest specific wear rate followed by semisolid A356 composite containing 1.5 wt.% TiB_2 and 2.63 wt.% TiB_2. The specific wear rate decrease due to the addition of reinforcement in the alloy. This similar result has been reported by Wang et al. [8] on decreasing wear rate for hypoeutectic Al–Si composite reinforced with TiCN nanoparticles. Ramesh et al. [9] also have reported by increasing content of reinforcement leads to lower wear rates of composites. Others author claims that higher hardness of composites, the wear rate will lowered. This due to the increases in hardness will increase the resistance to sliding wear [10]. All the samples have the specific wear rate in the range of 7.86 x 10^{-5} to 25.3 x10^{-5} mm^3 N^{-1} m^{-1}. According to Rabinowicz [11], there were four types of wear mechanism which are adhesive wear, abrasive wear, corrosion wear and surface fractures wear. In adhesive wear, there were three types of wear regime; severe wear which in the range 10^{-2} - 10^{-4}, moderate wear 10^{-4} - 10^{-6} and mild wear 10^{-6} - 10^{-8}. All the samples are fall under moderate wear regime.

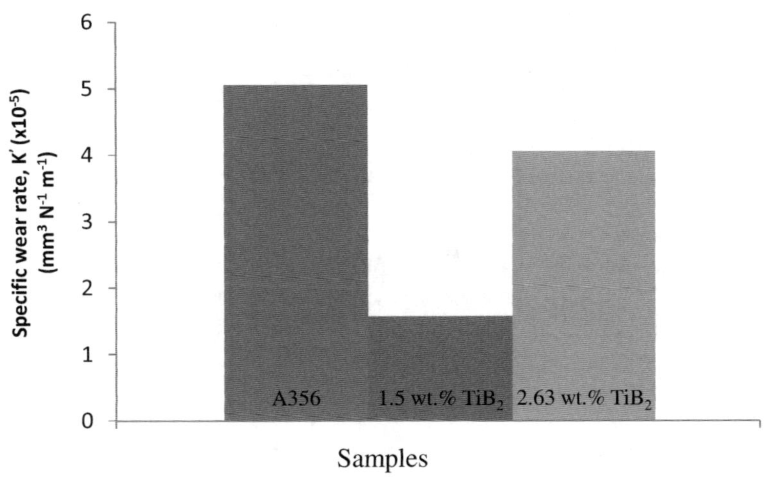

Figure 6. Specific wear rate of semisolid A356 alloy and semisolid A356 composite with various amount of TiB_2 at applied load of 70N

3.3.3. Worn surface morphology. Figure 7 shows the worn surface of semisolid A356 alloy and semisolid A356-TiB_2 composite at applied load of 70 N. The worn surface showed a present of groove, abrasive scoring marks and craters. Deeper and wider crater is observed on surface of semisolid A356 alloy than semisolid A356-TiB_2 composite. This is consistent with the higher volume loss observed for semisolid A356 alloy. According to Deuis et al. [12], the adhesive wear is normally characterised by transfer of material from one surface to the other, resulting in a tribolayer. This layer generally consists of plastically deformed and oxidised particles that reduce the wear rate of the material. Moreover, hardness of the wear surface also influences the adhesive wear rate. Semisolid A356 alloy with lower hardness has produced more debris compared to semisolid A356-TiB_2 composite. The formation of debris is resulted from the delamination wear. The formation of debris is influenced by the properties of materials such as hardness, fracture toughness and microstructure. Among the samples, semisolid A356-TiB_2 composite with 1.5 wt.% showed shallow crater and less abrasive scoring marks.

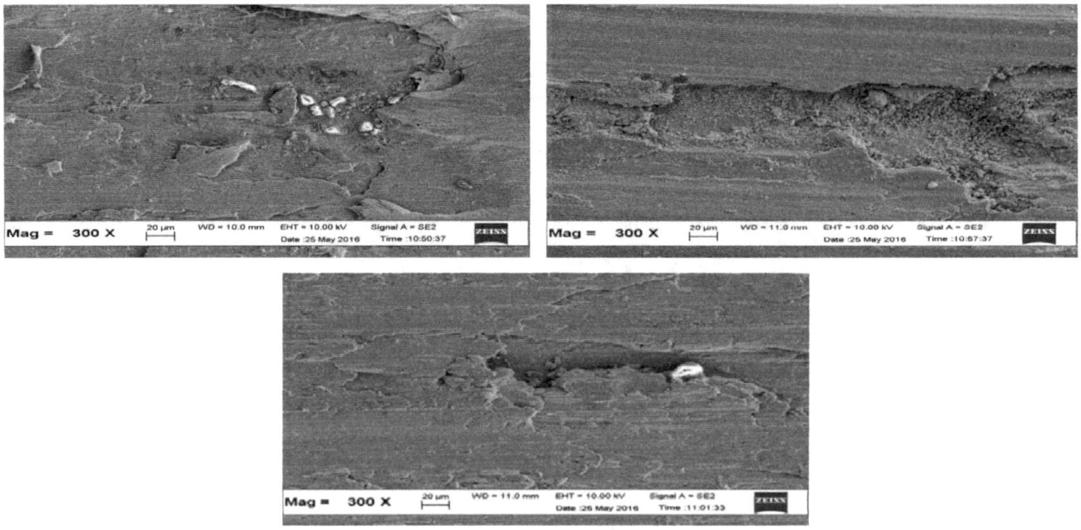

Figure 7. Worn morphology of (a) semisolid A356 alloy, (b) semisolid A356-TiB_2 composite with 1.5 wt.% and (c) semisolid A356- TiB_2 composite with 2.63 wt.%

4. Conclusion

Microstructure of semisolid A356 composites has been refined with the addition of TiB_2 and T6 heat treatment. Higher amount of TiB_2 reduced the size of α-aluminum while T6 heat treatment refined and distribute Si particles homogeneously in the semisolid A356 composite. Wear surface morphology revealed that the adhesive and abrasive wear mechanism present and fall under moderate wear regime.

Acknowledgements

The authors would like to acknowledge the grant provided by the Ministry of Education through the FRGS Grant No. 203/PBAHAN/6071231.

References

[1] Cavaliere P, Cerri E and Leo P 2004 *Journal of Materials Science* 39 1653.
[2] Wu S, Lu S, An P and Nakae H 2012 *Materials Letters* 73 150.
[3] Deepak K S, Mandal A and Chakraborty, M 2015 *Materials Science and Engineering: A* 636 254.
[4] Menargues S, Martin E, Baile M T and Picas J A 2015 *Materials Science and Engineering A* 621 236.
[5] Nieto A, Yang H, Jiang L, Schoenung J M 2017 Wear 390–391 228.
[6] Kori S A and Chandrashekharaiah T M 2007 *Wear* 263 745.
[7] Yasmin T, Khalid A A and Haque, M M 2004 J*ournal of Materials Processing Technology* 153-154(1-3) 833.
[8] Wang K, Jiang H Y, Wang Y X, Wang Q D, Ye B, Ding W J *Materials and Design* 2016 95 545.
[9] Ramesh C S, Ahmed R N, Mujeebu M A and Abdullah M Z 2009 *Materials & Design* 30(5) 1632.
[10] Dinaharan I, Nelson R, Vijay S J and Akinlabi E T 2016 *Materials Characterization* 118 149.
[11] Rabinowicz E 1984 *Wear* 100(1-3) 533.
[12] Deuis R L, Subramanian C and Yellup J M 1997 *Composites Science and Technology* 57(4) 415.

IOP Conf. Series: Journal of Physics: Conf. Series **1082** (2018) 012066 doi:10.1088/1742-6596/1082/1/012066

Preliminary Study on Potential of EAF Slag as Nutrients Source for Mangrove Seedling

KONG Ern Hun[1,2], Nurulakmal M.S. [1]*

[1]School of Materials and Mineral Resources Engineering, Universiti Sains Malaysia, Engineering Campus, Penang, Malaysia

[2]Research and Development Department, Southern Steel Berhad, Penang, Malaysia

Email: kong.ernhun@southsteel.com. srnurul@usm.my*

Abstract. Electric Arc Furnace (EAF) slag generated from steel making process mainly consist of calcium oxide (CaO), iron oxide (FeO), silicon oxide (SiO_2), manganese oxide (MnO), aluminium oxide (Al_2O_3) and magnesium oxide (MgO) in different phases. Since the elements are matching some of the nutrients or micro-nutrients that are needed in plant growth, this study was carried out with the purpose of finding the potential of EAF Slag as nutrients source for mangrove seedling in nursery centre. EAF slag taken from steel making plant during normal production process was crushed and analysed with XRF to obtain chemical composition result and XRD for phase analysis. ICP-OES elemental test was done on leachate from kinetic leaching experiment to assess the leaching trend for each elements form EAF slag in distilled water. Result obtained shows Ca and Si are the major elements that leached from EAF slag with uptrend line, and achieved about 25 mg/l for Ca and 7.5 mg/l for Si after 10 days in distilled water with L/S ratio as main affecting factor. Fe and Mn leaching will be enhanced by stirring at highest speed of 700 rpm. Both elements show uptrend with highest leaching at 0.27 mg/l for Fe and 0.14 mg/l for Mn at day 10, and this shows good potential of EAF slag as slow release element. Mg as an important element in chlorophyll shows uptrend leaching to about 1.5 mg/l at day 10, with L/S ratio as main influential factor. Potassium (K) and sodium (Na) leaching fluctuated below 0.50 mg/l for K and below 0.08 mg/l for Na. Al is the only element that achieved highest level in between day 3 to 7, and then drop gradually to day 10, at about 1.0 mg/l for sample with 700 rpm stirring speed. The results showed good potential for EAF slag to be used as slow release nutrients source for mangrove seedling. Further studies are currently underway to evaluate the effect of EAF slag in altering pH of the soil, as well as efficiency of EAF slag as fertilizer for mangrove plant by observing the growth rate of seedling planted in mangrove nursery site.

1. Introduction

The Tsunami of 26 December 2004 had caused at least 167,000 of lives and 500,000 of homes in Aceh, Indonesia. Besides, it swept away 800 km of mangrove forest along the coastal of the province of NanggrÖe Aceh Darussalam [1]. Since then, mangrove forest protection and restoration become one of the major activities in building up protection to community along the coastal area, especially in east

Asia. Thousands of people and millions of dollars were involved in mangrove forest planting and restoration in tropical and subtropical coastline.

1.1 Mangrove in Malaysia. There are 641,886 ha of mangrove forests in Malaysia, which is less than 2% of the total land area in Malaysia, at about 329,758 km^2. 57% of the mangrove forests are found in Sabah and 26% in Sarawak, with remaining 17% in Peninsular Malaysia, which is 106,554 ha. From this, 69% (441,092 ha) of the mangrove forests have been gazetted as forest reserves [2].

The oldest silvicultural systems for systematic mangrove forests management were developed in Malaysia since 1902 in Matang Mangrove Forest Reserve (MMFR), Peninsular Malaysia, which was still considered as the best managed mangrove forest in the world.

But, forest conversion to agriculture have intensified beginning in the early 1970s. Over a period of 22 years from 1970 to 1992, natural forest in Malaysia was reduced by 19.3% due to conversion of forest land to agricultural land especially for agricultural crops, oil palm and rubber. Lowland dipterocarp forests were affected the most. Peat swamp forests, freshwater swamp forest and mangrove forests were affected as well [3].

There are a lot of mangrove forest related activities being promoted in order to create awareness to people, especially mangrove planting activities. Mangrove seeds were collected and grew in a plastic bag filled with muddy soil at nursery center. After about 5 to 6 months, the seedlings were adopted by participants and were planted at mangrove forest area. But, not all planted seedling survived after a year, which may due to damage by animal in the forest, abnormal dry season, or unhealthy seedling.

1.2 EAF Slag. EAF slag was generated from steel making during refining process in electric arc furnace. The slag waste was unavoidable, in fact it is good for the process in reducing energy and refractory protection. The amount of generation depends on the scrap quality and process efficiency. It is about 8 to 15% of the total raw material that is being charged into the furnace. This means that in a steel mill with one-million-ton capacity, the mill will generate about 100 thousand ton of slag yearly. Currently, there is recycling plant to handle all the slag. The slag will be treated and used to replace aggregate and sand for concrete, road base, and brick making. However, the use of such aggregate is rather limited and could not successfully re-use all of the generated slag. Instead of just recycled the EAF slag as normal waste, the slag which is rich in Ca, Si, Fe, Al, Mn and Mg has created and attracted the interest of researchers to create value for EAF slag in new or advance applications.

Therefore, this study was initiated to find out the potential of re-using EAF slag as nutrients source for seedling growth of mangrove plant.

2. Experimental study

2.1 EAF slag sample.

EAF slag sample was taken from electric arc furnace slag pit, in dry form during normal steel making production. It was analyzed by SPECTRO XEPOS XRF to obtain chemical composition; inspected under EVO MA 10 Zeiss SEM and analyzed with INCA Penta FETx3 EDX to obtain chemical composition.

The raw slag sample was ground by disc mill to size 0.15-0.25, 0.25-0.5 and 0.5-1.0mm, and then used in kinetic leaching test for 10 days with liquid to solid (L/S) ratio of 20:1, 60:1 and 100:1 in distilled water, with stirring speed at 0, 350 and 700 rpm by Simarec Magnetic stirrer. The leachates were analyzed by Induced Coupled Plasma - Optical Emission Spectrometry (ICP-OES), PerkinElmer Optima 7300DV, to determine the quantity of Ca, Si, Fe, Al, Mn, Mg, K and Na.

2.2 Rhizophora mangrove leaves sample.

Mangrove leaves were taken from mangrove forest at Kampung Batas Ban, Nibong Tebal, P. Pinang, Malaysia. 10 fresh leaves were sampled from each of the 5 mangrove trees. The fresh leaves were dried in an oven at 105°C for 24 hours and analyzed by SPECTRO XEPOS XRF to determine the nutrients element content.

IOP Conf. Series: Journal of Physics: Conf. Series **1082** (2018) 012066 doi:10.1088/1742-6596/1082/1/012066

3. Results and discussion

Results of all the experiment were obtained and compiled for subsequent discussion.

3.1 Chemical composition, physical structure and kinetic leaching characteristic of EAF slag

Table 1 shows the EAF slag analysis result by XRF. Result shows that the EAF slag has Ca as the highest element and followed by Fe, Si, Mn, Al and Mg.

Table 1. XRF analysis result of EAF slag (weight %).

Sample	Ca	Fe	Si	Mn	Al	Mg
EAF Slag	29.8	20.2	17.3	8.17	6.13	5.61

Based on the result obtained from SEM-EDX in Figure 1, the EAF slag appeared in irregular shape and based on the EDX analysis, consisted mainly of iron oxide and calcium silicate.

Element	Weight%	Atomic%
O K	35.77	58.72
Mg K	2.21	2.39
Al K	3.71	3.61
Si K	8.38	7.84
Ca K	21.39	14.02
Ti K	0.69	0.38
Cr K	1.16	0.58
Mn K	3.49	1.67
Fe L	21.56	10.14
Zn L	1.64	0.66
Totals	100.00	

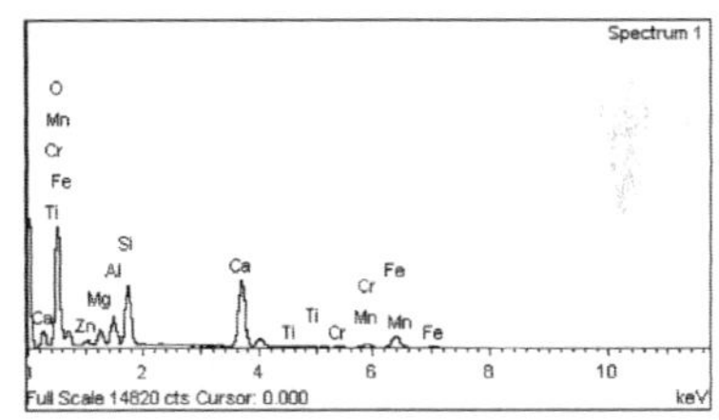

Figure 1: SEM-EDX inspection result of physical structure and element content in EAF slag

Result from Figure 2 shows the 10 days kinetic leaching for each test elements. Ca is the main element that leached out from slag in kinetic leaching test. It achieved about 25 mg/l after 10 days. Si is the second most leached element at about 8 mg/l followed with Mg at about 1.5 mg/l. Three elements show the highest leaching with L/S ratio of 20:1, without significantly affected by stirring speed. Leaching of Fe and Mn at stirring speed 700 rpm showed uptrend line. At day 10 of the experiment, Fe achieved highest at 0.27 mg/l and Mn at 0.14 mg/l. Low leaching concentration was detected for K and Na. Leaching concentration of K fluctuated below 0.5 mg/l and Na fluctuated below 0.08 mg/l. Al is the only element that achieved highest level in between day 3 to 7, and then drop gradually to day 10, at about 1.0 mg/l for sample with 700 rpm stirring speed.

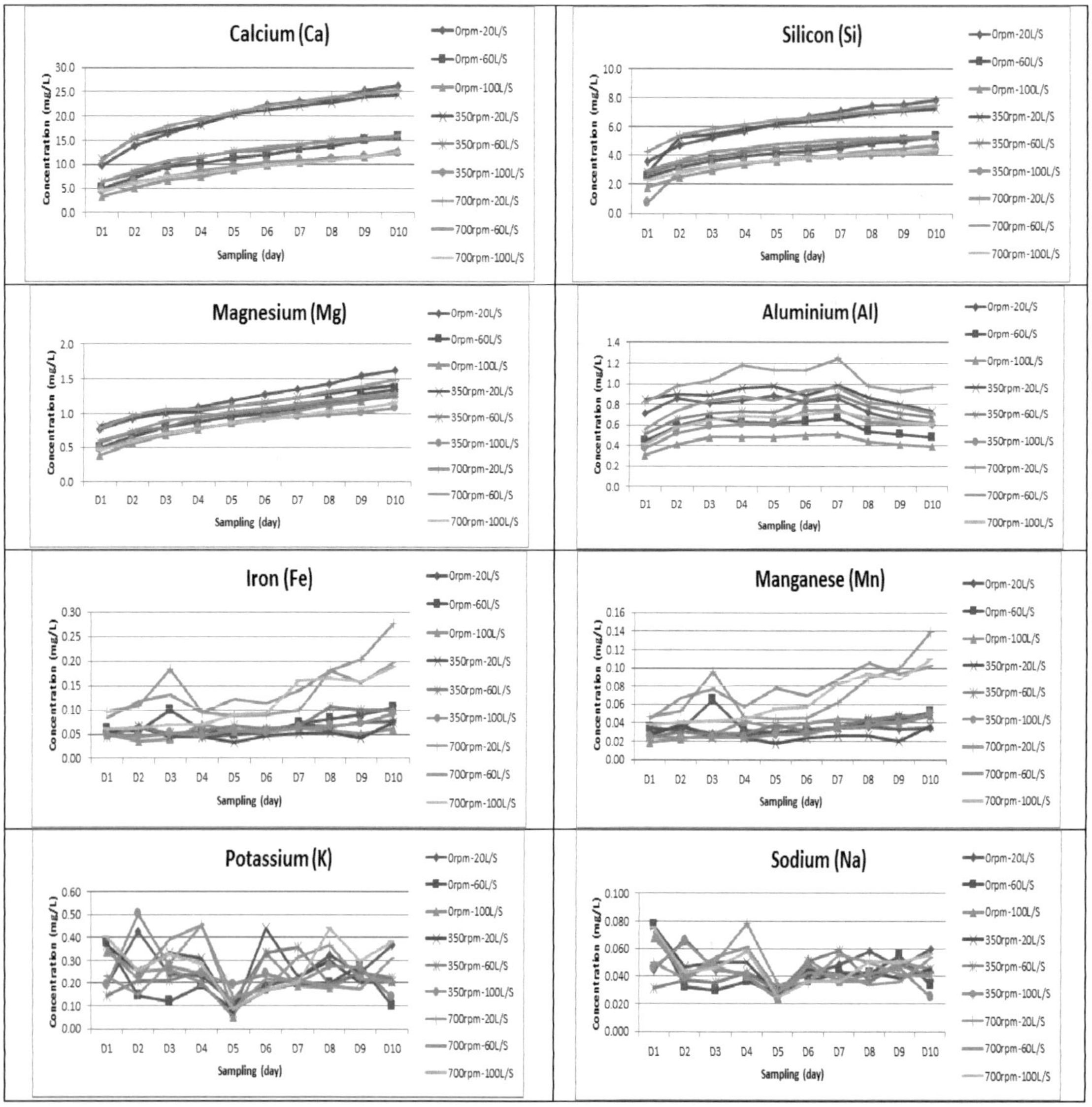

Figure 2: (a): Cacium, and (b): Silicon, (c): Magnesium, (d): Aluminium, (e): Iron, (f): Manganese, (g): Potassium, (h): Sodium, leaching result at L/S ration of 20, 60 & 100 and stirring speed of 0, 350 and 700rpm.

3.2 Nutrients element concentration in Rhizophora mangrove leaves

Figure 3 shows the nutrients element of mangrove in a kg of dry leave weight.

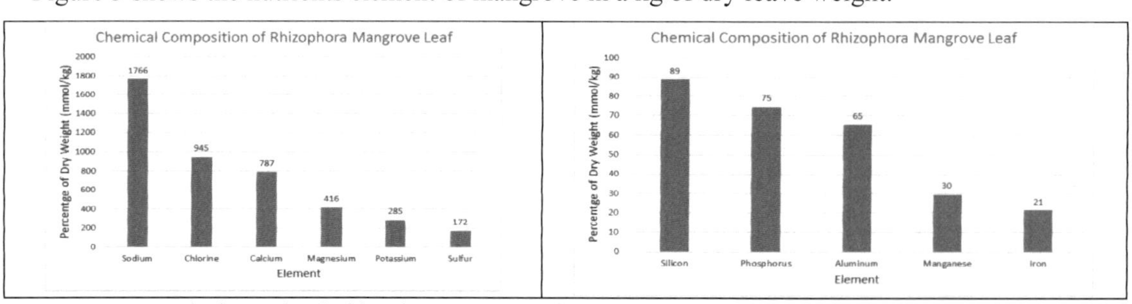

Figure 3: XRF analysis result of Rhizophora mangrove leaves (mmol/kg of dry Weight).

3.2.1 Method of analysis for nutrient elements concentration. In the study of nutrients in sediment and mangrove leaves by Bernini et al. [4], nutrient content was determined using methods described by Sarruge & Haag (1974) and Silva (1981), which is the use of nitric perchloric acid digestion to extract Ca, Cu, Fe, K, Mg, Mn, P, S and Zn. The determination of total P was done by colorimetry (725nm), S was determined by gravimetric determination based on the precipitation of sulphur by barium chloride, in the form of barium sulphate, and other elements by conventional atomic absorption spectrophotometry.

Most of the plant nutrient elements were analysed by Induced Coupled Plasma - Optical Emission Spectrometry (ICP-OES) [5] or Atomic Absorption Spectrophotometry (AAS) [4] [6]. But XRF was also being used in research work to determine the nutrients element in plant [7] [8]. Compared with conventional ICP and AAS methods, XRF has the advantages of simple sample preparation, sample in small quantity, less time spend for analysis test and less cost involved. The studies suggested that XRF technique could be an alternative analysis method for nutrient analysis [7].

Even though XRF has the limitation of unable to analyze low atomic number elements like carbon (C), hydrogen (H), oxygen (O) and nitrogen (N), but in this study, main study elements are not C, H, O and N. Thus, with the facility and reference that available, SPECTRO XEPOS XRF was used to analyze the elements in mangrove leaves.

3.2.2 Nutrients element in mangrove leaves. In mangrove leaves, nutrients followed the sequence of N> Ca> K> Mg> S> P> Fe> Mn> Zn> Cu and there was significant difference between mangrove species and sites of the studies [4].

Table 2 shows the comparison of leave nutrients element in this study with 2 results from Khafaji et al. [6] and Medina et al. [5]. The concentration results are different from each other. It may be due to different species and mangrove forest that the leaves were sampled as highlighted by Bernini et al. [4]. But the sequence of nutrients concentration was similar in all the study as presented in table 2.

Table 2: Rhizophora mangrove leave nutrients content.

	Rhizophora apiculata		Rhizophora mucronata		Rhizophora mangle
	Result in this study		(Khafaji et al. [6])		(Medina et al. [5])
Element	%	mmol/kg	%	mmol/kg*	mmol/kg old leaf **
Sodium (Na)	4.06	1766	1.81	787	~ 500
Chlorine (Cl)	3.35	945	-	-	-
Calcium (Ca)	3.16	787	1.26	314	~ 440
Magnesium (Mg)	1.01	416	0.62	255	~ 190
Potassium (K)	1.12	285	0.69	176	~ 120
Sulphur(S)	0.552	172	-	-	-
Silicon (Si)	0.249	89	-	-	-
Phosphorus (P)	0.231	75	0.14	45	-
Aluminium (Al)	0.177	65	-	-	-
Manganese (Mn)	0.163	30	0.0216	4	~ 10
Iron (Fe)	0.119	21	0.0109	2	~ 1.5
Titanium (Ti)	0.0132	2.76	-	-	-
Vanadium (V)	0.0061	1.19	-	-	-
Chromium (Cr)	0.0076	1.46	-	-	-
Zinc (Zn)	0.0017	0.25	0.0034	0.52	-
Copper (Cu)	0.0009	0.15	0.00054	0.08	-
Leaf water g/g dry mass	2.27	-	-	-	2-2.3

Remark: * Calculated from % result. **Figure estimated from graph (Medina et al. 2015)

3.2.3 Micronutrients. Micronutrients such as molybdenum (Mo), iron (Fe), magnesium (Mg), manganese (Mn), sodium (Na) and copper (Cu) are required in very small quantity for synthesizing chlorophyll and some enzymes. These micronutrients are important for metabolic and physiological

IOP Conf. Series: Journal of Physics: Conf. Series **1082** (2018) 012066 doi:10.1088/1742-6596/1082/1/012066

process. Deficiency of these micronutrients may severely affect the efficiency of photosynthesis, carbon balance, and rate of water use in plants [9]. In the study of mangroves' response to dissolved iron supply rate by Alongi [9], all experimental results showed that mangrove growth rate was enhanced by increasing iron supply. Maximum grow rate of Rhizophora apiculate was at 50-60 mmol Fe per meter square area per day, which is about 50% higher than control sample with 0 supply rate. In another study by Almahasheer et al. [10], result showed that the iron addition consistently led to a fast growth rate in mangrove seedling. The concentration of metallic elements with relative leaf age was presented with one of the conclusion concluded that Mn and Fe concentration were several times larger in Rhizophora mangle at all leaf stages compared with other type of mangrove, Laguncularia racemasa [5].. This shows that the Rhizophora mangle accumulated Mn and Fe more than other species.

In this work, it is observed that Ca and Si are the main leached elements from EAF slag. Ca is important in plant to prevent plant from disorder which may cause curling of leaves, stunt plant growth and death of root tips. Si on the hand, may strengthen the plant cell and prevent invading by bacteria or insect. Even though Fe and Mn which are important for plant growth, leached at low level from EAF slag, but the requirement is in minute quantity by plant. On top of that, it was observed that the leaching of Fe and Mn showed uptrend over time, and this provides the opportunity of EAF slag being used as slow release nutrients source for mangrove seedling.

4. Conclusion

At the end of current studies, it could be concluded that main elements in EAF slag, which are Ca, Fe, Si, Mn, Al, and Mg, are matching the main nutrient (>1% dry weight) of Ca and Mg; minor nutrient (0.1 – 1.0%) of Si and Al; and the micronutrient (in hundreds ppm) of Fe and Mn in mangrove leaves. High leaching rate of Ca, Si and Mg from EAF slag will improve nutrients supply to mangrove soil. Even though EAF slag shows low leaching rate of Fe and Mn, which are important for plant growth rate, but the leaching rate is in uptrend over time, creating a great potential and opportunities of EAF slag to be used as slow release nutrients source to enhance mangrove seedling growth.

References

[1] UNEP 2007 *United Nations Environment Programme*

[2] Abd. Shukor, A.H. 2004 *Philippines: Aquaculture Department, Southeast Asian Fisheries Development Center* 136

[3] Ministry of Science, Technology and the Environment Malaysia (MOSTE) Malaysia 1998 *Malaysia's National Biodiversity Policy (MNBP)*

[4] Bernini E., Maria A.B. da Silva, Tania M.S. do Carmo and Geraldo R.F. Cuzzuol 2010 *Braz. J. Plant Physiology* 22 177

[5] Medina E., Fernandez W. and Barboza F. 2015 *Web Ecol.* 15 1

[6] Khafji A.K., El-Hassan A.M., Mandura A.S., Al-Hachim G.H. and El-Nakkadi A.M.N. 1993 *J.K.A.U.: Mar. Sci.* 4 181

[7] Kimsamran P., Wonprasaid S., Tancharakorn S. and Tantanuch W. 2016 *Int'l Journal of Advances in Agricultural & Environmental Engg. (IJAAEE)* 3 82

[8] Tezotto T., Favarin J.L., Neto A.P., Gratão P.L., Azevedo R.A., and Mazzafera P. 2013 *Scintia Agricola* 70 263

[9] Alongi D.M. 2010 *Ecology* 91 3229

[10] Almahasheer H., Duarte C.M. and Irigoien X 2016 *Frontiers in Marine Science* 3 1

Influent of soaking time in liquid nitrogen on microstructure and mechanical properties of AISI 1018 low carbon steel processed by cryorolling

K.Motham[1], S.Khamsuk[2], H.Miura[3], A.S.Anasyida[1]

[1]Structural Materials Niche Area Group, School of Materials and Mineral Resources Engineering, Engineering Campus, Universiti Sains Malaysia, 14300 Nibong Tebal, Penang, Malaysia

[2]Department of Industrial Engineering, Burapha University, Chonburi, Thailand

[3]Department of Mechanical Engineering, Toyohashi University of Technology, Aichi, Japan

Email: anasyida@usm.my

Abstract. Cryorolling was identified as a new potential route to fabricate ultrafine grained of AISI 1018 low carbon steel plate with less plastic deformation. This work aims to study the influence of soaking time in liquid nitrogen on microstructure and mechanical properties of AISI 1018 carbon steel. AISI 1018 low carbon steel was processed by soaking in liquid nitrogen at various soaking times (5, 10, 20, 30 minutes) and then followed by rolling up to 50% thickness reduction. Microstructure and mechanical properties were investigated using an optical microscope, Vickers microhardness, and tensile test. The hardness and strength were increased up to 10 mins of soaking time and then gradually dropped as the soaking time increases. The highest tensile strength and hardness were 501.75 MPa and 176.04 HV achieved at 10 minutes soaking time.

1. Introduction

Nowadays, low carbon steel has been one of the most used metallic materials in the automotive, railway and construction due to the combination of properties in strength, ductility, good formability and low cost [1, 2]. Although, it has a good mechanical property as mentioned above, its strength is not enough for some engineering application uses. Hence, the improvement in mechanical properties is required through microstructural refinement. Microstructural refinement of steel is usually achieved by alloying and/or thermomechanical treatments accompanies by various types of phase transformation. Recently, severe plastic deformation (SPD) techniques provide another efficient access for grain refinement of metals and alloys. The fabrication of bulk materials with ultrafine grain sizes has attracted a great attention over the past two decades because of the materials' enhanced properties [3].

IOP Conf. Series: Journal of Physics: Conf. Series **1082** (2018) 012067 doi:10.1088/1742-6596/1082/1/012067

The common SPD processes used for bulk materials include such as ARB (Accumulative Roll Bonding), ECAP (Equal Channel Angular Pressing), and simple shear processes (torsion) [4]. However, the current SPD processes may be difficult to apply to practical manufacturing, because of their complicated procedures and/or inapplicability to bulky workpieces. Moreover, the current SPD processes required high strain to achieve ultrafine grained (UFG). To overcome this drawback, cryorolling process was developed to achieve ultrafine grained. Cryorolling is a simple low-processing route which requires a relatively lower load to induce severe strain for producing the ultrafine grain structural features in materials [5]. The cryorolling process involved by soaking the sample in the liquid nitrogen prior to rolling process. During soaking in liquids nitrogen, the material cools, atomic contraction of matrix occurred due to the extremely cold temperature and released great compressive deformation energy that served as the driving force for the enhancement and movement of dislocations during cryrolling. However, for low carbon steel, the extensive research on development of ultrafine grained was investigated using severe plastic deformation (SPD) [6,7] and no work was reported on cryorolling. Therefore, this study was initiated to evaluate the influence of soaking time in liquid nitrogen on the microstructure and mechanical properties of AISI 1018 low carbon steel prior rolling process.

2. Experimental work

Plates of commercial plain AISI 1018 low carbon steel (Fe-0.06wt.% C, 0.14wt.% Mn, 0.01wt.% P, 0.01wt.%) was used. The steel plate was 20 mm in width, 100 mm in length and 5 mm in thickness. The plates were austenized at 900ºC for 1 h then followed by cryorolling. Cryorolling was performed by soaking the plates in liquid nitrogen for 5, 10, 20, and 30 min before each rolling process. The plates then were rolled by 50% thickness reduction (equivalent strain, $\varepsilon = 0.75$). In each rolling pass, the thickness of the samples was reduced by approximately 15%.

The microstructure was observed under optical microscope (OM) attached with image analyzer on the reduction cross section side. Before observed under OM, the samples were prepared by ground with SiC paper up to 2000 grit then followed by diamond suspension polishing of 9 μm to 1 μm. The samples then were etched in 2% nital for 25 seconds. The length and width of the grains were measured and the grain aspect ratio can be achieved by calculating the ratio of the length, l to the side width, w. The hardness of sample was performed using Microhardness Vickers at ten different location at cross section with load of 300 g for 10 s dwell time. Tensile test was conducted at room temperature with crosshead velocity of 1 mm/min using Instron 5982 digital control testing machine. The strength value was an average from 3 individual tests. The crystallite sizes and lattice strain of the samples were calculated using Scherrer equation as shown in Equation 1 and 2 [8].

$$\text{Crystallite size (nm)}, \tau = k\lambda / \beta (\cos \theta) \quad (1)$$

$$\text{Lattice strain}, \eta = \beta / 4 \tan \theta \quad (2)$$

3. Results and discussion

3.1 Microstructure

Microstructure of as-received material contains ferrite and pearlite phase. The grains are in the regular shape with the average size of 11.42 μm (ASTM E1382) as shown in Figure 1a. The grain increases after underwent annealing at 900 ºC with average size of 83.48 μm as shown in Figure 1b. The grain increases due to annealing temperature fall in the grain growth region. Figure 2 shows the microstructure of the samples after cryorolled at various soaking times in liquid nitrogen. The cryorolled samples exhibited deformed grains which are elongated along the rolling direction at all the soaking times.

The grain aspect ratio increases as soaking time increases up to 10 mins and starts to decrease beyond 20 mins as indicated in Table 1. The optimum soaking time maintain the samples in the cryogenic temperature, ensuring the contraction occurred within structural [9-11]. The contraction enables a greater generation of compression stress and deformation energy in the microstructure. The deformation energy would serve as the driving force for the enhancement and movement of dislocations and increased the deformation rate of the samples. From Table 1, sample soaked at 10 mins showed the highest grain aspect ratio compared to other samples. Thus, sample soaked at 10 mins in liquid has the highest deformation rate during the cryorolling process. Thus, sample with higher soaking time causes higher deformation compared to lower soaking time. However, as the soaking time increases up to 20 mins, the sample might have over shrinkage and the volume of deformation energy might not enough to achieve high deformation rate and not much changes observed in the grain aspect ratio of AISI 1018 low carbon steel.

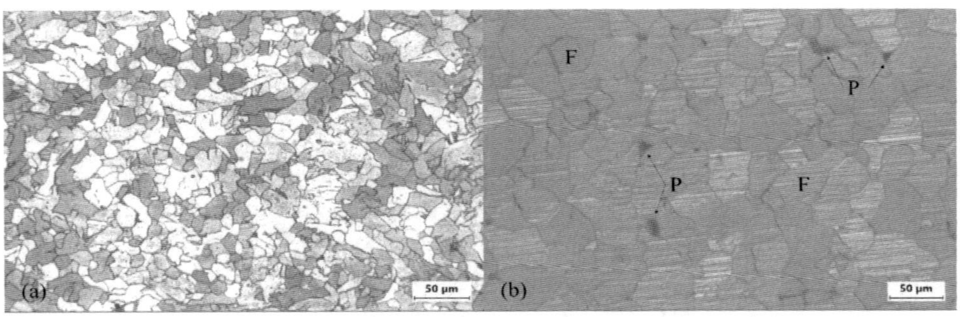

Figure 1. Microstructure of low carbon steel AISI 1018. (a) as-received and (b) annealing.

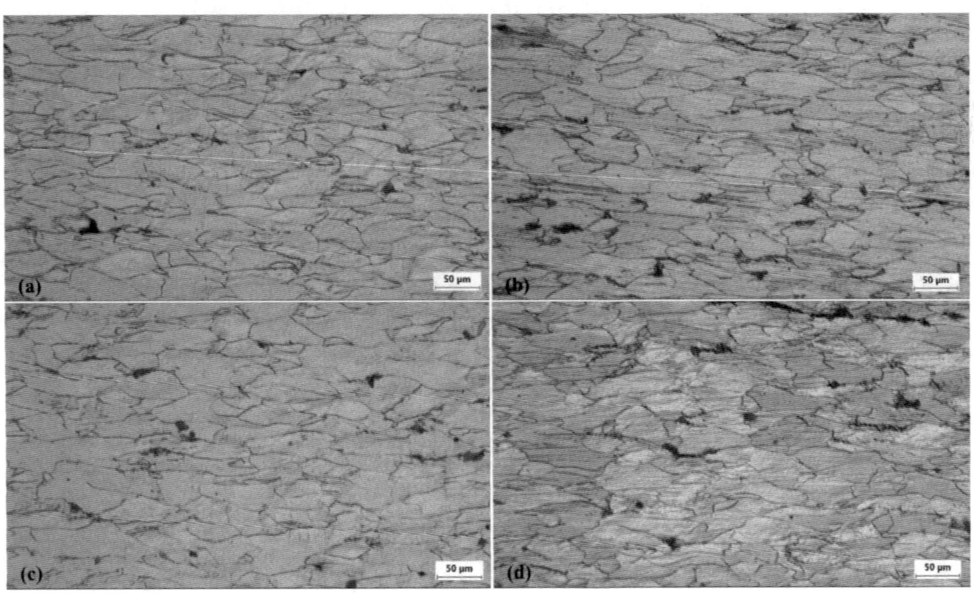

Figure 2. Microstructure of cryorolled low carbon steel AISI 1018 at various soaking times. (a) 5, (b) 10, (c) 20, and (d) 30 minutes.

Table 1. The average grain aspect ratio at various soaking time

Soaking time (min)	Grain Aspect Ratio
5	4.53 ± 1.47
10	5.27 ± 1.94
20	4.33 ± 0.89
30	4.13 ± 1.97

3.2 Crystallite Size and Lattice Strain

The variations of lattice strain ($\langle \varepsilon^2 \rangle^{1/2}$) and crystallite size against soaking time were determined qualitatively from the XRD peak broadening using Scherrer equations and is shown in Table 2. The result showed that the lattice strain for all the samples slightly increased meanwhile the crystallite size decreased as the soaking time was prolonged to 10 minutes. As reported by [12], the cryorolled Al–Cu alloy samples exhibited reduced crystallite size along with enhanced lattice strain and dislocation density compared to room temperature rolled samples. The factors that contribute to the high value of lattice strain are due to the high amount of large deformation strain achieved and the maximum contraction of Feum matrix. The soaking time of 10 minutes able to introduce contraction in Fe matrix which suppressed the atoms inside the sample.

Table 2. Variation of crystallite size and lattice strain of cryorolled AISI 1018 at different soaking times

Soaking time (min)	Crystallite size (nm)	Lattice strain
5	42.1	6.14×10^{-4}
10	30.9	9.67×10^{-4}
20	50.9	5.34×10^{-4}
30	34.0	8.56×10^{-4}

3.3 Mechanical Properties

The hardness and tensile strength of as-received material and cryorolled samples at different dipping time is presented in Figure 3. The as-received material recorded hardness, yield strength and tensile strength of 132.27 HV and 380.43 MPa. The highest hardness (176.04 HV) achieved for sample soaked at 10 mins with increment of 33.09% compared to as-received. The hardness increases as the soaking time increases and then dropped beyond 20 mins. The yield strength and the tensile strength also displayed the same trend as hardness. The highest tensile strength was 501.75 MPa at 10 minutes soaking time. The increase in both hardness and strength are due to increment of plastic deformation as can be observed in high grain aspect ratio.

The cryogenic temperature enables the suppression of dynamic recovery. The suppression of dynamic recovery retains the high strain (high dislocation density) in the sample and thus increased the hardness and strength of the samples. The high dislocation density impedes the propagation of dislocation in the slip system which lead to increase in hardness and strength.

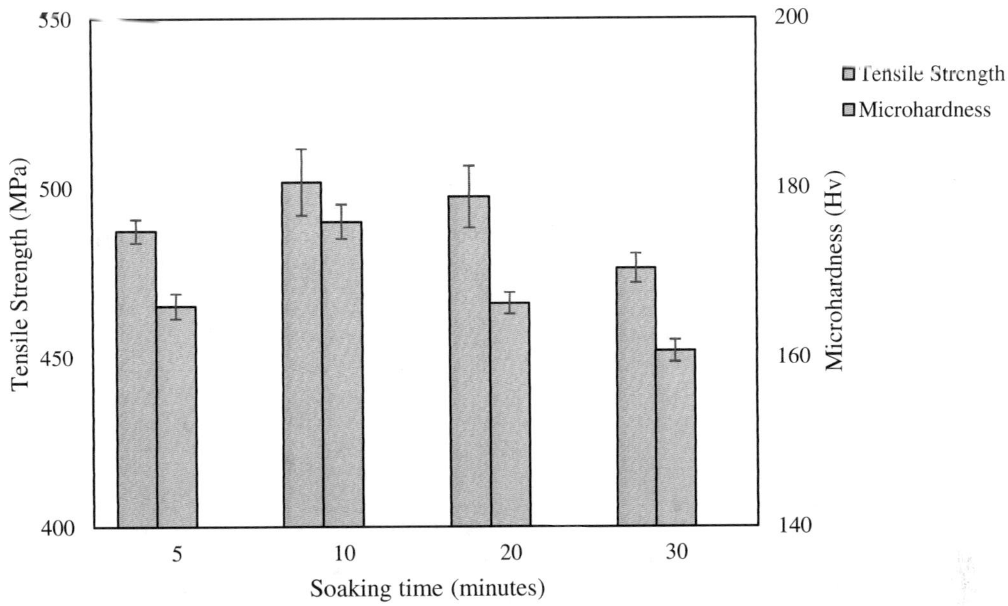

Figure 3. Strength and hardness of cryorolled low carbon AISI 1018 at various soaking times

4. Conclusion

The variation of soaking times affect the microstructure, hardness and strength of cryorolled AISI 1018 low carbon steel. The microstructure showed the highest elongated grain at 10 minutes soaking times with the highest grain aspect ratio of 3.16 μm. The hardness and tensile strength of low carbon steel showed improvements of 33.09% and 31.89% at 10 minutes soaking time compared to as-received. Thus, the suitable soaking time in liquid nitrogen prior rolling for AISI 1018 prior is 10 minutes due to the excellent mechanical properties.

5. Acknowledgment

The authors gratefully acknowledged Universiti Sains Malaysia for providing the facilities and AUN/SEED-NET for funding this study under AUN/SEED-NET grant (304/PBahan/6050352).

References

[1] Phoumiphon N, Radzali O, and Ismail A B 2015 *Procedia Chemistry* 19 822.
[2] Meng Q, Li J, and Zheng H 2014 *journal of Materials & Design* 58 194.
[3] Zrnik J, Dobatkin S V, Raab G, Fujda M and Kraus L 2010 *Revista Matéria* 15 240.
[4] Zrnika J, Dobatkin S V and Ondrej S 2008 *Metal* 13 1.
[5]Yuan C, Wang Y, Sang D, Li Y, Jing L, Fu R and Zhang X 2015 *Journal of Alloys and Compounds,* 619 513.

IOP Conf. Series: Journal of Physics: Conf. Series **1082** (2018) 012067 doi:10.1088/1742-6596/1082/1/012067

[6] Shin D H, Kim B C, Kim Y S and Park K T 2000 *Acta Materialia* 48 2247.
[7] Park K T, Kim Y S and Shin D H 2001 *Metallurgical and Materials Transactions A: Physical Metallurgy and Materials Science* 32 2373.
[8] Mahato DN 2009 State University of New York at Albany.
[9] Shi J T, Hou L G, Zuo J R, Zhuang L Z, and Zhang J S 2017 *Journal of Minerals, Metallurgy and Materials* 24 638.
[10] Siti A Z, Zuhailawati H, and Anasyida A S 2007 *AIP Conference Proceedings* 1865.
[11] Anas N M, Quah W L, Zuhailawati H and Anasyida A S 2015 *Procedia Chemistry* 19 241.
[12] Krishna N, Tejas R, Sivaprasad K and Venkateswarlu K 2013 *Materials and Design* 52 785.

IOP Conf. Series: Journal of Physics: Conf. Series **1082** (2018) 012068 doi:10.1088/1742-6596/1082/1/012068

Surface Roughness, Hydrophilicity and Encapsulation Efficiency of Gentamicin Loaded Surface Engineered PLA Microspheres

Tran Thi Thu Trang [1], **Mariatti Jaafar** [1], **Badrul Hisham Yahaya** [1], **Masakazu Kawashita** [2], **Nguyen Xuan Thanh Tram** [3] **and Zuratul Ain Abdul Hamid** [1*]

[1]Biomaterials Research Niche Group, School of Materials and Mineral Resources Engineering, Engineering Campus, Universiti Sains Malaysia, 14300 Nibong Tebal, Penang, Malaysia, Tel. +604 5996153, Fax. +604 5996907.

[2]Graduate School of Biomedical Engineering, Tohoku University, 6-6-12, Aramaki Aza Aoba Aoba-ku, Sendai, Miyagi 980-8579, Japan, Tel. +81 22 795 7491.

[3]Department of Silicate Materials, Faculty of Materials Technology, Ho Chi Minh City University of Technology, VNU-HCM, 268 Ly Thuong Kiet St, District 10, HCM city, Vietnam, Tel.+ 848 38 647 256.

Corresponding author: *srzuratulain@usm.my

Abstract. Polylactic acid (PLA) is one of the common biodegradable polymers utilised as therapeutic drug vehicles in drug delivery system (DDS). PLA has several desired properties for drug delivery including biocompatible, biodegradable and good mechanical properties. However, the poor hydrophilicity of PLA leads to low cell adhesion, disruption in the human body and causing inflammatory to the biological environment. Hence, the modification of PLA bulk and surface properties has become crucial to increase its potential. This research investigated the effect of surface modification with sodium hydroxide (NaOH) at different concentrations on the PLA microspheres properties which loaded with gentamicin at different loadings (0.25 ml, 0.5 ml, and 1 ml). The average size of the microspheres obtained was in the range of 1 μm to 50 μm; within the range of acceptable size in the parenteral injection. Confirmation of increased hydrophilicity has been evaluated using Fourier transform infrared (FTIR) and contact angle. Gentamicin was encapsulated within PLA microspheres and the PLA microspheres obtained the highest encapsulation efficiency (42%) for the highest volume drug loaded. Morphological changes of modified PLA microspheres have been confirmed via scanning electron microscopy (SEM) imaging.

1. Introduction

Biodegradable polymers are highly desirable in the field of drug delivery because they degrade in the body to biologically inert and compatible molecules. Besides that, by incorporating drugs in biodegradable polymers, this allow for no post-surgery since the remaining polymer will degrade and get cleared by the body. Thus, biodegradable polymers offer a versatile approach for developing sustained release drug delivery systems that are simple and convenient to the patient. One of the most commonly used for drug delivery system is poly(lactic acid) (PLA) since PLA comes from renewable resources [1]. In fact, PLA has outstanding advantages over other polymers such as biodegradable, biocompatible, non-toxic and good mechanical properties [2, 3]. It is being used a potential candidate for biomedical applications [4] and has been approved by the Food and Drug Administration (FDA) for certain clinical applications including drug delivery systems [5]. The solvent evaporation method is the most frequently used technique to prepare polymeric microspheres for drug delivery. The double emulsion method has been used by several researchers to entrap highly hydrophilic drugs such as gentamicin within biodegradable hydrophobic microspheres [6, 7]. Gentamicin, a highly water-soluble drug, is an aminoglycoside that has been used against a wide range of Gram-positive and Gram-negative bacteria [7]. The most disadvantage of PLA is poor its hydrophilicity due to it's unstable property in wet condition, which can undergo chain disruption in the human body, lead to low cell adhesion inflammatory, etc [3]. Thus, PLA needs to be surface modified in order to introduce hydrophilicity so that the biocompatibility property will be enhanced and therefore it is suitable to be used as drug delivery vehicles. There are several surface modification methods such as physical method, including surface coating, entrapment, plasma treatment, and chemical methods [3]. Among all methods, the conventional alkaline hydrolysis technique using sodium hydroxide (NaOH) is chosen to modify the PLA microspheres beacause they are simple, cheap and promising in improving the hydrophilicity property of the micropsheres. A low concentration of NaOH could be applied to avoid severve bulk degradation, and residual alkali was easy to remove. After surface hydrolysis, hydrophilic carboxyl and hydroxyl can be produced by cleaving the ester bond [8]. In this paper, PLA microspheres containing gentamicin was confirmed through UV-Vis and microspheres size was performed by particle size analysis. And the surface structure of PLA microspheres was also examined by using FTIR to determine the presence of hydroxyl groups. While for the surface morphology was investigated by using SEM. Other than that, dynamic water contact angle was done to assess the hydrophilicity of the microspheres after the modification.

2. Experimental study

Poly(lactic acid) (PLA, Nature Work), dichloromethane (DCM, ACS Grade, Merck), poly(vinyl alcohol) (PVA, 80% hydrolyzed, Mw 9-10K, Aldrich Chemistry), sodium hydroxide (NaOH, M=40 g/mol, Merck), hydrochloric acid (HCl, 37%, fuming acid, Merck) and gentamicin (reagent solution, Gibco). All chemicals were used as received. PLA microspheres prepared through emulsion and solvent evaporation (ESE) technique. The primary emulsion was obtained by dispersing at the different volume of gentamicin in the solution of PLA pellets were dissolved in DCM by magnetic stirring for 3 minutes. Then the mixture was immediately with PVA solution to form secondary emulsion at 250 ± 5 rpm continuously stirring for 24 hours. PLA microspheres were washed, filtered and dried in a desiccator (Table 1).

Table 1. The formula used for the synthesis of PLA microspheres

Sample	PLA Concentration (%, g/ml)	PVA Concentration (%, g/ml)	Volume of gentamicin at 10000ppm (ml)
G0	7.5	3	0
G1	7.5	3	0.25
G2	7.5	3	0.5
G3	7.5	3	1

After preparation of PLA microspheres was done, surface hydrolysis treatment was performed to modify surface and introduced functionally on PLA. PLA microspheres were immersed in both NaOH and HCl solutions adopted from previous studies [9]. The encapsulation efficiency was obtained using following equation 1: % EE = [(Total drug – Free drug in supernatant)/Total drug]*100 %. The obtained PLA microspheres were characterized by Malvern Mastersizer particle size analyzer (Worcestershire, UK), the transmission Fourier-Transform Infrared spectroscopy (FTIR) spectra, Dynamic Contact Angle Meter and Tensiometer (DCAT 11, Dataphysic), Cary 50-UV-Vis spectrometer (Varian Inc., Agilent Technology) and Zeiss Supra Gemini 35 VP field emission scanning electron microscope (SEM).

3. Results and Discussion

3.1 Particle size distribution

The particle size distribution curve of PLA microspheres is presented in Figure 1. It was found that size range of the measured particles was around 9-35 µm, where 90 % of the measured particles had a mean particle size ranging from 22-32 µm and below 50 µm. The homogeneity was determined based on the span values for each of formulations. The span value was calculated based on the formulation as described as follow: Span value = $(D_{90} - D_{10})/ D_{50}$. Whereby D_{90}, D_{10}, and D_{50} are the particle distribution at 90 %, 10 %, and 50 % respectively. The span lower than 1 indicates that the size distribution can be considered as homogeneously dispersed [10]. Based on the results, the particle distribution of microspheres within the range of acceptable size in the parenteral injection.

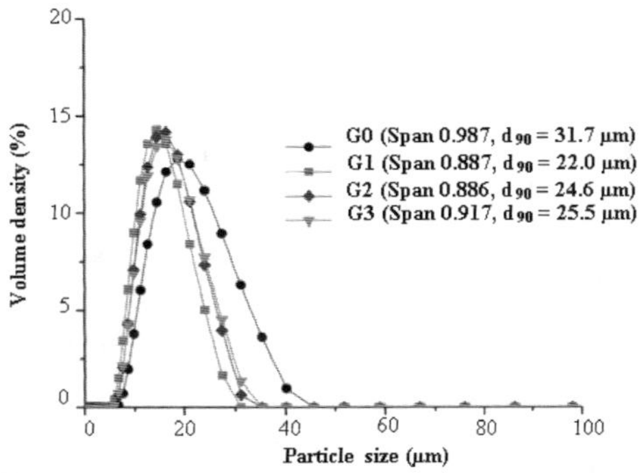

Figure 1. The particle size distribution of PLA microspheres

3.2 FTIR Analysis

Figure 2 (a-c) show the FTIR spectra of series of surface modified PLA microspheres at difference concentration of NaOH. Alkaline hydrolysis will produce hydroxyl (-OH) through cleaving the ester bond in PLA. The appearance of (-OH) groups theoretically was observed at wavelength 3300 to 3610 cm^{-1} and intensity of -OH peak increased as the concentration of NaOH used increases [9]. Besides that, the FTIR also showed that the PLA microspheres treated with 0.5 M NaOH showed the highest intensity because higher concentration NaOH was having more ability to cleave the ester bond. This confirmed the cleavage of ester bond when surfaces microspheres treated with NaOH has successfully occurred. Figure 2 d) shows the FTIR spectra of neat gentamicin, neat PLA (G0) and PLA loaded with 0.25 ml (G1), 0.5 ml (G2) and 1ml (G3) of gentamicin. It is known that amine group exist in gentamicin can be represented by peak in the range of 3100 - 3600 cm^{-1} [11]. From Figure 2 d), it was found that significant peak appearance within this range. Therefore, it can be concluded that the gentamicin was loaded within the PLA microspheres after ESE process. Based on these results, the encapsulation of gentamicin within PLA microspheres was achieved and hydrolysis treatment has resulted in increased of hydrophilicity property of surface engineered PLA microspheres.

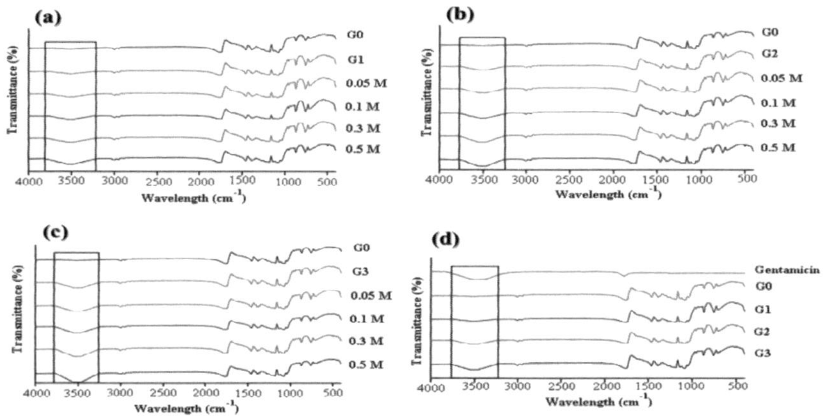

Figure 2. Comparison on chemical bonding Neat PLA (G0) with:
(a) PLA loaded 0.25 ml (G1) treated with 0.05 M, 0.1 M, 0.3 M and 0.5 M NaOH
(b) PLA loaded 0.5 ml (G2) treated with 0.05 M, 0.1 M, 0.3 M and 0.5 M NaOH
(c) PLA loaded 1 ml (G3) treated with 0.05 M, 0.1 M, 0.3 M and 0.5 M NaOH
(d) PLA loaded 0.25 ml (G1), 0.5 ml (G2) and 1ml (G3).

3.3 Hydrophilicity Testing

Figure 3 shows that the neat PLA (G0) has the highest hydrophobic behavior because the contact angle value obtained was more than 90° (103.91°). While PLA loaded gentamicin (G1, G2, and G3) undergo changes into hydrophilic behavior which was below 90°. This is due to the existance of loaded gentamicin within PLA microspheres. Furthermore, the contact angle values were lower when a higher concentration of NaOH. For PLA loaded 0.25ml (G1) was decreased from 72.8° – 58.59°, for (G2) also was decreased from 68.98° – 51.54°. The lowest degree of contact angle was observed for PLA loaded 1ml gentamicin (G3) from 57.78° – 46.33°. From the result, higher the concentration of NaOH used for surface treatment, the more hydrophilic the PLA microspheres. Therefore, PLA treatment with NaOH at difference concentration has successfully modified the surface hydrophilicity property of PLA microspheres.

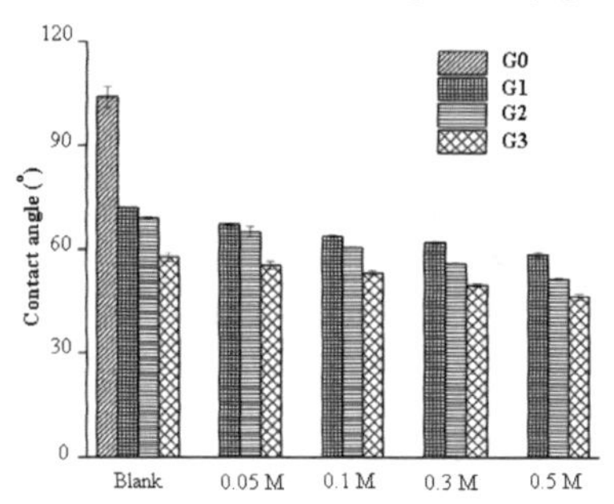

Figure 3. The contact angle of PLA microspheres before and after treated with NaOH.

3.4 Encapsulation Efficiency

The absorbance of gentamicin was measured using UV-Vis at 250 nm. The standard curve of different concentration gentamicin was plotted, as illustrated in Figure 4. The concentration of free drug in supernatant was determined from this standard curve and encapsulation efficiency was calculated based on the equation 1.

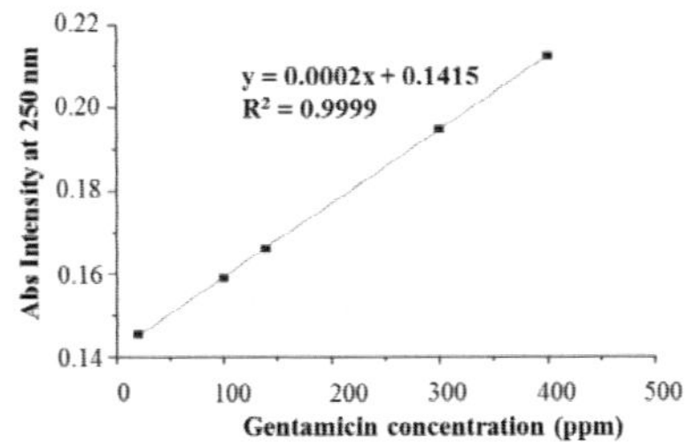

Figure 4. Standard curve obtained from absorbance intensity of Gentamicin at different concentration in 3% PVA solution.

The results of the encapsulated efficiency (EE) indicate that drug loading obtained was increased from 8.88 ± 2.05 % (G1), 18.36 ± 4.45 % (G2) to 42 ± 1.40 % (G3) when the volume of drug increased. This is due to the increased the volume of drug in the formulation has resulted in higher encapsulation efficiency obtained [12, 13].

3.5 Morphology Characterization via SEM

From Figure 5, the surface structure of PLA microspheres was observed at the same magnification at 1800x. In Figure 5 (a) the neat PLA (G0), (b), (c), and (d) PLA loaded gentamicin (G1, G2 and G3), the surface of microspheres is smooth. But Figure 5 (e), (i) and (m) treated with 0.05 M NaOH, showed the change on the surface as it becomes rough compared with PLA microspheres before treatment. This is due to the hydrolysed surfaces which resulted in the change from smooth to rough surfaces. When the concentration of NaOH increased (Figure 5 (f), (j) and (n) of 0.1 M; (g), (k) and (o) of 0.3 M; (h), (l) and (p) of 0.5 M), the surface morphology showed that PLA microspheres become rougher and the introduction of porous structure was observed. This is due to higher concentration of NaOH used for the surface hydrolysis causes more breakage ester bonds occurred. Therefore, more –OH groups being introduced into PLA although the different volume of the drug has been used, the roughness of each sample almost showed the same pattern due to the same concentration of NaOH has been used.

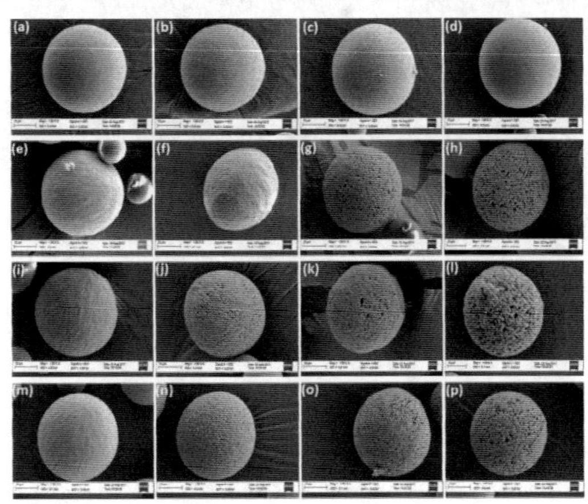

Figure 5. PLA microspheres surface morphology of (a) Neat PLA (G0) in comparison with:
(b) PLA loaded 0.25 ml (G1) treated with (e) 0.05, (f) 0.1, (g) 0.3 and (h) 0.5 M NaOH
(c) PLA loaded 0.5 ml (G2) treated with (i) 0.05, (j) 0.1, (k) 0.3 and (l) 0.5 M NaOH
(d) PLA loaded 1 ml (G3) treated with (m) 0.05, (n) 0.1, (o) 0.3 and (p) 0.5 M NaOH

4. Conclusion

The surface engineered PLA microspheres loaded with gentamicin was successfully developed. An optimized encapsulation efficiency of gentamicin at 42 % was achieved. The hydrolysis treatment of PLA microspheres with NaOH has resulted in the improvement of hydrophilicity property. This has been achieved as shown in FTIR, SEM and contact angle. Hydrophilic functional groups (-OH) that being introduced during surface treatment enhanced hydrophilic behavior 46.33^{o} at 0.5 M NaOH. This hydrophilicity would be expected to improve the cell-biomaterials interaction during drug delivery. The gentamicin loaded surface engineered PLA microspheres has great potential carriers for drug delivery particularly for bone regeneration application.

5. Acknowledgment

The authors would like to thank AUN/SEED-NET for the Collaborative Research Program Grant (Grant no. 304/PBAHAN/6050353) and gratefully acknowledge School of Materials and Mineral Resources Engineering, Universiti Sains Malaysia for the facilities provided.

References

[1] I. Manavitehrani, A. Fathi, H. Badr, S. Daly, A. Negahi Shirazi, and F. Dehghani 2016 *Polymers* 8 20
[2] R. E. Drumright, P. R. Gruber, and D. E. Henton 2000 *Advanced materials* 12 1841
[3] L. Xiao, B. Wang, G. Yang, and M. Gauthier 2012 *Biomedical science, engineering and technology*
[4] M. Jamshidian, E. A. Tehrany, M. Imran, M. Jacquot, and S. Desobry 2010 *Comprehensive Reviews in Food Science and Food Safety* 9 552
[5] Q. Xu, A. Crossley, and J. Czernuszka 2009 *Journal of pharmaceutical sciences* 98 2377
[6] Y.-Y. Yang, T.-S. Chung, and N. P. Ng 2001 *Biomaterials* 22 231
[7] C. Lecároz, M. A. Campanero, C. Gamazo, and M. J. Blanco-Prieto 2006 *Journal of Antimicrobial Chemotherapy* 58 557
[8] C. Tham, Z. A. A. Hamid, Z. Ahmad, and H. Ismail 2014 *Surface Engineered Poly (lactic acid)(PLA) Microspheres by Chemical Treatment for Drug Delivery System* 594
[9] M. M. Sabee, N. Kamalaldin, B. Yahaya, and Z. A. Hamid 2016 *Journal of Polymer Materials* 33 191
[10] A. F. H. Ismail, F. Mohamed, L. M. M. Rosli, M. A. M. Shafri, M. S. Haris, and A. B. Adina 2016 *Spectrophotometric Determination of Gentamicin Loaded PLGA Microparticles and Method Validation via Ninhydrin-Gentamicin Complex as a Rapid Quantification Approach*
[11] A. Rapacz-Kmita, M. Bućko, E. Stodolak-Zych, M. Mikołajczyk, P. Dudek, and M. Trybus 2017*Materials Science and Engineering* 70 471
[12] F. Boury, H. Marchais, J. Proust, and J. Benoit 1997 *Journal of controlled release* 45 75
[13] R. C. Mehta, B. Thanoo, and P. P. Deluca 1996 *Journal of Controlled Release* 41 249

Micro-CT evaluation on osteoconductivity of DCPD-coated β-TCP granule using experimental rats

Khairul Anuar Shariff[1], Kanji Tsuru[2], and Kunio Ishikawa[2]

[1] School of Materials and Mineral Resources Engineering, Engineering Campus, Universiti Sains Malaysia, 14300, Nibong Tebal, Penang, Malaysia

[2] Department of Biomaterials, Faculty of Dental Science, Kyushu University, 3-1-1 Maidashi, Higashi-ku, Fukuoka, 812-8582, Japan

Email: biokhairul@usm.my

Abstract. β-tricalcium phosphate (β-TCP) bone substitutes have been widely used because of its excellent tissue response and good osteoconductivity. However, recent study reported that the osteoconductivity of β-TCP could be enhanced by treating β-TCP surfaces with saturated acidic calcium phosphate solution. In this study, the osteoconductivity of treated β-TCP granule were evaluated using micro-computed tomography (micro-CT) scan technique. The granule specimens were obtained by exposing β-TCP granules with saturated acidic calcium phosphate solution for 10, 30 and 60 mins, respectively. Then, critical bone defect size with dimension of 9 mm were made at rat calvarial, and the defect was reconstructed with 10, 30 and 60 mins treated β-TCP granules and β-TCP granules without treatment as a control. After 2 and 4 weeks implantation, the specimens with the surrounding bone tissue were taken out and examined using micro-CT machine. Based on the cross section of micro-CT images, new bone formations were observed from the edge of the defect towards the center of the bone defect area in all specimens. At 2 weeks, new bone formations were observed for 10 min and 30 min treated β-TCP granules when compared with other specimens. The 30 min treated β-TCP granules showed faster new bone formation rate after 4 weeks implantation. It can be concluded that 30 min treated β-TCP granules revealed the highest osteoconductivity in comparison to the other specimens.

Keywords: Osteoconductivity, Dicalcium Phosphate Dihydrate, β-Tricalcium Phosphate, Microtomography, Acidic Calcium Phosphate Solution.

1. Introduction

Numerous studies have been reported that the β-TCP bone substitute shows an osteoconductivity [1, 2–4]. For example, comprehensive studies by Ogose et al. have been reported that the β-TCP bone substitute shows greater osteoconductivity than hydroxyapatite (HAp) bone substitute after implanted in human femur [1, 3, 4]. Although these findings demonstrated that the β-TCP bone substitute has a potential to be used as an osteoconductive material, its osteoconductivity is still inferior to the natural bone [5, 6]. In fact, Shinomiya et al. reported that if bone defect size becomes larger, the amount of new bone formation becomes low [7]. Therefore, if the ability of new bone formation for β-TCP bone

IOP Conf. Series: Journal of Physics: Conf. Series **1082** (2018) 012069 doi:10.1088/1742-6596/1082/1/012069

substitute can be enhanced, its range of clinical application is expected to be wider. Based on the histological study by Khairul et al. [8] reported that the osteoconductivity of β-TCP granule were enhanced by treating β-TCP granule surfaces with saturated acidic calcium phosphate solution. However, till now no study has been done using micro-CT scan technique. Therefore, the objective of this study is to investigate the osteoconductivity of treated-β-TCP granules using micro-CT scan technique.

2. Experimental study

2.1 Preparation of granular specimens

The granular specimens used in this study were 10 mins, 30 mins, 60 mins-treated β-TCP granule and DCPD free β-TCP granule. These granular specimens were prepared in the same manner as described by Khairul et. al [8]. The granular size used in this experiment was regulated between 300-600 μm using a sieve. Before implantation, these granular specimens were sterilized using gamma ray (2.27 kGy/hour, 11 hours, total dosage 25 kGy at 25°C) [9].

2.2 Animal surgical procedures

Animal experiment was performed according to Animal Care and Use Committee of Kyushu University (Reference number: A25-231-0). Twelve Wistar male twelve-weeks-old rats (Kyudo, Saga, Japan) with an average weight of 350 g were used. The rats were anesthetized with intraperitoneal injection of ketamine and 2% xylazine hydrochloride by using liquid to body weight ratio of 0.2 ml/100 g. The fur on the skin of rat calvarial was shaved and disinfected with 10% povidone iodine solution (Meiji Seika Pharma, Tokyo, Japan). An infiltration of localized anesthesia with 0.4 ml of 2% lidocaine-epinephrine solution (Dentsply Sankin, Tokyo, Japan) was applied over the surgery area in rat calvarial to arrest the bleeding and to control early post-operative pain. Then, the skin was incised sagittally with a surgical blade, and the calvarial bone area was carefully opened. A circular bone defect was created in each rat using 9 mm trephine burr under irrigation of saline solution (Otsuka, Osaka, Japan). The resected bone was gently removed to avoid injury the underlying tissue. Furthermore, the bone defect was irrigated with saline solution to remove debris that occurred during the drilling process. After wiping with gauge, the bone defect was reconstructed with 0.04-0.05 g of the granular specimens. Next, the periosteum layer and skin was sutured to close the wound. Then, the sutured area was disinfected with 10% povidone iodine solution (Meiji Seika Pharma). Figure 1 show the granular specimen condition reconstructed with the bone defect. After 2 and 4 weeks implantation, these rats were sacrificed by using overdosage of anesthesia solution. Then, the block section of rat calvarial bone containing granular specimens with surrounding tissue was taken out for micro-CT analysis.

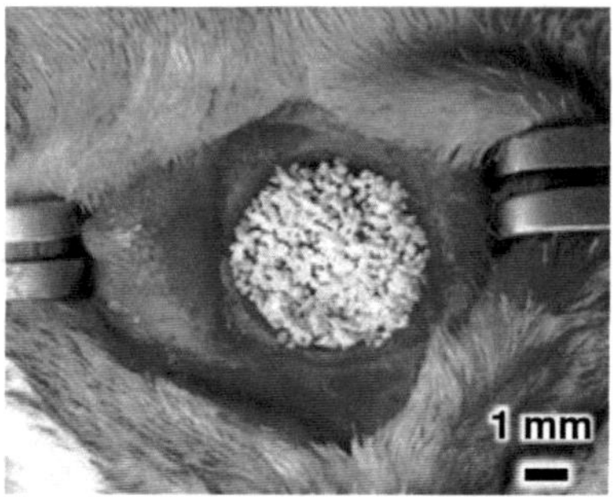

Figure 1. Treated β-TCP granule specimens (300-600 μm) was reconstructed with 9 mm in diameter of bone defect size at rat calvarial.

IOP Conf. Series: Journal of Physics: Conf. Series **1082** (2018) 012069 doi:10.1088/1742-6596/1082/1/012069

2.3 Microcomputed tomography analysis

Morphology of bone contained implant specimens was analysed using microcomputed tomography (μ-CT: Skyscan 1075 KHS, Kontich, Belgium) at a source of 59 kV and source content of 169 μA using a 0.5 mm aluminum filter [10]. The samples were scanned using the high-resolution mode (9 μm voxel resolution) image.

3. Results and discussion

3.1 Appearance of granular specimens

Figure 2 shows the appearance of the β-TCP granules (a) before and after reacted with saturated acidic calcium phosphate solution for (b) 10 min, (c) 30 min and (d) 60 min at 25°C. The physical appearance of β-TCP granule reacted with saturated acidic calcium phosphate solution is similar with the unreacted β-TCP granule. The 10 mins, 30 mins and 60 mins treated β-TCP granules were successfully fabricated after exposing β-TCP granule to 50 mmol/L MCPM-25 mmol/L H_3PO_4 solution at 25°C for 10, 30 and 60 min, respectively. The precipitation of DCPD crystals on the surface of β-TCP granule occurs through dissolution-precipitation reaction. [8]. Therefore, prolong the exposure time of β-TCP granule to acidic calcium phosphate solution results in the larger amount of DCPD coated on β-TCP granular surface. The 1 mass%, 5 mass% and 15 mass% of DCPD were produced after exposing β-TCP granule to saturated acidic calcium phosphate solution for 10 mins, 30 mins and 60 mins, respectively.

Figure 2. Appearance of the β-TCP granules (a) before and after reacted with acidic calcium phosphate solution for (b) 10 min, (c) 30 min and (d) 60 min at 25°C.

3.2 Micro-computed tomography analysis

Figure 3 and 4 show the cross-sectional view of the bone defect filled with (a) β-TCP specimen, (b) 10 min-treated β-TCP, (c) 30 min-treated β-TCP and (d) 60 min-treated β-TCP specimen at 2 and 4 weeks after implantation. These granular specimens remain in the bone defect in all specimens. At 2 week, the new bone formation was observed at the edge of bone defect area for 10 mins and 30 mins-treated β-TCP specimens as indicated by black arrows. At 4 week, the new bone formation was observed underneath the implanted granules in all specimens. Also, the new bone was regenerated from the surface of the original bone in all specimens at opposite direction. The new bone formation results obtained in this study indicated that the osteoconductivity of β-TCP granule could be enhanced by coating its surface with appropriate amount of DCPD. Interestingly, effect of the amount of coated DCPD on bone formation was not simple. Based on micro-CT results, new bone formation rate for 10 mins-treated β-TCP was faster compared to DCPD free β-TCP granule after 2 weeks implantation, indicating even 10 min of treated time on β-TCP granular surface is effective to stimulate new bone formation. However, no significant difference was observed between 10 mins-treated β-TCP and DCPD free β-TCP granule at 4 weeks implantation. In the case of 30 mins-treated β-TCP, the new bone

IOP Conf. Series: Journal of Physics: Conf. Series **1082** (2018) 012069 doi:10.1088/1742-6596/1082/1/012069

formation rate was faster after 4 weeks implantation. Although, detailed mechanism of the improvement of osteoconductivity of β-TCP granules coated with DCPD has not been clarified in the present study, it seems that the amount of coated DCPD is useful to enhance the osteoconductivity of β-TCP granule. Based on the reported by Khairul et al [8] mentioned that the enhancement of the new bone formation of DCPD-coated β-TCP granule may be explained by the release of Ca^{2+} and PO_4^{3-} from the specimens. The concentration amount of Ca^{2+} and PO_4^{3-} release from the specimens was increased with increasing the treatment time on the β-TCP granule surfaces. The 30 mins-treated β-TCP granules may release appropriate amount of Ca^{2+} and PO_4^{3-} that promoted the proliferation and differentiation of osteoblasts [8]. In the case of 60 mins-treated β-TCP granules, the large amount of Ca^{2+} release might decreased the proliferation and differentiation of osteoblasts cell [8].

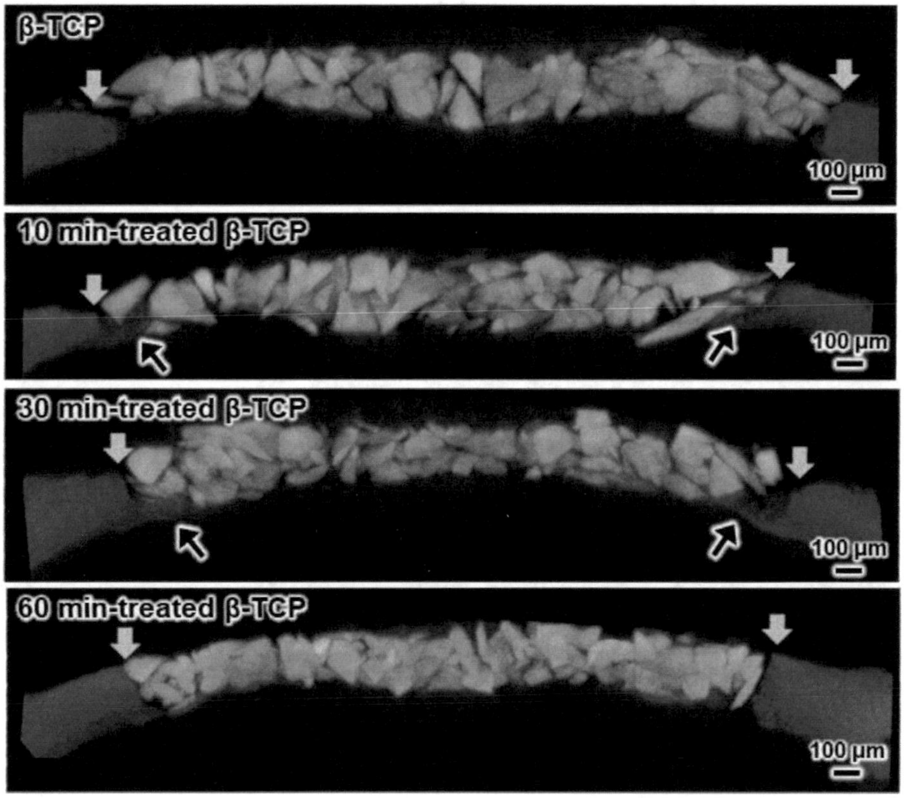

Figure 3. Cross-sectional view of the bone defect filled with (a) β-TCP specimen, (b) 10 min-treated β-TCP, (c) 30 min-treated β-TCP and (d) 60 min-treated β-TCP specimen after 2 weeks implantation. Yellow arrow = defect margin; Black arrow: new bone formation

IOP Conf. Series: Journal of Physics: Conf. Series **1082** (2018) 012069 doi:10.1088/1742-6596/1082/1/012069

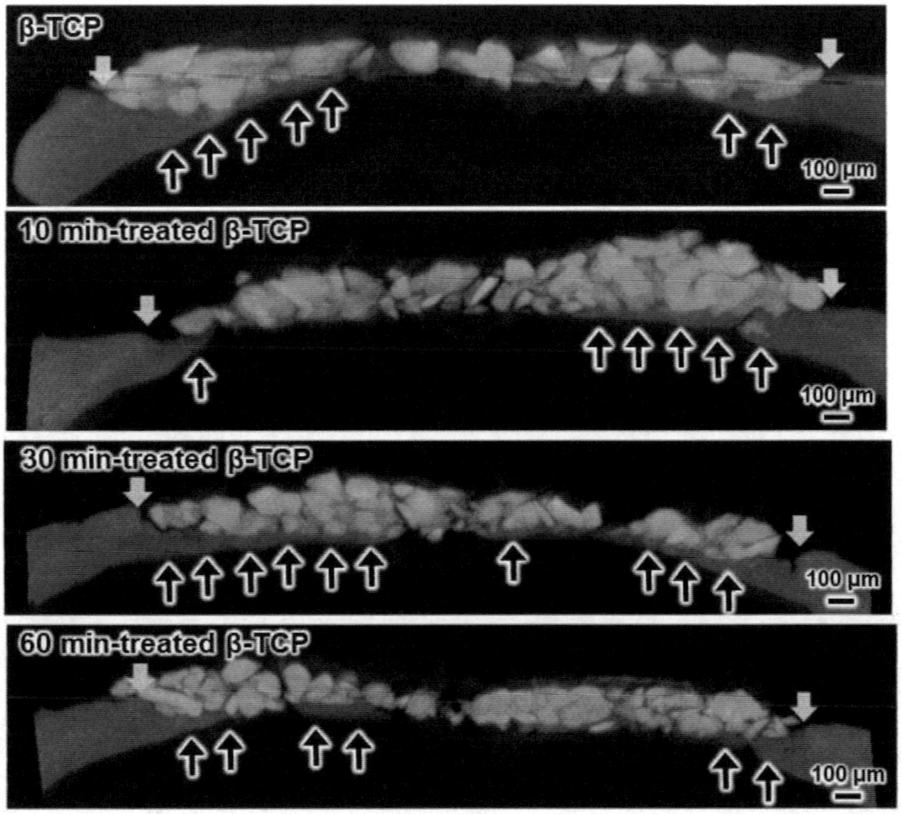

Figure 4. Cross-sectional view of the bone defect filled with (a) β-TCP specimen, (b) 10 min-treated β-TCP, (c) 30 min-treated β-TCP and (d) 60 min-treated β-TCP specimen after 4 weeks implantation. Yellow arrow = defect margin; Black arrow: new bone formation

4. Conclusion

It is found that 30 mins-treated β-TCP granule showed faster new bone formation rate than DCPD-free β-TCP granule for both 2 and 4 weeks after implantation. However, 60 mins-treated β-TCP granules showed lower new bone formation rate than 30 mins-treated β-TCP granule with similar implantation period. This results proved that appropriate treatment time of β-TCP granule with acidic calcium phosphate solution could improve osteoconductivity of β-TCP granule.

Acknowledgement

This study was supported by Short-Term Research Grant Universiti Sains Malaysia (Grant no: 304/PBAHAN/60313046), in part, by the Strategic Promotion of Innovative Research and Development Program (15im0502004h0004) from Japan Agency for Medical Research and Development. The first author would like to thank Department of Biomaterials, Faculty of Dental Science, Kyushu University for micro-CT analytical service.

IOP Conf. Series: Journal of Physics: Conf. Series **1082** (2018) 012069 doi:10.1088/1742-6596/1082/1/012069

References

[1] Ogose A, Hotta T, Kawashima H, Kondo N, Gu W, Kamura T and Endo N 2005 *Journal Biomedical Materials Research - Part B Applied Biomaterials* 72 94
[2] Onodera J, Kondo E, Omizu N, Ueda D, Yagi T and Yasuda K 2014 *Knee Surg, Sports Traumatology. Arthroscopy* 22 2763
[3] Ogose A, Hotta T, Hatano H, Hatano H, Kawashima H, Tokunaga K, Endo N and Umezu H 2002 *Journal Biomedical Materials Research - Part B Applied Biomaterials* 63 601
[4] Ogose A, Kondo N, Umezu H, Hotta T, Kawashima H, Tokunaga K, Ito T, Kudo N, Hoshino M, Gu W and Endo N 2006 *Biomaterials* 27 1542
[5] Frota R, Silva-Junior VAD, Teixeira M, Veras Sobral AP, Oliveira e Silva ED, Fonseca Da Silveria MM and Aragao-Neto AC 2011 *Med Oral Patol Oral Cir Bucal* 16 190
[6] Khatiblou F 2011 *Journal Oral Implantology* 37 727
[7] Shinomiya K, Ishizuki M, Morioka H, Matasumoto S, Nakamura T, Abe S and Beppu Y 2012 *Seikei Geka* 63 921
[8] Khairul A S, Tsuru K and Ishikawa K 2017 *Materials Science and Engineering C* 75 1411
[9] De Moraes M A, Weska R F and Beppu M M 2013 *Journal Biomedical Materials Research.-Part B Applied Biomaterials* 102 869
[10] Shariff K A, Tsuru K and Ishikawa K 2016 *Journal Biomaterials Applications* 30 838

IOP Conf. Series: Journal of Physics: Conf. Series **1082** (2018) 012070 doi:10.1088/1742-6596/1082/1/012070

Synthesis and Characterization of Bioactive Quaternary Silicate Gel-glasses

T.T.Swe[1], H.Mohamad[1], K.A. Shariff[1], A.F.M.Noor[1], K.Ishikawa[2], A.A.Thant[3]

[1] School of Materials and Mineral Resources Engineering, Universiti Sains Malaysia, Nibong Tebal, Penang, Malaysia

[2] Faculty of Dental Science, Kyushu University, Japan

[3] Department of Physics, University of Yangon, Yangon, Myanmar

Email: hasmaliza@usm.my

Abstract. Bioactive glasses and glass-ceramics were applied in bone fix applications, dental applications as well as in tissue engineering. In the current study, a new quaternary silicate bioglass S50P4 (50%SiO_2–24.5%Na_2O–21.5%CaO–4%P_2O_5) with different aging time (3, 5 and 7 days) was prepared by sol-gel method. These synthesized glasses were analyzed using X-ray powder diffraction (XRD), Scanning electron microscopy (SEM) and Fourier transform infrared spectroscopy (FTIR). Whereas the bioactivity of the optimized samples were observed through in-vitro test using Hank's Balanced Salt Solution (HBSS) at different soaking time (7 and 14 days). XRD results indicate that the crystalline patterns in gel-derived glass aged for 3 days exhibit stronger peaks compare to samples aged for 5 and 7 days. This could be attributed to the difference of water content inside the gels. Moreover, the peak at $2\theta \sim 33°$ became sharper and more intense for longer soaking times corresponds to the apatite structure. Besides, in-vitro test shows, the formation of hydroxyl carbonated apatite layer (HCA) with finer grains.

1. Introduction

Bioactive silicate glasses can be used as bone substitutes in orthopedic and dental applications as well as in the field of bone tissue engineering, in the form of granules, scaffolds, porous or dense powders, and coatings[1]. Bioactive glasses are synthetic materials, which show highly positive interactions with hard and soft tissues [2]. The first bioactive glass 45S5, known worldwide for its trademark Bioglass®, was invented by Professor L.L.Hench at the University of Florida in 1969 [3].

The production of 45S5 bioactive glass has traditionally been carried out through melting technology. In the melting process, precursors were mixed homogeneously and melted at a high temperature (>1300°C). The product was then quenched to form amorphous glass. However, the high temperature can induce the volatilization of phosphate components, which may change the composition of the glass. However, the sol-gel method can be used for the fabrication of both glassy and ceramic materials [4].The primary advantage of the sol-gel technique is the potential of gaining high purity, homogeneity materials and lower processing temperatures than the traditional melting method [5].

IOP Conf. Series: Journal of Physics: Conf. Series **1082** (2018) 012070 doi:10.1088/1742-6596/1082/1/012070

Generally, the sol-gel processes include hydrolysis, polycondensation, gelation, aging, drying, and stabilization. In the sol-gel processes, the precursors are catalyzed and dissolved in the solvent to form a sol. The sol gradually becomes a gel containing both a liquid and a solid phase. To strengthen the network, an aging process is necessary. Removal of the remaining liquid (solvent) phase requires a drying process. Finally, a thermal treatment (stabilization) is often carried out to enhance the mechanical properties and to improve the structural stability [5]. In this study, the new glass composition S50P4 is designed based on the assumption of L.L Hench [3] and synthesized by so-gel method in tend to get greater surface area for the purpose of better bioactivity. The effect of aging time on the structure of gel-derived S50P4 bioglass was studied by Fourier-transform infrared spectroscopy (FTIR) and X-ray diffraction (XRD) techniques. The bioactivity of the glass was evaluated by immersing the samples in HBSS for 7 and 14 days to study the formation of apatite layer on the glass surface.

2. Materials and Method

2.1 Gel Preparation

The S50P4 bioglass with the composition (50%SiO_2–24.5%Na_2O–21.5%CaO–4%P_2O_5) in wt. % was prepared by sol-gel method. The main raw materials used were tetraethyl orthosilicate (TEOS, $Si(OC_2H_5)$, Fluka), triethyl phosphate (TEP, $(OP(OC_2H_5)$, Merck), sodium nitrate (($NaNO_3$), Emsure) and calcium nitrate tetrahydrate (($Ca(NO_3)_2.4H_2O$), Emsure) as precursors of SiO_2, P_2O_5, Na_2O and CaO respectively. Samples with different aging times (3, 5 and 7 days) were prepared according to the ratio of SiO_2, P_2O_5, Na_2O and CaO stated for the composition.

Briefly, the required amount of TEOS (37 mL) was poured into 1M dilute nitric acid solution and stirred for 1 h at room temperature for acid hydrolysis. Then TEP (2 mL), calcium nitrate (18.09 g) and sodium nitrate (13 g) were added into the TEOS with constant stirring for 45 min for each reagent to react completely. The mixture was then continuously stirred for another 1 h to complete the hydrolysis reaction. The solution was kept and sealed at the ambient condition with different aging periods (3, 5 and 7 days) to homogenize and re-precipitation. The resulting gel was aged at 60°C for 2 days and dried at 100°C for 2 days to eliminate the adsorbed water from the pores and finally, calcined at 700°C for 5 h to eliminate the residual nitrate and organic substances.

2.1.1. X-ray Diffraction (XRD). X-ray diffraction was used to analyse the glass powders using an XRD Bruker DX 8 operating with CuKα radiation (λ = 1.5406 Å). The diffraction patterns were obtained in the 2θ range from 20° to 80° in a fixed counting time of 1 second. The obtained data were analysed using X'Pert HighScore Plus Software for phase analyzation.

2.1.2. Fourier Transform Infrared Spectroscopy (FTIR). Surface structural of glass powders was determined using FTIR (PERKINELMER SPECTRUM ONE). Transmission technique was implied with KBr powder used as a reference. The weight ratio of sample to KBr was set at 1:100. The mixture of KBr and glass powder was pressed into pellet form using hand press. Sample with lower thickness has good transparency which provides a relatively short path length for the beam. Spectra were collected in the region between wavenumber 4000 and 400 cm^{-1} with a resolution 4 cm^{-1}.

2.1.3. Field Emission Scanning Electron Microscope (FESEM). The microstructure of glass powders was characterized using thermal field emission gun scanning electron microscope (FESEM). FESEM was used to study the surface morphology and topography of the glass before and after soaking in Hank's Balanced Salt Solution (HBSS). In this study, Zeiss Supra 35VP Field Emission Scanning Electron Microscope (FESEM) was used. Because of the poor electrical conductivity of the samples, their surfaces were coated with a thin layer of gold before the test.

IOP Conf. Series: Journal of Physics: Conf. Series **1082** (2018) 012070 doi:10.1088/1742-6596/1082/1/012070

3. Results and Discussion

3.1 X-ray Diffraction (XRD)

The diffraction patterns of gel-derived S50P4 bioglass aged under different periods and stabilized at 700°C are shown in figure 1. Results show the formation of several crystalline sodium calcium silicate phases together with an amorphous structure of the bioactive glass. Among them, the most interesting phase is combeite ($Na_2Ca_2Si_3O_9$) because it was significantly influence the bioactivity. As pointed out by Peitl et al., combeite ($Na_2Ca_2Si_3O_9$) is highly bioactive [6]. The formation of the combeite phase was associated with the heat treatment performed at 700 °C during the drying step in sol-gel synthesis [7]. However, the structure of the glass still exhibit amorphous up to 7 days aged (figure 1). It can be seen that the S50P4 bioglass aged 3 days was crystalline, however by increasing of aging time, the formation of amorphous structure increased where the sharpness of peaks decreased. This phenomenon was believed due to the dehydration of the mixture during [8]. The formation of amorphous structure proved that the sol-gel method could prepare glasses with aging time up to 7 days. However, in the current study, combeite phase was only detected in glass powder aged 5 days which is optimized for in-vitro analysis by soaking in HBSS.

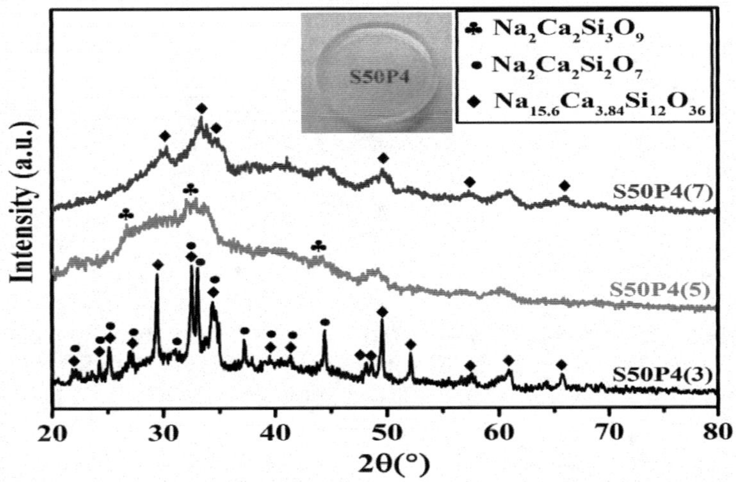

Figure 1. XRD patterns of the glass samples produced at different aging time (3, 5 and 7 days)

The XRD patterns of the 5 days aged bioglass after immersion in HBSS for 7 days and 14 days was shown in figure 2. The formation of an apatite layer during the immersion in HBSS was proved by the XRD analysis. As discussed in the previous part, before immersion in HBSS, the XRD spectrum shows the partially amorphous structure of $Na_2Ca_2Si_3O_9$ (Combeite) by a broad peak between 30° and 35°. After 7 days of soaking in HBSS, the formation of apatite layer on the glass surface can be detected as shown in figure 2. With increasing soaking times up to 14 days, the new peaks can be observed at around 26°, 33° and 49° assigned to the apatite reflections according to the standard JCPDS (09-0506). The peak at 2θ~33° became sharper and more intense with increasing the soaking times up to 14 days, it happens due to the reflection of (2 1 1) plane of apatite.

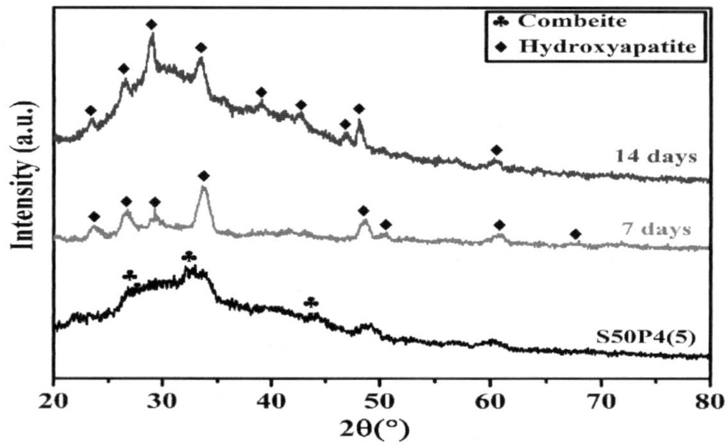

Figure 2. XRD patterns for S50P4 5 days aged and after soaking in HBSS at different periods (7 and 14 days)

3.2 Fourier Transform Infrared Spectroscopy (FTIR)

The FTIR spectra of the bioglass prepared at different aging times (3, 5 and 7 days) were compared in figure 3. The main characteristic bands ranging from 400 to 1400 cm^{-1} were related to the silicate network group vibrations with different bonding arrangements silicate. While the spectra from 1400 to 4000 cm^{-1} clearly consists of vibrations refer to water or hydroxyl groups [9]. No obvious distinction was seen among these samples, which means their compositions were not affected by increasing the aging time. The FTIR spectra reveal Si-O-Si bending (467 and 526 cm^{-1}), Si-O-Si stretching (symmetric) (884 cm^{-1}) and Si-O-Si stretching (asymmetric) (1040 cm^{-1}) bands, which were known and accepted to be mainly characteristic of silicate network [9]. This might be attributed to the presence of major SiO_2 as a basic building constituent. The wavenumbers 619 cm^{-1} was due to the presence of sodium calcium silicate ($Na_2Ca_2Si_3O_9$) crystalline phase [10]. The peak at 1384 cm^{-1} was due to the vibration of nitrate $(NO_3)^-$ [11] which was appeared only in 3 and 7 days aged. The existence of the peak at 1648 cm^{-1} band indicates there is water absorption reaction take placed at the glass interface [12]. The bands at the 800 - 1200 cm^{-1} were assigned to the stretching vibration of the SiO_4 tetrahedral with the different number of bridging oxygen atoms and the peak at 1400-1530 cm^{-1} corresponds to carbonate $(CO_3)^{2-}$ groups. The presence of carbonate was attributed to a carbonation process of the material due to the atmospheric CO_2 as a consequence of the high calcium content [11].

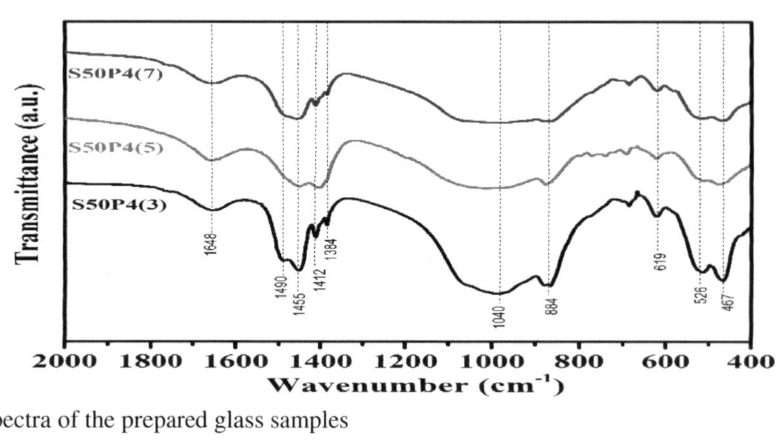

Figure 3. FTIR spectra of the prepared glass samples

The Fourier transform infrared (FTIR) transmittance spectra of bioactive glasses after soaking in HBSS for different period of time (7 days and 14 days) are shown in figure 4. Before soaking, the FTIR pattern shows Si-O-Si bending (467 and 526 cm^{-1}), Si-O-Si stretching (symmetric) (884 cm^{-1}) and Si-O-Si stretching (asymmetric) (1040 cm^{-1}) bands. After HBSS treatment, it was noticed that some peaks at 526 cm^{-1} (Si-O-Si bending) and 1412 cm^{-1} (C-O stretch) are disappeared. The peaks of O-H bending at 1643 and 3454 cm^{-1} became more intense. Evolving of P - O bending (610 - 600 cm^{-1}) bands indicates the formation of hydroxyl carbonate apatite (HCA) layer on the glass surface. However, the C - O stretching (~1429 cm^{-1}) and P - O stretching (1200 - 910 cm^{-1}) bands were attributed due to hydroxyl carbonated apatite (HCA) layer. These results confirmed that the formation of apatite layer on the surface of glass powder was carbonated apatite, which is similar in composition and structure to the bone apatite and was also found on bioactive glasses [13].

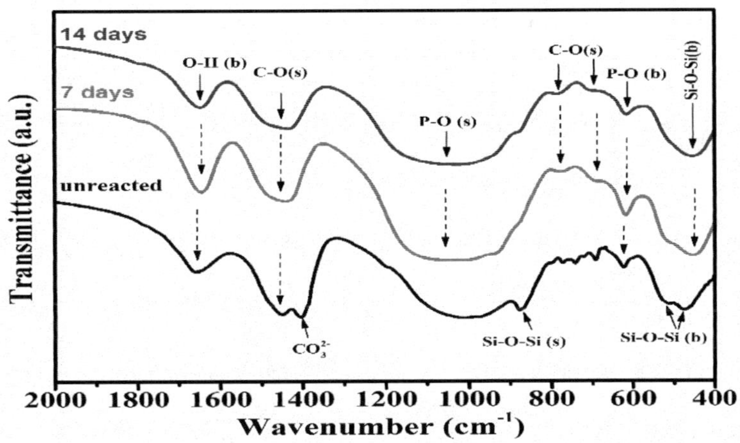

Figure 4. FTIR patterns for S50P4 before and after soaking at different periods in HBSS

3.3 Scanning Electron Microscopy (SEM)

The microstructure of 5 days aged gel-derived S50P4 glasses before and after soaking in HBSS were illustrated in figure 5. The gel-derived glasses were exhibited porous structure on the surface of all glasses. Pores are necessary for tissue formation because they allow migration and proliferation of cells [14]. The porous structure shows the typical morphology of gel-derived glasses fabricated in acid-catalysed conditions [14]. Figure 5 (b) and (c) shows the surface morphology of the glass specimen after soaking in HBSS for 7 days and 14 days respectively. Comparison with the unreacted sample figure 5(a), there was an evidence of apatite layer appeared on the glass surface in figure 5(b) and (c). The images displayed the formation of the apatite particles over the entire surface of the samples, as a result of the action and reaction between the HBSS and glass particles. This pattern was typical of HCA precipitated on the glass surface. In figure 5(c), the precipitation of apatite crystals can be seen as homogenously covered the glass particles. The formation of apatite layer was confirmed by not only XRD results but also FTIR patterns.

Figure 5. SEM images for 5 days aged S50P4 (a) before and after soaking in HBSS at different periods (b) 7 days and (c) 14 days

4. Conclusion

In the present study, the effect of aging time on the structure of gel-derived S50P4 glass and in-vitro bioactivity test with HBSS were investigated. The crystallinity of the bioglass was reduced with increasing aging time but combeite ($Na_2Ca_2Si_3O_9$) phase was only detected after 5 days aged. Before soaking in HBSS, the glass powder was agglomerated as the particle size is reduced in tend to get greater surface area for the purpose of better bioactivity. The precipitation of apatite crystals was homogeneously covered the glass particles after soaked for 14 days.

Acknowledgements. This research work is funded by AUN/SEED-Net under grant number 6050368/A119.

References

[1] Pishbin F, Simchi A, Ryan M and Boccaccini A 2011 *Surface and Coatings Technology* 205 5260
[2] Miguez-Pacheco V, Hench L L and Boccaccini A R 2015 *Acta biomaterialia* 13 1
[3] Hench L L 2006 *Journal of Materials Science: Materials in Medicine* 17 967
[4] Hench L L and West J K 1990 *Chemical reviews* 90 33
[5] Brinker C J and Scherer G W 2013 *Sol-gel science: the physics and chemistry of sol-gel processing*: Academic press
[6] Peitl Filho O, Latorre G P and Hench L 1996 *J Biomed Mater Res* 30 509
[7] Faure J, Drevet R, Lemelle A, Jaber N B, Tara A, El Btaouri H and Benhayoune H 2015 *Materials Science and Engineering: C* 47 407
[8] Zheng K, Solodovnyk A, Li W, Goudouri O M, Stähli C, Nazhat S N and Boccaccini A R 2015 *Journal of the American Ceramic Society* 98 30
[9] Wang X, Li X, Ito A and Sogo Y 2011 *Acta Biomaterialia* 7 3638
[10] ElBatal H, Azooz M, Khalil E, Monem A S and Hamdy Y 2003 *Materials Chemistry and Physics* 80 599
[11] Martinez A, Izquierdo-Barba I and Vallet-Regi M 2000 *Chemistry of Materials* 12 3080
[12] Zhao Y, Song M and Liu J 2008 *Surface and Interface Analysis* 40 1463
[13] Kontonasaki E, Zorba T, Papadopoulou L, Pavlidou E, Chatzistavrou X, Paraskevopoulos K and Koidis P 2002 *Crystal Research and Technology* 37 1165
[14] Bellantone M, Coleman N J and Hench L L 2002 *Antimicrobial agents and chemotherapy* 46 194

IOP Conf. Series: Journal of Physics: Conf. Series **1082** (2018) 012071 doi:10.1088/1742-6596/1082/1/012071

Fabrication and Characterisation of Gelatine/Hydroxyapatite Porous Scaffold

Mazlam. M.I[1], Ho. H.H.[1], and A. Nurazreena[1]*

[1]Niche Biomaterials, School of Materials and Mineral Resources Engineering, Engineering, Campus, Universiti Sains Malaysia, 14300 Nibong Tebal, Pulau Pinang, Malaysia

Email: nurazreena@usm.my

Abstract. Hydroxyapatite (HAp) has been used as bone substitute materials due to its bone compatibility after implantation. However, major drawbacks of HAp typical of ceramic materials are its brittleness which limits its applications. Therefore, to introduce toughness in HAp scaffold, gelatine layer was coated on the surface of the HAp scaffold. In this research, porous HAp scaffolds coated with gelatine are fabricated via polymer reticulate method using polyurethane foam (PU) template. Initially, HAp powders were calcined at 1100°C to reduce its viscosity and agglomeration effect when formed into slurry. Preparation of HAp slurry with 40wt% of calcined HAp powders and 3 wt% of polyvinyl alcohol (PVA) powder were prepared before immersing the PU foams. The HAp coated PU foams were then heat treated at 650°C for an hour to burnout all PU, then sintering was done at 1100°C. Finally, the sintered HAp scaffold was coated with 10, 15 and 20wt% of gelatine. Additionally, the gelatine was crosslinked at 130°C and 160°C. Compressive strength analysis shows that as the wt% of gelatine increases the compressive strength of the gelatine/HAp scaffold increases. Highest compressive strength of 71.0 kPa was obtained for 20wt% of gelatine coated on the HAp scaffold. Higher crosslinking temperature also increases the compressive strength. The average compressive strength for gelatine/HAp scaffold crosslinked at 130 °C and 160 °C are 66.4 kPa and 72.3 kPa respectively. XRD analysis shows that 66% of HAp transformed into beta-tricalcium phosphate (β-TCP) after calcination while 71% transformed into β-TCP after sintering. FTIR and EDX analysis showed that the higher amount of gelatine content found in scaffolds with higher crosslinking temperature correlate to the increased in compressive strength. SEM analysis shows that the increased in the compressive strength is attributed to the covering of cracks on the strut of the HAp scaffolds by the gelatine.

1. Introduction

Bone scaffolds are intended to be used for implantation in critical size bone defects[1]. Traditionally, autografts and allografts has been used to correct bone defects or fracture problem. However, autografts and allografts has its disadvantages such as limited quantity of bone to harvest, massive blood sepsis, extra surgery, donor-site morbidity and risk of disease transmission[2]. Therefore, 3-D synthetic bone substitute scaffold is an alternative for bone remodelling. To mimic a scaffold, a number of key considerations are important when designing or determining the suitability of a scaffold for use in tissue

IOP Conf. Series: Journal of Physics: Conf. Series **1082** (2018) 012071 doi:10.1088/1742-6596/1082/1/012071

engineering. Those considerations include biocompatibility, biodegradability, mechanical properties, scaffold architecture and its manufacturing technology[3]. Successful scaffold design should stimulate new bone growth leaving only native bone tissue with no trace of the implanted scaffold at end state of success bone implantation process[4]. Scaffolds play an important character in new tissue regeneration and solving bone defects problems. Thus, this research focuses on the fabrication of the scaffold with enhancement of natural polymer which is gelatine.

2. Experimental study

2.1. Preparation of HAp slurry

HAp slurry was prepared by mixing 40wt% pre-calcined (1100 °C) HAp powder, 60wt% of water and additional of 3wt% of PVA as a binder. PVA binder is first dissolve in water at around 40°C using stirring hotplate and a magnetic bar. HAp powders are then added into the PVA solution and the stirring continued at 500 rpm for 4 hours using Mechanical Stirrer (RW20D, MAXIMA, IKA, Germany).

2.2. Impregnation of HAp slurry onto PU foam

HAp scaffolds were prepared by polymer reticulate method using polyurethane (PU) foam template cut into dimension of 1.5cm x 1.5cm x 2.54cm. PU foam were then squeezed to remove air trapped by fingers before immersing into the HAp slurry. HAp slurry should cover the whole PU foam to allow better impregnation of HAp slurry. PU foam is then taken out and rolled for 5 times using corrugating machine. The same method of immersion of foam into slurry and rolling were repeated twice to ensure full penetration. Air bubbles created during rolling and excess HAp slurry were removed by using air pressure generated by air gun.

2.3. Heat Treatment

The HAp impregnated PU foams are then put into oven (Memmert, UN30) at 80°C for 24 hours. Then the coated HAp/PU foam are sintered at 650 °C for one hour soaking time for PU foam burnout follow by sintering of the scaffold at 1100°C for one hour with temperature increment of 5 °C.

2.4. Gelatine coating process

10 wt%, 15 wt% and 20 wt% of gelatine concentrations were prepared by dissolving and stirring gelatine in water Beaker containing the gelatine solution were covered with aluminium foil to prevent excessive water loss through evaporation. Once a clear solution of gelatine is obtained, the sintered HAp was immersed into the solution at 35°C to enable good flow ability of gelatine to completely impregnate the sintered HAp scaffold. Next excessive gelatine was allowed to flow out by leaving in an oven at 80 °C for one minute. Lastly, the scaffolds were placed into vacuum drying oven at 130°C and 160°C for 24 hours.

2.5. Characterization and analytical techniques

The starting material and finished scaffolds were characterized by X-ray diffraction (XRD) analysis and Fourier transform infrared spectroscopy (FTIR) analysis. The XRD pattern were recorded with a diffractometer system (X-ray Diffractometer D8 Advance BRUKER-binary V3) using CuKα (λ of K-A1 = 1.54060 Å) radiation generated at 40 kV. The FTIR spectra were measured using a FTIR spectrometer (PERKIN ELMER Inc., MA, USA) in the wavelength range of 450-4000 cm^{-1}.

Universal Testing Machine (Series 5982, Instron, USA) was used to determine the mechanical properties of HAp scaffold incorporated with gelatine using 5 kN load cell with a cross head speed of 1 mm per min. Five samples for each experimental group were tested and their mean value was calculated.

Morphology of the scaffold were observed using Scanning Electron Microscope (SEM) (Hitachi TM3000, Japan).

IOP Conf. Series: Journal of Physics: Conf. Series **1082** (2018) 012071 doi:10.1088/1742-6596/1082/1/012071

3. Results and discussion

3.1. X-Ray Diffraction (XRD) and Fourier Transform Infra-Red (FTIR) Analysis

Figure 1 shows the XRD pattern of as received commercial HAp powder, calcined HAp powder at 1100 °C and sintered HAp scaffold at 1100 °C. The as-received raw HAp gives diffuse intensities indicated that as-received raw HAp has lower crystallinity than the standard reference which has higher intensity resembling its high crystallinity HAp due to the processing route used. The XRD peak of calcined and sintered HAp in Figure 1 shown significantly different characteristics of sharper and narrow diffraction peaks as compared to as-received raw HAp powders before the heat treatment process. This changed revealed that calcination and sintering process also altered the crystal size of HAp powder. This finding is parallel with statement claimed by Ahmed et al. (2015) that the physical properties of HAp powder are largely influenced by the heat treatment process[5]. The crystallinity of the HAp powder was found to be largely enhanced with the increase in calcination temperatures.

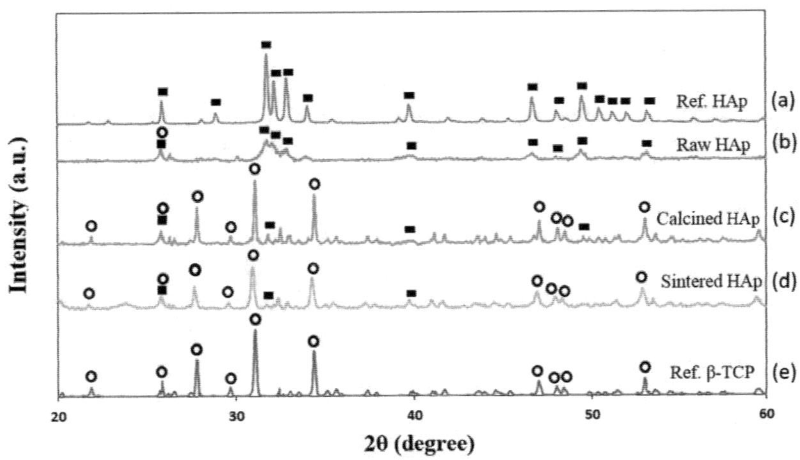

Figure 1. XRD patterns of (a)standard reference of commercial HAp; (b) raw commercial Hap used; (c) calcined HAp at 1100°C; (d) sintered HAp at 1100°C; (e) standard reference of commercial β-TCP

Nonetheless, all the XRD peaks of raw commercial HAp powders are found to match with the XRD peaks of the reference standard of commercial HAp as shown in the 1. This matching is done by using X'Pert High Score Plus software and the standard ICDD code number 98-004-0506. From the XRD analysis, it is found that HAp phase changes to β-TCP phase after calcination and sintering. Most of the HAp have been transformed into β-TCP as shown in Figure 1 with peaks under rectangular box represent HAp while circle representing β-TCP. Percentage of HAp being transformed during calcination and sintering are 66 % and 71 % This is supported by Rao et. al that stated above 600 °C, HAp structure starts to transfer to β-TCP explaining the XRD results obtained in this study[6].

Figure 2 shows the FTIR spectra obtained for HAp scaffold, with 10 wt%, 15 wt% and 20 wt% gelatine concentration coated. In this FTIR analysis, the functional group of gelatine and HAp are found and labelled in the Figure 2. According to Lawrie et al., 2007, broad band that ranges from 3700 cm^{-1} to 3000 cm^{-1} is assigned to O-H stretch[7]. Besides, peaks appear at 650 cm^{-1} are assigned to the O-H stretch[6]. Presence of O-H can are predicted to the presence of HAp. This have been confirmed with the XRD analysis shown in Figure 1. FTIR are also used to detect the presence of gelatine. Amide I with band range from 1630 cm^{-1} to 1695 cm^{-1} are matched with the peak detected in 2 confirms the existence of gelatine. Peaks detected as COO- with band range from 1596 cm^{-1} to 1412 cm^{-1} also represents the gelatine structure[7]. Thus, it is confirmed that the scaffold fabricated consist of HAp impregnated with gelatine.

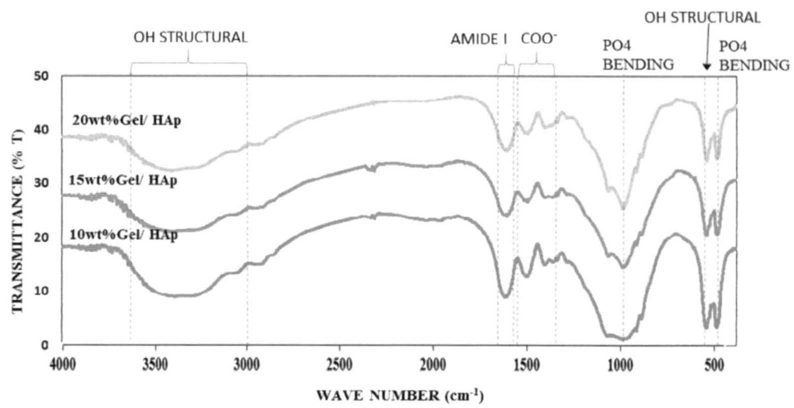

Figure 2. FTIR spectra of HAp scaffold with different concentration of gelatine coated.

3.2. Compressive Strength Analysis

The results from Figure 3 shows the trend of increasing compressive strength as the gelatine concentration increases. HAp coated with 20wt% of gelatine concentration has the highest compressive strength of 71.0 kPa followed by 22.8 kPa and 7.8 kPa for HAp coated with 15 wt% and 10 wt% gelatine concentrations respectively. The crosslinked gelatine after de-hydrothermal (DHT) process act as a strong adhesive to join the sphere-like shape HAp particles together from spreading away easily under stress thus increasing the toughness of this brittle HAp scaffold[8]. The effect of crosslinking temperature on the compressive strength is relatively not significant as compared to the effect of gelatine concentration on the compressive strength as shown in Figure 3. The compressive strength of pure HAp scaffold without gelatine coating are not shown in graph due to it brittleness characteristics.

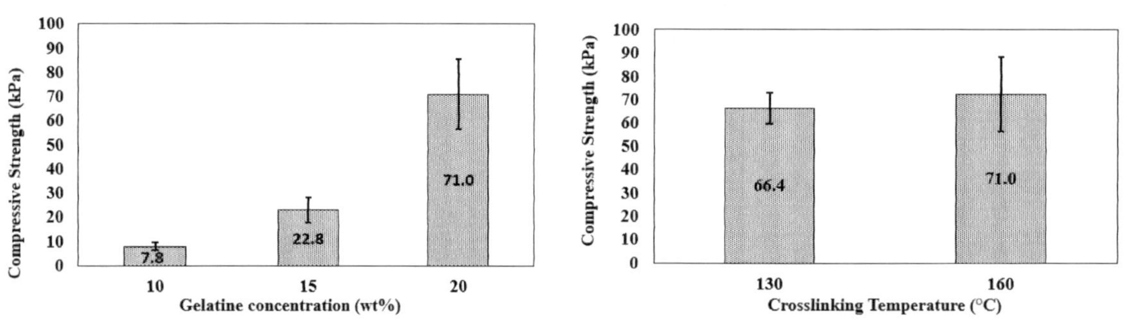

Figure 3. Compressive strength of 10 wt%, 15 wt% and 20 wt% of gelatine/HAp scaffold and compressive strength of gelatine/Hap scaffold coated under two crosslinking temperature.

3.3. Scanning Electron Microscope (SEM) Analysis

The morphology of the sintered HAp scaffolds is shown in Figure 4 (a) – (d). From these figures it can be observed the existence of missing part of strut as the supporting structure. Thus, this explains the brittleness of HAp scaffold which lowers the compressive strength of the HAp scaffold to undetected value during the compression testing. Moreover, Figure 4 (b) shows the pores existed on the strut. The reason of regular pores presented are unknown. Figure 4 (c) shows a relatively deep scratch line found on strut which may due to the organic type of impurities burn out during the high temperature firing leaving a structure look similar to scratch line. Figure 4 (d) shows the empty core of the scaffold strut. This is believed to be due to the PU foam burn out at 650 °C leaving this empty core behind. All these contribute towards the low compressive strength of HAp scaffold. Figure 4 (e) shows the cracking in strut linked to the pores with approximate 150 μm long in length. The initial point of cracking is unknown. However, the reason of crack presented was believe to be due to the volumetric expansion behaviour of unit cells during transformation of HAp to β-TCP[9].

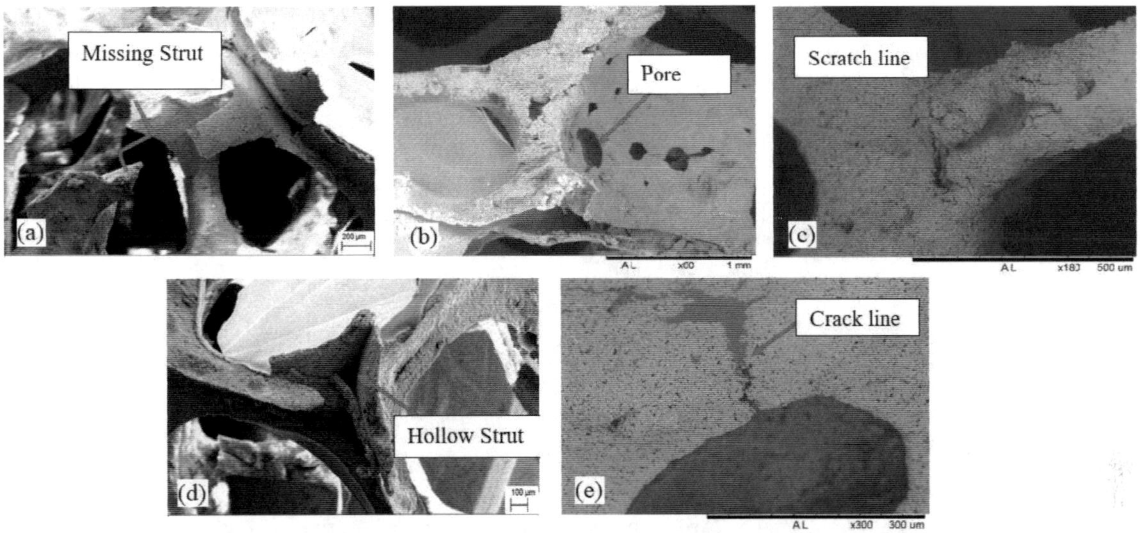

Figure 4 (a) – (e). Morphology of fabricated uncoated HAp porous scaffold.

Coating of the gelatine was found to contribute to the increases in compressive strength of HAp scaffold as the gelatine penetrates and closes the cracks and micropores on the surface of the HAp scaffold. Figure 5 (a) (b) shows a distinct gelatine layer at concentration 10 wt% and 15 wt% coated on HAp scaffold which was successfully observed and captured using SEM Tabletop machine. It can be observed that the gelatine layer is around 10 μm and 20 μm for 10 wt% and 15 wt% gelatine/HAp scaffold. The increment in gelatine concentration was proven to increase the strength of the scaffold as shown in Figure 3. This is because as the gelatine concentration increased, less water content is lost during the DHT crosslinking process. The higher gelatine concentration resembles the more bonding and crosslink achievable and thus strengthen the scaffold structure and avoiding it from collapsing during service. Gelatine not only shows excellent biocompatibility and biodegradability, but also exhibits superior plasticity and adhesiveness that is essential for developing mechanical strength in the scaffold. Besides, gelatine also acts as a binding agent in medical application because it is derived from collagen and contains biological functional groups, such as amino acids within its backbone which can enhance cell growth and proliferation[8].

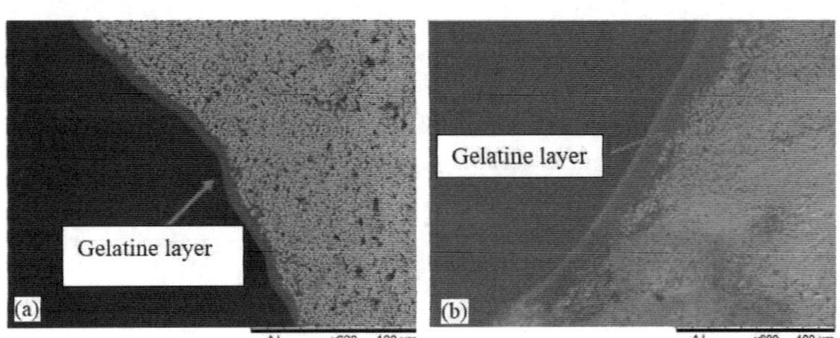

Figure 5. Morphology of fabricated gelatine/HAp porous scaffold at (a)10 wt% and (b)15 wt%.

4. Conclusion

Gelatine/HAp porous scaffold with 40 wt% of solid loading via polymer reticulate method was successfully fabricated in this study. The calcination treatment on HAp powder was proven to facilitate the slurry infiltration process. The HAp powder calcined at 1100 °C introduces more negative charges on the HAp particle surface to enable a well-stabilized HAp suspension of high solid loading with constantly low viscosity being prepared. In this study, the gelatine/HAp scaffold with average compressive strength of 7.8 kPa, 22.8 kPa and 71.0 kPa was achieved for HAp coated with 10 wt%, 15 wt% and 20 wt% gelatine concentrations respectively. On the other hand, the average compressive strength of scaffolds underwent crosslinking at temperature of 130 °C and 160 °C recorded as 66.4 kPa and 71.0 kPa. Increase in gelatine concentration and crosslinking temperature will increase the HAp scaffold compressive strength. However, the effect of crosslinking temperature on compressive strength is less significant compared to the gelatine concentration.

Acknowledgement

This work is supported by Trans Disciplinary Research Grant Scheme (TRGS) No: 203/PBahan/6761003

References

[1] M. Liebschner and M. A. Wettergreen 2003 *Optimization of bone scaffold engineering for load bearing applications*
[2] Ljupcho Efremov, Tatjana Kanjevac, Dušica Ciric, and D. Bosnakovsk 2014 *Serbian J. Exp. Clin. Res.* 14 145
[3] F. J. O'Brien 2011 *Mater. Today* 14 88
[4] Y. Ikada 2011 *Tissue Engineering: Fundamentals and Applications*
[5] Y. M. Z. AHMED, S. M. EL-SHEIKH, and Z. I. ZAKI 2015 *Bull. Mater. Sci.* 38 1807
[6] R. R. Rao, H. N. Roopa, and T. S. Kannan 1998 *Trans. Indian Ceram. Soc.* 57 73
[7] G. Lawrie *et al.* 2007 *Biomacromolecules* 8 2533
[8] K. Maji and S. Dasgupta 2014 *Trans. Indian Ceram. Soc.* 73 110
[9] D.-H. Kim, K.-H. Hwang, J. D. Lee, H.-C. Park, and S.-Y. Yoon 2015 *Biomater. Res.* 19 14

IOP Conf. Series: Journal of Physics: Conf. Series **1082** (2018) 012072 doi:10.1088/1742-6596/1082/1/012072

The Influence of Scaffold Designs on Cell Seeding Efficiency in Establishing A Three-Dimensional Culture

Yanny Marliana Baba Ismail[1,2*], Yvonne Reinwald[2], Ian Wimpenny[2], Oana Bretcanu[3], Kenneth Dalgarno[3], Alicia J El Haj[2]

[1]School of Materials and Mineral Resources Engineering, Universiti Sains Malaysia, Engineering Campus, 14300, Nibong Tebal, Penang, Malaysia.

[2] Institute for Science and Technology in Medicine, Keele University, Stoke-on-Trent, ST47QB, UK

[3]School of Mechanical and Systems Engineering, Newcastle University, Newcastle-upon-tyne, NE17RU, UK.

Email: yannymarliana@usm.my

Abstract. The concept of Bone Tissue Engineering (BTE) based on cell-seeded scaffold holds great promise for the treatment of bone defects. However, much optimization is required in establishing a three-dimensional (3D) culture before BTE can be successfully applied clinically. In this study, we have assessed the influence of scaffold designs on the initial cell seeding efficiency of anchorage cells. The rapid prototyping designed polylactic acid (PLA) scaffolds fabricated using fused deposition modelling (FDM) designated by 2C and 4C scaffolds were compared to the commercial hydroxyapatite (HA) scaffold. Prior to cell seeding, 2C and 4C scaffolds were deposited with silicon carbonated HA nanopowders/ Hyaluronic acid/ Collagen Type I by Polyelectrolyte Multilayers (PEMs) coating and sterilized under ultra-violet (UV) radiation. Human Bone Marrow derived-Mesenchymal Stem Cells (hMSCs) were then cultured on the scaffolds using different volume of cell suspension, seeding method (one-side and two-sides; coated and non-coated well plates) and culture condition (static and rotary bioreactor). Regardless of scaffold designs, cell-seeded scaffold with small volume of cell suspension and two-sides seeding using coated well plate resulted in better cell attachment and distribution across the surface of the scaffolds than one-sided seeding. In any case, commercial HA exhibited lower percentages of cell attachment. Culturing these scaffolds in different media but culturing these scaffolds in rotary bioreactor has pronounced impact on the cell survival as compared to static condition.

1. Introduction

In tissue engineering (TE) particularly scaffold-based TE, two critical challenges remained are the homogenous seeding of cells throughout a porous scaffold and providing sufficient oxygen as well as nutrients to the seeded cells throughout the culture[1]. Cell seeding of scaffolds involves the distribution of a cell suspension across the surface of a scaffold and potentially growth within the pores. It is considered as a determining factor for the development of tissue formation as it is the first step involved in establishing a three-dimensional (3D) culture[2]. Therefore, it has been highlighted that homogenous cell seeding can be considered as the building block to uniform tissue formation. However, even for a small scaffold (e.g. 5 mm diameter X 2 mm thick), it can be major challenge to distribute a high density of cells efficiently and homogenously throughout the entire scaffold[2,3].

Several methods have been proposed for the cell seeding into the microporous scaffolds such as static and dynamic seeding. Although static seeding is still the most commonly used, it is characterized by low seeding efficiencies and non-uniform distribution[1]. In order to improve cell distribution as well as providing sufficient oxygen and nutrients for cell growth and survival after seeding, dynamic culture system have been developed, for example, spinner flasks, perfusion, rotating vessel, compression and magnetic force bioreactors[4]. Besides seeding method, the scaffolds architecture (i.e. channel size, porosity, surface roughness and surface area) and properties of the scaffolding material itself are responsible for the cell activities, which subsequently leads to tissue formation[1,5]. Numerous kinds of scaffolds have been fabricated using polymer, ceramic or natural based materials for example poly (lactic acid) (PLA), poly (glycolic acid) (PLGA), hydrogel, chitosan, hydroxyapatite (HA), and so forth by various fabrication techniques[2,6]. Recently, Rapid Prototyping (RP) techniques such as Fused Deposition Modelling (FDM) has become an attractive alternative to fabricate an intricate scaffolds design for bone TE, as it allows customization of scaffold designs to treat the variable needs of patients. For instance, a section of bone defect from the patient can be imaged using Magnetic Resonance Imaging (MRI) or X-Ray computed tomography (CT scan), which can then be converted to the file format for RP manufacturing (usually in stereolithographic or .STL format). The final stage is to print a customized 3D scaffolds for that individual patient[7]. In addition, fabricating scaffolds by FDM also allows the pore/ channel networks of that particular scaffold to be designed such as to reduce resistance of mass transport by shortening diffusion paths[1].

To gain further insight into the influence of cell seeding efficiency on 3D scaffolds, we firstly fabricated two different designs of three-dimensional printed (3DP) hybrid scaffolds via FDM method followed by polyelectrolyte multilayers (PEMs) coating as described in our previous report[5]. We cultured human bone marrow derived mesenchymal stem cells (hMSCs) on the fabricated scaffolds to demonstrate the optimal cell seeding efficiency by investigating the effect of culture method (i.e., one-sided and two-sided seeding, coated and non-coated plate and cell volume suspension) and culture conditions (static and rotary bioreactor) on cell adhesion and viability.

2. Experimental

In this study, two different scaffold designs were fabricated by FDM technique followed by PEMs in order to produce 3DP hybrid scaffolds. Commercially available scaffolds of pure sintered HA scaffolds were used as the experimental control. The basic properties of commercial HA, two channels (designated as 2C) and four channels (designated as 4C) hybrid scaffolds are summarized in Table 1.

Table 1. Summary of basic properties of commercial HA, 2C and 4C hybrid scaffolds

Scaffolds/ Properties	Commercial HA	4C	2C
Diameter (mm)	10	10	10
Thickness (mm)	2	2	2
Pores/ Channels size (μm)	200-300	1500	1500
Porosity (%)	60	40	20
Nature of materials	100% HA	SiCHA/ Collagen type I/ Hyaluronic acid coated on PLA	SiCHA/ Collagen type I/ Hyaluronic acid coated on PLA

IOP Conf. Series: Journal of Physics: Conf. Series **1082** (2018) 012072 doi:10.1088/1742-6596/1082/1/012072

2.1 Scaffolds fabrication

The 3DP scaffolds were initially fabricated by 3DP technique via FDM method using a synthetic polymer, Poly (lactic acid) (PLA) resin (Product code: 4032 D) purchased from NatureWorks® LLC (United States). Scaffolds were printed using Ultimaker 2 from Ultimaker (United Kingdom). Two different structural designs were fabricated i.e., the two- and four channel scaffolds and the basic properties of the fabricated scaffolds are shown in Table 1. The fabricated scaffolds were coated using Polyelectrolyte Multilayers (PEMs) technique using silicon carbonated hydroxyapatite nanopowders (SiCHA), collagen type I and hyaluronic acid as described in our previous report[5]. SiCHA nanopowders were firstly synthesized based on nanoemulsion method followed by calcination at 500°C, which we have developed earlier[8].

2.2 Cell culture and seeding

Human bone marrow derived-mesenchymal stem cells (hMSCs) obtained from a 24-year old male (Lonza, United States) were expanded until passage two when the required cell number was obtained. Cells were cultured and incubated at humidified environment at 37°C with 5% CO_2 in proliferation media (PM) consists of 4.5 g/L Dulbecco's Modified Eagle Medium (DMEM), L-glutamine, Penicilin-Streptomycin and Fetal Bovive Serum (FBS; Biosera labtech, United Kingdom). All other reagents were purchased from Lonza, United Kingdom.

Prior to cell seeding, the fabricated 3DP hybrid scaffolds were initially sterilized for three times in the UV Chamber for 90 seconds each cycle followed by pre-wetting in PM for three hours before cell seeding. Commercial hydroxyapatite (HA) scaffolds were used as control samples and these scaffolds required longer soaking in PM (72 hours) as recommended by the manufacturer (Ceramisys, Sheffield, United Kingdom). Pre-wetting is a crucial step to improve the seeding efficiency by promoting a formation of thin layer of proteins which are required for cell adhesion to the scaffolds. Scaffold seeding was carried out in 24 well of coated and uncoated cell culture well plates. The coated well plate used was a 24 well plate coated with 1% w/v Pluronic F-127 (Sigma-Aldrich, United States), to prevent the cell from attached to the well plate. As control, the uncoated well plate (ordinary 24 well culture plate) was used. After cell counting, 100 000 cells were seeded onto each scaffold in the wells. To choose the right seeding method, cells were seeded by two ways i.e. (1) seeded the total amount of cells only one side; (2) seeded half of the amount on one side, incubate for 3 hours and repeat these steps on the other side before toping up with proliferation media. The cell volume suspensions were also optimized using two different volume suspensions namely, 20 and 40 μL. The cellular scaffolds were then cultured for 24 hours at 37°C and 5% CO_2 in humidified atmosphere.

2.3 Cell seeding efficiency

Cell seeding efficiency of hMSCs on three different designs of the scaffolds was determined through the observation of their adhesion and viability capability.

2.3.1. Cell adhesion. Seeding efficiency was calculated using the following formulation:

$$Cell\ seeding\ efficiency\ (\%) = \frac{n_i - n_f}{n_i} * 100$$

where, n_i is the cell number seeded and n_f is the cell number adhered to the scaffolds after 24 hours incubation at 37°C and 5% CO_2 in humidified atmosphere, respectively.

The cell adhesion throughout the scaffolds was observed using MTT, (3-(4, 5-Dimethylthiazol-2-yl)-2, 5-diphenyltetrazolium bromide. The reagent is reduced to a crystalline, purple salt by viable cells. After 24 hours of incubation, media was removed from samples and they were washed in PBS. Samples were then stained with 500 μL of MTT reagents and incubated for 1.5hours, washed with 1.0 mL of PBS once and immediately imaged using a bright field microscope AMG-EVOS X1 CORE at magnification of 10X.

IOP Conf. Series: Journal of Physics: Conf. Series **1082** (2018) 012072 doi:10.1088/1742-6596/1082/1/012072

2.3.2. Cell viability. The cell viability was observed using Confocal Laser Scanning Microscope (CLSM) Olympus Fluoview FV 1200 with Fluoview Version 4.1 software (Olympus, UK). The viability of the cells was assessed after 24 hours using the Live/Dead Assay Kit (Invitrogen, United Kingdom) according to the manufacturer's instructions. Calcein-AM ester was used to fluorescently label viable cells (green); the nucleus of dead cells is labelled with Propidium Iodide (red). Briefly, cell culture media was removed from samples. They were washed with PBS then immersed in a PBS staining solution containing 10 μM Calcein-AM and 1 μM Propidium Iodide and incubated at 37°C for 20 minutes in the dark. The samples were then washed once with 1.0 mL of PBS and immediately imaged using CLSM.

3. Results and discussion

The concept of Bone tissue engineering (BTE) holds great promise for the treatment of clinical defects. However, much optimization is required before BTE can be broadly applied clinically. This study evaluated various cell seeding methods on the different designs of 3D scaffolds i.e. the HA, 2C, and 4C seeded with different volume of cell suspensions (20 and 40 μL) in coated and non-coated well plates (Table 2).

Table 2. Percentages of cell attachment onto the scaffolds with different cell volume suspensions on different scaffold designs

Scaffolds	Coated well plates		Non-coated well plates	
	20 μL	40 μL	20 μL	40 μL
HA	55%	47%	46%	45%
2C	87%	80%	56%	48%
4C	85%	77%	60%	52%

The results suggested that scaffolds seeded with small cell volume suspension of 20 μL have higher percentages of cell attachment onto the scaffolds as compared to 40 μL cell volume suspension. Regardless of the scaffold designs, lower cell attachment was found on scaffolds seeded in the non-coated well plates. In order to further optimize the seeding efficiency, scaffolds were then seeded in two different methods using 20 μL volume suspension and coated well plates. Using small volume cell suspension allows the cell to retain on the surface of the 3DP hybrid scaffolds and in the porous structure of the HA scaffolds. While, larger volume of cell suspension could easily flow through the channels/pores of the scaffolds and cells would then rather adhere to the well plate than to the scaffolds. Thus, smaller volume suspension was found to be better on a small scaffold in order to avoid overflowing of the cell suspension around the scaffold, which could lead to the loss of cells that should attached to the scaffolds.

The percentages of cell attachment was found to be higher on 3DP hybrid scaffolds as compared to the HA scaffolds. This showed that the coating materials used to fabricate the 3DP hybrid scaffolds were more suitable for cell attachment than pure HA scaffolds. This proved the role of collagen as one of the elements in the coating materials as demonstrated in our previous study[5]. The presence of collagen on the 3DP hybrid scaffolds has effectively contributed towards better support for cell attachment[9]. While, pure HA scaffolds needed longer pre-wetting time with higher concentration of serum (20% FBS) before seeding compared to the 3DP hybrid scaffolds. This was to create a thin layer of protein on the surface as to provide better cell attachment.

IOP Conf. Series: Journal of Physics: Conf. Series **1082** (2018) 012072 doi:10.1088/1742-6596/1082/1/012072

To further improve the seeding efficiency, scaffolds were seeded with hMSCs on different culture substrates. Pluronic F-127 has been used as biological coating in particular to prevent cells adherent to the culture well plate[10]. Pluronic F-127 is a polypropylene oxide-polyethylene oxide (PPO-PEO) tri-block copolymer with two hydrophilic ethylene oxide chains and a hydrophobic propylene oxide chain in the middle, which provides Pluronic F-127 with very flexible molecular chains and high capacity to hydration[11]. Studies have shown that coating substrates with Pluronic F-127 reduces protein and hence cell adhesion, which prevents seeded cells from adhering to the substrate. The results obtained shows that hMSCs seeded on all scaffold designs in the coated well plates resulted in higher percentages of cell attachment compared to the non-coated well plates. This finding highlighted the beneficial use of Pluronic F-127 in improving seeding efficiency.

Seeding method is another crucial key towards efficient seeding. The distribution of cells on each scaffold seeded by one-sided and two-sided method was analysed using MTT stains (Figure 1). It was found that cells were more homogenously spread over the entire scaffolds when hMSCs were seeded on the scaffolds by the two ways method. The use of two ways seeding method is indirectly helping the cells to migrate faster throughout the entire scaffolds. While, the one way seeding method requires longer time for the cells to migrate and infiltrate from one side of the scaffold to the other[12].

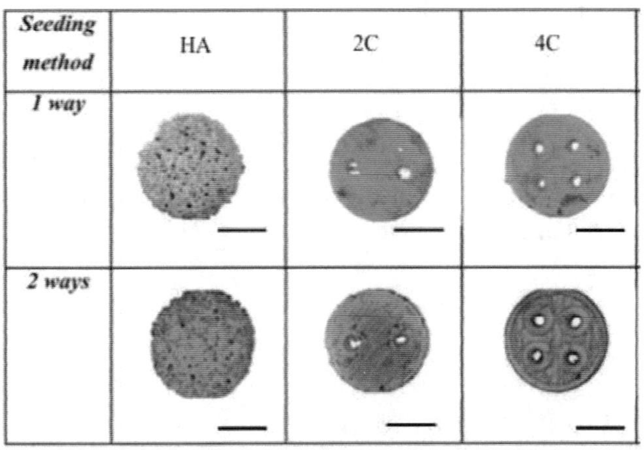

Figure 1. Cell distribution on different scaffold designs seeded with one-sided and two-sided seeding method. Scale bar = 1 mm.

The final part of our study investigated on the effect of culture condition on cell seeding efficiency. Rotary bioreactor was used in this study to investigate the impact of a microgravity environment on cells fate upon cultured on different scaffold designs. The aims of using this type of bioreactor were to minimize shear force and maximize the fluid flow throughout the scaffolds to provide enhanced mass transfer across large scaffolds for BTE[4]. Live/dead staining observation revealed that higher proportions of viable cells were found when scaffolds were cultured at 20 rpm compared to 40 rpm. This indicates that the lower rotational speed allows for a better cell attachment across the scaffolds whilst the cells tend to detach from the scaffolds when they were rotated at higher speed. At a higher speed of 40 rpm, scaffolds started to collide with each other and the wall of the chamber, resulted in some loss of cells.

IOP Conf. Series: Journal of Physics: Conf. Series **1082** (2018) 012072 doi:10.1088/1742-6596/1082/1/012072

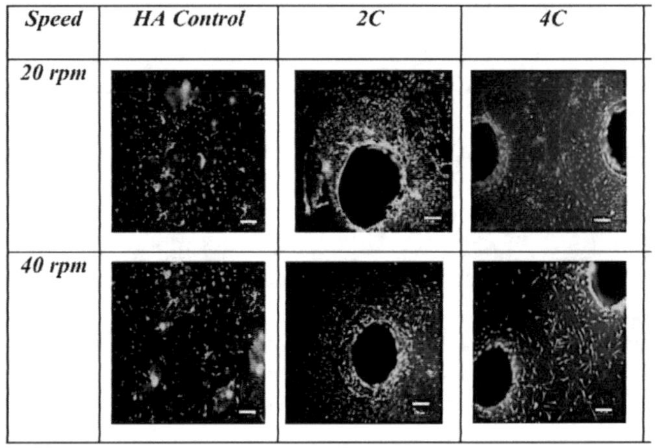

Figure 2. Cell viability on different structural designs of 3DP hybrid scaffolds at 20 and 40 rpm, respectively. Results suggested more viable cells were found as cells were cultured at 20 rpm where the scaffolds were free floating in the rotary bioreactor. Scale bar= 500 μm.

4. Conclusion

Five main parameters improving cell seeding efficiency were established in this study: (1) suitable composition of the tested scaffolds mimicking bone mineral content, (2) small volume of cell suspension, (3) using coated well plates during cell seeding, (4) seed cells by the two-sided method and (5) controlled speed of rotary bioreactor.

References

[1] Melchels, F. P. W. *et al.* 2010 *Acta Biomater* 6 4208
[2] Wang, P. *et al.* 2014 *Bone Res.* 2 14017
[3] Luo, Y. *et al.* 2015 *ACS Appl. Mater. Interfaces*
[4] El Haj, A. J. & Cartmell, S. H. 2010 *Proc. Inst. Mech. Eng. Part H J. Eng. Med.* 224 1523
[5] Baba Ismail, Y. M., Ferreira, A. M., Bretcanu, O., Dalgarno, K. & El Haj, A. J. 2017 *Colloids Surfaces B Biointerfaces* 159 445
[6] Barry, M., Pearce, H., Cross, L., Tatullo, M. & Gaharwar, A. K. 2016 *Curr. Osteoporos. Rep.* 14 87
[7] Subia, B., Kundu, J. & Kundu, S. 2010 *Tissue Eng.* 141
[8] Baba Ismail, Y. M., Wimpenny, I., Bretcanu, O., Dalgarno, K. & El Haj, A. J. 2017 *J. Biomed. Mater. Res. - Part A* 1775
[9] Somaiah, C. *et al.* 2015 *PLoS One* 10 1
[10] Dang, S. M., Kyba, M., Perlingeiro, R., Daley, G. Q. & Zandstra, P. W. 2002 *Biotechnol. Bioeng.* 78 442
[11] Deegan, A. J. *et al.* 2014 *Biomed. Eng. Online* 13 136
[12] Brennan, M. Á. *et al.* 2015 *Biomed. Mater.* 10 45019

IOP Conf. Series: Journal of Physics: Conf. Series **1082** (2018) 012073 doi:10.1088/1742-6596/1082/1/012073

Effect of chitosan concentration on the properties of PLA/β-TCP scaffold produced by combination of solvent casting and salt leaching techniques

Rosaniza Md Isa[1], Mariatti Jaafar[1], Zulkifli Ahmad[1], Badrul Hisham Yahya[2]

[1]Biomaterials Niche Area Group, School of Materials and Mineral Resources Engineering, Engineering Campus, Universiti Sains Malaysia, 14300 NibongTebal, Pulau Pinang, Malaysia

[2]Advanced Medical & Dental Institute (AMDI), Universiti Sains Malaysia Bertam, 13200 Kepala Batas, Penang

Email: mariatti@usm.my

Abstract. Polylactic acid (PLA)/beta-tricalcium phosphate (β-TCP) porous scaffold was fabricated using solvent casting and salt leaching. This scaffold is facing problems such as relatively low compressive strength and high brittleness. In order to overcome this problem, natural polymer coated porous scaffold was introduced. In this study, PLA/β-TCP scaffold was coated with four different concentration of chitosan (0.5%, 1.0%, 1.5% and 2.0% w/v). It was found chitosan coated scaffold has higher compressive strength compared to that of uncoated scaffold. The compressive strength of PLA/β-TCP coated with 1.5% w/v of chitosan indicates the highest compressive strength (0.49 MPa), which is 35% increament compared to the uncoated scaffold. Morphological observation by scanning electron microscopy (SEM) revealed that highly porous three-dimensional scaffold with well interconnected porous structure could be achieved by coated and uncoated scaffolds. The results reveal that the scaffold coated with chitosan layer achieves a highly interpenetrated polymer and scaffold network, potentially leading to a porous scaffold with higher mechanical strength and toughness.

1. Introduction

Scaffold is a porous three-dimensional structure that acts as a mechanical support for the cell growth, differentiation, proliferation, and other cellular functions. The current materials used in tissue engineering fields are metals, ceramics, composites, and natural/synthetic polymers [1]. These materials are used for replacement and repair of damaged or traumatized bone tissues. Polymer-based composite scaffolds are widely applied in tissue engineering. These composites improved the biological and mechanical performances of the scaffolds compared with other materials. Commonly, tissue engineering scaffolds are mainly made of biodegradable polymers such as poly(lactic acid) (PLA), poly(glycolic acid), and their copolymer poly- (lactic-co-glycolic acid) (PLGA) [2]. These polymers have the advantages of demonstrated host compatibility, variable physiochemical properties and predictable resorption rates. However, these polymers also have poor cell–material interactions because of their hydrophobic nature. Calcium phosphates (i.e., hydroxyapatite (HA) and beta-tricalcium phosphate (β-TCP)) were used due to their bone-like composition. β-TCP is a bioceramic materials that allows it to be resorbed and replaced by bone during the healing process. β-TCP has a good biocompatible, non-toxic, non-inflammatory and osteoconductive, but this material also facing the limitation of the biocompatibility, biodegradability and low mechanical properties [3]. In the present study, the purpose of combination of PLA and β-TCP is to achieve optimum mechanical and biological performance for bone tissue application.

There are several methods to fabricate highly porous biodegradable scaffolds including fiber-bonding, 3D-printing, replication, direct foaming, phase-separation, emulsion freeze drying, gas-foaming and solvent casting and particle leaching. The solvent casting and salt leaching are chosen as a fabrication method according to the simplicity of operation and producing scaffolds with a controlled composition, good pore interconnectivity degree, porosity and pore size [4]. Due to the limitation on the properties of PLA/ β-TCP scaffolds, coating of the scaffold has been introduced. Coating method is adopted in fabricating composite scaffolds since it can improve tissue interaction, mechanical scaffold stability, and toughness [5]. Chitosan is used as a bioactive coating according to biocompatible properties, nontoxic, biodegradable and hydrophilic polysaccharide [6]. In the present study, the effects of different concentration of chitosan layer (0.5%, 1.0%, 1.5% and 2.0% w/v) with silane treatment processes on the mechanical properties of the PLA/ β-TCP scaffolds were determined.

2. Materials and Methodology

2.1 Materials

The raw materials used for in the study were poly(lactic acid) (PLA), dichloromethane (DCM), chloroform, sodium chloride (NaCl) and β-tricalcium phosphate (β-TCP), chitosan, 3-(Trimethoxysilyl)propyl methacrylate (A174) and glutaraldehyde (GA). PLA was supplied by Nature Works in pellet form with a specified molecular weight average of 74,00000. DCM with > 99.5% purity, chloroform and NaCl were supplied by Merck Millipore. β-TCP, medium molecular weight of chitosan, GA and 3-trimethoxysiyl propyl methacrylate were supplied by Sigma Aldrich. β-TCP used in powder form with the weight average of 310.18 g/mol.

PLA/NaCl and β-TCP ($Ca_3(PO_4)_2$) were used as the main component in the fabrication of composite porous scaffolds. NaCl act as porogen agent while, 3-(Trimethoxysilyl) propyl methacrylate was used as silane coupling agent to produce chemical link between PLA/β-TCP scaffolds and chitosan.

2.2 Methodology

PLA/β-TCP porous scaffolds were prepared by using solvent casting and particulate (salt) leaching technique. In order to fabricate porous scaffolds, PLA pellets were dissolved in DCM/chloroform (v:v= 1:1) for two hours. About 40 wt% of β-TCP and NaCl powder with a mass ratio of polymer to NaCl of 1:5 (wt/wt) was added into the solution to form a paste-like gel. After mixing, the gel-like mixture was then cast into the Teflon mold (diameter=1.5 cm and height=2.5 cm). The scaffolds were air dried for 24 hours and de-aired inside a vacuum desiccators for 4 hours. Next, the scaffolds was dipped in ethanol about 2 hours at room temperature and immersed in distilled water for another 4 days before placed the samples in the oven at 80 °C for 24 hours to leach out of the water.

For treatment, 2% (v/v%) of silane (3-trimethoxysiyl propyl methacrylate) coupling agent was added to the 95% (v/v%) methanol and 5% (v/v%) distilled water solution. The scaffolds were soaked in this solution for 2 hours and continuously rinsed with deionized water and ethanol. After that, the scaffolds were placed in the oven at 80 °C for 2 hours. For coating process, the chitosan solution was

IOP Conf. Series: Journal of Physics: Conf. Series **1082** (2018) 012073 doi:10.1088/1742-6596/1082/1/012073

prepared by dissolving the powder in 99 ml of distilled water, 1ml of acetic acid and 0.025% of glutaraldehyde in order to produce chitosan solutions (0.5%, 1.0%, 1.5% and 2.0% w/v). The composite scaffolds were then dipped into the chitosan solution; next, vacuum infiltration was carried out 24 hours to infiltrate the chitosan into the porous structure of scaffolds and finally, the scaffolds were dried in vacuum oven for 24 hours to remove any residual solvent.

The AccuPyc II 1340 Series Pycnometers were applied to identify the density and porosity of the scaffolds. The morphology of porous composite scaffolds was carried out by using field emission scanning electronmicroscopy (FESEM), model HitachiTM3000 TableTop. The compression test was conducted using Instron Universal Testing Machine; model Instron3369 according to ASTM-F-1250-13. The crosshead speed of testing was at 1 mm/min with dimensions 25 x 15 mm were used.

3. Results and discussion

Figure 1 (a-e) present the fracture surface after compressive test of coated and uncoated PLA/ β-TCP scaffolds with different concentration of chitosan. All the scaffolds were treated with silane prior to chitosan coating. SEM result shows the coral-reef like structure and the distribution of interconnected macropores size in the range of 150 to 500 μm of the PLA/β-TCP scaffolds for uncoated and coated with chitosan. From Figure 1 (b-e), chitosan layer can be seen clearly seen on the cross-sectional surface of the scaffolds after coating with different concentrations of chitosan. The morphology of porous scaffolds coated with 1.5 w/v% (Figure 1(d)) indicates the chitosan layer on the surface of the struts compared other coated scaffolds. However, 2.0 w/v% of the chitosan shows a thick layer of chitosan on the cross-sectional surface of the scaffolds. This indirectly proved that the chitosan is stick on the top of the surface and failed to penetrate throughout the internal micropores of the scaffolds. Zairani et al. [7] reported that the β-TCP scaffolds coated at higher concentration (6 wt%) of alginate produce rough surface of the porous structure. Penetration of alginate into the inner scaffold structure is difficult due to high alginate concentration which relates to high viscosity.

Figure 2 shows the compression strength and total porosity of uncoated and coated PLA/ β-TCP scaffolds with different concentration of chitosan (0, 0.5, 1.0, 1.5, and 2.0 w/v%). The concentration of chitosan at 1.5 w/v % shows the highest compressive strength (0.49 MPa) compared to uncoated and other coated scaffolds. Concentration of 1.5 w/v% of chitosan was sufficient to permit the chitosan to infiltrate the entire structure of the porous PLA/β-TCP scaffold. Infiltration of the chitosan into the scaffold improves the brittleness and biocompatibility of the porous scaffolds [8]. Besides, higher porosity of the uncoated scaffold (in Figure 2) which is 78.5% compared to other coated scaffolds which are in the range of 73 to77.6%. Theoretically, the porosity of at least 50–90% is required for the bone ingrowth process [9]. This concludes that all the scaffolds have the minimum porosity which is more than 70% and suitable for bone tissues application.

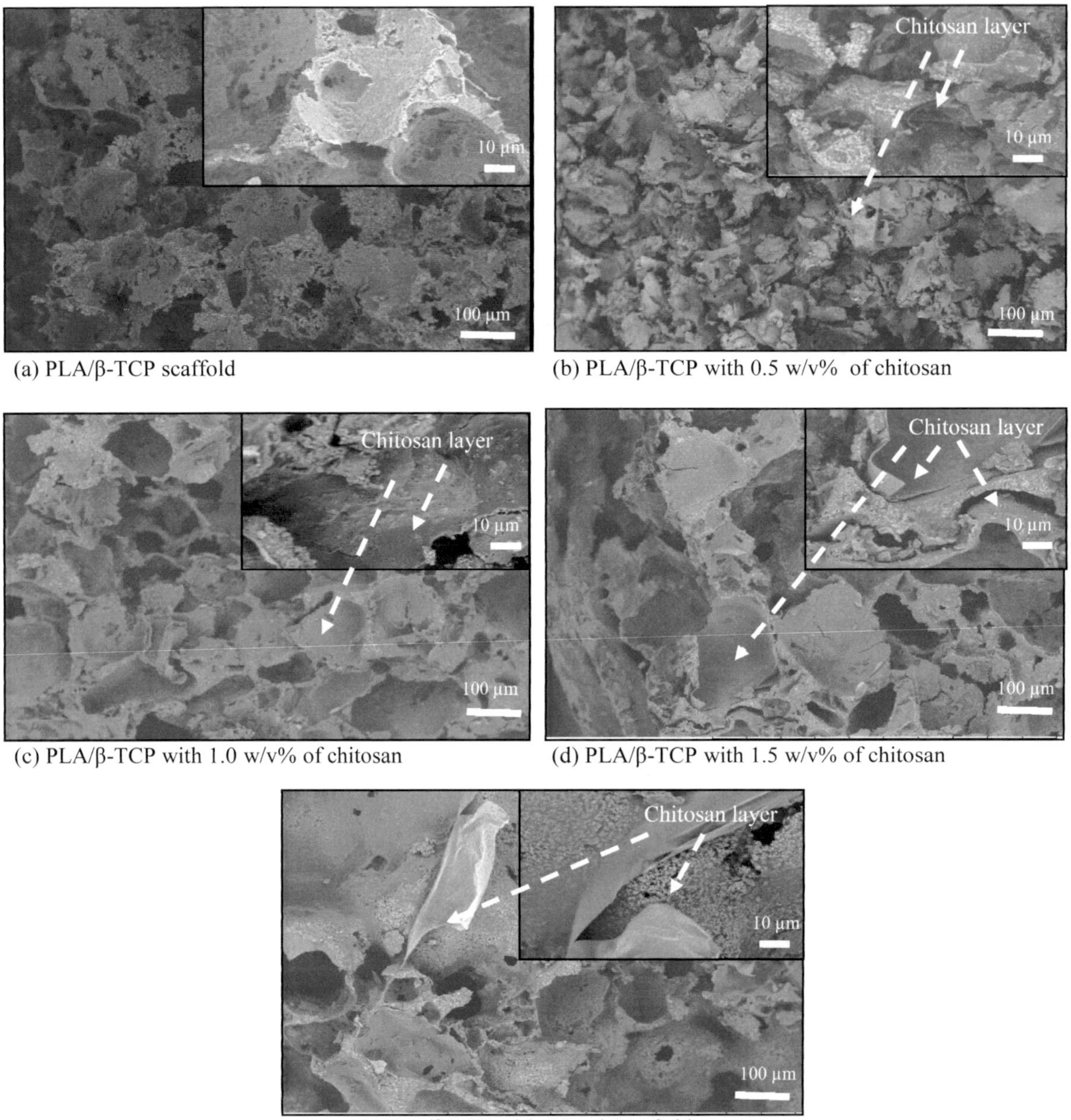

(a) PLA/β-TCP scaffold

(b) PLA/β-TCP with 0.5 w/v% of chitosan

(c) PLA/β-TCP with 1.0 w/v% of chitosan

(d) PLA/β-TCP with 1.5 w/v% of chitosan

(e) PLA/β-TCP with 2.0 w/v% of chitosan

Figure 1. SEM images of the (a) PLA/β-TCP scaffold and PLA/β-TCP prepared from different concentrations of chitosan; (b) 0.5 w/v, (c) 1.0 w/v %, (d) 1.5 w/v % and (e) 2.0 w/v % of chitosan at different magnification (100x magnification and 500x magnification (for insert), respectively).

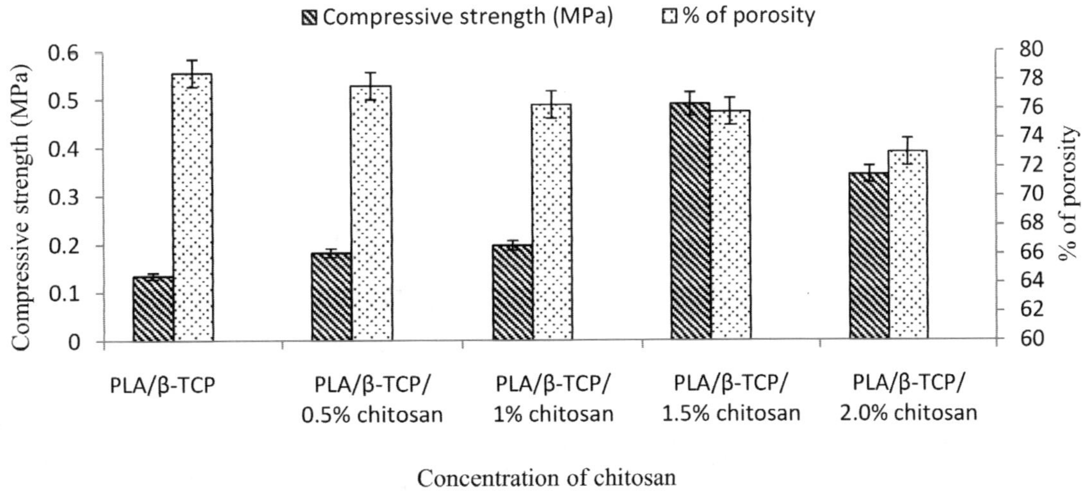

Figure 2. The compressive strength and percentage of porosity for PLA/ β-TCP with different concentration of chitosan

Figure 3 indicates the functional group for chitosan, silane, uncoated PLA/β-TCP scaffolds and treated-coated PLA/β-TCP scaffold. It is observed the presence of the chitosan and silane functional groups on the PLA/ β-TCP scaffold structure. The silane solution used as a reference to show the presence of hydrocarbon chain (-CH) vibrations at the peak around 3000-2800 cm^{-1}. The silanol group Si-O can be seen in silanization spectrum at 1122 cm^{-1} and the peak Si-O-Si linkage at around 500 cm^{-1}. This is because of the existence of covalent bonding between the silane coupling agent and PLA/ β-TCP when coated with chitosan. The band at 1173 cm^{-1} indicates of an amino group also indicates the possible interaction between the chitosan and PLA/ β-TCP during the coating process. The IR spectrum of the scaffolds beads showed a band at 1632cm^{-1} which corresponds to imine bond (C=N) which might be formed between chitosan and glutaraldehyde [10].

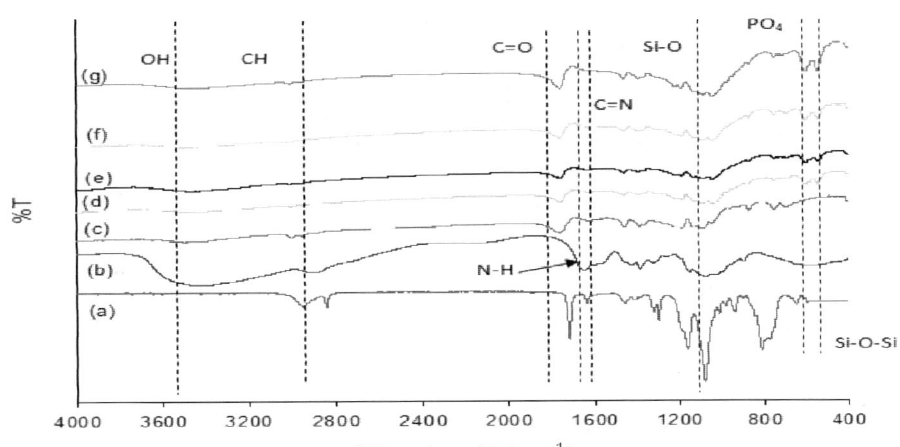

Figure 3. FTIR spectra of (a) Pure chitosan, (b) silane solution, (c) uncoated PLA/ β-TCP scaffold and treated-coated PLA/ β-TCP scaffolds with different concentration of chitosan ; (d) 0.5wt% of chitosan , (e) 1.0wt% of chitosan , (d) 1.5wt% of chitosan and (d) 2.0wt% of chitosan.

Table 1. FTIR spectra that exists from the PLA/ β-TCP scaffolds, treated and coated with chitosan

Bonding type	Wavenumber (cm^{-1})	Sample
OH bond	622, 3800-3600	Chitosan
C=N	1632	Chitosan/ glutaraldehyde
PO_4 bond	607-554	β-TCP
Si–O, Si-O-Si	1122, 500	Silane
C-H	1452-1353	Silane
NH	3300, 1654	Silane, secondary amide (chitosan)
C=O	1598,1728	Chitosan, PLA
C-H	1452-1353	Silane

4. Conclusion

The effect of different concentrations of chitosan on the morphology, compressive strength, pore size and porosity of the PLA/β-TCP scaffolds were investigated. The PLA/ β-TCP scaffold coated with 1.5 w/v% of chitosan exhibited the highest compressive strength with 75.8 % of porosity compared to uncoated and other coated scaffolds. The compressive strength shows the ascending order from 0.5 until 1.5 w/v% and the compressive strength decreased at 2.0 w/v%. That means the chitosan solution at higher concentration (2.0 w/v%) is not able to penetrate into the porous structure. FTIR analysis shows the presence of chitosan and silane functional groups on the surface of the PLA/β-TCP scaffolds.

5. Acknowledgement

The authors gratefully acknowledge support from the Universiti Sains Malaysia and the School of Materials and Mineral Resources Engineering Campus. This project was funded by Ministry of Education Malaysia through the Trans Disciplinary Research Grant Scheme (TRGS; Grant no. 6761004).

References

[1] Hadjipanayi E, Brown R A and Mudera V 2009 *Journal of Tissue Engineering and Regenerative Medicine* 3 3230
[2] Gunatillake P A, Adhikari R and Gadegaard N 2003 *European Cells and Materials* 5 1
[3] Cao Z, Dou C and Dong S 2014 *Journal of Nanomaterials* 2014 4
[4] Smita Mohanty M C, Mohanty S and Nayak S 2015 *Journal of Tissue Science Engineering* 6 1
[5] Kang Y, Scully A, Young D A, Kim S, Tsao H, Sen M and Yang Y 2011 *European Polymer Journal* 47 1569
[6] Liu S X, Kim J T and Kim S 2008 *Journal of Food Science* 73 3
[7] Zairani N A S, Jaafar M, Ahmad N and Abdul Razak K 2016 *Ceramics International* 42 5141
[8] Roohani-Esfahani S I, Nouri-Khorasani S, Lu Z F, Appleyard R C and Zreiqat H 2011 *Acta Biomaterialia* 7 31307
[9] Yeo A, Wong W J and Teoh S H 2010 *Journal of Biomedical Materials Research - Part A* 93 1358
[10] Sharma C, Dinda A K and Mishra N C 2012 *Journal of Biomaterials and Tissue Engineering* 2 133

IOP Conf. Series: Journal of Physics: Conf. Series **1082** (2018) 012074 doi:10.1088/1742-6596/1082/1/012074

Effects of Milling Speed and Sintering on the Formation of Akermanite ($Ca_2MgSi_2O_7$) Bioceramics

Ahmad-Fauzi Mohd Noor[1*], Yanny-Marliana Baba Ismail[1], Masakazu Kawashita[2], Aye Aye Thant[3], Myat Myat-Htun[1,3]

*[1]School of Materials and Mineral Resources Engineering, Engineering Campus, Universiti Sains Malaysia, 14300 Nibong Tebal, Penang, Malaysia

[2]Graduate School of Biomedical Engineering, Tohoku University, Sendai 980-8579, Japan

[3]Department of Physics, University of Yangon, 11041 Kamayut, Yangon, Myanmar

*E-mail: srafauzi@usm.my

Abstract. The aim of the present work is to study the effect of different milling speeds on the preparation of akermanite ceramics. In this study, akermanite ($Ca_2MgSi_2O_7$) was synthesized by high speed planetary ball milling and mixing method. Calcium oxide (CaO), magnesium oxide (MgO) and silicon dioxide (SiO_2) were used as initial reactants. The starting materials were milled with molar ratio of 2:1:2, respectively, using ball-to-powder ratio 10:1 at different milling speeds of 300 rpm, 400 rpm and 500 rpm. After synthesis, the powders were pelletized to form pellets (13 mm) with 150 MPa by uniaxial pressing. The pellets were then sintered at 1250°C for 3h. Phase analysis by X-ray diffraction (XRD), surface morphology by Field Emission Scanning Electron Microscopy (FESEM), Brunauer-Emmet-Teller (BET) specific surface area analysis, linear shrinkage, porosity and density measurement were performed on the milled and sintered samples. It was found that milling at 500 rpm resulted in powder with largest surface area (108 m^2/g), indicating the finest size of powder. FESEM observations showed that the sintered-akermanite ceramics from 500 rpm milling at 1250°C were the densest with microcrystalline appearance and clear grain boundaries, while milling at 300 rpm and 400 rpm showed some pores in the sintered samples.

1. Introduction

Bioactive materials, especially calcium phosphate (CaP) and calcium silicate (CS) materials, have been investigated due to their excellent biocompatibility and osteoconductivity [1,2]. Among CS biomaterials, akermanite ($Ca_2MgSi_2O_7$), a mineral containing Ca, Mg and Si, has recently received research interest due to its better mechanical properties compared to hydroxyapatite (HA) [3] and it has common degradation rate compared with other silicate bioceramics [4].

Wu. et al. produced akermanite bioceramics in sol-gel method at sintering temperature at 1370°C for 6h. Their results demonstrated that akermanite samples have apatite-forming ability on their surface when soaked in simulated body fluid (SBF) for 10 days, and possess improved mechanical properties compared to HA [5].

Many methods have been proposed for the synthesis of nano-sized calcium silicate materials such as sol gel, chemical precipitation, hydrothermal and solid reaction methods [6-9]. Among them, mechanical activation is a technique used for a wide range of preparation of nano powders and composites with reduced particle sizes [10-13].

The objectives of high energy mechanical ball milling and mixing are for reducing the particle size and simultaneously mixing the raw materials. Besides low cost process, the advantages of this technique are simplicity of experiment, and rapid and easy to synthesize the products [14]. Most of the researchers have focused on the influence of milling time on particle size and densitification parameters [15,16]. The current study is therefore designed to study the effect of different milling speeds via high energy wet ball milling and mixing as a processing technique and subsequent sintering process on the formation of akermanite.

2. Materials and Methods

2.1 Samples preparation

In this study, akermanite powders were synthesized by using calcium oxide (CaO), magnesium oxide (MgO) and silicon dioxide (SiO_2) as the starting materials with molar ratio of 2:1:2, respectively. These materials were then ball-milled with milling speeds of 300, 400 and 500 rpm for 4h (based on earlier study) in a planetary ball mill (Retsch PM 100). Zirconia balls were used as the milling media to assist the milling-mixing process. The ball-to-powder weight ratio was 10:1 and the deionized water-to-powder weight ratio was set to 4:1. Then, the resultant slurry was dried in an electric oven at 100°C for 24h, and subsequently grinded and sieved. Some as-milled powders were used for BET specific surface area analysis and some were uniaxially pressed at 150 MPa to obtain disc-shaped green bodies with diameter 13 mm. The pellets were then sintered in an electric furnace at 1250°C with a heating rate of 5°C/min for soaking time 3h.

2.2 X-ray diffraction (XRD) analysis

The phase identification, lattice parameter and crystallite size of as-milled and sintered powders were obtained with Cu-$K_{\alpha 1}$ radiation on X-ray diffraction analysis (XRD, Bruker D2 Phaser, Germany). The obtained experimental patterns were compared with the standards complied by the Joint Committee on Powder Diffraction and Standards (JCPDS). The average crystallite size of as-milled samples was obtained from the experimental XRD patterns using the Debye-Scherrer's relation [17], $D = k\,\lambda/\beta\cos\theta$, where D is the crystallite size for the (hkl) plane, λ is the wavelength of the incident X-ray radiation (Cu-$K_{\alpha 1}$ radiation, λ = 0.154 nm), β is the full width at half maximum (FWHM) in radiations, and θ is the corresponding angle of Bragg diffraction.

2.3 Field Emission Scanning Electron Microscopy (FESEM) analysis

Morphology and grain sizes of samples were investigated by using FESEM (Zeiss Supra Gemini 35 VP). A thin layer of gold (Au) samples was coated onto the sample. The average grain size of the sintered samples was evaluated using the image analysis software (Image J 1.51k).

2.4 Linear shrinkage measurement

Diameter (D_g) and thickness (T_g) of the green bodies were measured by digital vernier caliper (Mitutoyo, Japan). After sintering the pellets at 1250°C, the measurements of diameter (D_s) and thickness (T_s) of the samples were done again to evaluate the linear shrinkage of sintered samples. Changes in linear shrinkage of all samples during sintering was calculated by using the following equations [18]:

$$\text{Diameter shrinkage (\%)} = \frac{D_g - D_s}{D_g} \times 100\ \%$$

$$\text{Thickness shrinkage (\%)} = \frac{T_g - T_s}{T_g} \times 100\ \%$$

IOP Conf. Series: Journal of Physics: Conf. Series **1082** (2018) 012074 doi:10.1088/1742-6596/1082/1/012074

2.5 Porosity and density measurements

The apparent porosity P measurement of the sintered akermanite samples was carried out by Sartorius CP2245 Analytical balance set and calculated by the Archimedes' method [19]. P = (W-D)/(W-S) × 100%, where D is the weight of the sample in air, W is the weight of sample with water, and S is the weight of sample suspended in water.

The apparent density D_a of the ceramics was measured according to the Archimedes' formula [20]: $D_a = D/(W-S)$, where D is the weight of sample in air, W is the weight of sample with water, and S is the weight of sample suspended in water. The relative density R_d was calculated using the formula $R_d = D_a/T_d$, where D_a is the apparent density and T_d is the theoretic density of the ceramics.

2.6 Brunauer-Emmet-Teller (BET) analysis

The specific surface areas of akermanite ceramics were determined from the N_2 gas adsorption isotherms using Brunauer-Emmet-Teller (BET) technique on a Micromeritics ASAP-2020 surface area and porosity analyzer. BET equivalent particle size, under a spherical particle assumption, was calculated by the following equation [21]. $D_{BET} = \frac{6}{\rho \cdot S_{BET}}$, where S_{BET} is the specific surface area of as-milled powder ($m^2\ g^{-1}$), and ρ is density of akermanite (g/cm^3).

3. Results and discussion

3.1 X-ray diffraction (XRD) analysis

The XRD patterns of akermanite powders sintered at 1250°C is shown in Fig. 1. The sharpness of peaks shows high crystalline order of akermanite as evidenced by the highest peak at ~31.1°. The refined lattice parameter values of tetragonal akermanite were a = b = 7.83 Å , c = 5.01 Å, α = β = γ = 90° are matched well with the standard lattice parameters (a = b = 7.8340 Å , c = 5.0080 Å, α = β = γ = 90°) which reflects the formation of akermanite (JCPDS no. 35-592). The XRD patterns of akermanite milled with 300, 400 and 500 rpm showed impurity such as diopside ($CaMgSi_2O_6$, JCPDS no. 41-1370).

XRD average crystallite sizes were estimated by selecting highest 6 peaks of as-milled powders. The crystallite size values for as-milled samples are listed in Table (1). According to the results obtained from Scherrer equation revealed that increasing in milling speed from 300 rpm to 500 rpm reduced the crystallite size. This is due to the increase in particle fracturing induced by high mechanical energy ball impacts and the number of impacts between material particles and milling balls collision during ball mill process [22-24]. Therefore, ball milling-mixing synthesis gradually reduces the particle sizes of the powders by activating mechanically.

Table 1. Comparison of average values of crystallite size of as-milled powders

Milling speed	Average crystallite size (nm)
300 rpm, 4h	45.5
400 rpm, 4h	44.1
500 rpm, 4h	43.3

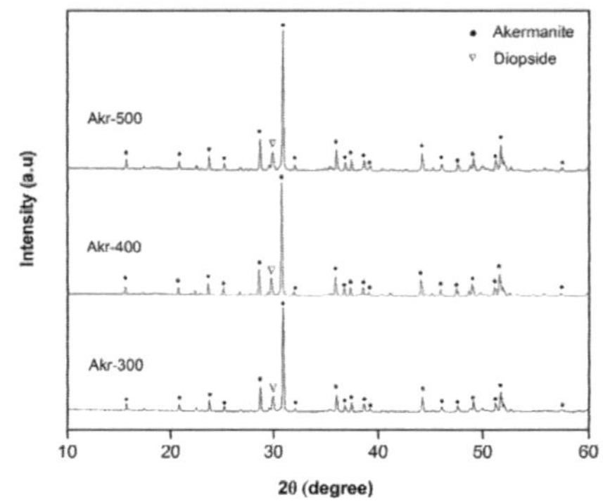

Figure 1. XRD patterns of akermanite powders at different milling speeds and sintered at 1250°C

3.2 Field Emission Scanning Electron Microscopy (FESEM) analysis

The micrographs of sintered akermanite pellets at different milling speeds are illustrated in Fig. 2 (a-c). As can be seen in the FESEM images, the grain boundaries of all samples are very clear after sintering at 1250°C for 3h that indicated the sample surfaces were well crystallized. The grains are irregular in shape and non-uniform in size in all cases. The number of grain decreases as the characteristic grain size increases.

The average grain sizes of the sintered samples with milling speeds 300, 400 and 500 rpm are 1.785, 2.143 and 2.314 μm, respectively. According to the result, akermanite ceramics milled with 500 rpm sintered at 1250°C is observed with the largest grain size. This suggested that the reduction in particle size leads to the increase in particle surface area which drives overall microstructural grain growth in nanocrystalline materials. The growth of apparently larger grains is attributed by the consumption of smaller grains. Subsequently, the grain growth positively influences the densification of material ceramics [25].

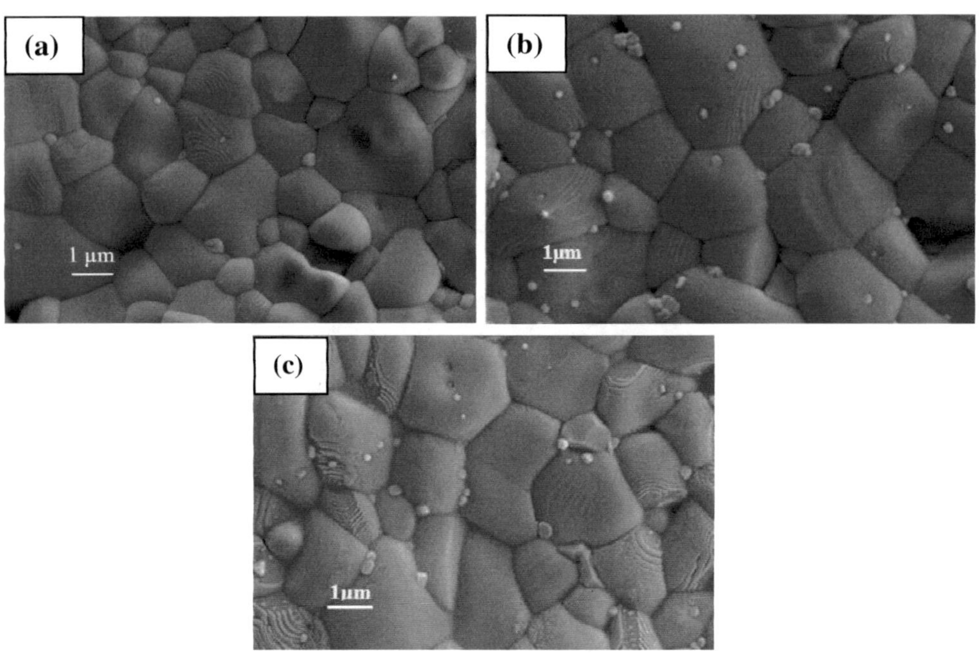

Figure 2. FESEM micrographs of the sintered akermanite pellets milled with (a) 300 rpm, (b) 400 rpm, and (c) 500 rpm

3.3 Linear shrinkage measurement

The changes of the linear shrinkage of the sintered ceramics at 1250°C are described in Table 2. The values presented in the table suggest that the extent of shrinkage is mainly attributed to the specific surface area of particles. Higher specific surface area, as well as, higher surface energy can give rise to more shrinkage during the thermal treatment and consequently improved densification by reducing in porosity. I. Cristofolini et al studied the effect of sintering temperature on the shrinkage and reported that sintering leads to the progressive reduction of the total pore surface and pore volume [26]. Therefore, particle size and sintering influenced the final product dimension.

3.4 Porosity and density measurements

The apparent porosity and relative density of sintered samples with different milling speeds are listed in Table 2. Based on the result, low residual porosity could be found in all ceramic samples [27]. The sintered samples milled with 500 rpm exhibited the highest relative density in comparison with 300 and 400 rpm. This reveals that the higher milling speed produces the finer powders which could be sintered to a higher relative density [28] due to the increase in contact area between the particles. Therefore, increasing the milling speed has a good effect on relative density and apparent porosity.

Table 2. Linear shrinkage, apparent porosity and density of sintered samples at 1250°C for 3h

Milling speed	Diameter Shrinkage (%)	Thickness Shrinkage (%)	Apparent Porosity (%)	Relative Density (%)
300 rpm	25.465	21.892	3.762	93.717
400 rpm	27.409	25.459	3.569	96.807
500 rpm	27.542	26.021	3.485	96.829

3.5 Brunauer-Emmet-Teller (BET) analysis

BET measurements (Table 3) revealed that the specific surface area S_{BET} of as-milled powders shows an increasing trend with the milling speed, while the BET equivalent particle size D_{BET} shows a decreasing trend. It means that particle size decreases because of the particle fracture induced by mechanical energy of ball impacts [29]. The highest specific surface area the finest particle size of the as-milled powders was reached to 108 m²/g at 500 rpm. This surface area result coincides again with results of XRD analysis and FESEM analysis.

Table 3. The characterization of BET specific surface area and particle size of as-milled powders

Milling speed	BET method	
	S_{BET} (m^2/g)	D_{BET} (μm)
300 rpm	77	0.026
400 rpm	88	0.023
500 rpm	108	0.019

4. Conclusions

Akermanite ceramics were mechanically activated by using high energy planetary ball milling and mixing. Based on the XRD results, the calculated crystallite size obtained from Scherrer method indicated that crystallite size was reduced with increasing the milling speed. According to BET analysis and SEM micrographs observation, synthesizing the starting materials by increasing milling speed produces the formation of smaller sized particles, showing the increase in the BET specific surface area of particles and leading to larger grain size due to mechanical activation and sintering temperature. The apparent porosity of sintered akermanite ceramics prepared with 500 rpm was

3.485% and relative density was 96.829%. Therefore, the results from the study showed the densest akermanite ceramics were produced at 500 rpm at sintering temperature 1250°C.

Acknowledgements

The authors sincerely acknowledge for the financial support (Grant no. 304.PBAHAN.6050369) provided by Japan International Cooperation Agency (JICA) and ASEAN University Network/Southeast Asia Engineering Education Development Network (AUN/SEED-Net) to accomplish this project under its collaborative research. The authors are grateful for the facilities support by Universiti Sains Malaysia (USM).

References

[1] YJ No, JJ Li, and H Zreiqa 2017 *Materials Journal* 10 1
[2] P. Feng, C. Gao, C. Shuai, and S. Peng 2015 *Journal of RCS Advances* 5 3498
[3] E. Bernardo, J.F. Carlotti, P.M. Dias and L. Fiocco 2014 *Ceramics International* 40 1029
[4] C. Wu, J. Chang 2013 *Biomedical Materials* 8 1
[5] C. Wu, J. Chang 2006 *Journal of biomaterials applications* 119
[6] H. Gheisari Dehsheikh, E. Karamian 2016 *International Journal of Bio-Inorganic Hybrid Nanomaterials* 5 223
[7] Noriyuki Y. Iwata, Geun-Hyoung Lee, Yoshikazu Tokuoka, Norimichi Kawashima 2004 *Colloids and Surfaces B: Biointerfaces* 34 239
[8] A. Yazdani, H.R. Rezaie, H. Ghassai and M. Mahmoudian 2013 *Journal of Ceramic Processing Research* 14 12
[9] K. Mondal, J. Manam 2016 *Journal of Molecular Structure* 1125 503
[10] M. Ramezani, T. Neitzert 2012 *Journal of achievements in materials and manufacturing engineering* 22 797
[11] E. Karamian, H. Gheisari 2015 *Nano Studies* 11 87
[12] A. Kazemi, M. Abdellahi, A. Khajeh-Sharafabadi, A. Khandan, N. Ozada 2017 *Materials Science and Engineering: C* 71 604
[13] TP Yadav, RM Yadav, DP Singh 2012 *Nanoscience and Nanotechnology* 2 22
[14] H. Shin, S. Lee, H. S. Jung, J. B. Kim 2013 *Ceramics International* 39 8963
[15] K. Marzban 2016 *Nanomedical research journal* 1 79
[16] F. Tavangarian, R. Emadi, 2010 *Materials Research Bulletin* 45 388
[17] L. Ghorbaniana, R. Emadi, M. Razavib, H. Shinc, A. Teimourid 2012 *Journal of nanostructures* 357
[18] Majid Muhi Shukur, Elham Abd Al-Majeed, Mohammed M. Obeid 2014 *International Journal of Engineering and Technology* 4 426
[19] A. Zare-Haroftehetal, S. Saber-Samandari, S. Saber-Samandari 2016 *Ceramics International* 42 17781
[20] S. Ramezani, R. Emadi, M. Kharaziha, T. Tavangarian 2016 *Materials Chemistry and Physics* 1
[21] H. Ohsato, M. Terada, K. Kawamura 2012 *Japanese Journal of Applied Physics* 51 1
[22] J. Kano, F. Saito 1988 *Powder Technology* 98 166
[23] Mio H, Kano J, Saito F 2004 *Chemical Engineering Science* 59 5909
[24] H. Ashrafizadeh, M. Ashrafizaadeh 2012 *Advanced Powder Technology* 23 708
[25] W. Liu, Z. Ren, G. Chen 2013 *Thermoelectric Nanomaterials, Springer Series in Materials Science* 182 259
[26] I. Cristofolini, A. Rao, C. Menapace, A. Molinari 2010 *Journal of Materials Processing Technology* 210 1716
[27] A. L. Yurkov, T.A. Sarkisyan, D.A. Ivanov, R.C. Bradt 1997 *Ceramic international* 23 389
[28] Mohamed N. Rahaman 2017 *Ceramic processing* 9
[29] Bill Lee, Rainer Gadow, Vojislav Mitic 2017 *Proceedings of the IV Advanced Ceramics and Applications Conference* 140

IOP Conf. Series: Journal of Physics: Conf. Series **1082** (2018) 012075 doi:10.1088/1742-6596/1082/1/012075

Effect of Different Reactive Diluents on PDMS-PEG Hydrogels Properties

Z.A.A. Hamid[1*], T.G. Lock[1], and B. Hilmi[1]

[1]Biomaterials Research Niche Group, School of Materials and Mineral Resources Engineering, Universiti Sains Malaysia, Engineering Campus, 14300 Nibong Tebal, Penang, Malaysia, Tel. +604 5996153, Fax. +604 5996907.

Email: *srzuratulain@usm.my

Abstract. Polydimethylsiloxane (PDMS) is commonly used biomaterials for tissue engineering due to inert and nontoxic properties. However, there is the main drawback of PDMS which it is hydrophobic in nature. To overcome this limitation, PDMS was crosslinked with polyethylene glycol (PEG) in order to enhance it hydrophilic property. In this study, the effect of allyl methacrylate (AMA) and PDMS-MA 1K as reactive diluents in the hydrogel fabrication was investigated. Firstly, PDMS-SiH and PDMS-MA that was synthesized and had been confirmed with Fourier transform infrared spectroscopy (FTIR) and nuclear magnetic resonance (NMR). While for UV-crosslinked hydrogel, contact angle and swelling study were conducted. From the analysis, it was found that increasing in reactive diluents content increased the swelling with insignificant trend for contact angle result. Importantly, from this work the biocompatibility of produced PDMS-based hydrogel was enhanced with the addition of reactive diluents.

1. Introduction

PDMS is an inorganic polymer which is biocompatible, low glass transition temperature, hydrophobic, exhibits excellent gas permeability, and exceptional elasticity when lightly crosslinked. The stability, lack of toxicity, and excellent biocompatibility of PDMS make these materials well suited for use in personal care, pharmaceutical, and medical device applications. (Ratner et al, 2004). Other than that, it is chemically inert, thermally stable, simple to handle and manipulate, exhibits isotropic and homogeneous properties as well as lower cost than silicon (McDonald and Whitesides, 2002). PDMS also shows some drawbacks. One the main drawbacks for cell biology is that PDMS can absorb small hydrophobic molecules like biomolecules and drugs from the solution. Also, many researchers noticed adsorption of proteins on the PDMS surface which has been identified as a major problem for molecular biology. PDMS's hydrophobicity also has been a noted stumbling block. Although the surface can be made hydrophilic, this is not a natural act for the polymer, and it will revert back to its hydrophobic self in air. The relapse can be a major problem because unwanted phenomena such as adsorption of proteins to the surface start to happen (Guilhem et al, 2010). To overcome this problem, numbers of PDMS treatments have been developed depending on the application. Hence, in this project, PDMS will be blend with other hydrophilic polymer which is PEGDA to overcome its drawback.

Poly(ethylene glycol) diacrylate (PEG-DA) based hydrogels have been extensively utilized as scaffolds for the regeneration of tissues including bone, cartilage, nerve, and vascular tissue. PEG

hydrogels are particularly useful for biomaterial study because of their intrinsic resistance to protein absorption and cell adhesion. PEG chains of any length can be easily synthesized by the controlled polymerization of ethylene oxide or ethylene glycol in aqueous solution. PEG is highly biocompatible and well-suited for use in hydrogels for biological studies. PEG is also non-immunogenic and resistant to protein adsorption, making it suitable for in vivo as well as in vitro studies.
In this project, the chemical and physical properties of polydimethylsiloxane (PDMS-MA) hydrogel were tuned by the introduction of PEG-DA. The effect of AMA and PDMS-MAS 1K as reactive diluents on the chemical properties of the PDMS-PEG hydrogels was also evaluated. In addition, the effect of hydrogel composition on physical properties, including, equilibrium swelling, contact angle and biocompatibility was examined.

2. Experimental study

Octamethylcyclotetrasiloxane (98%, Aldrich), 1,1,3,3-Tetramethyldisiloxane (97%, Aldrich), Trifluoromethanesulfonic acid (Aldrich), Platinum (0)-1,3-divinyl-1,1,3,3- tetramethyldisiloxane complex solution (Karstedt's catalyst, 2 wt % in xylene, Aldrich), Poly(ethylene glycol) diacrylate (average Mn 700, Aldrich), photoinitiator 2,2-dimethyl-2-phenylacetophenone (Aldrich), Anhydrous magnesium sulphate (Aldrich), Toluene for analysis (EMSURE® ACS) and Allyl Methacrylate were purchased and used as received. The hydrogels were prepared by photo-polymerization of aqueous mixtures of PEGDA and PDMS-MA solutions. Different composition of PDMS-MA 12k, PEG and reactive diluent were mixed according to the parameters in Table 1 and 2.

Table 1: PDMS with allyl methacrylate (AMA) as reactive diluent (A) with PEG-DA phase

Item	Weight ratio (wt%)		
	PDMS-MA	PEG-DA	AMA
PDMS/0A/0PEG	100	0	0
PDMS/7.5A/0PEG	92.5	0	7.5
PDMS/7.5A/2.5PEG	90.2	2.3	7.5
PDMS/7.5A/5.0PEG	87.9	4.6	7.5
PDMS/7.5A/10.0PEG	83.3	9.3	7.5
PDMS/15.0A/0PEG	85.0	0	15.0
PDMS/15.0A/2.5PEG	82.9	2.1	15.0
PDMS/15.0A/5.0PEG	80.8	4.3	15.0
PDMS/15.0A/10.0PEG	76.5	8.5	15.0

Table 2: PDMS with PDMS-MA (1K) as reactive diluent (B) with PEG-DA phase

Item	**Weight ratio (wt%)**		
	PDMS-MA	**PEG-DA**	**PDMS-MA (1K)**
PDMS/0B/0PEG	100	0	0
PDMS/25.0B/0PEG	75.0	0	25.0
PDMS/25.0B/2.5PEG	73.1	1.9	25.0
PDMS/25.0B/5.0PEG	71.2	3.8	25.0
PDMS/25.0B/10.0PEG	67.5	7.5	25.0
PDMS/50.0B/0PEG	50.0	0	50.0
PDMS/50.0B/2.5PEG	48.7	1.3	50.0
PDMS/50.0B/5.0PEG	47.5	2.5	50.0
PDMS/50.0B/10.0PEG	45.0	5.0	50.0

A total 4 wt% photoinitiator DMPA was added into the mixture. Then, the mixture was subjected with magnetic bar stirring for 15 minutes. Planar hydrogel samples (1.00 mm thick) were prepared by pipetting the precursor solution between two microscope slides (25.4x76.2 mm^2) separated by glass spacers. These hydrogels were synthesized via chemical crosslinking method using a UV curing system (Philip UVA). This UV chamber is a controlled radiation source with 6 UV-tubes that provides a spectral range between 315-400 nm at average intensity of 40 W/cm^2. Further characterization were conducted including FTIR, NMR, swelling, contact angle and biocompatibility test.

3. Results and Discussion

3.1 Fourier-Transform Infrared spectroscopy (FTIR) of PDMS-SiH and PDMS-MA

To determine the polymerisation of PDMS-SiH had completed, the present of Si-H peak was determined from the spectrum. The Si-H group is readily identified by a strong band in the range 2280-2080. As it shown in Figure 1(left), there was a small peak in the range of 2280-2080 for both PDMS-SiH 12k and PDMS-SiH 1k spectrum. The Si-H peak was not obvious in PDMS-SiH 12k and it is due to the very large molecular weight of PDMS 12k with 12000g/mol. Si-H is the functional group of the polymer chain which each chain only contains two Si-H group and make it relatively small and hard to be detected. After synthesized of PDMS-SiH, it needs to be further process to change the functional group from Si-H to MA. Figure 1 (right) show the FTIR spectrum of PDMS-MA 12k and PDMS-MA 1k. Si-H terminal groups were subsequently converted to photo-sensitive methacrylate groups by Pt-catalyzed hydrosilylation with allyl methacrylate. The chemical reaction was confirmed by the disappearance of the Si-H peak (~ 4.5 ppm) in the 1H NMR spectra and Si-H absorbance (~2125 cm−1) in the FTIR spectra to determine the process of changing functional group had completed. C=O is present in methacrylate groups hence after the reaction had completed, C=O peak which is at 1720 had appeared.

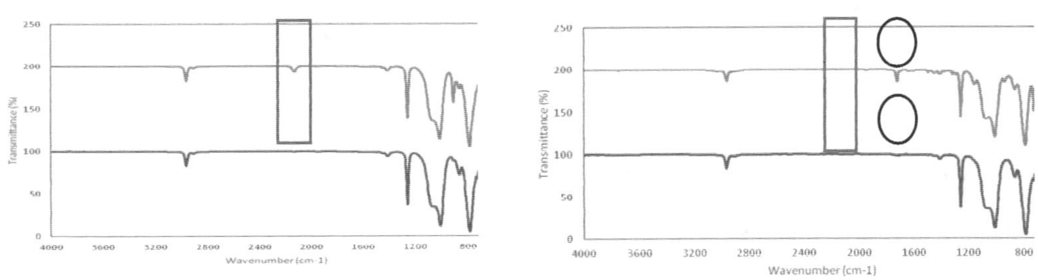

Figure 1: FTIR spectrum of PDMS-SiH 12k and PDMS-SiH 1k (left) and FTIR spectrum of PDMS-MA 12k and PDMS-MA 1k (right)

3.2 Nuclear magnetic resonance spectroscopy (NMR) of PDMS-SiH and PDMS-MA

Figure 2 show the NMR spectrum for PDMS-SiH 1k and PDMS-SiH 12k respectively. From the NMR spectrum of PDMS-SiH 1k, it can be seen a little tiny peak at 4.5ppm which representing Si-H peak. In the NMR spectrum of PDMS-SiH 12k, a Si-H peak was not present. It was due to the extremely low Si-H content in the polymer chain. By comparing PDMS 1k and PDMS 12k, the molecular weight is 12 times larger. With just a tiny peak in PDMS 1k, so the probability of not being seen in PDMS 12k is high.

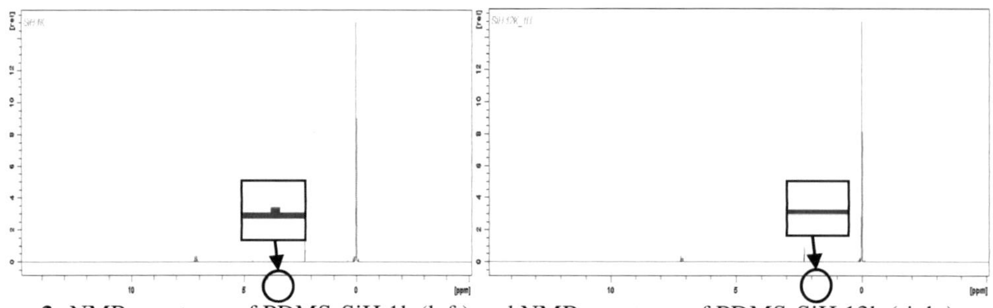

Figure 2: NMR spectrum of PDMS-SiH 1k (left) and NMR spectrum of PDMS-SiH 12k (right)

After synthesized of PDMS-SiH, it is then further processed into PDMS-MA. Figure 3 show the NMR spectrum for PDMS-MA 1k and PDMS-MA 12k. The purpose of NMR spectrum was to determine whether PDMS-SiH had been completely changing its functional group to MA. From the disappearance of peak 4.5 ppm in PDMS-SiH 1k and the arising of others AMA peak, it can conclude that AMA had been attached to the PDMS.

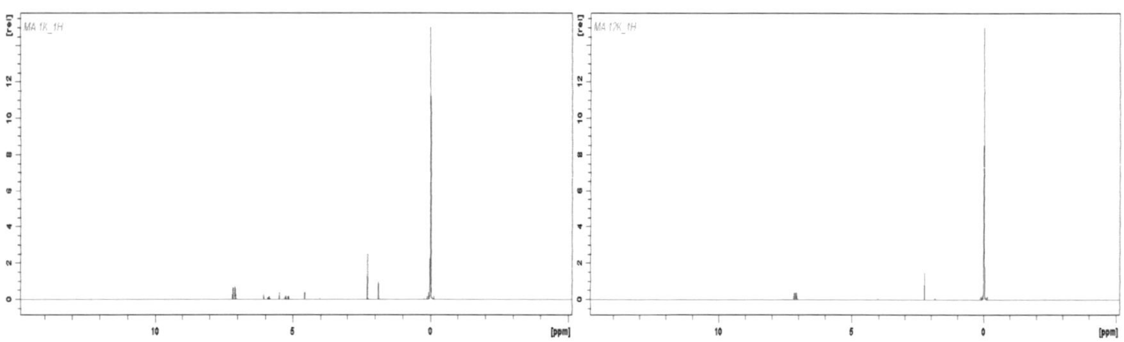

Figure 3: NMR spectrum for PDMS-MA 1k (left) and PDMS-MA 12k (right)

3.3 Contact Angle of PDMS-PEG Hydrogels

Figure 4 (left) shows the contact angle of PDMS-PEG hydrogel in various composition of PEG with 7.5 and 15 wt% AMA. From the Figure 4, it showed that AMA had the lowest contact angle with 77°. AMA is hydrophilic in natural which show in its ester bond with can form hydrogen bond with water. Hydrophilic properties also increase the wettability of the hydrogel which explains by the low contact angle of the result. PEG which is also hydrophilic in natural due to its OH functional group which readily for hydrogen bond with water which expected the contact angle to be lower than 90° but in the experiment, the contact angle that obtains is 102°. The value obtained here can be explained by the different surface roughness on PEG surface which due to air entrapment. In the UV crosslinking of PEG, air entrapment had happened between polymer chains as agreed by the surface roughness of PEG throughout the surface. Air entrapment had increased the contact angle of PEG. PDMS-MA 12k is hydrophobic which theoretically had a higher contact angle as compared to PDMS-MA 1k. The end chain of PDMS-MA is hydrophilic but due to steric hindrance effect, the end chain is sealed due to the high molecular weight of PDMS. In term of the concentration of functional group, PDMS-MA 12k had a lower concentration of functional group as compare to PDMS-MA 1k which cause it had a higher contact angle.

Increasing the concentration of AMA and PEG in hydrogel did not give trending effect on the contact angle. Theoretically, increasing in AMA and PEG concentration will decrease the contact angle of hydrogel. AMA and PEG are hydrophilic. Increasing the concentration will increase the hydrophilic properties of hydrogel and hence increase its wettability and decrease in contact angle. But in the experiment, the increased AMA and PEG content show fluctuated result on the contact angle of hydrogel. The results may be also associated to the better biocompatibility property of the PDMS-PEG hydrogels due to the addition of reactive diluents. In which, the PDMS hydrophobicity overshadowed the AMA and PEG hydrophilicity. Thus, decreased of contact angle results were achieved. The enhanced biocompatibility of PDMS-PEG hydrogels was reflected in the one phase occurrence from visual observation (data is not shown). However, at higher ratio of PDMS-PEG, the occurrence of opaque hydrogel was achieved due to the phase separation existed in the hydrogels.

Figure 4 (right) shows the contact angle for PEG, Pure PDMS-MA 1k and hydrogel at different PDMS 1k and PEG composition. PDMS 1k showed a very high contact angle at 114° which shows it hydrophobicity. The increase of PDMS 1k had an increase in contact angle for all different composition. It is because PDMS 1k is hydrophobic and increasing its content in hydrogel will increase the hydrogel hydrophobicity and thus decrease the wettability and decrease contact angle. Besides, increasing PEG content suppose had a decrease in contact angle. But due to uneven surface, increasing PEG had a fluctuating effect on contact angle.

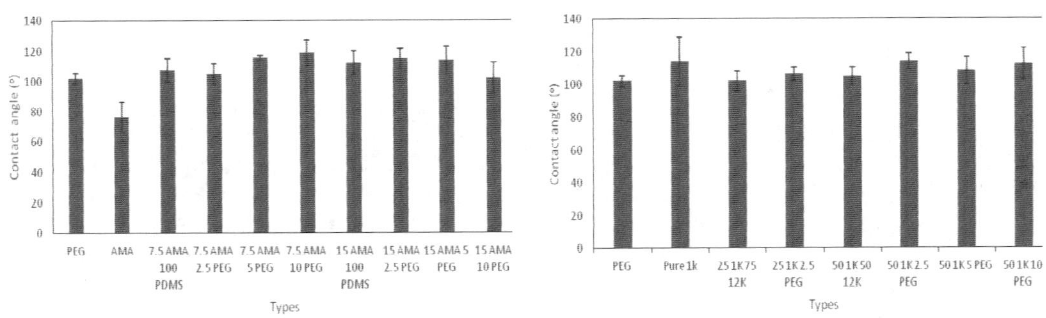

Figure 4: Contact angle of AMA hydrogel (left) and contact angle of 25 wt% and 50 wt% PDMS 1k at increasing PEG content (right)

3.4 Swelling Percentage of PDMS-PEG Hydrogels

Figure 5 (left) shows the swelling percentage of AMA, PEG and the mixture of AMA, PDMS-PEG hydrogels at a different ratio. PEG is hydrophilic which can show in the Figure 5 that it swells at 49 %. Although AMA is hydrophilic in natural due to its ester bond that can form hydrogen bond easily with water, it just swells for 1.4 % which can be explained by UV crosslinked AMA had high crosslink density which shields the water from the ester bond and makes it swelling percentage to decrease. PDMS is hydrophobic in nature due to its does not have any functional group that can react with water which shows by its limited swelling percentage in distilled water. As the PEG content is increasing, the swelling percentage also increased. This is due to the increasing PEG content which is hydrophilic that act as water absorber. 7.5 wt% AMA with 100 wt% PDMS shows a minimum swelling value in hydrogel due to its composition which is made up of mainly PDMS 12k only which are hydrophobic in natural.

Figure 5 (right) shows the swelling percentage of PDMS 1k and different mixing ratio of PDMS 1k, and PEG. From this figure, it shows that PDMS 1k had a very minimum value in swelling which is 0.3%. While PEG is a very hydrophilic material, the trend is still the same as AMA hydrogel which increasing in PEG content will increase the swelling percentage. It is true for both 25 wt% PDMS 1k and 50 wt% PDMS 1k. At constant 10 wt% PEG, 25 wt% 1k had higher swelling percentage than 501k. It is due to higher PEG content in 25 wt% 1k 10 wt% PEG. For 25 wt% PDMS 1k and 50 wt% PDMS 1k hydrogel, the swelling percentage did not have significant different to each other as indicate by the overlapping of error bar. This can be explained by both PDMS 1k and PDMS 12k are hydrophobic in nature. Thus, it only had a very minimum swelling effect in distilled water.

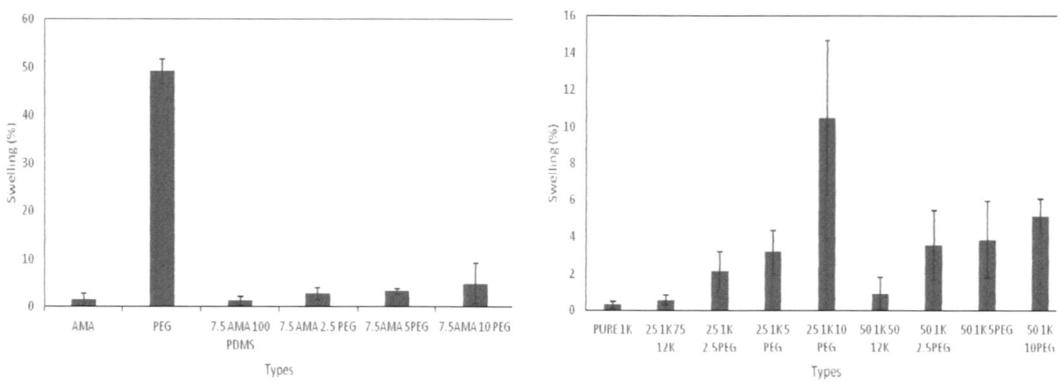

Figure 5: Swelling percentage of AMA, PEG and hydrogel at 7.5 wt% AMA (left) and swelling percentage of PDMS 1k and different mixing ratio of PDMS 1k, PDMS 12k and PEG

4. Conclusion

All hydrogel was fabricated and characterized to determine the properties of hydrogel at different formulations. The chemical structure of PDMS-SiH and PDMS-MA were confirmed by FTIR and NMR. Different types of reactive diluent which are AMA and PDMS-MA 1k had different effect to the properties of hydrogel. For contact angle of hydrogel, increasing of AMA should decrease the contact angle due to AMA is hydrophilic but from the testing, increasing AMA content had a fluctuating effect and this deviation can be explained by uneven surface of hydrogel which deviated the result. Besides, increasing PDMS 1k had an increasing trend to the contact angle of hydrogel. This is because PDMS 1k is hydrophobic and it decreases the wettability of hydrogel. Furthermore, in swelling test, different reactive diluent which are AMA and PDMS-MA 1k had no significant effect on swelling percentage. This is due to the same hydrophobic nature in crosslinked AMA and PDMS-MA 1k.

5. Acknowledgments

The authors would like to thank the Ministry of Higher Education (MOHE) for the financial support under Transdisciplinary Research Grant Scheme (TRGS) (203/PBAHAN/6761001).

References

[1] Ratner B D, Hoffman A S, Schoen F J and Lemons J E 2004 *Biomaterials science: An introduction to materials in medicine*
[2] McDonald JC1, Whitesides GM. 2002 *Acc Chem Res.* 35 491
[3] Guilhem G, Cornu C and Guevel A 2010 *Ann. Phys. Rehabil. Med.* 53 319
[4] Stoyanova, Kuesel A C and Brown T R 1995 *Journal of Magnetic Resonance, Series A* 115 265
[5] Ekwall P, Groth K and Runnström V 1964 *Surface Chemistry Proceedings of Second Scandinavian Symposium on Surface Activity* 18

IOP Conf. Series: Journal of Physics: Conf. Series **1082** (2018) 012076 doi:10.1088/1742-6596/1082/1/012076

Co-Sr doped carbonated hydroxyapatite: A biomaterial with enhanced mechanical and bioactivity properties

Muhammad Syazwan M.N., Ahmad-Fauzi M.N. and Yanny Marliana B.I*

School of Materials & Mineral Resources Engineering, Engineering Campus, Universiti Sains Malaysia, 14300, Nibong Tebal, Penang, Malaysia.

*E-mail address: yannymarliana@usm.my

Abstract: The present study is aimed at investigating the role of two biologically important cations, cobalt (Co^{2+}) and strontium (Sr^{2+}) doped into carbonated hydroxyapatite (CHA) crystal structure. Four different compositions of Co-Sr CHA powders were synthesized via nanoemulsion method at ambient temperature. Among the as-synthesized powders, Co-Sr CHA 1 (0.0082wt% Co, 0.0547wt% Sr) and Co-Sr CHA 2 (0.0085wt% Co, 0.0261wt% Sr) were chosen as the optimum compositions and used for fabrication of dense products. Sintering was then performed on the Co-Sr CHA compacted samples at 900°C in air, followed by cooled down in dry CO_2 atmosphere at temperature of 200°C. Regardless of their compositions, the sintered samples remained as single phase B-type CHA. Co-Sr CHA 1 has the optimum mechanical properties where the relative density and diametral tensile strength (DTS) values obtained were 89.15% and 7.67 MPa, respectively. *In vitro* bioactivity test revealed that the formation of apatite layer were detected on the surface of all sintered samples after soaking in SBF solution for 7 days, with Co-Sr CHA 1 showing the fastest apatite formation as compared to CHA and Co-Sr CHA 2. Results obtained indicate that simultaneous doping of Co^{2+} and Sr^{2+} into CHA structure gives rise to improved mechanical strength as well as enhanced the bioactivity of the apatite.

1. Introduction

Calcium phosphate-based materials are widely used for biomedical applications as bone implants to substitute damaged or ailing bones[1,2]. Hydroxyapatite (HA) with a chemical formula of $Ca_{10}(PO_4)_6(OH)_2$ is one of illustration that has received wide attention among the researchers in the previous decades. This is due to its affinity to bone minerals because of their ability as biocompatibility, bioactivity and osteoconductivity properties[1,3]. However, HA has few drawbacks that contrasted to the biogenic apatite, particularly, its resorption rate *in vivo* which has been reported to be too sluggish in inducing a massive new bone formation[3]. In term of mechanical properties, HA ceramic is known to be less suitable to be used in a load-bearing application. Some studies found that dense HA ceramic has potential to fail by fatigue after a period of clinical utility[4,5].

Biological and mechanical properties of HA can be enhanced by introducing new elements in their structure that are generally present in natural apatites of bone. Carbonated hydroxyapatite (CHA) has been reported to have a much closer compositional similarity to the mineral in natural bone compared to stoichiometric HA[6,7]. The amount of carbonate ions present in human bone is in the range of 4-8 wt% depending on the age of the individual[1,7]. The introduction of carbonate ions into HA structure was found to increase the dissolution rate in solution and enhance its osteointegration rate[8,9]. Hence, CHA is seem as better choice of material relative to HA due to its enhanced properties biocompatibility, bioactivity, and resorbability[9,10].

Over the past couple of decades, many researchers have demonstrated that variety of ionic substitutions can be incorporated into synthetic HA. This approach is seem as one of the effective ways to improve the properties of HA in order to produce a mineral composition closely mimic to the natural bone tissue[11]. Interestingly, the high flexibility, and stability of the apatite crystal structure allow the wide selection for replacement of cations (mainly Ca^{2+})and anions (OH^- and $(PO_4)^{3-}$) according to charge type, charge size and ionic radii[12,13]. The introduction of trace metal ions such as Ag^+, Na^+, Cu^{2+}, Mg^{2+}, Mn^{2+}, Zn^{2+}, Sr^{2+} and Co^{2+} into HA crystal structure can affect the surface charge, lattice parameters, crystallinity and morphology, which resulting in changes in thermal stability, mechanical properties, solubility and bioactivity of HA[2,14]. For example, the substitution of Co^{2+} was reported to improve the angiogenesis properties of HA[15,16]. The same goes for Sr^{2+} into HA structure which has been reported enhance the solubility and stimulated faster bone formation. Besides, the present of Sr ions also encouraged better cell attachment and proliferation[12,17].

The main aim of this work is to investigate the influences of the ionic substitutions by the incorporation of cobalt (Co^{2+}) and strontium (Sr^{2+}) ions simultaneously into the carbonated hydroxyapatite (CHA) structure. Several characterization techniques were used to analyze the as-synthesized powders and sintered multi-doped CHA such as X-ray Diffraction (XRD), Fourier transform infrared spectroscopy (FTIR), Field Emisson Scanning Electron Microscopy (FESEM) and Energy dispersive X-ray analysis (EDX). Besides that, Diametral Tensile Strength (DTS) was used to determine the compressive strength of the sintered multi-doped CHA. For bioactivity testing, the sintered multi-doped CHA pellets were immersed in Simulated Body Fluid (SBF) for 7 days followed by surface analysis using FESEM to detect the formation of apatite layer.

2. Experimental

2.1 Preparation of Powders

Co-Sr doped Carbonated Hydroxyapatite (Co-Sr CHA) powders were chemically synthesized via nanoemulsion method according to the previous studies reported in the literature[1,18]. The chemicals used were ammonium bicarbonate ($(NH_4)HCO_3$), di-ammonium hydrogen phosphate ($(NH_4)_2HPO_4$), calcium nitrate tetrahydrate ($Ca(NO_3)_2.4H_2O$), sodium hydroxide (NaOH), strontium nitrate ($Sr(NO_3)_2$) and cobalt nitrate hexahydrate ($Co(NO_3)_2.6H_2O$. Initially, aqueous solutions were prepared by dissolving 7.942g of ($(NH_4)_2HPO_4$) and 3.162g of ($(NH_4)HCO_3$) in 60 ml of deionized water separately. For the doping precursor, 0.14g of $Co(NO_3)_2.6H_2O$ was dissolved in deionized water until achieve the desired concentration (7.41×10^{-7}g) and directly poured into carbonate-phosphate mixture (solution 1). 0.012g Sr(NO3)2 was weighed and directly added to solution 1 and continuously stirred until well dissolved (solution 2). Meanwhile, a solution 3 was prepared by dissolving 23.409g of $Ca(NO_3)_2.4H_2O$) in 100 ml of acetone, which was then poured into solution 2 to create emulsion. In this work, CHA powders was used as a standard reference for Co-Sr doped CHA powders. The amount of raw materials used in the synthesis is shown in Table 1.

Table 1: The amount of raw materials used

Samples	Mass (g)				
	Ca^{2+}	PO_4^{3-}	CO_3^{2-}	Co^{2+}	Sr^{2+}
CHA	23.409	7.942	3.162	-	-
Co-Sr CHA 1	23.409	7.942	3.162	7.41×10^{-7}	0.012
Co-Sr CHA 2	23.409	7.942	3.162	7.41×10^{-7}	0.006

IOP Conf. Series: Journal of Physics: Conf. Series **1082** (2018) 012076 doi:10.1088/1742-6596/1082/1/012076

2.2 Fabrication of Dense Co-Sr CHA

0.5 g of Co-Sr doped CHA powders were pelletized using uniaxial hydraulic hand press with 50 MPa to produce compact green body. Sintering was then performed on the compacted powders at 900°C followed by cooled down under a wet CO_2 atmosphere at temperature of 200°C.

2.3 Characterizations Techniques

2.3.1. X-ray diffraction. X-ray diffraction (XRD) was carried out using a Bruker D8 XRD. The diffractometer was operated at 40 kV and 110 mA at 2θ range from 0° to 90° with step size of 0.02° using the Cu $K\alpha$ radiation ($\lambda = 1.542$ Å), operating with PANalytical X'pert PRO diffractometer. Stoichiometric HA with ICDD file number of 09-0432 was used as the reference pattern.

2.3.2. Fourier Transform Infrared Spectroscopy. Fourier transform infrared spectroscopy (FTIR) was performed to determine the chemical constituent presence in the as-synthesized powders and dense multi-doped CHA produced. Samples were ground and mixed with potassium bromide (KBr) and pelletized using hand press at a pressure of 5 MPa and hold for 2 minutes. The wave range was set up to be in the range of 4000 to 400 cm^{-1}. The samples was scanned for four times in transmittance mode.

2.3.3. Field Emission Scanning Electron Microscopy. The morphology of the as-synthesized powders and sintered samples before and after bioactivity study were observed using Zeiss Supra 55VP Field Emission Scanning Electron Microsope (FESEM). The as-synthesized Co-Sr CHA powders and sintered samples were initially coated with gold before observed under FESEM. Besides that, Energy Dispersive X-Ray (EDX) was used to analyze the elemental of each sample. This technique detects x-rays emitted from the sample during the bombardment by an electron beam to characterize the elemental composition of the analyzed sample.

2.3.4. Density and Porosity Measurement. Bulk density and porosity of the sintered multi-doped CHA pellets were determined using Standard Archimedes method in ASTM C830-00[10]. The dry weight of the samples (*D*) were measured using microbalance followed by submerged in deionized water by placing them in the desiccator in order to vacuum the samples for 120 minutes to allow water filled the open pores of the samples. Then, the suspended mass (S) and saturated mass (W) of the samples were weighed using Sartorius Balance and Bulk Density Apparatus.

2.3.5. Diametral Tensile Strength. Diametral Tensile Strength (DTS) was performed using INSTRON 3367 universal tensile machine in order to determine the tensile strength of the sintered samples. Prior to testing, the diameter and thickness of each sintered sample was measured. The sintered sample was then placed vertically between the two stainless steel plates. Compression load was then applied at the rate of 0.5 mm/min until the sample was broken.

2.3.6. Evaluation of Bioactivity. The bioactivity characteristic of dense multi-doped CHA samples was evaluated by observing the ability of the apatite to form on its surface when soaked in a simulated body fluid (SBF) for a certain period of time. Prior to bioactivity test, SBF solution was prepared according to the recipe proposed by Kokubo & Takadama (2006)[19,20]. After soaking for 7 days, the samples were gently taken out from the SBF solution, washed carefully using deionized water and finally observed under FESEM. The formation of apatite layer on the sample was confirmed by perfoming EDX to detect the element present on the surface.

3. Results and discussion

3.1 Physico-chemical Properties

XRD results of the sintered samples of CHA, Co-Sr CHA 1 and Co-Sr CHA 2 is shown in Figure 1. Regardless of the compositions, all sintered samples produced corresponds to typical HA pattern (ICSD 09-0432). The main peaks were detected at 31°, 32°, 33° and 34° in 2θ which represents (211), (112), (300) and (202) respectively. Other planes such as (002), (130), (222), (213) and (004) matched with HA pattern were also found. It is widely accepted that substitution of doping elements into apatite structure causes a change of lattice parameters and induces the contraction in *a*-axis and expansion in *c*-axis. This is due to the large radii atom of Sr^{2+} substitution for the small radii atom of Ca^{2+} while replacement of large radii atom Ca^{2+} by small radii atom of Co^{2+} where the atomic radius of : Sr^{2+}= 1.13 Å, Co^{2+}= 0.7 Å and Ca^{2+} = 1.00 Å[17,21]. However, due to small amount of Co^{2+} and Sr^{2+} ions substituted in this study, no significant change in the crystallographic apatite structure was observed[21].

The obtained FTIR spectra (Figure 2) showed that all sintered samples exhibited the characteristic bands of carbonate groups (CO_3) of the apatite structure at 1461.30 and 1419.22 cm^{-1} which corresponds to B-type CHA[22]. The main characteristic bands of PO_4^{3-} groups (v_4) vibrations of sintered Co-Sr CHA detected at 568.31 cm^{-1} and 601.04 cm^{-1}. While v_1, v_2 and v_3 detected at 474.81, 961.04 and 1040.52 cm^{-1}, respectively of of PO_4^{3-} groups. The bands about 3443.55 cm^{-1} and 1638.71 cm^{-1} correspond to absorbed water. The band at 3571.77 cm^{-1} were assigned to stretching vibration of OH^- groups of the apatite structure[3,9].

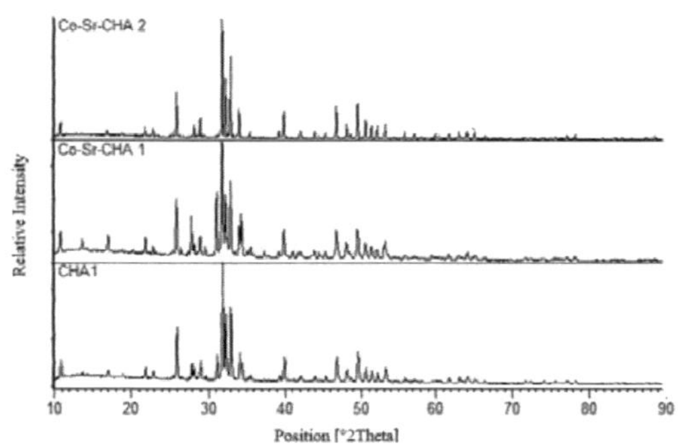

Figure 1: XRD pattern of sintered CHA and multi-doped Co-Sr CHA at 900°C

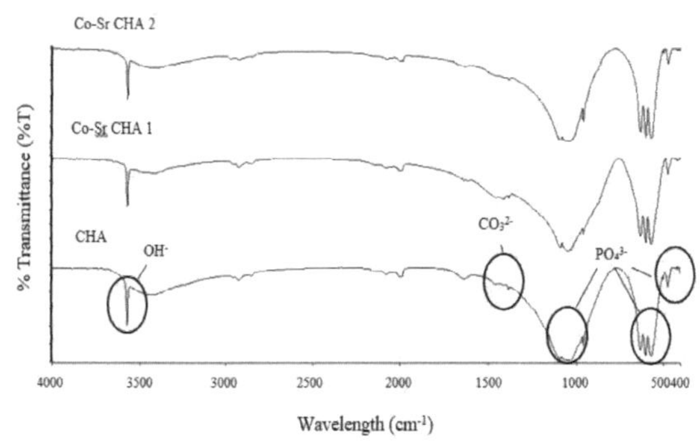

Figure 2: FTIR spectra for sintered of CHA and multi-doped Co-Sr CHA

3.2 Mechanical Study of the Sintered Samples

The mechanical properties of sintered multi-doped CHA were evaluated in terms of relative density and diametral tensile strength (DTS). It is observed that the relative densities of Co-Sr CHA 1 and Co-Sr CHA 2 achieved higher if compared to the sintered CHA which is 89.15% and 72.14%, respectively. For DTS, Co-Sr CHA 1 was showed the highest value between other two sintered samples which are 7.67 MPa while Co-Sr CHA 2 and CHA about 5.13 MPa and 1.95 MPa. These results reveal a fact that incorporating Sr^{2+} ions into the apatite structure helps to improve the mechanical properties especially in terms of density relative and compressive strength relative to pure CHA[23,24]. The present of Co^{2+} ions in the HA structure was reported to have potential effect on enhancing the biological performances to a great extent particularly in inducing angiogenesis properties of the apatite structure[25]. However, these ions does not showing significant impact on the mechanical properties of Co-Sr CHA as in agreement with the literature. Besides that, the presence of CO_2 atmosphere during cooling stage not only had assisted in compensating the carbonate loss from the sintered samples during sintering but also aids in densification (Table 2)[10].

Table 2: Comparison for relative density and diametral tensile strength of the sintered Co-Sr CHA

Samples	Relative density (%)	Diametral tensile strength (MPa)
CHA	56.80	1.95
Co-Sr CHA 1	89.15	7.67
Co-Sr CHA 2	72.14	5.13

3.3 Biological Evaluation of the Sintered Samples

Bioactivity of the sintered samples evaluated *in vitro* by immersed in SBF solution for 7 days. The formation of apatite layer on the surface of Co-Sr CHA 1 sample was fastest as compared to CHA and Co-Sr CHA 2 sintered samples (Figure 3). The present of layer apatite on the surface of the sample was confirmed by the detection of calcium and phosphate ions[5]. The EDX analysis was performed to show the presence of Ca and P by carried out the Ca/P ratio[26]. The values of Ca/P ratio for CHA, Co-Sr CHA 1 and Co-Sr CHA 2 are 1.93, 1.87 and 1.89, respectively. This finding prove that besides improved the mechanical properties, the presence of Sr^{2+} and Co^{2+} into CHA structure also enhanced the biological activity by encouraging the apatite layer formation.

Figure 3: Surface morphologies of the sintered samples after immersed in SBF solution. (a) CHA, (b) Co-Sr CHA 1 and (c) Co-Sr CHA 2 (dotted line: formation of thin apatite layer in granular-like form on the surface of the sample)

4. Conclusion

The new elements such as of strontium (Sr) and cobalt (Co) as ion substitutions into the apatite structure play an important role to enhance the physical, chemical, mechanical, and biological properties of HA. In terms of mechanical properties, the amount of ion substitutions that incorporated into HA structure were influenced the mechanical properties of final product. As indicated by biological study results, the amount of ion substitutions also influenced biological properties such as more soluble and bioactive material if compared to stoichiometric HA due to the ability to encourage the apatite formation. At this stage, it can be concluded that Co-Sr CHA 1 has the best properties as compared Co-Sr CHA 2 and CHA alone for both bioactivity and mechanical properties in order to create a product which is closer to the cancellous bone.

Acknowledgments:
The authors would like to thank Universiti Sains Malaysia (Short term, Grant No.: 304/ PBAHAN/ 6315041) for supporting this work.

References

[1] Kee, C. C., Ismail, H. & Mohd Noor, A. F. 2013 E *J. Mater. Sci. Technol.* 29 761
[2] Bang, L. T., Long, B. D. & Othman, R. 2014 *Sci. World J.* 2014
[3] Kovaleva, E. *et al.* 2009 *Open Chem.* 7 168
[4] Teraoka, K. *et al.* 1998 *J. Dent. Res.* 77 1560
[5] González Ocampo, J. I., Escobar Sierra, D. M. & Ossa Orozco, C. P. 2016 *J. Adv. Res.* 7 297
[6] Othman, R., Mustafa, Z., Loon, C. & Noor, A. 2016 *Procedia Chem.* 19 539
[7] Wong, W. Y. & Noor, A.-F. M. 2016 *Procedia Chem.* 19 98
[8] Kovaleva, E. S. *et al.* 2008 *Materwiss. Werksttech.* 39 822
[9] Mohammad, N., Othman, R., Abdullah, N. & Yeoh, F. 2016 *Procedia Chem.* 19 259

[10] Baba Ismail, Y.-M. & Mohd Noor, A.-F. 2011 *J. Mater. Sci. Eng. B* 1 157
[11] Ratnayake, J. T. B., Mucalo, M. & Dias, G. J. 2016 *J. Biomed. Mater. Res. - Part B Appl. Biomater.* 1
[12] Kumar, G. S., Thamizhavel, A., Yokogawa, Y., Kalkura, S. N. & Girija, E. K. 2012 *Mater. Chem. Phys.* 134 1127
[13] Uysal, I., Severcan, F., Tezcaner, A. & Evis, Z. 2014 *Prog. Nat. Sci. Mater. Int.* 24 340
[14] Bose, S., Fielding, G., Tarafder, S. & Bandyopadhyay, A. 2013 *Trends Biotechnol.* 31 594
[15] Tank, K. P., Chudasama, K. S., Thaker, V. S. & Joshi, M. J. 2013 *J. Nanoparticle Res.* 15
[16] Narwade, V. N. & Khairnar, R. S. 2017 65
[17] Aina, V. *et al.* 2012 *J. Mater. Sci. Mater. Med.* 23 2867
[18] Baba Ismail, Y. M., Wimpenny, I., Bretcanu, O., Dalgarno, K. & El Haj, A. J. 2017 *J. Biomed. Mater. Res. - Part A* 1775
[19] Kokubo, T. & Takadama, H. 2006 *Biomaterials* 27 2907
[20] Landi, E. *et al.* 2005 *Biomaterials* 26 2835
[21] Zyman, Z. Z. & Tkachenko, M. V. 2013 *Process. Appl. Ceram.* 7 153
[22] Youness, R. A., Taha, M. A., Elhaes, H. & Ibrahim, M. 2017 *Mater. Chem. Phys.* 190 209
[23] Curran, D. J., Fleming, T. J., Towler, M. R. & Hampshire, S. 2011 *J. Mech. Behav. Biomed. Mater.* 4 2063
[24] Guo, D., Xu, K., Zhao, X. & Han, Y. 2005 *Biomaterials* 26 4073
[25] Kulanthaivel, S. *et al.* 2016 *Mater. Sci. Eng. C* 58 648
[26] Mohamed, K. R., Beherei, H. H. & El-Rashidy, Z. M. 2014 *J. Adv. Res.* 5 201

IOP Conf. Series: Journal of Physics: Conf. Series **1082** (2018) 012077 doi:10.1088/1742-6596/1082/1/012077

Effect of Hydrothermal Reaction Temperature on Properties of Bismuth Nanoparticles and Its Properties as Modified Electrode for Pb Sensors

Khairunisak Abdul Razak[1,2,*], Teh Jin Jian[1], Zulfa Aiza Zulkifli[1], Noorhashimah Mohamad Nor[1], Nur Syafinaz Ridhuan[1], Nor Dyana Zakaria[2], Toshiaki Hattori[3]

[1]School of Materials and Mineral Resources Engineering Universiti Sains Malaysia, 14300 Nibong Tebal, Penang, Malaysia

[2]NanoBiotechnology Research and Innovation, INFORMM, Universiti Sains Malaysia, 11800 USM, Penang, Malaysia

[3] Department of Electrical and Electronic Engineering, Faculty of Engineering, Toyohashi University of Technology, 1-1 Hibarigaoka, Tempakucho, Toyohashi-shi, Aichi 441-8580, Japan

Email: khairunisak@usm.my

Abstract. Heavy metal pollution is a common environmental problems. Therefore a versatile method that can be used to detect heavy metal ions on field is sought after. Electrochemical sensor is a promising method for heavy metal detection, however, electrodes modification is of interest to enhance sensitivity and specificity of sensors. In this work bismuth nanoparticles (BiNPs) was produced using hydrothermal method with varying hydrothermal reaction temperature. X-ray diffraction analysis showed pure-phase bismuth with a Rhombohedral structure was successfully produced. BiNPs/ITO had good CV properties in which BiNPs produced at 170°C has the highest respond current. However, BiNPs synthesized at 160°C showed the highest DPASV stripping current response due to high surface area and less agglomeration of particles. The BiNPs modified ITO was successfully fabricated as electrochemical Pb sensor. The modified BiNPs/ITO has LOD of 2.5 μg/L Pb with good specificity.

1. Introduction

Heavy metal elements such as lead (Pb), cadmium (Cd), Mercury (Hg) and Arsenic (As) are typical hazardous pollution due to industrial and agricultural activities. Pb^{2+} is the most serious environmental pollutant that is highly toxic to human immune, nervous, reproductive and gastrointestinal systems [1]. Besides, Pb non-degradable and could be accumulated in ecological system in a long term, which leads to severe pollutions for people's health [2]. Therefore, it is necessary to develop highly sensitive, rapid and simple methods for detection of these kinds of heavy metal pollutants.

Bi has been chosen as viable replacement of mercury based electrodes for electrochemical sensors. Nanoparticles modified electrode surface is also known to enhance the electroanalytical properties of the electrode, primarily due to increases surface area to volume ratio and couple with the enhance mass transfer effect [3,4,5]. As a result, BiNPs modified electrodes exhibits high sensitivity and selectivity towards trace heavy metal ions in the analyte solution, which is comparable or even superior to that of Bi film coated electrodes [6]. Recently, several approaches that involves additional processing steps have been reported to synthesize BiNPs due to the high surface area of BiNPs, thereby resulting in enhanced performance of sensing electrodes in heavy metal detection. BiNPs can be synthesized using gas condensation method, chemical synthesis method, thermal plasma method, aerosol quenching, electron beam irradiation and laser ablation techniques [7,8]. However, these methods are inappropriate and impractical for mass production of nanoparticles towards the electrochemical approaches.

In this work BiNPs was synthesized using hydrothermal method with varying hydrothermal reaction temperature. Hydrothermal method is simple, cost effective, low growth temperature and energy-efficient technique to synthesis nanoparticles. The produced BiNPs was immobilized on substrates and tested using Cyclic Voltammetry (CV) and Differential Pulse Anodic Stripping Voltammetry (DPASV). Sensitivity, selectivity and limit of detection (LOD) of the analytical signal for the heavy metal ions interaction were systematically studied.

2. Experimental study

In this work, BiNPs was synthesized using hydrothermal method with varying hydrothermal reaction temperature: 150, 160, 170 and 180°C. Firstly, 2.94 g of polysodium 4-styrene-sulfonate (PSS), M_W = 70000, Aldrich) and 2.5g of hydrazine hydrate (80% water solution) were dissolved in 50 ml distilled water. Next, 3.6 g of bismuth nitrate ($Bi(NO_3)_3$) was added and stirred for 30 minutes to obtain a white suspension. The suspension was then transferred into a Teflon-lined stainless steel autoclave with capacity of 80 ml. Hydrothermal reaction was carried out in a preheated oven with varying temperature for 6 hours. After cooled to room temperature, black powder produced were centrifuged at 10000 rpm and rinsed several times with ethanol.

The produced BiNPs were then drop-casted on the pre-cleaned ITO electrodes, and dried overnight at room temperature. The modified electrodes were used for electrochemical analysis. An electrochemical analyzer was used to perform electrochemical measurement of cyclic voltammetry (CV) and differential pulse anodic stripping voltammetry (DPASV) by using BiNPs/ITO modified electrode as a working electrode, a saturated calomel electrode (SCE) as the reference electrode (3 M of KCl), and a platinum wire as the counter electrode. CV analysis of BiNPs/ITO modified electrodes was performed in 0.1 mol/L of KCl solution (pH 7.0) containing 2.0×10^{-3} mol/L of potassium ferrocyanide, $[Fe(CN)_6]^{3-}$. Differential Pulse Anodic Stripping Voltammetry (DPASV) measurement of BiNPs/ITO electrodes was performed using Pb(II) standard solution. All measurements were carried out at room temperature in 2.0×10^{-3} mol/L of acetic acid–sodium acetate (NaAc-HAc) buffer solution at pH 4.5 as supporting electrolyte. In this work, pH value of acetate buffer solution was changed using acetic acid (CH_3CO_2H) and sodium acetate (CH_3COONa). The concentration of Pb(II) standard solution was diluted accordingly in acetate buffer analyte solution in order to obtain limit of detection (LOD) of heavy metal ions for the corresponding sensor electrode. The interference study was performed by using ICP multi-element standard solution. 20 μg/L of 2.0×10^{-3} mol/L potassium ferrocyanide, $[Fe(CN)_6]^{3-}$ was added into the solution in order to act as masking agents. Then, 2.0×10^{-3} mol/L of acetic acid–sodium acetate (NaAc-HAc) buffer solution at pH 4.5 was mixed with 100 μg/L of the multi-element standard solution.

3. Results and discussion

Phase composition of the synthesized samples were determined using X-ray diffraction (XRD). Figure 1 shows the XRD patterns of the BiNPs synthesized with varying hydrothermal reaction temperature. The increase in hydrothermal reaction temperature increased the intensity of the peaks indicates the increase of crystalline structure with increasing amount of crystalline BiNPs. All XRD patterns can be indexed to a pure-phase bismuth with a Rhombohedral structure belonging to the R-3m

IOP Conf. Series: Journal of Physics: Conf. Series **1082** (2018) 012077 doi:10.1088/1742-6596/1082/1/012077

space group, (ICSD Reference Code: 98-002-1658). This is because no unidentified peaks in the diffractogram are noticed within the detection limit of the XRD method, suggesting that the pureness of the final product of BiNPs and also free from impurities. When the hydrothermal temperature is above 160°C, there are continuous sharpening and intensifying of the diffraction peaks for BiNPs with increasing hydrothermal temperature, thereby indicated that the crystallite size of BiNPs increases with the increasing temperature. This observation revealed that particle size increased with increasing reaction temperature. It is known that high reaction temperature contributed to faster growth rate due to faster diffusion of ions on nuclei of BiNPs [7].

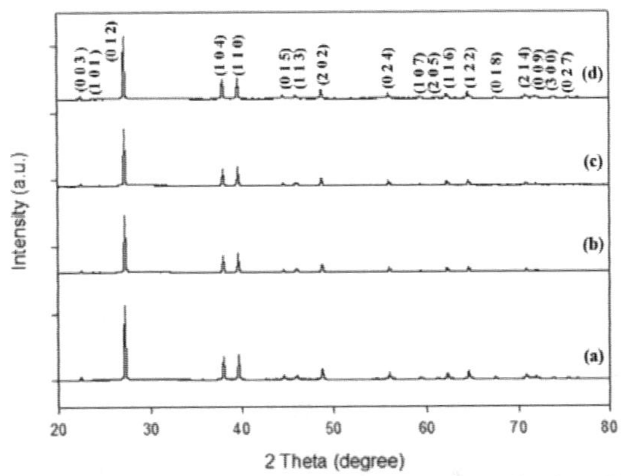

Figure 1: XRD patterns of BiNPs synthesized using hydrothermal method at 6 hours and 2.5 ml of hydrazine hydrate with varying hydrothermal reaction temperature: (a) 150°C, (b) 160°C, (c) 170°C, and (d) 180°C

Figure 2 shows the changes of reaction temperature contributed to successful hydrothermal reaction to form BiNPs with spherical shapes. However, Figure 2 clearly reveals the detail morphologies of the BiNPs, indicating that the agglomeration of BiNPs synthesized with varying reaction temperature which could later decrease its surface area for electrochemical measurement towards heavy metal ions. As reaction temperature increases, the obtained BiNPs shows only the approximate spherical particles with increasing particle size and with more agglomeration.

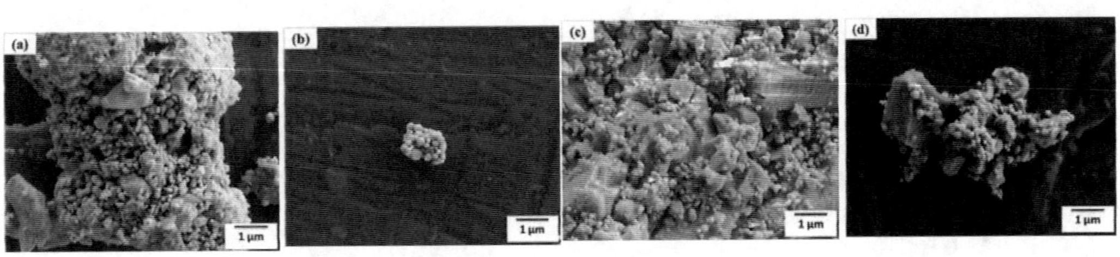

Figure 2: SEM images of BiNPs synthesized using hydrothermal method for 6 hours and 2.5 ml of hydrazine hydrate with varying hydrothermal reaction temperature (a) 150°C, (b) 160°C, (c) 170°C, and (d) 180°C

During hydrothermal reaction, the reaction temperature is crucial for the morphology and particle size of the synthesized BiNPs. The results revealed that the reaction temperature influenced the particle size, whereby particle size of the BiNPs increased gradually with the increasing reaction temperature. More specifically, the roughness of the particles increased, yielding less spherical particles. Therefore, in response to an increase in reaction temperature, size distribution increased. Nucleation rate of the Bi_2O_3 precursor is slower than the growth rate under the condition of the higher reaction temperature; thereby the formation of precursor Bi_2O_3 tends to grow into larger particle sizes [8]. Table 1 shows the temperature changes produce various sizes of BiNPs which also vary their surface areas later towards sensitivity of heavy metal ions.

Table 1: Average particle size of BiNPs synthesized using hydrothermal method for 6 hours and 2.5 ml of hydrazine hydrate with varying hydrothermal reaction temperature: 150°C, 160°C, 170°C, and 180°C

Hydrothermal reaction temperature (°C)	Average particle size of BiNPs (nm)
150	127.00
160	91.79
170	110.73
180	124.57

The Cyclic Voltammetry (CV) curves of bare ITO electrode and BiNPs/ITO electrodes synthesized using hydrothermal method with varying reaction temperature analyzed in 0.002 mol/L of $K_3Fe(CN)_6$ electrolyte are shown in Figure 3. The result of the anodic peak current with BiNPs clearly shows that BiNPs/ITO electrodes had higher current respond activity for the electrochemical behavior of $[Fe(CN)_6]^{3-}$ ions. This is because BiNPs increase the surface area of BiNPs/ITO electrodes, thus there are larger surface area for electrochemical detection of $[Fe(CN)_6]^{3-}$ ions. High electron transfer capability of BiNPs/ITO electrodes promotes faster electron transfer through BiNPs. This result prove that BiNPs/ITO electrodes have better sensitivity than bare ITO electrode. Figure 3 also shows the peak current responses of BiNPs/ITO electrodes increase from 150°C > 170°C > 180°C > 160°C. Sample synthesized at 160°C shows the smallest particle size (average size ≅ 91 nm), thereby faster electron transfer kinetic energy between ITO electrodes and $[Fe(CN)_6]^{3-}$ ions. The decrease of the peak current responses at CV properties could be due to agglomeration or larger particle size of BiNPs that blocked the connection between $[Fe(CN)_6]^{3-}$ and ITO glass substrate which then obstructs the electrons transfer between $[Fe(CN)_6]^{3-}$ ions.

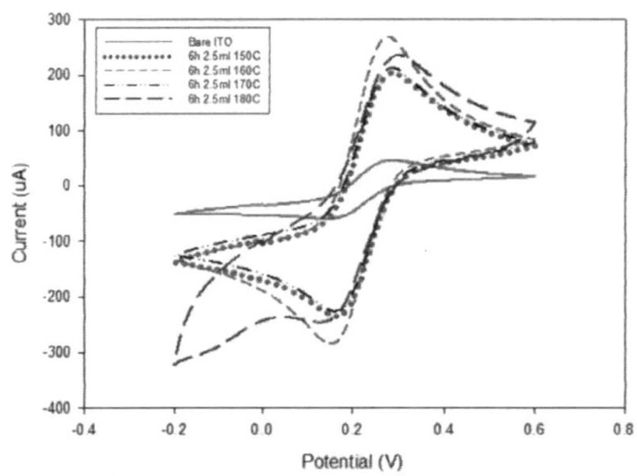

Figure 3: CV curves of modified ITO electrodes with BiNPs synthesized using hydrothermal method for 6 hours and 2.5 ml of hydrazine hydrate with varying hydrothermal reaction temperature: 150°C, 160°C, 170°C, and 180°C in 0.002 mol/L of $K_3Fe(CN)_6$

Figure 4 shows the stripping current responses of the DPASV measurements recorded for the determination of Pb(II) after a 300 seconds deposition time at −1.2 V in 0.002 mol/L of acetate buffer solution at pH 4.5. High current response of the electrode in this work indicates that drop casting of BiNPs on ITO electrodes has great potential to be used as heavy metal sensors. This is because the large surface areas between spherical BiNPs and ITO electrodes favor the reduction action of heavy metal ions thereby increases the sensitivity of heavy metal ions sensor. Large area of direct surface contact between BiNPs and ITO ease the electron transfer to be detected by the DPASV measurement. Moreover, Figure 4 shows the stripping signal of Pb(II) on the BiNPs/ITO modified electrodes are weaken due to the aggregation of the larger nanoparticle sizes of BiNPs materials which hinder the electron transfer for deposition of target heavy metal cations on ITO electrodes. Hence, BiNPs/ITO electrodes with larger particle size is insufficient to fully functions; thereby the electroanalytical activity produced is poor. Theoretically, small size of nanoparticles has larger surface area. However, in this case, BiNPs synthesized at 170°C did not obey this rule and it did not perform the higher peak current than 180°C. This is could be due to the agglomeration of BiNPs synthesized at 170°C has decreased its surface area.

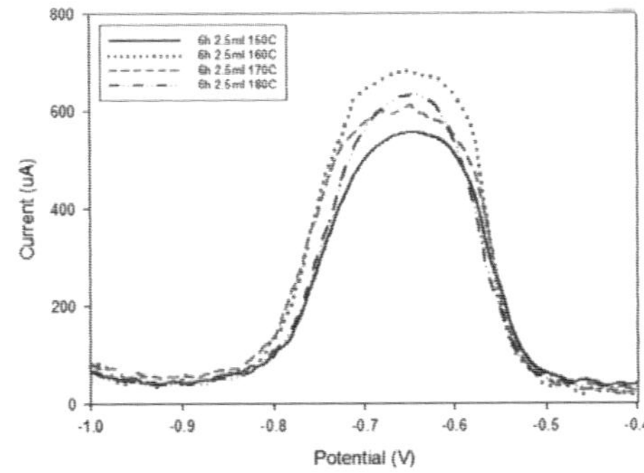

Figure 4: The DPASV stripping current response measured with the BiNPs/ITO electrodes synthesized using hydrothermal method for 6 hours and 2.5 ml of hydrazine hydrate with varying hydrothermal reaction temperature: 150°C, 160°C, 170°C, and 180°C in electrolyte containing 100 µg/L of Pb^{2+} standard solution with acetate buffer solution

The sensitivity response of the electrode upon changes in heavy metal ions concentration was studied. As shown in Figure 5, for the detection of Pb(II), the exponentially decreasing relationships between BiNPs signal changes and ion concentrations are achieved in a range from 2.5 to 100 µg/L. Throughout this study, the resulting BiNPs/ITO modified electrodes performed adequately in heavy metal ions detection. Herein, based on Figure 5, the limit of detection (LOD) for Pb(II) is 2.5 µg/L.

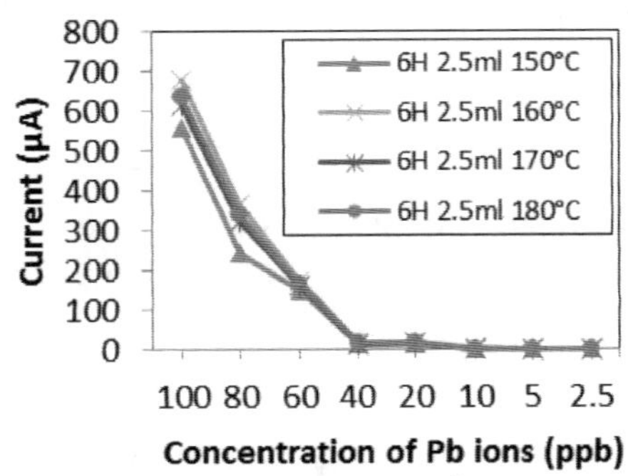

Figure 5: Graph of current (µA) against concentrations of Pb^{2+} standard solutions (µg/L) measured with the BiNPs/ITO electrodes synthesized using hydrothermal method for 6 hours and 2.5 ml of hydrazine hydrate with varying hydrothermal reaction temperature: 150°C, 160°C, 170°C, and 180°C

Figures 6 shows the stripping current responses of the DPASV measurements recorded for simultaneous determination of Pb(II) in 100 µg/L of multi-element standard solution added with 0.002 mol/L of potassium ferrocyanide $K_3Fe(CN)_6$. $K_3Fe(CN)_6$ was added to alleviate the interference of deposition competition among Cu(II), Pb(II) and Cd(II) on the electrode surface. This is because $[Fe(CN)_6]^{3-}$ is able to form relatively stable compounds with Cu^{2+} ions and therefore aid in alleviate the interference effect on the analyses of Pb(II) [9]. As shown in Figure 6, the recorded peak currents were observed, but the detection currents appeared to be well below than the 100 µg/L of Pb(II) standard solution. This suppression effect was probably due to the electrodeposition competition between the analytes, Pb(II) for the active sites on the BiNPs/ITO electrodes, as well as the formation of intermetallic compound between Cu(II) with Pb(II). Besides, from the electrochemical point of view, other heavy metals ions could also simultaneously accumulated on the BiNPs/ITO electrode surfaces and the corresponding stripping signals were recorded at potentials lower than those of Pb(II). The results from this research shows that BiNPs modified electrode has promising properties for electrochemical Pb sensor.

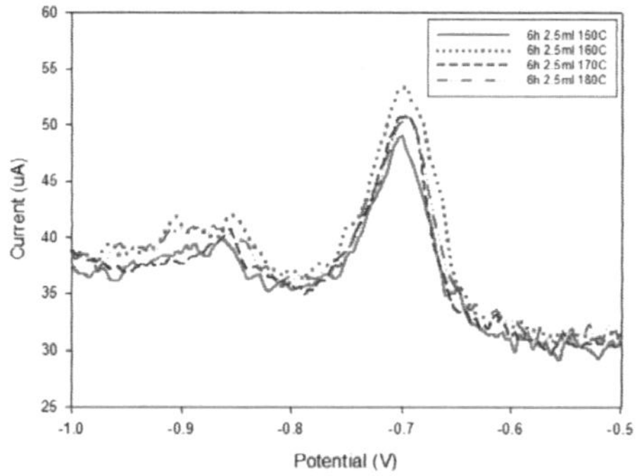

Figure 6: The DPASV stripping current response measured for interference study in electrolyte containing 100 µg/L of multi-element standard solution by using BiNPs synthesized using hydrothermal method for 6 hours and 2.5 ml of hydrazine hydrate with varying hydrothermal reaction temperature: 150°C, 160°C, 170°C, and 180°C

4. Conclusion

In this study pure-phase bismuth powders with a Rhombohedral structure were successfully produced using hydrothermal method with varying hydrothermal reaction temperature. BiNPs/ITO had good CV properties in which BiNPs produced at 170°C has the highest current response. However, BiNPs synthesized at 160°C showed the highest DPASV stripping current response due to high surface area and less agglomeration of particles. The BiNPs modified ITO was successfully fabricated as electrochemical Pb sensor. The modified BiNPs/ITO has LOD of 2.5 µg/L Pb with good specificity towards Pb.

Acknowledgment

The authors appreciate the technical supports from the School of Materials and Mineral Resources, and Institute for Research in Molecular Medicine (INFORMM), Universiti Sains Malaysia. This research was funded by RU grant 1001/Pbahan/870049.

References

[1] Nisar Ullah, Muhammad Mansha, Ibrahim Khan, and Ahsanulhaq Qurashi 2018 *TrAC Trends in Analytical Chemistry* 100 155
[2] Huangfu, C., Fu, L., Li, Y., Li, X., Du, H. and Ye, J. 2013 *Electroanalysis* 25 2238
[3] Cui, L., Wu, J. and Ju, H. 2015 *European Journal of Chemistry* 21 525
[4] Sureshkumar Kempahanumakkagari, Akash Deep, Ki-Hyun Kim, Suresh Kumar Kailasa and Hye-On Yoon 2017 *Biosensors and Bioelectronics* 95 106
[5] Metters J.P. and Banks C.E., 2014 *Nanoparticle modified electrodes for trace metal ion analysis, In Nanosensors for Chemical and Biological Applications* Woodhead Publishing, 54
[6] Lee, G. J., Kim, C. K., Lee, M. K. and Rhee, C. K. *Electroanalysis* 22 530
[7] Hwang, G. H., Han, W. K., Kim, S. J., Hong, S. J., Park, J. S., Park, H. J. and Kang, S. G. 2009 *Journal of Ceramic Processing Research* 10 190
[8] Hajra, P., Shyamal, S., Mandal, H., Fageria, P., Pande, S. and Bhattacharya, C. 2014 *Electrochimica Acta* 123, 494
[9] Niu, P., Fernández-Sánchez, C., Gich, M., Ayora, C. and Roig, A. 2015 *Electrochimica Acta* 165 155

The Effect of Alumina and Hydroxyapatite Content on Morphology and Mechanical Properties of Mg Hybrid Composite for Biodegradable Implants Materials

R. Nazirah[1], H. Zuhailawati [1]

[1]Universiti Sains Malaysia, Engineering Campus, Penang, Malaysia

[1]Biomaterials Niche areas, Institute of Research, Penang, Malaysia

Email: zuhaila@usm.my

Abstract. In biomedical applications, Magnesium (Mg) as a biodegradable implant material becomes one of the most potential candidates for temporary implants and has attracted much people's attraction. It is because Mg is a light weight, non-toxic and able to degrade in human body. The aim of this research is to investigate the effect of hydroxyapatite(HA) and alumina (Al_2O_3) content on morphology and mechanical properties of Mg hybrid composite for biodegradable implants materials. Several of Al_2O_3 contents (i.e., 0, 5, 10, and 15 wt. %) were introduced in forming the Mg-Zn/HA/Al_2O_3 hybrid composite. The powder mixture of hybrid composite was milled separately in a planetary mill under gas argon atmosphere using a stainless steel container and balls. The powder was milled at 220 rpm for 1 hours and 3% of n-heptane solution was added prior to milling process to avoid the excessive cold welding of the powder. The powder was compacted under 400MPa and sintered with sintering temperature 300°C in a tube furnace for 1hour under the flow of argon. The results show that the sintered compacts were dispersed by uniform Al_2O_3 powder corresponding to white distribution of particles, as shown in the micrograph. As the Al_2O_3 content was increased to 15 wt. %, the mechanical properties such as hardness and compression test was improved.

1. Introduction

Biomedical metals and biodegradable polymers are generally used as implants. However, medical metals such as titanium alloy and stainless steel may result in 'stress shielding' of the bone and these permanent implants must to be eliminate by a secondary surgery after curing [1]. Metals such as Mg and its alloys also exhibit a bright expectation as a potential candidate for bio-implants such as orthopedic and craniofacial repair. This is because Mg and its alloys demonstrate a unique characteristic such as can degrade in vivo and have comparable physical properties as natural bones. The mechanical properties of Mg and its alloys such as Young's Modulus of elasticity (E=41-45GPa) and density (1.74-1.84g/cm^3) are recognized to be comparable that of bone (E=15-20GPa and density = 1.8-2.1g/cm^3) [2]. There are only small numbers of element that can be tolerated in human body and can also retard biodegradable of Mg alloy including calcium (Ca), manganese (Mn), zinc (Zn) and very small amount of low toxicity rare earth (RE) elements such as niobium (Nb) and tantalum (Ta). Yang et al. 2014, reported that the addition of Zn mechanically improves the yield strength and ultimate tensile strength of Mg. Zn also provided the lowest hydrogen evolution rate in comparison to Al, Si, Sn and Zr alloying element [3]. Thus addition of Zn into Mg matrix beneficial improves its mechanical properties, corrosion resistance without giving harsh effect to human body [4].

Bioceramics have good osteoconductivity and corrosion resistance but they are brittle and consequently are not suitable used as implant materials alone. Therefore, researchers mix the metals and ceramic forming composites which have the qualities from both metals and ceramics. Hydroxyapatite (HAP) is a biodegradable and bioceramic materials are commonly used in bone replacement application due to its close similarity to bone mineral [5]. Alumina (Al_2O_3) ceramics are known as 'bioinert' (although a material should never be considered as totally inert), since no direct bone material interface is created. Al_2O_3 has two main benefits over other materials compared to metals and polymers such as low concentrations and low wear rates of inert wear particles (debris) in the surrounding tissue. The corrosion resistance of Al_2O_3 ceramics is also very high (rate of corrosion 10^{-4} mm/y corresponding to a maximum corrosion rate of 1 mm in 10 years) [2].

Therefore, current studies have proved the workableness of it's used in composite form which encourages the idea to add two reinforcements such as HAP and Al_2O_3 into Mg-Zn biodegradable alloy creating a new hybrid composite, Mg-Zn/HAP/ Al_2O_3 hybrid composite.

There were many Mg-Zn based alloys and other metal based composited reinforced by HAP via powder metallurgy method. However, fabrication method of hybrid composite Mg-Zn/HAP/Al_2O_3 is not well discovered and the rightful parameters to fabricate the hybrid composite are yet to be presented. Therefore in the current work, alumina content was investigated on mechanical and microstructure studies of Mg-Zn/HAP/ Al_2O_3 composite prepared by powder metallurgy.

2. Experimental study

Different amount of HAP and Al_2O_3 and balanced Mg and Zn powders were mixed in weight proportion. Pure powder were milled to form Mg-Zn/x HAP/x-1 Al_2O_3 (x=0-15 wt. %) composites in a Fritch planetary ball mill, while confined in sealed 250 ml steel containers rotated at 220 rpm for 1 h. The container was loaded with a blended of balls (θ=10, 20mm). The total weight of the powder was about 20g and the ball to powder mass ratio about 1:8.75. n-heptane (3 wt. %) was used as the process control agent to prevent excessive cold welding of powder particles. Then the milled powders were uniaxially cold pressed under 400 MPa for 2 min at room temperature to produce 10 mm diameter of green Mg-Zn/HAP/ Al_2O_3 hybrid composite. The sample was sintered at 300°C under argon flow at 10°C/min for both heating and cooling rate for an hour in order to form solid bodies.

3. Results and discussion

The result of compressive strength of hybrid composite can be represented in Figure 1 of Mg/0-15wt. % Al_2O_3. Ultimate compressive strength of 0 wt. % Al_2O_3 was found to be as low as 127.42 MPa compared to 15 wt. % of Al_2O_3 (171.09 MPa) in Mg-Zn/HAP/ Al_2O_3 hybrid composite. Thus, when adding Al_2O_3 from 5wt. % to 15wt. % of to the Mg composites, the values was again increased. Mechanical properties of Mg have been extensively investigated by many researches. Varied result have been reported and observed. More or less, these properties are influenced by on how the structures were processed. In addition, sintering temperature, time and presences of additives and their amount has played major roles in mechanical properties of sintered Mg. The values of compressive strength increased from 0 to 15 wt. % were mainly due to uniform dispersion of Al_2O_3 into Mg matrix. This is due to the hard particles of Al_2O_3 can hinder the mobility of dislocation in the Mg-Zn alloy matrix [5]. Thus, compressive strength is enhanced with higher amount of Al_2O_3. From previous research by

Rahimian et al. 2006, the addition of Al_2O_3 into Al matrix from 0 to 15 wt. % increased the value of compressive strength however with addition of 20 wt. % Al_2O_3 , the compressive strength value decrease. The decreasing value in compressive strength at 20 wt. % of Al_2O_3 and above was caused by deterioration and agglomeration of Al_2O_3 [6]. According to Khalil et al. 2012, addition of 10 wt. % HAP and above can caused agglomeration of HAP particles that lead to degradation of interfacial bonding strength between matrix and reinforcement thus leading to a lower compressive strength [7]. The result with higher HAP content caused the value of compressive strength decrease. It also agreed by Loy et al. 2015, in Mg-Zn/HAP composite finding that the addition of higher HAP content can decrease the value of compressive strength [8]. From this result it can see that the reinforcement constituents are one of the important factors controlling the strength of Mg matrix. Al_2O_3 particles being brittle and hard lead to dispersion hardening of matrix. These particles acts as second phase in the matrix and resist the movements of dislocations and hence the value of composites.

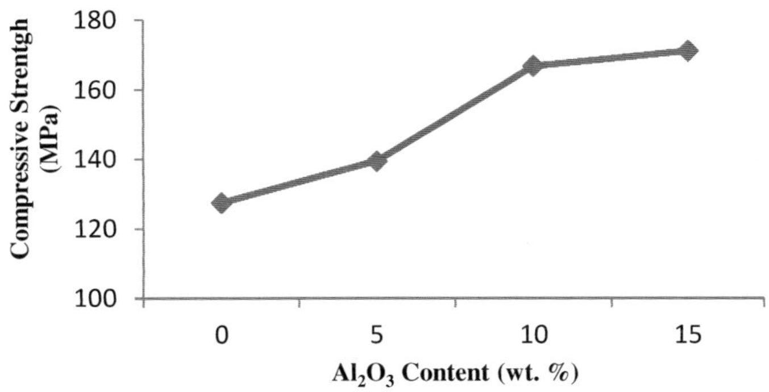

Figure 1. Compressive strength of Mg-Zn/HAP/ Al_2O_3 hybrid composite.

3.1 Microhardness Analysis of Sintered Hybrid Composite

As shown in Figure 2, microhardness of hybrid composite of Mg-Zn/HAP/Al_2O_3 increased from 70.35 HV to 79.85HV when Al_2O_3 content increased from 0 wt. % to 15 wt. %. The increases in microhardness value of the hybrid composite were mainly attributed to the dispersion effect of Al_2O_3 within Zn, and HAP into Mg matrix. The harder Al_2O_3 particles provide sufficient strength to the soft matrices and hence the microhardness values increases due to the dispersion strengthening effect [7]. Based on preview research by Chabri et al. 2013, said the addition of Al_2O_3 from 5 to 15wt. % into Al/Mg/Cu/Ti matrix increased the hardness values but the addition with 20 and 25 wt. % of Al_2O_3 caused the hardness values fall gradually due to agglomeration of the Al_2O_3 particles within the matrices. It is clearly seen that the composite samples dispersed with 15 wt. % of Al_2O_3 display a maximum hardness value [8]. The result agreed with Khalil et al. 2015, who found that addition of 10 wt. % of HAP and above gives significant drop in compressive strength and hardness value due to agglomeration of HAP particles. It leads to the degradation of interfacial bonding strength between matrix and reinforcement [5]. In this result, the maximum values of Al_2O_3 added in Mg composite is 15 wt. % and the results become increased trend without a drop value in hardness when Al_2O_3 added from 0 wt. % to 15 wt. %. The reason for the increase in the microhardness is probably due to the presence of HAP fragmented at the grain boundary regions. Besides that, the agglomeration of HAP not achieve yet or little occur compared to the previous research said that when added 20 wt. % Al_2O_3 and above into Mg the agglomeration occur and the result become decrease. Furthermore, with the increase of the content of Al_2O_3, the microhardness is strictly increased which demonstrates that the fine Al_2O_3 particles have an effect of strengthening. Al_2O_3 particles in mixed powder are beneficial to obtaining the smaller powder and can effectively refine grains of composite powders by inhibiting their growth [11].

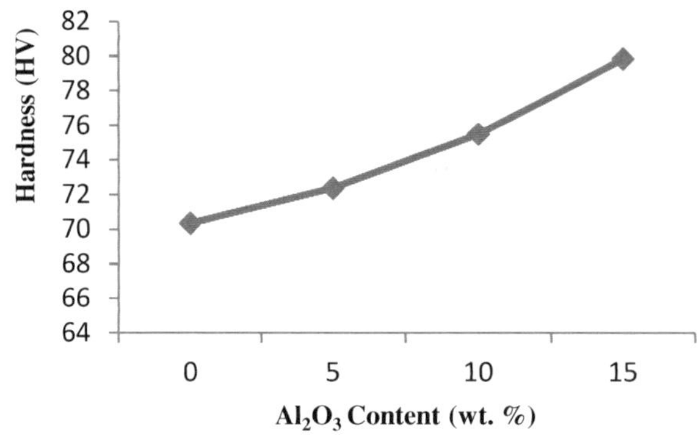

Figure 2. Microhardness of Mg-Zn/HAP/Al_2O_3 hybrid composite.

3.2 Microstructure Observation

The effect of alumina particle content on microstructure is depicted in optical micrographs. Figure 3 (a-d) shows the optical micrograph with magnification 5x from 0 to 15 wt. % of Al_2O_3 dispersed in hybrid composite. From the micrograph, the grey color is Mg matrix while the white color is Al_2O_3 powder and the dark color is porosity. From the Figure 3 (a-d) shows the presence of Al_2O_3 becomes increased and more white color presence in 15 wt. % of Al_2O_3 compared to the others. It reveals that the distributions of Al_2O_3 particles are well distributed throughout the matrices. The addition of Al_2O_3 into Mg composite can enhance the compressive strength and hardness vales due to dispersion strengthening effect.

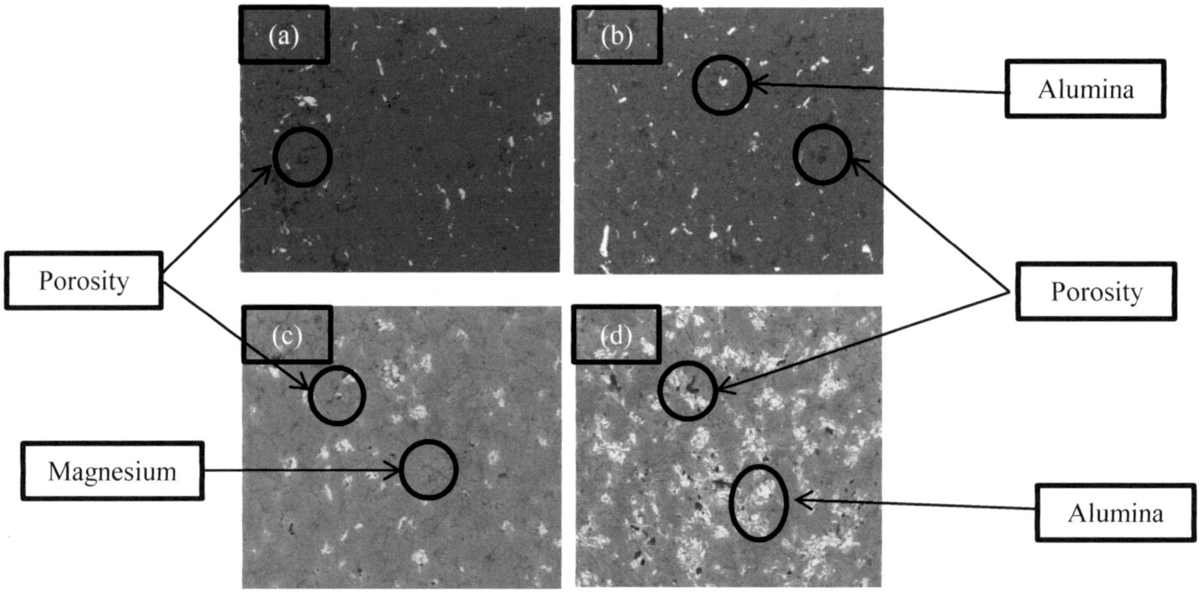

Figure 3. Optical Microscope of hybrid composite with magnification 5x (a) 0 wt. % Al_2O_3 (b) 5 wt. % Al_2O_3 (c) 10 wt .% Al_2O_3 and 15 wt. % Al_2O_3

4 Conclusion

The addition of two reinforcements such as HAP and Al_2O_3 content on Mg-Zn/HAP/ Al_2O_3 hybrid composite increase the values of hardness from 70.35 HV at 0 wt. % of Al_2O_3 to 79.85HV at 15 wt. % of Al_2O_3. Meanwhile for compressive strength the value become increased too from 127.42 MPa to 171.09 MPa. It shows that the addition of two reinforcements can enhance the mechanical properties and for microstructure analysis the Al_2O_3 and HAP content are dispersion well in the matrix of Mg.

5. Acknowledgement

The authors would like to thank to Universiti Sains Malaysia for FRGS Grant No.203/PBAHAN/6071304 and sponsorship scheme of Ministry of Higher Education (Malaysia)

References

[1] S.Shadanbaz and G.J.Dias, 2012 *Acta Biomater*
[2] M. Niinomi. M. Nakai, M. Emamy and A. Zolriasatein 2011 *Acta Biomaterialia*
[3] M.Staiger, A. Pietak, J . Huadmai and G. Dias 2006 *Biomaterials*
[4] F. Witte, V. Kaese and H. Haferkamp 2005 *Biomaterials*
[5] S. Sankaranarayanan, S. Jayalakshmi, and M. Gupta 2012 *Mater Charact*
[6] M. Rahimian, N. Parvin, and N. Ehsani 2011 *Mater Des*
[7] K.Khalil, 2012 *Int. J.Electrochem.Sci.* 7 10698
[8] L.L.Soon, H, Zuhailawati, I, Suhaina, B.K, Dhindaw 2016 *Acta Metall*
[9] M. Rashad, F. Pan, W. Guo, H. Lin, M. Asif, and M. Irfan 2015 *Mater Charact*
[10] S. Chabri, S. Chatterjee, S. Pattanayak, H. Chakraborty, N. Bhowmik, and A. Sinha 2013 *J Mater Sci Technol*
[11] F . Witte, F . Feyerabend, P. Maier, J .Fischer, M. Stormer, C .Blawert, W . Dietzel and N .Hort 2007 *Biomaterials*

IOP Conf. Series: Journal of Physics: Conf. Series **1082** (2018) 012079 doi:10.1088/1742-6596/1082/1/012079

Gel casting of β-TCP scaffolds: Effects of foaming time and polymer coating

S. Y.Cheng[1], M. Mariatti[1*], and Z. A. A. Hamid[1]

[1]Biomaterials Niche Area Group, School of Materials & Minerals Resources Engineering, Engineering Campus, Universiti Sains Malaysia, 14300 Nibong Tebal, Pulau Pinang, Malaysia

*Email: mariatti@usm.my

Abstract. Three-dimensional interconnected porous structure and ideal mechanical strength are essential for biological scaffold for bone tissue engineering application. However, ceramic scaffold is very brittle and exhibit low strength. The main aim of this study is to investigate the effect of foaming time on the microstructure and mechanical properties of porous beta tricalcium phosphate (β-TCP) scaffold. The effect of polymer coating on the β-TCP scaffold was also investigated. Porous β-TCP scaffolds were fabricated using the gel casting method by varying the foaming time (60s and 90s) followed by sintering at 1100°C. Porous β-TCP scaffold was coated with polymer coating by using dipping method. Results indicate that increased foaming time results in higher porosity and lower compression strength. It is observed that polymer coating improved around 116% compressive strength in comparison to that of uncoated scaffold. Fourier transform infrared (FTIR) confirm the presence of gelatin on scaffold walls. Based on the morphology observed by scanning electron microscope (SEM), it is found that the scaffold fabricated by gel casting method showed open pores characteristic.

1. Introduction

Biomaterial-based scaffold plays an imperative function in bone tissue engineering since it can affect the matrices of tissue formation [1]. There are some requirements need to be achieved during the fabrication of scaffold which include mechanical strength, pore size, porosity and interconnectivity. There are numerous methods to fabricate porous ceramic scaffold such as polymer sponge replication, direct foaming, sacrificial templating and gel casting [2, 3]. Gel casting is a favourable method because it can produce dense and high strength scaffold, near net shape forming and easy to produce. Gel casting method used to prepare porous ceramic structure by using gelling agents along with ceramic dispersed slurry which cause rapid setting of porous structure [4]. Gel casting method has been applied to various ceramic materials in the past decade such as alumina, silicon nitride, zirconia and silicon carbide [5]–[7]. The mechanical properties of the scaffold are very vital in handling prior to and throughout the surgical implantation. However, the main disadvantages of the bioceramic scaffold are very brittle and low strength [8]. For the purpose of improving the mechanical properties of bioceramic scaffolds, polymer coatings is found to be one of the approaches [9]. The polymer coatings can strengthen and toughened the scaffold by the activation of a micron scale crack-bridging

IOP Conf. Series: Journal of Physics: Conf. Series **1082** (2018) 012079 doi:10.1088/1742-6596/1082/1/012079

mechanism. The polymeric coating film will undergo continuous deformation and form thin fibrils which are able to bridge the cracks edges, this crack-bridging mechanism are opposing to the crack opening process and prevent the catastrophic fracture of the entire structure of the scaffold. Consequently, it can increase the strength and crack resistance of the bioceramic scaffolds [10]. In this study, gel casting method was employed to fabricate scaffolds. The effect of stirring time and polymer coating on pore characteristics and mechanical properties of the scaffolds was investigated.

2. Experimental Study

β-TCP scaffold was fabricated using gel casting process. In gel casting method, ceramic suspension was prepared using polyethyleneimine as the dispersant. Sodium dodecyl sulfate (SDS) and sorbitol act as foaming agent and gelling agent, respectively. Initially, Polyvinyl alcohol, polyethylene glycol, polyethyleneimine, β-TCP powder and distilled water were mixed together to prepare the ceramic suspension. After that, Sodium dodecyl sulfate was added into the mixture, it was vigorously stirred and foamed. Lastly, sorbitol was added into the mixture. Then, the mixture was transferred into mold and dried at 50°C for 24 hours. After the drying, the green samples were heated at a rate of 1°C/min to 400°C and then sintered at a rate of 5°C/min to 1100°C for a dwell time of 2h. The fabrication process of scaffold was repeated with different foaming time (60s and 90s). In this study, the foaming time was manipulated to investigate the effect of scaffold properties. Next, the sintered porous scaffolds were infiltrated with 5% w/v gelatin solution mixed with 0.25% glutaraldehyde for 24hours at 50°C. The coated scaffold was then dried at 40°C in vacuum oven.

3. Results and Discussions

3.1 Effects of different stirring time

The impact of stirring time on morphology of the scaffolds was characterized by scanning electron microscopy. With the increasing of stirring time from 60s to 90s at a constant stirring speed of 900 rpm, more porous structure was obtained. The number of pores was increased significantly with the increasing of foaming time, from 66±6 pores per image for 60s of foaming to 134±10 pores per image for 90s of foaming. This is because longer stirring time caused much more air bubbles were entrapped in the ceramic suspension. As can be seen from Figures 1 (a) and (b), both different foaming times facilitated the formation of interconnected porous structure. The foaming time has the significant effect on the porosity and compressive strength of scaffold. Figure 2 showed with longer period of foaming, scaffold with high porosity was produced. In addition, it showed the decrement in the compressive strength from 1.204MPa to 0.470MPa. This is because the mechanical properties of porous materials are strongly influence by stress concentration around the pores [11]. As a result, the load bearing area was reduced with increasing the porosity, which causes stress concentration around the pores and initiated the failure of scaffold [12].

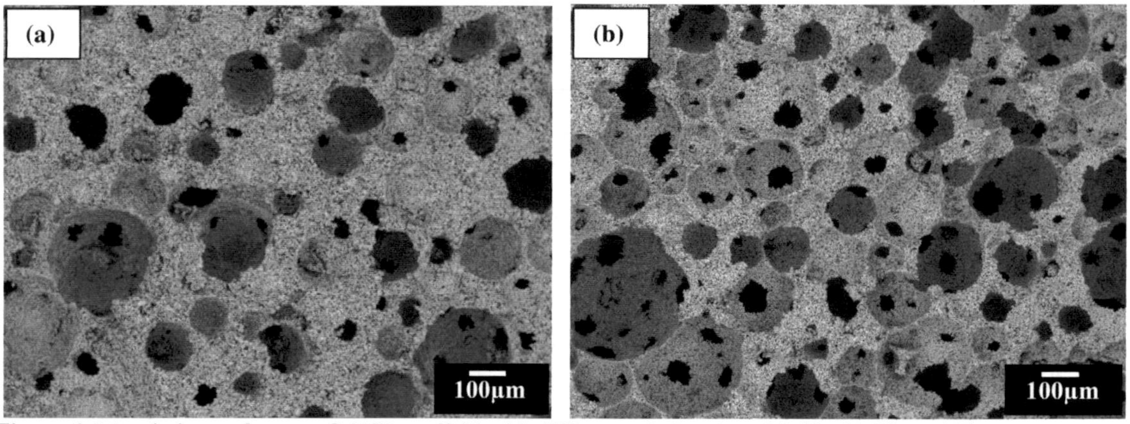

Figure 1. Morphology of porous β-TCP scaffold with different stirring time (a) 60s (b) 90s at 100x magnification.

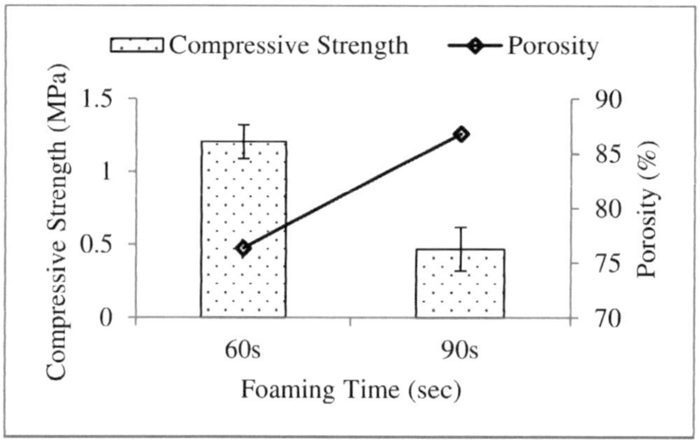

Figure 2. Compressive strength and porosity for different foaming time.

3.2 Effects of polymer coating

Figure 3 (a) showed the surface morpholgy of uncoated β-TCP scaffolds fabricated with 60s foaming time. It was noticed that the scaffolds fabricated by gel casting method had micro-size pores and dense surface. Figure 3 (b) showed SEM micrograph of gelatin deposited on the microstructure of scaffold. FTIR-ATR analysis was performed to confirm the presence of gelatin coating on the scaffold walls. Fig. 4 compares the chemical structure of gelatin coated and uncoated scaffold. The characteristic of the OH bands in gelatin was observed at 3435cm^{-1}. The spectrum for gelatin coated showed the characteristic of gelatin bands; amide I and II bands at 1657cm^{-1} and 1545cm^{-1}, respectively [13]. As indicated in Fig. 5, gelatin coated scaffold showed the increment of compressive strength from 1.204MPa to 2.602 MPa. In contrast, the porosity of gelatin coated scaffold decreased about 5% compared to that of uncoated scaffold. This result had good agreement with the previous study by Li et. al [9].

IOP Conf. Series: Journal of Physics: Conf. Series **1082** (2018) 012079 doi:10.1088/1742-6596/1082/1/012079

Figure 3. Morphology of porous β-TCP scaffold (a) uncoated scaffold fabricated with 60s foaming time (b) gelatin coated scaffold at 3000x magnification.

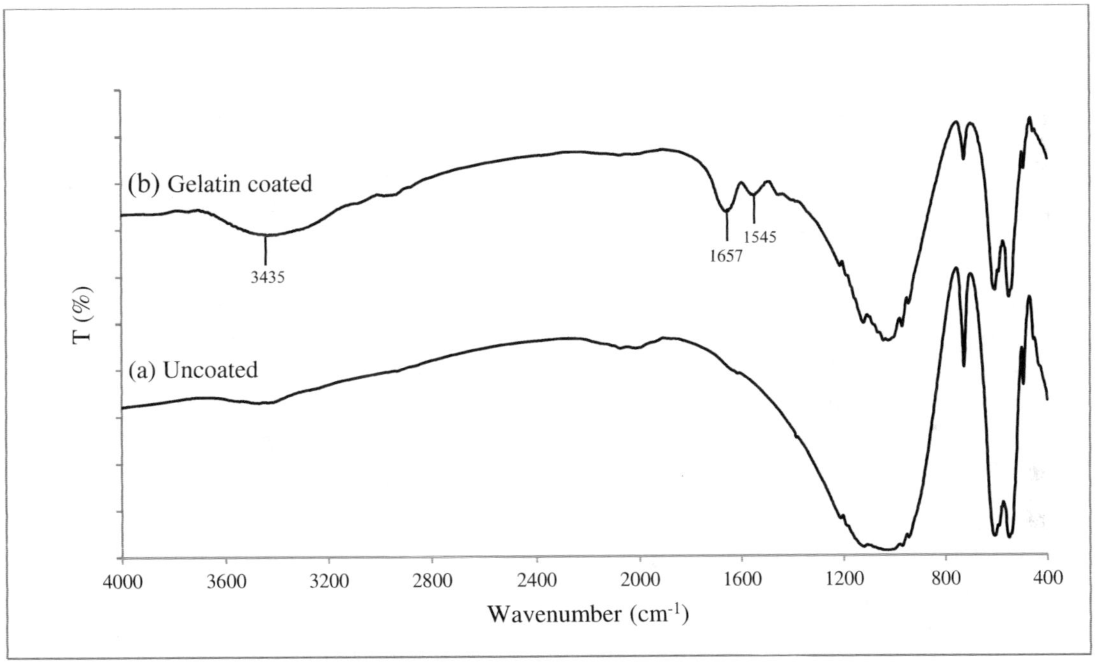

Figure 4. FTIR spectra for (a) uncoated scaffold and (b) gelatin coated scaffold.

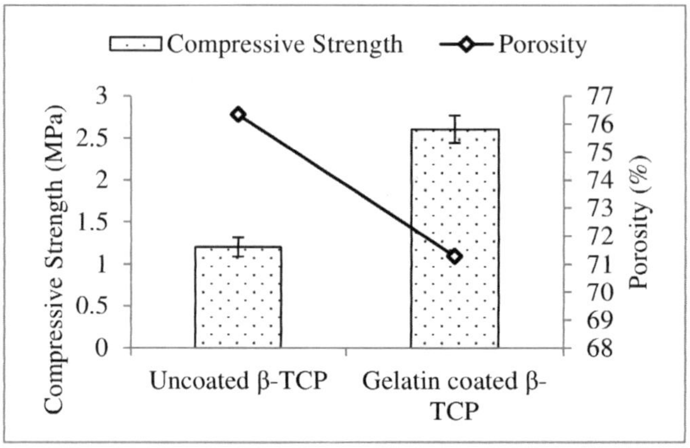

Figure 5. Compressive strength and porosity for uncoated scaffold and gelatin coated scaffold.

4. Conclusions

The effects of different foaming time and polymer coating on the scaffold properties were investigated. In this study, 60s of stirring time at 900 rpm yielded scaffold with higher 60.96% of compressive strength compared to 90s of stirring time. Gelatin coating improved around 116% compressive strength in comparison to that of uncoated scaffold (60s foaming time) while the porosity is observed to be in the range of required porosity for bone tissue engineering applications.

5. Acknowledgement

The authors gratefully acknowledge support from the Universiti Sains Malaysia and the School of Materials and Mineral Resources Engineering. This project was funded by Ministry of Education Malaysia through the Trans Disciplinary Research Grant Scheme (TRGS; Grant no. 6761004)

References

[1] J. Elango, J. Zhang, B. Bao, K. Palaniyandi, S. Wang, W. Wenhui, and J. S. Robinson 2016 *Int. J. Biol. Macromol.* 91 51
[2] A. R. Jamaludin, S. R. Kasim, A. K. Ismail, M. Z. Abdullah, and Z. A. Ahmad 2015 *J. Eur. Ceram. Soc.* 35 1905
[3] I. Sopyan and J. Kaur 2009 *Ceram. Int.* 35 3161
[4] S. R. Dash, R. Sarkar, and S. Bhattacharyya 2015 *Ceram. Int.* 41 3775
[5] C. Jiang, X. Gan, D. Zhang, R. Xie, and K. Zhou 2013 *Ceram. Int.* 39 9429
[6] E. Adolfsson 2006 *Journal of the American Ceramic Society* 89 1897
[7] M. Dong, X. Mao, Z. Zhang, and Q. Liu 2009 *Ceram. Int.* 35 1363
[8] A. Motealleh, S. Eqtesadi, F. H. Perera, A. Pajares, F. Guiberteau, and P. Miranda 2016 *J. Mech. Behav. Biomed. Mater.* 64 253
[9] W. Li, M. I. Pastrama, Y. Ding, K. Zheng, C. Hellmich, and A. R. Boccaccini 2014 *J. Mech. Behav. Biomed. Mater.* 40 85
[10] L. Bertolla, I. Dlouhý, and A. R. Boccaccini 2014 *J. Eur. Ceram. Soc.* 34 3379
[11] R. Paskaramoorthy, S. Bugarin, and R. G. Reid 2011 *Int. J. Solids Struct.* 48 2255
[12] D. Hoey and D. Taylor 2009 *Acta Biomater.* 5 719
[13] D. Bellucci, A. Sola, P. Gentile, G. Ciardelli, and V. Cannillo 2012 *J. Biomed. Mater. Res. Part A* 100 3259

IOP Conf. Series: Journal of Physics: Conf. Series **1082** (2018) 012080 doi:10.1088/1742-6596/1082/1/012080

A Review of Mechanical Properties of Scaffold in Tissue Engineering: Aloe Vera Composites

T.T. Tran[1], Z.A. Hamid[1], and K.Y. Cheong[2]

[1] Biomaterials Research Niche Group, School of Materials and Mineral Resources Engineering, Engineering Campus, Universiti Sains Malaysia, Penang, Malaysia

[2] Electronic Materials Research Group, School of Materials and Mineral Resources Engineering, Engineering Campus, Universiti Sains Malaysia, Penang, Malaysia

Email: tranthanhtam@student.usm.my

Abstract: Aloe Vera (AV) is a well-known pharmaceutical herb and traditional biomaterial from thousand years ago. AV is also a great potential material in tissue engineering because of its excellent properties such as biocompatible, biodegradable, antioxidant, anti-inflammatory, anti-diabetic, and antimicrobial. In tissue regeneration, scaffold plays an important role to form template for cell growth or tissue repair. There are some requirements in fabricating scaffold to mimic the structure of native parts of the body such as skin, cardiac, nerve, or bone. Here, mechanical properties of AV aiming to be used as scaffold are reviewed. Tensile strength as the main mechanical property of scaffold made from AV and incorporated with other natural materials (chitosan, alginate, collagen, etc.) and synthetic polymers (PCL, PVA, PLGA, etc.) will be discussed. The methods to measure mechanical properties are also being presented in this brief review article. Finally, some of the perspectives will be given for the future development of AV in tissue engineering.

1. Introduction

To date, tissue engineering has gone through remarkable developments and advancements. It proposes excellent ways to repair or replace lost tissues and damaged organs of the human body, such as skin, cartilage, bone, etc. Generally, the working principles of tissue engineering includes three important elements: (i) living cells/tissue, (ii) biomaterials scaffold, and (iii) growth factors/bioreactor [1]. In tissue engineering, researchers focused on methods and materials that can be used to fabricate porous scaffold that mimics the native extracellular matrix (ECM) of the cell that meet the requirements of tissue regeneration, as well as the suitable technique to produce scaffold for each type of tissue. There are certain criteria that scaffold materials must meet for development and designation in tissue engineering or medicine applications [2].

Among these requirements, mechanical properties are essential factors to consider in fabrication of scaffolds. Mechanical strength is identified by the impact resistance of final products in order to maintain their integrity of scaffold during implantation [3]. Most common mechanical tests to evaluate scaffold include tensile and compressive tests. Each application during tissue remodelling requires a different acceptable range for mechanical properties, because it is desirable for the scaffolds to have mechanical properties close to those of the native tissue or organ, and to be able to avoid "possible side effects" resulting from the stress-shielding mechanism. For example, the native skin has tensile strength values approximately between 5.0 to 30.0 MPa, Young's modulus in the range of 4.6-20.0 MPa and elongation at break about 35.0-115.0% [4,5]. Therefore, scaffold to be used in native sk n must exhibit good mechanical strength in order to ensure their stability for angiogenesis, the lymphatic system, nerve bundles, and other structures to generate [4,6,7].

Synthetic polymers and natural materials are two types of materials suitable to fabricate scaffold structure and have been used in last decade. However, every material has limitations to be used in tissue engineering. Synthetic polymers have show several advantages especially in processability and good mechanical properties but exhibit insufficient bioactivity in scaffold products; low potential in cell attachment, hydrophobic, therefore its limit in surface cell recognition sites [8-10]. Natural materials, on the other hand, have excellent biocompatibility, highly hydrophilic, and efficient biological activity when interacting with cells. However, they have low mechanical strength and fast degradation rate [11,12]. To overcome the limitations of each individual type of materials, they have been combined to exploit their favourable properties [13,14].

Aloe Vera (AV) has been known as an ancient therapeutic herb belonging to the Liliaceae family [15]. Its compositions possess numerous bio-actives including antioxidants, anti-inflammators, immunomodulators, hepatoprotective, anticancer, antiulcer and antidiabetic agents [15-17]. AV is also well known as a great candidate for tissue regeneration because of its ability to promote cell migration, proliferation and growth, especially in wound healing. Moreover, AV has attracted research attention due to its biocompatibility, biodegradability, and low toxicity [18]. AV contains over 75 compounds, including polysaccharides, vitamins, amino acids, enzymes, and other substances [19,20]. Lim Zhe Xi et al. showed the structure of the AV plant and components of the AV leaf is [21]. Most valuable bioactivity comes from the inner gel of Aloe Vera leaf [22].

In the recent years, AV has been used in combination with other materials to form scaffolds for many applications because of its unique bioactivity [23]. When compared to other natural materials used in tissue engineering such as chitosan, alginate or silk, AV showed superior ability in accelerating wound healing. Sheikh Rahman et al reviewed the applications of AV use for tissue engineering, as well as the techniques to fabricate scaffolds from AV, including electrospinning, freeze – drying, phase separation and self – assembly [24]. This review summarizes mechanical properties of scaffolds made from compounds of AV with synthetic polymers and natural materials and discusses testing methods to measure those properties.

2. Experimental study

As mentioned above, when scaffolds is implanted into the human body, it must have adequate mechanical strength to efficiently support the growth of cells. For each application, different requirements of mechanical properties are needed and thus, different testings have been utilized. For example, a scaffold applied in bone requires a compression test [25] while a tensile test is more important for a scaffold to be used in skin or cartilage [26,27]. Furthermore, sample shape and size are also need to be considered when choosing of testing methods. This paper mainly discusses tensile test (which is usually) used in evaluation of composites of AV incorporated with other materials. Generally, the tensile properties are described by the stress-strain curve. Stress and strain are defined by equations (1) and (2):

$$Stress\ (\sigma) = \frac{F}{A} \quad (1)$$

$$Strain\ (\varepsilon) = \frac{\Delta l}{L} \quad (2)$$

where F (N) is the force applied to the sample, A (m^2) is the cross-sectional area of the sample, Δl (mm) is the extension in length after deformation and L (mm) is the initial length between clamps.

There are several standards for sample preparation and testing method depending on the type of materials and sample sizes. Table 1 shows list of standards for tensile testing for scaffolds. Most of scaffold samples are prepared as or cut into films or thin sheets for tensile test. Sónia P. Miguel et al. determined the tensile strength of scaffold made from polycaprolactone, aloe vera and chitosan under wet and dry state according to the guidelines of ASTM D3039/A3039M [28]. The test sample was in the form of membrane with thickness in the range of 0.41 to 0.52 mm. Sadegh Khoshgozaran-Abras et al. measured the tensile strength of a film fabricated from chitosan and aloe vera using ASTM D 882 – 02 in thickness ranging from 0.163 to 0.213 mm [29]. Similarly, Ong-ard Saibuatong et al. also employed ASTM D 882 – 02 to evaluate aloe aera – bacterial cellulose composite films with 0.03 mm thickness [30].

On the other hand, in several situations where it is impossible or difficult to prepare the samples according to the standard, for instance, the length of the sample is too short to grip with the machine, certains have to be made in order to carry out the test. For example, Princeton Carter et al. fixed the film on a paper frame because the sample size was too small and the length of the sample was only 16 mm [31]. The force was zeroed after clamping and the sides of the frame were cut to release the tension of paper.

Table 1: List of standards for the tensile strength of scaffold.

Standard	Name of standard	Thickness
ASTM D3039/ D3039 M	Standard Test Method for Tensile Properties of Polymer Matrix Composite Materials	Up to 2.5 mm
ASTM D 638 - 02a	Standard Test Method for Tensile Properties of Plastics	Up to 14 mm
ASTM D 882 – 02	Standard Test Method for Tensile Properties of Thin Plastic Sheeting	Less than 1 mm

3. Results and discussion

3.1 Aloe Vera/synthetic polymers

Biopolymers are a source of inexpensive materials that possess excellent mechanical properties and are easy to process, making them the first consideration to use for tissue engineering [32]. When combined with natural materials, especially aloe vera, the biological properties of scaffolds fabricated from synthetic polymers are enhanced, which allow the scaffold more to apply in tissue regeneration and greater potential in this field. Table 2 show some aloe vera/ synthetic polymers systems used and their mechanical properties.

Table 2: Composites of aloe vera/synthetic polymers scaffold in tissue engineering.

Year	Aloe vera model	Synthetic polymer	Application	Technique	Tensile strength (MPa)	Reference
2017	Powder	PLGA	Wound healing	Electrospinning	PLGA-AV (1:1) – 4.66 ± 0.90	[27]
2016	Powder	PCL	Barrier membrane	Electrospinning	PCL/AV (80/20) – 1.35	[31]
2015	Powder	PCL	Wound dressing	Electrospinning	PCL/AV 5% - 2.00	[33]
2014	Powder	PCL	Skin	Electrospinning	PCL/AV 10% - 6.28	[16]

Until now, several synthetic polymers have been studied to incorporate with aloe vera, such as polyvinyl alcohol (PVA), polycaprolactone (PCL), or polylactic-co-glycolic acid (PLGA) [27,31,34]. Most of the scaffolds from these combinations have been fabricated by electrospinning technique because it is a simple way to achieve the porous structure. Aloe Vera has mostly been used in powder form and small concentrations. Aloe Vera gel contains nearly 55% polysaccharide in dry matter and the rest of components are amino acids, salicylic acid, ascorbic acid, vitamin A and vitamin E [35]. The functional groups – OH, - COOH and – NH in AV pulp have potential to combine with other groups in synthetic polymers [27,36,37]. However, the mechanism of bonding between AV and other materials is still in discussion. It depends on the source of the materials, fabricated techniques, or the solvents used to dissolve components of aloe vera [38,39].

As seen in Table 3, the scaffolds from AV/synthetic polymers have shown good tensile strength that meets the requirements for tissue regeneration. Recently, Itxaso Garcia-Orue et al. obtained uniform nanofibrous membranes based on PLGA and AV with fraction 1:1 [27]. This was the highest amount of AV in powder form in the literature. In this case, AV increased the tensile strength of pure PLGA from 3.06 ± 0.35 MPa to 4.66 ± 0.90 MPa, which was adequate mechanical property for use in tissue growth processing. Similarly, S. Suganya et al. synthesized nanofibrous scaffold from PCL and AV at low concentration (5 and 10 %) by electrospinning [16]. At 10% AV in the complex scaffold, the tensile strength gained the maximum value of 6.28 MPa compared to PCL (2.43 MPa) and PCL/AV (5%) (2.96 MPa). Furthermore, a PCL/AV blend also displayed better mechanical properties compared to a PCL/collagen blend. S. Agnes Mary and V.R. Giri Dev achieved the same outcome for scaffold with PCL and 5 % AV [33]. However, according to Princeton Carter et al., when the concentration of AV exceeds 20 %, tensile strength decreases [31]. This is a limitation of AV when incorporated into synthetic polymers. The mechanical strength of this scaffolds is depended on the bonding of functional groups in the matrix. PLGA and PCL have mainly C=O groups on the chains can be connected with functional groups existed in AV powder (-OH, -COOH). In low concentration, it was assumed that all the number of functional groups in AV got the bonding with functional group of synthetic polymers. However, the high concentration of AV in the composite may have interrupted the bonding of functional groups in the matrix with excess functional groups. This issue can be resolved by using suitable solvent or cross-linker. Kyoung Ran Park, Young Chang Nho also reported biopolymers such as PVA and PVP to be excellent candidates to prepare hydrogel with AV [40]. Compression test results presented increased strength along with increasing concentration of AV gel up to 10%.

3.2 Aloe Vera/natural materials

Natural materials have attracted remarkable attention in tissue engineering. These type of materials exhibit excellent behaviour in tissue regeneration because of their unique biological properties, biodegradability, biocompatibility, low toxicity and suitable structure for live organism [2,41,42]. However, the main disadvantages of natural materials are weak mechanical properties and difficulties in processing. Several biomaterials have been studied to combine with AV in tissue repairing such as chitosan, gelatin, alginate, collagen, bacterial cellulose supplied in Table 3 [26,29,30,43-46].

Table 3: Composites of aloe vera/natural materials scaffolds.

Year	Aloe Vera model	Natural materials	Technique	Application	Tensile strength MPa	Reference
2017	Powder	Chitosan/ Collagen	Freeze-drying	Burn skin tissue	At 0.1% AV – 0.17	[26]
2016	Gel	Gelatin/ Chitosan	Freeze-drying	Skin	At 0.07% AV – 0.11	[43]
2013	Powder	Alginate	Solvent casting	Wound healing	Maximum force at break: 54 N	[45]
2013	Powder	Alginate	Solvent casting	Wound healing	In dry films in non-cross-linked state: 21.44 to 40.44 At dry films after crosslinking: 42.36 to 50.91	[46]
2012	Gel	Chitosan	Solvent casting	Wound healing	At 20% AV: 8.49 ± 0.7	[29]
2010	Gel	Bacterial cellulose	Solvent casting	Tissue engineering	At 30% AV – 8.67	[30]

Uniform scaffold fabricated from comprising of AV and other natural materials (chitosan and gelatin) using freeze-drying technique have shown excellent porosity and biological properties [26,43]. However, adding AV also decreases the strength of film by using this method. Andini Isfandiari et al. mixed AV powder with chitosan/collagen to form a novel scaffold [26]. It was found that tensile strength decreased when increasing AV concentration from 0.1 % to 0.25 %. The highest tensile strength value was only about 0.17 MPa with 0.1% AV powder. This limited tensile strength can be explained by the presentation of AV which interfered with bonding between collagen and chitosan. Similarly, tensile strength of a gelatin/chitosan blending decreased from 0.11 MPa to 0.04 MPa when AV gel concentration was increased from 0.07 to 0.15%, although other characteristics showed good results [43]. With the in vitro and in vivo assay results, these combinations are worth investigating in tissue engineering.

Another technique, solvent casting, displayed more favourable peculiars in fabrication of scaffolds with strong mechanical properties from a combination of natural materials. By that method, Pereira et al. successfully produced scaffolds from alginate and 5, 15 and 25% AV concentration [45,46]. Although the increasing of AV proportion caused reduction of tensile strength, the change in the values was unremarkable in both dry and wet test conditions. Scaffold gained tensile strength in the range of 21.44 to 40.44 MPa in the dry state without crosslinker and 42.36 to 50.91 MPa after crosslinked by $CaCl_2$. The role of crosslinking agent was clear in this case and is an important factor to consider in scaffold processing. Similarly, chitosan and bacterial cellulose were mixed with AV gel up to 50% amount by solvent casting [29,30]. AV gel improved the tensile strength of scaffold at low concentration to 8.49 ± 0.7 MPa at 20% incorporate with chitosan and 8.67 MPa at 30% for bacterial cellulose mixing. The higher amount of AV caused a decrease in mechanical properties. However, the resultant product still had adequate properties for tissue regeneration. In short, when AV was incorporated with other individual natural materials, the bonding occurred mostly between functional groups of them and mechanical strength presented good results at low concentration of AV.

3.3 Composite aloe vera/natural/synthetic polymers scaffolds

In composite scaffolds of synthetic polymers and natural materials, these two types of materials are able to balance out each other's limitations in tissue applications. To enhance the characteristics of this type of composites, AV was added to improve the bioactivity of scaffolds by electrospinning, as shown in table 4 [28,47-49]. All the experiments listed used the electrospinning method to construct scaffolds. This novel group of composites is still in development and has great potential for future work.

Table 4: Composite aloe vera/natural/synthetic polymers scaffolds.

Year	Amodel	Incorporate compound	Technique	Application	Tensile strength (MPa)	Reference
2017	Gel	PCL/Chitosan/PEO	Electrospun	Skin	In dry state: 6.39 ± 2.09 In wet state: 6.23 ± 0.33	[28]
2017	Powder	PCL/ gum tragacanth	Electrospun	Skin	PCL/GT/AV (5%): 0.75 PCL/AV: 0.32	[47]
2014	Powder	PCL/silk fiber/HA	Electrospun	Bone	PCL/SF/AV: 4.80 PCL/SF/HA/AV: 10.89	[48]
2014	Powder	PLACL/silk fiber	Electrospun	Cardiac	PLACL/AV: 4.7 PLACL/SF/AV: 3.7	[49]

In order to improve biological properties of the scaffold, AV has been considered for use in applications as skin, bone, and cardiac. Miguel et al. established new scaffolds with PCL on the top and chitosan/PEO/AV at the bottom [28]. Mechanical properties of membranes were examined under dry and wet conditions. Although adding 40% AV gel reduced tensile strength from 9.48 MPa to 6.39 MPa (compared to PCL/CS_PEO) in dry state, the achieved result was sufficient to apply in skin repairing [4,5]. Ranjbar-Mohammadi used AV powder mixed with PCL and gum tragacanth (GT) to determine the effect of AV addition [47]. Tensile strength of nanofibers increased from 0.21 to 0.75 MPa by adding 5% AV into PCL/GT. Although there was an increase in tensile strength, the application of new scaffolds in skin regeneration had many limitations. Shanmugavel et al. fabricated scaffolds to apply in bone tissue engineering from a combination of PCL, silk fibre (SF), hydroxyapatite (HA) and AV powder [48]. The role of HA can be seen in increasing tensile strength of PCL/SF/AV from 4.80 to 10.89 MPa. This implied that AV powder was suitable to combine with HA in bone scaffolds, which is a noteworthy point. Furthermore, tensile strength of PCL/SF/AV had higher values compared to PLACL/SF/AV with 3.70 MPa [49]. The AV gel has the variety of functional groups when comparing to powder model. It shows the better benefit in combination with both synthetic polymer and other natural materials.

4. Conclusion

Aloe Vera is an excellent natural material and suitable for incorporation with other biomaterials in tissue engineering for improved biological properties; enhanced scaffolds' biodegradability, porosity, biological properties for the growth of new tissue implanted in the human body. According to certain researches, Aloe Vera showed different mechanical properties behaviours when combined with synthetic polymers or natural materials or both. In most cases, Aloe Vera can enhance mechanical properties at low concentrations as well as improve the bioactivity of scaffolds, especially, in mixture of synthetic polymers (PLGA, PCL, and PVA/PVP) and other natural materials (chitosan, alginate, bacterial cellulose). It depends mainly on the bonding of functional group between AV model and other materials through specific synthetic technique. However, increasing AV concentration over 30 % weakens the bonding of scaffold and decreases the mechanical properties. In other cases, combination of AV and compounds of both synthetic polymers and other natural materials exhibited great results. This is a new approach for scaffolds in future time. Crosslinking agents are another important role in enhancement of mechanical properties of AV scaffolds with other materials. This should be put into consideration when blending different materials.

Acknowledgement

The authors acknowledge financial supported from AUN/SEED-Net scholarship organization (Grant Number: 304/PBAHAN/6050351).

References

[1] Nerem R M 1992 *Med. & Biol. Eng. & Comput.* 30 8
[2] O'Brien F J 2011 *Materials Today* 14 88
[3] Song A et al. 2012 *Acta Biomater* 8 41
[4] Elsner J J et al. 2010 *Tissue Viability* 19 54
[5] Zahouani H et al. 2009 *Skin Research and Technology* 15 68
[6] Zahedi P et al. 2009 *Polymers for Advanced Technologies* n/a
[7] Elsner J J et al. 2010 *Journal of Biomedical Materials Research B: Applied Bimaterials* 93 425
[8] Nair L S et al. 2007 *Progress in Polymer Science* 32 762
[9] Ramakrishna S et al. 2001 *Composites Science and Technology* 61 1189
[10] Gunatillake P et al. 2006 *Biotechnology* 12 301
[11] Gunn J et al. 2010 *Trends Biotechnol* 28 189
[12] Ratner; B D et al. 2004 *Biomaterials science : an introduction to materials in medicine*
[13] Chong E J et al. 2007 *Acta Biomater* 3 321
[14] Roether; J A et al. 2002 *Biomaterials* 23 3871
[15] Reynolds T et al. 1999 *Journal of Ethnopharmacology* 68 3
[16] Suganya S et al. 2014 *Iranian Polymer Journal* 23 237
[17] Maenthaisong R et al. 2007 *Burns* 33 713
[18] Choi S et al. 2003 *Seminars in Integrative Medicine* 1 53
[19] Hamman J H 2008 *Molecules* 13 1599
[20] Li X et al. 2013 *Colloids and Surfaces B: Biointerfaces* 103 182
[21] Lim Z X et al. 2015 *Physical Chemistry Chemical Physics* 17 26833
[22] Ni Y et al. 2004 *Int Immunopharmacol* 4 1745
[23] Hashemi S A et al. 2015 *Biomed Res Int* 2015 714216
[24] Rahman S et al. 2017 *Functional Biomaterials* 8
[25] Balagangadharan K et al. 2017 *Int J Biol Macromol* 104 1372
[26] Isfandiary A et al. 2017 *Journal of Biomimetics, Biomaterials and Biomedical Engineering* 32 82
[27] Garcia-Orue I et al. 2017 *Int J Pharm* 523 556
[28] Miguel S et al. 2017 *Polymers* 9 183
[29] Khoshgozaran-Abras S et al. 2012 *Carbohydrate Polymers* 87 2058
[30] Saibuatong O-a et al. 2010 *Carbohydrate Polymers* 79 455
[31] Carter P et al. 2016 *Journal of Biomaterials Science, Polymer Edition* 27 692
[32] Dhandayuthapani B et al. 2011 *International Journal of Polymer Science* 2011 1
[33] Agnes Mary S et al. 2015 *The Journal of The Textile Institute* 106 886
[34] Isfahani F R et al. 2017 *Fibers and Polymers* 18 264
[35] Kang M C et al. 2014 *Carbohydr Polym* 99 365
[36] Torres-Giner S et al. 2017 *J Agric Food Chem* 65 4439
[37] Fox L T et al. 2017 *J Ethnopharmacol* 200 1
[38] Femenia A et al. 1999 *Carbohydrate Polymers* 39 109
[39] Ray A et al. 2015 *Industrial Crops and Products* 76 298
[40] Park K R et al. 2003 *Journal of Applied Polymer Science* 90 1477
[41] Topman G et al. 2013 *J Biomech* 46 170
[42] Malafaya P B et al. 2007 *Adv Drug Deliv Rev* 59 207
[43] Lopez Angulo D E et al. 2016 *Int J Biol Macromol* 92 645
[44] Silva S S et al. 2013 *Carbohydrate Polymers* 98 581
[45] Pereira R F et al. 2013 *Carbohydrate Polymers* 98 311
[46] Pereira R et al. 2013 *Int J Biol Macromol* 52 221
[47] Ranjbar-Mohammadi et al. 2017 *Journal of Industrial Textiles* 152808371769259
[48] Shanmugavel S et al. 2014 *J Biomater Appl* 29 46
[49] Bhaarathy V et al. 2014 *Mater Sci Eng C Mater Biol Appl* 44 268

IOP Conf. Series: Journal of Physics: Conf. Series **1082** (2018) 012081 doi:10.1088/1742-6596/1082/1/012081

Uniform Coating of Polyaniline on Porous Carbon Nanofibers as Efficient Electrodes for Supercapacitors

Chau D Tran*, A.Q Vu, and Thang Van Le

Faculty of Materials Technology Ho Chi Minh City University of Technology, Vietnam National University Ho Chi Minh City Vietnam

Email: chau_tran@hcmut.edu.vn

Abstract. Porous activated carbon nanofibers (A-CNFs) are fabricated and considered as promising electrode materials for electrochemical double layer capacitors (EDLCs) due to their free-standing and hierarchical pore structure. They exhibit high specific capacitances in 1 M H_2SO_4 without any significant loss at high scan rates. To further enhance charge storage, we incorporated polyaniline onto A-CNFs with the aim to combine the benefits of both EDLCs (high power density) and pseudocapacitors (high energy density). By using a simple galvanostatic deposition, we achieve a uniform coating of polyaniline on A-CNFs. Combination of two different storage mechanisms results in an excellent specific capacitance of 290 F g^{-1}. Cyclic voltammetric measurements show no significant change at scan rates up to 100 mV s^{-1}.

1. Introduction

Supercapacitors are widely known as high-power energy storage devices that can be utilized for many practical applications. Depending on working mechanism, supercapacitors are divided into two types, namely electrical double layer capacitors (EDLCs) and pseudocapacitors. Carbon-based materials with a high surface area are typically used for EDLCs while fast faradic redox exhibiting materials, such as conducting polymer and metal oxides, are used for pseudocapacitors. Conducting polyaniline (PANI) has been shown as a promising material for pseudocapacitors due to its high theoretical pseudocapacitance (750 F g^{-1}) [1]. PANI exists in three different oxidation states: leucoemeraldine (fully reduced), emeraldine (half oxidized), and pernigraniline (fully oxidized). At low pH values, these states can get partially or fully protonated. Under electrochemically oxidizing condition, PANI transforms from leucoemeraldine to emeraldine, and finally to pernigraniline through the loss of electron. These reactions are also associated with a gain or loss of protons and anions. Depending on pKa value of electrolyte condition, transformation pathways of PANI will be different. Most of studies in the past have been carried out in an acidic solution, which is often linked with a high activity of polyaniline. Several reports [2-4] also tried to study in non-aqueous electrolytes to increase voltage window, but often associated with instability of PANI.

Polyaniline often faces a short cycle life due to the swelling and shrinking during the doping/dedoping process [5]. As an example, PANI expands due to anion insertion during transformation from leucoemeraldine to emeraldine. A thick PANI film will suffer an inefficient utilization at high power due to the long diffusion path length of ions. PANI with various nanostructures have been fabricated to reduce the diffusion path length; however, the reported sizes are still in the range

IOP Conf. Series: Journal of Physics: Conf. Series **1082** (2018) 012081 doi:10.1088/1742-6596/1082/1/012081

of 50-300 nm [6]. Growth of nanostructured PANI on carbon based materials is often reported and provides an extremely high capacitance and a good cycle life. An example is the growth of PANI on porous carbon monolith [7]. The obtained specific capacitance with respect to PANI weight was 1580 F g^{-1}, which is much higher than the theoretical value of PANI. However, the specific capacitance with respect to the total mass of PANI and carbon support was only about 300 F g^{-1}. These composites are often impractical for commercialization due to either limited scalability or insignificant capacitive contribution from the carbon substrate [8]. Moreover, most works in the past synthesize powder-based electrodes. In this work, we demonstrate that conformal growth of nanostructured PANI on porous carbon nanofibers offers both high specific capacitance as well as stability upon cycling.

PANI can be deposited on carbon materials by either chemical or electrochemical method. Powder based electrodes are often obtained through the chemical method. As an example, previous studies have obtained a whisker-like or needle-like structure of PANI on graphitized carbon nanofibers (CNFs) [5, 9] by using oxidative polymerization. Before polymerization, nanofiber mats were ultrasonicated to obtain a uniform suspension of nanofibers. This resulted in loss of the free-standing morphology of nanofibers. Subsequently, the obtained PANI/CNFs powder needed to be mixed with a binder and conductive filler before coating on the current collector, resulting in reduced performance compared to free-standing materials. In contrast to chemical deposition, electrodeposition can be carried out on fabricated porous CNFs, while preserving their free-standing morphology. In this study, PANI will be coated on non-porous and porous activated CNFs which are both free-standing substrates using electrochemically galvanostatic technique. The non-porous carbon nanofibers (NP-CNFs) serve as a control substrate for initial studies on PANI electrodeposition. The porous carbon nanofibers (A-CNFs) are then utilized for PANI deposition to combine both aspects: double layer capacitance from porous carbon nanofibers and pseudocapacitance from polyaniline. We achieve a composite capacitance of 290 F g^{-1}. The CVs show a near-rectangular shape with well-defined redox peaks up to 100 mV s^{-1}.

2. Experimental

2.1 Fabrication of non-porous and activated porous carbon nanofibers

Polyacrylonitrile (Mw = 150000 g mol^{-1}, purchased from Sigma-Aldrich) were dissolved in N,N-dimethylformamide (purchased from Sigma-Aldrich) at 10% w/w under gentle heating and stirring for 1h. Fibers were electrospun at room temperature under controlled humidity. The distance between the tip of the needle and the grounded collector was 5 inches, and the applied voltage of 12 kV was used to obtain a stable Taylor cone. The flow rate was kept at a constant of 1 mL h^{-1}. The electrospun nanofibers were placed in a horizontal tube furnace and stabilized by heating to 280^{0}C at a rate of 5°C min^{-1} under an air environment. The stabilized nanofibers were then carbonized by heating to 800°C at a rate of 5°C min^{-1} and held at 800°C for 1h under nitrogen atmosphere to obtain NP-CNFs. Then NP-CNFs was submerged in a 30% w/w KOH solution for 1h and activated at 800°C for 30 minutes to obtain A-CNFs.

2.2 Polyaniline deposition on non-porous and activated porous carbon nanofibers

Aniline (purchased from Alfa Aesars) was dissolved in 1M H_2SO_4. The electrochemical polymerization of PANI on these substrates was carried out using a three-electrode cell containing Ag/AgCl as the reference electrode and a platinum mesh as the counter electrode. All reported potentials were versus Ag/AgCl. A T-type Swagelok cell was employed as shown in Figure 1. In this setup, a graphite rod of ½" diameter was used as the current collector. A ½" CNFs mat was placed between the graphite rod and a porous stopper and the assembly was compressed to minimize the contact resistance. The reference electrode was positioned such that it is sufficiently close to CNFs for lower solution resistance. The electropolymerization was carried out by holding a constant current density at 6 mA (galvanostatic). After deposition, samples were thoroughly washed with deionized water and dried at 60°C.

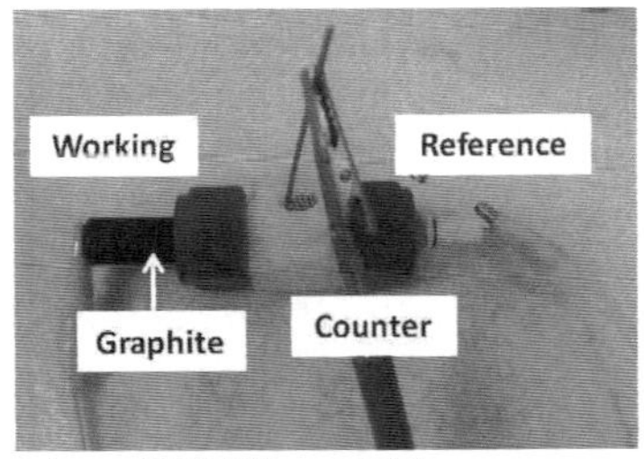

Figure 1. Photo of setup for electrodeposition of PANI on CNFs.

2.3 Electrochemical and structural characterization

The electrode materials are electrochemically tested by using cyclic voltammetry. The capacitance, C, is calculated by using the following equation:

$$C = \frac{1}{2}\frac{Q}{m \times V}$$

where m, V, and υ are mass of composite, voltage window and scan rate. The integral was done by using quick integrate feature of Echem Analyst software. The specific capacitance of polyaniline, C_{PANI} (F g_{PANI}^{-1}), was calculated by the following equation:

$$C_{PANI} = \frac{\frac{1}{2}Q - C_{substrate} \times m_{substrate} \times V}{m_{PANI}\, V}$$

where $C_{substrate}$ and $m_{subtrate}$ are specific capacitance and mass of NP-CNFs or A-CNFs repsectively, and m_{PANI} is mass of PANI. The external morphology of electrospun nanofiber mats was characterized using scanning electron microscopy (SEM, Zeiss Supra 50VP). To characterize internal morphology, nanofibers were embedded in epoxy and microtomed into thin sections. The sections were then placed on lacey carbon copper grids and characterized using transmission electron microscope (TEM).

3. Results and discussion

The morphologies of free-standing carbon nanofibers were observed under SEM (Figure 2). Both NP-CNFs and A-CNFs show smooth surfaces with no visible porous structure. However, the TEM images of cross section of A-CNFs show clear intra-fiber pores. The appearance of porous structure inside A-CNFs was due to KOH which chemically etched carbon during heating at high temperature. The severely activating behaviour of KOH lead to a broad pore size distribution consisting of mesopores and micropores within nanofibers, which in turn resulted in high surface area carbon nanofibers.

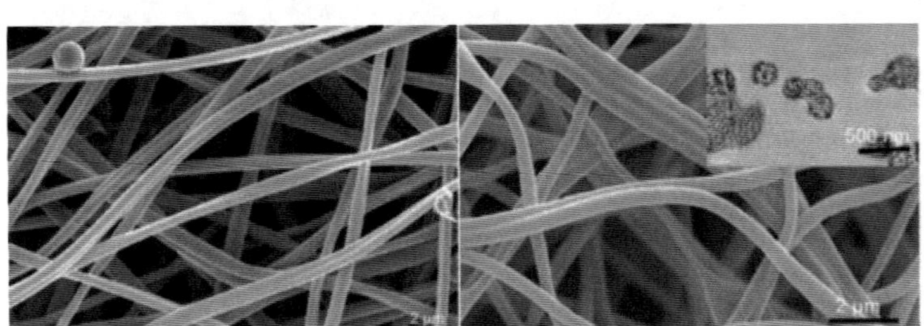

Figure 2. SEM images of the NP-CNFs (left) and A-CNFs (right) used for PANI deposition. The inset is the TEM micrograph of the microtomed section of A-CNFs.

Free-standing carbon nanofibers were tested as electrode materials by cyclic voltammetry (CV) in a two-electrode cell. No binder or conductive filler was added to the free-standing films. As shown in Figure 3, the specific capacitance of A-CNFs was 120 F g^{-1} while P-CNFs showed minor capacitance. CV curve of A-CNFs exhibited a near-ideal rectangular shape indicating an excellent rate handling capability at 100 mV s^{-1}. This was not only due to micro-textures of A-CNFs but also their inherent free-standing property which eliminated the usage of any insulating binder.

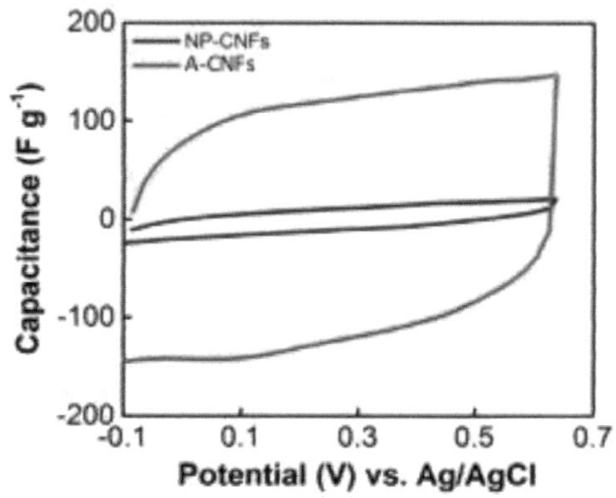

Figure 3. Cyclic voltammetry of NP-CNFs and A-CNFs at 100 mV s^{-1}.

To further increasing the performance of these electrode materials, electropolymerization of PANI was carried out on these substrates. Electropolymerization of PANI depends on many factors: i) concentration of aniline, ii) concentration of electrolytes, iii) nature of the technique: potentiodynamic, potentiostatic or galvanostatic. In general, electropolymerization begins with aniline oxidation to form a radical cation which then reacts with a second radical cation to give a dimer while eliminating two protons [10]. The dimer or oligomer will then oxidize and react with another radical cation to build up the polymer chain.

Our initial investigation of electropolymerization using potential sweep technique between -0.1 and 1.0 V at scan rate of 20 mV s^{-1} in a solution containing 0.1 M aniline and 1 M H_2SO_4 resulted in formation of degraded products instead of polyaniline. These degraded products were easily washed away with DI water indicating poor adhesion of the deposited materials. We increased aniline concentration to 0.5 M and reduced potential sweep between -0.1 and 0.9 V to minimize the formation of degraded products. Figure 4 shows SEM image of PANI deposition on NP-CNFs after 10 cycles of potential sweep. Non-uniform and non-conformal coating of PANI on outer surface of NP-CNFs can be easily seen. This result is similar to previous works on other carbon-based substrates [11-12]. The reason was probably due to the use of potentiodynamic technique which often associated with high current density and fast growth of polymer. These in turn affected the morphology of the deposited PANI.

Figure 4. SEM after depositing PANI on NP-CNFs using potential sweeping at 20 mV s^{-1} in 0.5 M aniline.

We switched to galvanostatic method by keeping the current density constant at 6 mA for 900 seconds for electropolymerization. Figure 5 shows a uniform and conformal coating of PANI on NP-CNFs. The SEM image of cross-section of PANI coated NP-CNFs revealed an approximated thickness of PANI around 100 nm. The deposited PANI on NP-CNFs was washed and then studied the electrochemical behaviour in 1 M H_2SO_4. Figure 6 shows the CV curve of PANI on NP-CNFs. There is a high intensity of redox peak at 0.04 V and 0.2 V which corresponds to conversion between leucoemeraldine and emeraldine. The specific capacitance of PANI was approximately 500 F g_{PANI}^{-1} with an assumption of negligible capacitive contribution of NP-CNFs. This value is comparable with several earlier reports [13-15]. However, the specific capacitance of the whole composite was reduced to 205 F g^{-1} if we also counted the mass of NP-CNFs. This was understandable since NP-CNFs contributed a negligible amount of capacitance to the composite. In order to achieve higher capacitance of whole composite, we electropolymerized PANI on A-CNFs with the aim of integrating double layer capacitance of A-CNFs and pseudocapacitance of PANI.

Figure 5. SEM after depositing PANI on NP-CNFs by keeping current density at 6 mA for 900 seconds.

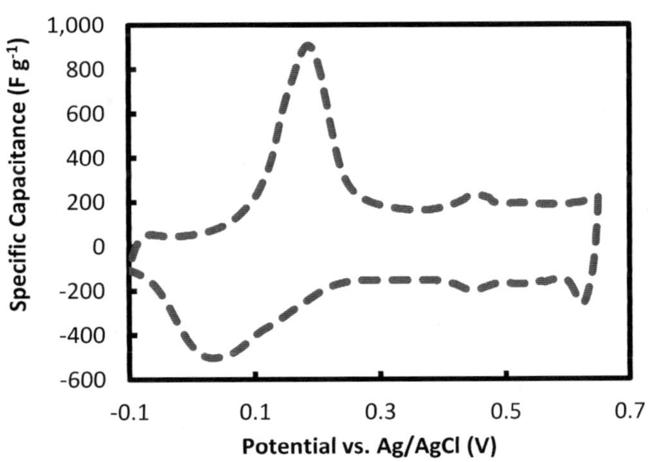

Figure 6. CV curve at 100 mV s^{-1} after depositing PANI on NP-CNFs by keeping current density at 6 mA for 900 seconds.

Figure 7 shows the CV curve after depositing PANI on A-CNFs. The specific capacitance for the whole composite increased to 290 F g_{PANI}^{-1}. At a similar loading of PANI, the PANI/A-CNFs composite capacitance is much higher than that of PANI/NP-CNFs sample. This is due to the higher capacitance contribution of A-CNFs (120 F g^{-1}) compared to NP-CNFs (negligible). This work demonstrated the fabrication of a true hybrid device integrating electric double layer and pseudo capacitance. Cheng et al. [16] deposited PANI on etched carbon fiber cloth by combining two different techniques: potentiostatic for initial nucleation and galvanostatic for a uniform deposition. They obtained a PANI nanowire coating, which showed a specific capacitance of 673 F g^{-1}. However, the specific capacitance of the whole composite was only 210 F g^{-1} due to low capacitance contribution from the etched carbon fiber cloth. We achieved a composite capacitance of 290 F g^{-1}. Yang et al. [17] reported vein-like nanostructured PANI on activated CNFs using potentiodynamic approach. The CV curve is completely distorted at a scan rate of only 10 mV s^{-1} indicating high resistance possibly due to larger PANI particle size. Their capacitance values decreased quickly with increasing scan rates to 70 mV s^{-1}.

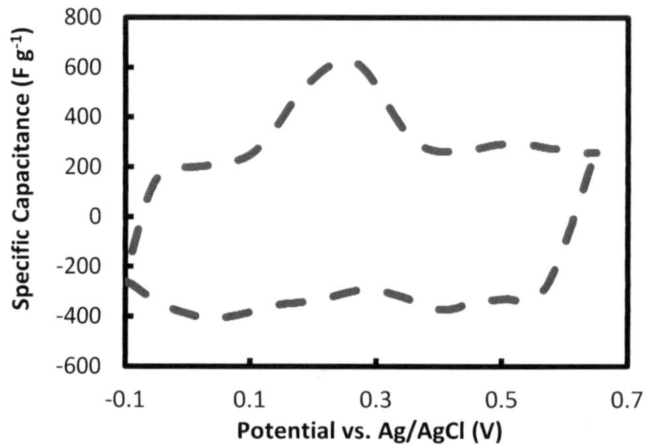

Figure 7. CV curve at 100 mV s^{-1} after depositing PANI on A-CNFs.

4. Conclusion

We studied the deposition of pseudocapacitive polyaniline on freestanding carbon nanofiber mats. By using galvanostatic electropolymerization method, we achieved a uniform and thin conformal coating of polyaniline on both NP-CNFs and A-CNFs. At the same loading of PANI, A-CNFs exhibited much higher capacitance than NP-CNFs due to the contribution from double layer capacitance of A-CNFs. The combination of two different storage mechanisms resulted in excellent specific capacitance of 290 F g^{-1} at 100 mV s^{-1}. The thin conformal coating of PANI reduces the diffusion length of ions during doping/dedoping process of polyaniline for stability and high energy density at high power rate. This is evident from good capacitance retention even at a high scan rate of 100 mV s^{-1}.

References

[1] G.A. Snook, P. Kao, A.S. Best 2011 *Journal of Power Sources* 196 1.
[2] F. Fusalba, P. Gouerec, D. Villers, D. Belanger 2001 *Journal of Electrochemical Socciety* 148 A1.
[3] J. Desilvestro, W. Scheifele, O. Haas 1992 *Journal of Electrochemical Society* 139 2727.
[4] W. Lu 2002 *Science* 297 983.
[5] J. Wei, J. Zhang, Y. Liu, G. Xu, Z. Chen, Q. Xu 2013 *RSC Advances* 3 3957.
[6] Y.G. Wang, H.Q. Li, Y.Y. Xia 2006 *Adv. Mater.* 18 2619.
[7] L.Z. Fan, Y.S. Hu, J. Maier, P. Adelhelm, B. Smarsly, M. Antonietti 2007 *Advanced Functional Materials* 17 3083.
[8] Y. Li, X. Zhao, P. Yu, Q. Zhang 2013 *Langmuir* 29 493.
[9] S. He, X. Hu, S. Chen, H. Hu, M. Hanif, H. Hou 2012 *Journal of Materials Chemistry* 22 5114.
[10] A. A. Syed, M.K. Dinesan 1991 *Talanta* 38 815.
[11] V. Gupta, N. Miura 2006 *Journal of Power Sources* 157 616.
[12] H.-F. Jiang, X.-X. Liu 2010 *Electrochimica Acta* 55 7175.
[13] D. Dhawale, A. Vinu, C. Lokhande 2011 *Electrochimica Acta* 56 9482.
[14] H. Khdary, M.E. Abdesalam, G. EL Enany 2014 *Journal of Electrochemical Socciety* 161 G63.
[15] H.-P. Cong, X.-C. Ren, P. Wang, S.-H. Yu 2013 *Energy and Enviromental Science* 6 1185.
[16] D. He, C. Zeng, C. Xu, N. Cheng, H. Li, S. Mu, M. Pan 2011 *Langmuir* 27 5582.
[17] J.E. Yang, I. Jang, M. Kim, S.H. Baeck, S. Hwang, S.E. Shim 2013 *Electrochimica Acta* 111 136.

IOP Conf. Series: Journal of Physics: Conf. Series **1082** (2018) 012082 doi:10.1088/1742-6596/1082/1/012082

Effect of Freezing Temperature on The Pore Formation of Beta Tricalcium Phosphate Scaffold

M. Muhammad Iqbal[1], I. Nurul Fatin Nabila[1], and A. Nurazreena[1]*

[1]Niche Biomaterials, School of Materials and Mineral Resources Engineering, Engineering, Campus, Universiti Sains Malaysia, 14300 Nibong Tebal, Pulau Pinang, Malaysia

Email: nurazreena@usm.my

Abstract. Beta-tricalcium phosphate (β-TCP) has been widely used for biomedical application due to its excellent mechanical strength, good tissue compatibility and outstanding chemical stability. There are several methods to fabricate porous scaffolds including solvent casting, phase separation, rapid prototyping and electrospinning methods. In the present study, β-TCP scaffolds were fabricated using freeze drying method. β-TCP slurry was frozen and freeze dried for 24 hours followed by sintering process to form β-TCP scaffolds. The porous β-TCP scaffold were fabricated at different percentages of β-TCP powder used (10 wt. %, 20 wt. % and 30 wt. %) and froze at different freezing temperature (-10°C, -20°C, -40°C and -80°C) in order to study the effect of these parameters on the pore formation of the scaffolds. Scanning electron microscopy (SEM) analysis for 30 wt. % β-TCP scaffold frozen at -10°C revealed acceptable pore size distribution with majority pore size within the range of 102.65 μm to 372.90 μm. Furthermore, this scaffold also possessed highest compressive strength of 0.31 MPa.

1. Introduction

Bone tissue is a living organ which mainly composed of both organic and inorganic component. Generally, bone consists about 10% to 20% of water and approximately about 60% to 70% bone mineral. The remaining essential material existed in living bone is collagen with small trace amounts of proteins and inorganic salts. Typically, living bone is widely recognized organ that capable of forming highly functionalized connective tissue [1]. To ensure the bone receives enough supply of minerals that able to prevent breakage of bone, this connective tissue also serve as mineral reservoir for essential minerals such as calcium and phosphorus. All these functions demand for a balance and healthy bone system [2]. Over the years, there are various type of different clinical alternatives used for treating bone defects including autologous and allogeneic transplantations using autografts and allografts. Despite numerous benefits, the disadvantages of allografts and autografts comprised of possibility of disease transmission such as human immunodeficiency virus (HIV) and Hepatitis, limited supply of donor tissue, complication to the donor site and formation of scar after surgery [3]. Decades ago, a noval approach to bone reconstruction has been widely discovered and developed. Which today known as bone tissue engineering. Among the tissue engineering technique, a three-dimensional (3D) scaffold offer temporary structural support during bone reconstruction for cell infiltration and act as a physical support to lead the cell activity into the targeted tissues or organs seems to be the most promising technique for the treatment of bone defects [4]. In this research, beta Tricalcium phosphate (β-TCP) porous scaffold were prepared via freeze drying technique to investigate the effect of β-TCP powder loading and freezing temperature on the pore formation. β-TCP was chosen as the bone substitute material because β-TCP are known for its ability to bond with living tissues and it has been reported that β-TCP shows excellent biological performance with surrounding host tissues.

2. Experimental study

2.1 Preparation of β-TCP scaffold

The porous bioceramic β-TCP (Sigma Aldrich, Germany) scaffolds were fabricated by freezing of tricalcium phosphate (TCP) slurries. The slurries were first prepared by mixing 5 wt% of gelatine (Halagel, Malaysia) with distilled water. The mixture was stirred using magnetic stirrer to ensure homogenous mixing until the temperature reached 60°C. The slurries were then mixed together with β-TCP powders and 0.25 wt. % of polyvinyl alcohol (PVA) (Merck, Germany) which act as a binder. After the addition of PVA and β-TCP powder, the suspension was homogenously mixed by using mechanical stirrer for 15 minutes at a speed of 300 rpm. Three different slurries were prepared respectively with 10 wt. %, 20 wt. % and 30 wt. % β-TCP powder for each different freezing temperature. Afterwards, the homogeneous slurries were de-aired using a vacuum desiccator for 10 minutes with pressure supplied of 0.01 MPa to remove trapped air bubbles. The mixture was then stirred by using mechanical stirrer at a speed of 200 rpm and the gelatine cross-linking agent, 0.25 wt. % of glutaraldehyde solution (Merck, Germany) was added into the solution and continuously stirred for one minutes. In order to study the effect of freezing temperature on the pore formation of the scaffold, the slurries were casted into cylindrical polyvinyl chloride (PVC) mold and froze in a freezer at different temperature of -10°C, -20°C, -40°C and -80°C for 24 hours. The frozen samples were then freeze dried using LYOQUEST-55 freeze dryer at very low vacuum pressure which was 0.05 mBar and at low temperature of -50°C for 24 hours.

2.2 Sintering of the β-TCP scaffold

After freeze drying process was completed, the porous of β-TCP scaffold was sintered in an air furnace. The sintering process involved two stages, the first stage involved low initial heating rate which was 5°C/min to reach temperature of 500°C. Heating at 500°C was held for 2 hours to ensure complete burn out of polymer substances. The second sintering stage was performed at 1100°C for 6 hours. After 6 hours, the temperature cooled down to room temperature at rate of 3°C/min.

2.3 Characterization and analytical techniques

The starting material and the produced scaffolds were characterized by X-ray diffraction (XRD) analysis. The XRD pattern were recorded with a diffractometer system (X-ray Diffractometer D8 Advance BRUKER-binary V3) using 1.5406Å at 40 kV on all phase of sintered β-TCP powders. Meanwhile, the diffraction angles (2°) were set starting from 10° to 90°.

Morphology of the scaffold were observed using Scanning Electron Microscope (SEM) (Hitachi TM3000, Japan). As these specimens are non-conductive, a conductive layer coating was coated on each specimen prior to observation.

Universal Testing Machine (Series 5982, Instron, USA) was used to determine the mechanical properties of HAp scaffold incorporated with gelatine using 5 kN load cell with a cross head speed of 1 mm per min.

Porosity of the sintered β-TCP scaffolds were measured by using Archimedes' principle according to the Equation 1.

$$Porosity\ (\%) = \frac{(Mw\text{-}Md)}{(Mw\text{-}Ms)}\ x\ 100\% \qquad Equation\ 1$$

3. Results and discussion

3.1 X-Ray Diffraction (XRD) Analysis

The standard pattern of XRD was obtained from the Joint Committee on Powder Diffraction Standard (JCPDS) database and β-TCP in accordance with ASTM file number of 09-0169. Figure 1 compares the results obtained from the primary XRD analysis of non-sintered β-TCP powder, β-TCP powders sintered at 1100°C and reference β-TCP powder. The high and narrow peaks shown in the spectrum of each XRD patterns imply the presence of high crystallinity β-TCP phases. From the comparison of the patterns, it can be seen that all the diffraction peaks positions match well with the reference XRD pattern of β-TCP. Based on this analysis it is confirmed that single phase β–TCP was formed without presence of any secondary phase. Therefore, these results suggest that the crystallographic structure of β-TCP was not affected by the sintering process.

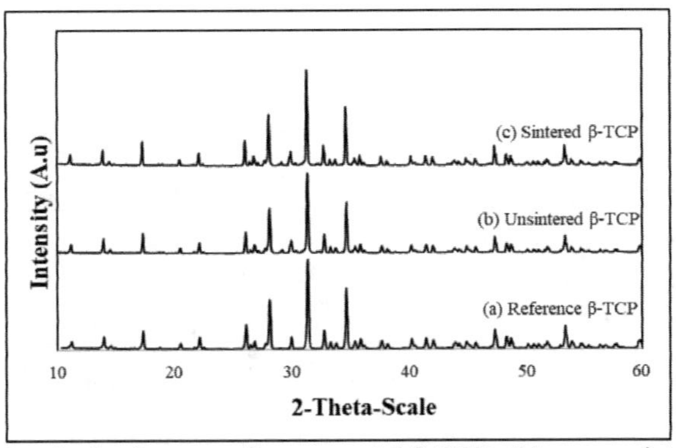

Figure 1. XRD patterns of (a)standard reference of commercial β-TCP; (b) raw unsintered β-TCP; (c) sintered β-TCP 1100°C

3.2 Morphological analysis of β-TCP scaffold

The results of the SEM analysis can be categorized into two parts: 20 wt. % and 30 wt. % of β-TCP powder. β-TCP scaffold with 10 wt. % of β-TCP powder were unable to proceed with the following characterization due to its brittleness. This is because the composition of β-TCP used was minimal and thus resulted in significant reduction in mechanical properties as shown in Figure 2. From the SEM images in Figure 3, the pore size of each scaffolds was calculated and the pore size obtained are shown in Table 1. From Table 1, 20 wt. % of β-TCP powder scaffolds has bigger average pore size than 30 wt. % of β-TCP powder. The pore size of both scaffolds is within the preferably pore size ranges for cell and nutrients transportation in bone tissue engineering (BTE) applications. It can be observed that a significant difference between these two groups of scaffolds whereby for β-TCP scaffold with 30 wt. % of β-TCP powder were having smaller pore when compared to β-TCP scaffold with 20 wt. % of β-TCP powder as shown in Figure 2.

Another important finding is that, from Table 1 it is clearly shows a decreasing trend in pore size from higher freezing temperature (-10°C) to lower freezing temperature (-80°C) regardless the amount of β-TCP powder used. This result may be explained by the fact that lower freezing temperature cause rapid nucleation rate of ice crystal and poor rate of growth throughout freezing stage. These factors contribute to speedy crystallization of solvent into smaller crystal. Subsequently, the small porous sizes were left behind during sublimation of ice crystal during freeze drying [5].

Another possible reason for variation in pore size formation is due to the difference in the fabrication temperature is the difference proximity of void formation during freezing stage and final pore size. As can be seen from Table 1, smaller pores size is obtained at lower freezing temperature (-80°C) and larger pore size is obtained at higher freezing temperature (-10°C). It can also be observed that large void space forms at -80°C and less void space are formed at -10°C as shown in Figure 4.

Figure 2. β-TCP scaffold with 10 wt. % of β-TCP powder

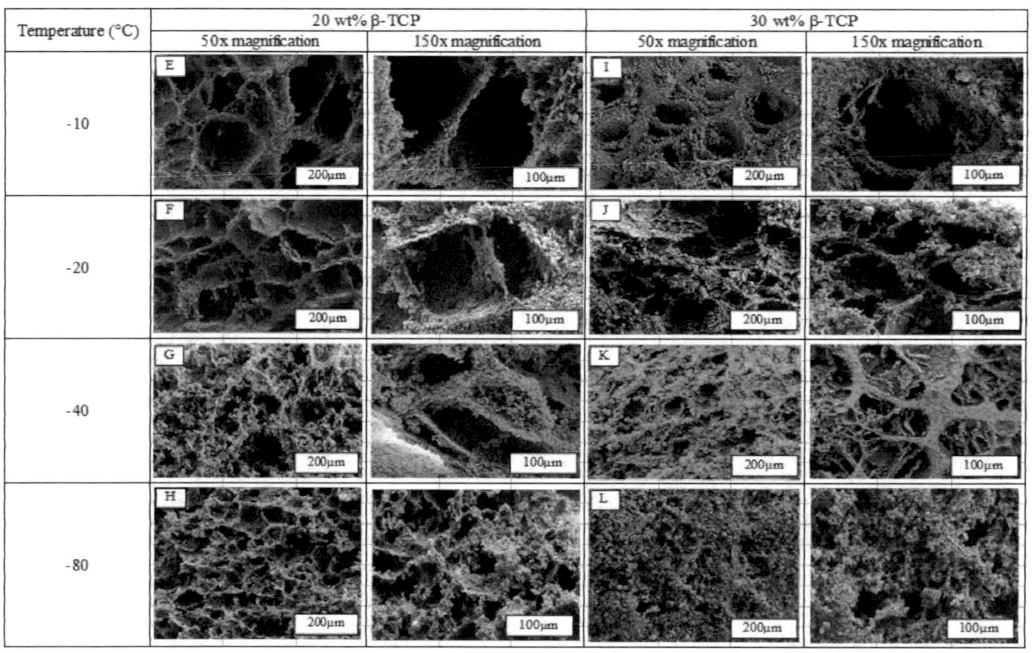

Figure 3. SEM images of β-TCP scaffolds at difference freezing temperature and percentage of β-TCP contents

Table 1. Average pore size of 20 wt. % β-TCP and 30 wt. % β-TCP

Composition of β-TCP (wt. %)	Freezing Temperature (°C)	Pore size (µm)
20	-10	570.76
	-20	343.52
	-40	267.98
	-80	171.97
30	-10	372.90
	-20	210.45
	-40	136.97
	-80	102.65

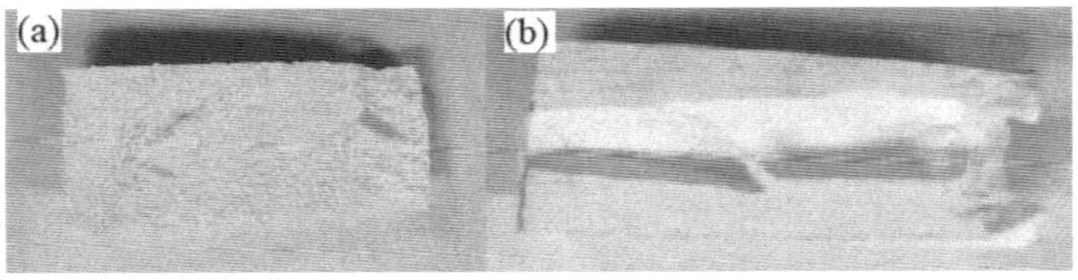

Figure 4. Void formation in scaffold at difference freezing temperature (a) -10°C and (b) -80°C

3.3 Compressive strength of β-TCP scaffolds

From Figure 5 it can be observed that at both β-TCP powder content, similar decreasing trend in the strength of the scaffolds as the freezing temperature decreases. Although it can be noticed that β-TCP scaffolds with 30 wt. % of β-TCP powder has higher compressive strength varying from 0.14 MPa to 0.31 MPa when compared to β-TCP scaffolds with 20 wt. % of β-TCP powder content. The range of compressive strength for β-TCP scaffolds with 20 wt. % of β-TCP powder are from 0.08 MPa to 0.13 MPa. These differences in mechanical strength for both scaffolds may be due to variation in percentage of porosity. The porosity measurement for scaffolds with 20 wt. % of β-TCP powder were 83% while for β-TCP scaffolds with 30 wt. % of β-TCP powder has only 72% porosity as reported in Table 2. It is known that porosity of scaffolds is inversely proportional to the mechanical properties. Therefore, as the number of pores increases, the strength of the scaffold decreases. This finding are consistent with other researchers which found that increased in porosity cause severe reduction in mechanical properties [6].

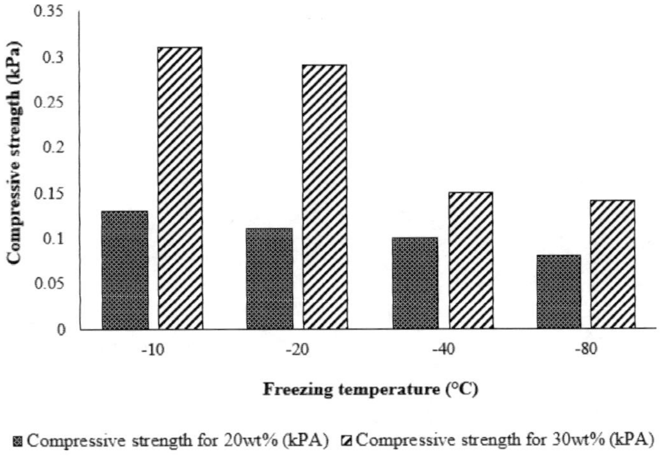

Figure 5. Compressive strength of porous β-TCP scaffolds at different freezing temperature

The observed decreases in compressive strength could be attributed to void formation. This study shows that at lower temperature of -80°C, large central voids are identified. These large voids formation act as stress concentration points that have higher tendency to initiate failure of the scaffolds upon subjected to additional loading [7]. It is therefore likely that such void formations exist within microstructure of scaffold can alter the mechanical properties.

Table 2. Average porosity of β-TCP scaffolds

Composition of β-TCP (wt. %)	Freezing Temperature (°C)	Average porosity (%)
20	-10	82.77
	-20	82.50
	-40	83.45
	-80	83.47
30	-10	71.56
	-20	72.45
	-40	71.86
	-80	73.14

4. Conclusions

Characterization and analysis of 10 wt. % of β-TCP scaffold cannot be conducted since it was too fragile and brittle for handling. In term of pore size of the scaffold and their corresponding effect on both porosity and mechanical properties, SEM analysis performed revealed that smaller average pore size from 30 wt. % of β-TCP powder content is better for bone regeneration compared to 20 wt. % of β-TCP powder content. The SEM results also indicate that the pore size gradually increases as the freezing temperature increased. Compressive strength analysis showed that mechanical properties for 30 wt. % of β-TCP powder content were better than for the 20 wt. % of β-TCP powder content. The compressive strength value increases significantly as freezing temperature decreased. Therefore, scaffold with 30 wt. % of β-TCP powder content fabricated at -10°C freezing temperature yields highest mechanical properties of 0.31 kPa.

5. Acknowledgement

This work was supported by Fundamental Research Grant Scheme (FRGS) No: 203/PBahan/6071302.

References

[1] G. A. Rodan 1992 *Bone* 13 S3
[2] R. Florencio-Silva, G. R. da S. Sasso, E. Sasso-Cerri, M. J. Simoes, and P. S. Cerri 2015 *Biomed Res. Int.* 2015 421746
[3] P. Vang 2006 *Significance* 82
[4] K. Ikeda and S. Takeshita 2014 *J. Bone Metab.* 21 163
[5] L. Jiang, Y. Li, C. Xiong, S. Su, and H. Ding 2017 *ACS Appl. Mater. Interfaces* 9 4890
[6] G. Hannink and J. J. C. Arts 2011 *Injury* 42 S22
[7] A. Zhang, D. Li, H. Lu, and D. Zhang 2011 *Mater. Des.* 32 4803

IOP Conf. Series: Journal of Physics: Conf. Series **1082** (2018) 012083 doi:10.1088/1742-6596/1082/1/012083

Morphology and mechanical properties fabricated from Ti, Nb and HA by powder metallurgy method

Farrah Noor Ahmad[1,2], Zuhailawati Hussain[2]

[1]Faculty of Mechanical Engineering, Universiti Teknologi Mara (UiTM), Penang Campus, Jalan Permatang Pauh, 13500 Permatang Pauh, Penang, Malaysia

[2]School of Materials and Mineral Resources Engineering, Engineering Campus, Universiti Sains Malaysia, 14300 Nibong Tebal, Penang, Malaysia

Email: farra728@ppinang.uitm.edu.my, zuhaila@usm.my

Abstract. Hydroxyapatite (HA) is a bioactive ceramic material with poor mechanical properties which hinders its application in bone implant. However, HA has a similar structure to bone and can promote the growth of natural tissues. Combining the low modulus β-type Titanium (Ti) based alloy (e.g:Ti-40Nb) which has excellent mechanical properties and non-toxic composition with HA is considered as a good candidate to expand its biomedical applications. In this current work, Ti, Ti-40Nb and Ti-40Nb-15HA (wt.%) was compacted under 500 MPa using powder metallurgy process. The mixtures were performed by high energy milling and sintering in a tube furnace with temperature of 1200°C/2 h under argon flow. The aim of this research is to investigate the morphology and the mechanical properties of metal alloys. Images obtained by SEM show a homogeneous morphology of voids with smooth surface for Ti and Ti-40Nb while Ti-40Nb-15HA show the presence of pores located at grain boundary and exhibit rough surface appearance. Ti-40Nb-15HA alloy was found to give a favourable compression strength and low density at 171.93 MPa and 5.086 g/cm^3, respectively. Thus, considering its mechanical properties, Ti-40Nb-15HA was found ideal for orthopaedic implant materials as it has properties closed to human bone.

1. Introduction

Titanium (Ti) and its alloys have been widely used for biomedical implants. However, the mismatch between Ti and the surrounding natural bones make the bone insufficiently loaded and becomes stress shielded resulting in bone resorption and lead to implant failure. Thus, a material with an excellent combination of high strength and low modulus closer to the bone has to be used for implantation to avoid loosening of implants and give higher service period [1]. The ultimate compressive strength, yield strength and Young's modulus of human cortical (dense) bone are ~ 130-180 MPa, 104-121 MPa and 10-30 GPa, respectively [2]. This mechanical properties of Ti alloys can be achieved by selecting an alloy composition with low Young's modulus (e.g:β -type alloy). The elastic modulus of pure Ti is 100

IOP Conf. Series: Journal of Physics: Conf. Series **1082** (2018) 012083 doi:10.1088/1742-6596/1082/1/012083

GPa while Ti-6Al-V alloy (α + β type) which is the most used Ti-based alloy in bone implant has 110 GPa and β-Ti alloys has 55 GPa [3]. Thus, the β-Ti alloys display more compatible elastic behaviour with human bone. In terms of binary Ti-Nb alloys, the stability of β phase is enhanced with the increase of Nb content to 40 wt% which lead to the more stable β-Ti phase, minimizing the precipitation of ω phase and lowering the elastic modulus [4]. Therefore, the composition of β-type Ti-40Nb was chosen for the present work due to its unique combination of mechanical properties, especially its low elastic modulus, non-toxic and excellent corrosion resistance [5]. Among the factors that affect the mechanical properties of bone is its material properties [6]. In this study, our approach is to incorporate Hydroxyapatite (HA) into β-type Ti-Nb to form a new metallic biomaterials. HA has been widely used as a bone replacement material due to its chemical and crystallographic structure being similar to the bone mineral which result in excellent biocompability [7]. Furthermore, it has a good bioactivity and osteoconductivity [8-9]. However, HA cannot be used for load bearing implants due to its low strength and brittle nature [10]. Combining HA with the excellent mechanical properties of low modulus β-type Ti-Nb (non toxic metal) is considered to be a promising approach to fabricate most suitable biomaterial for medical implant applications. This comparative study is about the morphological and mechanical properties of Ti, Ti-40Nb and Ti-40Nb-HA15 (wt.%) which was produced by powder metallurgical approach. The aim of this study is to reduce the mechanical properties (e.g. : compressive strength) of Ti of the implants to the level approaching human bones which may decrease the stress shielding and reduce the failure of implants. Compressive strength is necessary for the development and growth of the bone which must be taken into account in producing new biomaterial. Therefore, in this study, mechanical properties under compressive stress has been mainly considered and investigated.

2. Experimental study

The elemental powder of Ti, Nb and HA was weighed with an electronic balance according to nominal compositions of Ti, Ti-40Nb and Ti-40Nb-15HA (wt.%). The mechanical milling was performed in a high-energy Fritsch Pulveristte P-5 planetary ball mill with stainless steel jar under argon atmosphere. To obtain a complete uniform mixture, the powders were milling for 2 hours at a weight ratio (balls to powder) of 10:1 at a rotation speed of 200 rpm. 3% n-heptane solution was added prior to milling process to avoid excessive cold welding of the elemental alloy powders. The mixture was uniaxially pressed at a pressure of 500 MPa into green compacts. Pellets of approximately 10 mm in diameter were fabricated. Finally, the green compact samples were sintered ina Lenton tube furnace at 1200°C for 2 hours in an argon atmosphere. These samples were mechanically polished using SiC abrasive papers from grade 100 up to 2000 grits, followed by a 0.05 μm diamond finish. The morphology of metal alloys was performed by VPFESEM SUPRA 35VP. Several magnifications weres focused to obtain a good viewing point on the samples. The density of the sintered samples were determined by Archimedes method. Compression test was carried out to investigate the compressive strength of samples which was prepared using Universal Testing Machine (UTM) INSTRON 5982 at a crosshead speed of 1 mm/min at room temperature. Samples were subjected to 100 kN.

3. Results and discussion

3.1 Morphology Observation

It can be seen from Figure 1 that the morphology of Ti, Ti-40Nb and Ti-40Nb-15HA (wt.%) after milling was varies due to its different particle size and properties. The raw materials of the starting materials were broken down and reduced to smaller particles. Ti and Ti-40Nb powder appears as an irregular particle shape with different size. With the addition of HA content, the morphology of Ti-40Nb-15HA was changed and appeared as flower-like particles. The presence of small agglomerates of HA particles can also be observed (label at figure).

The thermal treatment at 1200°C leads to a sintering process of the mixture. Figure 2 shows SEM micrographs of the sintered Ti, Ti-40Nb and Ti-40Nb-15HA (wt.%) at 1200°C. After sintering, porosity was observed for Ti sample while for Ti-40Nb sample, the interconnected porosity with irregular shapes and various sizes was seen. Furthermore, there was a lack of fusion regions observed in Ti-40Nb which caused by its fusion temperature. The melting temperature of Ti and Nb was 1668°C and 2477°C, respectively. The energy needed to melt the metal powder increased, especially with Nb content increased. Due to incomplete melting of the metal powders, this lead to a lack of fusion of the sample which produce porosity. The pores also may result from entrapment of gas during the melting process [11]. When adding HA (e.g: 15 wt.%) into Ti-Nb40, the agglomeration phenomenon of HA powders occurred. This is due to the high surface activity and this phenomenon is more obvious, especially with the increase of HA content [12]. For Ti-40Nb-15HA, the agglomeration of the particle create porous structure. Different heights of grain, which exhibit high irregularity of sample surface was observed. The implant surface should be rough to enhance load bearing capacity and osseointegration which could allow for bone ingrowth [13].

Figure 1. SEM images of the particle morphology obtained after 2 hours of milling: Ti (Figure 1a), Ti-40Nb (Figure 1b) and Ti-40Nb-15HA (Figure 1c).

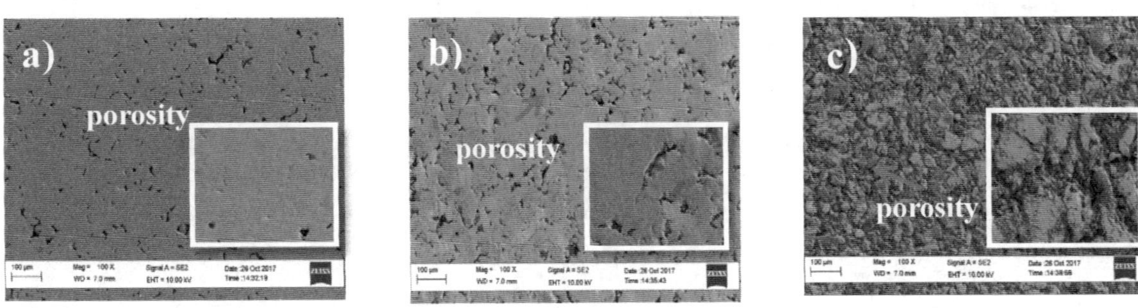

Figure 2. SEM images of metal alloys sintered at 1200°C with different magnification: Ti (Figure 2a), Ti-40Nb (Figure 2b) and Ti-40Nb-15HA (Figure 2c).

3.2 Density Measurement

As seen in Table 1, the density of Ti increase with the addition of Nb. The density of Nb is high which is around 8.6 g/cm^3. Therefore, the addition of Nb into Ti-Nb40 give a significant increase to the density value. However, the addition of HA into Ti-40Nb-15HA which has a density of 3.156 g/cm^3 has significantly decreased the density of metal alloy due to its low density. The bulk density achieved for the samples are considered light and close to natural bones (1.8-2.1 g/cm^3), compared to other similar metallic biomaterials such as stainless steel (8 g/cm^3) or cobalt-based alloys (8.5 g/cm^3).

Table 1. Theoretical density (g/cm^3) and bulk density (g/cm^3) of Ti, Ti-40Nb and Ti-40Nb-15HA

	Theoretical Density (g/cm^3)	Bulk Density (g/cm^3)
Ti	4.510	4.265
Ti-40Nb	5.569	5.281
Ti-40Nb-15HA	5.160	5.086

3.3 Compressive Strength

The mechanical behaviour of metal alloys was studied by compression tests using cylindrical samples are shown in Table 2. It can be observed that the strength of Ti increase when added 40% Nb due to a solid solution hardening effect. This is consistent with observations by other author working with Ti-Nb alloys [14]. Ti-40Nb-15HA give the lowest strength compared to other sample. When the HA incorporated into Ti-40Nb, the strength of the sample decline drastically which related to the brittleness and agglomeration of HA. Besides, microcracks and pores on Ti-40Nb-15HA also affect its strength and fracture pattern. The lower compressive strength can supported by the fracture surface. The microstructure appears to be less dense and brittle in Ti-40Nb-15HA as compared to Ti and Ti-40Nb, which show ductile behaviour. This is shown on the morphology of fracture surface in Figure 3.

Table 2. Compressive strength of Ti, Ti-40Nb and Ti-40Nb-15HA

	Compressive Strength (MPa)
Ti	1186.66
Ti-40Nb	1353.00
Ti-40Nb-15HA	171.93

Figure 3. Fracture morphology of the samples after compression test: Ti (Figure 3a), Ti-40Nb (Figure 3b) and Ti-40Nb-15HA (Figure 3c).

IOP Conf. Series: Journal of Physics: Conf. Series **1082** (2018) 012083 doi:10.1088/1742-6596/1082/1/012083

4. Conclusion

In this paper, the metallic biomaterials were synthesized by mechanical milling of powder consisting of Ti, Nb and HA forming Ti, Ti-40Nb and Ti-40Nb-15HA. Then, the samples were fabricated by powder metallurgy method. The effect of Nb and HA content on microstructure and mechanical property were investigated. The main conclusions can be summarized as follows.

i. Ti and Ti-40Nb show a smooth surface morphology and ductile behaviour while Ti-Nb-HA15 exhibited rough surface and brittle behaviour.

ii. With the addition of Nb content, the strength of Ti-Nb increase due to solid solution strengthening effect. HA is brittle and this result in a reduction of mechanical strength for Ti-40Nb-15HA. Besides, excessive HA content would also reduce the mechanical properties even further.

Ti-Nb-HA15 not only has a low density of metal alloy but also display better results in terms of its mechanical properties. By establishing a balance between strength and elasticity with human bone, a successful performance of a load-bearing bone implant can be achieved.

References

[1] Mitsuo Niinomi 1998 *Materials Science and Engineering : A* 243 231
[2] Havaldar R, Pilli S C and Putti B B 2014 *Advanced Biomedical Research* 3 101
[3] Li Y, Yang C, Zhao H, Qu S, Li X and Li Y 2014 *Materials* 7 1709-1800
[4] Welsch G, Boyer R and Collings E W 1993 *ASM International*, Ohio 1169
[5] Mitsuo Niinomi 2016 *Regenerative Biomaterials* 3 173
[6] Vikas Tomar 2009 *Journal of Engineering Fracture Mechanics* 76 884
[7] Zreiqat H, Valenzuela S M, Nissan B B, Roest R, Knabe C, Rradlanski R J, Renz H and Evans PJ 2005 *Biomaterials* 26 7579
[8] Woodard J A Hilldore S L, Park C J, Morgan A W, Eurell J C, Clark S G, Wheeler M B, Jamison R, and Wagoner Johnson A J 2007 *Biomaterials* 28 45
[9] Saber-Samandari S, Ghonjizade-Samani F, Aghazadeh J, and Sadeghi A 2016 *Ceramics International* 42 11055
[10] Prokopiev O and Sevostianov I 2006 *Materials Science and Engineering* A 431 218
[11] Kasperovich G, Haubrich J, Gussone J and Requena G 2016 *Materials and Design* 105 160
[12] Han C, Wang Q, Song B, Li W, Wei Q, Wen S, Liu J and Shi Y 2017 *Journal of the Mechanical Behavior of Biomedical Materials* 71 85
[13] Halldin A, Ander M, Jacobsson M and Hansson S 2015 *BioMedical Engineering OnLine* 1-13
[14] Wang Q, Han C, Choma T, Wei Q, Yan C, Song B and Shi Y 2017 *Materials & Design* 126 268–277

Elemental distribution of the greenish-Thai decorative glass

K. won-in[1,*], P. Dararutana[2]

[1]Department of Earth Sciences, Faculty of Science, Kasetsart University, Bangkok 10900 Thailand

[2]Royal Thai Army Chemical Department, Bangkok 10900 Thailand

Email: kritwonin@yahoo.com

Abstract. Glasses have been used in ornaments and decorations in Thailand for several hundred years, especially various colors of decorative glass which called Ancient Thai Glass (ATG) that made by Thais were found as architecture components of historical materials. Nowadays, the ATG has been almost broken and no products. Restoration with glass for conservation can be found with the modern imported glass (MIG) which bought from the market that brought from the foreign countries. However, it was shown that the colors of ATG were unchanged for hundreds of years, while as those of MIG were mostly changed within ten years. In this work, scanning electron microscope coupled with energy dispersive X-ray fluorescence spectrometer (SEM-EDX) was used to characterize the structure and composition of the greenish-ATG compared with those of MIG. It was found that the glass substrate of the ATG was lead-based glass, while as that of the MIG was soda-based one. It was revealed that the green coloration came from the Cu, Fe and Cr ions that doped into the glass matrix that confirmed with either the EDX or line scan. The results from this work are used as information that understanding the characteristics and how their production. X-ray fluorescence spectroscopy can be applied successfully in the historical glass study. It can be concluded that this technique can be applied successfully in the field of archaeology.

1. Introduction

Glasses have been used in ornaments and decorations in Thailand for several hundred years as seen by archaeological evidences, such as, decorative glass and glass beads discovered in regions throughout the country. Decorative glasses called Ancient Thai Glass (ATG) can generally be found as architectural components in places, such as, ancient palaces and Buddhist temples which have been center points among Thai commoners for centuries. Various colors of glass ranging from transparent to blue, green, red, and amber have been found. Under the tropical sun, it gives out an artistic harmony of flamboyance and serenity [1-4]. Figure 1 show an artifact decorated with 0.5-2.0 mm sheets of glasses coated with a thick reflecting metal film similar to an ancient glass mirror. As the glasses have kept deteriorating over time, restorations must be performed routinely to preserve these cultural crafts for later generations to appreciate. The most recent restoration work was done about 150 years ago.

However, presently there are no glass makers producing this type of glasses. At that time, they used conventional method which the melting was processed by adding small amount of colorant into the glass mixture. It was known that they were more shade of color from slightly to dark green. They may be produced at difference factories and the use of chemically raw materials. However, it was possible that they were made in the same one, slight compositional differences between glass samples may be a result of variables in the glass production or the use of chemically similar raw materials with different batches [5-6]. Only modern, imported, difficult-to-cut glasses, called modern imported glass (MIG) with greater thickness and shorter lifetime are available in the market.

The previous studied using scanning electron microscopy and Raman spectroscopy confirmed that the ATG was lead-based glass and MIG was soda-based glass. It was also because of the presence of lead in the composition that made G-ATG samples color unchanged [4-7].

X-ray spectroscopy that is one of non-destructive techniques is widely used to study in archaeological science. Scanning electron microscope coupled with energy dispersive X-ray fluorescence spectrometer (SEM-EDX) is well-known as an appropriate technique for investigation of archaeological or historical objects. Scanning electron microscope (SEM) was used to analyze a line scan on cross-sections of samples. The elemental composition of the sample was characterized using energy dispersive X-ray fluorescence spectrometer (EDX) [8-10].

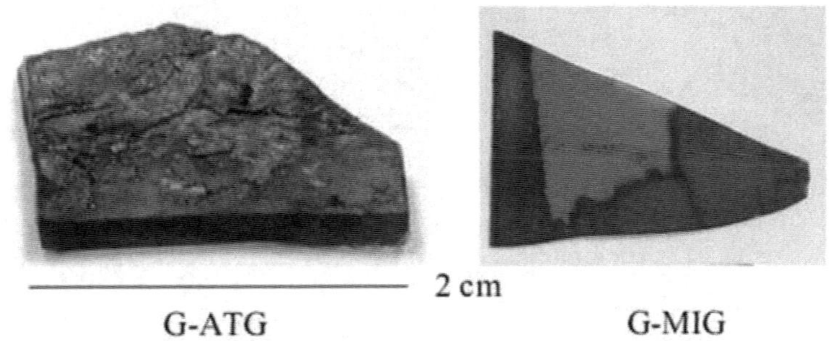

Figure 1. The greenish-Thai decorative glass samples.

2. Experimental study

In this work, two greenish-Thai decorative glass samples namely G-ATG and G-MIG were used to analyze. They were collected to study because the green colored glass must be used for conservation and restoration the ancient cultural artifacts. The shade of color demanded was slightly green that could not buy in the market. Its property was less studied while other types of coloring glass were already studied [4-7 and 11-13].

The ATG sample was collected from the Bureau of the Royal Household at the Grand Palace (Bangkok), while the MIG one was bought from the market. They were coated with a thin layer of graphite to prevent charging of the sample surface prior to the scanning electron microscope observations.

The cross-sections were investigated using a Jeol JSM-6335F field emission scanning electron microscope (FESEM), operated at 15 kV. The compositional elements of these samples were analyzed using the Oxford Instrument INCAx-sight energy dispersive X-ray fluorescence spectrometer (EDX), operated with Si(Li) detector. Line scans were carried out with 20 runs each sample and at a live time of 100s for each run.

3. Results and discussion

Figure 2 showed the EDX line scan and concentration profiles in the samples. The scan over a line resulted information about the element relative distribution on the selected line. It was found that the glass substrate of the G-ATG sample was lead-based glass, while as that of the G-MIG one was soda-based glass. The line scan showed the compositions of the metal backing of G-ATG was contained with majority of Pb and minority of Al, Fe, Cu, Cr and Mn, while as that of G-MIG was Mg, Ti, Fe, Cr, Al and Cu. It was also showed that the green color was affected from Cu, Fe and Cr. The line scan indicated the same metallic ions. However, the previous and advanced studied found that the colors of the ATG was colorant melted together with glass, but those of the MIG was coloring mixture coated.

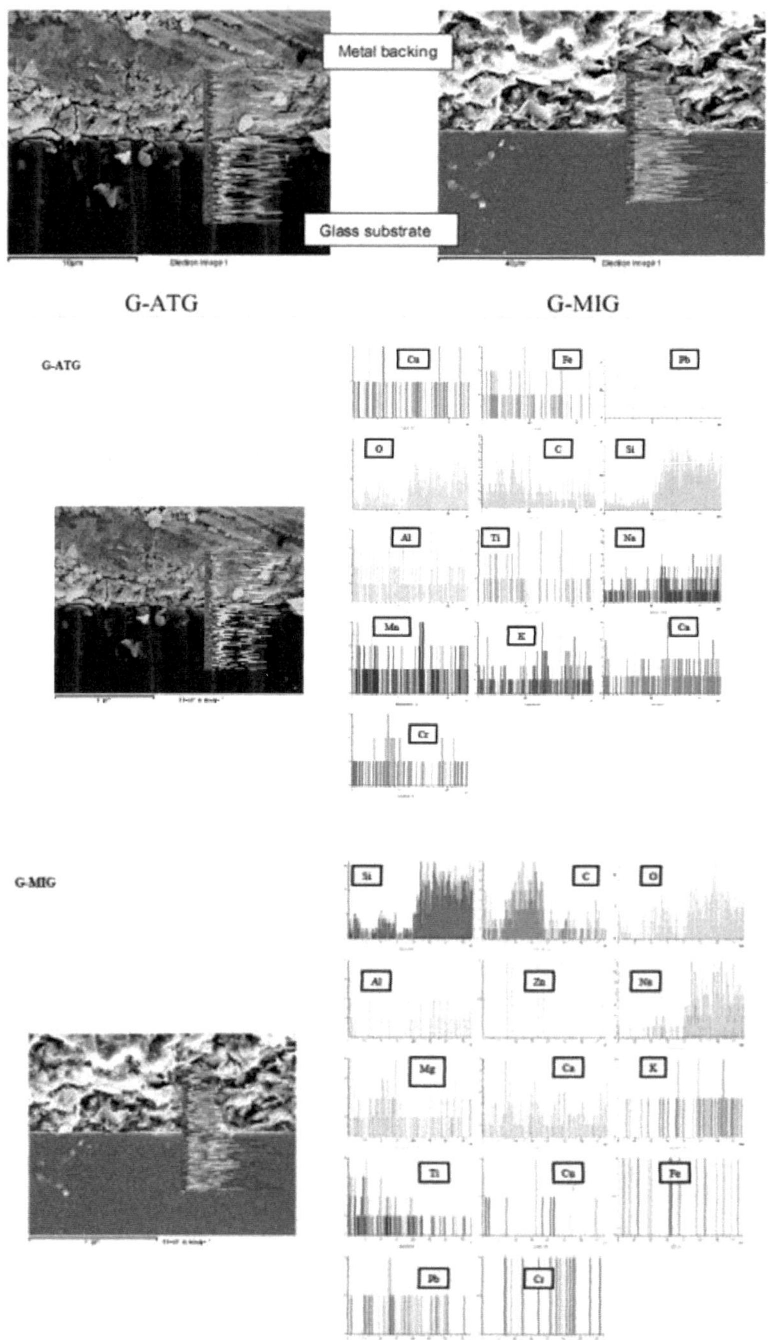

Figure 2. The line scans of the greenish-Thai decorative glass samples.

4. Conclusion

Scanning electron microscopy and X-ray fluorescence spectroscopy that are non-destructive methods were suitable to investigate the surface coating of Thai decorative glasses. Obtained images showed the morphological structure, while EDX analysis allowed details on compositions of both the glass substrate and the metal backing.
It was revealed that the green coloration came from the Cu, Fe and Cr ions that doped into the glass matrix. It was confirmed either the EDX or line scan.
It can be concluded that these equipment can be applied successfully in the field of archaeology

Acknowledgments

The Bureau of the Royal Household is thanked for the samples supporting. The Faculty of Science at Chiang Mai University is also thanked for providing SEM-EDX apparatus for analysis. The Department of Earth Sciences at Faculty of Science of Kasetsart University (Bangkok) provided team expertise.

References

[1] Chotichutrakul K 1982 *Thai Art and Culture* 6
[2] Punjabhan N and Na Nakhonphanom S 1992 *The Art of Thai Wood Carving: Sukhothai, Ayutthaya, Ratanakosin, Rerngrom Publishing Co.Ltd., Bangkok*
[3] Punjabhan N, Wichienkeeo A and Na Nakhonphanom S 1994 *The Charm of Lanna Wood Carving, Rerngrom Publishing Co.Ltd., Bangkok*
[4] Dararutana P and Sirikulrat N 2007 *Modern Research and Educational Topics in Microscopy*
[5] Won-in K, et.al. 2011 *Advanced Materials Research* 324 501
[6] Won-in K et.al. 2011 *Spectrochimica Acta Part A* 83 231
[7] Dararutana P 2016 *Journal of Microscopy Society of Thailand* 29 3
[8] Mirti P 2000 *X-ray Spectrometry* 29 63
[9] Jembrih-Simburger D, Neelmeijer C, Schalm O, Fredrickx P, Schreiner M, De Vis K, Mader M, Schryvers D and Caen J 2002 *JAAS* 17 321
[10] Sakurada T, Hashimoto S, Tsuchiya Y, Tachibana S, Suzuki M and Shimizu K 2005 *Journal of Surface Analysis* 12 118
[11] Dararutana P et.al. 2008 *Advanced Materials Research* 55 601
[12] Won-in K, et.al. 2011 *PIXE Applications on Archaeological Materials in Thailand, Thai Journal of Physics* 7
[13] Dararutana P et.al. 2012 *Current Microscopy Contributions to Advances in Science and Technology*

IOP Conf. Series: Journal of Physics: Conf. Series **1082** (2018) 012085 doi:10.1088/1742-6596/1082/1/012085

Photocatalytic activity of ZnO nanodisks in degradation of Rhodamine B and Bromocresol Green under UV light exposure

Yuet Lee Ying[1], Swee Yong Pung[1*], Ming Thong Ong[2] and Yuh Fen Pung[3]

[1]School of Materials and Mineral Resources Engineering, Engineering Campus, Universiti Sains Malaysia, 14300 Nibong Tebal, Pulau Pinang, Malaysia

[2]Molecular Medicine Research (INFORMM), Universiti Sains Malaysia, Jalan Inovasi, 11800 Gelugor, Pulau Pinang, Malaysia

[3]Department of Biomedical Sciences, University of Nottingham (Malaysia Campus), Jalan Broga, 43500 Semenyih, Selangor Darul Ehsan, Malaysia

E-mail*: sypung@usm.my

Abstract. Semiconductor-based photocatalyst is an alternative approach for wastewater treatment. The effectiveness of semiconductor nanoparticles to eliminate textile dye substances in the contaminant water has been widely studied in the last decade. This work discusses the photocatalytic activities of ZnO nanodisks by using different dye substance (rhodamine B, RhB and bromocresol green, BCG) under UV exposure. ZnO nanodisks were successfully synthesized through solution precipitation method. The formation of ZnO in hexagonal nanodisks structure was confirmed by XRD, FESEM and EDX analysis. The result showed that the photodegradation efficiency of BCG dye is higher than RhB dye under 75 min of UV irradiation. This could be explained by the dye molecular structure as BCG dyes containing more hydroxyl groups (-OH) than RhB dyes.

1. Introduction

The waste water containing organic compounds has raised serious issue on human health and affected the ecological system [1]. Particularly, organic dyes in water effluent of textile industry are harmful to environment even though present in low concentration [2]. If this water effluent from textile industry is discharged into river without treatment, it will cause water pollutant. The emerging waste water treatment technique that based on semiconductor photocatalysts provides alternative route to photodegrade the organic pollutants [3]. In fact, the effectiveness of using these semiconductor nanoparticles has been widely studied due to its simplicity; low operation cost and decomposes the organic dye into less harmful by-product such as water and carbon dioxide [4].

Metal oxide, such as zinc oxide (ZnO), is well known as multifunctional II-IV compound semiconductor with high photocatalytic activity [5]. The unique properties of ZnO such as large binding energy (60 meV) wide band gap (3.37 eV) at room temperature [6-7], making it a potential material for various applications e.g. LEDs [8], light emitters [9] and lasers [10]. In addition, the good antibacterial activity [11] and photocatalytic performance in organic compounds removal [12] are useful in waste water treatment plants, textile and paint industry [13]. It is also noted that the physical properties of ZnO nanostructures could be varied by tailoring their morphology into nanoflowers [14], nanoneedles [15], nanowires [16], nanorods [17] and nanodisks [18].

In this study, ZnO nanodisks (NDs) were synthesized by using zinc acetate dehydrate (Zn $(CH_3COO)_2{\bullet}2H_2O$ and ammonium hydroxide (NH_4OH) as precursors. Whereas, aluminium sulfate ($Al_2(SO_4)_3$) was used as complexing agent to customize as nanodisks shape. The particles were synthesized via solution precipitation method as it is one of the most commonly used techniques [19]. The advantages of solution precipitation methods are low raw materials cost, low synthesis temperature [20] and the composition of precursors could be adjusted easily [21].

2. Material and Methods

The synthesis of ZnO NDs is described in details in reference [22]. In a typical synthesis, an equal molar of precursors were prepared in two separate beakers, i.e. 6.07 mM of zinc acetate dehydrate (Zn $(CH_3COO)_2{\bullet}2H_2O$) was dissolved in 300 ml distilled water and 6.07 mM of aluminium sulphate ($Al_2(SO_4)_3$) was dissolved in 30 ml distilled water. The solutions were stirred separately while heating at 60 °C for 2 h in water bath. Subsequently, 2 ml of ammonium hydroxide (NH_4OH) was added drop by drop into the zinc acetate dehydrate solution. Then, aluminium sulphate solution was added and continuously stirrer for 1 h by maintaining the temperature at 60 °C. The solution was left overnight to wait for the nanoparticles to be formed and settled down. The precipitate was washed several times by deionized water and ethanol. Finally, the precipitate was dried at 100 °C for 8 h.

The sample was characterized by X-Ray diffraction technique (XRD, Bruker Advanced X-ray Solution D8 Diffractometer (Cu K_α, $\lambda = 0.154$ nm)) to determine the type of crystal structure and phases; field-emission scanning electron microscope (FESEM, Zeiss Supra 35VP) to reveal its morphology and energy dispersive X-Ray spectroscope (EDX) to identify the elements of the sample.

In photocatalytic study, $1\text{x}10^{-5}$ M of dye aqueous solutions of rhodamine B (RhB) and bromocresol green (BCG), were prepared. A fixed amount of nanoparticles, i.e. 0.10 g, was added into 250 ml of dye aqueous solution and continues to stir in dark condition for 30 min. The mixed solution was irradiated under UV light ($\lambda = 254$ nm). Subsequently, a 2.5 ml of photodegraded dye solution was sampled at every 15 min interval (0, 15, 30, 45, 60 and 75 min). The absorbance of photodegraded dye solutions in the range of 400-700 nm was recorded using UV-Vis spectrophotometer (Cray 50). By measuring the characteristic absorbance peak of organic dyes, i.e. 553 nm for RhB and 620 nm for BCG, the photodegradation efficiency (%) of ZnO NDs was calculated by $\left(1 - \frac{A_t}{A_0}\right) x\ 100\ \%$, where A_t = initial absorbance and A_0 = absorbance at time 't'.

3. Results and Discussions

Figure 1 (a) shows the XRD patterns of as-synthesized particles. The characteristic diffraction peaks at 2θ = 31.7°, 34.4°, 36.2°, 47.5°, 56.5°, 62.8°, 66.3°, 67.9°, 69.0°, 72.5° and 76.9° are corresponding to the crystal planes of (100), (002), (101), (102), (110), (103), (200), (112), (210), (004) and (202) respectively. All diffraction peaks can be assigned to the wurtzite structure of hexagonal ZnO by comparing with ICDD reference code database (98-008-1294), with lattice constants of a = b = 3.25 Å, c = 5.207 Å, $\alpha = \beta = 90°$, $\gamma = 120°$. Additional peaks at 38.8°, 46.4°, 52.6°, 59.8°, 61.2° and 73.85° are assigned to Al_2O_3, which could be contributed from the surface directing agent, aluminium sulphate (98-006-2983 and 98-006-6546) as labelled in Figure 1(a). The strong and sharp diffraction peaks indicate that the particles were highly crystallized.

Figure 1(b) reveals that the particles consist of large ZnO NDs and some small flower-like ZnO particles. The ZnO hexagonal disk is highlighted in yellow dotted line. The ZnO NDs have average diameter of 12.301 ± 1.050 μm and thickness of 0.067 ± 0.020 μm. The elemental analysis of ZnO NDs was performed using EDX. As shown in Figure 1(c), the particles composed of Zn, O and Al elements. The atomic percentages of Zn, O and Al are 46.69%, 47.6% and 5.71%, respectively. The small atomic percentage of Al was come from the surface directing agent, $Al_2 (SO_4)_3$. The stoichiometric ratio of Zn:O of the particles is closed to the theoretical value, i.e. ~1:1.

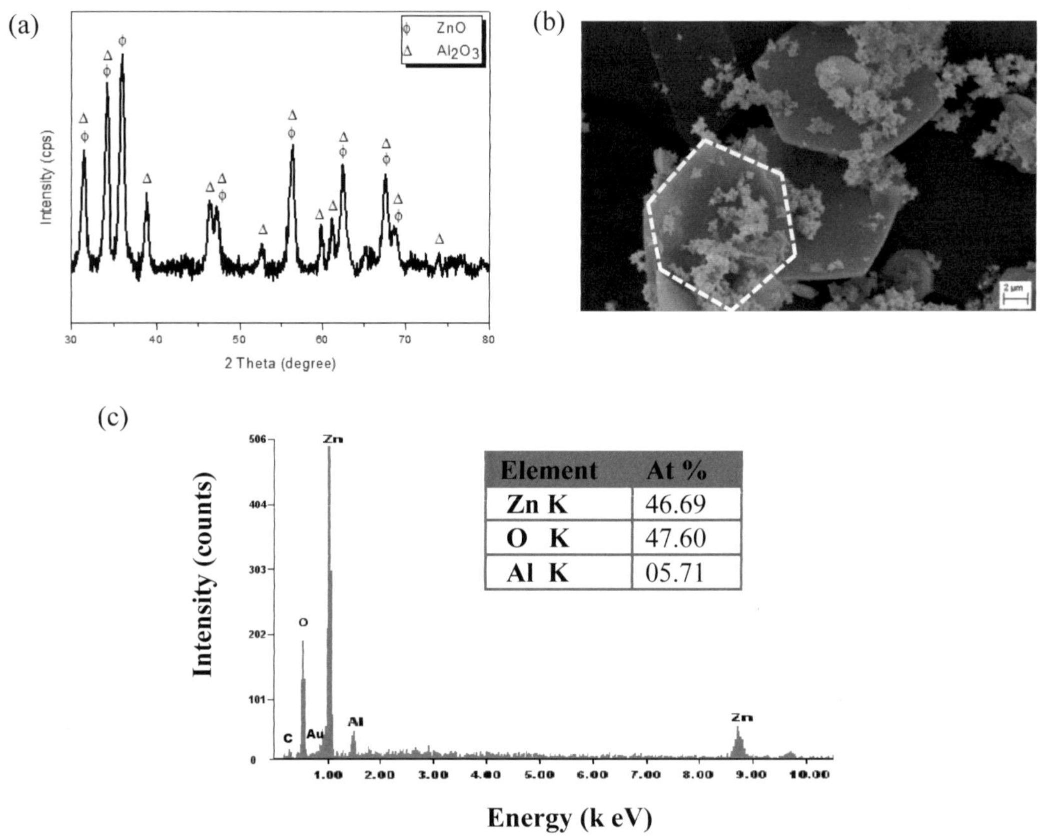

Figure 1. (a) XRD pattern, (b) FESEM image and (c) EDX analysis of ZnO NDs.

The photocatalytic activity of ZnO NDs under UV light irradiation was evaluated by using two kinds of dye solution, i.e. RhB and BCG dyes. Figure 2(a) and (b) display the absorption spectra in degradation of RhB and BCG dye respectively. The intensity of characteristic adsorption peak of RhB (553 nm) and BCG (620 nm) reduced slowly and the colour of dye solution was faded over time. The photodegradation efficiency (PE) of ZnO NDs for RhB and BCG dye solution is illustrated in Figure 2 (c). The BCG dye solution (PE = 44.39%) degraded faster than RhB dye (PE = 21.14%) after 75 min of UV irradiation.

The difference in PE of dyes by ZnO NDs could be explained from their molecular structures. As shown in Figure 3, the partial conjugated π–π* system (benzenic ring) and n–π* bonding (hydroxyl group (–OH) in BCG dye is more easy to destruct [24]. Another fact is the BCG dye containing hydroxyl group (–OH), in which hydrogen/electrostatic bonding with the surface hydroxyl group of ZnO NRs may easy to construct [25]. The •OH radicals generated from ZnO could attack the π–π* system, causing destruction of the conjugated π systems in benzenic rings of BCG dyes. In addition, the BCG dye with more –OH group will have a higher amount of dye absorbance rate with maximum adsorption capacity (q_{max}). Therefore, ZnO NDs could adsorb and decolourize the BCG dyes better than RhB dyes.

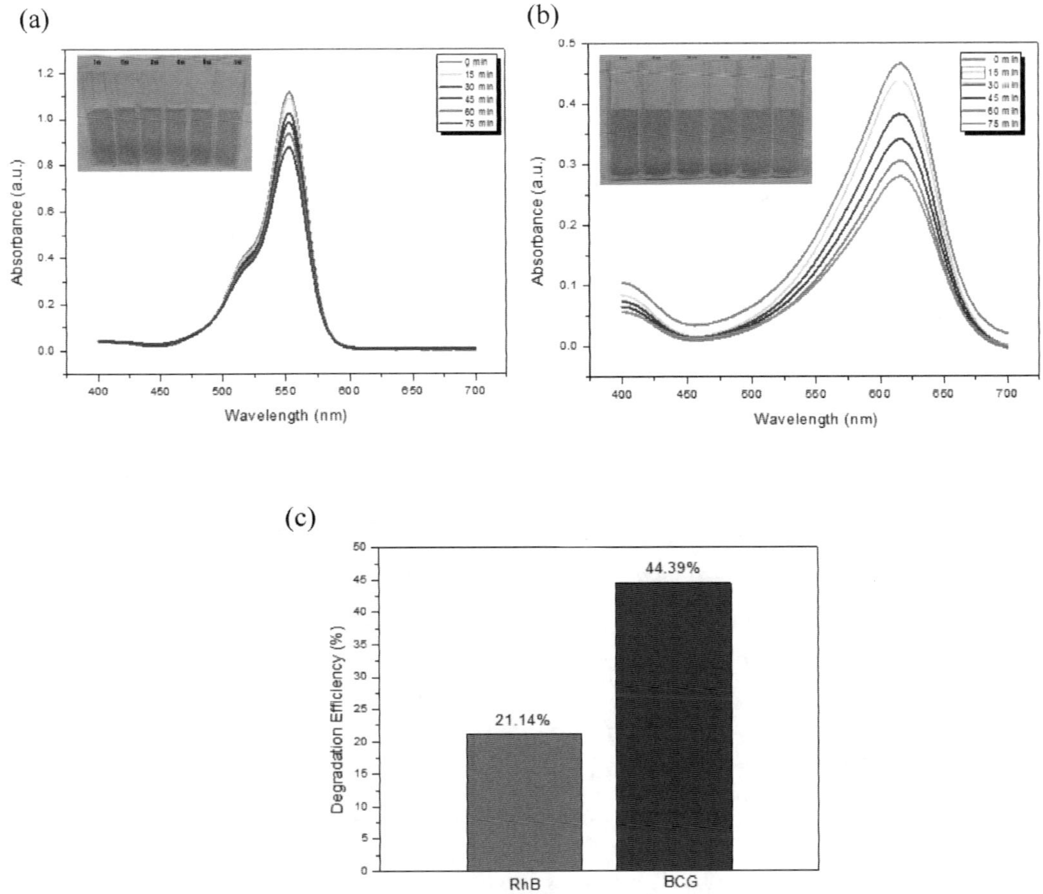

Figure 2. UV-Visible absorption spectra of (a) RhB solution and (b) BCG solution photodegraded and (c) their photodegradation efficiency by ZnO NDs under UV irradation.

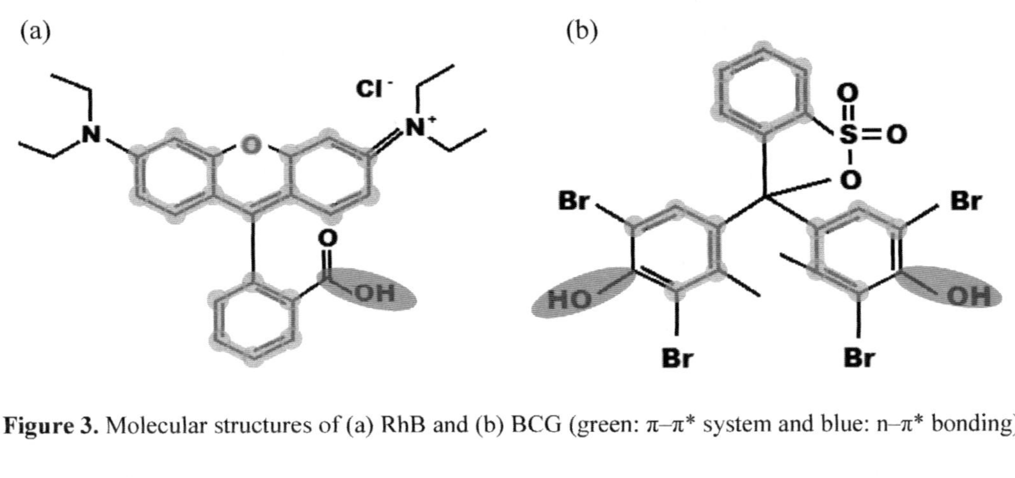

Figure 3. Molecular structures of (a) RhB and (b) BCG (green: π–π* system and blue: n–π* bonding).

4. Conclusions

ZnO NDs were synthesized successfully through solution precipitation methods at room temperature. It is clearly shows that ZnO NDs are in hexagonal wurtzite crystal structure. The ZnO NDs have an average diameter of 12.301 ± 1.050 μm and thickness of 0.067 ± 0.020 μm. The EDX analysis shows particles composed of Zn and O elements with ratio of ~1:1. A small amount of Al, i.e. 5.71 % which originated from the $Al_2(SO_4)_3$ was detected. The BCG dye solution degraded faster than RhB dye by using ZnO NDs after 75 min of UV irradiation. This could explained by the presence of more hydroxyl group (–OH) in BCG dye than RhB dye.

Acknowledgements

The authors are grateful to Fundamental Research Grant Scheme (FRGS, 2031/PBAHAN/6071327) Ministry of Higher Education, Malaysia for their financial support throughout the research study.

References

[1] Golzad-Nonakaran B *et al* 2016 *Materials Chemistry and Physics*, 210-221.
[2] Yaqoob U *et al* 2015 *Sensors and Actuators B: Chemical*, 221, 760–768.
[3] Kumar S G *et al* 2015 *Applied Surface Science*, 355, 939–958.
[4] Claro E M T *et al* 2016 *Journal of Environmental Management*, 177, 264–270.
[5] Shah A H *et al* 2014 *Journal of Alloys and Compounds*, 606, 164–170.
[6] Karunakaran C *et al* 2011 *Materials Science in Semiconductor Processing*, 14(2), 133–138.
[7] Podporska-Carroll J *et al* 2017 *Journal of Hazardous Materials*, 324, 37-47.
[8] Shabannia R 2016 *Journal of Molecular Structure*, 1118, 157–160.
[9] Ambika S *et al* 2015 *Journal of Photochemistry and Photobiology. B, Biology*, 146, 52–57.
[10] Rekha K *et al* 2010 *Physica B: Condensed Matter*, 405(15), 3180–3185.
[11] Ann L C *et al* 2014 *Ceramics International*, 40(2), 2993–3001.
[12] Georgekutty R *et al* 2008 *Journal of Phyical Chemistry*, 112, 13563–13570.
[13] Akhil K *et al* 2016 *Journal of Photochemistry and Photobiology B: Biology*, 160, 32–42.
[14] Sharma R K *et al* 2015 *Ceramics International*, 41(1), 967–975.
[15] Baskar G *et al* 2013 *Asian Journal of Pharmacy and Technology*, 3(4), 142–146.
[16] Hatamie A *et al* 2015 *Langmuir*, 31(39), 10913–10921.
[17] Amornpitoksuk P *et al* 2012 *Superlattices and Microstructures*, 51(1), 103–113.
[18] Ghayempour S *et al* 2016 *Carbohydrate Polymers*, 136, 232–41.
[19] Zak A K *et al* 2011 *Ceramics International*, 37, 393–398.
[20] Amornpitoksuk P *et al* 2012 *Superlattices and Microstructures*, 51(1), 103–113.
[21] Poongodi G *et al* 2015 *Spectrochimica Acta. Part A, Molecular and Biomolecular Spectroscopy*, 148, 237–43.
[22] Thein M T *et al* 2015 *Journal of Experimental Nanoscience*, 10(14), 1068–1081.
[23] Fassi S *et al* 2012 *Journal of Materials and Environmental Science*, 3(4), 732–743.
[24] Saha B *et al* 2011 *Journal of Physical Chemistry C*, 115, 8024–8033.
[25] Saharan *et al* 2014 *J. Nanosci. Nanotechnol*, 14(1), 627–643.

Study on Bamboo Sawdust Structure Breakage on Manganese Leaching Application

N. Muthalib[1], N.S. Abdullah and S. Ismail[1*]

[1]Strategic Mineral Niche Area, Universiti Sains Malaysia, Engineering Campus, Penang, Malaysia

Email: suhaina@usm.my

Abstract. The purpose of this work is to study Bamboo Sawdust (BSD) structure breakage to assess its use as potential reducing agent in manganese leaching application. The determination of holocellulose, cellulose and hemicellulose was carried out using BSD reaction towards acetic acid and sodium chlorite ($NaClO_2$). Afterwards, the residue reacted with acetic and nitric acid. Result shows BSD contents 28.60% hemicellulose, 84.71% holocellulose and 56.11% cellulose. During the holocellulose analysis formation of pores was observed. This indicated degradation of holocellulose into reducing sugars. Meanwhile, cellulose analysis caused BSD bundles isolated into individual component. The treatment generates degradation of cellulose into reducing sugars. In addition, the suitability of BSD as a reducing agent is also investigated. In this part, raw BSD undergone leaching reaction and the residue structure was determined using Field Emission Scanning Electron Microscope (FESEM). Two type of manganese (Mn) ores were analysed which were Low Grade Manganese Ore (LGMO) and Synthetic Manganese Ore (SMO). As a result, leaching with manganese ores in sulphuric acid solution develop pores on BSD structure illustrate degradation of BSD into reducing sugars. In addition, the recovery reveals Mn was extracted up to 70% for LGMO and 95% for SMO within 120 minutes of reaction Accordingly, BSD capable released cellulose as reducing agent hence suggests BSD is suitable reducing agent to extract manganese ore.

1. Introduction

Bamboo is a renewable woody grass and often used to replace wood [1]. It is easily cultured in tropical and subtropical zone [1-2]. Some of the bamboo species are *Gigantochloa Levis, Gigantochloa Wrayi and Gigantochloa Brag*. The most common species is known as Buluh Semantan *(Gigantochloa Scortechinii)* that grows abundantly in Peninsular Malaysia [4-5].

Bamboo cell wall composed of 90-98% cellulose, hemicellulose and lignin. The remaining 2-10% of bamboo cell wall composed of extractive. Cellulose presents in bamboo is a long chain polymer of glucose and the configuration is differs from starch [5]. The degradation of cellulose produces oligosaccharides which are cellotetraose, cellotriose and cellobiose and monomeric sugars which are glucose and fructose [1]. Meanwhile, amorphous hemicellulose configuration is branched polymers of five carbon sugars (pentoses). Generally, hemicellulose degrades into pentose sugars such as xylose and arabinose [1]. These sugars are normally helps in reducing metals particularly manganese dioxide ore (MnO_2). They function as a reducing agent to reduce metals into lower form [9–12].

Previous researches had studied acid treatment to identify cellulose, hemicellulose and holocellulose of bamboo sawdust [3–5], [10–13]. However, most of the research works focused on bamboo composites, cellulose fibres and pulp production. For example, Cao et al., 2014 in his works studied characterization of bamboo to produce pulp production and Parnia et al., 2014 studied preparation of bamboo fibre reinforced composites [10, 12]. Besides that, there are some works described on characterization of bamboo and bamboo in leaching application [3–5], [14]. However, none of these researchers studied on the structure breakage of bamboo sawdust on leaching application.

Therefore, this work aims to study BSD structure breakage to assess its use as potential reducing agent in manganese leaching application. Thus, FESEM was utilised to study the BSD structures before and after chemical treatments. The structures are observed and described. Amount of cellulose, hemicellulose and holocellulose in the BSD are calculated. In addition, the suitability of BSD as a reducing agent is investigated. BSD structures changes after leaching is described. The reduction of Mn ore using BSD as a reducing agent is discussed.

2. Materials and Methods

2.1 Raw materials and chemicals

Bamboo sawdust was obtained from Forest Research Institute Malaysia (FRIM). The type of BSD used was *Gigantochloa Scortechinii.* The bamboo part used was culm. The bamboo was washed, dried, cut and grinded until -75µm size was achieved. The ash content and extractive analysis of BSD were 3.37% and .

LGMO was obtained from Pahang, Malaysia. The ore was dried, grinded and sieve to achieve -75µm. Elemental content of the ore consist of 15.26% Mn, 4.05% Fe, 2.74% Si and 2.71% Al. Phases present in the ore were pyrolusite, MnO_2 (ICDD card no. 98-006-2906), goethite, FeO(OH) (ICDD card no. 98–003-4797) and quartz, SiO_2 (ICDD card no. 98-010-7202) [15].

SMO was obtained from Acros Organics (99%, -10 mesh). Ore was sampling and grinded into -75µm. The elemental content of SMO was 34.33% Mn and phase present in this ore was pyrolusite, MnO_2 (ICDD card no. 98 003 5167).

Chemicals used include 95%- 97% sulphuric acid, 100% acetic acid, 65% nitric acid, 95% ethanol, acetone, 99% EDTA, thmolphatalexone indicator and silicone oil. All chemicals used are analytical grade. Dilution was made using deionized water.

2.2 Hemicellulose, holocellulose and cellulose content of BSD

Holocellulose identification was performed in 100mL conical flask. Approximately 1g of BSD samples were hydrolyzed with 30mL deionized water, 0.08mL acetic acid and 0.8g $NaClO_2$ for an hour at 75°C. After an hour, 0.08mL acetic acid and 0.4g $NaClO_2$ were added. Same amount of acetic acid and $NaClO_2$ were added every hour for 3 hours respectively. Then, residue was filtered, washed using deionized water and acetone and dried at 60°C. The final weight of holocellulose was taken.

Cellulose identification was carried out using dry solid residue from previous final holocellulose analysis. The residue was mixed with 3mL of reagents mixture (80% acetic acid and 70% nitric acid). The reagent ratio was 10:1. The mixture was heat at 100°C for 2 hours in oil bath. Later, mixture was cooled down and 60mL deionized water was added. The mixture was filtered, washed with deionized water and dried at 60°C. The final weight of cellulose content was recorded. Equation 1 to Equation 3 indicates hemicellulose, holocellulose and cellulose calculation.

$$\text{Hemicelluloses (\%)} = \text{Holocellulose (\%)} - \text{Cellulose (\%)} \quad (1)$$

$$\text{Holocellulose (\%)} = [\frac{\text{Weight initial - Weight final}}{\text{Weight final}} \text{ x (100- extractives) \%}] \quad (2)$$

$$\text{Cellulose (\%)} = [\frac{\text{Weight initial - Weight final}}{\text{Weight initial}} \text{ x (100- extractives) \%}] \quad (3)$$

2.3 Leaching of LGMO and SMO using BSD as a reducing agent

In this work, BSD application as a reducing agent to extract Mn ore was analyzed. Two types of ores were used which were LGMO and SMO. Reductive leaching was performed in 500mL three neck flask immersed in silicon oil bath. The reactor was connected with condenser and thermometer was placed inside the glass reactor and at silicon oil bath to measure the temperature.

25g of Mn ore (LGMO or SMO) size -75µm was added into 4M of 250mL sulphuric acid at 100°C. Later, 7.5g of raw BSD (-75µm) was poured in reaction flask. The solution was stirred at 300rpm and leaching was continued until 360min.

2mL sample was withdrawn during leaching process at 15min, 60 min, 120 min, 240 min and 360 min. These samples were filtrate using Whatman, Kent, UK, 0.45 um (GMF- PTFE). Mn determination was determined using complexometric titration with EDTA as a titrant. Meanwhile solid residue was undergone SEM observation to identify BSD structure breakage.

3. Results and Discussion

3.1 Hemicellulose, holocellulose and cellulose content of BSD

The function of this analysis was to determine holocellulose, cellulose and hemicellulose content in BSD. The finding indicated BSD contents 28.60 % hemicelluloses, 84.71 % holocellulose, and 56.11 % cellulose. These contents were coherent with previous researchers which reported holocellulose content in BSD was in the range 65.91% to 82.3% and cellulose was between 38.96% to 64.6% [3–5]. This high holocellulose, cellulose and hemicellulose content have high tendency to degrade into reducing sugar to extract Mn ore.

3.2 Structure identification of BSD using FESEM analysis

3.2.1 Structures of raw BSD. The observation of BSD under FESEM was performed to understand the structure surface of BSD. Figure 1 shows the structure of raw untreated BSD observed by FESEM.

Figure 1. Hollow cylinder structure of BSD

Figure 1 display the structure of BSD was in hollow cylindrical shape. The cell wall was in rigid structure. Besides that, the structured showed the culm was rough and uneven.

In addition, bamboo composed of several vascular bundles. The function of these vascular bundles was to provide strength to the culm [10]. Generally, vascular bundles consist of fibre strand. This fibre strand contents lignin and hemi-cellulose. Figure 2A and Figure 2B display the vascular bundle meanwhile Figure 2C and Figure 2D display the fibre strand of BSD.

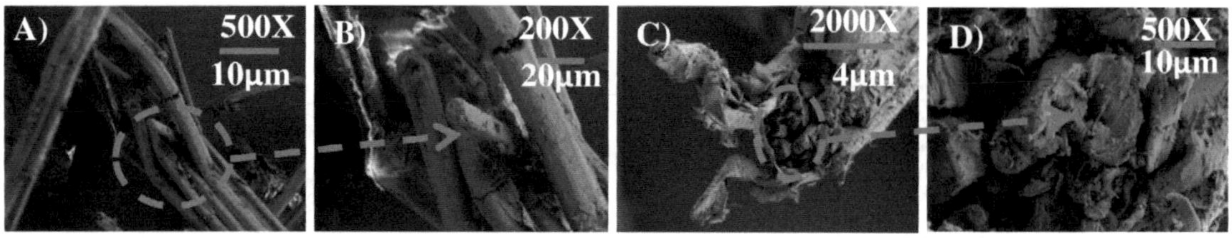

Figure 2. Raw BSD structure A) Vascular bundle 200X magnification, B) Vascular bundle 500X magnification, C) Fibre strand 500X magnification D) Fibre strand 2000X magnification

3.2.2 Structures of holocellulose and cellulose analysis of BSD. Meanwhile, holocellulose analysis produces rougher BSD structures. The BSD was no longer rigid. The study developed formation of pores in the BSD structure. These pores indicate the holocellulose had been degraded into reducing sugars. A lot of pores present in the BSD after the reaction between acetic acid and sodium chlorite analysis. The generation of numerous pores supported with the result of high holocellulose content in BSD. High yield of holocellulose degraded into reducing sugars. Figure 3A and Figure 3B reveals the formation of pores in BSD structure after pretreatment with acetic acid and sodium chlorite at 75°C. Meanwhile, Figure 3C and Figure 3D expressed the FESEM of BSD structures after cellulose analysis.

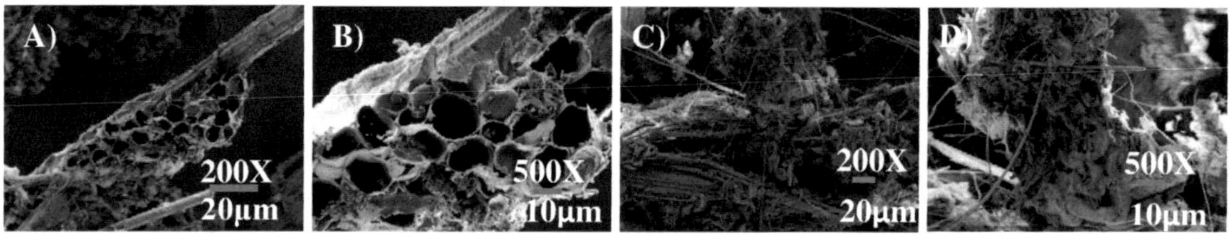

Figure 3. BSD structures; A) after holocellulose extraction analysis 200X magnification, B) after holocellulose extraction analysis 500X magnification, C) after cellulose analysis extraction 200X magnification, D) after cellulose analysis extraction 500X magnification

Further analysis of cellulose extraction using acetic and nitic acid caused some part of vascular bundles separated into individual components. The rigid structures of BSD start to dissociate and binders that protect the structures were isolated. Some structures were also loosening. Therefore, further treatment of BSD with 80% acetic and 70% nitric acid shows cellulose had been degraded into reducing sugars.

3.3 Application: Leaching of LGMO and SMO using BSD as a reducing agent

Generally, Mn does not exist as a free element in nature. It often combines with other elements such as oxygen and iron. The common ore of Mn is pyrolusite (MnO_2) [16]. This Mn (IV) oxide has limited value in the industry [17]. Hence, it needs to reduce into Mn (II) or Mn metal to make it commercial in industry like iron and steel making.

Previous research had studied reduction of Mn (IV) in the absent of reducing agent. The research found out Mn was unable to reduce without the presence of reducing agent [18]. Therefore, this work utilized raw BSD as a reducing agent to reduce Mn ore. The leaching was made using diluted sulphuric acid (4M) instead of high acid concentration. Raw BSD was used to determine the BSD structure changes without pretreatment analysis.

It was found out Mn was successfully reduce into Mn (II) in the presence of BSD as a reducing agent. Figure 4 illustrates FESEM of BSD structures after leaching process. Figure 4A indicates the cell wall of BSD becomes rougher compared to raw BSD. Besides that, the images shown revealed BSD was not fully breakout but there were some formation of pores develops in the BSD structures.

The formation of pores develops in both leaching of LGMO ore and SMO which was seen in Figure 4B and Figure 4C. This expressed the diluted acid reaction occurred during leaching process able to degrade BSD into reducing sugars. In addition, based on the observation of FESEM analysis the ore was accumulated around BSD structures. These shows there were reaction occurred between ore and BSD in sulphuric acid.

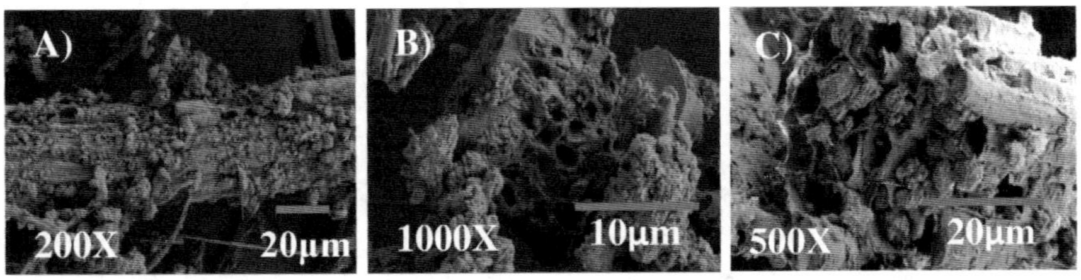

Figure 4. BSD structure after leaching

Further, the recovery reveals Mn was extracted up to 70% for LGMO and 95% for SMO within 120 minutes of reaction. This indicates the suitability of BSD as a reducing agent to reduce Mn ore using diluted sulphuric acid. Figure 5 shows the reduction of Mn ore using BSD as a reducing agent in sulphuric acid.

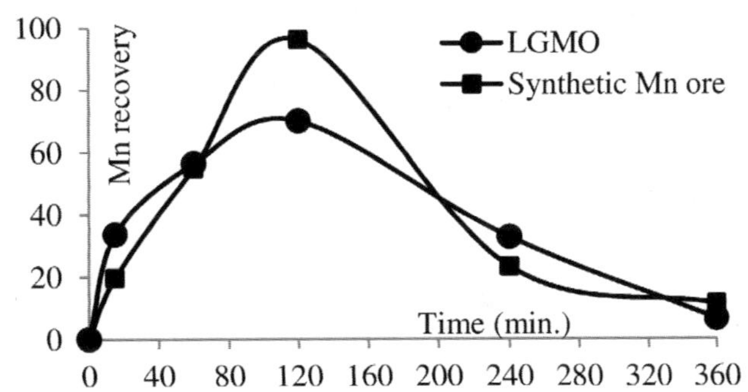

Figure 5. Mn extraction using BSD as a reducing agent

4. Conclusion

This work aims to determine BSD structure breakage to assess its use as potential reducing agent in manganese leaching application. The treatment of BSD with acetic acid and sodium chlorite at 75°C developed formation of pores in the BSD structure. These pores indicate the holocellulose had been degraded into reducing sugars. In addition, BSD structure becomes rougher.

Further treatment analysis using acetic and nitic acid caused some part of vascular bundles separated into individual components. This indicates cellulose had been degraded into reducing sugars. The treatment also indicates BSD contents 28.60 % hemicelluloses, 84.71 % holocellulose, and 56.11 % cellulose.

In addition, the suitability of BSD as a reducing agent is also investigated. LGMO and SMO were reduced using BSD as a reducing agent in H_2SO_4. The reaction shows formation of pores develops in both leaching of LGMO ore and SMO. This revealed acid reaction occurred during leaching process able to degrade BSD into reducing sugars.

The result was supported with Mn was extracted up to 70% for LGMO and 95% for SMO within 120 minutes of reaction. This indicates the suitability of BSD as a reducing agent especially to reduce Mn ore.

5. Acknowledgement
The authors are thankful to School of Materials and Mineral Resources Engineering, USM and Fundamental Research Grant Scheme (FRGS) (203.PBAHAN.6071359).

References
[1] M. Mohan, T. Banerjee, and V. V. Goud 2015 *Bioresource Technology* 191 244
[2] C. Xie, L. Xu, T. Peng, K. Chen, and J. Zhao 2013 *Chinese Journal Geochemistry* 32 222
[3] S. Ismail, N. S. Abdullah, H. Hussin, and S. F. Saiyid Hashim 2012 *Advance Materials Research* 620 405
[4] H. N. Hisham, S. Othman, H. Rokiah, M. A. Latif, S. Ani, and M. M. Tamizi 2006 *Jornal Tropical Forest Science* 18 236
[5] R. Wahab, M. T. Mustafa, M. A. Salam, M. Sudin, H. W. Samsi, and M. S. M. Rasat 2013 *Journal Agriculture Science* 5 66
[6] F. Beolchini, M. P. Papini, L. Toro, M. Trifoni, and F. Vegliò 2001 *Mineral Engineering* 14 175
[7] F. Wu, H. Zhong, S. Wang, Lai, J. Li, and Y. Liu 2014 *Journal Central South University* 15
[8] F. Vegliò, M. Trifoni, F. Pagnanelli, and L. Toro 2001 *Hydrometallurgy* 60 167
[9] S. Ismail, H. Hussin, S. F. Saiyid Hashim, and N. S. Abdullah 2016 *Advance Materials Research* 1133 629
[10] P. Zakikhani, R. Zahari, S. M.T.H, and D. L. Majid 2014 *Materials and Design*
[11] P. P. Zhang, D. S. Tong, C. X. Lin, H. M. Yang, Z. K. Zhong, W. H. Yu, H. Wang, C. H. Zhou 2014 *Asia Pacific Journal Chemistry Engineering* 689
[12] S. Cao, X. Ma, L. Lin, F. Huang, L. Huang, and L. Chen 2014 *BioResources* 9 4528
[13] Y. Teramoto, S. H. Lee, and T. Endo 2009 *Bioresource Technology* 100 4783
[14] D. Hariprasad, B. Dash, M. K. Ghosh, and S. Anand 2007 *Minerals Engineering* 20 1293
[15] N. Muthalib, N. Abdullah, H. Hussin, and S. Ismail 2017 *Materials Science Forum* 888 MSF 453
[16] A. Abdallah, Y. Darmane, and K. Said 2015 *Journal Mining and Metallurgy* 1 29
[17] A. a Ismail, E. a Ali, A. Ibrahim, and M. S. Ahmed 2004 *Canadian Journal Chemical Engineering* 1296
[18] M. F. H. Mohd Rozali, N. Nasir, S. Ismail, and N. S. Abdullah 2016 *Materials Science Forum* 840 364

IOP Conf. Series: Journal of Physics: Conf. Series **1082** (2018) 012087 doi:10.1088/1742-6596/1082/1/012087

Significant of Gold Mineralogical Study for The Chemical Separation of Weathered Ore in Malaysia

W.N.Syafiqah Anuar[1], Norlia Baharun[1] and Suhaina Ismail[1]

[1]Universiti Sains Malaysia, Engineering Campus, Penang, Malaysia

Email: wansyafiqahanuar@gmail.com

Abstract. This paper provides a mineralogical study on the gold ore sample obtained from a local gold mine in Malaysia that underwent the lateritisation process. The mineralogical characteristic of the secondary gold was investigated using the optical microscope study. Then, the presence of other minerals was identified by X-ray fluorescence, X-ray diffraction and scanning electron microscope/ energy dispersive X-ray. X-ray mapping was also employed for the determination of the elemental distribution in the gold ore samples. From the optical microscopy and SEM/EDX study, the gold grain was found embedded in the quartz and also was spotted interlocked in iron oxide and arsenopyrite as well within size from sub-microscopic (1<μm) to microscopic (≤10μm) with a flaky type of surface morphology as well as the presence of element of Si, O and Fe. XRD shows the mineral phases of quartz, muscovite and magnetite. In XRM, the gold was also found disseminated on the silicate minerals surface like quartz (SiO_2), chlorite ($(Mg_5Al)(AlSi_3)O_{10}(OH)_8$) and clay minerals, as also confirmed by the chemical composition presents in the XRF analysis. Hence, from this mineralogical studies, the characteristic of the gold is unsuitable to be processed by physical separation as the submicroscopic gold will tend to be lost during processing and a sustainable processing method such as non-cyanide leaching can be proposed to overcome the problem.

1. Introduction

In the Peninsular Malaysia, there is a large volume of published studies describing the role of the hydrothermal fluid as a conduit which carries the metallic minerals and resulting in the gold mineralization. The precipitation of this hot fluid originated from the metamorphism and magmatic events causing the deposition of the gold in the extensive deformation (brittle-ductile and shearing zone) and develops the gold deposits which mostly mined from quartz lode as well as in the stockwork deposits.

Gold is a native metal that may happen as primary or secondary deposit where the secondary type of gold sometimes comes with very high purity of gold grains instead of an alloy of gold, silver and copper that often formed during the gold occurrences. These secondary gold grains are formed from the weathering process that most likely attacking either through chemically beneath the rock by the help of the supergene solution or mechanically transported that often resulted in depletion of silver content. The lateritic weathering profiles are created as the gold deposit is exposed to the tropical weathering [1] and this prolonged weathering is due to the arid environment where Malaysia is one of the countries having

IOP Conf. Series: Journal of Physics: Conf. Series **1082** (2018) 012087 doi:10.1088/1742-6596/1082/1/012087

tropical rainforest climate with an average annual rainfall of 2500 mm with hold an average temperature of 27°C that is hot and humid all year.

Several studies have revealed that the secondary gold deposit is characterized by several features particularly in morphology and their grain sizes. The gold content also decreases towards the top of the soil profile as it experiences the chemical alteration. This type of gold is mostly displayed by the flat rounded shape resembling the botryoidal type with etch pits as well as corrosion cavities that appeared on the grain surface. Generally, the gold forming in the secondary environment is mostly very fine (<10μm) in size and also found scattered and surface-bound onto the surface of other minerals [2] such as stained quartz vein, iron hydroxides and others.

This paper will focus on the mineralogical characterization of the secondary gold from weathered ore as it is compulsory before the gold deposit is brought to the beneficiation process. Prior to separation and extraction process of the gold, the waste mineral needs to be identified and separated physically and chemically as gangue minerals would affect the quality of final product. With a better understanding on the mineralogy of the gold, this would effectively minimize the cost of gold processing. In this research work, the methods employed to characterize the gold ore are optical microscopy, X-ray diffraction and fluorescence, SEM/EDX analysis and also X-ray mapping.

2. Experimental study

In this research study, the gold ore samples were obtained from a local gold mine in Malaysia. This gold ore sample was collected from the gold-bearing vein that been exposed to the lateritic weathering front. The samples were crushed and ground until passing 80% of 75 μm sieve. The ground ore was blended homogeneously before the analysis while for the ore microscopic study, representative rock samples were prepared as polished mounts and characterized using Scanning Electron Microscope–Energy Dispersive X-ray (SEM/EDX) as well as X-ray Mapping (XRM) analysis. Then, X-ray Fluorescence (XRF) which is a semi-quantitative analysis tool used to determine the chemical composition whereas identification the major mineral phase were done using X-ray Diffraction (XRD) technique.

3. Results and discussion

3.1 Semi-quantitative and Phase Identification Analysis

The elemental contents are analysed by the XRF analysis which is tabulated in Table 1. It shows that silica and alumina were the major compositions present in the gold ore sample with the weight percentage of 75.22% and 16.10% respectively. While other oxide compounds were detected like K_2O and Fe_2O with the weight percentages of 3.23% and 2.80% respectively. This led to the conclusion that silicate minerals such as quartz, feldspar, clay and other were the major gangue minerals as well as iron oxide present in the gold ore sample.

Table 1. XRF analysis of the major elements presents in the bulk ore sample

Component	Result (%)	Component	Result (%)	Component	Result (%)
SiO_2	75.22	SO_3	0.29	PbO	0.02
Al_2O_3	16.10	MgO	0.22	CaO	0.02
K_2O	3.23	TiO_2	0.21	CuO	0.01
Fe_2O	2.80	ZnO	0.06	MnO	0.01
As_2O_3	0.34	Na_2O	0.04	Au_2O	Traces

In the XRD analysis, the peaks which represent the mineral phases had been plotted and shown in Figure 1. There were three phases that had been identified in the sample where the major phase was quartz mineral, SiO_2 with 81.3% matched with the reference data (98-001-7203). Meanwhile, other peaks identified were muscovite, $KAl_2(AlSiO_2)(OH)_2$ with 18.1% (98-001-1927) and the iron present as Fe_3O_4 with 0.6% (98-004-0552). The phases reported were referred to the Inorganic Crystal Structure Database (ICSD) as a standard reference. These findings were tally to the XRF result and optical microscopy studies that showing the presence of silicate minerals and iron oxide as gangue minerals.

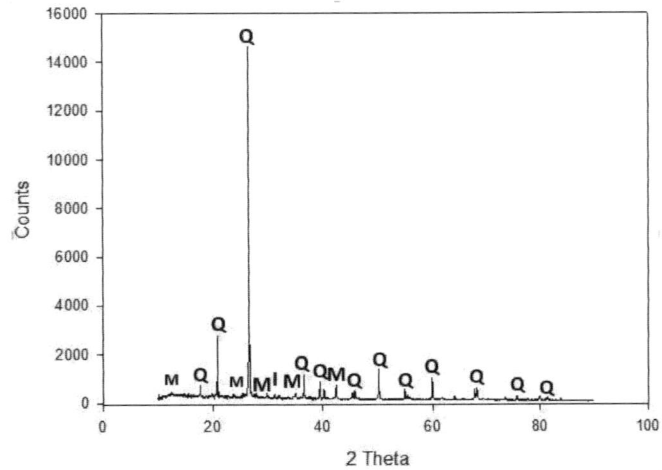

Figure 1. XRD diffractogram of the gold ore sample (Q: quartz, M: muscovite and I: magnetite)

3.2 Optical Microscopic Study

The images displayed in Figure 2 are the optical micrographs of the thin section obtained from the stained vein and goethite rock sample. The microscopy images were taken in the cross-polarised light. Figure 2A shows elongated biotite was found in the clay groundmass where another silicate mineral; orthoclase feldspar also was spotted. From the modal analysis in Figure 2B, about 90% of quartz is shown and other 10% was iron oxide (FeOx) that show the yellowish-brown colour of iron oxide which outlines the micro-crystalline quartz having grainy medium size with the first order birefringence colour; white to grey. Then, in Figure 2C reveals iron oxides were surrounded by quartz mineral where the quartz at the above region having grey interference colour while at the lower part of quartz was yellowish interference colour indicates the thin section was slightly thick.

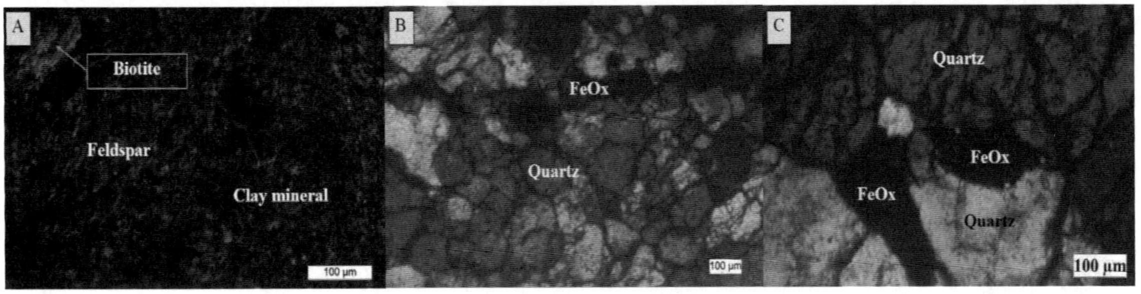

Figure 2. Optical microscopy images of thin section samples in cross-polarized light obtained from the (A and B) stained vein and (C) goethite rock (Qz: quartz, FeOx: iron oxide)

In Figure 3, microscopy images of the polished section analysis observed by polarized microscope. The fine size gold grain was observed to be embedded in the quartz and also was spotted interlocked in brownish iron oxide and arsenopyrite. The presence of iron oxide minerals pseudomorphs after sulphide minerals is an indication that the ore had been exposed to weathering. Description of minerals in the photomicrograph of thin section study was compared to the reference images obtained from Mackenzie et.al, (1980), Raith et al. (2012) and Goeke (2015). For the optical microscopic study of the polished section, the images were referred to the Reflected Light Microscopy by Francis, (2013), Atlas of Ore Minerals by Picot and John, (1982) as well as literature by Berrezueta et al. (2016).

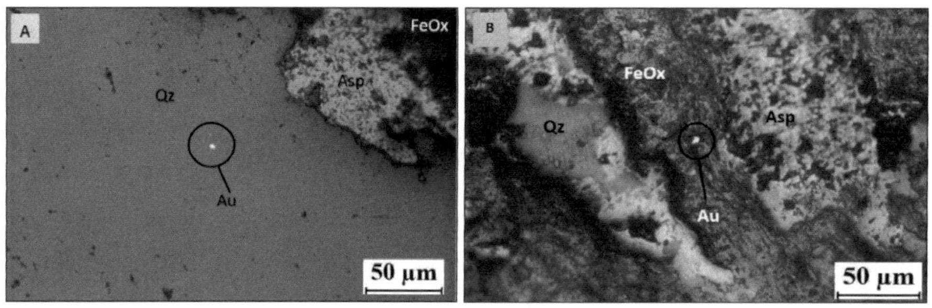

Figure 3. Optical microscopy images of polished section samples observed by polarizing microscope in (A) quartz vein (B) goethite (Qz: quartz, Asp: arsenopyrite, FeOx: iron oxide and Au: gold)

3.3 Scanning Electron Micrograph Study

Figure 4A portrayed a very fine gold that was found embedded in the quartz at the spot A while Au also was detected together with Mg, Al, Si and O (chlorite) in the spot B. Then, in Figure 4B reveals the presence of the particle of Au and O with a flaky morphology around 0.5 μm found in the region containing the abundant presence of Si and O (quartz mineral). Further, in Figure 4C, a gold grain with an estimated size of 20 μm was found associated together with the elements of Fe and O at spot C while there also a gold that still interlocked in arsenopyrite in spot D. From here, it can be said that this iron oxide (spot C) was resulting from the oxidation of sulphide mineral (spot D) that has been undergone weathering environment. These findings are similar in the book written by Brinck, 1956, who reported that the secondary gold grain present in the iron oxide (as inclusion or embedded) has a flaky morphology and fine in size.

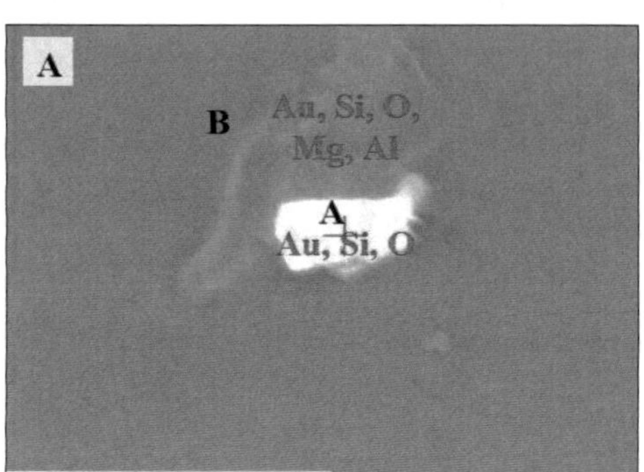

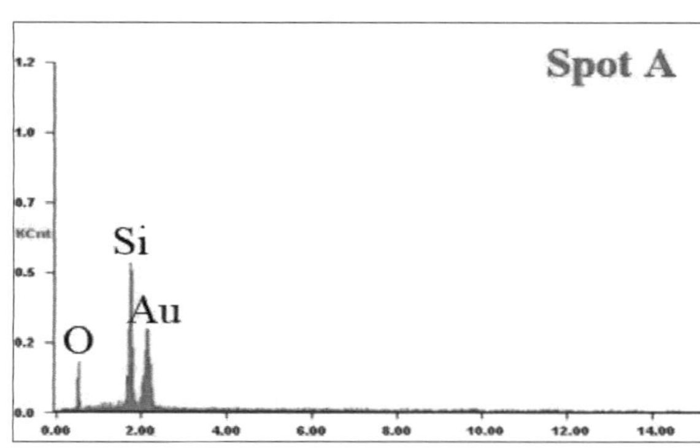

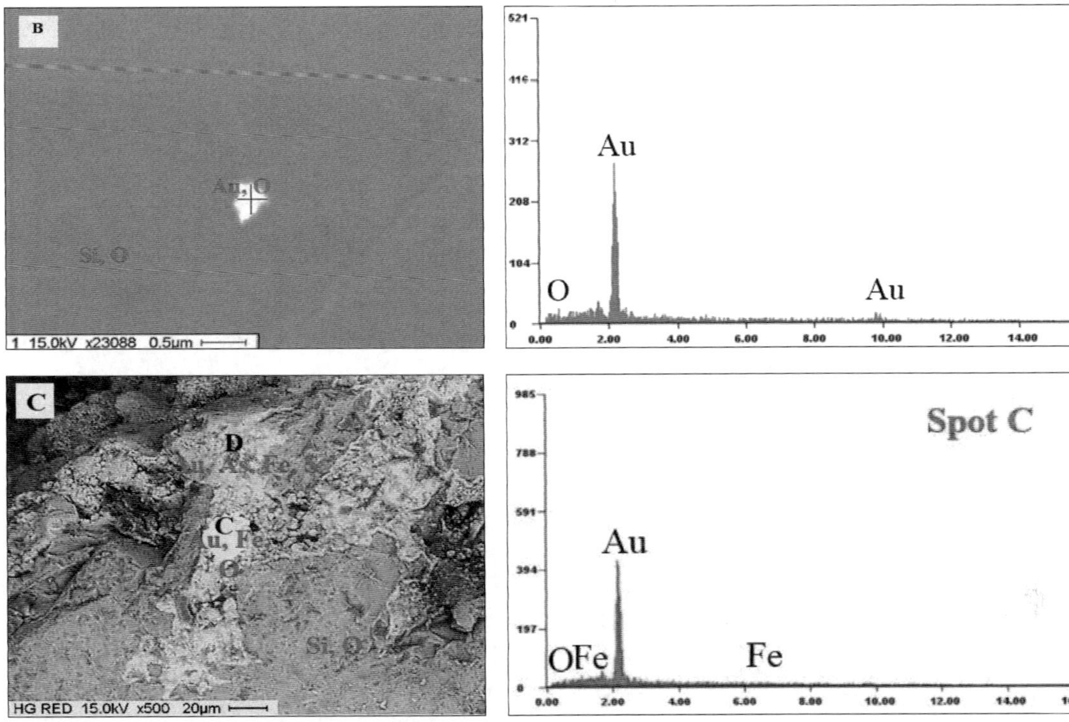

Figure 4. SEM Back-scattered electron micrograph and EDX diffractogram of gold grains in (A) quartz mineral and (B and C) goethite

3.4 X-ray Mapping Study

X-ray mapping is a tool to study the elemental distribution in the interest area of the sample. The distribution of the elements; O, Mg, Al, Si and Au are presented in Figure 5A – 5E while in the Figure 5F shows the overlay minerals association and distribution in the gold ore sample. From the photomicrograph shown in Figure 5E, the gold grains were scattered and surface-bound onto the silicate minerals like quartz, chlorite and clay minerals as the Si and O were majorly found in the sample. Hence, the gold ore samples that being studied was concluded as gold colloids and occur as submicroscopic grains from this mineralogical studies.

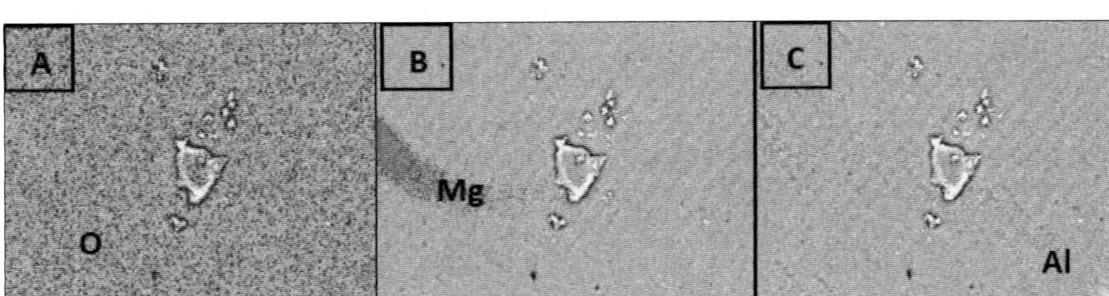

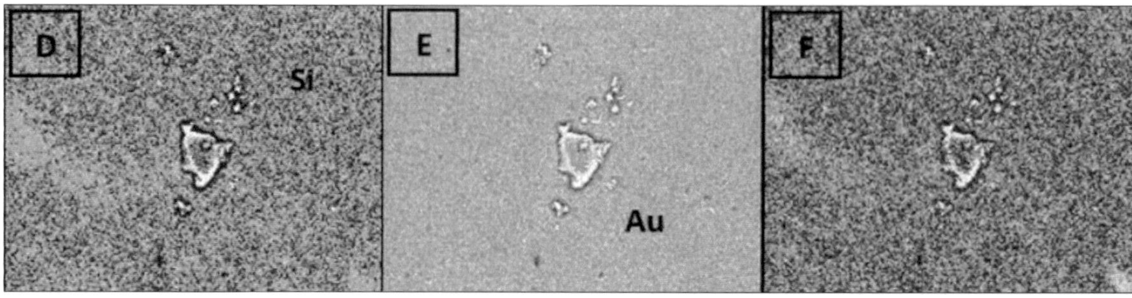

Figure 5. X-ray mapping photomicrographs of elemental distribution of O, Mg, Al, Si, and Au presence in the quartz vein of the gold ore sample

4. Conclusion

From the X-ray analysis, three mineral phases had been identified which are quartz, muscovite as well as iron present as magnetite. The associated oxides present were SiO_2, Al_2O_3, K_2O, Fe_2O_3 and As_2O_3. From optical microscopy and SEM/EDX study, the gold grain was found interlocking in quartz, iron oxide and arsenopyrite as well as spotted as oxide of about 5μm in size with a flaky surface morphology. In XRM, the gold was also found disseminated on the silicate minerals surface like quartz (SiO_2), chlorite ($(Mg_5Al)(AlSi_3)O_{10}(OH)_8$) and clay minerals, as also confirmed by the chemical composition presents in the XRF and in SEM/EDX analysis displayed in Figure 4A at spot B where Au was detected together with chlorite mineral. The presence of sulphide minerals associated with iron oxide proof that the deposit has undergone the lateritisation process resulting in the formation of iron oxide minerals (pseudomorphs after sulphides). This lead to the staining of the nearest quartz from the oxidation of iron. Hence, this mineralogical studies led to the conclusion that the very fine gold found in the sample being studied is the result of lateritic weathering attacking the outcropped of the gold ore deposit. Thus, this type of gold ore is unsuited to be recovered by physical separation as the submicroscopic to microscopic size of gold grain will tend to be loss during the process.

5. Acknowledgement

This research study is part of author's MSc thesis and credited to School of Materials and Minerals Resources Engineering (SMMRE) and Universiti Sains Malaysia (USM) for the lab facilities and technical assistance. This work was done with the financial support rendered by USM RUI grant (Account No.: 1001/ PBAHAN/ 814263), USM FRGS grant (Account No.: 203/ PBAHAN/ 6071359) and Ministry of Higher Education Malaysia (MyBrain15). Thanks also go to all fellow researchers and technical staffs for their contribution and numerous support.

References

[1] Roberts, P.A. et al., 2010 *Minerals & Metallurgical Processing* 27 72
[2] Shuster, J. et al., 2017 Secondary gold structures: Relics of past biogeochemical transformations and implications for colloidal gold dispersion in subtropical environments
[3] Berrezueta, E., Ordóñez-Casado, B., Bonilla, W., Banda, R., Castroviejo, R., Carrión, P., Puglla, S. 2016 Ore Petrography Using Optical Image Analysis: Application to Zaruma-Portovelo Deposit (Ecuador), Geosciences, 6 30
[4] Raith, M.M., Raase, P., Reinhardt, J. 2012 Guide to Thin Section Microscopy Second Edition
[5] MacKenzie, W.S. and Guilford, C. 2014 Atlas of the Rock-Forming Minerals in Thin Section. Routledge
[6] Picot, P. and Johan, Z. 1982 Atlas of ore minerals BRGM

IOP Conf. Series: Journal of Physics: Conf. Series **1082** (2018) 012088 doi:10.1088/1742-6596/1082/1/012088

Mineralogical Characterization and Determination of Refractoriness of Malaysian Gold Ore

Wahab, S.A.R[1*], Baharun, N.[1], and Ariffin, K.S.[1]

[1] Strategic Mineral Niche Area, School of Materials and Mineral Resources Engineering, Engineering Campus, Universiti Sains Malaysia, 14300 Nibong Tebal, Pulau Pinang, Malaysia

*Email: kaimu92@gmail.com

Abstract. This study aims to conduct mineralogical characterization and determine the refractoriness of gold ore from Eastern Coast of Peninsular Malaysia. Elements presence in the sample were investigated using X-ray fluorescence (XRF) analysis while X-ray diffraction (XRD) method was used for the identification of mineral phases. Polished sections samples were studied by optical microscopy (OM), scanning electron microscopy and energy-dispersive X-ray (SEM/EDX). XRD phase analysis showed the presence of mineral phases of quartz, iron sulphide and silicate minerals, which were confirmed by XRF with the elements detected as Si, O, Al, Fe, As and S. From SEM/EDX analysis, gold was found mostly interlocked in sulphide minerals such as arsenopyrite, galena and others. Fire assay and cyanide bottle roll test were employed on the bulk samples to assay the gold content and free gold, respectively. Atomic Absorption Spectrometry (AAS) revealed the gold content to be 31 gram per tonne in the sample. The cyanide bottle roll test determined the free gold assay of 16 gram per tonne. These findings imply that the samples contained free gold of about 52% along with gold present interlocked with minerals unfavourable to cyanidation, which further provide valuable information in terms of the refractoriness of the gold for further processing.

1. Introduction

Gold has been mined for centuries for its unique properties and abundant applications. From jewelleries to electronic applications [1], gold has been utilized worldwide. Its price is relatively higher compared to other metals since it requires more complex and tedious processing methods. However, this depends on the type of ore that bears the gold. In general, gold ore can be divided into two types, namely "free milling" ore and "refractory" ore [2]. Free milling ore refers to ore in which gold can be liberated easily from other minerals by physical means (crushing and grinding) and then by direct cyanidation. This type of ore requires less investment to be processed. As for refractory ore, it refers to ore which requires pre-treatment, usually by chemical means and it cannot be processed by direct cyanidation. Cyanide is used mainly to liberate free gold particles. Refractory gold usually co-exist with preg-robbing minerals or "preg-robbers" such as sulphide minerals which are very difficult to be processed [3].

In comparison to the previous decades, free milling gold ores are currently found to be less abundant in nature. This is due to their easy accessibility and simple processing route which led to their maximum utilization and in return brought them to the verge of depletion. Thus, currently most gold processing plants took the initiative to process refractory gold ores. However, there is no established processing routes for treating and processing refractory gold since the refractoriness of the ores are dependent largely on their mineralogical and geological data, which are variable from place to place. In Malaysia, almost all gold processing plants are using cyanide leaching as their method for gold recovery and this situation is quite irrelevant to the processing of refractory gold ores. This is because the ores contained preg-robbers which tend to deplete cyanide and resulted in the inefficient consumption of cyanide for gold leaching purposes. Therefore, the purposes of this research are to characterize the gold ore obtained from the Eastern Coast of Peninsular Malaysia and to determine the refractoriness of the ore. These information are very crucial and important for geologists and engineers to upgrade the processing routes of Malaysian refractory gold ores, since cyanide utilization alone is not efficient in treating this type of ores. Some physical or chemical processing stages can be added into the established cyanidation processes to treat the preg-robbers before being treated with cyanide.

2. Experimental study

Rock sample from the Eastern Coast of Peninsular Malaysia was crushed by using laboratory scale jaw crusher and cone crusher to produce rock sample of size $\leq$4 mm. The crushed rock sample was then ground to particle size of 80% passing -75 µm by using ring mill. Cone and quartering sampling method was used on the ground rock sample to produce two well-distributed sample portions [4][5].

For mineralogical characterization, one sample portion from the cone and quartering method was used. The sample was analyzed qualitatively using X-ray fluorescence (XRF; Rigaku Rix-3000) and the mineral phase identification was analyzed by X-ray diffraction (XRD; Bruker D8-advance). Besides powdered sample, some selected fresh rock samples were also used in studying the phases and morphology of the ore. Polished sections of the rock sample were prepared by using mixture of resin and epoxy [5]. Morphology of the rock sample was identified using optical microscopy (OM; Olympus BX51M), scanning electron microscopy and energy-dispersive X-ray analysis (SEM/EDX; Leo Supra 35VP).

Fire assay and cyanide bottle roll test were employed on the other portion from the cone and quartering method. Fire assay aims to determine the grade or assay of gold in the ore by using Carbolite HTF 1800 furnace. Fire assay is the most effective digestion technique in determining the quantitative grade of gold in an ore [4] [6]. Cyanide bottle roll test, on the other hand determines the free gold content in the ore [4]. Sodium cyanide (NaCN) solution was used as the lixiviant in leaching the free gold from the ore into the solution. 24 hour cyanide leaching test was conducted on the sample to determine its free gold content [1] [7]. For both fire assay and cyanide bottle roll test, the gold content was determined quantitatively using the Atomic Absorption Spectrometry (AAS; AAnalyst 700, Perkin Elmer). The fire assay and cyanide bottle roll test were repeated three times and the average gold content were obtained for each method.

3. Results and discussion

3.1 Mineralogical Characterization of Gold Ore Sample

The mineral characterization of the gold ore sample was conducted by using X-ray fluorescence (XRF), X-ray diffraction (XRD), optical microscopy (OM), scanning electron microscopy (SEM) and energy dispersive X-ray (EDX). The XRF result is presented in Table 1. As shown in Table 1, the most abundant compound in the sample are silica and feldspar, due to the significant amount of Al_2O_3 and K_2O. The amount of iron oxide in the sample is also quite significant, which is about 2.17%.

Table 1. Mineral composition of ground ore sample from XRF analysis

Compound	wt%	Compound	wt%	Compound	wt%
SiO_2	81.14	MgO	0.21	ZnO	0.01
$Al2O_3$	11.18	Na_2O	0.03	Ga_2O_3	0.01
Fe_2O_3	2.17	P_2O_5	0.02	Rb_2O_3	0.01
K_2O	2.10	CaO	0.01	SrO	0.01
SO_3	0.78	MnO	0.01	ZrO_2	0.01
As_2O_3	0.59	NiO	0.01	Au_2O	0.01
TiO_2	0.21	CuO	0.01	PbO	0.01
Loss on Ignition (LOI) = 1.50 %					

In determining the main and trace minerals and their interlocking in the sample, X-ray diffraction (XRD) studies were performed [5]. The major phases of minerals identified were quartz, iron sulphide and silicate minerals as shown in Figure 1. This findings were then further confirmed by microscopic analysis. From the optical microscope (OM) studies of the rock sample, it was found that the sample contained gold (Au), pyrite (FeS_2), silica (SiO_2) and arsenopyrite (FeAsS) [8]. Pyrite and arsenopyrite were not detected in the XRD analysis due to their low amount within the sample. Images of the OM were shown in Figure 2. Gold has a distinct bright yellow colour and its shape is irregular with soft edges. Pyrite on the other hand has hard edges with euhedral cubes and triangles shapes, while arsenopyrite has euhedral rhombs shape. It was shown that gold occurred both free and as inclusions in arsenopyrite [9].

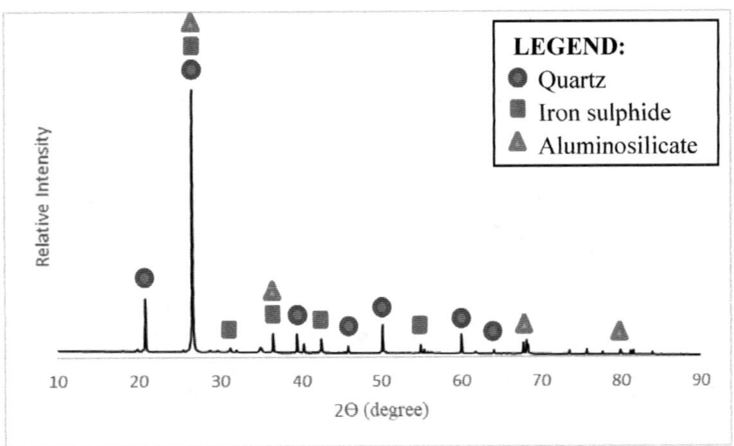

Figure 1. XRD diffractogram showing the major phases of minerals identification in the ground ore sample.

IOP Conf. Series: Journal of Physics: Conf. Series **1082** (2018) 012088 doi:10.1088/1742-6596/1082/1/012088

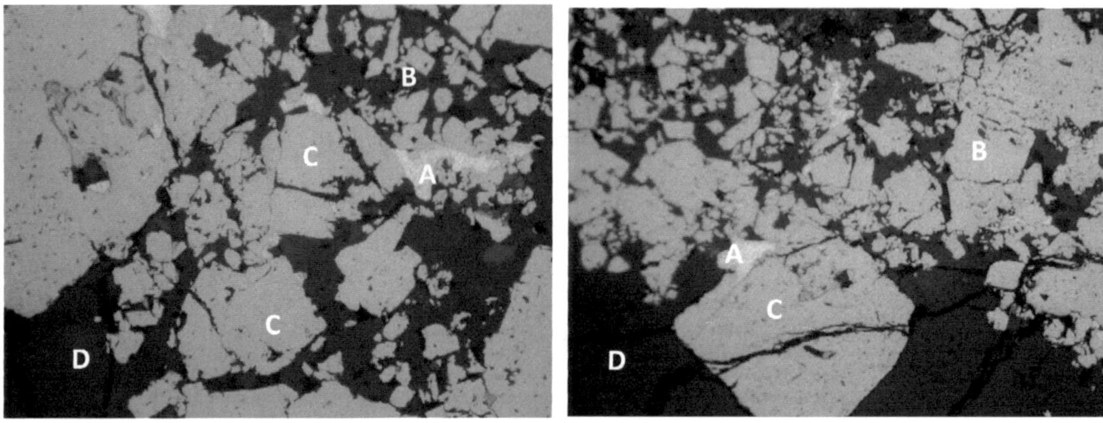

Figure 2. Optical images of polished rock samples under reflected light (100x magnification).
A: Au; **B**: FeS_2; **C**: FeAsS; **D**: SiO_2

For further identification of sample characterization, scanning electron microscopy (SEM) and electron dispersive X-ray (EDX) methods were used and the results were presented in Figure 3. As it can be seen, gold was found interlocking in the sulphide mineral with major components presence such as As, Pb, Fe and S [8]. This shows that the refractoriness of the gold ore sample could be due to the presence of gold associated with sulphide minerals [10]. These findings complimented with analysis results from XRD, XRF and OM. Besides, from SEM micrographs, the average liberation size of gold is 20 μm.

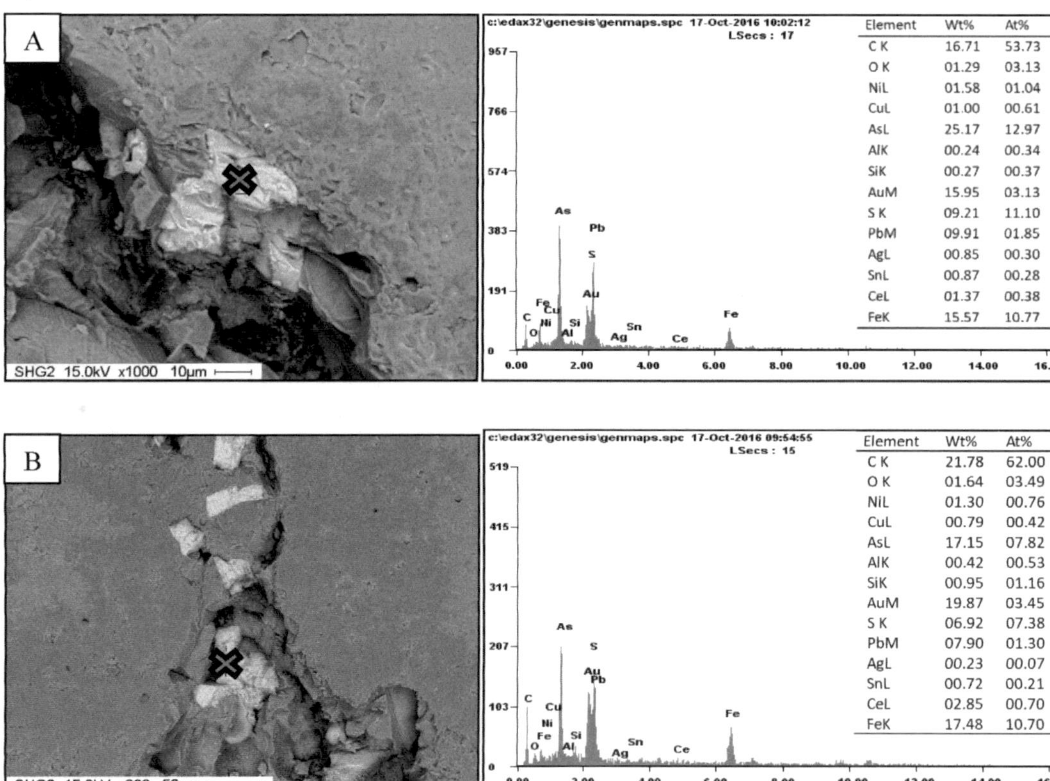

Figure 3. SEM micrographs/EDX diffractograms of the polished sections of the rock ore samples at spots label X (in A and B).

3.2 Gold Content Determination and Refractoriness Identification

Fire assay and cyanide bottle roll test were performed on the ground ore samples. Based on the atomic absorption spectrometry (AAS) analysis, gold content of the ore sample were determined. Fire assay method gave an average gold content of 31 gram per tonne. This result shows the total gold in the ore sample, since almost all impurities were removed as slag during fire assay [6].

Cyanide bottle roll test is a method in which cyanide solution was used to recover free gold in ore [1] [11]. From this test, it was found that the amount of free gold in the sample to be of average 16 gram per tonne. This finding shows that there are about 15 gram per tonne of gold interlocked within the ore, as compared to the amount of total gold obtained by fire assay method. As discussed from the mineralogical characterization, gold in this ore were found mainly interlocked in sulphide minerals.

Therefore, we can say that gold recovery using direct cyanidation method will not provide high gold recovery due to unliberated gold [3]. The ore sample contained gold associated with other components such as sulphide minerals and silicate minerals which reduced the effectiveness of gold cyanidation process. Thus, it is concluded that the ore sample is a refractory gold ore type and this information is very useful in determining the next step in gold recovery process.

4. Conclusion

Several major findings were obtained from this research. It can be concluded that the gold ore sample contain the mineral phases of quartz, iron sulphides and silicate minerals based on the XRD analysis, which were also confirmed by XRF, OM and SEM/EDX analysis. Other minerals to be found associated with gold were galena, pyrite and arsenopyrite. From the mineralogical characterization of the ore sample, it was found that gold was interlocked in the sulphide and silicate minerals. Fire assay of the ground sample gave an average total gold of 31 gram per tonne. For cyanide bottle roll test, the amount of free gold within the sample was found to be of average 16 gram per tonne. These findings imply that the samples contained free gold of about 52%, leaving the remaining unliberated gold still interlocked in the sulphide and silicate minerals. Thus this ore sample can be classified as a refractory gold ore type, as proven by the mineralogical characterization, fire assay and cyanide bottle roll test. This information will be very useful in gold recovery since it will provide the possible processing route for this type of gold ore. However, the selection of processes will also depend on the cost of operating rather than only high percentage of gold recovery. From the mineralogical characterization of the gold ore sample, a diagnostic leaching approach could be used as an analytical tool to determine gold deportment in gold ore sample.

Acknowledgement

The authors would like to express their gratitude and appreciation to the ASEAN University Network/Southeast Asia Engineering Education Development Network (AUN/SEED-Net) for the financial support through a collaborative research grant with grant number 304/PBAHAN/6050347/A119.

IOP Conf. Series: Journal of Physics: Conf. Series **1082** (2018) 012088 doi:10.1088/1742-6596/1082/1/012088

References

[1] S. Syed 2012 *Hydrometallurgy* 115 30
[2] S. Keskinen 2013 "Comparison of cyanide and thiosulphate leaching (a literature review)"
[3] O. Celep and V. Serbest 2015 *Trans. Nonferrous Met. Soc. China English Ed.* 25 1286
[4] Labtium Ltd 2016 *Sample Prep. Anal. Methods Geochemistry, Explor. Min.* 41
[5] S. Ismail, S. F. S. Hashim, H. Hussin, and N. S. Abdullah 2016 *Period. di Mineral.* 85 277
[6] S. A. Oke, A. F. Abimbola, and D. Rammlmair 2014 "Mineralogical and Geochemical Characterization of Gold Bearing Quartz Veins and Soils in Parts of Maru Schist Belt Area, Northwestern Nigeria" 2014
[7] T. Chingosho, O. T. Johnson, and M. Mashingaidze 2014 "Factors affecting Gold Recovery from Secondary Ore" II 4
[8] Monument Mining Limited 2016 "Malaysia's Geology," *http://www.monumentmining.com/ : Malaysia*
[9] B. Sayiner 2013 "Influence of Lead Nitrate on Cyanide Leaching of Gold and Silver from Turkish Gold Ores" 50 507
[10] J. Zhang, S. Shen, Y. Cheng, H. Lan, X. Hu, and F. Wang 2014 *Hydrometallurgy* 144 114
[11] B. Aydin, H. Basturkcu, and A. Gul 2015 *Physicochem. Probl. Miner. Process.* 51 647–660

IOP Conf. Series: Journal of Physics: Conf. Series **1082** (2018) 012089 doi:10.1088/1742-6596/1082/1/012089

Characterization and Leaching Kinetics of Ilmenite in Hydrochloric Acid solution for Titanium Dioxide Production

N.A Jabit[1*], G. Senanayake[2]

[1]Strategic Mineral Niche Area, Universiti Sains Malaysia, Engineering Campus, Penang, Malaysia

[2]School of Engineering and Information Technology, Murdoch University, Western Australia

Email: ainjabit@usm.my

Abstract. A kinetic study of ilmenite in hydrochloric acid (HCl) leaching in the presence of reducing agent has been investigated. The effect of temperature ranging from 60 °C to 80 °C and HCl concentrations (4-9 M) on titanium and iron dissolution has been reported. The chemical and mineralogical studies of ilmenite samples was also performed using scanning electron microscopy with an energy disperse X-Ray (SEM-EDX), X-ray diffraction (XRD) and X-ray fluorescence techniques (XRF). From XRD analysis, ilmenite was found to be the main mineral phase in the presence of associated minerals such as rutile and hematite. SEM micrograph indicated that ilmenite before leaching has a homogenous and angular shape. After leaching for 3 hours, a few particles have reduced its particle sizes due to structural disruption during chemical attack by HCl, but some still retain their shapes. From chemical analysis of leached product, it is indicated that 85.6% of titanium dioxide with low iron content. The dissolution of ilmenite leaching in HCl is reasonably agreed well with a shrinking core kinetic model. The apparent activation energy for iron and titanium is 71.9 kJ mol^{-1} and 90.1 kJ mol^{-1} respectively.

1. Introduction

Ilmenite ($FeTiO_3$) has been one of the major sources of titanium dioxide (TiO_2) which is extensively used in paints, pigments, papers, porcelain, plastics and fibres [1]. The forms of high grade massive ilmenite minerals of greatest importance are relatively rare. Most deposits are often found in association with oxides such as hematite (Fe_2O_3) or magnetite (Fe_3O_4) and the more widely occurring titaniferous beach sand, which have resulted from natural erosion and concentration processes [2].

White titanium dioxide pigment is produced by two processes, namely the sulfate process and the chloride process. The two processes differ in both their chemistry and raw material requirement. The sulfate process, which utilizes ilmenite as a raw material, is well known and widely used but unfavorable. This is due to environmental concerns and the production of a large volume of by-products (ferrous sulfate) that require high capital cost for treatment. Alternatively, the chloride process, which utilizes rutile as a raw material, offers a more economical flowsheet and generates less waste materials. Nowadays, about 60% of the world's TiO_2 pigments are manufactured using the chloride process, in which natural or synthetic rutile (SR), or titanium-rich slag are used as feedstocks [3]-[4]. The shortage of rutile feedstock has prompted the upgrading of ilmenite to synthetic rutile. The use of ore with greater TiO_2 content allows a pigment producer to reduce the amount of iron sulfate by-product produced in the process [5]. In this work, characterization study of ilmenite has been done to really understand the composition and mineralogy of the sample and to study the effect of leaching parameters on ilmenite dissolution in hydrochloric acid.

2. Experimental study

2.1 Ore Characterization study

The chemical composition of samples was determined using X-ray fluorescence analysis (XRF). Their phase compositions were obtained using GBC EMMA (Enhanced Multi Materials Analyzer) diffractometer (CuKα) with subsequent match search by current Visual XRD EMMA software. The morphological studies of ilmenite and leached samples were done via Scanning Electron Microscopy (SEM) (Philips XL 20).

2.2 Leaching and Kinetic Study

2.2.1 Materials and Procedures. The ilmenite ore is heavy mineral sand originated from Western Australia, Australia which provided by School of Engineering, Murdoch University. They were mixed homogeneously, split and sampled to obtain a representative sample prior to the analyses. All samples were stored at room temperature. About 100 g of samples was ground in an iron-ring mill for 30 seconds. The crushed samples were then subjected for screening and sieving to achieve the required size fractions for leaching study. Hydrochloric acid (HCl) as leaching reagents with 32% purity and milipore water was used throughout the experiments. Batch leaching test was carried out using 500 cm^3 glass reactor provided with a reflux condenser and a mechanical agitator. The desired volume of HCl of the required concentration was poured in the reactor and about 20 g of ilmenite ore was added. The reactor was heated to the desired temperature using a thermostatically controlled water bath. The leachate were taken from the reactor using syringe at various time intervals and filtered through Whatman Celulose Nitrate membrane filters with pore size of 0.45 μm. Leachate samples was analysed using Atomic Absoprtion Spectrometer (AAS) and UV/Visible Spectrophotometry to determine iron and titanium content respectively. After the leaching process completed, the slurry was filtered under vacuum and the residue was washed with 100 cm^3 of 3% HCl. The washed residue was dried at 110 °C, calcined at 900 °C, and weighed. The iron and titanium contents were determined in the filtrate, wash liquor, and solid product.

2.2.2 Leaching Kinetic Study. The well-known kinetic models such as the shrinking sphere and the shrinking core models for the dissolution of sized particles under the assumption that the particles are homogeneous and spherical [6] is used to rationalise the kinetics of ilmenite dissolution in HCl solution. The models can be expressed as follows in the form of integrated rate equations for the reaction with H^+ ions for the reaction A(aq) + bB(s) = products:

For the shrinking particle model:

$$1-(1-X)^{1/3} = b[H^+]_{bulk}kr^{-1}\rho^{-1}t = k_{ss}t \quad (1)$$

For the shrinking core model:

$$1 - 3(1-X)^{2/3} + 2(1-X) = 6\,b\,[H^+]_{bulk}\,D_{H+}\,r^{-2}\rho^{-1}t = k_{pl}t \quad (2)$$

$$1 - 3(1-X)^{2/3} + 2(1-X) = 6\,b\,[H^+]_{bulk}\,D_{H+}\,r^{-2}\,\{(1-\varepsilon)\,\rho\}^{-1}t = k_{pl}t \quad (3)$$

where X = the fraction dissolved at time t, k_{ss}, k_{pl} = apparent rate constant, $[H^+]$ = concentration of acid in bulk solution (mol cm^{-3}), D_{H+}= diffusion coefficient of H^+ through the product layer ($cm^2\,s^{-1}$), r = initial particle radius (cm), ρ= molar density of the dissolving metal in the initial particle (mol cm^{-3}), ε = particle porosity, b = stoichiometric factor.

IOP Conf. Series: Journal of Physics: Conf. Series **1082** (2018) 012089 doi:10.1088/1742-6596/1082/1/012089

3. Results and discussion

3.1 Particles grain shape, mineral phase and chemical composition

The Capel area of south-west Western Australia has extensive buried beach sand deposits and this sample is a typical commercial concentrate from the region. The visual assessment of the samples showed that the sand is lighter in colour but still retains the submetallic lustre of ilmenite. From the surface morphology observation in Figure 1, it can be observed where some of the particles shapes have a rounded and angular shape but most of the particles have a very low and medium in sphericity. The chemical analysis indicated that this sample contained 53.9% TiO_2 and showed some alteration, with 45.0% of the iron being in the ferric form. Examination of the XRD pattern of sample, ilmenite phase are the main minerals present in the sample followed by minor traces of hematite and rutile phase, as shown in Figure 2. The polished section of ilmenite sample is shown in Figure 3 showed the inclusion of other minerals by the patchy pattern with dark and grey appearance under high magnification. Further EDX analysis identified the presence of other elements such as vanadium, manganese and iron along with titanium. Further investigation can be seen by the optical micrograph in Figure 4 which also shows the weathering effect on the mineral.

Figure 1. SEM micrograph of ilmenite particles grain.

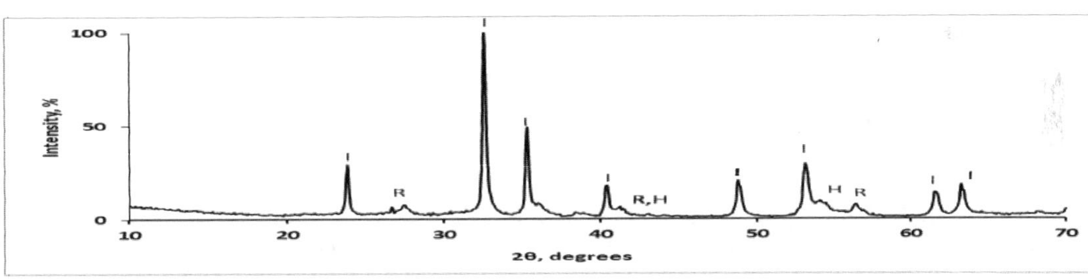

Figure 2. XRD patterns for ilmenite ore from Western Australia ($K_\alpha Co$, λ = 1.788965) (I: ilmenite; R: rutile; H: hematite)

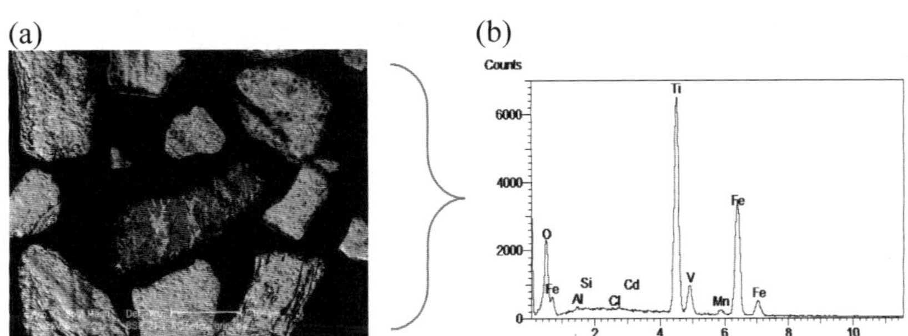

Figure 3 (a) Backscattered image of ilmenite polished section sample and (b) EDX analysis on selected area, at 200x magnification.

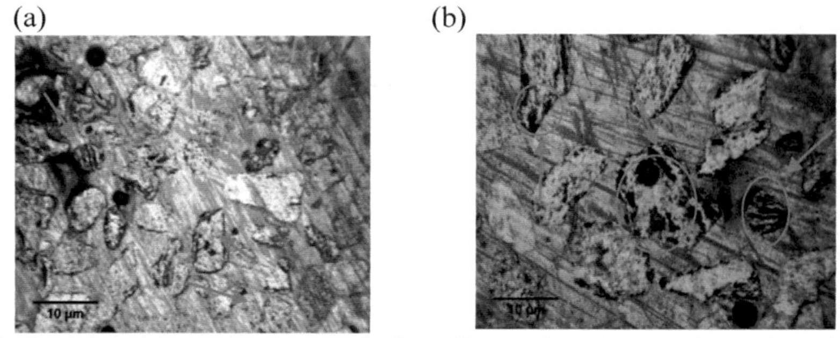

Figure 4. Optical micrograph showing the weathering effect on ilmenite sample (a) altered particles (b) particles with patchy pattern showed by red arrows.

3.2 Ilmenite leaching and kinetic study

3.2.1 Effect of HCl and leaching temperature. The degree of oxidation of the ferrous iron is expressed by the mole ratio Fe^{3+}/Fe while the degree of leaching of iron during weathering is expressed by the mole ratio Ti/Fe. To express an overall degree of alteration of ilmenite, a quantitative method described by Equation 4 [7]:

$$M = (\frac{Fe^{3+}}{Fe}) \text{ x } (\frac{Ti}{Fe}) \qquad (4)$$

where M is defined as the alteration factor. The value of M falls between the theoretical limits of zero, for unweathered ilmenite (i.e. theoretical $FeTiO_3$), to an infinitely large number, for completely weathered material containing little or no iron (e.g. leucoxene, TiO_2). Based on Equation 5, this ilmenite sample have alteration factor (M) of 1.2 which is moderately altered ilmenite.

During ilmenite leaching in HCl solution, Fe and Ti will dissolved in solution according to chemical reaction which can be represented by Equation 5 below:

$$FeTiO_3 + HCl \longrightarrow Fe^{2+} + TiO^{2+} + 2H_2O \qquad (5)$$

For the effect of different HCl concentration ([HCl] = 4 M- 11 M; leaching temperature was remained constant at 70 °C for 5 hours) it shows that the initial leaching of Fe was greater than that of Ti but it did not remain so at longer times (Table 1). As the HCl concentration increased, the percentage dissolution of Ti became higher than that of Fe. This behaviour can be a result of the presence of hematite (Fe_2O_3) and rutile (TiO_2) in the sample which influence the initial and final dissolution behaviour. The alteration factor showed that this sample was slightly altered ilmenite, which contains rutile and fine grained hematite (in the altered region) as discussed in section 3.1. The results listed in Table 1 show that in 4 M HCl and 7 M HCl the iron dissolution predominated over that of titanium for up to 60 minutes. This was followed by a time at which some rates were similar. However, after 60 minutes Ti dissolution became greater than that of Fe, until it reached complete leaching at higher concentrations of HCl.

For the study of effect of different temperature ([HCl] = 7 M was kept constant for 5 hours), it can be seen that as the leaching time progressed up to 1 hour, the extractions of Fe and Ti increased with increasing temperatures from 80 °C to 110 °C. After 4 hours of leaching at 90 °C, 95% of Fe and Ti was dissolved in solution and less than 2 hours at 100 °C and 110 °C.

Table 1. Effect of HCl concentration and temperature on the leaching efficiency (%) of Fe and Ti for ilmenite leaching in HCl solution at various parameters shown in the table.

Leaching	1h		2 h		3 h		4 h	
conditions	Fe	Ti	Fe	Ti	Fe	Ti	Fe	Ti
[HCl]/ (M)								
4	35	29	45	46	63	65	67	79
7	40	43	55	60	67	74	79	83
9	86	95	97	98	97	98	97	98
10	90	98	97	98	97	98	97	98
11	94	100	97	100	97	100	97	100
T/ (°C)								
80	35	43	55	60	67	74	67	83
90	64	69	81	88	94	90	95	91
100	85	95	93	98	96	99	97	100
110	92	99	99	99	100	100	100	100

3.2.2 Evaluation of Activation energy. The activation energy can be derived from the Arrhenius plot of Ln dX/dt as a function of -1000/RT. The activation energy for ilmenite reaction in HCl obtained in this study is 90.0 kJ mol^{-1} for Fe and 71.9 kJ mol^{-1} for Ti. A high values of activation energy of ilmenite ranging from 62 to 92 kJ mol^{-1} in HCl or H_2SO_4 was also reported suggesting that the mixed chemical-diffusion controlled reaction [8]– [11].

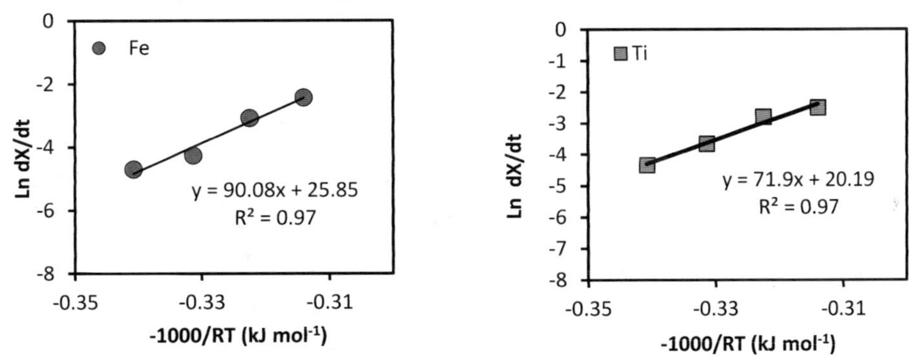

Figure 5. Arrhenius plot for Fe and Ti dissolution on ilmenite samples Arrhenius plot for Fe and Ti dissolution on ilmenite samples ([HCl] = 7 M, pulp density = 4 g/L, T = 80 °C, 90 °C, 100 °C, 110 °C, time = 5 h, size fraction = +53-63 µm).

3.2.3 Kinetic reactions. The reaction scheme of ilmenite leaching in HCl solutions is complex because of the high acid concentration used and the complex mineralogy of ilmenite samples. The reactions between ilmenite and HCl change the concentration of reagents with time and affect the kinetics of ilmenite dissolution. Examination of plots based on the equations and the two expressions on the left hand side of Equations 1 and 2 as a function of time for ilmenite sample showed that only Equation 2 (shrinking core model) gave a straight line (for the time period of 0 to 30 min) with a higher correlation ($R^2 \approx 1$) compared to Equation 1 (shrinking particles) as shown in Figure 6, both for Fe and Ti dissolution. Thus, the reaction kinetics of ilmenite is controlled by diffusion of the mass transport through a thickening product or solid layer on the surface. This assumptions also can be validated with SEM analysis on leaching residue after 5 hours of leaching as shown in Figure 7 below. The observation on surface of leaching ilmenite residue showed that the solid layer forms on the surface of particles after 5 hours of leaching and this solid layer inhibit further dissolution of ilmenite. Further investigation should be done on solid layer that form around the ilmenite grains

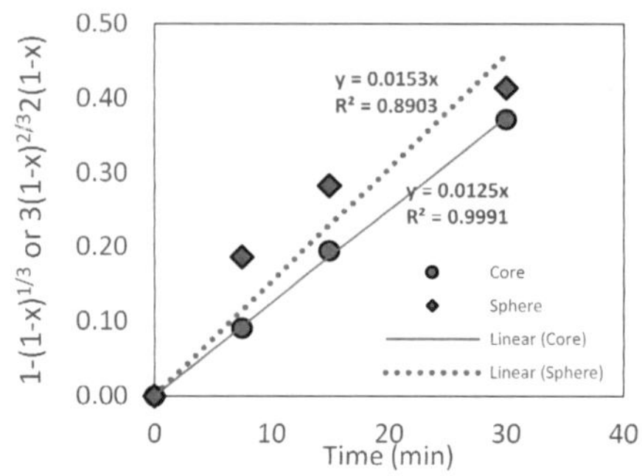

Figure 6. Validity of kinetic models the leaching results (Data was plotted from Fe leaching rate in 7 M HCl at 80 °C with pulp density of 4 g/L based on Equation 1 and 2).

(a) (b)

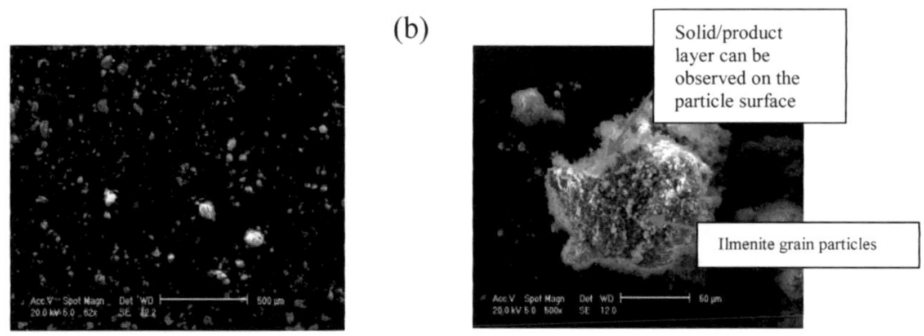

Figure 7. SEM images of particles obtained after leaching in 7 M HCl at 40.6 g/L pulp density, at 80 °C (a) after leaching (2 h) and (b) after leaching (5 h) at higher magnification.

4. Conclusion

The direct leaching of ilmenite in HCl to obtain synthetic rutile can be achieved without the pre-treatment of the raw materials which consume large energy and cost. However, it is still depend on the mineralogical nature of ilmenite samples. The more altered ilmenite become more difficult to directly leach ilmenite in acid without oxidation or reduction process. Further investigation on using reducing agent should be done to enhance the leaching rate of ilmenite. In this work, ilmenite samples obtained can be categorized as moderately altered ilmenite had resulted high leaching efficiency of Fe and Ti in HCl solution. Therefore, direct leaching method could be new potential methods for production of titanium dioxide from synthetic rutile.

Acknowledgement

The authors are thankful to School of Materials and Minerals Engineering, Universiti Sains Malaysia for financial assistance and Murdoch University, Western Australia where this work was conducted.

References

[1] Zhang W, Zhu Z and Chang CY 2011 *Hydrometallurgy* 108 177
[2] Pownceby MI, Sparrow GJ and Fisher-White MJ 2008 *Minerals engineering* 21 587
[3] Xue T, Wang L, Qi T, Chu J, Qu J and Liu C, 2009 *Hydrometallurgy* 95 22
[4] Egerton TA, 2000 *Kirk-Othmer Encyclopedia of Chemical Technology*
[5] Crundwell FK, 2016 *Hydrometallurgy* 161 34
[6] Levenspiel O, 1972 *John Wiley & Son: New York USA* 668
[7] McConnel SR, 1978 *Thesis Murdoch University*
[8] Olanipekun E, 1999 *Hydrometallurgy* 53 1
[9] Tsuchida H, Narita E, Takeuichi H, Adachi M and Okabe T, 1982 *The chemical society of Japan* 55
[10] Jackson JS, Wadsworth ME, 1976 *Light Metals* 481
[11] Zhang S and Nicol MJ, 2010 *Hydrometallurgy* 103 19

IOP Conf. Series: Journal of Physics: Conf. Series **1082** (2018) 012090 doi:10.1088/1742-6596/1082/1/012090

Characterization of Malaysian Monazite Concentrate for The Recovery of Thorium Dioxide

Sanjith Udayakumar[1], Sheikh Abdul Rezan[1], Ahmad Fauzi Mohd Noor[1], Teuku Andika Rama Putra[1*], Khaironie Mohamed Takip[2]and Roshasnorlyza Hazan[2]

[1]School of Materials & Mineral Resources Engineering,Universiti Sains Malaysia, Engineering Campus, Penang, Malaysia.

[2]Material Technology Group, Industrial Technology Division, Malaysian Nuclear Agency, Bangi, Selangor, Malaysia.

[*]Email: teukuandika@usm.my

Abstract. Malaysia has many potential mineral resources including some heavy and light rare earth element (HREE and LREE's) minerals, such as Monazite and Xenotime. Their occurrence is in association with ores containing phosphates and oxides of Cerium (Ce), Lanthanum (La), Neodymium (Nd) and Thorium (Th). Mineralogical characterization of Malaysian Monazite concentrate for the recovery of Thorium dioxide (ThO_2) has been carried out in this study. Scanning Electron Microscope (SEM) equipped with an Energy Dispersive X-ray Analyser (EDX) was used to assess the morphology and elemental composition of the mineral. The latter analysis showed that EDX surface analysis detected P-K_α Th-M_α, La-L_α, Ce-L_α and Nd-L_α as the major elements. The SEM micrographs with the EDX analysis revealed the morphological characteristics of Monazite which showed angled and irregular structures with dark coloured patches of Al, Si and Mn based inclusions. XRF analysis confirmed the presence of Th (8.52 wt.%) and U (0.25 wt.%). XRF also showed that the major rare earth elements (REE) of interest (> 5 wt.%) such as Ce, Nd and La were abundant in the mineral. XRD analysis identified $LaPO_4$, $CePO_4$ and $NdPO_4$ as the major phases. Thorium phosphate ($Th_4(PO_4)_4$) phase was not detected via XRD due to stronger X-ray reflection from the rare earth phosphates. The information collected from this study will help in developing a suitable process of treatment for the local Monazite, thereby facilitating the mineral to be used a domestic source of Th and other associated REE's.

1. Introduction

The potential of rare earth elements (REEs) comprising the fifteen elements of the lanthanide series have been recently explored in different applications such as high-strength permanent magnets, catalysts for petroleum refining, metal and glass additives and phosphors used in electronic displays [1]. The solution for a sustainable energy source has gathered widespread attention on minerals that contain Naturally Occurring Radioactive elements namely Monazite and Xenotime in the recent times. Thorium (Th) and uranium (U) represent the heaviest naturally occurring elements on Earth however, the abundance of Th over U in nature is very limited. Added to this, Th is fissile in nature and has replaced U in the nuclear energy industry as a sustainable and safe alternative [2]. For many years, it has been a tantalizing prospect to use Th as a new primary source of nuclear energy. However, a considerable research and development investment was required to realize the challenge of extracting the latent energy of this element in a cost effective manner [3]. In Malaysia, Th exists in minerals and REE production residue. Malaysia has several minerals (e.g. monazite, zircon, xenotime and ilmenite) which are classified as strategic minerals as it contains a significant amount of Th and U (above 500 ppm) which requires a regulatory control. Malaysian monazite and xenotime minerals contains an average range of about 70,000 and 15,000 ppm respectively. Although Th can become a major radiological problem to our environment if not processed, its prospect as a future alternative fuel in nuclear technology is very promising [4].

Monazite concentrate are found with a variety of other minerals, including silica, magnetite, ilmenite, zircon and garnet [5]. Th is mainly obtained from monazite concentrate as a by-product of extracting rare earth oxides and metals. Although the mineral has a similar combination of elements, they exist with unique mineralogical composition different from another mineral deposit. In this case, a thorough understanding of mineralogy, chemical composition, size, morphology and elemental association with other minerals, by means of in-situ characterization, will provide a basis on the mineral's behaviour during the separation of the particular REE's of interest. In this paper, preliminary characterization of the Malaysian monazite concentrate was carried out to understand the morphology, elemental composition and mineral phases present to improve the prospects of using it as a source of Th and other associated REE's.

2. Materials and Methods

Malaysian Monazite concentrate from Ipoh silica sands were supplied by Ooi Cheng Huat Private Limited. Cone and quartering sampling method was used on the concentrate to have a representative sample for the characterization tests. Polished sections of the samples were prepared using a mixture of epoxy resin and hardener(1:1 ratio) prior to mounting for morphological analyses. The polished sections were coated with Au coating via sputtering for improved conductivity during SEM and mounted. High definition examination of surface morphology and elemental analysis were carried out with a Scanning Electron Microscope equipped with Energy Dispersive X-ray Analyser (Hitachi S-3400N). The identification of mineral phases present in the sample was done using Bruker D8 Advanced X-ray diffractometer. The bulk chemical composition of the mineral was determined by X-ray Fluorescence (XRF) analysis using a Panalytical Axios-Max Spectrometer. The samples were homogenized and hydraulically pressed to form pellets for the XRF analysis.

3. Results and Discussions

3.1 SEM/EDX Analysis

Figures 1 (a)-(b) present the SEM images with EDX analyses taken at specific points of the mineral. The surface morphology showed an angular and irregular shape of the grains due to the weathered nature of the concentrate. The elemental EDX analysis shows that the major elements present are Ce, La, Th, Nd as phosphates. Based on morphological and optical studies in the literature [12], the darker grain in Fig.1(b) indicate traces elements of Al, Si and Mn which exhibit different shades of grey under SEM imaging depending upon their association with other elements in the mineral. The elements of Ce and La in Fig. 1(b) exhibit higher weight concentration which was the characteristic of monazite mineral. Added to this, U can also be observed in trace levels (wt. % < 1) for this micrograph. The high carbon concentration detected was because of the carbon tape in the poly-section. The variations in the elemental combination of the minerals in the SEM/EDX analysis can be attributed to non-uniformity of the texture and particle distribution in the mineral.

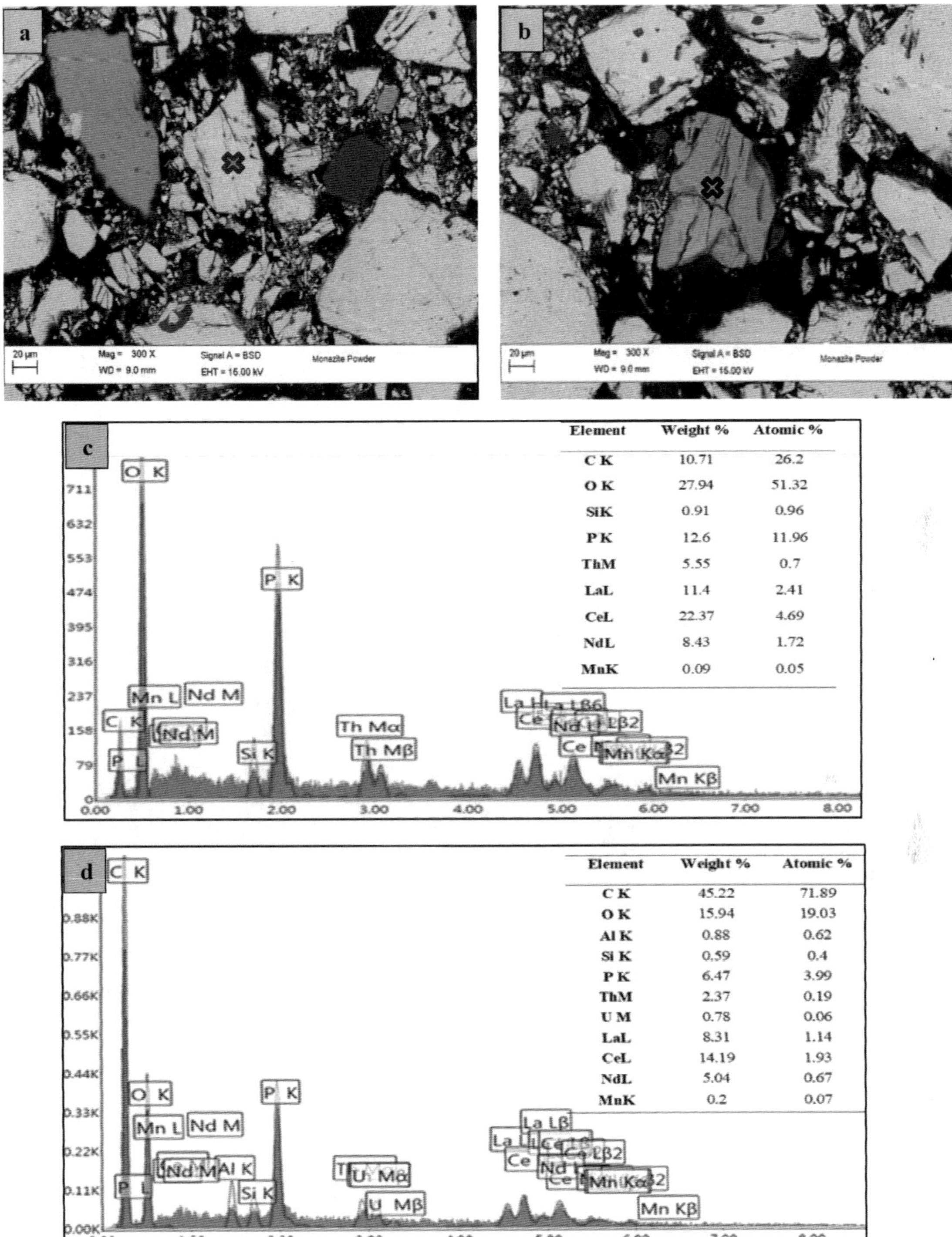

Element	Weight %	Atomic %
C K	10.71	26.2
O K	27.94	51.32
SiK	0.91	0.96
P K	12.6	11.96
ThM	5.55	0.7
LaL	11.4	2.41
CeL	22.37	4.69
NdL	8.43	1.72
MnK	0.09	0.05

Element	Weight %	Atomic %
C K	45.22	71.89
O K	15.94	19.03
Al K	0.88	0.62
Si K	0.59	0.4
P K	6.47	3.99
ThM	2.37	0.19
U M	0.78	0.06
LaL	8.31	1.14
CeL	14.19	1.93
NdL	5.04	0.67
MnK	0.2	0.07

Figure 1(a)-(d). SEM/EDX micrographs of Malaysian monazite mineral. Figures 1(c) and (d) correspond to the points marked in SEM micrographs of Figures 1(a) and (b) respectively.

3.2 XRF Analysis

The results of the X-ray Fluorescence Spectroscopy analysis conducted for the monazite was tabulated in Table 1. The analysis indicated that it contained significant amounts of REE elements like Ce, La, Nd and about 8.5 wt.% of Thorium (NORM). Traces of Y, Sm and Gm and Si can also be observed in the mineral with the overall composition below 5 wt.%. A considerable amount Mn of about 8.5 wt.% was present and could be extracted for industrial use in steelmaking [6]. The XRF results show that Th was a chemically bonded along with a wide range of light REE's which requires distinctive processing and separation techniques. It is due to the presence of numerous elements in the Malaysian monazite mineral which adds to the complexity in separation and recovery of Th. For example, to recover Th, the phosphates composition should be digested [7, 8] or desphosphorized [9, 10].

Table 1. Bulk chemical composition of Malaysian monazite mineral by XRF

Composition	Content (wt.%)	Composition	Content (wt.%)
Al_2O_3	0.94	SnO_2	0.301
SiO_2	2.56	La_2O_3	11.569
P_2O_5	19.391	CeO_2	21.528
SO_3	0.118	Pr_2O_3	2.797
K_2O	0.056	Nd_2O_3	10.634
CaO	0.782	Sm_2O_3	1.756
TiO_2	0.267	Gd_2O_3	1.276
MnO	8.585	ThO_2	8.525
Fe_2O_3	0.529	U	0.253
Y_2O_3	1.718	Others	~5.65
ZrO_2	0.764		

3.3 X-ray diffraction (XRD) Analysis

The elements of Th and REOs usually exist together as they have similar atomic radii [1, 11]. As the monazite bearing of the rare earths was fine in terms of grain size, their identification cannot be done by visual inspection. XRD analysis is an important technique in identifying the mineral phases present in monazite wherein there were chances for multiple phases of the elements to be present in co-existence with the other REE's. From XRF analysis and literature review, Malaysian monazite comprises of phosphates of Ce, La, Nd, Th predominantly and lesser amounts of Mn, Y, Sm, Gd, Al, Si based compounds [12, 13]. The diffraction pattern of the bulk sample was shown in Figure 2. Phases of $LaPO_4$, $CePO_4$ and $NdPO_4$ are very prominent and are assumed to be present together as their peaks are overlapping with each other [14]. It can be noticed that the analysis identifies $CePO_4$ in almost every peak because the composition of the mineral is dominated with Ce. However, thorium phosphate is not identified in the diffraction peaks which could be due to the stronger reflection of the X-ray due to the heavy atomic nucleus of Thorium.

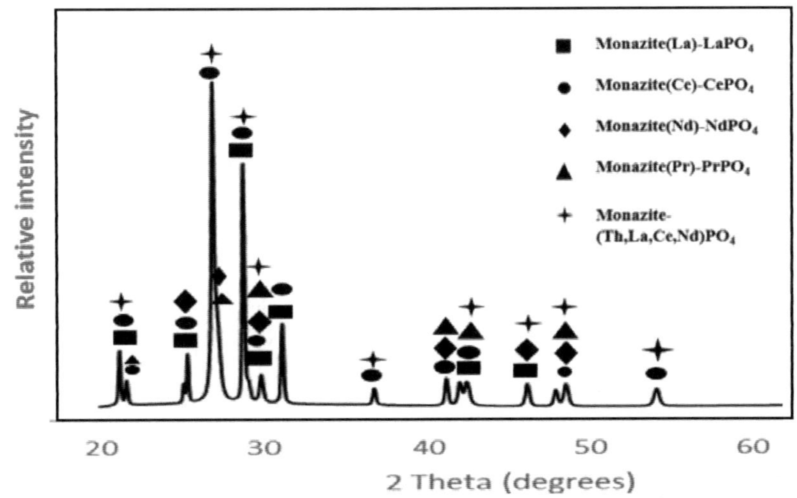

Figure 2. X-ray diffractogram of powder monazite.

It is quite interesting to note that a lot of elements identified in the elemental and bulk analysis like Si, Al, Mn, have not been detected by the XRD analysis. One reason is that the mineral was pre-concentrated such that a lot of impurities have been removed at an earlier stage. Also, the intensity of reflected XRD from the rare earths is very high compared to that of the energy of the reflected rays from the lighter elements which makes their detection difficult. Some of the peak positions and their intensities may vary somewhat for specimens of an individual mineral collected from different localities [15]. These differences are due mainly to variations in the elemental content of a particular mineral. It is also possible to identify the minor phases in the mineral, however, it requires partitioning of the diffraction pattern into multiple figures and analysing the figures separately corresponding to one particular mineral and its related phases.

4. Conclusion

The mineralogical characterization on a preliminary level was carried out for Malaysian monazite in this study. The morphological and elemental analysis confirm the presence of Thorium in the concentrate indicating the potential of Malaysian monazite as a source of Thorium. The various REE phosphates (Ce, La, Nd, Pr) present in the mineral were identified by diffraction studies and their composition was determined by X-ray Fluorescence spectroscopy. The knowledge on the mineralogy of Malaysian monazite can be applied to study and improve the available methods of separation for the REE's and increase the production efficiency of Thorium. Further characterization studies will aid in understanding the chemistry of the composing elements, and will be helpful to propose specialized selective extraction techniques based on the element of interest.

Acknowledgements
The authors gratefully acknowledge the support from University Sains Malaysia (USM) Fellowship Award. This work was financially supported by USM and Ministry of Higher Education (MOHE) of Malaysia through the following research grants: Fundamental Research Grant Scheme (FRGS) (No. 203/PBAHAN/6071364), USM Research University Individual (RUI) grant (No. 1001/PBAHAN/814273) and Nippon Sheet Glass Research Grant (NSGRG) (No. 304/PBAHAN/650360/N120).

IOP Conf. Series: Journal of Physics: Conf. Series **1082** (2018) 012090 doi:10.1088/1742-6596/1082/1/012090

References

[1] Jordens A, Cheng YP, Waters KE. 2013 *Minerals Engineering* 41 97

[2] IAEA NFC. 2005 International Atomic Energy Agency

[3] Lainetti PE. 2015 Thorium and Its Future Importance for Nuclear Energy Generation. Brazil: 2015. Report No.: INIS-BR--15811.

[4] Sulaiman MY, editor. 1991 An overview of the rare-earth mineral processing industry in Malaysia. Materials Science Forum: Trans Tech Publ.

[5] Mohanty AK, Sengupta D, Das SK, Saha SK, Van KV. 2004 *Journal of Environmental Radioactivity* 75 15

[6] Beer Jd, Worrell E, Blok K. 1998 *Annual Review of Energy and the Environment* 23 123

[7] Al-Areqi WM, Ab. Majid A, Sarmani S. 2014 *Malaysian Journal of Analytical Sciences* 18 221

[8] Bahri CNACZ, Al-Areqi WM, Majid AA, Ruf MIFM. 2016 *Malaysian Journal of Analytical Sciences* 20 44

[9] Panda R, Kumari A, Jha MK, Hait J, Kumar V, Kumar JR, et al. 2014 *Journal of Industrial and Engineering Chemistry* 20 2035

[10] Xing P-f, Zhuang Y-x, Tu G-f, Guo J. 2010 *Transactions of Nonferrous Metals Society of China* 20 2392

[11] Gupta CK, Krishnamurthy N. 1992 *International Materials Reviews* 37 197

[12] Sadri F, Rashchi F, Amini A. 2017 *International Journal of Mineral Processing* 159(Supplement C) 7

[13] Ito A, Otake T, Shin K-C, Ariffin KS, Yeoh F-Y, Sato T. 2017 *Applied Geochemistry* 82(Supplement C) 89

[14] Panda R, Kumari A, Jha MK, Hait J, Kumar V, Rajesh Kumar J, et al. 2014 *Journal of Industrial and Engineering Chemistry* 20 2035

[15] Staatz MH, Brownfield IK. 1982 *US Geological Survey*

IOP Conf. Series: Journal of Physics: Conf. Series **1082** (2018) 012091 doi:10.1088/1742-6596/1082/1/012091

Effect of Grinding Aids in Cement Grinding

Syed Fuad S.Hashim and **Hashim Hussin**

Strategic Mineral Group, School of Material and Mineral Resources Engineering
Engineering Campus, Universiti Sains Malaysia, 14300 Nibong Tebal, Penang, Malaysia

Email: *mrsyfuad@usm.my*

Abstract In fine grinding of cement in a ball mill, it is sometimes impractical to grind finer in a dry state. Even though the chemical reactivity requires the material to be dry ground, it is sometimes necessary to use different breakage machines to obtain the product, but it is usually more expensive, requires more energy and reduces capacity. The economic alternative is to use a grinding aid. Grinding aid or grinding additives refer to substances which when mixed into the mill contents cause an increase in rate of size reduction and flowability. Grinding process of clinker was carried on with a laboratory scale ball mill by varying different type of grinding additives and dosage while the operating conditions of the mill was kept constant. These additives were added into mineral in certain ratio based on the mineral weight and the grinding has been done for a definite time at the same condition. Analysis were then conducted for their size distribution and compaction factor. Results shows that there were significant improvement in size distribution as the dosage and types of grinding additives changes. Different types of grinding additives were found only suitable for clinker grinding while other give no significant results. The results obtained showed that with the addition of grinding additives can improve the grindability by decreasing agglomeration and increase breakage and hence, reducing ball coating Industrials application of grinding additives also give significant improvement in terms of throughput and size distributions.

1. Introduction

Grinding aid or grinding additives refer to substances which when mixed into the mill contents cause an increase in rate of size reduction [1]. These terms are very common in cement industries where it is used to increase the throughput to the mill. Grinding aid also affect the cement flowabilitiy throughout the circuits. The grinding aid facilitates size reduction so that the mill has to apply less grinding power without having adverse effect on any of the properties of the resulting cement. Another beneficial effect of grinding aids used today is to decrease compaction in storage. It is generally improves grinding efficiency, by bringing down the limit of grinding.

In dry ball milling of fine powders such as cement grinding, a common phenomenon is the slowing down of breakage as the powder becomes finer, i.e the grinding efficiency decreases with the accumulation of fines in the mill content. This might be due to the regrowth or rebuilding of particles from smaller particles by either agglomeration involving van de Waal's force or by direct coating of balls to give soft surfaces [1].

However, the action mechanism of grinding aids, has not been understood perfectly despite large amounts of useful information [2,3]. But from industrial experience, there is significant effect in terms of throughput and flow of material in the mill and throughout the circuits.

Cement clinker is difficult to grind and the fine grinding of this material is one of the major problems of cement industry. The benefits of any grinding aid must outweigh its cost and the grinding aid should have no detrimental effect on the downstream process of the finished product.

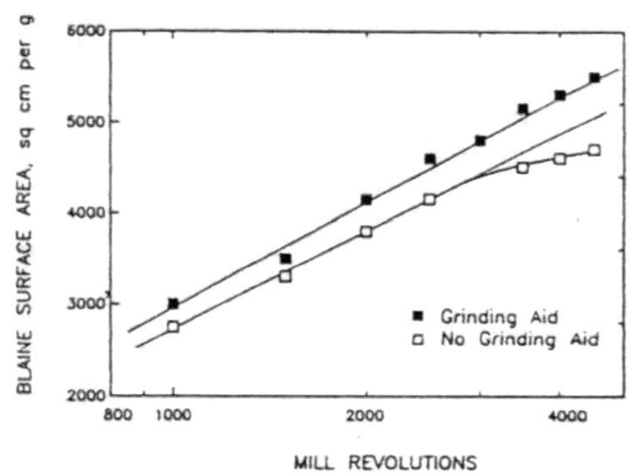

Figure 1. Comparison between grinding of cement clinker with and without grinding aid (after Fuerstenau, 1990)

Figure 1 shows the typical comparison between grinding of cement clinker with and without a grinding aid in a laboratory mill. As discussed in the literature [1,2,4], there was significant effect of using grinding aid in cement grinding. Grinding aids are active surface agents that counteract the agglomeration of very fine particles on the grinding media, thus inhibiting the grinding effect and therefore reducing production rates. A number of grinding aids are available on the market. The most suitable grinding aid for a certain grinding process has to be determined by full scale test runs. The use of grinding aids in both open and closed circuit systems will improve the control of the grinding process, increasing the Blaine number (finer product). These advantages usually justify the cost of a grinding aid. But some cement manufacturers refuse to use grinding aids for economic reason.

Locher and von Seebach [5] investigated the use of organics in cement dry grinding and found that the use of chemical additives does not affect the breakage of coarse material and become a factor when fine material builds up in the mill. This conclusion was in parallel to literature [4]. In their work found that grinding aids work effectively in the range of very fine grinding. The use of an additive also increases flowability of the powder and reduces the mean residence time at a given feed rate [1].

2. Experimental Work

Experimental work was divided in two parts: industrial survey and laboratory work. Industrial survey was conducted from a cement plant in Perak. The milling facilities consist of a two-compartment air-swept ball mill arranged in close circuit. The first compartment of the open circuit mill contains balls ranges from 90mm to 30mm in diameter, while for the second compartment contain 20mm ball. Samples are taken from the two-compartment ball mill in close circuit the cement grinding with and without grinding aid.

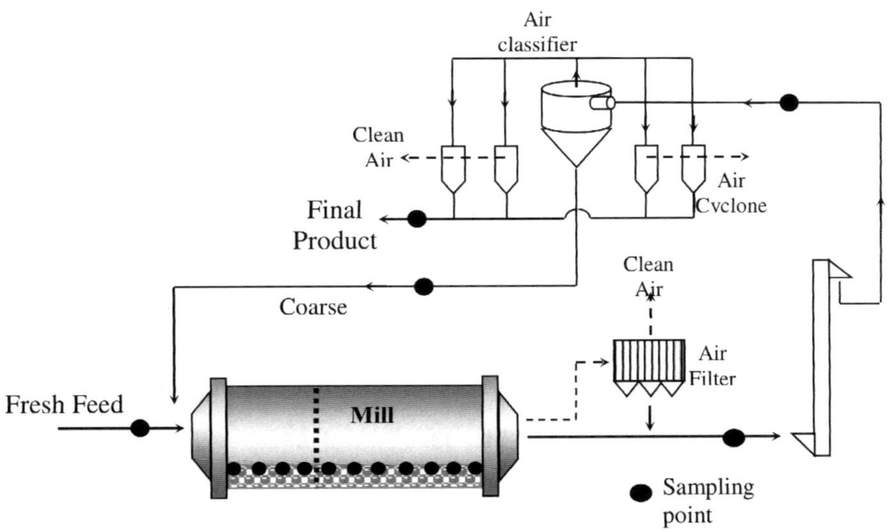

Figure 2. Sampling point for the close circuit mill

Table 1. Operating condition and product type for close circuit

	Survey 7	Survey 8	Survey 9	Survey 10	Survey 11	Survey 12
Clinker Feed (TPH)	87	82	86	100	104	95
Gypsum Feed (TPH)	4.8	4.9	5.3	6.2	5.9	5.3
Mineral Add (TPH)	4.1	4.1	4.3	5.4	5.6	4.8
Combined Feed (TPH)	95.9	91	95.6	111.6	115.5	105.1
Grinding Aid	No	No	No	Yes	Yes	Yes

Laboratory experimental work was also conducted to study the effect of various types of grinding additives on product fineness and compressability. Triethanolamine (TEA), Ethylene Glycol (EG) and CEMEX (commercial grinding additives) was tested at various dosage in ball grinding mill at fixed grinding condition (ball size, speed and filling ratio). The product was tested for their size distribution using laser sizer while compressibility test (Carr's Index) was used for this study.

3. Result and Discussion

3.1 Industrial experimental work

Figure 2 shows the size distribution of the surveys conducted. The was a sudden decrease from the feed to the first point. The effect of recycle material was obvious and indicated by Point 1 and Point 2 having finer size distribution. The classification of material due to middle grate was not obvious. Only the coarse fraction indicate clear finer size along the mill. The important conclusion from these observations was that the use of grinding aids gave finer product, therefore increased the throughput (Table 1).Product from surveys 10,

IOP Conf. Series: Journal of Physics: Conf. Series **1082** (2018) 012091 doi:10.1088/1742-6596/1082/1/012091

11 and 12 gave finer product compared with the one without grinding aid. Survey 7 and 12 are also plotted in the graph for comparison (Figure 2)

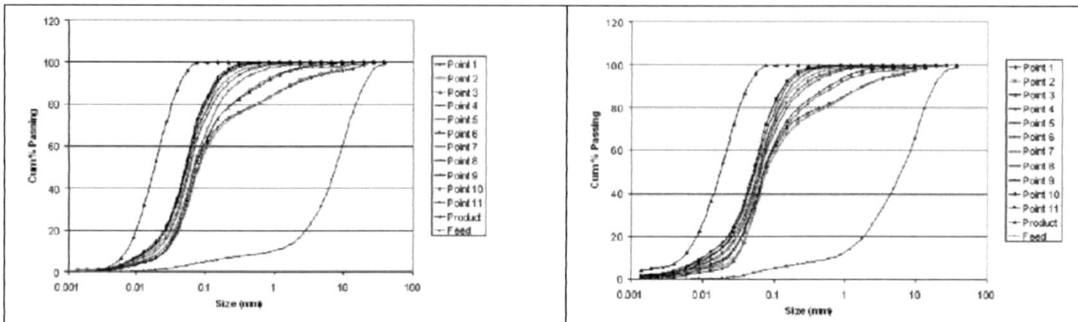

Size distribution without grinding additives

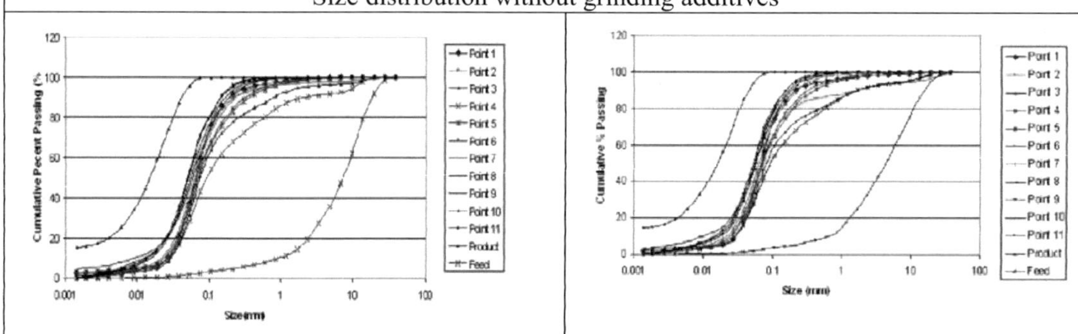

Size distribution with grinding additives

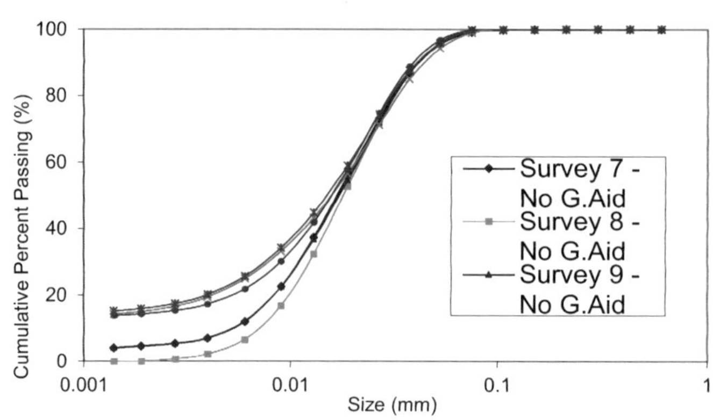

Figure 3. The effect of grinding aid in clinker grinding

Figure 3 shows the size distribution analysis for various grinding additives at 0.1% dosage. Although the particle size analysis has resulted in the increasing of coarser particles with the increasing concentration of TEA, it shows that the powder is free flowing with higher concentration of TEA. The increasing of coarser size is because of the agglomeration of particles during the grinding process and most probably because of the concentrations of the TEA used in this experiment is too high until it can't meet its optimum percentage usage. However, from the observation of ball coating, it has also shows that with the addition of grinding additive, it has results in less ball coating and mill lining coating and thus, proving that the flowability of the clinker powder has increase with the addition of grinding additive

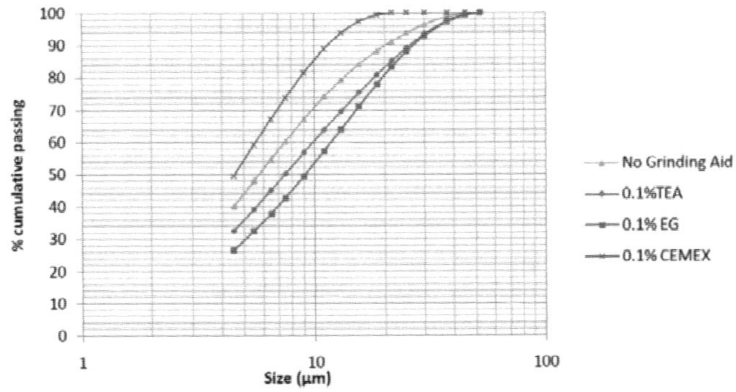

Figure 4. Size distribution analysis for various grinding additives at 0.1% dosage.

The particle size distribution with the addition of CEMEX results in the movement of curve towards finer size region with the increasing percentage of CEMEX except for 0.02% addition of CEMEX. This is because at the 0.2% concentration, the particles tend to agglomerate and make the size coarser but with the increasing of CEMEX concentration, it has results in finer size of particles. The adsorption of the grinding additive to the surface of particles has decrease the cohesive force and thus, prevents the agglomeration of particles. The increasing in the flowability can be proved when there is less ball coating (Figure 5), mill lining coating and less agglomeration occurred that can be showed with the movement of curve towards the finer particle size.

The particle size distribution with the addition of EG has results in the movement of curve towards a coarser size region with the increasing percentage of EG. The agglomeration of particles with the addition of EG has results in the less breakage of cement clinker. Although the particle size analysis has resulted in the increasing of coarser particles with the increasing concentration of EG, the compressibility index has shows that the flowability increase with the increasing of percent grinding additive used. It increasing of flowability can be proved by looking at the ball coating and mill lining coating observation. It has shows that with the addition of EG it has results in less ball coating and mill lining coating.

Figure 5. Ball and lining coating without grinding additives

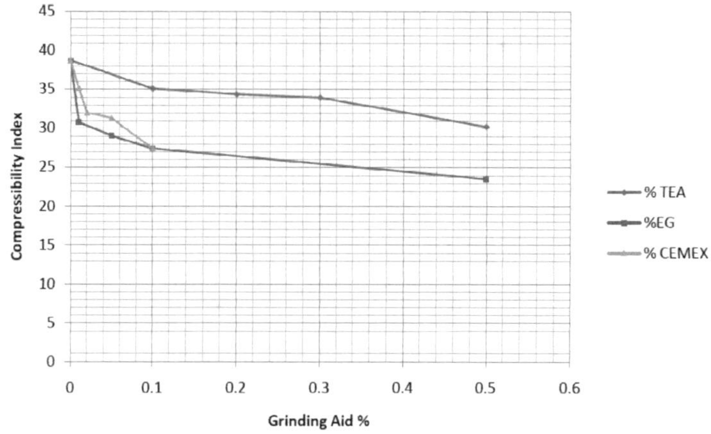

Figure 6: Compressibility index for various grinding additives and dosage

Figure 6 shows that at 0.1% TEA, EG and CEMEX, EG and CEMEX give the lower compressibility index compared to TEA. Thus, we can conclude that EG and CEMEX has higher flowability properties compared to TEA. The flowability of CEMEX and EG are almost similar but in term of particle size distribution, the cement clinker product with CEMEX addition gives finer size than EG.

4. Conclusions

In conclusion, with the addition of correct type of grinding additive in a small amount to the cement clinker, it will results in a finer ground product compared to grinding without an additive. For the current work, CEMEX was found to be the most suitable. The decreasing in the compressibility index number with the increasing of additive concentration may possibly due to the reduced tendency of compaction and the agglomeration of the ground cement clinker. The flowability of the cement clinker powder will also increase with the increasing of grinding additives concentration. It is hoped that the outcome of the work will help the cement industry to justify the use of grinding aids in their operations

References

[1] Austin, L.G., Klimpel,R.R., Luckie,P.T., 1984. *Society of Mining Engineers, New York*, 561
[2] Hasegawa,M., Kimata,M., Shimane,M., Shoji,T and Tsuruta,M., 2001 *Powder Technology*, 114 15 Fuerstenau, D.W., Abouzeid,A.-Z.M, 2002. *International Journal of Mineral Processing*, 67 161
[3] Rajendran Nair, P.B.a.P., R., 1999 Powder Technology 10131
[4] Sohoni, S., Sridhar,R. and Mandal,G., 1991 Powder Technology 67 277
[5] Locher W.F., Seebach H.M. (1972) *Indus. Eng. Chem. Process Des. Develop* 2 190

IOP Conf. Series: Journal of Physics: Conf. Series **1082** (2018) 012092 doi:10.1088/1742-6596/1082/1/012092

Remote Sensing Technique for Lineament Extraction in Association with Mineralization Pattern in Central Belt Peninsular Malaysia

T.G. Mathew[1,2], K.S. Ariffin[2]

[1,2]Universiti Sains Malaysia, Engineering Campus, Penang, Malaysia

Email: kamarsha@usm.my

Abstract. Geographical information system and remote sensing data are now more remarkable for the extraction of lineaments. Lineaments studies provide the tectonic history of geological settings. It is the objective of this study to demonstrate the extraction of lineaments from Landsat 8 Operational Land Imager (OLI) of Kelantan area. The current approach applied different image processing techniques on Landsat 8 imagery to achieve its gold. The image processing techniques include; principal component analysis (1,2,3), Band ratios (7/5,6/4, 4/2) and Minimum Noise Fraction bands (7,5,3). Gamm stretching and directional filtering facilitate the extraction of lineaments. The results of this study indicate that the area is affected by several structural trends (N-S, NW–SE, and NE–SW). The structural trends sever as conduits for gold mineralization fluids. The methodology demonstrates capability in extracting lineaments from Landsat 8 imagery. The methods locate known gold deposits and also highlight new locations for new ones.

1. Introduction

The interest in Geographical Information System (GIS) and remote sensing technology had increased its application in various fields of scientific research. Scientific research fields such as geology and mineral exploration considered GIS and remote sensing data as an excellent resource for structural mapping. Its importance in structural geological studies is remarkable. Remote sensing technology is capable of capturing and storing geological features from the earth surface. Lineaments are straight lines surface element that can be extracted directly from satellite imagery [1, 2]. Remote sensing research had reported the successful application of Landsat 8 imagery in arid and semi-arid regions in pieces of literature including [3,4, 5].

Remote sensing techniques such as PCA (principal component analysis), MNF (Minimum Noise Fraction), and Band rationing, improve and enhance satellites images. The enhanced images facilitate the differentiation and characterization of the structural geological elements [6,7, 8]. This study reviews the application of Landsat 8 imagery with regard to climate conditions. The review shows a more successful application in arid to semi-arid region compared to tropical environments[9,10,5]. Thick vegetation cover characterizes tropical environments. Landsat 8 application studies are challenging by vegetation cover in a tropical environment. Most times the techniques produced images without much clarity. The solutions to this hurdle lie in the selection of bands for the image analysis. This research provides the appropriate band combination for Landsat 8 application in Kelantan. The study extract lineaments from Kelantan area. And further, established the relationship between lineaments and gold deposits distribution within the study area. The idea of this work applied different image enhancement techniques on Landsat 8 imagery to achieve its gold.

IOP Conf. Series: Journal of Physics: Conf. Series **1082** (2018) 012092 doi:10.1088/1742-6596/1082/1/012092

1.1 Study Area

The study area is located between longitudes E 101°48′30″-102° 0′ 10″ and Latitudes N 4°38′50‘′-4°54′0″ in the southern part of Kelantan state Malaysia (Figure 1). Kelantan is located at the north-eastern corner of Malaysia Peninsular. The area covers approximately 570km^2. The region is characterized by tropical monsoon climate, with a significant temperature, which ranges from 21°C to 32 °C from January to December. An average yearly rainfall range of 2030 mm to 2540 mm characterizes the study area. The wettest months begins from November through January. The topography is mountainous with extensive deep valley. In this selected area of study, the altitude ranges between 80.0 m to 780 m (Figure 2). The high mountainous region is located North-central and the South-eastern part of the study area. An extensive drainage network with other small tributaries cut the study area and conditioned the morphology. The drainage pattern often displayed a general orientation, i.e. N-S, to NW-SW and NE-SW, N-S to NW-SE, furthermore, these orientations are related to the past orogeny activity. The study area is partly accessible by vehicle. The region is known for its historic gold exploration and mining activities.

Peninsular Malaysia is an integrated part of the Southeast Asian continental core of Sundaland, and it is characterized by two tectonic blocks/terranes, the Sibumasu terrane in the west while the Sukhothai Arc in the east [11,12, 13]. The selected region of study is composed of a variety of rocks, Igneous, sedimentary and metamorphic. The rocks are evenly distributed in a north-south trend. Further classification of the rocks in the area of study highlight four classes, granitic rocks, sedimentary/metasediment rocks, extrusive rocks and unconsolidated sediment. Localized geological features also characterize the area comprises of faulting and jointing in the granitic rocks and folding, faulting and jointing in sedimentary rocks[13].

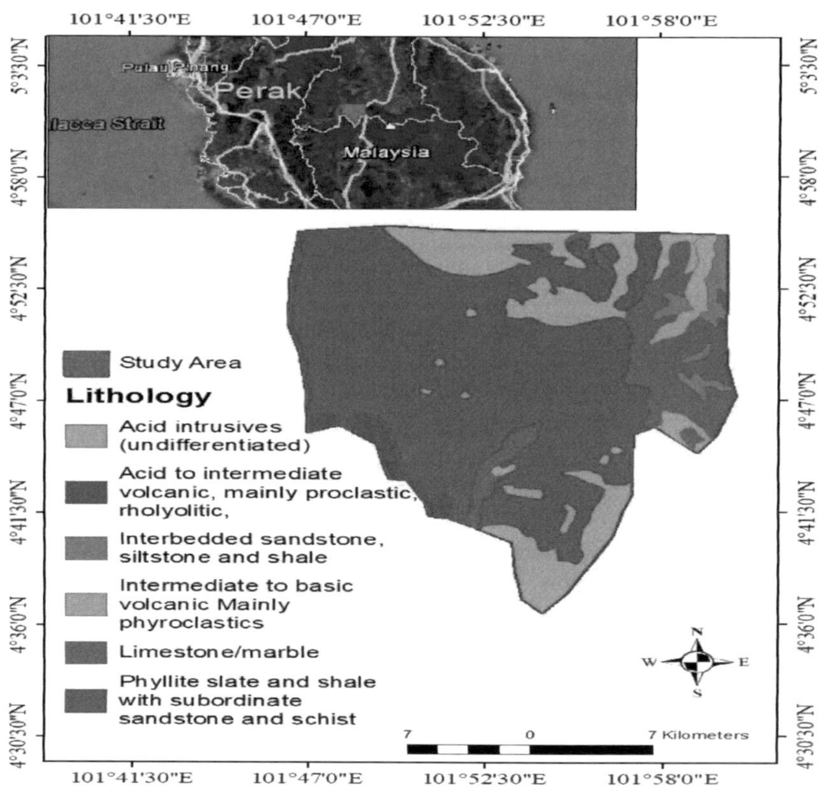

Figure 1. Geology and study area map (modified from Mineral and Geoscience Department Malaysia).

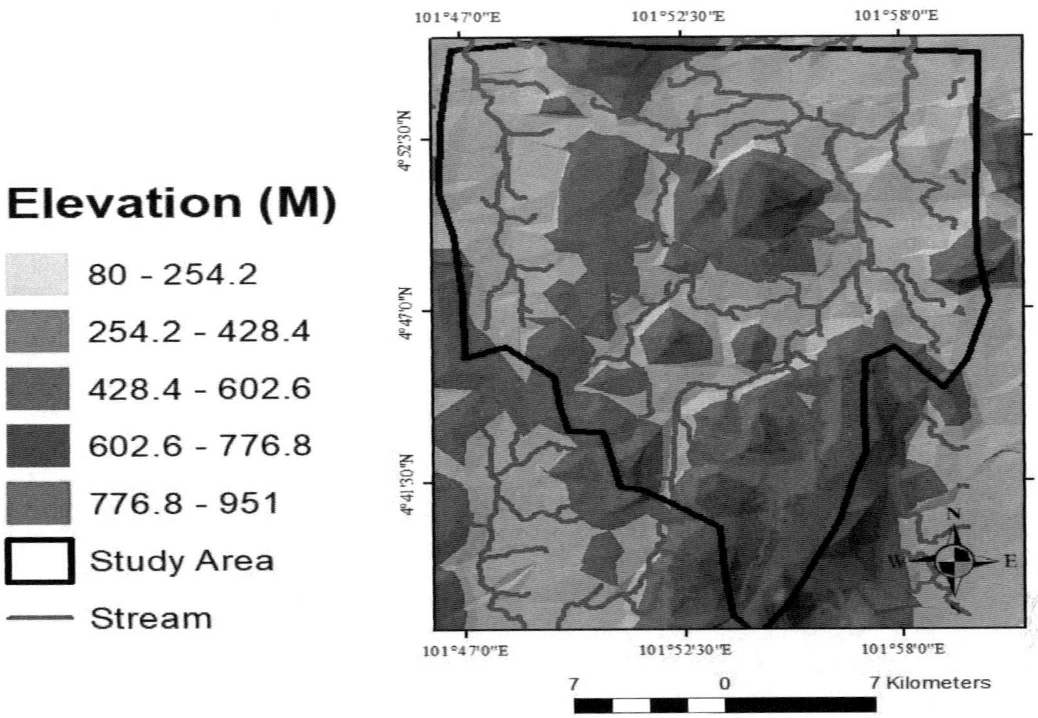

Figure 2. Elevation and stream pattern of study area

2. Methodology study

Focused on lineaments extraction have rendered many methods for their identification. Review in this regards has identified the application of the individual bands of Landsat 8 imagery. Band 5 is specific for soils and rock discrimination. While Band 7 is specific for geology, Band 3 is for soil discrimination from vegetation [10, 14].

2.1 Data Description and Image Enhancement

Data and equipment such as (i) Gold deposits locations provided by Mineral and Geoscience Departments of Malaysia (GJM) (ii) ArcGIS 10.3 and (iii) Landsat 8 OLI imagery (Path 127, Row 57) acquired on 3rd July 2016 free downloaded from USGS web page, were used in this study. The approach applied different enhancement techniques such as PCA, Band combination, and Band ratios on Landsat 8 imagery for lineaments extraction. PCA analysis considered bands (1-7). PC (1, 2, 3) revealed the structural information. Gamma stretch was applied to the RGB combination of PC 1, 2, 3 for better distinction of geological features. MNF analysis considered band (7, 5, 1). All the considered bands enhanced lithological features. All the resulted images were also compared with the band ratios RGB combination of the band (7/5, 6/4, 4/2) for lineaments extraction.

2.1.1. A Manual Lineament Extraction from enhanced Landsat 8 OLI imagery and Geological Map. As earlier defined, lineaments are the observable straight lines structures on satellite images. The definition excludes anthropogenic features such as road, railways, and electric grids. This study avoided all anthropogenic features [2, 15]. The Current study delineates and extracts lineaments from the geological map of Kelantan. The extractions were performed using on a scale of 1:200000 using ArcGIS.

The idea also delineates and extracts lineaments from enhanced Landsat 8 images of the study area. The number of lineaments extracts from satellite imagery depends on the analysis applied to the images. The images realized to bring out lineaments details includes PCA 123 as RGB (**Figure 3**a) and MNF 751 as RGB (Figure 3b) Lineament clarity relied on enhanced textural property observed in band ratios

RGB of 7/5, 6/4, and 4/2 (Figure 3c). Resolution of the satellite imagery data determined the quality of images and the information contents. This work performed a layer stacking of 1 to 7 bands combined with band 8 as resolution merge, the resultant image enhanced lithology and allow differentiation of natural lineaments from artificial ones.

3. Results and discussion

The combinations of bands from the visible, mid-infrared and shortwave infrared bands highlight geological features. Images resulting from RGB combination of PCA (PC 1,2,3) (Figure 3a), MNF (band 7,5,1) (Figure 3b) and Band ratios (7/5, 6/4, 4/2) (Figure 3c), were used as reference points to digitalized the lineaments in the selected area of study. The RGB band combination outline lithological boundaries. In this study, layer stacking of bands (1-7) combined with the high-resolution band (band 8) distinguished lithological boundaries.

The lineaments extractions and analysis recorded high counts of 58 lines from the enhanced satellite images compared to only 36 digitalized from the geological map (Table 1). Lineaments are aftermaths effects of tectonic plays on geological terrane, fewer counts of lineaments almost always suggests less effects of tectonic activity on surrounding rocks. The results of lineament analysis give us a proper interpretation of the main structural geology and tectonic forces (Figure 3(d)) and suggest the area is profoundly affected by several structural trends N-S, NW-SE and E-W directions.

The methodology of this work demonstrates the capability of lineaments extraction from Landsat 8 imagery. Band selection determines the outcome of structural mapping using satellite images. The selected band use in this research demonstrates efficiency in highlighting lineaments from Landsat 8 imagery. Directional filtering and Gamm stretching allow the orientation of lineaments to be determined. The rose diagram for both the lineament digitalized from the geological map and the enhanced images shows a dominant trending direction of NW-SE and NE-SW (Figure 4a and b) . These trending directions had been reported in the literature on this focus as the dominant trending direction in the selected area of study [16]. Comparing, the distribution of gold deposits and the lineaments pattern (Figure 3d), established the spatial relationship between the lineaments and the gold deposits. The lithologies hosting gold mineral deposits are highly structured compared to regions without any gold deposits. The current research is in conjunction with the gold mineralization zonation [17], which reported that two major directions (NE-SW and NW-SE) control gold mineralization in Kelantan. Heng et al. [17], focused only geochemical and geophysical data on unraveling the distribution pattern of gold deposits within the study area. Their results conclude that lineaments serve as channels for gold deposition [18, 19], hence the presence of lineaments may indicate gold deposits occurrence. This current study uses the satellite imagery to high and delineates the lineament within the area of study. Since the known gold deposits coincide with NE-SW and NW-SE lineament directions, areas dominated by lineaments trending NE-SW and NW-SE have the high potential of gold deposits occurrence.

Table 1. The Statistical comparison between the lineaments extracted from geological map and Lineaments from the enhanced Landsat 8 images

Criteria	Geological Map	Image process
Counts	36	58
Total Length of all Lineation (m)	174,507.1	230,292.73
Maximum Bin Population (m)	8.0	13
Mean bin Population (m)	3.0	4.14
Standard deviation of Bin Population (m)	2.13	3.48
Maximum Bin length (m)	21,458.99	27,1567.71
Mean Bin Length (m)	7,271.13	8,224.74
Standard deviation of Bin Length (m)	5,5169.07	7,414,48

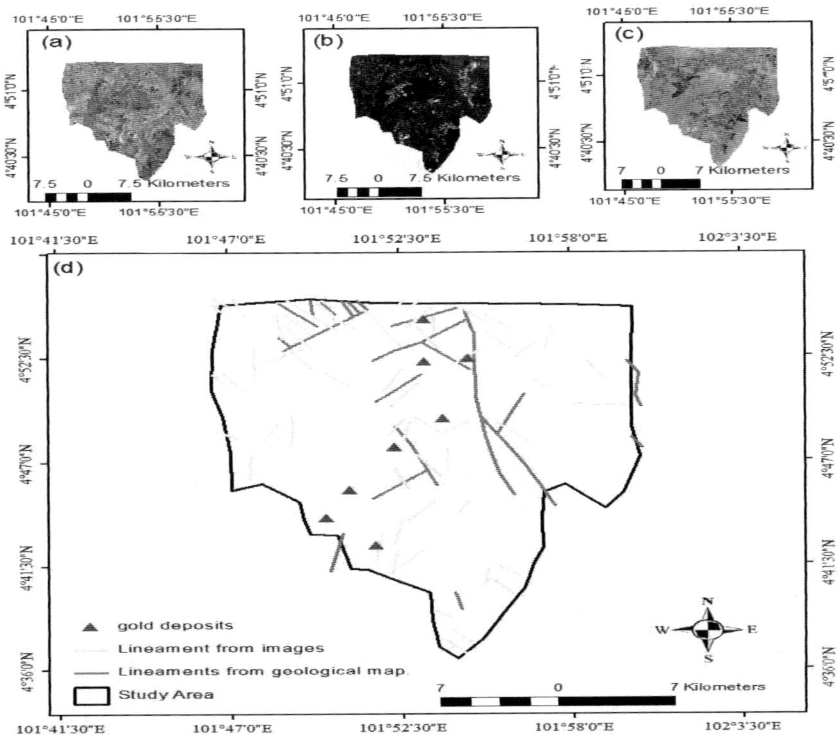

Figure 3. Structural maps; (a) PCA 1,2,3 (b) MNF 7,5,2 (c) Band ratio RGB 7/5,6/4,4/2 and (d) extracted lineament from Landsat 8 imagery and geological map.

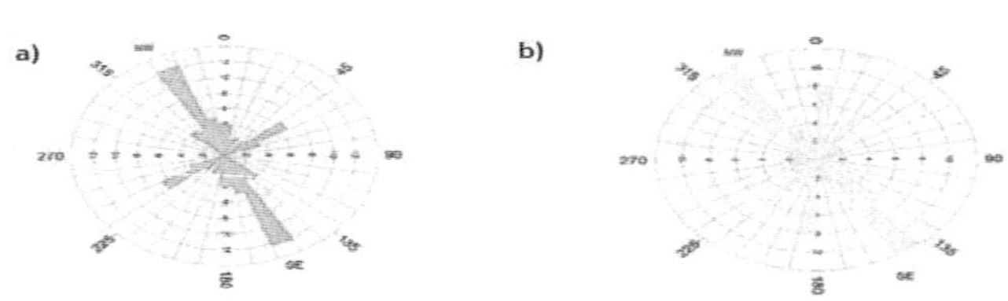

Figure 4. Rose diagram (a) lineaments from geological map (b) lineaments from Landsat 8 imagery.

3.1 Validation

The approach of this study compared the lineaments extracted from satellite images with the digitalized lineament from the geological map (**Figure 3d**). The synthesized structural map suggests a significant pattern of trend or structural direction compatible with the geology of the area. The structural pattern is also in conjunction with the published geological report on mineral resources of Kelantan region[12] Field visits to the selected area of study also confirmed most of the extracted lineaments.

4. Conclusion

Landsat 8 OLI is characterized by high spatial resolution SWIR (30 m) (band 1-7) and band 8 (15 m). Resolution merge facilitates lineament analysis. It plays a vital role in this current study. Enhancement of images is paramount in satellite imagery mapping. PCA, MNF, and Band ratios enhancements techniques improve satellite imagery interpretability of vegetated area. The further stretching and pan sharpening of enhanced images increased contrast on lithological features and allowed easy extraction of lineaments. This research demonstrates that band combination such as (i) RGB combination of PCA (1, 2,3) (ii) MNF Combination of (7,5,1) and (iii) Band ratios combination of 7/5, 6/4, and 4/2 enhanced geological features Landsat 8 imagery in a tropical environment. The approach demonstrated efficiency in extracting lineaments from Landsat 8 imagery of Kelantan area. The study area had suffered deformation from past tectonic events. The past tectonic activity resulted in various trending lineaments extracted (NW-SE, NE-SW, and N-S). Gold deposits distributions pattern shows a spatial relationship lineaments. Two significant lineaments (NW- SE, and NE-SW), control gold deposits occurrence in the area. The study locates not only the known gold deposits but also locate the new area of possible gold deposits occurrence. Highly structured areas are site for potential gold mineralization. The methodology of this study is adoptable for Greenfield exploration especially in regions with similar geological setting like Malaysia.

References

[1] Abdullah A, Nassr S and Ghaleeb A 2013 *Globa J. Hum. Soc. Sci. Geogr. Geo-Science, Environ. Disaster Manag.* 13 35
[2] Wang Arlegui L E and Soriano M A 1998 *Int. J. Remote Sens.* 19 3169
[3] Ali A and Pour A 2014 Sudan *Int. J. Basic Appl. Sci.* 3
[4] Tangestaninejad S, Moghadam M, Mirkhani V, Baltork I M and Salavati H 2008 *Ultrasonics Sonochemistry* 15 815
[5] Martins A and Lima C De 2015 Mapping hydrothermal gold mineralization using Landsat 8 data. A case of study in Chaves license, Portugal.
[6] Abdelsalam M G, Stern R J and Berhane W G 2000 *J. African Earth Sci.* 30 903
[7] Pour A B and Hashim M 2016 *IOP Conf. Ser. Earth Environ. Sci.* 37 012067
[8] Anderson F J and Anderson F J 2011 Lineament Mapping and Analysis in the Southern Williston Basin in Southwestern North Dakota.
[9] Pour A B, Hashim M, Makoundi C and Zaw K 2016 *Resour. Geol.* 66 368
[10] Sabins F F 1999 *Ore Geol. Rev.* 14 157
[11] Metcalfe I 2002 *J. Asian Earth Sci.* 20 551
[12] Hutchinson C S 1989 *Geological Evolution of South-East Asia*
[13] Batchelor D A F 1994 *J. Southeast Asian Earth Sci.* 10 101
[14] Pour A B and Hashim M 2017 *Nat. Hazards Earth Syst. Sci.* 17 1285
[15] Acharya T and Mallik S B 2012 *J. Earth Syst. Sci.* 121 453
[16] Beng Y E 2000 *Proc. Geol. Soc. Malaysia Annu. Conf.* 325
[17] Remedio E M 2006 W $ M $ 129
[18] Khamar Shah a. 2012 *Earth Sci.* 314
[19] Heng C L and Singh D S 1986 *GEOSEA V Proceedings, Vol. I. Publ. Bull. Geol. Soc. Malaysia* 19 431

Deterministic and Probabilistic Assessments of the Slope Stability of Kota Bunyih Dam, Pengkalan Hulu, Perak

M W Agam[1], M H M Hashim[1*] and H Zabidi[1]

[1]Strategic Mineral Niche, School of Materials and Mineral Resources Engineering Engineering Campus, Universiti Sains Malaysia, Penang, Malaysia

*Corresponding author's e-mail: mohd_hazizan@usm.my

Abstract. This paper analyses the stability of the design and construction of Kota Bunyih (KB) tailings dam of Rahman Hydraulic Tin Mine which is located at Kelian Intan, Pengkalan Hulu Perak. The stability assessment of KB tailings dam was conducted using 2D Limit Equilibrium Methods analysis software, Slide 6.0 from Rocscience Inc. using two different approaches – the deterministic and probability approaches. In the analyses, it was found that the upstream and downstream factors of safety (FOS) obtained through deterministic analysis are satisfactory as compared to the recommended minimum FOS for both normal operating freeboard and flood freeboard conditions. In addition, both of the normal and lognormal reliability index values obtained from overall slope surface analysis are twofold the minimum recommended value, which indicate the satisfactory level of safety of KB tailings dam slope.

1. Introduction

Mining produces a very large amount of tailings in its operation which might include slurry with high content of slimes, water and also coarse sandy tailings. In a proper tailings management, the tailings dam is divided into a several partitions called tailings ponds. Usually, at the very front side of tailings pond that is connected to processing plant tailings outlet is allocated for the settlement of coarse sandy tailings. Further distance from the outlet, the smaller size of tailings materials fill the other ponds as the particles flow under the influence of water flow current and gravity. This tailings flow happens in Rahman Hydraulic Tin Mine (RHT) Sdn Bhd, currently the largest producer of tin ore in Malaysia. RHT has been in operation for more than a century and recognised as the oldest hard rock tin mine in Malaysia [1, 2]. Thus, this paper presents the analysis of stability of Kota Bunyih (KB) tailings dam using different types of limit equilibrium methods (LEM) of slices in both deterministic and probabilistic approaches.

1.1 Site Geology

RHT is located at Kelian Intan which is a sub-district of Pengkalan Hulu at the northern part of Perak, Malaysia. Geologically, beneath the Kelian Intan is the Kroh formation which is one of Baling Group formations and represented by Early Silurian to Devonian marine sediments [2, 3]. These marine sediments consist of black shale, sub-mature arenite, calcareous shale and limestone [2, 3, 4].

The dominant lithofacies of Kroh Formation is argillaceous facies that composed of dark-grey black shale and mudstone which have experienced a local low-grade metamorphism by igneous rock intrusion underlying the formation and turned into metasedimentary rocks such as slate, phyllite, hornsfels, metamudstone and quartzmica schist [2, 3]. The KB tailings dam covered around 0.15 square kilometres and is located at the north east of RHT mining lease area. Based on previous subsurface investigation work [5], underlain the dam is highly fractured slate and layers of hard material with Standard Penetration Test value greater than 50. Slate dominantly made up the northern area of study which has heavily-jointed structure while at the southern area; residual soil mostly can be seen.

1.2 Design of Kota Bunyih Dam

Based on the general cross-section of KB tailings dam as shown in Figure 1, it can be seen that the dam is constructed on top of original ground that consists of weathered and highly fractured metasedimentary rocks. Here, the KB tailings dam embankments are made of three different layers of embankments – starter dam, existing tailings dam, and the new dam of which their construction is staged over the life of the impoundment.

By comparing types of construction method for raised embankment as classified by Vick [6], KB tailings dam is classified as the centreline embankment, as fill was placed onto the beach and also onto the downstream slope of the existing starter dam that makes the embankment rising vertically upward. In further detail, the downstream slope of KB tailings dam is divided into eight benches in which every bench has a 3 m width and 6 m height with a bench slope angle of 27 degrees. The design parameters of benches produce an overall slope angle of 22 degrees that span along 140 m width. On the other hand, the upstream slope consists of a single bench that inclined at 34 degrees from the horizontal with the top width of 10 m.

The stability of KB tailings dam is enhanced by constructing a few safety features such as rock toe with geotextile filter, impermeable blanket covering the upstream slope, impermeable soil covering the downstream slope and discharge spillway.

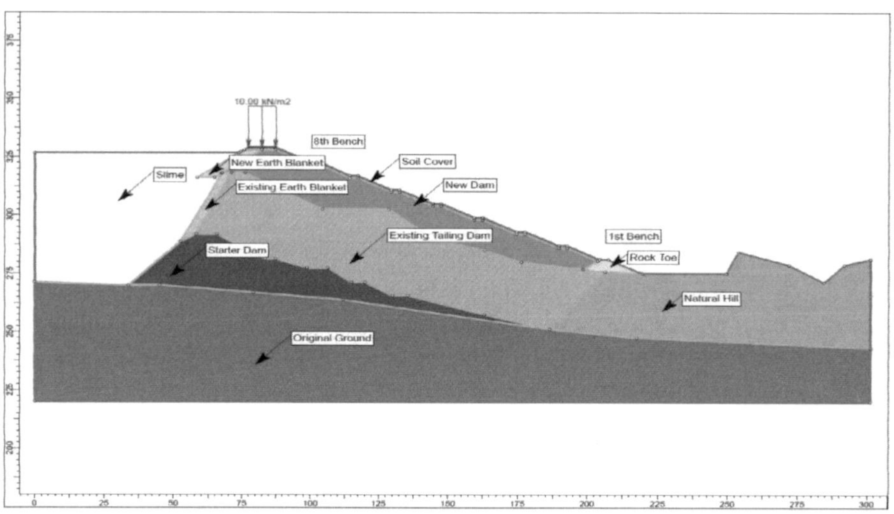

Figure 1. The general cross-section of KB tailings dam.

1.3 Limit Equilibrium Methods

Limit equilibrium method (LEM) of slices and strength reduction method (SRM) by finite element analysis are two of the slope stability analysis methods that are widely nowadays. In general, SRM produced a slightly higher factor of safety (FOS) as compared to those by LEM for most cases [7]. However, in most cases, both of the methods have produced similar critical slip surfaces (CSS), but this condition is only applied if the cohesive strength is low. Even though there are minor differences in both of the methods, both of them generally show a good agreement in the results produced which indicates that the use of either one of them is satisfactory [7].

Thus, LEM is chosen as a slope stability analysis method in this paper due to the low cohesive strength (soil) nature of the case study area. Along the slip surface in LEM, linear Mohr-Coulomb failure criterion is applied to determine the shear strength (τ_f) of soil. Once the soil mass is divided into several number of slices, free body is drawn for each of the slice to determine the FOS. The ratio of shear strength available (τ_f) to the shear stress mobilised (τ) is then taken to determine the FOS of a slope as shown in Eq. (1):

$$FOS = \frac{\tau_f}{\tau} = \frac{c' + \sigma' \tan \phi'}{\tau} \qquad (1)$$

Some assumptions need to be made in order to make the problem determinate and in deriving the FOS. Usually, the most common assumptions to make the problem determinate are the location of the acting normal force on the base of the slice and consideration on the inter-slice forces [8, 9].

2. Methodology

The analysis methods used were Ordinary, Bishop Simplified, Janbu Simplified, Spencer and Morgenstern-Price (MP) with half-sine inter-slice force function. It has to be noted that the implementation of General Limit Equilibrium (GLE) in Slide is essentially equivalent to the Morgenstern-Price method [10]. These analysis methods of slices were selected based on the reasons that they are widely used in slope stability analysis despite the fact that they possess different equilibrium conditions and assumptions.

In this paper, steady-state FEA is chosen as the groundwater analysis method. In applying this groundwater analysis method, the KB tailings dam model was meshed and discretized using 6-noded triangles as the mesh element type with 3000 approximate number of elements. A high number of elements were set to obtain an accurate solution especially in the case of a thin material boundary in the steady-state FEA [7].

For probabilistic analysis, it was set similarly to the deterministic analysis except that probabilistic analysis is conducted only for downstream slope under normal operating condition. In probabilistic analysis, Latin Hypercube was used as the sampling method of random variables with 1000 number of samples instead of Monte Carlo as the former is able to produce a comparable result to the latter with a lesser samples and a smoother sampling of probability distributions [10]. After defining the analysis type, cohesion, friction angle and unit weight of materials were selected as the random variables with normal statistical distribution.

3. Results and Discussion

3.1 Deterministic analysis

The deterministic analysis of KB tailings dam was conducted using the mean value of parameters which are shown in Table 1. This analysis is conducted to search for the FOS that related to CSS of the slope for every method of slices chosen. In this deterministic analysis, there are two cases whereby different freeboard levels of the ponded water were considered. Case A considered a normal operating freeboard level of 1.84 m while case B considered a flood freeboard level of 0.3 m. Both of these cases were considered to investigate the effect of freeboard level towards the FOS for both upstream and downstream slope. Based on Table 2, it can be seen that cases A and B, investigated using LEM gave FOS greater than 1.5. According to Duncan & Wright [11] and Jansen [12], these FOS are satisfactory as they recommended the minimum FOS for long-term steady seepage for upstream and downstream slope designs are both limited to 1.5.

By comparing the FOS produced by LEM in flood condition to the normal operating condition for both upstream and downstream slopes, the flood condition does not affecting the FOS of the downstream slope. This pattern was proven by the same CSS and FOS obtained for both cases. For the downstream CSS produced by the LEM for both cases, there is no significant difference can be seen in terms of their positions and sizes. All of the CSS produced are similar for which the left CSS endpoints are located around the top crest of the dam while the right endpoints are located around the lower part of seventh bench slope. This condition could be due to the effect of impermeable earth blankets that do not allow groundwater to penetrate across the existing tailings dam and new dam even though the level of ponded water is high. If there is no earth blanket covering the upstream slope of KB tailings dam, the chance for the downstream slope to fail is high. This is because groundwater seeps through the existing tailings dam and new dam which causing the rising of water table. If the CSS is in contact with water table, the existing pore pressure at the CSS will become higher and this pressure will separate the soil grains apart. This condition results in the reduction of normal forces and friction resistance at the base of slice which eventually fail the slope.

Table 1. The mean value of KB tailings dam materials properties used in deterministic analysis

Materials	Υ (kN/m3)	c' (kPa)	ϕ' (°)	K_s (m/s)	Model Properties (Soil Type)
Original Ground	18.5	0	35	1.00E-07	General
Starter Dam	18.5	5	31	1.00E-07	Silt
Existing Tailings Dam	18.5	0	33	1.00E-06	Sand
Existing Earth Blanket	13.0	1	16	1.00E-09	Clay
New Earth Blanket	18.0	5	28	1.00E-09	Clay
New Dam	18.5	0	33	1.00E-05	Sand
Natural Hill	18.5	0	33	1.00E-07	Clay
Slime	13.0	1	16	1.00E-09	Clay
Soil Cover	18.0	5	28	1.00E-09	Clay
Rock Toe	25.5	0	45	1.00	General

Table 2. The upstream and downstream slopes FOS obtained from both cases A and B

Conditions	**Limit Equilibrium Methods**				
	Ordinar y	**Bishop Simplified**	**Janbu Simplified**	**Spencer**	**GLE / MP**
Case A					
Downstream slope	1.63699	1.65722	1.63660	1.65222	1.65478
Upstream slope	1.72636	1.85739	1.67771	1.85103	1.85115
Case B					
Downstream slope	1.63699	1.65722	1.63660	1.65221	1.65499
Upstream slope	2.68439	2.83445	2.60868	2.82455	2.82476

In studying the effect of freeboard towards water table, it was found that both of the cases have resulted in very similar water tables which only differ in terms of water discharged at the rock toe. The water tables of both of the cases do not make their ways across the earth blankets to the middle body of KB tailings dam directly, but it only cross the dam horizontally at the lower part of the starter dam towards the rock toe at $3.0757e^{-9}$ m^3/s and $3.3852e^{-8}$ m^3/s respectively. This is due to the effect of impermeable earth blankets that do not permit the seeping of water which eventually lowers the level of water tables in the dam. In terms of water discharged, the rate is proportional to the level of ponded water, however, the discharges are very nominal.

3.2 Probabilistic analysis

The probabilistic analysis is only conducted to determine the reliability of the KB tailings dam downstream slope that operating at a normal condition. In details, there are two main CSS analyses that carried out in overall slope probabilistic analysis based on Rocscience [10]; the overall slope surface analysis and the critical probabilistic surface analysis. The critical deterministic surface analysis is also carried out in probabilistic analysis.

IOP Conf. Series: Journal of Physics: Conf. Series **1082** (2018) 012093 doi:10.1088/1742-6596/1082/1/012093

Based on the summary of probabilistic results of KB tailings dam in Table 3, all the mean FOS have comply the minimum recommended value required for long-term steady seepage downstream slope designs. It is also noticed that the critical probabilistic surfaces analysis of Janbu Simplified, Spencer and GLE/MP have produced two surfaces that correspond to the normal and lognormal FOS distributions. On the other hand, there are two FOS limit used in determining the probability of failure – the probability of failure when FOS < 1.0 and FOS < 1.5. In this stability assessment, the former is automatically computed for all of the surface analyses while the latter is determined by sampling the exact value of FOS at 1.5 from cumulative probability plot and only available for overall slope surface analysis. It is also can be seen that, there is zero probability of failure when FOS < 1.0 for all surface analyses. This means that from all of the number of samples, there is none of the sample has FOS < 1.0. However, for the case of probability of failure when FOS < 1.5, Bishop Simplified method gave the lowest by having 8.539 percent of its FOS samples less than 1.5. This probability of failure is relatively low and considered acceptable although there is no universal acceptable values implemented [10, 11].

There are two reliability indices in probabilistic analysis that are computed assuming normal and lognormal distribution of the FOS. According to Rocscience [10], FOS is often best fit by a lognormal instead of a normal distribution because lognormal distribution is best used with variables that have non-zero values. Generally, in order to ensure a safe slope design, a reliability index of at least 3 is recommended [10]. Based on the summary of probabilistic results of KB tailings dam in Table 3, it is noticed that only four reliability index values are less than 3. Moreover, these reliability index values are from normally-distributed FOS and were given by critical probabilistic surface analysis. In the meantime, all of the reliability indices for critical deterministic surface analysis exceed the minimum recommended value and also surpass the indices given by critical probabilistic surface analysis. Still, the indices given by critical deterministic surface analysis do not represent the KB tailings dam slope reliability. In a more representative overall slope surface analysis, it can be seen that both of the normal and lognormal reliability index values are twofold the minimum recommended value which indicate the high level of safety of KB tailings dam slope.

Table 3. The summary of probabilistic results of KB tailings dam.

LEM	Surface Analysis	Mean FOS	P_f at FOS < 1.0 (%)	P_f at FOS < 1.5 (%)	β	B_{LN}
Ordinary	Overall Slope	1.577	0.000	17.370	7.005	8.702
	Critical Probabilistic	105.174	0.000	N/A	0.910	4.823
	Critical Deterministic	1.640	0.000	N/A	5.388	6.803
Bishop Simplified	Overall Slope	1.643	0.000	8.539	6.146	7.774
	Critical Probabilistic	110.782	0.000	N/A	1.373	6.945
	Critical Deterministic	1.660	0.000	N/A	5.564	7.066
Janbu Simplified	Overall Slope	1.588	0.000	15.620	6.776	8.442
	Critical Probabilistic	43.957* / 1.658**	0.000	N/A	4.204	6.743
	Critical Deterministic	1.639	0.000	N/A	5.392	6.807
Spencer	Overall Slope	1.639	0.000	9.227	6.105	7.713
	Critical Probabilistic	71.516* / 1.669**	0.000	N/A	2.926	6.914
	Critical Deterministic	1.656	0.000	N/A	5.513	6.992
GLE/MP	Overall Slope	1.640	0.000	9.255	6.081	7.685
	Critical Probabilistic	79.340* / 1.670**	0.000	N/A	2.285	6.869
	Critical Deterministic	1.657	0.000	N/A	5.509	6.990

* Critical probabilistic surface related to normal distribution
** Critical probabilistic surface related to lognormal distribution
N/A Not available

4. Conclusion

This paper analysed the stability of the design and construction of KB tailings dam, using different types of limit equilibrium methods of slices in both deterministic and probabilistic analyses. It was found that the upstream and downstream FOS obtained through deterministic analysis are satisfactory as compared to the minimum recommended FOS under both normal operating freeboard and flood freeboard conditions. In terms of reliability indices, both of the normal and lognormal reliability index values from overall slope surface analysis are twofold the minimum recommended value, which indicate the high level of safety of KB tailings dam slope.

Acknowledgements

The first authors gratefully acknowledge the financial support of USM Fellowship and Short Term Grant no 60313015 in carrying out this study. This research would also not have been possible without the permission by RHT in providing the necessary data for the analysis purposes.

References

[1] Beng Y E 2000 *Geological Society of Malaysia Annual Geological Conference,* Pulau Pinang, Malaysia, 325
[2] MT-JGSC (Malaysia-Thailand Border Joint Geological Survey Committee) 2009 *Geology of the Pengkalan Hulu-Betong Transect Area along the Malaysia-Thailand Border,* Kuala Lumpur: Minerals and Geoscience Department Malaysia
[3] Ulfa Y, Yu E H R & Kit O C 2014 *Eksplorium, Buletin Pusat Teknologi Bahan Galian Nuklir,* 35 85
[4] Rosle Q A & Teh G H 1998 *Bulletin of the Geological Society Malaysia* 42 161
[5] Gue & Partners, Geotechnical Design Report of Kota Bunyih Tailings Dam at Rahman Hydraulic Tin Mine, Klian Intan, Pengkalan Hulu, Perak Darul Ridzuan, Unpublished, 2011.
[6] Vick S G 1983 *Planning, Design and Analysis of Tailings Dams*. NY: John Wiley & Sons Inc.
[7] Cheng Y M and Lau C K 2008 *Slope Stability Analysis and Stabilization – New Methods and Insight,* London: Routledge, Taylor & Francis Group
[8] Abramson L W, Thomas S L, Sharma S & Glenn M B2002. *Slope Stability and Stabilization Methods* (2nd ed.), New York: John Wiley & Sons Inc.
[9] Fredlund D G, Krahn J & Pufahl D E 1981 *Proc. 10th Conference of the International Society for Soil Mechanics and Foundation Engineering (ISSMFE)* 3 409
[10] Rocscience Inc., Slide Version 6.0 - 2D Limit Equilibrium Slope Stability Analysis, Retrieved December 11, 2014, from http://www.roscience.com, 2010.
[11] Duncan J M & Wright S G 2005 *Soil Strength & Slope Stability*, NY: John Wiley & Sons Inc.
[12] Jansen R B 1988 *Advanced Dam Engineering for Design, Construction, and Rehabilitation,* NY: Van Nostrand Reinhold

IOP Conf. Series: Journal of Physics: Conf. Series **1082** (2018) 012094 doi:10.1088/1742-6596/1082/1/012094

Reliability of Using Standard Penetration Test (SPT) in Predicting Properties of Soil

N.Q.A.M. Yusof[1], H.Zabidi[2]

[1,2]School of Materials and minerals Resources Engineering Campus, Universiti Sains Malaysia, Penang, Malaysia

Email: srhareyani.zabidi@usm.my

Abstract. The properties of soil play important role in many practices of geotechnical engineering. The soil classification tests such as Atterberg limits are quick and easy to perform and also cheaper compared to the test required to determine the soil parameter such as shear strength. It may take long time to get the result and difficult to perform. So, the prediction of these properties with the Standard Penetration Test (SPT) gives a chance to obtain these parameters without using of more laboratory test. Standard penetration test is used widespread due its simplicity and not expensive. The test is performed to measure the strength of the soil. In this study, the reliability of using standard penetration test (SPT) in predicting some properties such as Atterberg limits and shears strength parameters. The soil samples collected from 14 boreholes to depths between 1.5 to 19.5 m. The site of this study is at Ulu Jelai which is located in the State of Pahang; in the district of the Cameron Highlands have been investigated. Simple correlations between Standard Penetration Test and Atterberg limit is performed by using simple regression method. The results of the research show that the shear strength of the soil affects the standard penetration test number.

1. Introduction

Estimation of engineering characteristics of soil are important parts in geotechnical works. As part of design stage, an engineer needs to determine the average value and variability of soil properties. The importance of in-situ testing in geotechnical engineering is undeniably increasing because most of laboratory test are laborious, time consuming and costly [1]. However, sometimes the outcomes are not accurate due the poor laboratory condition [2]. Another long discussion is that the result of strength values determined from laboratory testing is not applicable to the in-situ soil mass due the scale effect.

Standard Penetration Test (SPT) is the one of the most common in-situ tests to identify soil type and stratigraphy along with being a relative measure of strength during site investigation. The standard penetration test consists of driving split barrel sampler with a distance 460 mm into the soil at the bottom of the boring. The number of blows is counting to drive the sampler the last two of 150 mm distances to obtain the SPT number. A 63.5 kg driving hammer is used to drive the sampler by falling free from a height of 760 mm [3].

The SPT number could provide useful in-situ test data [4]. The number of SPT might correlated with the consistency and density the soils encountered [5]. Many studies have been developed to investigate the correlations between the number of blows, N with different soil properties including the relative density of sands, consistency and unconfined compressive strength of clays and allowable bearing

IOP Conf. Series: Journal of Physics: Conf. Series **1082** (2018) 012094 doi:10.1088/1742-6596/1082/1/012094

pressure on sands and clays. According to Kulhawy and Mayne [6], SPT data have been used to correlate the shear strength parameter such as cohesion and angle of friction angle of soil but Bowles and Aggour [7], [8] recommend that the measured N value is standardised by multiplying it by the ratio between the measured energy transferred to the rod and 60 % of the theoretical free fall energy of the hammer. In this study, the relationships between SPT number and soil parameters such as Atterberg limit and shear strength parameter have been discussed to ensure that which of these parameters affect the SPT number.

2. Experimental study

The study area is located in the State of Pahang, in the district of Cameron Highlands about 140 km north of Kuala Lumpur and 80 km east of west coast of mainland Malaysia. Field work consisted of drilling and samplings of 14 boreholes to depths between 13.95 m to 50 m below ground surface were carried out. A standard penetration test (SPT) was carried out by Test Sdn Bhd using standard equipment in accordance with British Standard BS EN ISO 22476-3. An indication of the in-situ density of sands is given by driving of a standard 50 mm OD diameter split barrel sampler into the bottom of a drill hole using a 63.5 kg weight falling 0.76 m. The test also provides a rough indication of in-situ strength of cohesive materials (silts and clays).

The test involves the measurement of the number of blows for three successive penetration of 150 mm. The total blow count for the lower (final) 300 mm is recorded as the SPT number (N) values. The values that greater than 50 are commonly regarded as effective refusal. Some of the specimens with different SPT number have been take out in order to carry out for laboratory testing. The specimens were carefully transported according to standard classifications and techniques. Atterberg limits including liquid limit (LL), plastic limits (PL) and plasticity index (PI) were determined according BS 1377 Test 2B and 3 while to determine the shear strength parameters such as cohesion (c) and angle of friction (ϕ, triaxial compression (consolidated, undrained) is used according to BS 1377. These tests were performed by Test Sdn Bhd at their laboratories.

3. Results and discussion

The SPT results gave N values over the full range to 50 blows per 300 mm or until refusal but until a depth of more than 24 m that there was an absence of N values below 10. This is to indicate a significant variability in the residual soils and completely weathered (CW) to highly weathered (HW) rock. The weathering of granitic rock under tropical conditions produced zoned of very weak material where aggressive chemical alteration of the biotite and feldspar minerals results in a fabric that readily breaks down under dynamic loading.

The result from Atterberg limits show that the site is made of a typical of weathered granitic materials. The samples tested are classed according to the Unified Soil Classification System (USCS) as silts of low to high liquid limit (ML to MH). The results obtained from field and laboratory tests are presented in Table 1 and discussed in detail under this section

Table 1 Field and laboratory results

Sample No.	Depth of soil (m)	SPT number	Atterberg limits (%)			Cohesion (kPa)	Angle of friction (°)
			LL	PL	PI		
1	3.9	7	36	24	12	113	37
2	6.9	6	37	26	11	-	-
3	10.2	5	50	31	31	90	36
4	4.1	5	59	37	22	0	30
5	7.5	13	45	31	14	-	-
6	12	13	37	25	12	76	36
7	4	4	68	43	25	0	34
8	10.5	17	39	27	12	199	37
9	7	4	71	44	27	-	-
10	15	25	53	35	18	338	40
11	2.5	4	53	37	16	-	-
12	6	7	45	31	14	-	-
13	10.4	11	70	43	27	172	34
14	4.4	1	64	19	45	-	-
15	7.5	10	64	39	25	58	36
16	11.3	7	43	27	16	-	-
17	4.4	6	57	36	21	-	-
18	7.1	3	37	27	10	-	-
19	9.45	18	49	28	23	-	-
20	16.5	38	46	27	19	-	-
21	19.1	6	45	27	18	46	30
22	4.5	11	49	32	17	118	36
23	7.5	11	55	34	21	105	36
24	9	2	53	32	21	-	-
25	12.3	6	51	30	21	-	-
26	15	28	40	28	12	-	-
27	18	24	40	29	11	-	-
28	7.5	4	48	32	16	-	-
29	4.5	14	56	35	21	-	-
30	7.5	12	63	44	19	-	-
31	10.5	5	63	36	27	-	-
32	19.5	5	57	34	23	-	-

The relationships between field and experimental data included Standard Penetration Test number (N) at depth of sample taken from ground surface, Atterberg limits such as liquid limit (LL), plastic limit (PL) and plastic index (PI), shear strength parameters such as cohesion (c) and friction angle (ϕ) is plotted. The coefficient of correlation (R^2) between field and experimental data is determined. According to Taylor [9], coefficient of correlation (R^2) is representing the degree of linear association between two variables. The relationship between any parameters is classified as:

1. $R^2<0.3$ (no correlation)
2. $0.3<R^2<0.499$ (mild relationship)
3. $0.5<R^2<0.699$ (moderate relationship)
4. $0.7<R^2<1.0$ (strong relationship)

The relationship between these mentioned parameters have been performed as follows:

3.1 Effect of soil depth (D) below ground surface

Figures 1 to 4 show the relationships between depth of sample from the ground surface with different parameters such as SPT number, liquid limit (LL), plastic limit (PL) and plastic index (PI). As shown in Figure 1, the correlation coefficient ($R^2 = 0.34$) is considered to be mild relationship. The SPT number is increase as the depth of sample from the ground increase. It means that the density of soil is increase as the depth is increase. This is because the effect of stress imposed on the layer of soil by the weight of overlaying material also called as overburden pressure.

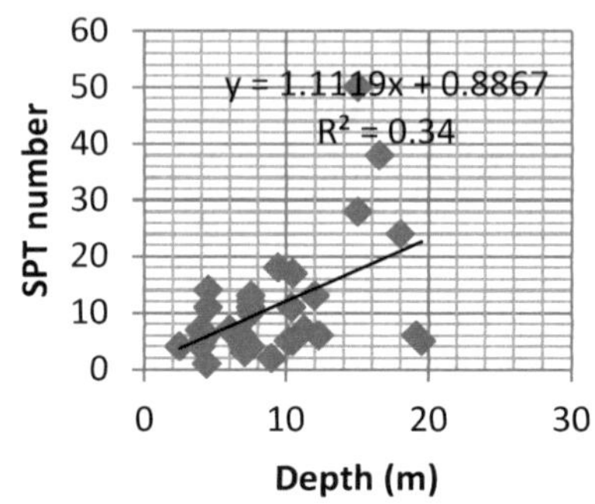

Figure 1. Depth of sample from the ground versus SPT number

Figure 2 (a)-(c) show the relationships between depth of soil and Atterberg limits (LL, PL and PI) respectively. From these figures, it shows that there is no correlation between Atterberg limit and depth of soil since the coefficient of correlation are less than 0.3. This means that the depth of soil do no effect the plasticity of soil because the plasticity of soil is depend on the physical and mechanical properties of soil particle.

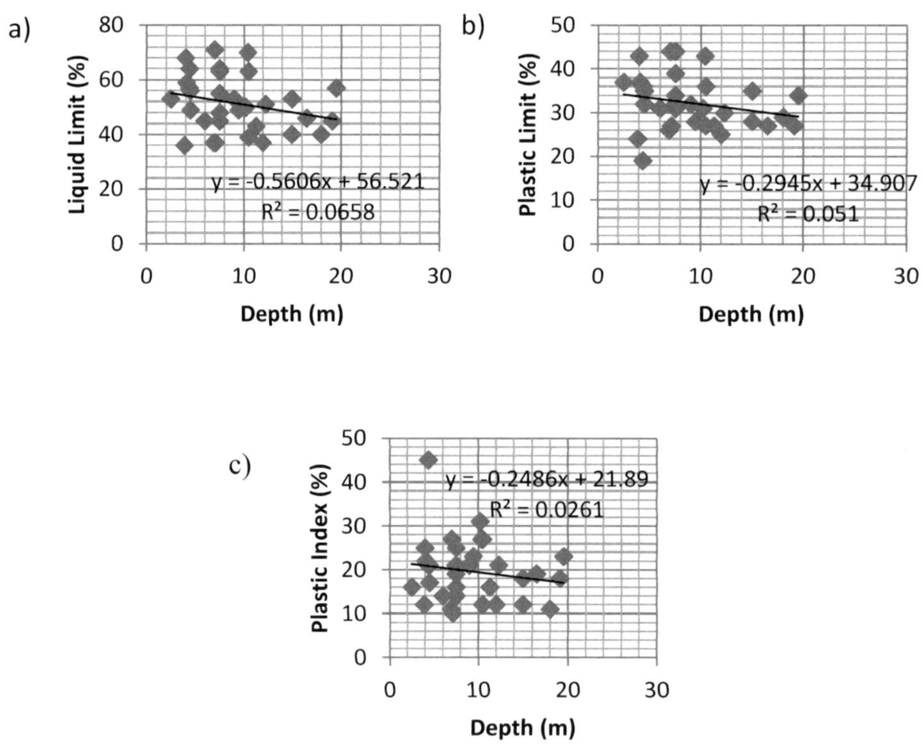

Figure 2. Depth of soil and Atterberg limits a) liquid limit (LL), b) plastic limit (PL) and c) plastic index (PI)

IOP Conf. Series: Journal of Physics: Conf. Series **1082** (2018) 012094 doi:10.1088/1742-6596/1082/1/012094

3.2 SPT versus Atterberg limits LL, PL and PI

The relationships between SPT number and Atterberg limit such as liquid limit (LL), plastic limit (PL) and plastic index (PI) are show in Figures 3 (a)-(c) respectively. Based on these figures, the Atterberg limits do not depend on SPT number since the values of correlation coefficient is less than 0.30. it is because the SPT number is depend on the relative density of layer as mentioned in previous section

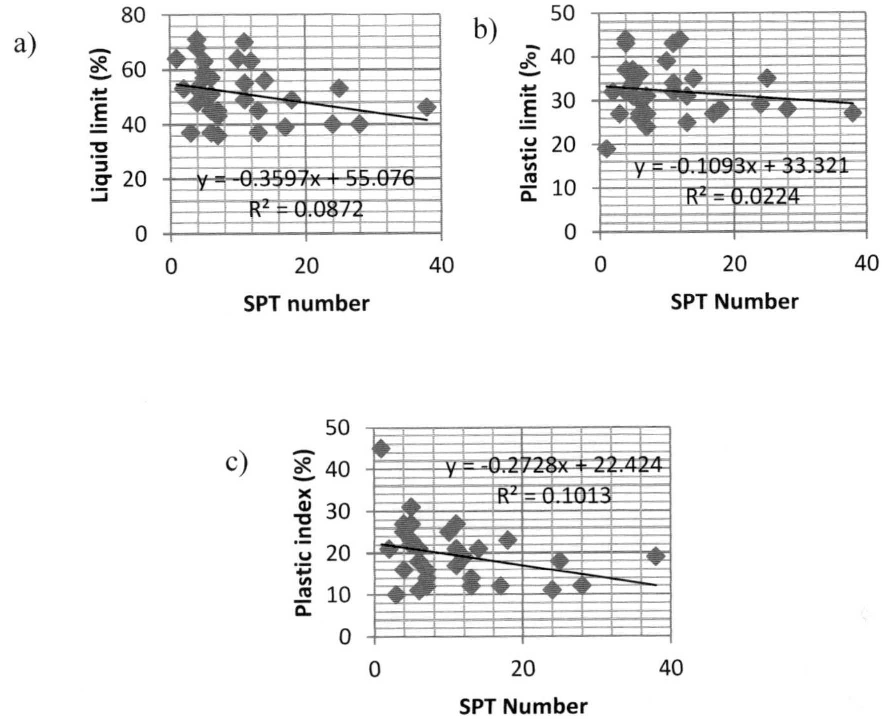

Figure 3. The relationships between SPT number and Atterberg limit a) liquid limit (LL), b) plastic limit (PL) and c) plastic index (PI)

3.3 SPT versus shear strength parameter (cohesion and friction angle)

Figures 4 (a) and (b) show the relationships between SPT number and shear strength parameters of soil (cohesion and angle of friction). The number of SPT is increases as the shear strength of the soil increase as shown in these figures. As the depth of soil increase, the relative density of soil increase due to the overburden stress effect. The density of soil increase may be due to the insertion of fine soil particles in the voids between sand particles to make the specimens are less voids. So, the friction surface is increases. As the friction surface increases, the shear strength also increases. In these figures, it show strong correlation between SPT number and shear strength parameters since the correlation coefficient is greater than 0.7. This means that the shear strength parameters have strongly affected the SPT number.

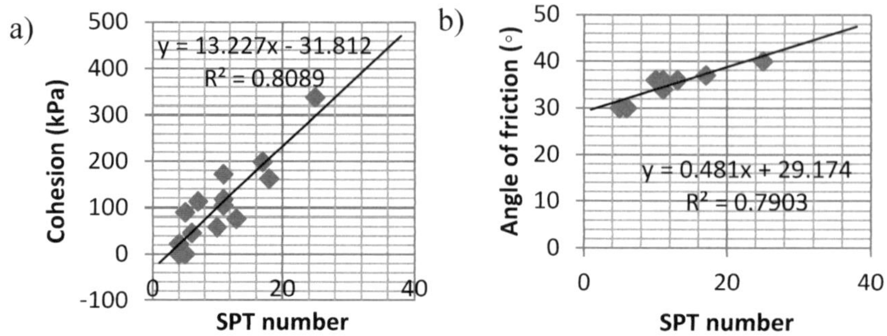

Figure 4. The relationships between SPT number and shear strength parameters of soil a) cohesion, and b) angle of friction.

4. Conclusion

Standard Penetration Test is the most popular tool for geotechnical characterisation at site due its simplicity and low cost besides it can provide useful and reliable data. Based on this study, the results are concluded as follows:

1. The depth of soil is significantly affected the SPT number.
2. The Atterberg limits have no effect on SPT number because it depends on the physical and mechanical properties of soil particles.
3. The shear strength parameter is strongly affected the SPT number.
4. Despite the fact that SPT is performed in the field on disturbed soil, it is adequate using SPT rather than using laboratory tests to determine shear strength parameter.
5. By considering standard penetration test, it is reliable to predict the shear strength parameter

Acknowledgement

This research project is financially supported by the Ministry of Higher Education (MOHE), Malaysia under the **Research University (RUI) research grant of 814229.** The permission to assess the sites and data are acknowledged for the support from the Tenaga Nasional Berhad.

References

[1] Mair R J and Wood D M 1987 Construction Industry Research and Information Association
[2] Venkatasubramanian C and Dhinakaran G 2011 *International Journal. Civil Structure. Engineneering* 2 614
[3] Bowles J E 2001 *Foundation Analysis Third Edition*
[4] Terzaghi K 1940 *Purdue Conference on Soil Mechanics and its Applications* 151
[5] Terzaghi K and Peck R B 1948 *Soil Mechanics In Engineering Practice*
[6] Kulhawy F H and Mayne P W 1990 *Manual on estimating soil properties for foundation design*
[7] Bowles J E 1997, *Foundation Analysis and Design Fifth Edition* 20 3
[8] Aggour M S and Radding W R 2001 *Maryland State Highway Administration Rep. No. MD02-007B48*
[9] Taylor R 1990 *Journal of diagnostic Medical Sonography* 6 35

IOP Conf. Series: Journal of Physics: Conf. Series **1082** (2018) 012095 doi:10.1088/1742-6596/1082/1/012095

Petrography and Geochemistry of Kinta Valley Palaeozoic Carbonate Rock

N.E.Q.M. Anua[1], H. Zabidi[1]

[1]School of Materials and Mineral Resources of Universiti Sains Malaysia, Engineering Campus, Penang, Malaysia

Email: srhareyani.zabidi@usm.my

Abstract. Petrographical and geochemical analyses of carbonates rocks collected from four different quarries around the Simpang Pulai areas were studied to determine the provenance and hydrothermal setting of the source region. Geologically, the selective quarries are predominantly underlain by the Kinta Limestone Formation. Locally, limestone at the quarry site predominantly made up of massive and thin bedded varieties, greyish white and with black carbonaceous patches/spots, fined grained limestone, frequently and in places intercalated or associated with carbonaceous shale or phyllite. Massive, of about few centimetres to 4 m thick, cream to pinkish white coloured, fined grained dolomite have been encountered running across the central part of the quarry are in N-S direction adjoining most of quarries. The carbonate rocks were analysed by performing microscopic study, mineralogical analysis of XRD and chemical analysis of XRF. The colour present by the carbonate rocks (limestone, dolomite, and marble) depends on its composition. White, semi-white, light-grey and beige colours can be attributed either to pure calcite or dolomite. All carbonate samples exhibit limited existence of major elements except for Ca and Mg. The particular characteristics of each rock depend highly upon the type of minerals present, their composition, grain size, and their geochemistry. Moreover, the integration of petrographic and geochemistry studies can lead to much better understanding on different secondary processes, including dissolution, karstification and dolomization.

1. Introduction

The Kinta Limestone Formation is part of Western Belt in the centre of Perak, and it is an area of interest for several reasons. The metamorphic and diagenetic alterations that have an impact on the stratigraphic complexity, the huge potential on the industrial usage of limestone, the stanniferous mineralization, and heritage values of caves are among the reasons, which raise the interest of researchers in the past [1]. The limestone is considered to be recrystallized due to metamorphism of granitic intrusions [2, 3] and other diagenesis phases (2, 4, 5] to change into marble and dolomite. Due to these alterations, attempts have been made to study and characterize further the Kinta Limestone Formation. For example, the Kinta Limestone Formation has been fully tested in detail for its geochemical characteristics to obtain supporting data for the paleontological dating via chemostratigraphy, except for a few oxide analysis carried out to optimize the industrial usage of Kinta Limestone [5].

IOP Conf. Series: Journal of Physics: Conf. Series **1082** (2018) 012095 doi:10.1088/1742-6596/1082/1/012095

In this paper, we had selected four quarries around the Kinta area to study their petrographic also geochemical data and to further correlate with the minerals present in the selected carbonate rocks that represent the quarries, the grain size of the samples as in photomicrograph, and the composition of oxides especially CaO and MgO that might influence the colour, strength that can also effect the fragmentation degree [6] and dissolution rate of carbonate rocks.

1.1 Geological setting

Kinta Valley is characterized by remnant limestone hills sandwiched by the Triassic granitic batholiths [7] of the Main Range in the east and the Kledang Range in the west (Figure 1). There are three main facies in Kinta Valley; carbonate facies, argillaceous facies and arenaceous facies [8]. The carbonate facies is known as Kinta Limestone Formation [4]. The Kinta Limestone is spanned from Silurian to Permian age in stratigraphy lexicon of Peninsular Malaysia [2, 9]. The Palaeozoic limestone was exposed for a very long period of time that only the limestone hills/karst are the remaining visible part [10]. Karst is a particular style of landscape that contain caves and underground water systems that develop specially on soluble rocks such as limestone, dolomite, marble, and gypsum. It is an open system which closely made up of two combined hydrological and geochemical subsystem [11].

At the selected quarry sites, the formation is predominantly made up of massive and thin bedded varieties, greyish white and with black carbonaceous patches/spots, fined grained limestone, frequently and in places intercalated or associated with carbonaceous shale or phyllite [12]. Massive, of about few centimetres to 4 m thick, cream to pinkish white coloured, fined grained dolomite ($CaMg(CO_3)_2$) have been encountered running across the central part of the quarry are in N - S direction adjoining most of quarries. Also, massive, medium to coarse grained and white to cream coloured marble.

2. Materials and methods

Samples of carbonate rock were taken from four different quarries around Simpang Pulai and Gopeng (Figure 1). The selected samples to undergo tests are varies from whitish to pinkish to dark grey. Petrographic thin section analysis was realized on 12 rock samples. Polarizing microscope Meiji model were used, under magnifications of 4x-40x and photomicrographs were taken by Canon camera. Nine carbonate samples were selected for geochemical study and these samples were powdered first in agate mortar. Major oxides composition were obtained by X-ray fluorescence (XRF) using an X-ray spectrometer Rigaku RIX 3000. Loss-on-Ignition (LOI) was measured by heating up the samples to 1000°C in porcelain crucibles. The powdered samples were also mineralogically characterized using X-ray diffraction (XRD). XRD spectra were acquired from 10° to 89° 2θ. All the method used were performed in School of Materials and Mineral Resources, USM Engineering Campus.

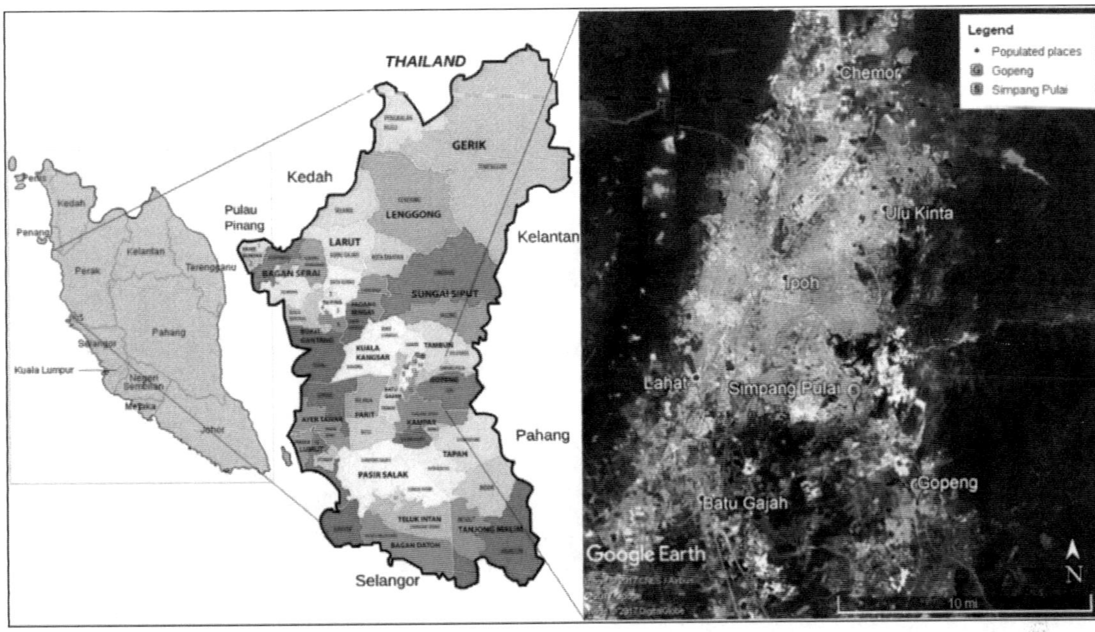

Figure 1. Location of Simpang Pulai and Gopeng, Perak, Malaysia from Google Earth (12.11.2017)

3. Results and discussion

3.1 Petrography

Petrographic studies were performed to identify the carbonate components in the samples. Three samples were studied from each of the quarries, where every thin section was selected beforehand so that it can shows variant characteristics of the quarry. Figure 2 shows the photomicrograph of the prepared thin sections and it is expected to have the existence of calcite (most abundance) and dolomite minerals.

Based on the figure, we can see the grain size are varies from fined to coarse grain, from 200μm up to 1200μm for calcite and dolomite minerals. L1, L2 and L3 have the finest grain of all and the other minerals are hard to be distinguished except for calcite as its characteristics like cleavage can still be seen under the microscope. Calcite and dolomite are easily been recognized as they possessed euhedral to subhedral shaped with perfect cleavage and visible twinning. A very clear cleavage (red arrow) can be seen in samples H1, I2-I4, T3 and T4 while sample I4 show calcite with lamellae twinning (blue arrow). The cleavages also appeared under plain polarized light. The colour of minerals under the polarized microscope are colourless to light brown and as the minerals extinct, it turns to dark grey. There are calcite veins found in L3 that are coarser than the host minerals and microfracture filled by iron found in H1. T1 are found to have opaque mineral that have inclusion of calcite and dolomite.

3.2 Geochemistry

Table 1 shows the elemental data from XRF analysis of 9 selected samples. The elemental concentration includes CaO, MgO, and other major elements including SiO_2, Al_2O_3, Fe_2O_3, MnO, K_2O, P_2O_5, and SrO concentrations. These concentrations give clues of the composition of minerals in the samples that can help in justified the petrographic analysis and XRD analysis. Geochemical contents of carbonate rocks are important to study as it is involved in dissolution process which can also impact the underground water.

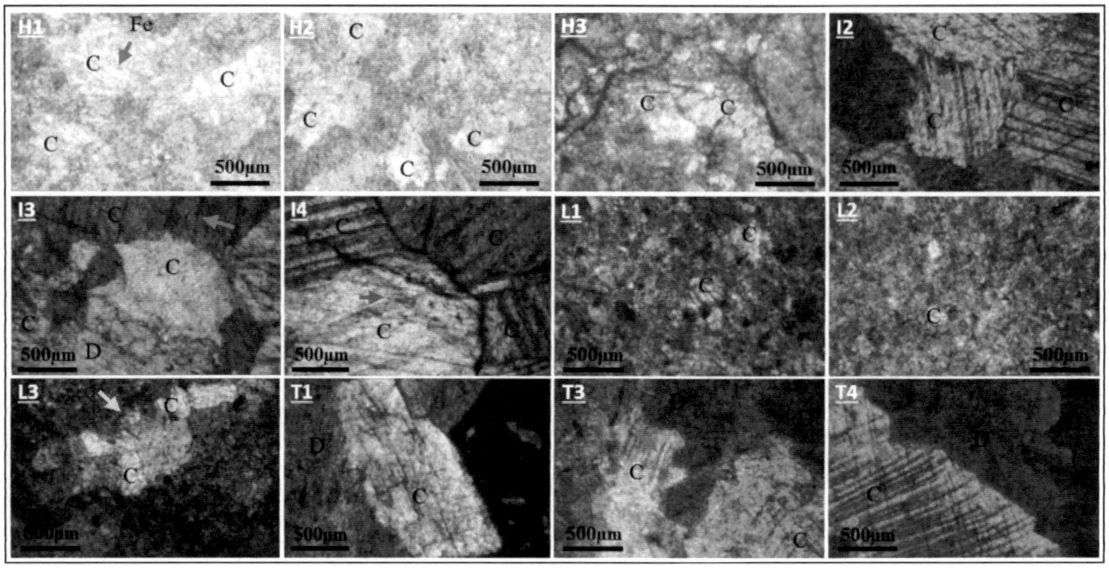

Figure 2. Photomicrograph of the carbonate rocks from the quarries in Kinta Valley under cross polarized light (XPL). H1, H2 & H3 shows calcite minerals with no real rhombohedra shape; calcite in I2, I3 & I4 appear to have a very clear cleavage and lamellae twinning; L1, L2 & L3 have fined visible grain. L3 has calcite vein (green arrow) that are coarser than the host; dolomite are clearly seen in T1, T3 & T4 together with calcite. Opaque mineral are found in T1. C: calcite; D: dolomite; Op: opaque mineral; Fe: iron

Based on the table, samples from Simpang Pulai and Gopeng reveal significant differences in their major compositions. CaO content dominate the concentration of each samples. It ranges from 35.48% to 68.55%, where the average is 56.84%. H2 has the highest amount of CaO while T1 has the lowest amount of CaO. T1 and T2 have the highest concentration of MgO, which are 18.19% and 18.64% respectively whilst for other samples, the concentration is less than 2.00%. As H2 has the highest content of CaO, the sample does not contain any MgO content and other major elements also contain no more than 1.00%, same goes for I4. It is noted that as MgO concentration increases, concentration of CaO decreases.

Al_2O_3 and Fe_2O_3 concentrations were significantly low (>1.00%) except for L1, where the concentrations are 2.11% and 1.12% respectively. L1 also contains high content of SiO_2 (6.78%) same goes to T2 with SiO_2 content of 1.32%. Besides of these major concentrations, there are also some other compositions existed in the samples like SO_3, NiO, CuO, Cl, Rb_2O, Y_2O_3, and ZrO_2 with concentration less than 1.00% that make the total of 100%. Loss-on-Ignition (LOI) of these carbonate rocks were dominantly high (range: 30.00-44.78%) due to the CO_3 content and the calcination process takes place during the heating process.

X-ray diffraction patterns that display peaks of possible mineral contents in the samples are shown in Figure 3. The patterns are between °2Theta (2θ) versus intensity (counts). The graph have the highest peak of calcite minerals for all samples except T1 and T2 at degree of about 29° and dolomite mineral for T1 and T2 with highest peak at about 31°. Based on the XRD pattern, it can be justified that the high MgO content of XRF in samples T1 and T2 and dolomite mineral in petrographic analysis. As for the other samples, the peaks of calcite justified the abundance of calcite grains in photomicrograph and the high content of CaO in XRF analysis.

IOP Conf. Series: Journal of Physics: Conf. Series **1082** (2018) 012095 doi:10.1088/1742-6596/1082/1/012095

Table 1. Some of the major elements composition of the selected Palaeozoic carbonate rocks of Kinta Valley

Sample/ Conc.(%)	SiO_2	Al_2O_3	Fe_2O_3	MnO	MgO	CaO	K_2O	P_2O_5	SrO	LOI	Total
H1	0.3733	0.2603	0.1370	0.0178	0.5195	62.7831	0.0103	0.0051	0.0308	35.8400	99.9772
H2	0.1055	0.0760	0.0417	0.0425	0.0000	68.5549	0.0108	0.0000	0.0298	31.1300	99.9912
I2	0.0756	0.0274	0.0565	0.0000	1.9014	67.8691	0.0099	0.0056	0.0481	30.0000	99.9936
I3	0.2489	0.1229	0.1908	0.0105	0.8018	63.1044	0.0433	0.0234	0.0380	35.4000	99.9840
I4	0.1086	0.6682	0.2613	0.0295	0.0000	62.5467	0.0000	0.0396	0.0277	36.2800	99.9616
L1	6.7842	2.1137	1.1272	0.1371	1.2841	50.5441	0.6450	0.0313	0.0278	37.2000	99.8945
L3	0.3840	0.3315	0.0740	0.0122	0.4870	64.3695	0.0077	0.0000	0.0282	34.2900	99.9841
T1	0.4018	0.2137	0.3327	0.4710	18.1953	35.4877	0.0035	0.0167	0.0095	44.7800	99.9119
T2	1.3299	0.7718	0.2399	0.6075	18.6470	36.3077	0.0040	0.0616	0.0086	41.9400	99.9180

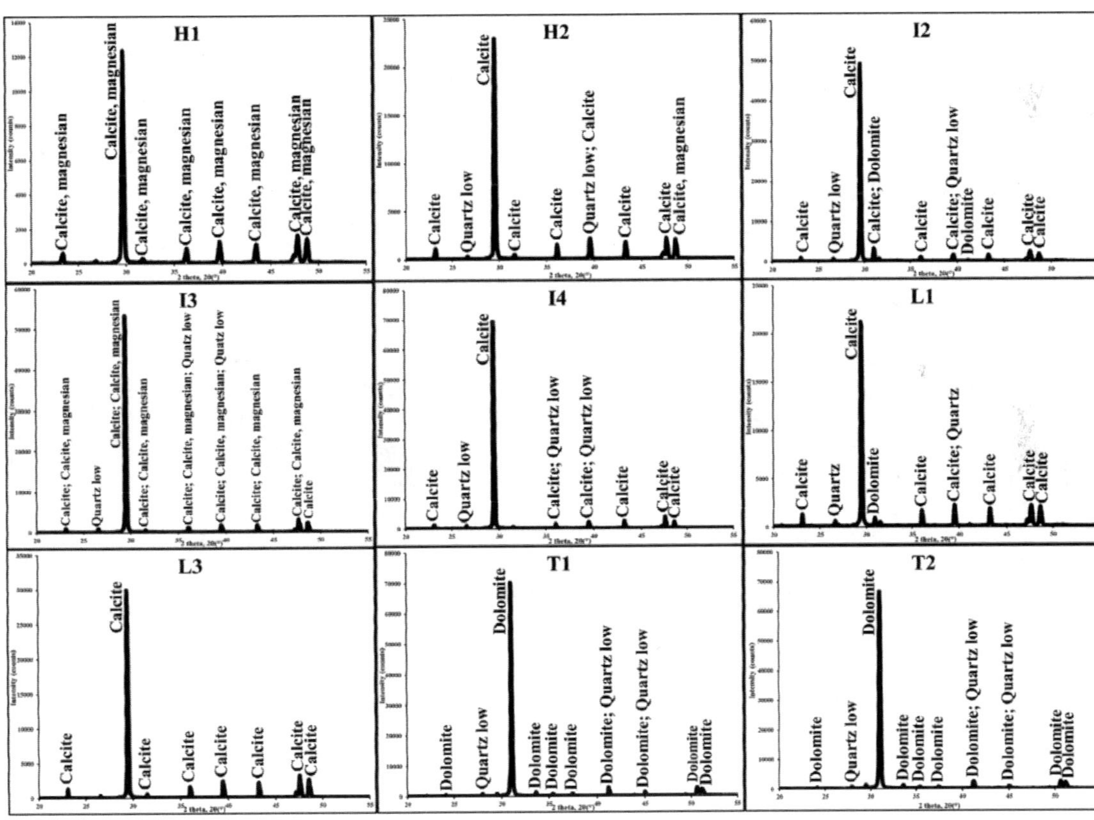

Figure 3. X-ray diffraction pattern of each selected carbonate rocks from quarries in Simpang Pulai and Gopeng. All samples showed a clearly highest peak that represent the most significant mineral in the sample. 2θ showed are from 20-55° with varies of counts

4. Conclusion

Carbonate rocks from selected quarries in Simpang Pulai and Gopeng have high content of calcite with low content of other minerals. Though, there are two samples that contains high concentration of MgO and it is proves by XRF analysis and XRD pattern graphs. This is good for industrial usage as they want the high quality of limestone and dolomite. Sample with the highest amount of Ca (calcite) is H2, the highest content of Mg (dolomite) is T2. The difference of mineral contents, grains size, and the compositions can help in justifying the particular characteristics of each of the carbonate rocks like colour, strength, and moisture content. The comprehensive of petrographic and geochemical of carbonate rocks in Kinta Valley can lead in understanding the secondary process including karstification, dolomization and dissolution processes. Petrographic, XRF and XRD are three analysis that compliment with each other. From photomicrograph, we can see the minerals and its characteristics that cannot be seen by naked eyes and the existence of the mineral were being approved by XRF and XRD analysis.

5. Acknowledgments

This research project is supported by the Ministry of Higher Education (MOHE), Malaysia under the Fundamental Research Grant Scheme (FRGS) of 6071361. The authors greatly acknowledge the quarries for permission to access the quarry sites and assistance during the study period.

References

[1] Gebretsadik H T, Sum C W and Hunter A W 2015 *Proceedings of the International Conference on Integrated Petroleum Engineering and Geosciences* 291-312
[2] Hutchison C S 1994 *Geologisches Rundschau* 82 388-405
[3] Khoo T T and Tan B K 1983 *The Geological Society of Thailand and the Geological Society of Malaysia* 253-290
[4] Foo K Y 1983 *Geological Society of Malaysia* 1 1-19
[5] Hutchison C S 1975 *Geological Society of America Bulletin* 86 797-806
[6] Lyana K N, Hareyani Z, Shah A K and Hazizan M M 2016 *Procedia Chemistry* 19 694-701
[7] Cobbing E J, Pitfeld P E, Derbyshire D P and Mallick D I 1992 *British Geological Survey of London* 78-86
[8] Ingham F T and Bradford E P 1960 *Federation of Malaya Geological Survey, Ipoh* 347
[9] Hutchison C S 2007 *Geological Society of Malaysia* 433
[10] Muhammad R F 2003 *University of Malaya* 443
[11] Suri S, Zabidi H and Ahmad F 2017 *AIP Conference Proceedings* 1892
[12] Gobbett D J and Hutchison C S 1973 *John Wiley & Sons* 438

IOP Conf. Series: Journal of Physics: Conf. Series **1082** (2018) 012096 doi:10.1088/1742-6596/1082/1/012096

Mechanical anisotropy of jointed rock mass under uniaxial compression and tensile loading

Hareyani Zabidi[1]*, Nurmaliana Mohd Khadri[1] and Muhammad Irfan Mohd Azmi[1]

[1]School of Material and Mineral Resources Engineering, Universiti Sains Malaysia, 14300, Nibong Tebal, Penang, Malaysia

Email: srhareyani.zabidi@usm.my

Abstract. Anisotropic in a rock mass are the properties which have directional mechanical behaviours. These anisotropic characteristics are originated from foliation of minerals in metamorphic rock, stratification in sedimentary rocks and discontinuities in the rock mass. The strength and deformability of rock mass are influenced by the orientation and distribution of fractured networks and the failure mode is varies significantly with the variation of fracture orientation. Therefore, the mechanical behaviour of rock mass becomes one of the most crucial topics in rock engineering fields. In this paper, stratified and jointed limestone samples from two different quarries of Ipoh areas (LafargeHolcim and Hume Cement) were studied to determine the mechanical strength and failure patterns under uniaxial compression and tensile loading. The results showed that the failure pattern and maximum strength of was correlated with relative layer orientations. It was observed that the higher the inclination angle the lower the failure strength. The maximum tensile strength also observed at 0° with 16.77 MPa and the lowest occurred at 75° which are 5.68 MPa. The highest compression loaded at 0° which to break at 104.81 MPa and the lowest occurred at 75° which are 40.38 MPa. It is suggested that the failure pattern analysis should be considered in mechanical study of jointed rock mass.

1. Introduction

Rock mass is a geological body consists of discontinuities, ranging from large scale fractures such as faults and bedding planes to small scales of joints and schistosity to create complex natures of rock like heterogeneous and anisotropic properties. Considering the geological history and repeated events of deformation experienced, a rock is considered to be an anisotropic. Rocks which to have properties that vary with direction are said as an anisotropic. It is also refer to the properties of a material that is dependent on the direction. In general, anisotropic is defined as the presence of different properties in different directions [1]. These anisotropic characteristics are originated from mineral foliation in metamorphic rock, discontinuities in any of rock mass and stratification in sedimentary rocks.

The anisotropic created by the discontinuities will have much greater engineering significance [2]. Anisotropy in rocks significantly affects many engineering aspects such as in civil and mining engineering where the stability of deeper foundation, underground and surface excavations can be questioned [3]. It also affects drilling, blasting and rock cutting in mining and quarrying activities [4]. Rock anisotropy is also a critical factor in petroleum engineering as in controlling the deviation, stability, deformation and failure of boreholes [5]. It also influences the propagation and fraction of cracks in shale gas extraction activities [6]. Therefore, the degree of mechanical behaviour of anisotropic rock masses need to be aware in order to decide whether it is relevant or not to consider the anisotropy before start any certain operations [7].

This paper focuses on the anisotropic strength and fracture pattern of laminated limestones. In order to understand the effects of the layer orientation on the strength and the fracture pattern, the anisotropic rocks were studied.

2. Methods

Intact laminated limestone blocks were collected from two limestone quarries in Kinta Valley areas, QA and QB (Fig. 1). For this purpose, 15 block samples were collected from each quarry to later be cored into different lengths with the same 54 mm diameter. The length varied with the ratio of thickness to diameter as it is two (2) in UCS and one (1) in Brazilian. The rock were cored at different direction with respect to the inclination angles (β) of 0, 30, 60 75 and 90°. In this study, the inclination angle (β) is the angle between the loading direction and lamination plane. The visual inspection of the inclination was indicated during the loading of force applied by the Universal Testing Machine (UTM) used. The UCS test on cored limestones was carried out in accordance with the test methods suggested by ISRM (1981). The stress rate was kept at constant value of 1kN/s so that the failure could occurred within 5 – 10 min of loading applied. The layer is adjusting based on the required inclination angle before the load is applied. When the load is applied, the sample fail and the fracture pattern occurred is observed and determine if it is a layer activation fracture, central fracture or non-central fracture.

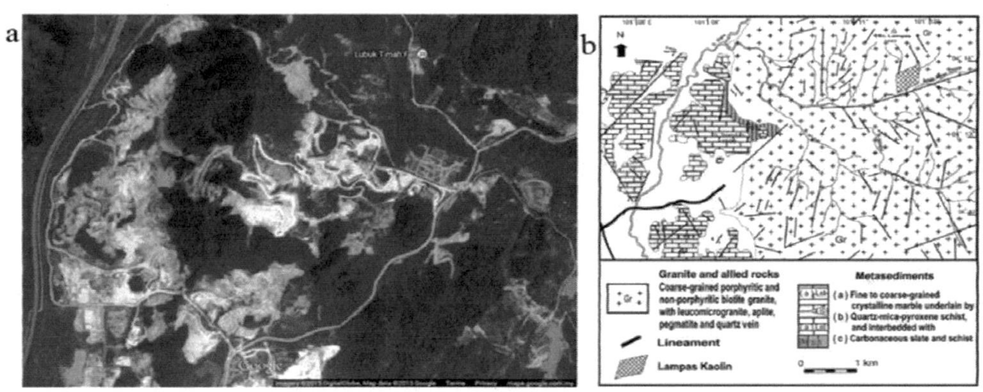

Fig. 1. (a) Part of Keramat Pulai industrial area; (b) Geology of Keramat Pulai area, Perak.

3. Results and discussion

3.1 Uniaxial Compressive Strength (UCS) Test

A total of 20 cored samples (10 from QA and 10 from QB) were tested under the uniaxial compression strength (UCS) test. For this test, five inclination angles were chosen, 0, 30, 60, 75 and 90°. Fig. 2 and Table 1 show the physical properties of samples, the failure pattern and ranges of uniaxial compressive strength with respect to the anisotropy orientations. Fig. 3 shows the variations of the UCS values (MPa) ranging from15 MPa to 106 MPa as a function of inclination angle. The results indicate the maximum UCS is in the inclination angle of 60° for QA at 51.18 MPa and 0° for

QB at 105.94 MPa. In contrast, the least strength recorded at 15.42 MPa of 90° angle of inclination for QA and at 31.79 MPa at 75° inclination for QB.

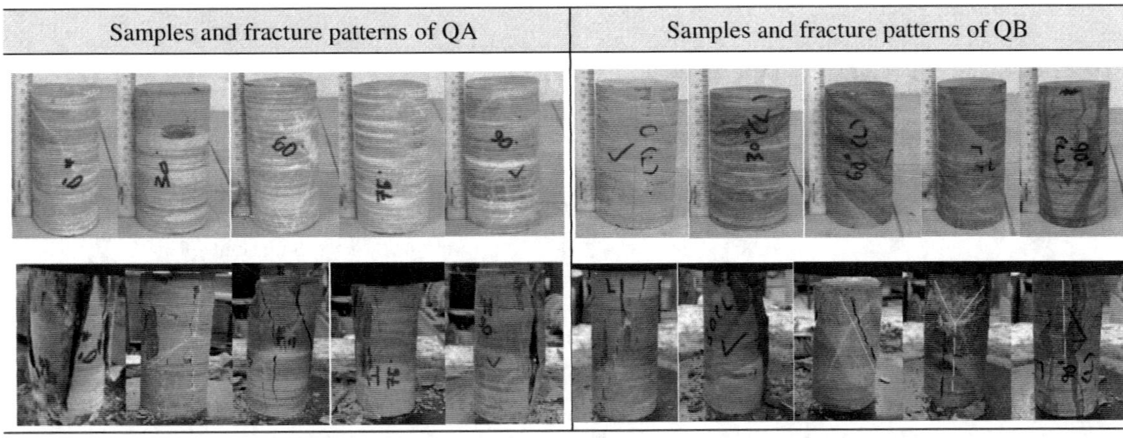

Figure 2. The observed condition of limestone samples of before and after the failure under different lamination-loading angles of UCS test.

Generally, the variant strength of intact rock in uniaxial and triaxial loading conditions with respect to the anisotropy orientation is defined as the "strength anisotropy" and its magnitude is representing the degree of anisotropy

$$R_c = \frac{\sigma_{c(90)}}{\sigma_{c(\min)}}$$

where R_c is the degree of anisotropy, $\sigma_{c(90)}$ or can use $\sigma_{c(\max)}$ is UCS perpendicular to the planes of anisotropy which is maximum value σ_c of and $\sigma_{c(min)}$ is the minimum value of σ_c commonly at β = 30°–45°. The strength anisotropy classification depends on the degree of anisotropy (R_c), which is the ratio of the maximum to the minimum values of uniaxial compressive strength (UCS) of anisotropic intact rock samples. The lower the value of R_c, the closer the rock nature to be isotropic. The higher value of R_c the higher anisotropy of rock class. Table 1 shows relationship between value of R_c and degree of anisotropy. In this case of QA and QB, both of limestones are valued at 3.32 and 3.33 respectively, thus to classify as medium anisotropy rock.

In reference to the previous studies, the minimum strength of laminated rock is at an orientation angle of β = 30° and the maximum failure strength is either at β = 0° or β = 90°. This is because at β = 30°, the weak plane is in the orientation in relation to major loading direction. The tests were carried out at orientation angles, β (the angle between the major principal stress direction and weakness plane), of 0°, 30°, 60°, 75° and 90°. So for QB, it was noted that it was in accordance to the theory as the maximum value of UCS (MPa) is at 0°. For QA, it does not follow the theory of 90° to record as the weakest angle measured from loading direction. The results are believed due to the moisture content of the specimens, and it thus should be noted. The UCS value of rock is highly dependent on the effective porosity, which in other term is the existence of water and applicable to the reduction of strength in the specimen. The inclined fissures, intrusions, and other anomalies will often cause premature failures on those planes.

Table 1. Anisotropic strength and failure patterns of Kinta Limestone samples in UCS test

No	Sample no	Inclination angle (β)	UCS (MPa)	Failure pattern
1	QA0°	0°	22.83	Axial splitting
2	QA0°	0°	33.39	Shearing along single plane
3	QA30°	30°	35.25	Along foliation
4	QA30°	30°	26.12	Axial splitting
5	QA60°	60°	30.77	Axial splitting
6	QA60°	60°	51.18	Along foliation
7	QA75°	75°	33.26	Shearing along single plane
8	QA75°	75°	28.29	Along foliation
9	QA90°	90°	33.28	Shearing along single plane
10	QA90°	90°	15.42	Shearing along single plane
11	QB0°	0°	104.81	Along foliation
12	QB0°	0°	105.94	Along foliation
13	QB30°	30°	71.51	Shearing along single plane
14	QB30°	30°	61.34	Shearing along single plane
15	QB60°	60°	40.38	Axial splitting
16	QB60°	60°	49.11	Axial splitting
17	QB75°	75°	65.08	Axial splitting
18	QB75°	75°	31.79	Shearing along single plane
19	QB90°	90°	46.54	Shearing along single plane
20	QB90°	90°	33.32	Axial splitting

3.2 Indirect Tensile test

A total of 25 cores were tested for Brazilian test, 15 for QA and 10 for QB. Based on Fig. 3 and Table 2, the maximum tensile strength for QA occurred at or near $\beta = 0^0$, at 13.23 MPa and the minimum tensile strength occurred at $\beta = 90^0$, at 2.35 MPa. Theoretically, the tensile strength should decrease when the inclination angle increase. However, when the average of tensile strength is calculated, it was observed that the tensile strength decreased from 0^0 to 60^0, suddenly increased at 75^0 and later dropped at 90^0. As in Tab. 2, the tensile strength for the inclination angle of 75^0 is quite high compared to the sample of inclination angle of 60^0.

For QB, the results are in contrast to the theory stated (Tab. 2). The maximum tensile strength is near or equal to $\beta = 0^0$ which is 16.77 MPa and the minimum tensile strength was at $\beta = 75^0$ which is 5.68 MPa. This may cause by certain factors that influence the fracture pattern. The existence of tight cracks can be observed in QB-75 in comparison to much intact core of QB-90. In general, the strength decreases from QB-0 to QB-75 and increase in QB-90. If there are no other factors that influence the crack, it can be concluding that the values of tensile strength are inversely proportional to the inclination angle. The lower the inclination angle, the higher the tensile strength.

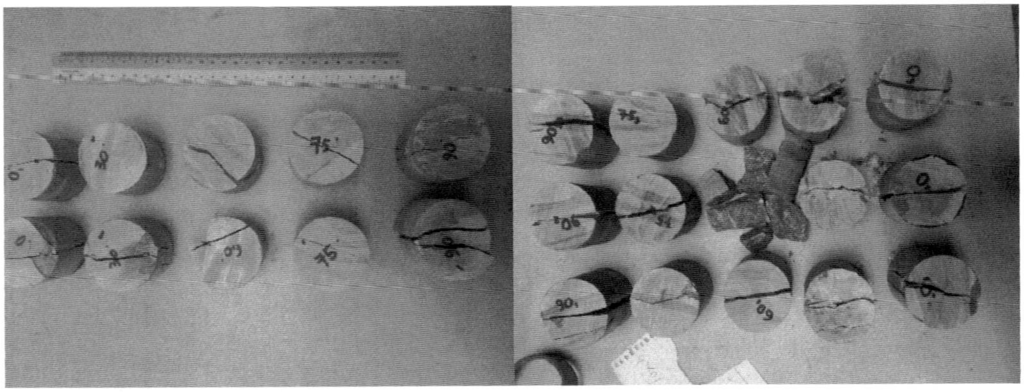

Figure 3. The observed condition of limestone samples from QA and QB after the failure under different lamination-loading angles of Brazilian test.

Table 2. Anisotropic strength and failure patterns of Kinta Limestone samples in Brazilian test

No	Sample no	UCS (MPa)	Failure pattern
1	QA0°	11.92	Central fracture
2	QA30°	8.26	Central fracture
3	QA60°	6.62	Central fracture
4	QA75°	9.22	Central fracture
5	QA90°	6.42	Central fracture
6	QB0°	16.77	Central fracture
7	QB30°	8.48	Central fracture
8	QB60°	9.77	Central fracture
9	QB75°	5.68	Central fracture
10	QB90°	10.01	Central fracture

The different strength can be observed in these two sites. As in both tables, the highest tensile strength is from QA site. Even though the rock from QA is more stratified compared to the QB, it is much stronger than QA. Compared the maximum compressive load between QA and QB for every inclination angle, it can be seen that the samples of QA is easily to break compare to the samples from QB.

The fracture pattern occurred are differently for every inclination angle. There are five inclination angle were tested ranging between 0^0 (perpendicular to loading direction) and 90^0 (parallel to loading direction). It should be highlighted that the orientation of the layers with respect to loading direction is evaluated. The loading direction in this test is always vertical while the layer orientation become the variable where it always change followed by the layer orientation required. Five different values are considered: 0^0, 30^0, 60^0, 75^0, and 90^0. There are three types of fracture pattern had occurred (Fig. 3). Normally, in most cases, there is two or three different fracture patterns will occurred in the same experiment. After conduct the test, three fractures are observed: (1) - layer activation occurred if some fractures are parallel to the isotropic layers; (2) - central fractures will occurred if some fracture are roughly parallel to the loading direction and it is located in the central part of the sample between the two loading lines; (3) - non-central fractures occurred if the fractured occurred outside the central and do not correspond to part of layer activation.

4. Conclusion

This paper studied the experimental characterization of the anisotropic behaviours of laminated limestone using UCS and Brazilian test. Observations were made on anisotropic strength and failure patterns at five different inclination angles of loading, 0^0, 30^0, 60^0, 75^0 and 90^0 in respect to lamination planes of limestones collected from two different quarries in Kinta Valley, Ipoh. The results show that the anisotropic strength and failure patterns of laminated limestones depend significantly on the inclination angle of planes and loading direction. From this study, it shows that the failure strength is correlated with the relative layer orientation (Fig. 4). But, nonetheless, due to the condition of rock itself which have crack cause the some failure strength are not correlated with the relatives' layer orientation as shown in the graph of the comparison between two sites where at 75^0 of sample QB, it is the lowest tensile strength while for the QA, 0^0 is the lowest tensile strength that is occurred. The different of this inclination angle cause by the condition of rock itself. From both sites, different fracture patterns were observed based on the inclination test. The fractured sample of QA and QB follow the trend where the fracture occur for the inclination angle of 0^0, 30^0, 60^0, 75^0, and 90^0 starts with the central fracture then followed by the combination of the central fracture and activation layer fracture and finally activation layer respectively. Based on the fracture pattern occurred, it can be conclude that the range between 30^0 and 60^0 is the range where the combination of the central fracture and activation layer fracture occurred.

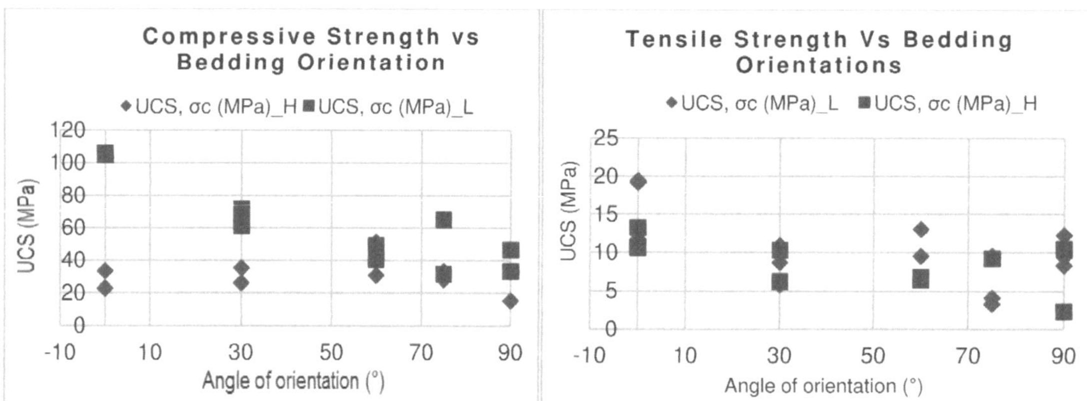

Figure 4. Variation of anisotropic strength (MPa) with the inclination angle (β) for each limestone samples (QA and QB)

Acknowledgements

The Ministry of Higher Education (MOHE), Malaysia under the Fundamental Research Grant Scheme (FRGS), No. 673044, financially supports this research project.

References

[1] Q. Gao, J. Tao, J. Hu, and X. Yu 2015 *J. Rock Mech. Geotech. Eng.* 7 213
[2] M. Trisugiwo, H. Zabidi, and F. Ahmad 2016 *Procedia Chem.* 19 751
[3] X. Tan, H. Konietzky, T. Frühwirt, and D. Q. Dan 2015 *Rock Mech. Rock Eng.* 48 1341
[4] S. Kahraman, M. Fener, and E. Kozman 2012 *J. South. African Inst. Min. Metall.* 112 331
[5] T. Yang *et al.* 2015 *Tunn. Undergr. Sp. Technol.* 48 129
[6] N. L. K, Z. Hareyani, K. S. A, and M. H. M. H. 2016 *Procedia Chem.* 19 694
[7] A. Delisio, J. Zhao, and H. H. Einstein 2013 *Tunn. Undergr. Sp. Technol.* 33 131

IOP Conf. Series: Journal of Physics: Conf. Series **1082** (2018) 012097 doi:10.1088/1742-6596/1082/1/012097

Synthesis and Characterization of Schwertmannite by Neutralization Method

Mohd Khairul Nizam Mahmad[1,2], Norlia Baharun[1], Mohd Remy Rozainy M.A. Z[2,3], Suhaina Ismail[1] and Teuku Andika Rama Putra[1]

[1]Srategic Mineral Niche Area, School of Materials and Mineral Resources Engineering, Engineering Campus, Universiti Sains Malaysia, 14300 Nibong Tebal, Penang, Malaysia

[2]School of Civil Engineering, Engineering Campus, Universiti Sains Malaysia, 14300 Nibong Tebal, Penang, Malaysia

[3]Center of Excellence Geopolymer and Green Technology (CEGeoGTech), School of Materials Engineering, University Malaysia Perlis (UniMAP), P.O Box 77, d/a Pejabat Pos Besar, 01000 Kangar, Perlis, Malaysia

E-mail: nizam.mahmad@gmail.com

Abstract. This research work involved the synthesis of Schwertmannite using neutralization method at room temperature and characterizations by X-ray Diffraction (XRD), Field Emission Scanning Electron Microscope (FESEM) and Transmission Electron Microscopy (TEM). The precipitate produced using neutralization method at room temperature and synthesis time of 27 mins was found to have the characteristic peaks position of Schwertmannite mineral from XRD analysis with D-spacing peaks values of 3.40, 2.55, 2.28, 1.96, 1.66, 1.51 and 1.46. FESEM photomicrograph shows that the Schwertmannite produced have a relatively rough surface, uneven and tunnel-like morphology. This structures are due to the existence of more cavities on the Schwertmannite surface which would play a significant role in enhancing the specific surface area of the Schwertmannite. The measured average diameter of cavities or tunnel is about 0.5-1μm. From TEM analysis, the precipitate is shown to have a compact Schwertmannite aggregates of > 200nm diameter. These surface properties are significant to the Schwertmannite produced and would be potential in the removal of arsenic from mining wastewater.

1. Introduction

Schwertmannite is a poorly crystalline and metastable ferric oxyhydroxy sulfate. It has an identical tunnel structure with Akaganéite and the ideal formula for Schwertmannite is $Fe_8O_8(OH)_6SO_4$ but may range to $Fe_80_8(OH)_{4.5}(S0_4)_{1.75}$ reliant on the degree to which tunnel and surface sites are saturated with sulfate. Additional details regarding the properties of Schwertmannite are given elsewhere [1], [2]. Schwertmannite has a large reactive surface [3] and has great potential in the scavenging significant quantities of Arsenic , Selenium, Chromate [4], Antimony[5], Phosphorus, Fluorine [6], Molybdenum [7], Vanadium, Caesium[8] and Uranium(VI) [9].

IOP Conf. Series: Journal of Physics: Conf. Series **1082** (2018) 012097 doi:10.1088/1742-6596/1082/1/012097

Schwertmannite actually forms in acidic condition (sulfate-rich) in range pH 3-4. Acid Mine drainage (AMD) and Acid Rock Drainage (ARD) is a common natural environment for Schwertmannite formation. Theoretically, the mobility of arsenic may be retarded by Schwertmannite. In point of fact, the Schwertmannite will decrease the arsenic solubility by sorption process [10].

This mineral also can be produced in the laboratory. There are several methods, the Schwertmannite can be synthesized such as hydrolyzation [11] [4], [12], ,oxidation [11], [12] and neutralization/precipitation [13], [14][15]

The objective of this research work was to synthesis Schwertmannite by using neutralization method and characterizations by X-ray Diffraction (XRD), Field Emission Scanning Electron Microscope (FESEM) and Transmission electron microscopy(TEM).

2. Experimental

The synthesis of Schwertmannite was done according to the procedure of Houng Aloune et al., (2015), this method is known as neutralization method. However, some modification has been done to study the influence of pH on Schwertmannite mineral. 0.1M Fe^{3+} solution was prepared with addition of 0.1M H_2SO_4 solution to make the solution rich with sulfate ion under magnetic stirring at 250 rpm and at room temperature. The solution was titrated with 1M of Na_2CO_3 to obtain the final pH 3-4. The precipitates formed and supernatant in the beaker were separated by filtration. The filtrate was collected for analysis by using Inductively Coupled Plasma-Optical Emission Spectrometer (ICP-OES), Model Varian 715-ES.

The residue was dried at 60 °C in a vacuum oven for 24 hours, and then were analysed using X-ray Diffraction (XRD), Model Bruker Advance D8, Field Emission Scanning Electron Microscope (FESEM), Model Zeiss Supra 35VP and Transmission Electron Microscopy (TEM), Model Philip CM12. The initial and final pH and redox potential of the solution before and after experiment were measured using pH meter, Model CyberScan pH 510.

3. Results and Discussions

3.1 Characterization of Schwertmannite

Fig. 1 shows the results obtained from the XRD analysis of the precipitate produced in the synthesis Schwertmannite. The major 7 references d-spacing's peaks values suggested by Bigham et al., (1996a) are attained for sample SCH-13. The pattern showed diffraction d-spacing peaks values at 3.40, 2.55, 2.28, 1.96, 1.66, 1.51 and 1.46 which can be indexed to Schwertmannite mineral.

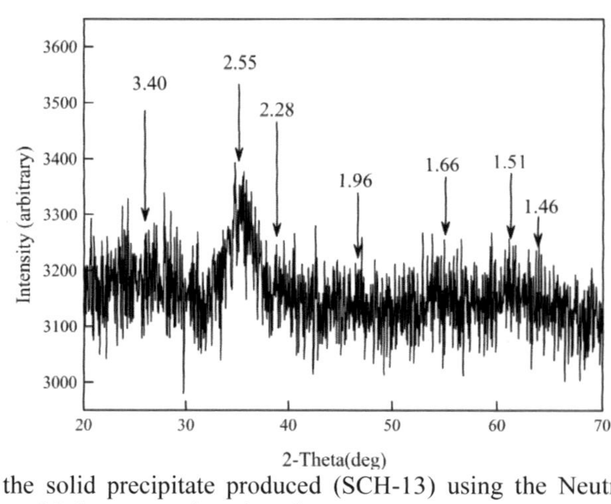

Fig. 1: XRD analysis of the solid precipitate produced (SCH-13) using the Neutralization Method showing characteristic peaks position of Schwertmannite mineral

Comparison made between the observed XRD peaks as seen in sample SCH-13 and the reference peaks position for Schwertmannite by Bigham et al., (1996), HoungAloune et al., (2014) and Loan et al., (2004) are shown in Table 1. From Table 1, this shows that the precipitate produced has almost similar reference peaks position with Schwertmannite mineral. This finding shows that Schwertmannite can be produced in the range of pH 3.0-3.2. This is also proven from the work carried out by Bigham et al., (1996a) and some other researchers. At higher pH (>3.2), other minerals will be formed during precipitation. At pH lower than 3.0 Jarosite mineral will be formed. Between the pH range between 3.3-3.6 goethite will be produced [16].

Table 1: Comparison between the observed XRD peaks for Sample SCH-13 and the reference peaks position for Schwertmannite mineral [16]–[18].

Schwertmannite Mineral		Sample SCH-13	
2-Theta(deg.)	D spacings	2-Theta(deg.)	D spacings
26	3.39	26.26	3.39
35	2.55	35.14	2.55
39	2.28	39.49	2.28
46	1.95	46.51	1.95
55	1.66	55.27	1.66
61	1.51	61.33	1.51
63	1.46	63.67	1.46

Fig. 2 shows the FESEM photomicrograph of Schwertmannite produced using neutralization method at room temperature. The FESEM analyses for sample SCH-13 was found to be similar to the previous studies of Schwertmannite [12], [17], [19]. From photomicrograph it shows that the Schwertmannite produced have a relatively rough surface, uneven and tunnel-like morphology. This structures are due to the existence of more cavities on the Schwertmannite surface and it has a significant role in enhancing the specific surface area of Schwertmannite.[20]. The average diameter of cavities or tunnel is about 0.5-1μm. According to Davidson et al., (2008), Schwertmannite has a high reactivity and surface area. These characteristics contributed to high sorption capacity for various dissolved trace elements.

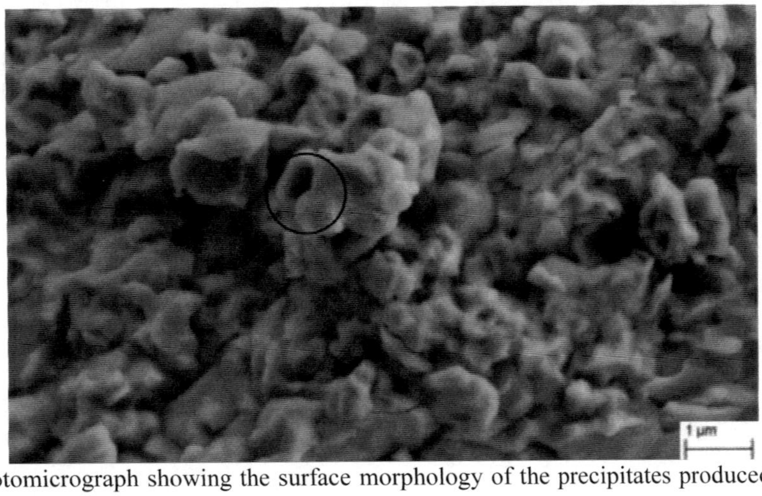

Fig. 2: FESEM photomicrograph showing the surface morphology of the precipitates produced at pH 3-3.2 for sample SCH-13 (Mag. 10.00Kv)

The precipitate produced from the neutralization method at room temperature (synthesis time of 27 mins) was further characterized using TEM analysis as shown in Fig.3 (a). The precipitate is shown to have a compact Schwertmannite aggregates of > 200nm diameter. A comparison is also made with the work done by French et al., (2013). Fig.3 (b) shows HRTEM image of the Schwertmannite aggregate produced by oxidation method at room temperature (synthesis time of 60 mins). She found that the bulk aggregates of the Schwertmannite particles have a size of > 100nm diameter with the distinctive 'pin-cushion' morphology of needles extending from a dense core [22]. From both findings, it is observed that different synthesis conditions would produce Schwertmannite mineral with different morphology and finest.

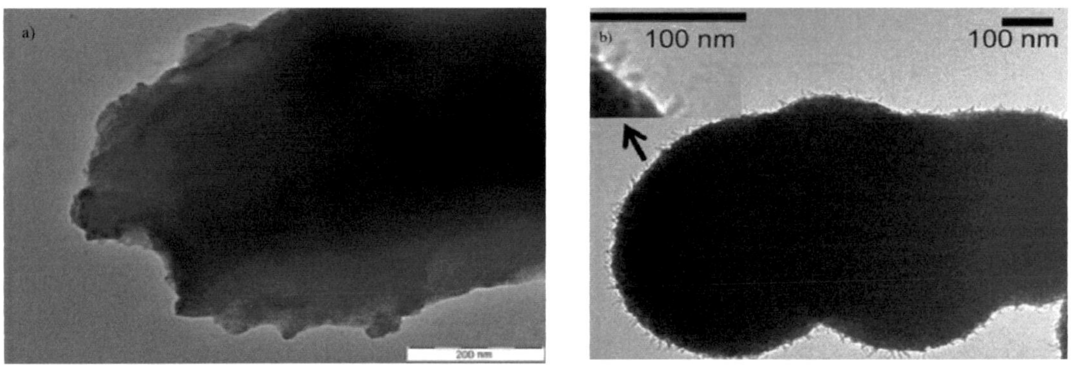

Fig. 3: (a) TEM image showing the Schwertmannite mineral produced (SCH-13) using neutralization method at room temperature (b) HRTEM image showing the Schwertmannite aggregate produced by using oxidation method at room temperature synthesized by French et al., (2013)

However, the best morphology would be the one with large reactive surface to enhance the sorption capacity for various dissolved trace elements. For the Schwertmannite produced in this work, arsenic removal experiments would be carried out to determine its potential to removal of arsenic from mining wastewater.

4. Conclusions

The precipitate produced using neutralization method at room temperature and synthesis time of 27mins was found to have the characteristic peaks position of Schwertmannite mineral from XRD analysis with D-spacing peaks values of 3.40, 2.55, 2.28, 1.96, 1.66, 1.51 and 1.46. FESEM photomicrograph shows that the Schwertmannite produced have a relatively rough surface, uneven and tunnel-like morphology. This structures are due to the existence of more cavities on the Schwertmannite surface which would play a significant role in enhancing the specific surface area of the Schwertmannite. The measured average diameter of cavities or tunnel is about 0.5-1μm. From TEM analysis, the precipitate is shown to have a compact Schwertmannite aggregates of > 200nm diameter. These surface properties are significant to the Schwertmannite produced and would be potential in the removal of arsenic from mining wastewater.

5. Acknowledgments

The authors thank to Universiti Sains Malaysia (USM) for use of X-ray Diffraction (XRD), Field Emission Scanning Electron Microscope (FESEM) and Transmission Electron Microscopy (TEM) and wish to thank Ministry of Higher Education Malaysia and MyBrain15 (Malaysia) for financial support.

IOP Conf. Series: Journal of Physics: Conf. Series **1082** (2018) 012097 doi:10.1088/1742-6596/1082/1/012097

References

[1] Y. Liao, L. Zhou, J. Liang, and H. Xiong 2009 *Mater. Sci. Eng. C* 29 211
[2] J. M. Bigham, U. Schwertmann, S. J. Traina, R. L. Winland, and M. Wolf 1996 *Geochim. Cosmochim. Acta* 60 2111
[3] E. D. Burton, R. T. Bush, L. a. Sullivan, and D. R. G. Mitchell 2008 *Geochim. Cosmochim. Acta* 72 4551
[4] S. Regenspurg and S. Peiffer 2005 *Appl. Geochemistry* 20 1226
[5] Y. Li, D. Mohan, C. U. Pittman, Y. S. Ok, and X. Dou 2016 *Desalin. Water Treat.* 1
[6] A. Eskandarpour, M. S. Onyango, A. Ochieng, and S. Asai 2008 *J. Hazard. Mater.* 152 571
[7] A. Eskandarpour, K. Sassa, Y. Bando, M. Okido, and S. Asai 2006 *Mater. Trans.* 47 1832
[8] T. Ishikawa, T. Harada, and F. Akahori 2016 50 235
[9] B. D. Stewart *et al.* 2003 *Environ. Sci. Technol.* 37 2898
[10] E. D. Burton *et al.* 2009 *Environ. Sci. Technol.* 43 9202
[11] J. Song, S. Y. Jia, H. T. Ren, S. H. Wu, and X. Han 2015 *Int. J. Environ. Sci. Technol.* 12 1559
[12] J. M. Blgham, U. Schwertmann, L. Carlson, and E. Murad 1990 *Geochim. Cosmochim. Acta* 54 2743
[13] W. D. Burgos *et al.* 2012 *Geochim. Cosmochim. Acta* 76 29
[14] S. Houng Aloune, N. Hiroyoshi, M. Ito, S. Houngaloune, N. Hiroyoshi, and M. Ito 2015 *Int. J. Chem. Eng. Appl.* 6 32
[15] S. HoungAloune 2015
[16] J. M. Bigham, U. Schwertmann, and G. Pfab 1996 *Appl. Geochemistry* 11 845
[17] M. Loan, J. M. Cowley, R. Hart, and G. M. Parkinson 2004 *Am. Mineral.* 89 1735
[18] S. HoungAloune, T. Kawaai, N. Hiroyoshi, and M. Ito 2014 *Hydrometallurgy* 147 30
[19] U. Schwertmann, J. M. Bigham, and E. Murad 1995 *Eur. J. Mineral.* 7 547
[20] F. Liu, J. Zhou, S. Zhang, L. Liu, L. Zhou, and W. Fan 2015 *PLoS One* 10 1
[21] L. E. Davidson, S. Shaw, and L. G. Benning 2008 *Am. Mineral.* 93 1326
[22] R. A. French, N. Monsegue, M. Murayama, and M. F. Hochella 2013 8 no. Iii

IOP Conf. Series: Journal of Physics: Conf. Series **1082** (2018) 012098 doi:10.1088/1742-6596/1082/1/012098

A study of performance of a diesel engine fueled with emulsified biofuel and its blends

M.F.Hamid[1], M.K. Abdullah[2*], M.Y. Idroas[1], Z.A. Zainal Alauddin[1], C.M. Sharzali[1], M.A.M. Naser[2], and S.R. Khimi[2]

[1]School of Mechanical Engineering, Universiti Sains Malaysia, Engineering Campus, Penang, Malaysia

[2]School of Materials and Mineral Resources Engineering, Universiti Sains Malaysia, Engineering Campus, Penang, Malaysia

Email: mkhalil@usm.my

Abstract. Diesel and natural gas are most popular fuels application in the natural aspirated (NA) compression ignition (CI) engine. However, burning fossil fuel results in serious environmental changes i.e. global warming. The direct use of liquid biofuel to the engine can harmful the engine system, engine performance, and emission. To mitigate this problem, it is necessary to change biofuel's properties. Emulsification techniques are one of the possibility to obtain the emulsified biofuel (EB). However, EB has an undesirable higher viscosity, larger and heavier molecules which lead to lower atomization and less mixing with fuel-air mixing. Therefore, a proper mixture of surfactant and co-surfactant is focused on this study. The combination of two different types of surfactant possible to reduce surface tension and improved atomization during injection. The 2% of surfactant agent mixture containing 60% of span-80 and 40% of Tween-80 is the best formulation. EB 5% shows the stable solution and produced the closest result to fossil diesel curve performance.

Keywords: *Compression ignition (CI) engine, Biofuel, Emulsified biofuel, Emulsifier, Diesel engine*

1. Introduction

Compression Ignition (CI) engines are become most popular power generation basis and widely used in both heavy and light duty transportations [1]. They are most reliable and powerful in term of fuel efficiency. Contrasting with spark-ignition (SI) or known as a gasoline engine, diesel engine technology used high compression ratios and combust via elevated temperature of the air in the cylinder due to mechanical compression (adiabatic compression) [2]. However, great achievement of diesel engine power generation are overshadowed by its emission poison like PM (particulate matters), Soot, THC (total hydrocarbons), NOx, SOx, and CO. These emissions effect have contribute to the adverse health and environmental issues [3]. Moreover, strict regulations imposed by many European countries have restricted many application in various field and reduce the potential of this engine to be used in a near future power generation. The extensive efforts have been made by engine manufacturers, researchers, and scientists to control emission emitted by this CI engine such as external devices and chemical, but only a few to a certain extent [4]. To mitigate this matter, exploring a new renewable energy source to replace fossil fuel are substantially being conducted globally. The used of biofuel, particularly made from palm oil is among the popular fuels to be discover on their potentials.

Burning the biofuel to the engine has improved emission indirectly due to the high oxygen content of biofuel [5]. However, straight application biofuel to the engine are strictly prohibited since it will harmful the engine such as carbon deposit, injector clogging, and incomplete combustion. This problem can be tackled fundamentally either need to improve via engine hardware system or improved the chemical properties of biofuel [6]. Improved the chemical properties of biofuel is seen feasible techniques and no modification required on the engine [7]. The most popular techniques is emulsification techniques which considering water in the fuel system [8]. Water present in the fuel system will reduce the emission of NOx, reduce the temperature in the combustion chamber and improve combustion as well via micro-explosion effect [9]. Introducing water in fuel system as such as thrust boosters or power supplements for fuel system. Azahari et al. [10] have studied the effect of water in diesel (W/O) emulsion with combination of surfactant type of Span 80 and Tween 80. They found that the 15% of water content will reduce almost 71% of reduction in PM emission. It is because of micro-explosion effect aids in faster fuel breakdown, allowing them for secondary atomization due to different boiling point of each fuel and improved injection penetration because of different fuels density.

To obtain an economic and environment-friendly engine fuel from biofuel, it is necessary to modify the biofuel chemical properties such as viscosity, surface tension, phospholipids, free fatty acids, sterols and moisture content [11]. Nevertheless, the high surface tension and viscosity of biofuel is the main obstacle confronted before any emulsification approach will be introduced [12]. Emulsification system consists of two immiscible and miscible fluids and form together to be a stable solution [13]. To stabilize emulsification in biofuel system, the combination of two types of surfactant agent has been refocused. A surfactant molecule has two parts of affinity; affinity for water and affinity for oil [14]. The concentration of surfactant and water quantity in the emulsion play a major role as well as the suitable combination of dual surfactants [15]. The combination of dual surfactants will specify the suitable value of HLB number. In this paper, the effect of formulation of surfactant agents on palm oil biofuel and its effect on the engine performance running with using emulsified biofuel as a primary fuel in diesel engine has been investigated.

2. Experimental study

2.1 Materials

The palm oil used in the present study for emulsified biofuel were collected from School of Mechanical Engineering, Universiti Sains Malaysia, and surfactant agents were purchase from CSI Labshop Malaysia. Two types of surfactant agents used were Tween 80 and Span 80.

2.2 Experiment engine setup

The setup as shown in Figure 1 consist of one cylinder, four-stroke Yanmar L70AE diesel engine connected to eddy current type dynamometer for loading (15kW). Provision is also made for interfacing airflow, fuel flow, temperatures and load measurements. Exhaust was assembled with ventilation system and the water cooling system was turned on. For initially, the fossil diesel was run, the torque data was collected and compared with standard manual performance given by manufacturer to validated 0purposed. The pure biofuel and EB was run then and the data was collected.

No.	Name
1.	Yanmar diesel engine model L70AE
2.	Dynamometer
3.	Exhaust ventilation
4.	Transaxle shaft
5.	Water cooling system

Figure 1: Experiment engine setup

3. Results and discussion

3.1 Physiochemical characteristic of emulsified biofuel

Emulsified biofuel (EB) properties is a basic study to determine the liquid characteristics. The properties of EB is a vital stage to experiment. The testing was based on the American Society of Testing and Materials (ASTM). It was observed that, the Increased of water percentage will increase the viscosity. It will benefit during injection due to different boiling point of auto ignitions [16].

Table 1. Physiochemical properties of EB

EB/Test	Specific weight (ASTM D854)	Kinematic Viscosity (at 40°C) (ASTM D445)	Calorific value (ASTM 2382)	Poor Point (ASTM D97)	Flash Point (ASTM D93)
Neat Biofuel	8.83	31.7	41.3	16	310
1% of EB	8.87	32	40.3	16.5	313
2% of EB	8.88	33.0	39.8	17	315
3% of EB	8.89	33.5	39.4	18.5	319
4% of EB	8.90	34.5	39.2	19	325
5% of EB	8.91	35	39	19.5	332

3.2 Engine torque

Figure 2 shows that diesel fuel produced higher torque performance curve compared to that of other biofuels. The higher torque was generated via diesel due to its lower viscosity and more complete combustion. Higher engine torque can be initially observed and it consequently decrease due to friction losses (negative torque). EB5 produced almost near to diesel and pure biofuel generated lower torque. EB produced a micro-explosion phenomenon and it's depend on the water droplet size or water volume. EB5 is closer to diesel performance is because it consist of bigger water dispersed phase size and providing additional energy from water explosion.

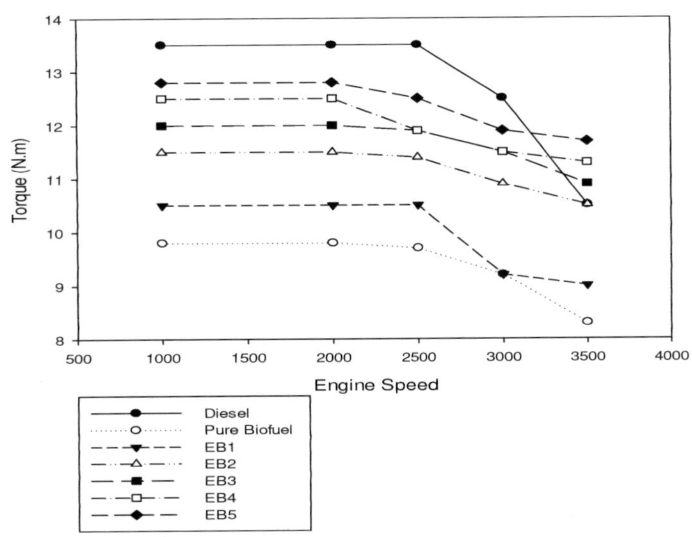

Figure 2. Engine Torque against engine speed

3.3 Brake power

The brake power produced by the engine using pure diesel, pure biofuel, and EB is presented in this section under variable conditions and speeds. Figure 3 shows that the brake power gradually increases with the increase of engine speed. Diesel fuel produced the maximum speed compared to EB. It was found that EB5 is much closer to diesel performance curve. This is because of the influence of the composition water volume in biofuel consequently improve combustion efficiency, improve the retardation of flame propagation, prolong the ignition delay and lower the rate of pressure in combustion.

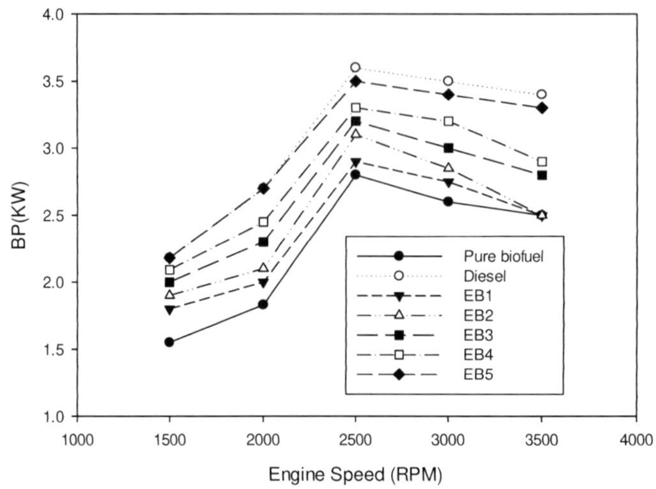

Figure 3. Engine brake power against engine speed.

4. Conclusion

The experiment investigation fueled with diesel, pure biofuel and EB has been conducted. The results of EB were compared with diesel and pure biofuel in order to validate and enhance the understanding of performance characteristic, influence of water disperse phase, water volume, and air-fuel mixing process of EB. EB is a mixture of two immiscible liquids. In order to ensure the mixture between them is homogeneous, surfactant agent is used to weakening the surface tension. 2% of surfactant agent mixture containing of 60% of span-80 and 40% of Tween 80 is the best formulation. EB5 is a more reliable solution and produced closest result to diesel curve performance, with an approximately 15% total difference in overall. This was effect from micro-explosion phenomenon and the water sized gave influence of results during combustion.

5. Acknowledgement

This study was funded by Research University Grant Scheme (Project code of 1001/PBAHAN/8014006).

References

[1] Hamid, M. F., Idroas, M. Y., Basha, M. H., Sa'Ad, S., Mat, S. C., Abdullah, M. K., & Zainal Alauddin, Z. A. 2016 *MATEC Web of Conferences* 90

[2] Heywood, J. B. 1988 *Internal Combustion Engine Fundamentals. McGrawHill series in mechanical engineering* 21

[3] Chen, S. M., & He, L. Y. 2014 *China Economic Review 31* 106

[4] Tartakovsky, L., Baibikov, V., Czerwinski, J., Gutman, M., Kasper, M., Popescu, D., Zvirin, Y. 2013 *Atmospheric Environment* 64 320

[5] Hamid, M. F., Idroas, M. Y., Ishak, M. Z., Zainal Alauddin, Z. A., Miskam, M. A., & Abdullah, M. K. 2016 *BioMed Research International 2016*

[6] Vallinayagam, R., Vedharaj, S., Yang, W. M., Lee, P. S., Chua, K. J. E., & Chou, S. K. 2014 *Applied Energy* 130 466

[7] Wong, K. I., Vong, C. M., Wong, P. K., & Luo, J. 2015 *Neurocomputing* (Part A) 397

[8] Mrad, N., Varuvel, E. G., Tazerout, M., & Aloui, F. 2012 *Energy* 44 955

IOP Conf. Series: Journal of Physics: Conf. Series **1082** (2018) 012098 doi:10.1088/1742-6596/1082/1/012098

[9] Demirbas, A. 2008 *Energy Conversion and Management* 49 2106
[10] Azahari, S. R., Salahuddin, B. B., Noh, N. A. M., Nizah, R., & Rashid, S. A. 2016 *Biofuels* 7 337
[11] Kannan, T. K., & Marappan, R. 2011 *Journal of Applied Sciences* 11 2961
[12] Koc, A. B., & Abdullah, M. 2013 *Fuel Processing Technology* 109 70
[13] Debnath, B. K., Sahoo, N., & Saha, U. K. 2013 *Energy Conversion and Management* 69 191
[14] Lin, Y. S., & Lin, H. P. 2011 *Renewable Energy* 36 3507
[15] Nadeem, M., Rangkuti, C., Anuar, K., Haq, M. R. U., Tan, I. B., & Shah, S. S. 2006 *Fuel* 85 2111
[16] Hamid, M. F., Idroas, M. Y., Sa'ad, S., Saiful Bahri, A. J., Sharzali, C. M., Abdullah, M. K., & Zainal, Z. A. 2018 *Renewable Energy* 127 84

IOP Conf. Series: Journal of Physics: Conf. Series **1082** (2018) 012099 doi:10.1088/1742-6596/1082/1/012099

Failure Mode of Epoxidised Natural Rubber - Microalumina Particle Composite (ENRAM) Using Electron Microscopy

M.M. Rosli[1], N. Mohamad[2], & C.H. Azhari[3]

[1,3]The Faculty of Engineering and Built Environment, Universiti Kebangsaan Malaysia, Selangor

[2]Faculty of Manufacturing Engineering, Universiti Teknikal Malaysia Melaka, Melaka

Email: m.mustaqim.89@live.com

Abstract. ENRAM at 50% ENR epoxidation at 5phr microalumina was evaluated for failure morphology in the tensile mode (tensile strength of) 19.9MPa at magnifications of 25x-3300x. The sample morphology before fracture showed uA particles with jagged edges due to agglomeration of finer particles. The mean particle size measured was 45μm. After fracture, the features observed on the fracture surfaces were i. ENR matrix ii. microalumina dispersed on the surfaces after being dislodged and iii. voids in the matrix. There was no matrix breakage observed, but predominantly pull-out of the microalumina agglomerated flakes leaving voids of 30μm-100μm. These values commiserated with the particle size range of the starting alumina particles. The voids showed an unusual surface with villi-like protrusions. The ENRAM particle pull-out failure can be explained by the failure model theory put forward by [1], where particle pull-out happened when the villi-microalumina agglomerated flakes adhesion was overcome. We would expect this mechanism to predominate as the particle sizes are greater than 10 μm, the critical size for pull-out to happen.

1. Introduction

Natural rubber is derived mainly from *Hevea brasiliensis*, a type of plant originated from the Amazonian Basin. On its own, rubber has high tensile strength. However, natural rubber tends to degrade when exposed to elevated temperature, ozone, and oxygen, due to its unsaturated nature as a polymer [2]. To improve natural rubber, epoxidation was conducted on NR to make it more resistant to ageing. As a result of the presence of epoxy ring in the NR molecular structure, epoxidised natural rubber (ENR) has excellent damping characteristic in increasing order of epoxidation [3].

Particulate alumina used as filler in rubber is nothing new. Addition of particle ranging from micro to nano size have been proven to improve mechanical property of polymer [4,5]. [1,6–10] used ENR together with alumina nanoparticle. Meanwhile in [11,12] NR was used with micro sized alumina particle. Another combination is nanoalumina with NR as exampled in [13]. [14] used alumina with NR with ENR as the compatibiliser. Various studies [15,16] have shown that smaller filler particle size resulted in composite with better mechanical properties. However, since nano sized alumina particle is significantly more expensive than its micro sized counterpart, the use of alumina microparticle as the filler will result in a material that will cost less to produce. Consequently, the cost of the total system will be driven down. Therefore, the effect of using a micro-sized alumina filler with ENR as its matrix on the characteristic of the tensile fracture surface of said composite (ENRAM) is the focus of this study as there is not enough body of work that can explain the fracture behaviour of this particular composite.

2. Experimental study

2.1 Materials

Table 1 shows the ingredients of the material used in this study. The epoxidised natural rubber (ENR) was supplied by Felda Rubber Industries Sdn Bhd. The ENR was bought under the trade name EKOPRENA 50 with mean epoxidation level of 49.5%. The Mooney viscosity measured at ML (1+4 @100°C) was 90. Alumina microparticle was obtained from Finetech Chemical Sdn. Bhd.

Table 1. Formulation of samples

Ingredients	LOADINGS (phr)[a]
ENR 50	100
Sulphur	1.6
Zinc oxide	2.0
Stearic acid	1.5
CBS[b]	1.9
TMTD[c]	0.9
6PPD[d]	2.0
Alumina	5

Note:
[a]Parts per hundred rubber, [b]N-cyclohexylbenthiazyl sulphenamide [c]Tetramethyldiuram disulphide, [d]N-(1,3-Dimethylbutyl)-N'-phenyl-p-phenylenediamine

2.2 ENRAM composite preparation

The ingredients were compounded together in an external mixer according to ASTM D-3192 at 90°C and a rotor speed of 60rpm for 6 min. All the ingredients were mixed for 4 minutes except for sulphur after the ENR was masticated for 1 minute. At about 1 minute before the mixture was dumped and cooled down to room temperature, sulphur was added. To cure the stock, a semi efficient vulcanization (EV) system was used. The material was pressed in a hot press at 150°C for about 3 minutes in a 300 mm x 300 mm x 1 mm mould.

2.3 Tensile testing

A sheet of the material was taken as a sample for tensile testing. 5 dumbbell shaped samples were taken along one of the sides of the sheet. Another 5 samples were taken from another side perpendicular to the first side. The tensile testing was run per ISO 37 standards at the Science & Technology Research Institute for Defence in Kajang on Hounsfield Testing Machine; model H25KS/06 with QMat 5.16 software. The nominal rate of traverse is 200 mm/min. The gauge length was determined to be 20 mm as indicated in the standard.

2.4 Scanning electron microscopy

Scanning electron microscope was used to view the characteristic of the raw material and to view the fracture surface of the composites. Two fracture surfaces from each direction of the sample which had undergone tensile testing were put under the microscope. The microscope used was JEOL JSM-7001F, a thermal Field Emission Scanning Electron Microscope (FESEM). The samples were coated with a thin layer (25 nm) of platinum-gold. The sample was scanned using EDS mapping technic to determine the location of different elements to help determine the location of the alumina microparticles and other substances.

3. Results and discussion

3.1 Tensile test

Tensile testing of the samples showed that the average value of tensile strength for the composite is 18.43 MPa in the first test direction and 21.36 MPa in the second one. The total average of the tensile strength of the composite is 19.9 MPa. This value is slightly higher than that reported in [8], which form the basis of this research. In the research, nano sized alumina was used as the filler. However, further study needs to be done before any conclusion can be made regarding effect of filler size on tensile strength of ENR – alumina composites.

3.2 Al_2O_3 microparticle FESEM microscopy

On 100x magnification, the alumina microparticle demonstrated wide polydispersity with diameters ranging from around 50 to 150 μm of nearly spherical individuals (Figure 1a). Upon closer inspection at 650x magnification, each particle was found to be an agglomerate of much smaller alumina particle (Figure 1b). The shape of the alumina particle was thin pieces instead of spherical. Two main shapes of particle were found out, rectangular (Figure 1c) and circular (Figure 1d) flake where diameter ranges from 4 to 6 μm with thickness around 0.3 μm.

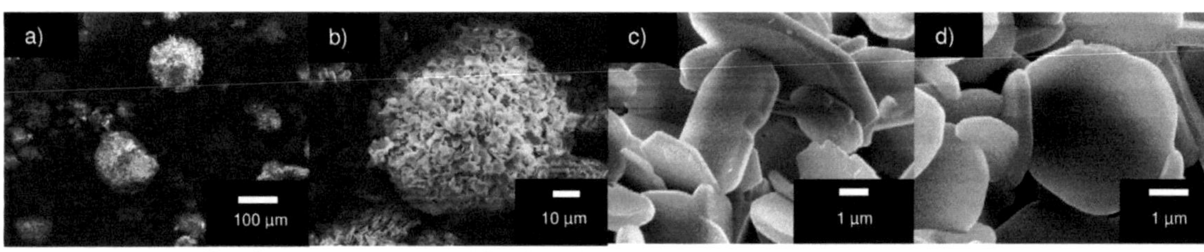

Figure 1. a) Magnification: 100x. b) Magnification: 650x. c) Magnification: 7500x. d) Magnification: 15000x.

3.3 ENRAM FESEM microscopy

For the study of fracture surface of the composites after tensile testing, random selection of fracture surface was then prepared as explained in the experimental method. Upon inspection at 200x magnification, the fracture surface exhibit areas of rough surface of the ENR-50 matrix (Figure 2a). Alumina filler can also be seen dispersed on the fracture surface. Based on the previous study in [8] it is found out that the rough surfaces were indicator of shear yielding.

As mentioned before, apart from rough surface, holes were another feature of the fracture surface similar to phenomenon observed in [6]. The micrograph in Figure 2a shows the matrix, ENR-50, with holes with varied sizes from around 30 to 100 μm. Upon closer inspection at 900x magnification, villus-like features can be seen on the surface inside the hole (Figure 2b). These villi are made of the ENR-50 matrix. A further inspection on the previous site at 2300x magnification reveals that some alumina particle was removed from the main agglomerate and sticks to the matrix (Figure 2c).

From the micrograph, it can be concluded that these villi were mere remnants of the interface between the matrix and alumina. The interface failed during tensile loading that resulted in particle pull out. This means that the ENR-50 matrix fully enveloped and wetted the alumina particle upon curing. These are strengthened by the fact that these features exhibit large surface area instead of a simple smooth surface.

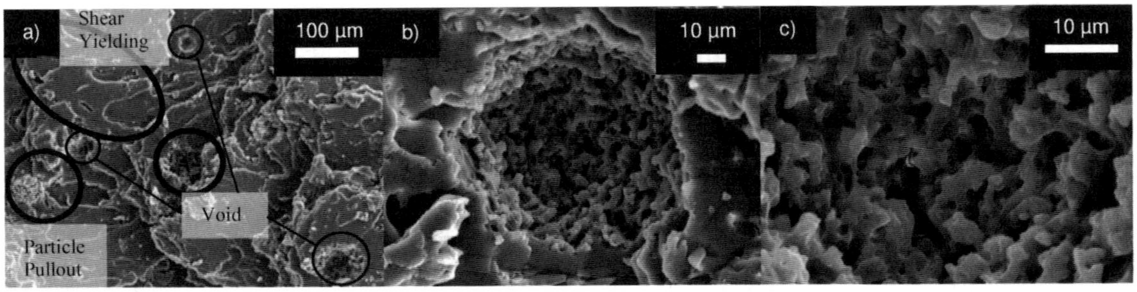

Figure 2: a) Magnification: 200x. b) Magnification 900x. c) Magnification 2300x.

Another interesting feature on the fracture surface can be seen in Figure 3a and 3b at 1200x and 3300x magnification respectively. The ENR-50 matrix can be seen in the background with alumina agglomerate sticking out of the matrix. The alumina agglomerate appears to be inside a void in the matrix. The size of the agglomerate is measured at about 30µm. It can be seen clearly that some matrix still bonds to the particle agglomerate. As such, the conclusion is that the agglomerate was pulled out from the other side of the ENR-50 matrix during tensile loading. The presence of ENR-50 residue further confirms that the alumina agglomerates were properly wetted during mixing and curing process. Apart from that we can see clearly from the formed void around the agglomerate that it is almost dislodged from the matrix because of debonding during the application of uniaxial force. This phenomenon of debonding and crack progression was studied by [17] by observing ongoing tensile loading of polymer composite under microscope. The model of debonding and crack progression for this study will be presented in the next part of the paper based on another study by [18] on failure of particulate polymer.

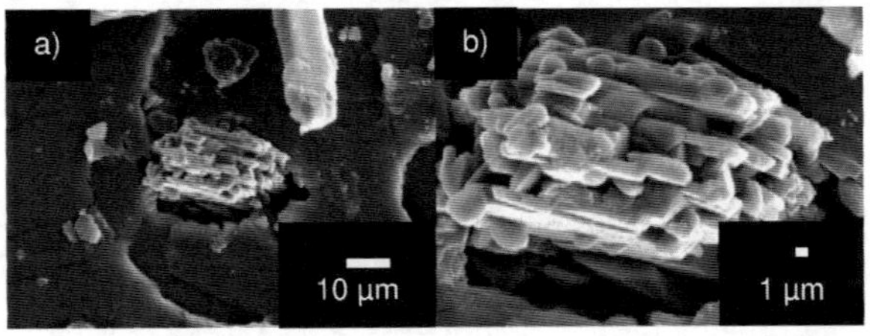

Figure 3. a) Magnification 1200x. b) Magnification 3300x.

Figure 4a shows an alumina agglomerate protruding out of the ENR matrix at 750x magnification. However, unlike Figure 3, the size of the particle is much bigger (70 µm), so there is little matrix residue seen left on the outermost surface of the agglomerate. Upon closer inspection, a crack line can be seen across the agglomerate vertically. This is also another example of particle pullout but at a larger scale than observed on Figure 3.

In Figure 4b, a larger agglomerate of filler around 100 µm can be seen at 500x magnification embedded in the ENR matrix. The agglomerate appears to be flat and does not protrude out from the matrix. The agglomerate surface also appears to be on the same plane as the fracture surface of the matrix. However, there is still ENR-50 matrix residue left on the surface of the agglomerate. It can be proposed that this phenomenon happens because the agglomerate have a flat surface and the debonding happened at said surface.

From observed fractographs, particle pullout and void were the dominant mechanism of failure found in this study. This may be explained by the size of the particle agglomerate (30 µm – 100 µm) which is greater than the 10 µm critical limit for particle pullout to happen in this type of composites as discovered in [1].

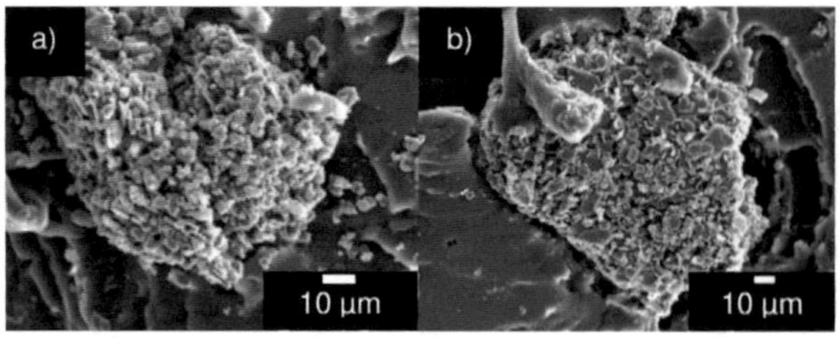

Figure 4. a) Magnification 750x b) Magnification 500x

3.4 Failure modes of particulate composite

From the SEM micrograph of the fracture surface of the material, two main modes of failure can be found out, which are the yielding of the ENR matrix and the pulling out of the alumina microparticle agglomerate, All the mechanism that leads to failure can be visualised in Figure 5. When no force is acting on the composites, the particle agglomerate is embedded within the matrix as shown in Figure 5a. Once uniaxial tension is applied, when the composites is approaching failure, voids begin to develop. If a void happens to occur on a site where particle agglomerate is present, the whole particle agglomerate may still stay whole and attached to only one side of the void as shown in figure 5b. Upon failure, the situation will develop into complete pull-out of the particle agglomerate from the matrix. Figure 5c shows the example from the specimen that corresponds to a specific type of failure modes. The propagation of crack seems to agree with the visual representation of void formation in [8,18] and similar to how microdamage evolve in [19], with the different being that the model in that study was intended for a solid particle instead of agglomerates in this study and the previous two.

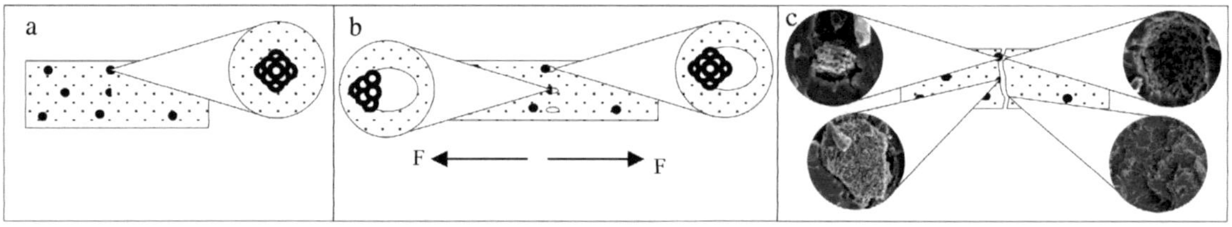

Figure 5. Fracture surface model of failure during the progression of crack. a) Alumina agglomerate as filler in ENR-50 matrix. b) Tensile loading along the horizontal axis. c) Features of the fracture surface. Adapted from [18]

4. Conclusion

Failure mechanism of ENR – alumina micro particle was studied by subjecting it to tensile failure. Then, the fracture surface was studied using FESEM. Among the features observed on the fracture surface were the ENR matrix, voids and pulled out particles. Crack propagating on the interfacial boundary of matrix and filler will result in particle pullout and voids due to adhesion force between the agglomerate and the matrix being overcame by the tensile force. The size of the void ranges from 30 to 100 µm, which match the observed size of the raw and pulled out microalumina particle agglomerates. It was also observed that the particle agglomerates were properly wetted by the matrix, suggested by matrix residue on pulled out particle agglomerate and the villi inside the void left by it. This type of failure mechanism predominates due to the critical diameter for the agglomerate (10µm) have been surpassed for agglomerate pull out to happen.

References

[1] Mohamad N, Muchtar A, Ghazali M J, Mohd D and Azhari C H 2011 *Glob. Eng. Technol. Rev.* 1 26
[2] Guimaraes Furtado C R and Amorim Moreira Leite M C 2014 *Natural Rubber Materials: Volume 1: Blends and IPNs* 1 550
[3] I.R.Gelling 1991 *J. Nat. Rubber Res.* 6 184
[4] Quaresimin M, Ricotta M, Martello L and Mian S 2013 *Compos. Part B Eng.* 44 133
[5] Sun L, Gibson R F, Gordaninejad F and Suhr J 2009 *Compos. Sci. Technol.* 69 2392
[6] Mohamad N, Muchtar A, Ghazali M J, Dahlan H M and Azhari C H 2009 *Solid State Sci. Technol.* 17 133
[7] Almaslow A, Ghazali M J, Talib R J, Ratnam C T and Azhari C H 2013 *Wear* 302 1392
[8] Mohamad N 2011 *Pembangunan komposit getah asli terepoksida-alumina (ENRAN) untuk penyerapan tenaga hentaman balistik dalam perisai jasad* (The National University of Malaysia)
[9] Mohamad N, Muchtar A, Ghazali M J, Mohd D H and Azhari C H 2010 *J. Appl. Polym. Sci.* 115 183
[10] Mohamad N, Muchtar A, Ghazali M J, Mohd D H and Azhari C H 2008 *Eur. J. Sci. Res.* 24 538
[11] Tangboriboon N, Chaisakrenon S, Banchong A, Kunanuruksapong R and Sirivat A 2013 *Plast. Rubber Compos.* 42 26
[12] Tangboriboon N, Chaisakrenon S, Banchong A, Kunanuruksapong R and Sirivat A 2012 *J. Elastomers Plast.* 44 21
[13] Thomas T, Ayswarya E P and Thachil E T 2013 *Int. J. Innov. Res. Sci. Eng. Technol.* 21 2365
[14] Konar B B, Roy S K and Pariya T K 2010 *J. Macromol. Sci. Part A* 47 416
[15] Fu S Y, Feng X Q, Lauke B and Mai Y W 2008 *Compos. Part B Eng.* 39 933
[16] Lauke B 2008 *Compos. Sci. Technol.* 68 3365
[17] Cho J, Joshi M S and Sun C T 2006 *Compos. Sci. Technol.* 66 1941
[18] Kausch H H, Beguelin P and Fischer M 2000 *Mech. Compos. Mater.* 36 177
[19] Moshev V V and Evlampieva S E 1997 *Polym. Eng. Sci.* 37 1348

The Effect of Sago Starch Filler Dispersion preparation on the Biodegradation Properties of Natural Rubber Latex Films

M.A. Misman[1,*], A.R. Azura[1], Y.S. Rohana[1]

[1]School of Materials and Mineral Resources Engineering, Universiti Sains Malaysia, Engineering Campus, 14300,Nibong Tebal, Penang, Malaysia.

[*]E-mail: afiqmisman@usm.my

Abstract. Sago starch dispersion was prepared by mechanical (ball milling) and chemical (reaction with alkali) processes in order to reduce the size of dispersed sago starch particles. The sago starch dispersion was compounded with natural rubber latex (NRL) as a filler and a NRL thin film was prepared by coagulant dipping technique. The ball milled sago starch (MSS) and alkali dispersed sago starch (CSS) filled NRL films were then subjected to soil burial test. The effect of the degradation process on weight loss percentages, surface morphologies, tensile strength behavior and Fourier Transform Infrared (FTIR) analysis of the starch filled NRL films were investigated. The utilization of sago starch dispersion prepared by alkali treatment showed the improvement in mechanical properties of NRL films compared to sago starch dispersion prepared by the ball milling method. From the results obtained, CSS/NRL films are more feasible to degrade compared to MSS/NRL films.

1. Introduction

The use of starch as bio-fillers in natural rubber latex compounding is a promising technique in producing the biodegradable NRL films. However, using starch in powdery form might result in latex destabilization. Thus, prior compounding, starch need to be dispersed in the form of the dispersion. Starch dispersion can be produced via chemical, mechanical, or temperature induced processes. During the process, starch gelatinization is expected to occur. The most common method to prepare filler dispersion before adding into NRL compound is a ball milling process. The method involves physical action without further alteration in chemical characteristic of starch [1]. The large particle size and hydrophilic in nature of starch causes the deterioration of final properties of NRL films. Therefore, starch treatment should be carried out to improve the compatibility between rubber and starch. Alkali treatment is a promising method to improve the starch dispersion. Alkali is generally used as a stabilizer in latex compounds as well as in preparing filler dispersion and it is able to gelatinize the native starch under room temperature. Alkali treatment for starch gelatinization also could prevent the retrogradation to occur in quite a long time [2]. A majority of hydroxyl groups in the anhydrous glucose units is ionised at high pH and therefore, the agglomeration of starch chains and paste retrogradation can be retarded [3]. Hence, it will improve the starch distribution in NRL matrix and enhance the final properties of NRL films. In this studies, different technique of preparing starch dispersion was conducted to observe the effect on NRL composite films.

2. Methodology

2.1 Preparation of ball milled sago starch (MSS) and Alkali dispersed sago starch (CSS)

Sago starch was mixed with anchoid (20% of filler weight), 10% potassium hydroxide (12% of filler weight) and distilled water before being stirred at 270 rpm of stirring speed for 15 minutes under room temperature. The mixture (pH 10) was then transferred into the milling jar with ceramic balls and subjected to rotary milled at 20 Hz for 24 hours under room temperature. The alkali treatment was carried out by mixing sago starch powder (NSS) with 100g of 5.35% KOH. The mixture was stirred at 200 rpm for 30 minutes before leaving it for two days under room temperature to gelatinize completely. Meanwhile, the CSS dispersion was stirred for 15 minutes at 200 rpm of stirring speed under room temperature. A faster preparation of alkali dispersed CSS mixture was due to the ability of the filler to dissolve in KOH instead of gelatinization. All pH of filler dispersion were maintained in the range of 12.60-12.80.

2.2 Preparation of MSS and CSS filled NRL composite films

The NRL mixture was prepared according to formulation as given in Table 1. The latex compound was diluted to 50% of total solid content (TSC) with distilled water.

Table 1: Compounding ingredients for MSS- and CSS/NRL

Ingredients	Control NRL, (phr)	NRL with filler, (phr)
NRL	100	100
KOH	0.3	0.3
ZnO	0.25	0.25
AnO	0.5	0.5
ZDEC	0.75	0.75
Sulfur	0.5	0.5
MSS/CSS	-	10

All the ingredients were mixed and stirred for two hours using Wise Stir HS-300 mechanical stirrer at stirring speed of 270 rpm and pre-vulcanized at temperature of 70°C. The chloroform number test was carried out until chloroform number 2 was obtained. The NRL compound was maturated overnight under room temperature before films preparation. The clean aluminum plates were dipped in coagulant (10% calcium nitrate) for 5 seconds and dried in the air oven for 10 minutes at 100°C before being cooled under room temperature for 5 minutes. The aluminum plates were then dipped into the NRL compound for 5 seconds and cured in the air oven for 20 minutes at 100°C before being cooled under room temperature for 24 hours. The films were stripped by the aids of calcium carbonate powder to prevent the films from sticking together and damaging the films surfaces.

2.3 Scanning Electron Microscopy

The morphologies of fillers and NRL films were observed using SEM model ZEISS Supra-35VP. The latex films test pieces were frosted with liquid nitrogen for 30 seconds before being fractured immediately. The freeze fractured surface of the film was attached on 45° holder. The samples were sputter coated with a thin layer of gold before analysis to avoid electrostatic charging during analysis. These coated samples were inserted into the microscope and the images were captured at 300x magnifications.

2.4 Fourier Transform Infrared Analysis of Fillers and NRL Films

The FTIR spectrometer used in this study was Perkin Elmer, Model Spectrum One. Analysis was carried out under room temperature within a typical wave number range of 4000 cm-1 to 550 cm-1 under transmission mode.

2.5 Soil burial test and determination of changes in the NRL properties

Soil burial test was carried out in accordance to Chuayjuljit [4]. The films were buried in comtrolled soil for 1 to 10 weeks at room temperature. The weekly withdrawn films were washed off and dried at room temperature overnight before the weight loss percentages calculation were done according to Equation 1.

$$Weight\ loss\ (\%) = \frac{W_i - W_b}{W_i} \times 100 \qquad \text{(Equation 1)}$$

where, *Wi* is an initial weight and *Wb* is the weight after soil burial of NRL films in gram. The appearance of biodegradation signs on the surface of films was observed using Dino Lite Optical Microscope at 100x magnification. The morphologies of film surface after soil burial test also was observed under SEM with magnification of 300x. Tensile strength retention and FTIR analysis were carried out to observe the changes in properties of NRL films. Tensile strength retention was conducted according to ASTM D 412-92 using tensometer Universal Testing Machine (Model Instron 3366) with a crosshead speed of 500 mm/min and for biodegraded specimens which is determined using Equation 2:

$$Tensile\ strength\ retention\ (\%) = \frac{\tau_t}{\tau_0} \times 100 \qquad \text{(Equation 2)}$$

where τ_t and τ_o are the tensile strength after biodegradation and the initial tensile strength, respectively.

3. Results and discussion

3.1 Weight Loss Measurement

Table 2 shows the comparison of weight loss percentages of NRL films with the addition of sago starch dispersions prepared through the ball milling method and alkali treatment, respectively. It was observed that the control NRL films showed the slow biodegradation process where only 20.9% of weight loss obtained after week 10 of the soil burial test.

Table 2: The percentages of tensile strength retention and weight loss with respect to weeks of biodegradation

NR Latex films	Properties / weeks	1	2	3	4	6	8	10
MSS	Tensile Retention, %	>80	>80	>80	>80	35	20	10
	Weight loss, %	< 5	< 5	< 5	< 5	7	18	25.1
CSS	Tensile Retention, %	>80	>80	>80	>80	7	2	0
	Weight loss, %	< 5	< 5	< 5	< 5	20	23	38.6

The MSS/NRL films showed comparable weight loss percentages to the unfilled NRL films up to week 4 of the soil burial test. After week 6 to week 10 exposed to soil environment, the MSS/NRL films showed the increment in weight loss percentages up to 25.1%. On the other hand, the CSS/NRL films showed a slight increase in weight loss percentages from week 1 to week 4 of soil burial compared to the control

IOP Conf. Series: Journal of Physics: Conf. Series **1082** (2018) 012100 doi:10.1088/1742-6596/1082/1/012100

NRL films and MSS/ NRL films. The increased in weight loss percentages of CSS/ NRL films were observed after week 6 and became more significant after week 8 and week 10 (38.6%) of soil burial. The present of KOH could change the large particle size of starch granules into the gel network formation. More amorphous regions in the gel network exist compared to starch granules could assist the biodegradation process of CSS/NRL films. The polysaccharide structures in starch generally consist of hydrolysable linkages along the polymer chain which are susceptible to biodegradation by microorganisms and hydrolytic enzymes [5-7].

3.2 Morphological Analysis

The morphological analysis of MSS and CSS/NRL films upon soil burial test was given in Figure 1, respectively. There were insignificant effect on the surface of MSS/NRL films after exposed to soil environment from week 1 to week 3 as observed in Figure 1 (a.ii- iv) due to the low filler loading and the ability of high molecular weight polyisoprene chains to withstand the biodegradation. For MSS/NRL films, it was observed that small blisters were detected after week 4 of the soil burial test (Figure Figure 1 a.v). The NRL film surface became brownish after week 6 (Figure 1 (a.vi)) before losing their structural integrity at week 8 (Figure 1 (a.vii)) of soil burial.

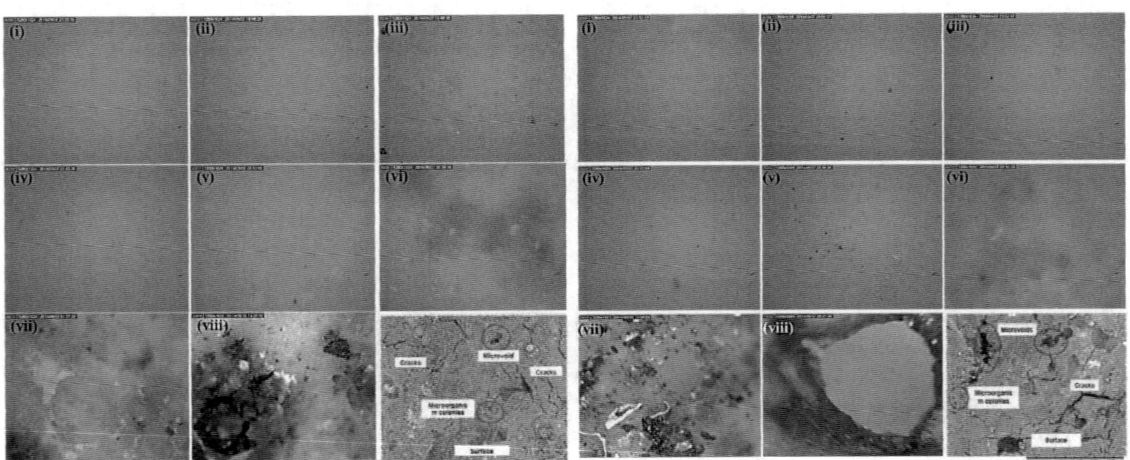

(a) (b)

Figure 1: The morphologies of (a) MSS/NRL films at (i) control week, (ii) week 1, (iii) week 2, (iv) week 3, (v) week 4, (vi) week 6, (vii) week 8, (viii) week 10 and (ix) SEM images, and (b) The morphologies of CSS/NRL films at (i) control week, (ii) week 1, (iii) week 2, (iv) week 3, (v) week 4, (vi) week 6, (vii) week 8, (viii) week 10 and (ix) SEM images.

At week 10 of the soil burial test (Figure 1 (a.viii)), the dark patch with notable cracks were observed on the MSS/ NRL films. The employment of alkali dispersed NSS showed insignificant changes on the surface of NRL films after 2 weeks of soil burial test (Figure 1 (b.i-iii)). The appearance of yellow patches was observed on the CSS/NRL films at week 3 (Figure 1 (b.iv)) before becoming more significant after week 4 and week 6 of soil burial (Figure 1 (b.v-vi)). The development of yellow colour on the films indicating that the microorganism was able to initiate the rubber degradation and decrease the average molecular weight of polymer [8]. Small void was observed on the film surface at week 8 (Figure 1 (b.vii)) and developed into a larger void at week 10 (Figure 1 (b.viii)) of CSS/NRL films. The incorporation of polysaccharides in the films could influence the ability of amylolytic microorganisms to attach to the

IOP Conf. Series: Journal of Physics: Conf. Series **1082** (2018) 012100 doi:10.1088/1742-6596/1082/1/012100

surface of the film for the enzymatic hydrolysis process [9]. Figure 1 (a.ix)-(b.ix) show the SEM observations on the biodegraded surface of MSS and CSS/NRL films after 8 weeks of soil burial test, respectively. As observed in Figure 1 (a.ix), MSS/NRL films showed the appearance of a number of cracks and microvoids along with surface erosion of samples owing to biodegradation activities during the soil burial test. The addition of alkali dispersed NSS could promote major surface erosion and microvoids formation on the NRL films upon disposal in soil as shown in Figure 1 (b.ix). The presence of microorganism colonies also were detected on the surface of NRL films. The well dispersed of alkali dispersed NSS, including with the gel network formation which able to consume by microorganisms could assist further biodegradation of NRL films in soil.

3.3 Tensile Strength Retention

The effect of the soil burial test as a function of time on tensile strength retention of MSS and CSS/NRL is shown in Table 2. Both the control and MSS/NRL films exhibited deliberate reduction up to week 4 before it showed drastically decreased in tensile strength retention after week 6 of the soil burial test. Further exposure to soil environment caused the decreased in tensile strength retention of NRL films up to week 10. Overall, the MSS/NRL films showed more significant decreased in tensile strength retention compared to the control NRL films. The starch is metabolized first and acts as growth substrate for the microorganism so that the microorganism population rises, thus increase the rate of degradation of NR [8]. For CSS/NRL films, the tensile strength retained up to week 2 before decreasing after week 3 and week 4 of the soil burial test. At week 6 of soil burial, the tensile strength retention of CSS/NRL films was reduced remarkably, and became very weak after buried for week 8 and week 10. These results can be correlated with weight loss (Table 2) and morphological studies (Figure 1). As the microorganisms consumed the starch, the NRL lose their structural integrity which led to the deterioration of the strength of the polymer [10]. The voids formation also was expected to decline the tensile strength of the films upon soil burial.

3.4 Fourier Transform Analysis

The FTIR analysis before and after week 4 and week 8 of the soil burial test of sago starch filled NRL films are shown in Figure 2. The addition of MSS and CSS in NRL films exhibited the decreased in bands after week 4 as observed in the region of 3420-3250 cm^{-1} which indicated the dehydration and depolymerisation of starch occurred upon composting. The MSS/NRL films showed an increase of the band in this region after week 8 of the soil burial test which attributed to the increased in biodegradation by-products. The significant increment of the band at 1641 cm^{-1} for both MSS and CSS/NRL films were observed in the spectra, especially after exposed to soil environment for week 8 which corresponded to tightly bound water present in amorphous region of fillers. At longer period of soil burial, the decrement of transmittance peaks at 1442 cm^{-1}, 1375 cm^{-1} and 875 cm^{-1} were detected in the FTIR spectra of MSS and CSS/NRL films. These results were signified the reduction of C-H deformation as well as decreased in C-O-C stretch in glucosidic linkages of starch. Both types of NRL films also exhibited the appearance of a new peak at 1544 cm^{-1} which was attributed to the present of amide groups after soil burial.

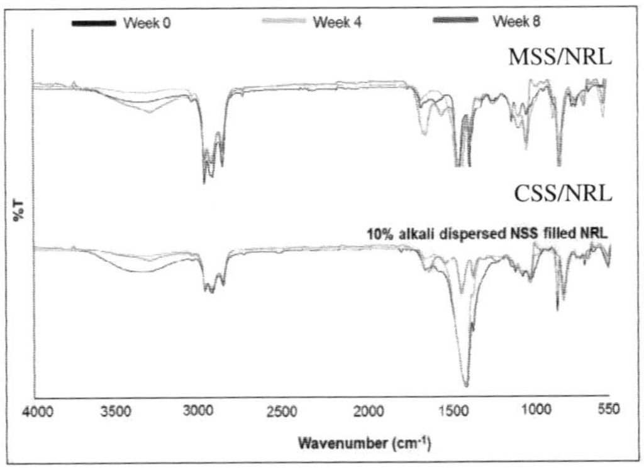

Figure 2: FTIR spectra of the control, MSS/NRL, and CSS/NRL films with respect to the week of biodegradation

4. Conclusion

Results indicate that CSS filled NRL (CSS/NRL) films shows the highest weight loss percentage at 10 week of soil burial. Optical analysis shows that CSS/NRL film undergo rapid development of void at week 10 compared to ball milled sago starch filled NRL (MSS/NRL) films. Further observation on the morphology of both films by Scanning Electron Microscopy (SEM) shows that, the CSS/NRL film undergoes a severe degradation process at week 8. Tensile retention obtained indicate the rapid deterioration of CSS/NRL films compared to MSS/NRL at week 6. The reduction of C-H (wavenumber 1442cm^{-1}) and C-O-C (wavenumber 1375cm^{-1}) bond in both starch filled films were recorded via FTIR analysis. From the results obtained, CSS/NRL films are more feasible to degrade compared to MSS/NRL films.

Acknowledgement

The authors would like to thank the Universiti Sains Malaysia for USM post-doctoral fellow scheme and RU grant (1001/ PBAHAN/814279).

References

[1] Huang Z Q, Xie X L, Chen Y, Lu J P, and Tong Z F 2008 *C R Chim* 11 73
[2] Karim A A, Nadiha M Z, Chen F K, Phuah Y P, Chui Y M, and Fazilah A 2008 *Food Hydrocolloid* 22 1044
[3] Han J A, and Lim S T, 2004 *Carbohyd Polym* 55 (2) 193
[4] Chuayjuljit S, Su-Uthai S, Tunwattanaseree C, Charuchinda S, 2009 *J Reinf Plast Compos* 28 1245
[5] Afiq M M, and Azura A R 2013 *J Chem Chem Eng* 7 137
[6] Abraham E, Pothen L A, and Thomas S 2012 *European Conference on Composite Materials,* Venice, Italy.
[7] Chandra R, Rustgi R, 1998 *Prog Polym Sci* 23 1273
[8] Watcharakul S, Umsakul K, Hodgson B, Chumeka W, and Tanrattanakul V, 2012 *Electron J Biotechnol* 15 (1) 1
[9] Imam S H, and Gould J M, 1990 *Appl Environ Microb* 56 872
[10] Shamala R, Hanafi I, and Yamuna M, 2014 *J Vinyl Addit Technol* 21(2) 128–133

IOP Conf. Series: Journal of Physics: Conf. Series **1082** (2018) 012101 doi:10.1088/1742-6596/1082/1/012101

CCTO thin film synthesis by radio frequency magnetron sputtering

M. Ahmadipour[1], N.A. Rejab[1], M.F. Ab Rahman[1], M.F. Ain[2], and Z.A. Ahmad[1*]

[1]Structural Materials Niche Area, School of Materials and Mineral Resources Engineering, Universiti Sains Malaysia, Engineering Campus, 14300 Nibong Tebal, Penang, Malaysia

[2]School of Electrical and Electronic Engineering, Universiti Sains Malaysia, Engineering Campus, 14300 Nibong Tebal, Penang, Malaysia

*E-mail: srzainal@usm.my

Abstract. In the present work CCTO thin film has successfully deposited on FTO substrate in a non-reactive (argon) atmosphere with frequency (RF) magnetron sputtering at 300 W. This work aims to determine the structural, morphological and optical properties of thin film using X-ray diffraction (XRD), atomic force microscopy (AFM) and Ultraviolet–visible spectroscopy (UV-Vis), respectively. The result shows that the nature of CCTO thin film is polycrystalline with cubic structure. XRD peaks confirmed the presence of CCTO phase with the crystal planes of (022), (013), (033), (024), and (224). The average crystallite size was 32 nm as measured by Willamson-Hall method. The surface roughness (Ra) and root mean square (RMS) of CCTO thin film were 26 nm and 33 nm, respectively. The maximum optical transmittance of CCTO thin film was more than 80% in the visible region (550-1000 nm) while calculated optical energy bandgaps was 3.2 eV.

1. Introduction

Recently, $CaCu_3Ti_4O_{12}$ (CCTO) perovskite-like compound has received increasing interest because of the attractive attribute like high dielectric constant (ε_r) at wide range of temperature (100 - 600 K). This special feature has increased the potential of CCTO as a useful material in the microelectronics field. In various CCTO studies, the chemical methods have applied to prepare bulk materials and the main aim its focusing on possible mechanism that account for the unusual CCTO properties [1]. The remarkable properties such as high ε_r, low dielectric loss (tan δ) [2,3], high electrical and thermal stability [4] can be achieved when CCTO is in the form of thin film owing to superior uniformity and porosity in contrast to bulk materials. With these properties, the CCTO thin film becomes very attractive to utilize in several applications such as capacitor [5], antennas [6], sensors [7], and solar cell [8]. Nevertheless, further comprehension of the structural characteristics of the CCTO thin films that is correlated to growth parameters, are essential prior to practical application as electronic devices.

There are various chemical and physical methods to deposit CCTO thin film such as, sol–gel process [9], pulsed laser deposition (PLD) [10], sputtering [11], and metal organic chemical vapor deposition (MOCVD) [12]. RF magnetron sputtering possesses huge beneficial as it's capable to control a deposition parameter, good adhesion and deposit uniformly in large surface area, among these methods. It is reported previously [13-15] the effect of thickness, annealing temperature, and RF power on the properties of CCTO thin films deposited on ITO substrates. All works show that the properties of CCTO thin film changes with changing the parameters. Thus far there is no scientific report to investigate the properties of CCTO thin film deposited on fluorine tin oxide (FTO) substrate by RF sputtering.

Therefore, in this study, 250 nm thickness of CCTO thin film has been deposited on FTO substrate by using RF magnetron sputtering and the crystallite structural, surface morphological, and optical properties of CCTO thin films were investigated for the first time.

2. Experimental Method

2.1 Preparation CCTO thin film

CCTO thin film with a 250 nm thickness has been deposited on FTO substrate by using RF magnetron sputtering (HHV Auto 500) technique. The FTO demonstrates to be a favorable electrode substrate in sensors and solar cell applications. CCTO target was used with in this experiment with 76 mm diameter and 5 mm thickness. In order to eliminate the surface impurities, the substrate (FTO) was cleaned with acetone, ethanol and deionized water before deposition process. Furthermore, the pressure in the deposition chamber was maintained at 30×10^{-5} mbar and the CCTO target was pre-sputtered for about 300 s at 30 W to eliminate surface contamination, prior to deposition process. The operating pressure during film deposition was 1.23×10^{-2} mbar, while an argon gas flow rate and RF power were10 sccm and 300 W, respectively at room temperature. The film thickness was measured by thin film controller of RF magnetron sputtering machine and was also confirmed by thin film analyzer (Film metric, F20). X-ray diffraction (XRD Bruker, D8 Advance system) and atomic force microscope (AFM, Nano Navi, SPA 400) were used to examine the crystalline phases and surface topography of the sputtered CCTO films. UV-Vis spectrophotometer (Varian, cary 50 conc) with the wavelengths range of 280 nm to 880 nm was used to determine the transmittance of CCTO thin film while energy bandgap had calculated from Tauc plot method.

3. Results and Discussion

3.1 X-ray diffraction analysis

Figure 1 depicts the X-ray diffraction (XRD) patterns of the CCTO thin film on a FTO substrate with 250 nm thickness. The diffraction peaks of the film are corresponds to cubic perovskite related structures (ICDD data card no. 095-005-8088) with space group of Im-3 [16]. It is visible that five peaks centered at 34.20°, 38.55°, 52.51°, 55.65°, and 61.42° which match to crystal planes (022), (013), (033), (024), and (224), respectively. The characteristic peak of FTO substrate can be seen in the pattern at $2\theta = 27.20$ ° which is related to the crystalline structure of FTO [17]. However, not all of the expected peaks for CCTO thin film are appeared in the spectrum. This phenomenon can be attributed to the small amount of CCTO being deposited on the FTO substrate. The average crystallite size was determined referring to Willamson-Hall method [18] and it was found to be 32 nm.

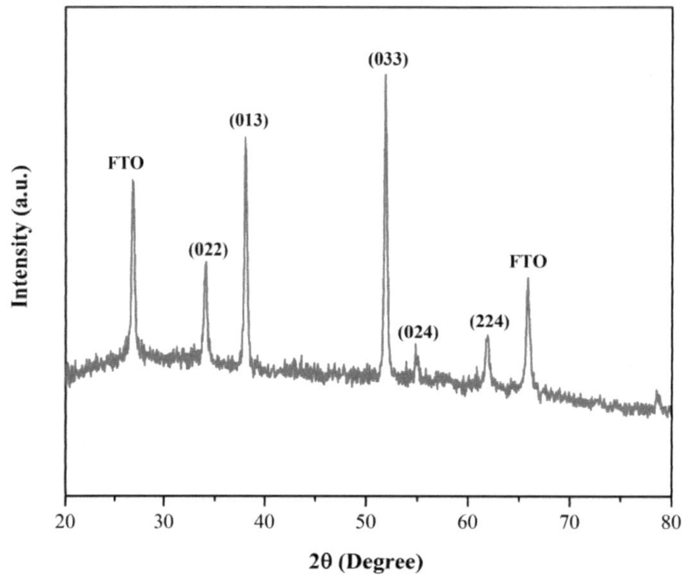

Figure 1. X-ray diffraction patterns of the sputtered CCTO thin films layer on FTO substrate

3.2 Morphology analysis (AFM)

The AFM analysis (tapping mode at ambient condition) was conducted to evaluate the surface morphology of the sputtered CCTO thin films. The 2D images of the sputtered CCTO thin film are presented in Figure 2. The roughness (Ra) and its root mean square (RMS) of a scan area (10 × 10 μm^2) were determined by the AFM. It can be seen (Figure 2 a) that the CCTO film illustrated a microstructure which contains a mixture of small and large grains. Furthermore, the recorded values of Ra and RMS are 26 nm and 33 nm, respectively. These values may have originated from the initial roughness of the substrate itself. Roughness values for FTO substrates before deposition were: RMS = 20 nm. It was also perceived from surface profile variations (corresponding average profile line) Figure 2 (b) that the pore size is about 85 nm. However, both surface roughness and pore size can be potential parameter to influence practical applications like optical and gas sensors.

IOP Conf. Series: Journal of Physics: Conf. Series **1082** (2018) 012101 doi:10.1088/1742-6596/1082/1/012101

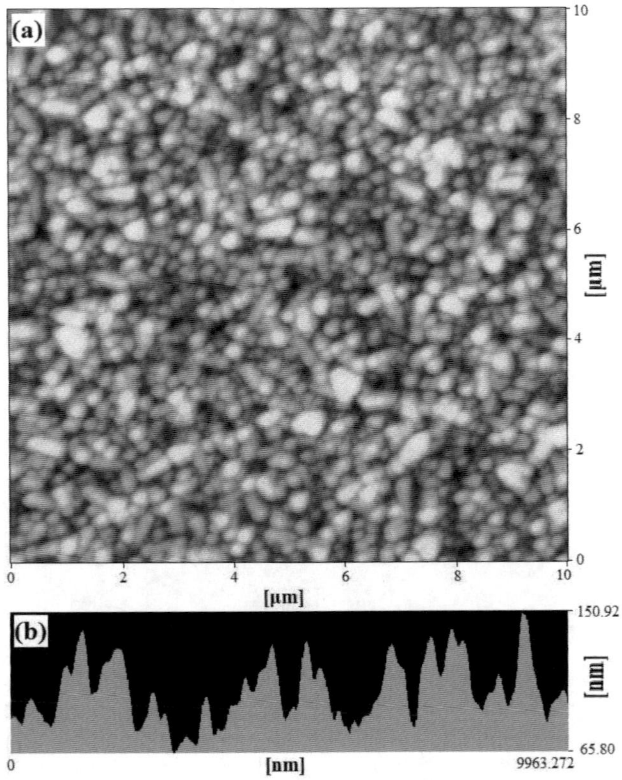

Figure 2. Atomic force micrographs of CCTO thin film (a) 2D image (b) the corresponding average profile line

3.3 Optical properties

The UV-Vis transmission spectra of CCTO thin films deposited on FTO substrates is shown in Figure. 3. The transmittance of samples was measured as a function of wavelength in the range of 300-1000 nm. The transmittance of CCTO thin films was detected in the range of 60% to 80% (Figure. 3inset) on FTO substrate in the visible region. The optical energy bandgap (Eg) was measured by using the Tauc model [19]. The Eg of CCTO thin films on FTO substrate was 3.2 eV as shown in Figure. 3. The wide bandgap is particularly important for allowing wide bandgap devices to operate at much higher temperatures. Additionally, most wide bandgap materials also have a much higher critical electrical field density. Combined, these properties allow them to operate at much higher voltages and currents, which makes them highly valuable in military, radio and energy conversion settings.

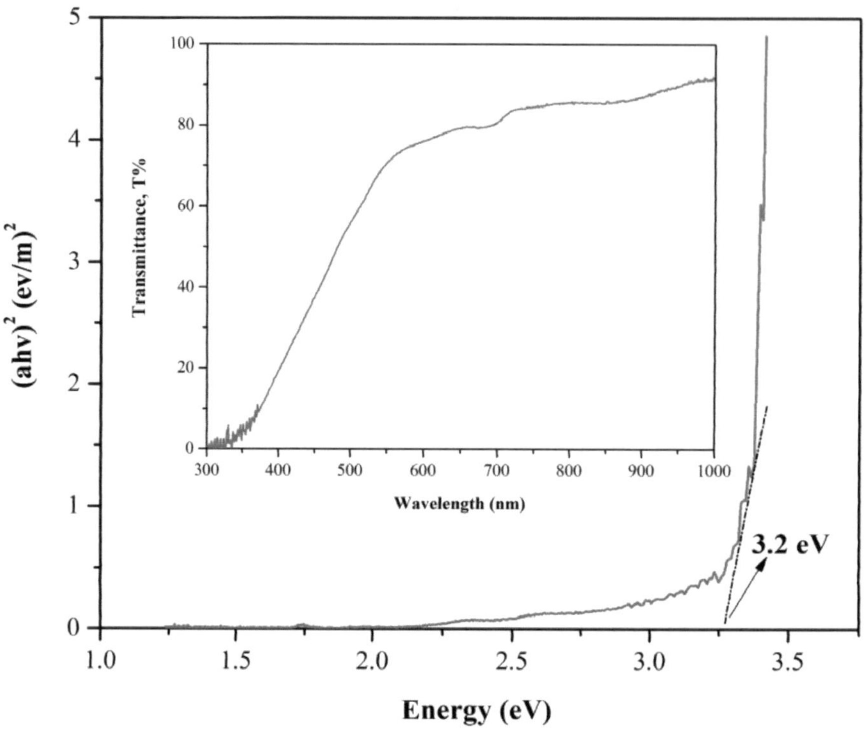

Figure 3. Optical transmittance spectra and Tauc's plot (inset) of CCTO films deposited on FTO substrate

4. Conclusion

CCTO thin film layer with 250 nm thicknesses was deposited on FTO substrate by using RF magnetron sputtering process at room temperature and power 300 W. The average crystallite size of CCTO thin films was obtained to be 32 nm while the surface roughness was 33 nm. CCTO film also showed good optical transparency (over 60%) in the visible region while the obtained energy bandgap was 3.2 eV. Therefore, having knowledge on properties of CCTO thin film on FTO substrate could be suitable for specific applications.

5. Acknowledgments

This research has been supported by the Universiti Sains Malaysia (USM) fellowship (APEX 91002/JHEA/ATSG4001) and fundamental research grant scheme (FRGS) under grant number of203/PBAHAN/6071263.

6. References

[1] Ramirez AP, Subramanian MA, Gardel M, Blumberg G, Li D, Vogt T and Shapiro SM 2000 *Solid State Commun.* 115 217
[2] Ahmadipour M, Ain MF and Ahmad ZA 2016 *Nano. Micro. Lett.* 8 291
[3] Shen YS, Chiou BS and Ho CC 2008 *Thin Solid Films*, 517 1209
[4] Singh L, Rai US, Mandal KD and Singh NB 2014 *Prog. Cryst. Growth Ch.* 60 15
[5] Homes CC, Vogt T, Shapiro SM, Wakimoto S and Ramirez AP 2001 *Science,* 293 673
[6] Kretly LC , Almeida AFL , Fechine PBA , de Oliveira RS and Sombra ASB 2004 *J. Mater. Sci-Mater. El.*, 15 657

[7] Ahmadipour M, Ain MF and Ahmad ZA, 2017 *IEEE Sensors Journal* 17 3224
[8] Kushwaha HS, Madhar NA, Ilahi B, Thomas P, Halder A and Vaish R 2016 *Scientific Reports* 6 18557
[9] Jime´nez R, Calzada ML, Bretos I, Goes JC and Sombra ASB 2007 *J. Eur. Ceram. Soc.* 27 3829
[10] Fang L and Shen MR 2003*Thin Solid Films,* 440 60
[11] Prakash BS, Varma K, Michau D and Maglione M 2008 *Thin Solid Films*, 516 2874
[12] Nigro RL, Toro RG, Malandrino G, Fragala IL, Fiorenza P and Raineri V 2007 *Surf. Coat. Technol.* 201 9243
[13] Ahmadipour M, Ain MF and Ahmad ZA 2016 *Appl. Surf. Sci.* 385 182
[14] Ahmadipour M, Ain MF and Ahmad ZA 2017 *J. Mater. Sci-Mater. El.* 28 12458
[15] Ahmadipour M, Ayub SN, Ain MF and Ahmad ZA 2017 *Mat. Sci. Semicon. Proc.* 66 157
[16] Ahmadipour M, Abu MJ, Rahman MFA, Ain MF and Ahmad ZA 2016 *Micro Nano Lett.* 11 147
[17] Bykkam S, Narsingam S, Ahmadipour M, Dayakar T, Rao KV, Chakra CS, and Kalakotla S 2015 *Superlattice Microst.* 83 776
[18] Nigro RL, Toro RG, Malandrino G, Fragala IL, Fiorenza P and Raineri V 2007 *Thin Solid Films*, 515 6470
[19] Ahmadipour M, Cheah WK, Ain MF, Rao KV and Ahmad ZA 2018 *Mater. Lett.* 210 4

IOP Conf. Series: Journal of Physics: Conf. Series **1082** (2018) 012102 doi:10.1088/1742-6596/1082/1/012102

Effects of Molar Ratios Between Surfactant and Aniline in Synthesis of Polyaniline

E. Netnapa[1], M. Mariatti[1,*], Z. Ahmad[1], N. Ohatake[2], H. Akasaka[2], L. Banhan[3]

[1]School of Materials and Mineral Resources Engineering, Universiti Sains Malaysia, Engineering Campus, Penang, Malaysia

[2]Department of Mechanical Sciences and Technology, Tokyo Institute of Technology, Tokyo, Japan

[3]Department of Industrial Engineering, Faculty of Engineering, Burapha University, Chonburi, Thailand

Email: mariatti@usm.my

Abstract. Conductive polymers are widely used in many applications such as sensors, solar cells, diodes, electrodes, and actuators. Among of conductive polymers, Polyaniline (PANI) has gained more interests due to high environmental stability, electrical conductivity, easy processability, and low cost. Synthesis of PANI with different molar ratios between Dodecylbenzenesulfonic acid (DBSA) and Aniline (An) ([DBSA]/[An]) at 0.5, 1.0, and 1.5 were investigated. The increased molar ratio of surfactant affects to improvement in higher electrical conductivity of 18.91×10^{-2} S/m. This is due to the increasing of doping in PANI-DBSA which can be obtained by Fourier Transform Infrared Spectroscopy (FTIR). At high molar ratio, thorn-liked structure and smooth surface of PANI are observed compared to irregular structure at low molar ratio. Moreover, the chemical structures of PANI-DBSA are investigated by Solid-state Nuclear magnetic resonance spectroscopy (NMR).

1. Introduction

Conductive polymer is an organic based polymer which can act as a conductor or a semiconductor due to their conjugated polymeric chains in form of polarons or bipolarons [1]. Among conductive polymers, Polyaniline (PANI) is widely studied because of ease of synthesis, low cost, good environmental stability, and good electrical conductivity [2, 3]. PANI has been used in many applications such as rechargeable batteries, shielding of electromagnetic interference, microwave and radar absorbing materials, sensors and indicators, membranes, electrochemical capacitors (supercapacitors), and solar cells [3, 4].

In general, conducting polymer is not dissolved by common organic solvents, this is because of their large intermolecular interfacial tension. Therefore, addition of functionalized protonic acids with long alkyl chain such as dedecylbenzene sulfonic acid (DBSA), sodiumdodecylsulfate (SDS), sodiumdodeclybenzenesulfonate (SDBS), and camphor sulfonic acid (CSA) could increase the solubility of PANI [1]. Commonly, emulsion polymerization is carried out with hydrophobic monomer

IOP Conf. Series: Journal of Physics: Conf. Series **1082** (2018) 012102 doi:10.1088/1742-6596/1082/1/012102

accumulated in the center of micelles with surrounding hydrophobic tails of emulsifiers, which providing suitable environment for para-directed coupling of monomer and leads to formation of conducting PANI [5, 6]. The aim of the present work is to investigate the effect of molar ratio between DBSA and aniline in PANI polymerization, effect of DBSA doped PANI at different molar ratio were carried out and characterized in this study.

2. Experimental study

2.1 Materials

Aniline ($C_6H_5NH_2$, 99% of purity) and Ammonium peroxodisulfate ($(NH_4)_2S_2O_8$, APS) were purchased from Wako, Japan. Dodecylbenzenesulfonic Acid (DBSA with >95% of purity) was purchased by Tokyo Chemical Industry Co., Ltd. APS and DBSA are used as oxidant and surfactant in the polymerization, respectively.

2.2 Synthesis of PANI

4 mmol of Aniline was mixed with DBSA and 90 ml of distilled water, solution was stirred 3 h. 5 mmol of APS was mixed with 10 ml of distilled water, solution was stirred 1 h. Next, APS solution was dropped with low speed into Aniline solution, the polymerization was carried out at 20 °C for 24 h. After polymerization is done, the dark green precipitates are obtained. 100 ml of Methanol was added into the solution to stop the polymerization, the precipitates were filtered, washed with methanol and distilled water until the pH of washing liquid reached pH 7.0. Finally, precipitates were dried in vacuum at 60 °C for 24 h, and the sample was pulverized into powder form.

2.3 Characterizations

Samples were characterized using Fourier Transform Infrared Spectroscopy (FTIR) (Perkin Elmer FT-IR Spectroscopy Frontier), 20 scans were used with scan range from 4000 cm^{-1} to 400 cm^{-1}. Nuclear Magnetic Resonance Spectroscopy (NMR) by JEOL (ECA400) was used to identify the chemical structure of the samples. PANI power was formed into pellet form with diameter of 13 mm at 700 MPa of pressure. The resistance of PANI pellets was tested by Prostat Resistance Meter (PRS-812), which the resistance was converted into electrical conductivity. Microstructures of PANI were observed by Scanning Electron Microscopy (SEM) (JCM-6000).

3. Results and discussion

3.1 Fourier Transform Infrared Spectroscopy (FTIR)

FTIR spectrum of PANI-DBSA at various molar ratio between DBSA and aniline is illustrated in Figure 1. Main absorbances are observed at 1567 cm^{-1}, 1493 cm^{-1}, 1445 cm^{-1}, 1298 cm^{-1}, and 1136 cm^{-1}, these peaks are assigned to C=N stretching vibration of quinoid ring, C=C stretching vibration of benzenoid ring, C=C stretching of aromatic ring, C-N stretching of secondary aromatic amnie, and NH+ stretching mode which indicates the degree of electron delocalization and its characteristic of conductivity, respectively [7-11]. In addition, relativly intensity of peak at 1136 cm^{-1} compared to peak at 1298 cm^{-1} is considerlably increased in doped of sample at high molar ratio, which result in higher conductivity of PANI-DBSA [11]. Besides, peaks at 1245 cm^{-1}, 1033 cm^{-1}, 1007 cm^{-1} are associated to C-N stretching in benzenoid unit, S=O stretching of DBSA, and C-H stretching of benzenoid ring in DBSA molecule, accordingly [10, 12, 13]. In addition, H-boned of NH vibration and aliphatic C-H stretching are located at 3233 cm^{-1}, and region of 2800 - 3000 cm^{-1}, which C-H stretching is depending on long alkyl of DBSA [10]. Molar ratio of 1.5 exhibits the obvious peaks in the region of 2800-3000

cm^{-1} due to the higher content of DBSA. Small peaks at 823 cm^{-1} and 667 cm^{-1} are assigned to C-H out of plane bending vibration of 1,4-disubtituted benzene ring and S-O stretching vibration of DBSA [8, 14].

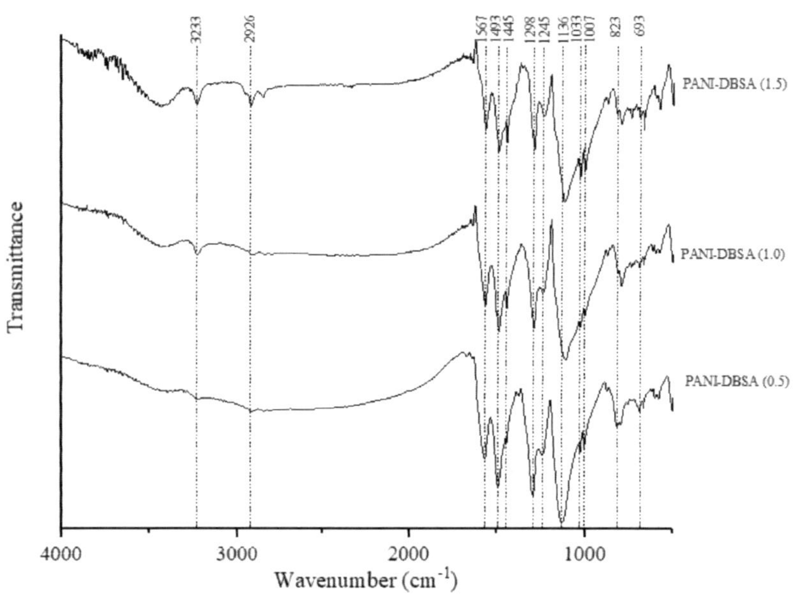

Figure 1. FTIR spectrum of PANI-DBSA at 0.5, 1.0, and 1.5 of molar ratio

3.2 Nuclear Magnetic Resonance Spectroscopy (NMR)

The chemical structures of PANI and DBSA, and NMR spectra of PANI-DBSA are illustrated in Figure 2 and 3, respectively. Sharpest peaks at 128 and 124 ppm are associated to C-2 and C-3 position in benzenoid structure, respectively. Besides, non-protonated carbon C-1, protonated carbon C-4 and C-5 in benzenoid part, and C-8 in quinoid part are observed at 148, 140, and 137 (shoulder), accordingly. Small peak at 163 ppm is assigned to C-7 non-protonated carbon in quinoid part, while peak at 114 (shoulder) ppm in the spectra is associated with C-6 in benzenoid part. Several peaks of aliphatic carbon region between 10-50 ppm is arised solely form aliphatic DBSA chains [15, 16]. In addition, presence of peaks at 180 and 96 ppm are referred to the cross-linking of well-ordered structures [17].

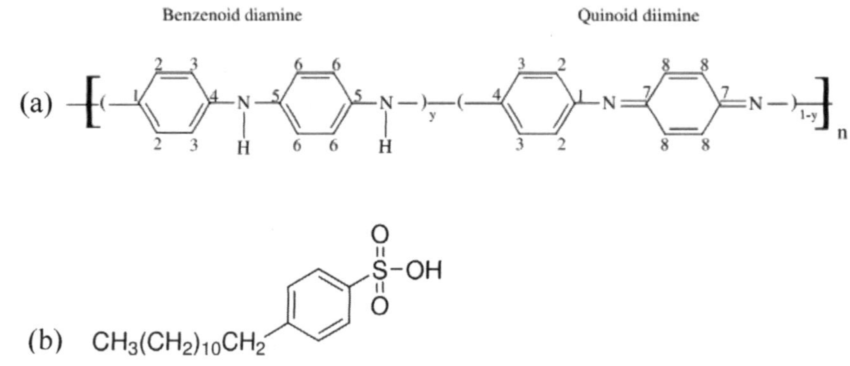

Figure 2. Chemical structure of (a) PANI and (b) DBSA

IOP Conf. Series: Journal of Physics: Conf. Series **1082** (2018) 012102 doi:10.1088/1742-6596/1082/1/012102

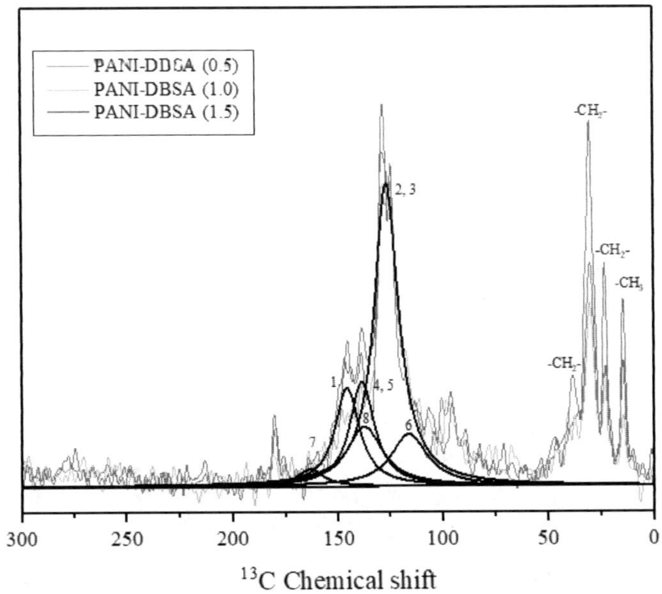

Figure 3. Comparison of 13C NMR spectrum of PANI-DBSA at 0.5, 1.0, and 1.5 of molar ratio

3.3 Electrical conductivity

Electrical conductivities of PANI-DBSA at different molar ratios are illustrated in Figure 4. Low molar ratio of [DBSA]/ [An] at 1.25:0.5 exhibit the lowest electrical conductivity of 1.167×10^{-6} S/m. However, high molar ratio of 1.5 shows the highest electrical conductivity of 18.91×10^{-2} S/m. It is observed that increasing of molar ratio between DBSA and aniline influences on electrical conductivity of PANI-DBSA. This is due to addition of protonic acid as dopant. This doping agent is source of charge compensation which increases the conjugation length without protonation [16, 18].

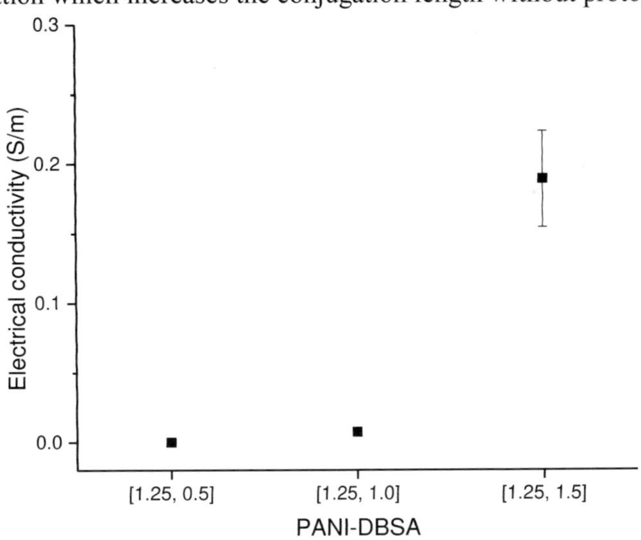

Figure 4. Electrical conductivity of PANI-DBSA at different molar ratio [DBSA]/ [An]

3.4 Scanning Electron Microscopy (SEM)

Figure 5 illustrates the microstructures of PANI-DBSA at different molar ratios between [DBSA]/ [An]. At molar ratios of 0.5 (Figure 4a) and 1.0 (Figure 4b), PANI-DBSA microstructures are agglomerated and formed irregular rock-like shape. However, at molar ratio of 1.5 (Figure 4c), thorn-like structure and smooth surface are observed in PANI-DBSA microstructure. This is due to the changes in molar ratio causes the transition of micellar shape [19].

Figure 5. SEM images of PANI-DBSA at (a) [1.25, 0.5], (b) [1.25, 1.0], and (c) [1.25, 1.5] of [DBSA]/ [An] molar ratios

4. Conclusion

DBSA doped PANI was polymerized at different molar ratios between DBSA and Aniline. High molar ratio of 1.5 exhibits the highest electrical conductivity of 18.91×10^{-2} S/m compared to the electrical conductivity at 05 low molar ratios of 1.167×10^{-6} S/m. This is due to increasing of doping in PANI-DBSA at high molar ratio of 1.5. Besides, the morphology of high molar ratio illustrates the smooth surface and thorn-like structure, while low molar ratio exhibits the irregular agglomerated structure.

Acknowledgement

We acknowledge the financial support by Japanese Government through Japan International Cooperation Agency (JICA) and Aun/Seed-Net scholarship (grant no. 304/PBAHAN/6050314/A119). This study has been supported for facilities by Universiti Sains Malaysia (USM).

References

[1] Campos, M., et al. 2014 *Journal of Applied Polymer Science* 131 40688
[2] Palaniappan, S. and A. John 2008 *Progress in Polymer Science* 33 732
[3] Ciric-Marjanovic, G. 2013 *Synthetic Metals* 177 1
[4] Lenhart, N., et al. 2011 *Thin Solid Films* 519 4351
[5] Wu, Y.-J., et al. 2013 *Polymer International* 62 581
[6] Streltsov, A.V., et al. 2009 *Journal of Applied Polymer Science* 114 928
[7] Shokry Hassan, H. 2015 *American Journal of Applied Chemistry* 3 54
[8] Babazadeh, M. 2009 *Journal of Applied Polymer Science* 113 3980
[9] Neelgund, G.M. and A. Oki 2011 *Polym Int* 60 1291
[10] Trchová, M. and J. Stejskal 2011 *Pure and Applied Chemistry* 83 1803
[11] Horta-Romarís, L., et al. 2017 *Materials & Design* 114 288
[12] Han, D., et al.2005 *Colloids and Surfaces A: Physicochemical and Engineering Aspects* 259 179
[13] Luo, C., et al. 2011 *Macromolecules* 44 6899
[14] Ansari, M.O. and F. Mohammad 2012 *Composites Part B: Engineering* 43 3541
[15] Kababya, S., et al. 1999 *Macromolecules* 32 5357

[16] Ho, K.S., et al. 2005 *Journal of Polymer Science Part a-Polymer Chemistry* 43 3116
[17] Zoran D. Zujovic, Marija Gizdavic-Nikolaidis, and a.G.A. Bowmaker 2013 *Chemistry in New Zealand* 87
[18] Ayad, M.M. and E.A. Zaki 2008 *European Polymer Journal* 44 3741
[19] Bhadra, J., et al. 2017 *Arabian Journal of Chemistry* 10 664

IOP Conf. Series: Journal of Physics: Conf. Series **1082** (2018) 012103 doi:10.1088/1742-6596/1082/1/012103

Synthesis and Characterisation of Bismuth Oxide Nanoparticles using Hydrothermal Method: The Effect of Reactant Concentrations and application in radiotherapy

Z.A. Zulkifli[1], K.A. Razak[1,2], W.N.W.A Rahman[3], and S.Z. Abidin[4]

[1] School of Materials and Mineral Resources Engineering, Engineering Campus, Universiti Sains Malaysia, 14300 Nibong Tebal, Penang, Malaysia.

[2] NanoBiotechnology Research and Innovation (NanoBRI), Institute for Research in Molecular Medicine, Universiti Sains Malaysia, 11800 Penang, Malaysia.

[3] School of Medical Sciences Health Campus, Universiti Sains Malaysia, 16150 Kubang Kerian, Kota Bharu, Kelantan, Malaysia

[4] Advanced Medical and Dental Institute (AMDI), Universiti Sains Malaysia, Bertam, Jln Tun Hamdan Sheikh Tahir, 13200 Kepala Batas, Penang

*E-mail: zulfa6254@gmail.com

Abstract. High atomic number (Z) of bismuth oxide nanoparticles (Bi_2O_3 NPs) has more cell penetration and less adverse effects than conventional radiosensitisers. In this work, 60 nm and 90 nm of Bi_2O_3 NPs were successfully synthesised using the hydrothermal method by varying the bismuth nitrate, $Bi(NO_3)_3$ concentration. The properties of Bi_2O_3 NPs were characterised to determine the phase presence, crystallinity, morphology, elements presence and size of nanoparticles. The as-synthesised Bi_2O_3 NPs was in monoclinic Bi_2O_3 phase (ICDD 98-008-5622). As the $Bi(NO_3)_3$ concentration increased, the particle size of Bi_2O_3 NPs decreased due to less ions diffusion per nuclei. The morphology observation showed that Bi_2O_3 NPs were in rods form. The produced Bi_2O_3 NPs were subjected to cytotoxicity analysis and radiotherapy. Bi_2O_3 NPs did not induce cytotoxicity in breast cancer (mcf-7) cell lines at concentration from 0.05 μM. The radiotherapy performance of the as-prepared Bi_2O_3 NPs was obtained by calculating the sensitisation enhancement ratio (SER). The optimum result was found for 60 nm Bi_2O_3 NPs with SER of 3.45.

1. Introduction

The National Cancer Registry of Malaysia (NCR) recorded 21,773 Malaysians being diagnosed with cancer and estimated that almost 10,000 cases are unregistered every year. One of the top 5 cancer in Malaysia is breast cancer [1]. In current clinical oncology practices, half of all cancer patients are treated with radiation therapy either alone or in combination with other treatments [2]. In radiotherapy, the central pathways to increase the ratio of treatment efficacy to side effects are reversal of radiation resistance in tumour tissue, where the enhancement of radioresistance in healthy tissue increasing radiosensitisation in tumour tissue and better confinement of the deposited dose to the tumour volume [3]. Recent advances in nanotechnology have led to the development of novel nanomaterials and integrated nanodevices for cancer detection and screening, *in vivo* molecular and cellular imaging, and the delivery of therapeutics such as cancer cell killing radio-isotopes. An increasing number of studies have shown that the selective delivery of therapeutic agents into a tumour mass using nanoparticle platforms could improve the bioavailability of cytotoxic agents and minimize toxicity to normal tissues [4].

The most common radiosensitising approach is to exploit the increase photon absorption of high atomic number (Z) materials at kilovoltage (kV) photon energies [5]. Adopting this approach, therapeutic nanoparticles have been produced using silver (Z = 47), gadolinium (Z = 64) and gold (Z = 79). Recently, Bi_2O_3 NPs has received wide attention in the field of radiotherapy research based on its remarkable radiation dose enhancement under kilovoltage X-ray beams, which is significantly higher than the well-known gold radiosensitiser [6]. However, the study on Bi_2O_3 NPs in radiotherapy is still new. Several approaches have been employed to synthesis Bi_2O_3 NPs in materials science and engineering. Bi_2O_3 NPs has been synthesised via metallorganic Chemical Vapor Deposition (CVD) [7], oxidative metal vapor-phase deposition [8], and slow oxidation of Bi nanowires at 750 °C [9], etc. Although these methods have been proven to be successful in the synthesis of Bi_2O_3 NPs, the methods normally require high temperature heat treatment, long synthesis period, and post treatment, which is far from low cost and apparently require sophisticated equipments. Hence, a more simple synthesis method, which is precipitant-free, additive-free, and low cost, is sought after for the synthesis of Bi_2O_3 NPs. A hydrothermal method is a powerful method in synthesis of Bi_2O_3 NPs at a considerably low temperature, energy saving and cost effective benefits [10]. The hydrothermal method also has the advantages such as controllable particle size, morphology and the degree of crystallinity by simply changing the $Bi(NO_3)_3$ concentration.

Hence, in this study, Bi_2O_3 NPs were produced using hydrothermal method. The effect of bismuth nitrate, $Bi(NO_3)_3$ concentration was studied. The effect of size of Bi_2O_3 NPs in cytotoxicity study and as a radiosensitiser also was studied systematically.

2. Experimental Study

2.1 Synthesis of Bi_2O_3 NPs

Bi_2O_3 was synthesised using the hydrothermal method. In this synthesis, 0.05 M and 0.3 M of $Bi(NO_3)_3.5H_2O$ was mixed with 3.0 mmol of Na_2SO_4 and dissolved in 40 ml distilled water, separately. The solution was stirred at room temperature for 45 min. Then, 18.0 mmol of NaOH dissolved in 40 ml of distilled water was added dropwise into the above solution with continuous stirring, which immediately resulted in the formation of yellowish precipitates in the container. The mixed solution was transferred into a Schott Bottle, sealed and subjected to hydrothermal reaction at 60˚C. The samples were then allowed to cool to room temperature naturally. The as-formed yellow precipitates were washed with deionized water and ethanol several times, and finally dried in air at 80˚C to obtain the final faint yellow Bi_2O_3 samples. The phase presence and crystallinity of the Bi_2O_3 were characterised by using XRD (Bruker AXS D8 Advance with Cu-Kα as a radiation source). The Bi_2O_3 morphology was examined by FESEM (ZEISS SUPRA 35VP), and analyzed using the ImageJ software (ImageJ, 2015) by counting 100 nanoparticles for particle size calculation.

2.2 Cytotoxicity analysis

For cytotoxicity analysis, firstly, 86 mM Bi_2O_3 NPs stock solution (each with 60 nm and 90 nm size) was diluted in media culture (DMEM) to a concentration of 0.05 µM. Secondly, DMEM containing mcf-7 cells in 25cm^2 flasks were rinsed with phosphate buffer saline (PBS) three times. 2 ml of trypsin was added into the flask for cell detachment from the surface of the flask and incubated for 5 minutes in 5% CO_2 incubator. 2 ml DMEM was added again to deactivate trypsin reaction. DMEM containing mcf-7 cells was transferred into conical tubes for centrifugation. After that, the cells were counted and 20µl of cells suspension containing 1 x 10^4 cells were seeded in a 96-well plate (Thermo, USA). 3 wells from 96-well plate were left without cell for control. The plate was left overnight at 37˚C in a humidified 5% CO_2 incubator. Next day, 100µl of prepared Bi_2O_3 with 500 µM concentration in DMEM was added into each well containing cells. Treated cells were incubated again in 5% CO_2 incubator for 24 hours. Then, the incubated treated cells were rinsed briefly with PBS three times. 90µl of DMEM was added again into each well followed by 10µl of PrestoBlue® reagent. The plate was incubated at 37 ˚C for 2 hours to increase sensitivity of Presto blue in order to detect live cells. Finally, the cell viability was measured using a microplate ELISA reader (Multiskan spectrum, USA). The work was performed in triplicate and the data was expressed as viability cell percentage using formula below.

$$\text{Cell viability \%} = \frac{\text{(Average absorbance sample} - \text{Absorbance blank)}}{\text{(Average absorbance sample} - \text{Absorbance blank)}} \times 100$$

2.3 Cell irradiation

60 nm and 90 nm of prepared Bi_2O_3 with 0.05 µM concentration were added into each micro-vial containing Mcf-7 cells. Then, the cells were exposed to 6 MV photon beams with varying radiation dose from 0 to 10Gy. The LINAC machine was set-up to 100 cm source above the surface distance (SSD) with 10 x 10 cm^2 field. The irradiated cells were sub-cultured in 6-well plates and incubated in 5% CO_2 incubator for 5 days for clonogenic assay. Medium in 6-well plate was removed and rinsed with PBS. The cold methanol was added and left for 30 minutes to fix the cells on the plate. The methanol was then discarded and crystal violet was inserted and stained for 30 minutes. The plates were rinsed with distilled water and left drying before counting colonies. The colonies of the cell were counted and plot the survival fraction. Finally, the sensitisation enhancement ratio (SER) was calculated based on the survival fraction graft to measure lethal dose for loss of proliferative capacity to 50 % in the presence of Bi_2O_3 NPs. The SER formula as shown below.

$$\text{Sensitiser enhancement ratio (SER)} = \frac{\text{Radiation dose alone (without NPs)}}{\text{Radiation dose with NPs}}$$

3. Results and discussion

XRD patterns of the samples prepared using different concentration of bismuth nitrate, $Bi(NO_3)_3$ of 0.05 M and 0.3 M are shown in Figure 1. All diffraction peaks can be indexed as a pure monoclinic Bi_2O_3 phase, which matched with ICDD 98-008-5622. The strong and sharp XRD peaks indicated that the as-prepared Bi_2O_3 powders are in crystalline state. From Figure 1, the intensity of Bi_2O_3 decreased with increasing

$Bi(NO_3)_3$ concentration. In hydrothermal reaction, nuclei formed first followed by growth. During hydrothermal reaction at 60°C for 10 min, the amount of $Bi(NO_3)_3$ concentration that was mixed with Na_2SO_4 and NaOH directly affected the number of nucle.After hydrothermally synthesised at 60°C, growth process would take place. Therefore, since number of nuclei increased with increasing $Bi(NO_3)_3$ concentration, less ions could diffuse per nuclei, thus particle size decreased. The average crystallite size of the as–synthesised samples were calculated using the Scherrer equation. From the calculation, the crystallite sizes are 29.6 and 23.2 nm for 0.05 M and 0.3 M of $Bi(NO_3)_3$, respectively.

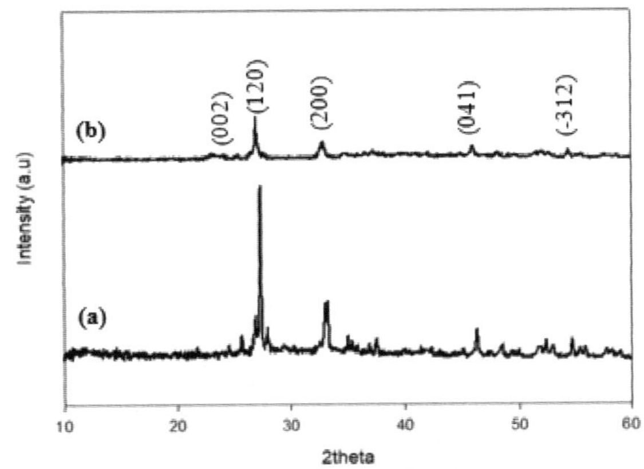

Figure 1. XRD patterns of Bi_2O_3 NPs synthesised using hydrothermal method at 60 °C and 10 min with varying $Bi(NO_3)_3$ concentration: (a) 0.05 M, and (b) 0.3 M.

The morphology of the synthesized Bi_2O_3NPs was observed using SEM as depicted in Figure 2. All samples are in rods form. When $Bi(NO_3)_3.5H_2O$ and Na_2SO_4 were mixed in distilled water at room temperature, $Bi_2O(OH)_2SO_4$ nanowires formed, which played a role as a template in the following process of the reaction. By adding NaOH solution into the reaction system, OH^- ions reacted with $Bi_2O(OH)_2SO_4$ gradually to form $Bi(OH)_3$. When the $Bi(OH)_3$ was subjected to hydrothermal reaction at 60°C and 10 min reaction time, the mixture transformed into Bi_2O_3 nanorods.This process is similar to the formation of ZnO nanorods [11].The diameters of particles measured usingthe ImageJ software are approximately 60 nm and 90 nm in diameter for samples prepared with 0.05 M and 0.3 M of $Bi(NO_3)_3$, respectively.

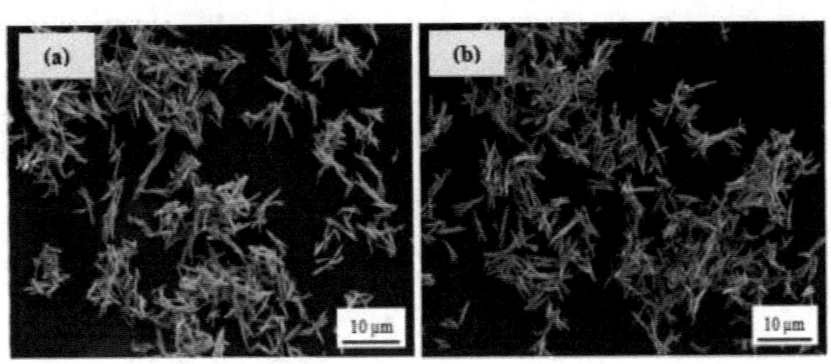

Figure 2. SEM images of Bi_2O_3 NPs synthesised using hydrothermal method at 60 °C and 10 min with varying $Bi(NO_3)_3$ concentration: (a) 0.05 M, and (b) 0.3 M.

The particle size dependence of Bi_2O_3 NPs on the cytotoxic response in mcf-7 cell lines is shown in Figure 3. A ratio between the nanoparticle-treated cell and untreated control cell was taken as the viabilities, assuming the viability of untreated control was 100 %, despite the fact that normal cell metabolism processes still produced a small number of dead cells. The result showed that cell viability decreased as the particle size increased. The calculated cell viabilities were 100 %, 118 %and 96 % for control, 60 nm and 90 nm, respectively. Both Bi_2O_3 NPs sizes did not affect the cell viability of mcf-7 cell lines, indicated that both size were non-toxic. The study on cytotoxicity analysis of Bi_2O_3 NPs on mcf-7 cells revealed that 90 nm Bi_2O_3 NPs was more toxic to the mcf-7 cells than 60 nm Bi_2O_3 NPs. This happened because of the size of nanoparticles was found to play a critical role in both the rate and extent of cellular uptake. Larger nanoparticles exhibited higher cellular uptake, suggesting that larger nanoparticles more strongly induced apoptosis [12]. The cellular uptake of nanoparticles is closely associated with cellular response which is apoptosis. In general, as cellular uptake of nanoparticles increased, cell viability declined and apoptosis increased. Hence, the apoptotic features in the treated mcf-7 breast cancer cells including cells shrinkage indicated that mcf-7 breast cancer cells underwent cells death via apoptosis pathway.

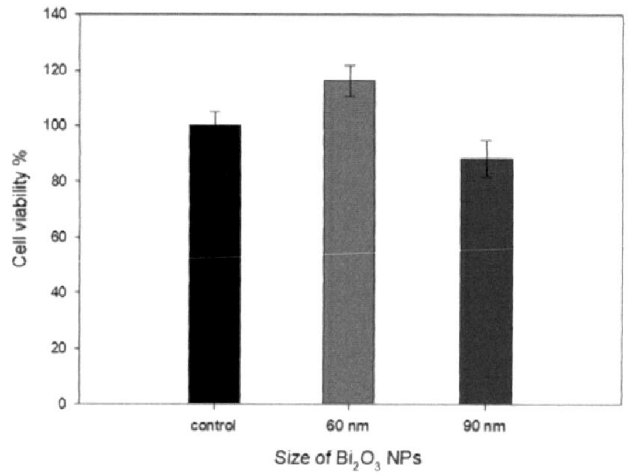

Figure 3. Cell viability of mcf-7 cells treated with different sizes of Bi_2O_3 NPs after 24 hours incubation.

The radiation enhancement effect of Bi_2O_3 NPs in mcf-7 cell lines was measured by clonogenic assay method (calculating the sensitisation enhancement ratio (SER)). Figure 4 shows the effect of Bi_2O_3 NPs on cell survival fraction at 6 MV photon energy with different radiation dose from 0 to 10 Gy. The calculated SER index are 4.36, 3.45 and 7.64 for control, 60 nm and 90 nm Bi_2O_3 NPs, respectively. This result indicated that the small size of nanoparticles is more internalised inside the cells and absorbed more ionizing radiation to damage the chromosomes. The internalization of Bi_2O_3 NPs could be via endocytosis that cross cell membranes and moved to vital component like DNA and mitochondria to generate 'free radicals' thus damage the cell. Several studies have demonstrated that AuNPs [13] and Bi_2S_3 NPs [14] were internalised by cells via endocytosis, and in some cases those nanoparticles have been found clustered around the membrane within the cytoplasm. Therefore, the NPs was expected to be internalized and engulfed inside the cells enhancing the radiation effect. This could be attributed to the fact that the inclusion of high electron density and high atomic number compounds in the cells was expected to generate high number of free radicals via both photoelectric and Compton interactions. These extra free radicals would increase the likelihood of DNA damage and subsequent cell death. The results suggested that 60 nm Bi_2O_3 NPs would be more beneficial as radiosensitiser.

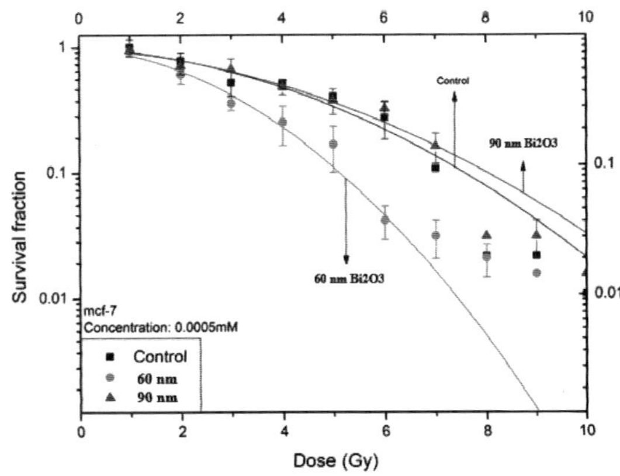

Figure 4. Survival curves of mcf-7 cell lines untreated and treated with Bi_2O_3 NPs and exposed to 6 MV radiation doses.

4. Conclusion

In this work, Bi_2O_3 NPs were successfully synthesised by a simple and cost-effective hydrothermal method. The concentration of $Bi(NO_3)_3$ influenced the size of synthesised powders. Bi_2O_3 rods in monoclinic phase with diameter of 60 nm and 90 nm, were produced with varying $Bi(NO_3)_3$ concentration of 0.05 M and 0.3 M, respectively. Bi_2O_3 NPs did not induce cytotoxicity in breast cancer (mcf-7) cell lines at concentration from 0.05 μM. The radiotherapy performance of the as-prepared Bi_2O_3 NPs was obtained by calculating the sensitisation enhancement ratio (SER). The optimum result was found for 60 nm Bi_2O_3 NPs produced using 0.05 M $Bi(NO_3)_3$ concentration with SER of 3.45.

References

[1] T. Watanabe, T.-T. Wu, P. J. Catalano, T. Ueki, R. Satriano, D. G. Haller, A. B. Benson and S. R. Hamilton, 2001 *New England Journal of Medicine* 344 1196
[2] L. Zhang, H. Chen, L. Wang, T. Liu, J. Yeh, G. Lu, L. Yang and H. Mao 2010 *Nanotechnology, science and applications* 3 159
[3] B. B. Ma, R. G. Bristow, J. Kim and L. L. Siu 2003 *Journal of clinical oncology* 21 2760
[4] X.-H. Peng, X. Qian, H. Mao and A. Y. Wang 2008 *International journal of nanomedicine* 3 311
[5] J. Coulter, W. Hyland, J. Nicol and F. Currell 2013*Clinical Oncology* 25 593
[6] M. Alqathami, A. Blencowe, U. Yeo, R. Franich, S. Doran, G. Qiao and M. Geso 2013 presented at the *Journal of Physics: Conference Series*
[7] H. W. Kim, J. H. Myung and S. H. Shim 2006 *Solid state communications* 137 196
[8] L. Kumari, J.-H. Lin and Y.-R. Ma, 2007 *Journal of Physics: Condensed Matter* 19 406204
[9] L. Li, Y. W. Yang, G. H. Li and L. D. Zhang 2006 *small* 2 548
[10] C. Wu, L. Shen, Q. Huang and Y.-C. Zhang 2011 *Materials Letters* 65 1134
[11] H. Li, Y. Zhang, H. Liu and J. Wang 2011 *Journal of the American Ceramic Society* 94 3267
[12] Y. Pan, S. Neuss, A. Leifert, M. Fischler, F. Wen, U. Simon, G. Schmid, W. Brandau and W. Jahnen-Dechent 2007 *Small* 3 1941
[13] S.-W. Tsai, Y.-Y. Chen and J.-W. Liaw 2008 *Sensors* 8 2306
[14] K. Ai, Y. Liu, J. Liu, Q. Yuan, Y. He and L. Lu 2011 *Advanced materials* 23 4886

IOP Conf. Series: Journal of Physics: Conf. Series **1082** (2018) 012104 doi:10.1088/1742-6596/1082/1/012104

Iron Oxide Magnetie-Polyethylene Glycol-Layered Double Hydroxide Core Shell Nanoparticles for Drug Delivery

Mona Ebadi[1], Saifullah Bullo[1,2], Mohd Zobir Hussein[1*]

[1]Materials Synthesis and Characterization Laboratory, Institute of Advanced Technology (ITMA), University Putra Malaysia, Selangor, Malaysia.

[2] Laboratory for vaccine and immunotherapeutic, institute of Biosciences, University Putra Malaysia, Selangor, Malaysia

E-mail: mzobir@upm.edu.my

Abstract. Recently, micro- and nanoparticles with magnetic core and functional shell structures have received much attention because of their potential applications in drug delivery, biosensors catalysis, selective separation and chromatography etc. Layered double hydroxides (LDHs) nanoparticles (NPs) are an efficient cellular delivery vehicle for many negatively charged chemical species. Recently, Mg and Zn/Al-layered double hydroxide (Mg/Al-LDH) and (Zn/Al-LDH) has emerged as a promising nanocarrier for drug delivery system due to its unique properties such as low cytotoxicity, pH-controlled release, good biocompatibility, tunable particle size, and protection of drugs in the interlayer. Various anionic biofunctional molecules, including DNA, SiRNA, drugs and vitamins have been successfully intercalated into LDH interlayers, protected from enzyme degradation and delivered into the cytoplasm in mammalian cells through the clathrin-mediated endocytosis pathway. In this study, a magnetite core shell nanodrug delivery formulation was designed based on Fe_3O_4-PEG-Zn/Al-LHDs or Mg/Al-LDHs loaded with anticancer drug 5-Fluoro-2,4(1H,3H)-pyrimidinedione commonly called 5-fluorouracil. Designed core shell magnetite nanoparticles formulation was characterized using XRD, zeta sizer and high performance liquid chromatography (HPLC).

1. Introduction

Nanomedicine is an advanced field of medical science where nanomaterials are designed and applied for various biomedical applications. Lately, nanomedicine is widely getting attention for theranostics applications i.e. simultaneous disease diagnosis and treatment. In addition, the use of magnetic nanomaterials is highly sought-after for nanomedicine applications particularly for targeted delivery, hyperthermia and biosensors. As important biomedical nanomaterials, magnetic nanomaterials need to have (1) superparamagnetic property which prevents residual magnetism induced aggregation;(2) sufficient magnetic responsiveness, which ensures the feasibility for external magnetic field induced separation and delivery;(3) monodispersing in size, which is critical for the electrical and magnetic properties of any nanoparticles due to their fundamental size dependent properties; (4) easily

IOP Conf. Series: Journal of Physics: Conf. Series **1082** (2018) 012104 doi:10.1088/1742-6596/1082/1/012104

functionalizable surfaces, which are important for further functionalization, such as the linkage of recognition molecules. Core–shell structural superparamagnetic iron oxide hybrid nanoparticles are exceptionally promising materials. Polymer coatings could stabilize magnetic nanoparticles leading to better dispersion and biocompatibility and could also be further modified with functional molecules of various descriptions for diverse applications. In this study we describe the development of theranostic core shell nanoparticles based on superparamagnetic of Fe_3O_4 core coated with PEG fabricated with Mg/Al-LDHs and Zn/Al-LDHs. The designed core shell nanocarriers were loaded with anti-cancer drug 5.FU.

2. Experimental study

2.1. Materials and Procedures

Distilled deionized water (18.2 MΩcm^{-1}) was used in all experiments. Iron (II) chloride tetra hydrate ($FeCl_2.4H_2O \geq 99\%$), iron (III) chloride hexahydrate ($FeCl_3.6H_2O$, 99%) were purchased from Merck, Germany. polyethylene glycol, average M.W. 6000 was purchased as a raw material from Acros Organics. Ammonia solution (25%) was obtained from Scharlau. Aluminum nitrate ($Al(NO_3)_3.9H_2O$) with 98.5% purity, Zinc nitrate hexahydrate $Zn(NO_3)_2{\cdot}6H_2O$ with 98% purity and magnesium nitrate ($Mg(NO_3)_2{\cdot}6H_2O$) with 99% purity were supplied by ChemAR. The drug used in this study is 5-fluorouracil ($C_4H_3FN_2O_2 \geq 98\%$) was obtained from AKSci. Dimethyl sulfoxide ($(CH_3)_2SO \geq 99\%$) was purchased from Sigma Aldrich, Merck.

2.2 Experimental

In order to prepare magnetite iron oxide coated with polyethylene glycol (FPEG), the mixture of 2.43 gr ferrous chloride tetra hydrate (FeCl2.4H2O), 0.99 gr ferric chloride hexahydrate (FeCl3.6H2O) and 80 mL deionized water in the presence of 6 mL ammonia hydroxide (25 % by mass) was exposed to ultrasonic irradiation for 1 h. The precipitates were centrifuged and washed 3 times and then the washed precipitates were dispersed in 100 mL deionized water and mixed with 2 % PEG. Then the nanoparticles were placed in an autoclave at 200 o^{C} for 24 h. The black precipitates were then collected by a permanent magnet and washed three times to remove the excess PEG which does not participate in the coating process. The washed precipitates were re-dispersed in 50 mL ethanol and added dropwise into a layered double hydroxides solution (Mg/Al) by stirring the mixture until the pH decreased to 9. Then the core-shell nanocomposite was centrifuged at 5000 rpm. The 3% of drug, 5-FU [4, 21] which was dissolved in dimethyl sulfoxide, was added into the nanocomposite (iron oxide-PEG-LDH (Mg/Al) and the mixture was stirred for 24h. Finally, the coated iron oxide was washed and dried in an oven. The same procedure was done to prepare iron oxide coated with LDH (Zn/Al). The structure and the composition of the iron oxide with poly ethylene glycol coating and LDH(Mg/Al) and LDH(Zn/Al) nanocarrier were characterized on an X-ray diffraction (XRD) in the 2θ range of 2°–80°, High performance liquid chromatography (HPLC) and particle size analysis (PSA).

3. RESULTS AND DISCUSSION

3.1. X-Ray Diffraction of the Nanoparticles

The XRD patterns of the Zn/Al-LDH and the Mg/Al-LDH nanocarrier are shown in Figure 1. The results display for pure Zn/Al-LDH layered structure characteristic with strong peaks. Corresponding to the diffraction planes of (003) and (006) at the 2θ positions of 10.9° and 22.1°, respectively. The XRD peaks observed for pure Mg/Al-LDH with high intensity in three main characteristic diffraction planes (003), (006) and (009) at the 2θ positions of11.5° and 23.2° and 34.8°, respectively.

The XRD patterns of the Mg/Al-LDH and the Zn/Al-LDH nanocarrier coating formed on iron oxide-PEG-drug nanocomposites are shown in Figure 1(a) and (b), respectively. All iron oxide nanoparticles have six characteristic peaks at $2\theta = 30.16^{o}$, 35.95^{o}, 43.34^{o}, 54.17^{o}, 57.27^{o} and 62.98^{o}. From the XRD patterns, it has also been found that these six diffraction peaks correspond to the pure magnetite nanoparticles with a cubic inverse spinal structure (Reference JCPDS Number 82-1533). Owing to observing these characteristic peaks in two nanocomposites (Figures 1(a) and (b)) it is evident that the coating process did not result in a phase change of the iron oxide nanoparticles. In addition, the XRD also indicates that PEG was well loaded onto the iron oxide nanoparticles. The peaks belong to LDH and drug are also observed on both samples, which means that the combination of nanocomposite and drug are well integrated into the LDH.

It is noted that obvious Zn/Al and Mg/Al nanocarrier peaks appeared on the coated samples. The XRD phases of iron oxide-PEG-drug composite coating remained the same as those of the Zn/Al-LDH and Mg/Al-LDH coating, but the crystallization intensity of iron oxide, drug and PEG the phases weakened. Obviously, the thick LDH coating lessens the penetration of the X-ray effectively.

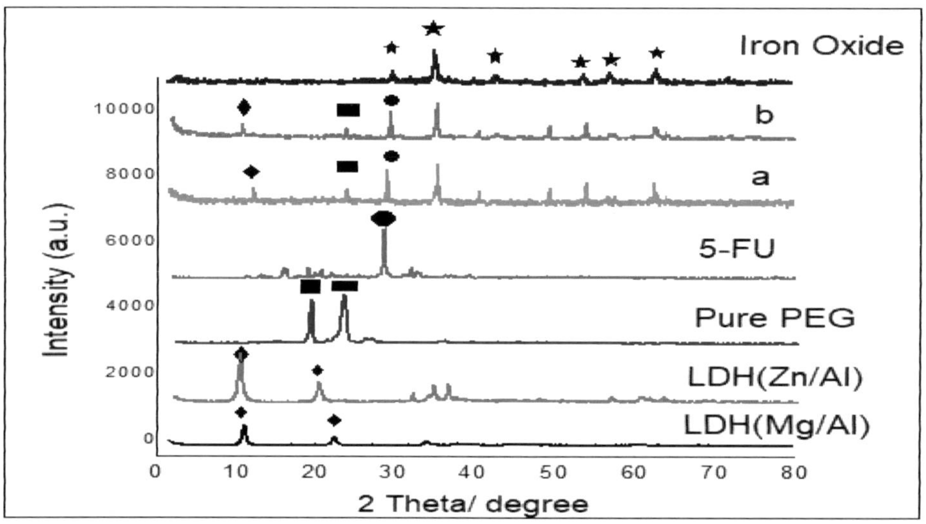

Figure 1. XRD patterns of pure LDH(Mg/Al), LDH(Zn/Al), Poly ethylene glycol, 5-FU, core shell nanoparticles with LDH-Mg /Al(a), core shell nanoparticles with LDH-Zn/Al(b) and Iron oxide nanoparticles.

3.2. Particle Distribution Measured by PSA with different nanocarrier

Particle size of nanocomposites with Mg/Al-LDHs and Zn/Al-LDHs as a nanocarrier (a and b) were determined with a dynamic light scattering technique by a zeta sizer. The samples were dispersed in methanol at once exposed to ultrasonic irradiation for 60 min at 30% power in 40 temperature and then they were analyzed with a particle size analysis (PSA) method. According to the cumulative distribution frequency, ~62% of particles in sample which coating with Mg/Al-LDHs as a nanocarrier were less than the 120 nm (Figure a) and the average hydro-particle size is 120 nm following the size distribution in the range of 90 to 150 nm. in addition, The particle size of core shell nanoparticles sample that coating with Zn/Al-LDH as a nanocarrier was distributed over a narrow range of 240–260 nm and ~63% of the particles were found to have the size less than 260 nm as shown in Figure 2.

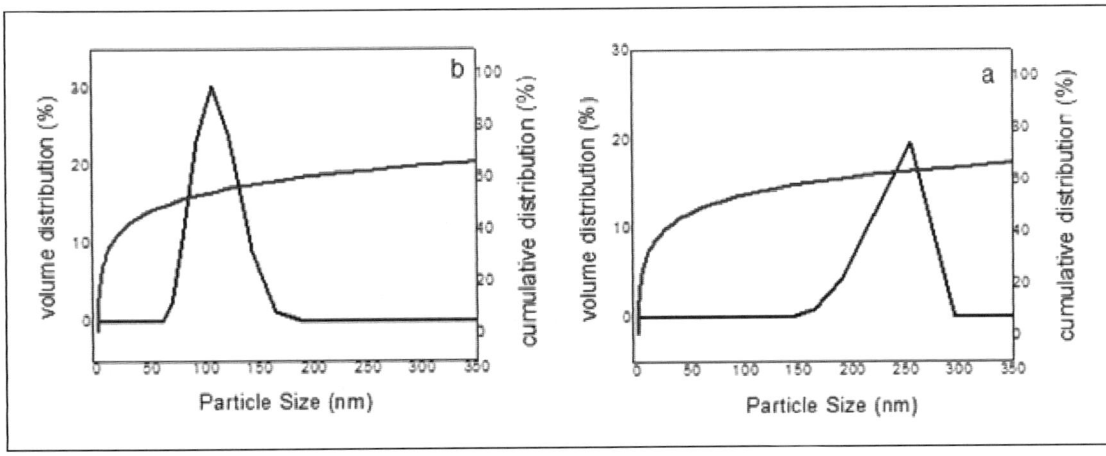

Figure 2. Hydrodynamic size of nanocomposites (a) with Mg/Al-LDHs and nanocomposites (b) with Zn/Al-LDHs.

3.3. high-performance liquid chromatography (HPLC) analysis

Figure 3 shows the calibration curve standard for the nanocomposites. The correlation coefficient, R2, for the standard solutions was 0.923. The percentage loading of drug in the nanocomposites determined by HPLC is given in figure 4. According to the results, the percentage of drug loading for iron oxide–PEG-Mg/Al-LDHs is higher than iron oxide–PEG-Zn/Al-LDHs, with 68.41% and 50.80% loading, respectively. Importantly, it was also observed that drug loading percentage in sample (a) that using Mg/Al-LDHs as the nanocarrier agent is more than when the using Zn/Al-LDHs for nanocarrie (sample b).

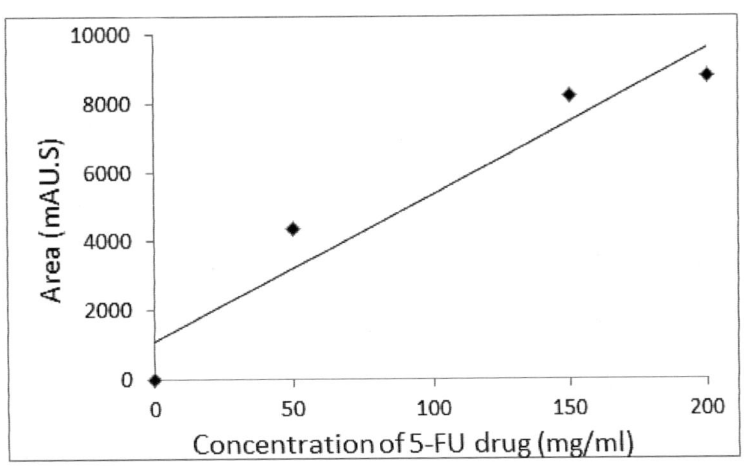

Figure 3. Calibration curve of 5-flourouracil drug determined using HPLC with standard concentrations of 0 mg/mL, 50 mg/mL, 150 mg/mL, and 200 mg/mL.

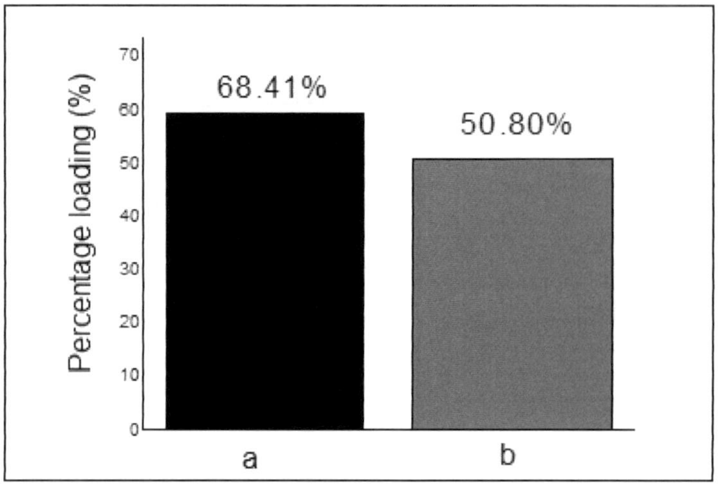

Figure 4. Percentage loading of nanocomposites a and b by HPLC.

4. Conclusion

X-ray diffraction results showed that both nanoparticles with cubic inverse spinal structures were successfully prepared using the co-precipitation method. High performance liquid chromatography indicated that when using of Mg/Al-LDHs as nanocarrier resulted in markedly enhanced percentage drug loading with smaller particle size and narrow size distribution. Further studies will be carried out such as sustained release, cytotoxicity and anticancer assays.

Acknowledgements

The authors would like to thank University Putra Malaysia and the Ministry of Higher Education of Malaysia (UPM-MOHE) for funding this project under NANOMITE programme Vot No 5489100.

References

[1] Dorniani, D. Kura, A. U. Hussein, M. Z. B. Fakurazi S. Shaari, A. H. 2014 *Journal of Materials Science*, 49 8487
[2] Hong, K. Khwaja, A. Liapi, E. Torbenson, M. S. Georgiades, C. S. 2006 *Clin Cancer Res,* 12 2563
[3] Mirza, A. Z. & Siddiqui, F. A. 2014 *International Nano Letters* 4 1
[4] Sophie Laurent, Delphine Forge, Marc Port, Alain Roch, Caroline Robic, Luce Vander Elst and Robert N. 2008 *Chem. Rev.* 108 2064

IOP Conf. Series: Journal of Physics: Conf. Series **1082** (2018) 012105 doi:10.1088/1742-6596/1082/1/012105

Synthesis of TiO_2 Nanotubes Decorated with Ag Nanoparticles (TNTs/AgNPs) For Visible Light Degradation of Methylene Blue

Norain Isa[1,2], Tan Wan Kian[3], Go Kawamura[3], Atsunori Matsuda[3], Zainovia Lockman[1*]

[1]Green Electronics NanoMaterials Group (GEMs), School of Materials and Mineral Resources Engineering, Universiti Sains Malaysia, Pulau Pinang, Malaysia

[2]Sensor and Environmental Research Group (SERG), Department of Applied Sciences, Universiti Teknologi MARA, Cawangan Pulau Pinang, Kampus Permatang Pauh, Pulau Pinang, Malaysia

[3]Center for International Education, Toyohashi University of Technology, Aichi, Toyohashi, Japan

Email: zainovia@usm.my

Abstract. TiO_2 nanotubes (TNTs) decorated with Ag nanoparticles (AgNPs) were fabricated first by anodisation of titanium wire to produce the TNTs then by dispersing AgNPs on the TNT arrays. The AgNPs were produced by reduction in *kyllinga brevifolia* extract. The AgNPs have diameter in the range of 10 – 30 nm measured from transmission electron microscope image and average particle size of 22.3 nm. They were dispersed on TNTs by wetness impregnation technique. The morphology of the fabricated TNTs/AgNPs was evaluated by scanning electron microscope (SEM). SEM images of the sample shows highly ordered TNTs with 7.30 ± 0.46 μm long with 74.59 ± 9.22 nm inner diameters and wall thickness 33.51 ± 4.56 nm. After impregnation, the dispersed AgNPs appear to be distributed uniformly across the surface of the TNTs. The TNTs/AgNPs composite were used to remove methylene blue (MB) dye from an artificial wastewater. On TNTs/AgNPs 63% degradation was recorded compares to pure TNTs with only 17% degradation after 1 hr reaction under visible light. The addition of AgNPs has contributed to the activation of the composite under visible light perhaps due to the surface plasmon of the AgNPs as well as electrons transfer process to the conduction band of TiO_2.

1. Introduction

TiO_2 is the most widely researched photo-catalysts semiconductor because it is highly photoactive, has a strong oxidizing potential, long term photo-stability and is also nontoxicity and economical [1]. However, the application of TiO_2 is limited by its high band gap energy (3.0–3.2 eV), it is only able to act as a photo-catalysts under ultraviolet (UV) light, which accounts for about 4-5% of sunlight, greatly limiting its practical applications [2]. From the view point of exploiting the solar light, visible light responsive TiO_2 photocatalysts are highly preferred.

One of the possible solutions for improving the photo-catalytic efficiency of TiO_2 photocatalysts in the visible region is to shift its optical absorption from the UV region to the visible light region by making a composite of TiO_2 with other materials [3]. In order to improve the photo-catalytic efficiency and extending the spectral response of TiO_2 to the visible spectrum, decoration of Ag on the surface of TiO_2 is employed in this study. The Ag enhances the photo-catalytic activity by reducing the band gap and/or reducing electron-hole recombination [2].

In this context, the TiO_2 nanotubes (TNTs) were prepared by anodizing method of titanium wire in mixture of ethylene glycol (EG) / anhydrous NH_4F. Throughout the anodisation process, air bubbles were introduced to the electrolyte. As to investigate the effect of alkaline salt, based on carbonate, we added in 0.1 M K_2CO_3 to the electrolyte half way through the anodisation process. It is acknowledged that alkaline species can increase the pH of electrolyte and may reduce the formation of TNTs as TNTs require acidic environment to induced the chemical etching for growth.

However as shown by Taib *et al* [4] and Nyein *et al* [5], the addition of alkaline species did not bring negative consequence to the growth of nanotubes. Alas, the TNTs formed were well crystalline and rather long. It was then concluded, that the growth process is not affected too much by the pH of the electrolyte. Nonetheless, TNTs formed were extremely large with diameter reaching 140 nm. It is believed that the large diameter is perhaps due to the increase in lateral growth of pore occurring at the beginning of the anodic process.

Here, anodic process was done in anhydrous electrolyte for 10 mins with air bubble injected to electrolyte as oxygen provider. After 10 mins, K_2CO_3 was added to electrolyte. First to investigate if the diameter can be reduced by performing this step and two is to see the effect of the carbonate to the adherence of the nanotubes on Ti. As reported by Bashirom *et al* [6], the presence of carbonate species can affect the adherence of anodic ZrO_2 on Zr.

In this work, *kyllinga brevifolia* silver nanoparticles (AgNPS) were synthesized by reduction of *K. brevifolia* extract. The decoration of AgNPs on surface of TNTs was done using wetness impregnation technique. In addition, catalytic activity of AgNPs decorated on TNTs was investigated and compared by degradation of methylene blue (MB). It was found that the presence of silver significantly improved the photo catalytic performance of samples to degrade MB under visible light.

2. Experimental study

TNTs were fabricated by anodizing method. Commercial pure Ti wire (0.50 mm thick, 4 cm length, 99.5%, NILACO, Japan) was used as the anode electrode and platinum as counter electrode in an anodisation bath comprising of fluoride EG electrolyte (97 mL EG with 0.3 wt% anhydrous NH_4F). Prior to anodization, Ti wire was degreased by sonicating in acetone and ethanol for 15 mins each then air dried. Anodisation bath was saturated with air, as to provide sufficient oxygen for oxide growth and 3.0 vol% 1.0 M K_2CO_3 was added at the middle of anodizing experiment. This was to investigate the influence of such salt on the morphology of anodic film as well as on investigating the adherence of the film. The anodization voltage was kept constant at 60 V for 20 mins anodizing time. The anodized Ti wires were then rinsed in deionized water and subsequently annealed at 400 ^{0}C in air for 2hrs to induce phase formation.

AgNPs were prepared by reduction of Ag salts in *K. brevifolia* extract (KBE) as reductant. 5 g of *K. brevifolia* dried powder was soaked in 100 mL deionized water at 70 °C for 1 hr. The extracts were filtered through Whatman No. 1 filter paper. This solution was considered as 100% *K. brevifolia* extract (KBE). 1 mL KBE was added to 9 mL 2 mM silver nitrate solution at room temperature for 90 mins reaction time. Once the reaction was completed, AgNPs were dispersed on TNTs by wet impregnation technique. The as-synthesized TNTs/AgNPs were dried in an oven at 60 ± 3 °C. The morphology of the fabricated TNTs/AgNPs was evaluated by field emission scanning electron microscopy (FESEM, Hitachi S3700N), energy-filtered transmission electron microscopy (EFTEM, PHILIPS CM12) and high-resolution transmission electron microscopy (HRTEM, FEI Tecnai™ G2 20 S TWIN) operated at 200 kV with a LaB_6 filament hybrid with selected area electron diffraction (SAED). The chemical

IOP Conf. Series: Journal of Physics: Conf. Series **1082** (2018) 012105 doi:10.1088/1742-6596/1082/1/012105

composition of the samples has been obtained with energy dispersive X-ray spectroscopy (EDX) using SEM instrument equipped with Bruker Quantax EDX system. The catalytic activity was studied by measuring decolouration of 30 ppm MB in the presence of TNTs/AgNPs and TNTs only. 1 mL 0.1M $NaBH_4$ was also added to the MB solution as to investigate its influence on the removal efficiency (%RE) of MB. 4.0 mL capacity quartz tube was used to contain the solution and catalysts and the tube was left in the dark for 60 mins. Then the quartz tube was exposed to visible light and time-dependent absorption data points of the peak at 664 nm were recorded after every 10 mins.

3. Results and discussion

Figures 1(a) and (b) are FESEM images (surface and cross section) of anodised Ti wire in the presence of air bubble and K_2CO_3. As can be seen similar to the case of KOH [4] or NaOH [5] addition, the presence of basic salt, K_2CO_3 did not hinder the formation of TNTs. As reported by Bashirom *et al* [6] the adherence of the anodic film to the underlying substrate was weakened and can peeled off when adding K_2CO_3 in electrolyte due to evolution of CO_2 gas during anodisation process. But in our case, the TNTs produced have better mechanical stability and were not peeled off from Ti wire. To increase the adherence of TNTs on Ti wire substrate as well as induce the phase formation, the TNTs were annealed at 400 °C in air for 2 hrs. Annealing causes the fluoride rich layer to sublimate and makes TNTS attached to the substrate through a newly formed thermal oxide layer in between [7]. TNTs grown in this condition have length of 7.30 ± 0.46 μm, 74.59 ± 9.22 nm inner diameters and wall thickness 33.51 ± 4.56 nm. Table 1 shows the comparison of length and diameter of TNTs using various conditions of anodisation. It can be seen clearly that the length and diameter of TNTs prepared from this experiment shows better length and diameter with shorter anodisation time.

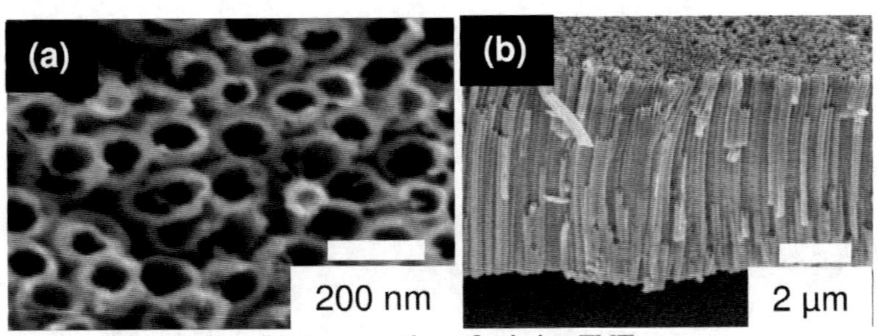

Figure 1. FESEM images of (a) surface (b) cross section of pristine TNTs

Table 1. Comparison of length and diameter of TNTs by various conditions

Electrolyte, voltage, substrate	Time, mins	Length, μm	Inner diameter, nm	Wall thickness, nm	Reference
Eg/NH_4F/KOH, 60 V, Ti foil	30	6.1	50	50	[4]
Eg/NH_4F/NaOH, 60 V, Ti foil	60	23	175	-	[5]
Eg/NH_4F, 20 V Ti wire	30	0.3	60	-	[8]
Eg/NH_4F/K_2CO_3, 60 V, Ti wire	20	7.30 ± 0.46	74.59 ± 9.22	33.51 ± 4.56	Present work

To prepare TNTs/AgNPs, we prepared the AgNPs via salt reduction technique utilising KBE as reductant. TEM image in Figure 2 (a) shows the appearance of the AgNPs produced by KBE. As seen from TEM image, the shape of colloidal AgNPs is quasi-spherical and the particles are highly dispersed. The particle size of AgNPs as obtained from TEM is in the range 10-30 nm with average particles of 22.3 nm as shown in Figure 2 (b). The AgNPs were then dispersed on TNTs by wet impregnation technique. After AgNPs deposition as seen from Figures 3 (a) and (b), the surface morphology and walls

IOP Conf. Series: Journal of Physics: Conf. Series **1082** (2018) 012105 doi:10.1088/1742-6596/1082/1/012105

features have not changed but the existence of AgNPs as spherical particles distributed around the walls of the nanotubes is rather obvious.

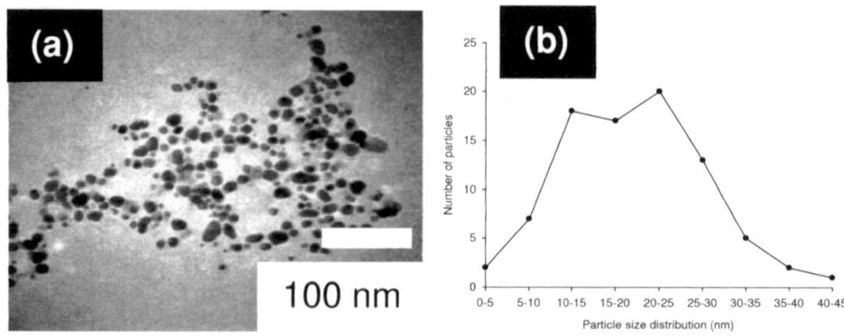

Figure 2. (a) EFTEM image of AgNPs (b) Histogram for particle size distribution of AgNPs

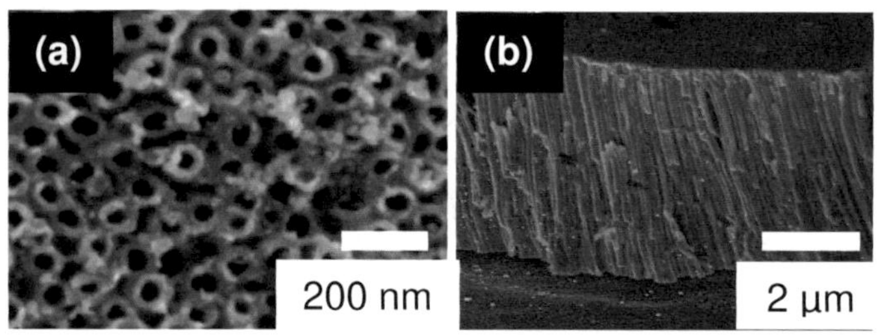

Figure 3. FESEM images of (a) surface (b) cross section of TNTs/AgNPs

In Figure 4, the morphology and crystallinity of TNTs/AgNPs were explored using HRTEM and SAED. Figure 4 (a) displays the HRTEM image of TNTs/AgNPs showing (101) lattice fringe of anatase TNTs (d = 0.35 nm) and AgNPs (d = 0.24 nm). The darker image on TNTs is AgNPs dispersed on the surface of TNTs. SAED pattern as shown in Figure 4 (b) further confirms this observation.

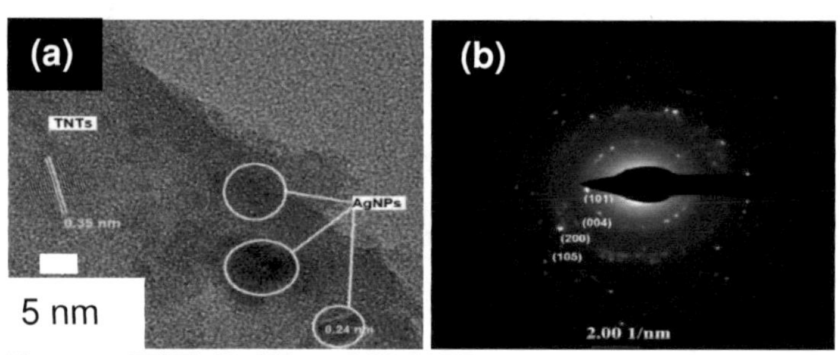

Figure 4. (a) HRTEM images of TNTs/AgNPs and (b) SAED

Figure 5 shows the EDX spectrum for TNTs/AgNPs sample. The peak at 0.277 eV is related to carbon species, whereas the peak at 0.525 eV and 0.677 eV are related to oxygen and fluorine species respectively. The common peak around 0.452 eV belongs to titanium species and peak at 2.984 eV is assigned to silver atom decoration on TNTs. The wt% each element is 2.49%, 39.72%, 21.71%, 1.62% and 34.47% for carbon, titanium, oxygen, fluorine and silver respectively as summarize in insect table in Figure 5.

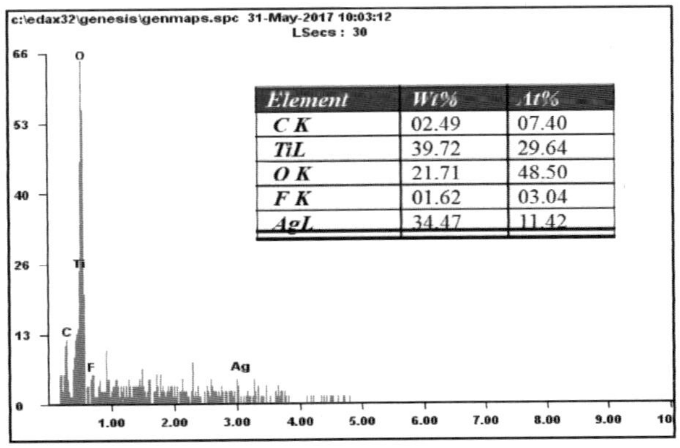

Figure 5. EDX spectrum and data for TNTs/AgNPs sample

For catalytic study, the MB degradation measured by %RE in aqueous solution with $NaBH_4$ was carried out in the presence of TNTs/AgNPs and TNTs (control) under visible light. MB has a main maximum peak at 664 nm due to the azo bond of the dye. As seen from Figure 6, for both samples, the %RE increased very slowly in dark. Thus, no obvious degradation of dye can be observed indicating extremely poor degradation of dye under dark despite the presence of catalysts. When the solution was placed under visible light for another 80 mins, degradation was seen to be much enhanced for the solution with TNTs/AgNPs. The maximum %RE for TNTs/AgNPs is 63% compared to pristine TNTs only around 17%. This indicates the success in forming visible light activated catalysts for dye degradation. The effect of $NaBH_4$ to degradation of MB is as sacrificial reagent (hole scavenger). The positive-holes at valence band (h_{VB}^{+}) of TNTs will consumed through the oxidation reaction with $NaBH_4$ as electron donor irreversibly. The remaining photo-generated electrons will then reduce MB to less harmful substances.

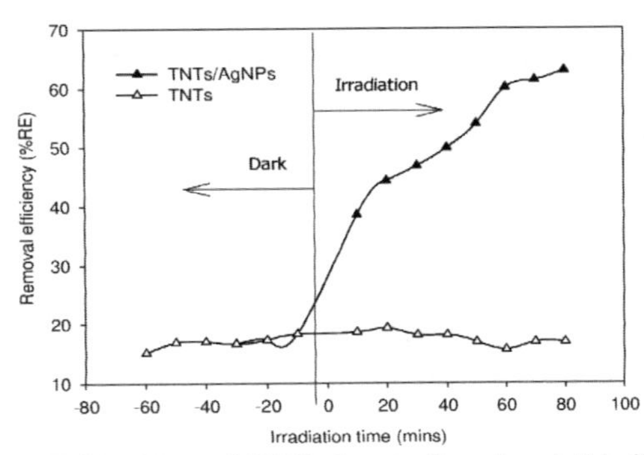

Figure 6. %RE of MB dye by TNTs/AgNPs and TNTs (control) under visible light.

Figure 7 shows the pseudo-first order and pseudo-second order kinetics study of TNTs/AgNPs and TNTs for degradation of MB respectively. The results imply that the photo-degradation of MB obeys the pseudo-first order kinetics. The corresponding linear relationships of ln (C_t/C_o) versus reaction time are shown in Figure 7 and the rate constant, *k*, is calculated from the slopes of the fitted straight lines. The rate constant obtained are 7.5×10^{-3} and 6.0×10^{-5} cm^{-1} for TNTs/AgNPs and TNTs (control) respectively. The improvement progress for TNTs/AgNPs compared to bare TNTs was attributed to the localized surface plasmon resonance (SPR) that enables AgNPs to absorb light in the visible region [9]. Similar findings were reported by Nyein *et al* [5] whereby deposited AgNPs were found to enhance photochemical properties of their nanotubes in visible region of sunlight. According to them, the influence of AgNPs on TNTs are to extend the absorption of light of TNTs in the visible region and help in reducing recombination of photo-generated charge carriers due to effect of SPR produced by oscillation of surface electrons. Table 2 shows the reaction rate constant (k), R^2 value and removal efficiency (%RE) for pseudo-first order and pseudo-second order kinetics of TNTs/AgNPs and pristine TNTs.

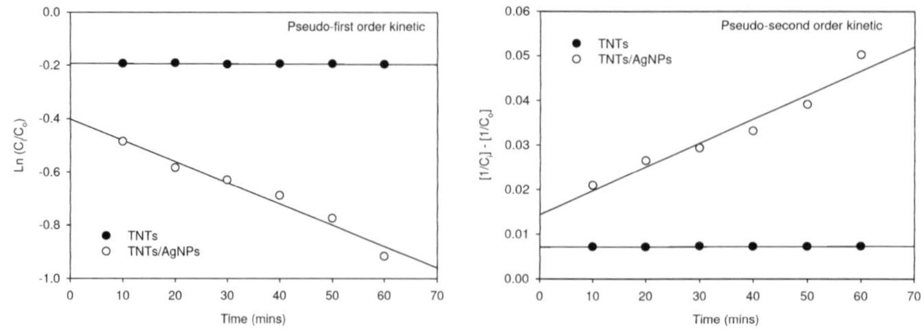

Figure 7. Kinetics study of TNTs/AgNPs and TNTs for degradation of MB under visible light.

Table 2. Pseudo-first, pseudo-second order kinetics and %RE for TNTs

Sample	Pseudo-first order		Pseudo-second order		% RE
	k	R^2	k	R^2	
TNTs/AgNPs	7.5 x 10^{-3}	0.9807	5.0 x 10^{-4}	0.9754	62.97
TNTs	6.0 x 10^{-5}	0.4002	3.0 x 10^{-6}	0.4000	17.77

4. Conclusion

TNTs/AgNPs have been successfully fabricated firstly by anodisation in anhydrous fluorinated/ethylene glycol/carbonated electrolyte and then dispersing AgNPs on the TNT arrays by wetness impregnation technique. The morphology of TNTs/AgNPs shows the TNTs produced highly ordered 7.30 ± 0.46 μm long with 74.59 ± 9.22 nm inner diameters and wall thickness 33.51 ± 4.56 nm. Photocatalytic study shows that TNTs/AgNPs are able to reduce MB with around 63% compared to bare TNTs (control) only 17% under visible light illumination. The reaction rate values for the degradation MB are 7.5 x 10^{-3} 1nd 6.0 x 10^{-5} min^{-1} for TNTs/AgNPs and TNTs respectively. AgNPs has contributed to the activation of the composite under visible light perhaps due to the SPR of the AgNPs as well as electrons transfer process to the conduction band of TiO_2.

5. Acknowledgements

TNTs and environmental application are supported by Research University Grant under TUT-USM collaboration on Heavy Metal Project (1001.PBAHAN.87004). AgNPs work is supported by 600-RMI/RAGS 5/3 (20/2013) and 600-RMI/RACE 16/6/2 (5/2013) grant. KPT-UiTM's SLAB Scholarship for postgraduate scholar is also acknowledged.

References

[1] Abdullah M, and Kamarudin SK, 2017 *Renew. Sustain. Energy Rev.* 76 212
[2] Zhao Y, Hoivik N, and Wang K, 2016 *Nano Energy J.*, 30 728
[3] Kong J, Song C, Zhang W, Xiong Y, Wan M, and Wang Y, 2017 *Superlattices Microstruct.*, 109 579
[4] Taib MAA, Razak KA, Jaafar M and Lockman Z, 2017 *Materials & Design*, 128 195
[5] Nyein N, Tan WK, Kawamura G, Matsuda A and Lockman Z, 2016 *Nanotechnology* 27 35
[6] Bashirom N, Razak KA, and Lockman Z 2016 *Surface & Coating Technol*, 320 371
[7] Miraghaei S, Ashrafizadeh F, Santamaria M, Di Quarto F and Shimizu K 2011 *J. Electrochem. Soc* 11 158
[8] Kar A, Smith YR, and Subramanian V 2009 *Environ. Sci. Technol*, 43 3260
[9] Momeni MM, Ghayeb Y, and Gheibee S, 2017 *Ceram. Int. J.*, 43 564

IOP Conf. Series: Journal of Physics: Conf. Series **1082** (2018) 012106 doi:10.1088/1742-6596/1082/1/012106

Improved Hydrophobicity of Polydimethylsiloxane (PDMS) Coating with ZnO Nanostructures

K.A Saharudin*, S. Sreekantan, N. Basiron and N.S.C. Endut

School of Materials & Mineral Resources Engineering, Universiti Sains Malaysia, Engineering Campus, 14300 Nibong Tebal, Penang, Malaysia

Email: khairularifah@gmail.com

Abstract. In this present work, an environmentally safe hydrophobic cotton surface comprising of PDMS and different morphology of Zinc oxide (ZnO) nanostructures via simple dip coating method was developed. ZnO nanosphere and ZnO nanoflakes were synthesized via sol gel method while ZnO nanorods were synthesized via solution precipitation method. The precursors used to prepared ZnO nanospheres, nanoflakes and nanorods are zinc acetate dehydrate, zinc chloride and zinc nitrate tetrahydrate respectively. ZnO nanostructures were analysed using FESEM and XRD. The hydrophobic cotton surface was characterized via FT-IR and WCA analysis. A WCA of 144.15° was achieved when coating the cotton with ZnO nanoflakes (ZnO-NF:PDMS).

1. Introduction

Hydrophobicity is the physical property that provides water repellence and non-wettability of a solid surface. The hydrophobic coating has been achieved in many ways with different kinds of materials by mimicking the surface roughness of lotus leaves and low surface energy. The surface structure of the leaves enables entrapment of air between water droplets and the surface [1]. Meanwhile, low surface energy attained from non-polar molecules. This contributes to less wetting and adherence between the water droplets and the surface, thus creating a highly water repellent surface.

The attractive features of hydrophobicity behaviour on solid surface clearly is very important and can be exploited in a variety of applications including in electronic, medical, paint coating, textiles and many other industries. Self cleaning glass, anti-corrosion coatings and waterproof textiles are some of the potential areas for its application[2]. Typically, fabricating hydrophobic textiles involve the treatment of the textile with selected low surface energy material purposely to prevent properties changes of the textiles[3].

Various research has been directed in several decades towards the creation of textiles which are water repellent, dust resistant and anti-bacterial. However, its application as a hydrophobic barrier to replace paraffin wax in batik making has not been explored yet. Research in substituting paraffin wax is essential to reduce large volume of solid suspended discharge from paraffin wax that cause clogging issue on batik waste water treatment system. Therefore, in this present work we attempt to study the effect of ZnO with different morphology to promote the bonding interactions. The solvated zinc acetate acts as a "bridge" between the textile substrate and PDMS by forming hydrogen bonds.

IOP Conf. Series: Journal of Physics: Conf. Series **1082** (2018) 012106 doi:10.1088/1742-6596/1082/1/012106

2. Experimental study

2.1 Materials

Zinc acetate dihydrate (183.48 g/mol, 99.8%), Zinc nitrate tetrahydrate (261.44 g/mol, 99.8%), Zinc chloride (136.315 g/mol, 99.8%), hydroxy terminated poly(dimethylsiloxane) (PDMS), Hexamethylenetetramine (HMT), Polyvinylpyrrolidone (PVP) were purchased from Sigma-Aldrich. Ethanol, Potassium hydroxide (KOH) and n-hexane were purchased from Merck Millipore. All the chemicals used were of AR grade and used as received without further purification. Textile/cloth was purchased from Ayu Fashion Sdn. Bhd.

2.2 Synthesis of ZnO spherical particles

In this experiment, 5.2 g of zinc acetate dihydrate [$Zn(CH_3COO)_2.2H_2O$ (99.5%)] was dissolved in 200 ml deionized water (DI) and heated at 65 °C under constant stirring. Simultaneously, 2.96 g of potassium hydroxide (KOH) solution was prepared in 100 ml deionized water. Then, appropriate amount of potassium hydroxide solution was added drop wise with this zinc acetate dihydrate solution under vigorous stirring at 1200 rpm at 65 °C. This mixtures solution was stirred continuously at 65°C for 1 h to achieve the white colour milky suspension. Now the product was centrifuged several times in ethanol for filtration and dried at 100 °C for 2 hours in normal oven for characterization.

2.3 Synthesis of ZnO rod particles

ZnO rods were synthesized via solution precipitation method using zinc nitrate tetrahydrate ($Zn(NO_3)_24H_2O$, Merck), hexamethylenetetramine (HMT, $(CH_2)6N_4$, Merck) and polyvinylpyrrolidone (PVP). In a typical synthesis, 3.14 g of $Zn(NO_3)_2.4H_2O$ in 100 ml of de-ionized water (solution I), 0.56 g of HMT in 100 ml of de-ionized water (solution II) and 1 g of PVP in 100 ml of de-ionized water (solution III) were prepared separately. The solution was stirred for 30 min at 90 ºC. Subsequently, solution I and solution II were added into the solution III under constant stirring rate at 90 ºC. The mixture of solution was heated to 90 ºC in a water bath. White precipitate was found in the solution. Centrifugation was carried out at 3000 rpm for 20 min in order to separate the precipitate from the solution. The precipitate was washed by de-ionized water for several times. The precipitate was dried in the oven for overnight at 100 ºC.

2.4 Synthesis of ZnO flake particles

In this experiment, 5.45 g of zinc chloride ($Zn(Cl_2)_2 \cdot 4H_2O$) was dissolved in 100 ml of ethanol and kept under constant stirring using magnetic for one hour. 4.49 g of Potassium hydroxide (KOH) was also prepared in 100ml with stirring of one hour. After complete dissolution of zinc chloride, KOH aqueous solutions were added under high speed constant stirring drop by drop for 45 min. The beaker was sealed at this condition and the solution was allowed to settle for overnight. Further, the supernatant solution was separated carefully. The remaining solution was centrifuged for 20 min and the precipitate was removed. Precipitated ZnO particles were cleaned three times with de-ionized water and ethanol then dried in oven for 8 hours at 60°C.

2.5 Preparation of hydrophobic coating

PDMS and hexane were used as the raw materials. The volume required of PDMS and hexane was measured. PDMS was dissolve in hexane that act as solvent. The PDMS-hexane dissolved with ratio of 1:3 (ml) using a magnetic stirrer for 30 minutes. Different amount of ZnO particles (0.02g, 0.05g, 0.1g, 0.30g, 0.50 g) was added in the PDMS solution and ultrasonicated separately for 10 minutes. Optimum amount of ZnO particles were varies with different morphology of spherical, rod and flake particles. Homogeneous solution typically occurred within 8 hours after stirred at heating temperature 60 ºC.

Before the coating, cloth of 3 cm x 3 cm was cleaned with ethanol and deionizes water. The pre-cleaning process is compulsory to remove possible impurities before drying for further processing. The cleaned cloth was then dip-coated into the hydrophobic solution at a controlled rate of 5 mm/s. The coatings were allowed to cure at room temperature for 1 hour and then dried in oven at 100 °C for 48 h.

2.6 Characterization

Fourier Transform Infrared (FTIR, Perkin Elmer) spectroscopy was used to investigate the functional groups of super-hydrophobic coating. The morphology was examined with a help of field emission scanning electron microscopy (FESEM-EDX, Zeiss, Supra 35VP). The crystalline phases of ZnO were identified using a Philips PW1729 powder X-ray diffractometer (XRD). The water contact angle (WCA) was measured by the sessile drop method with a droplet volume of 5 μL using a goniometer (Rame hart Instrument. Co, USA) on one side of the sample. Briefly, the samples were placed in a Contact Angle Goniometer, attached to an Image analyzer. Each sample was subjected to 10 measurements in each 4-angle position: vertical left, vertical right, horizontal left, horizontal right. DROPimage Advanced software was used to obtain the WCA.

3. Results and discussion

The surface morphologies of ZnO particles were investigated by FESEM. The ZnO particles which synthesized by using sol-gel method via zinc acetate dehydrate and potassium hydroxide, exhibited quasi-sphere structure as shows in Figure a. It is found that the particles are well dispersed and have a regular quasi-spherical with average diameters of 60.66 nm. Figure 1b show the SEM image of the synthesized ZnO flake particles by using zinc chloride as Zn^{2+} and potassium hydroxide as OH^- to growth the ZnO. The edge length of the nanoflakes are vary from 370 nm to 840 nm, with the average thickness of 61.91 nm. Meanwhile, Figure 1c shows the representative FESEM image of ZnO rod particles which synthesized by precipitation method using zinc nitrate tetrahydrate. The sample composed of rod-like structure with diameter ranging from 140 nm to 340 nm and average length of 2.51μm.

Figure 1. FESEM images of the as-synthesized samples in different morphology (a) ZnO spherical particles, (b) ZnO flake particles and (c) ZnO rod particles: inset showing the stacking rod

The crystal structure of various ZnO was characterized by XRD in order to determine the crystallinity of the as-synthesized ZnO particles. The diffraction peaks at 31.79°, 34.43°, 36.26°, 47.56°, 56.62°, 62.85°, 66.38°, 67.96°, 69.08°, 72.55°, and 76.95° are corresponded to (100), (002), (101), (102), (110), (103),(200), (112), (201), (004) and (202) crystal planes of hexagonal wurzite structure. No other diffraction peak from ZnO could be found, indicating high purity of ZnO formation. Although the obtained ZnO shows similar crystal structure of hexagonal ZnO, there are some differences with the lattice constant. For ZnO spherical, the wurzite structure match ICDS of 01-075-0576 with lattice constant of a = 3.2477 Å and c = 5.1948 Å, for ZnO flake particles match with ICDS of 01-079-0205 with lattice constant of a = 3.2499 Å and c = 5.2066 Å while for ZnO rod particles match with ICDS of 01-079-0205 with lattice constant of a = 3.2417 Å and c = 5.1800 Å.

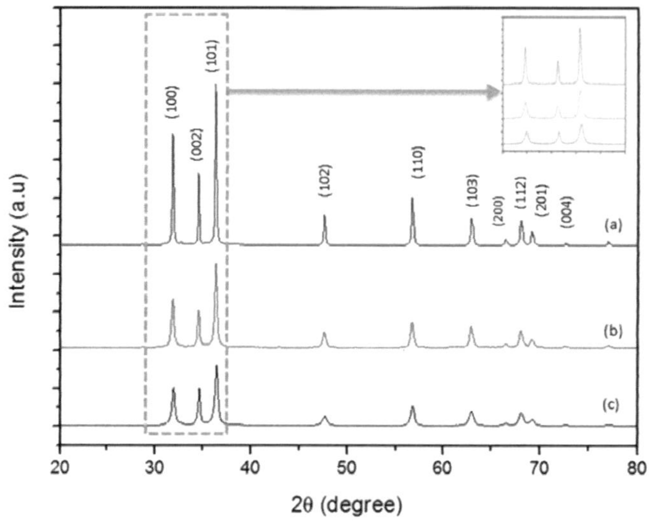

Figure 2: XRD patterns of the as-synthesized samples in different morphology (a) ZnO rod, (b) ZnO flakes and (c) ZnO spherical

As seen in Figure 2a, the peak intensity of rod structure is higher than the rest of the sample, indicating better degree of crystallinity. The (002) and (100) planes for rod appear to be stronger than the other sample, indicating rods grows along the direction of polar (002) that surrounded by six (100) planes[4], preferentially. This is in agreement with the FESEM image which shows the growth of the rod along *c*-axis. Besides (002) and (100) plane, (101) appear to be intense for rod probably due to non-uniform alignment of the rods perpendicular to the substrate (XRD sample holder). As demonstrated in Figure 2b, the interfaces are found mainly in the middle of the ZnO particles, indicating that each particles at least consists of two rods that stacked in their (002) polar surface. This result is in agreement with the work reported by Thuya and Pung, 2015[4], which demonstrated the role of Zinc nitrated tetrahydrate as zinc precursor as charged positively (Zn^{2+} terminated) and HMT which provides hydroxide ion (OH-) as charged negatively (O^{2-} terminated)[4]. In zinc-HMT growth process, these ions are tends to stacked together on another polar surface (002) plane in order to minimize the total surface energy. Such a process is repeated over time, leading to a fast growth of ZnO nuclei along the ±[0001] directions. As a result, ZnO rods will be continue synthesized if no complexing agent is added to alter the preferential growth direction of ZnO in [0001][5]. In order, to overcome the fast growth of ZnO, the PVP was adding into the solution as capping agent or stabilizer to minimize the issues of agglomeration and secondary nucleation on the surface of ZnO micronrods. As shown in Figure 2c, the peak intensities of (002) and (100) plane for spherical particles are similar while (101) plane are more intense, indicating preferential orientation of hexagonal ZnO[5].

Hydrophobic solution were analysed by FTIR to determine functional group of ZnO particles and ZnO/PDMS hydrophobic solution. Figure 3a, 3b, and 3c show the FTIR spectra of ZnO, PDMS and ZnO/PDMS hydrophobic solution in the range of 4000-400 cm^{-1} respectively.

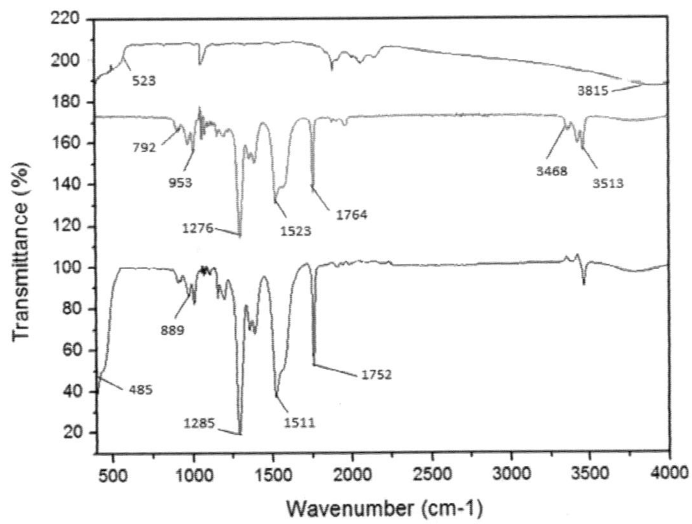

Figure 3: FTIR spectra of (a) ZnO particles, (b) PDMS solution and (c) PDMS/ZnO hydrophobic solution

From the FTIR spectrum, it was observed that ZnO particles peak at 3815 cm^{-1} is attributed to O-H stretching band vibration, assigned to small amount of adsorbed water on the ZnO surface[6]. A strong peak at 523 cm^{-1} is ascribed to Zn-O stretching vibration. For PDMS hydrophobic solution, strong peaks at 3468 $cm^{-1}$and 3513 cm^{-1} correspond to the intense anti-symmetric and symmetric –CH_3 group stretching confirming the presence of Si-$(CH_3)_3$ and Si-O-CH_3 groups. A series of absorption peaks from 1000 to 2000 cm^{-1} are found corresponding to vibration of various functional group of the silane compound such as Si-O in Si-O-Si back bone and Si-CH_3 vibration present on the PDMS surface. The absorbance peaks between 953 cm^{-1} to 792 cm^{-1} represent the vibration of the CH_3 at $Si(CH_3)_2$.

Meanwhile, PDMS/ZnO spectrum was observed peak at 2963 cm^{-1} is attributed to O-H stretching band vibration, assigned to small amount of adsorbed water on the ZnO surface. A series of absorption peaks from 1000 to 2000 cm^{-1} are found corresponding to vibration of various functional group of the silane compound such as Si-O in Si-O-Si back bone and Si-CH_3 vibration present on the PDMS surface. A strong peak at 485 cm^{-1} is ascribed to Zn-O stretching vibration. Other than that, a weak peak at 889 cm^{-1} was observed in PDMS-ZnO spectra, which could be originated from Zn-O-Si. This observation gives an idea that ZnO particles reacts also with some of the residual silanol (Si–OH) groups. The super-hydrophobic properties of a coating are greatly affected by the existence of these methyl group. From this spectra, it can be inferred that the PDMS/ZnO contain methyl group, thus improvement of hydrophobicity behaviour is observed.

Figure 4 shows the overall result of WCA with ZnO different morphologies. PDMS/ZnO-flake particles, PDMS/ZnO-rod particles and PDMS/ZnO-spherical particles coated on cotton cloth is more hydrophobic as the water contact angle increased to 144.15°, 139.65° and 141.15° respectively as compared to without ZnO particles. It has proved that ZnO morphology studied which are spherical, rods and flakes has influences wettability properties ZnO/PDMS coating. The differences in water contact angle of these coatings can be attributed to different topography induced by the different morphology of the ZnO particles. Morphology of ZnO particles has improved the wettability by following Wenzel state where the hydrophobicity was improved by increasing in surface roughness. However, due to non-perpendicular alignment of ZnO rod particles, the wettability has decrease followed the stable Wenzel state where the liquid fills between the rough structures. Remarkable WCA can be observed from results which show the incorporation of ZnO particles with flake shape produced better roughness surface, supporting Wenzel model, thus lead to an increase in static contact angle, 144.15° compared to other structures.

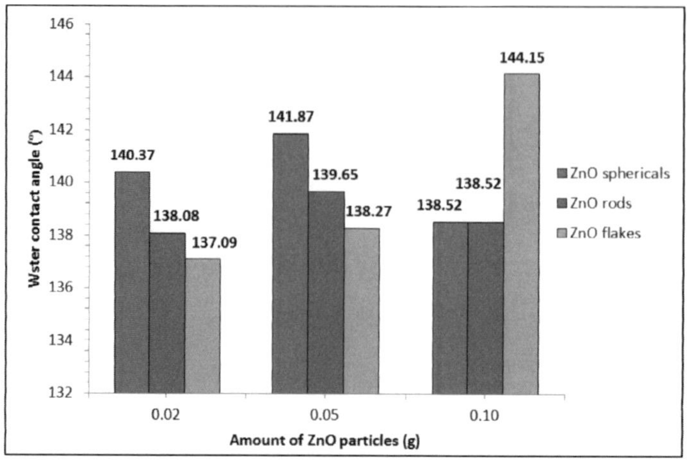

Figure 4: Overall WCA result with different morphology

4. Conclusion

In this project, ZnO particles with different morphology have been successfully synthesized using different precursor by sol-gel and solution-precipitation method. XRD and FESEM characterization shows that ZnO particles has good crystallinity and morphology structure of quasi-sphericals, rods and flakes. WCA of ZnO-NF:PDMS shows the highest WCA 144.15° due to rough surface morphology which followed Wenzel theory.

Acknowledgement
The authors are thankful to the Ministry of Education (MOE) Malaysia for funding this work under Fundamental Research Grant Scheme (FRGS) grant no. 6071319. The authors are very much grateful to Universiti Sains Malaysia (USM) for providing the necessary facilities to carry out the research work and financial support under Research University (RU) grant no. 814281.

References
[1] Ramaratnam K, Iyer SK, Kinnan MK, Chumanov G, Brown P, Luzinov I. 2008 *AATCC review* 8
[2] Yazdanshenas ME, Shateri-Khalilabad M. 2013 *Industrial & Engineering Chemistry Research* 52 12846
[3] Bhattacharyya R. 2009 Technological Applications of Superhydrophobic Coatings: Needs and Challenges: Indian Institute of Technology, Bombay
[4] Thein MT, Pung S-Y, Aziz A, Itoh M. 2015 *Journal of Sol-Gel Science and Technology* 74 260
[5] Xu S, Wang ZL. 2011 Nano Research 4 1013
[6] Chakradhar R, Kumar VD, Rao J, Basu BJ. 2011 *Applied Surface Science* 257 8569

IOP Conf. Series: Journal of Physics: Conf. Series **1082** (2018) 012107 doi:10.1088/1742-6596/1082/1/012107

Corrosion Behaviour of Hastelloy C22 in Alkali Salt Environment

A.K. Syamimi[1], L.S.Yeian[1], and T.K.Abdullah[1*]

[1] School of Material and Mineral Resources Engineering, Universiti Sains Malaysia, Nibong Tebal, Penang, Malaysia

Email: tutikatrina@usm.my

Abstract. Chloride-induced corrosion at elevated temperature is the major drawback of commercialization of biomass fired power generation. The incapability of the biomass-fired boilers to increase the temperature has resulted in a lower efficiency of the power plant. Hastelloy C22 is one of the potential materials for superheater due to the high content of Ni-Cr which provides an excellent corrosion resistance. The aim of this work is to obtain more information on corrosion behaviour of Hastelloy C22 at high temperature in alkali salt environment with the influence of air humidity, hence to justify the suitability of this alloy as a superheater material for higher operating temperature. Corrosion tests were performed in the synthetic salt containing 80 wt% KCl and 20 wt% K_2SO_4 under normal and dry air conditions. The test was conducted at temperature 650°C for 218 hours. The surface morphology, elemental distribution and the corrosion products were characterized using Field Emission Scanning Electron Microscope (FESEM) equipped with Energy Dispersive X-ray (EDX). The phase identification was carried out using X-ray Diffraction (XRD) analysis. The results revealed a significant effect of humidity on the corrosion behaviour of the alloy where the presence of air moisture reduces the protective capability of the chromium oxide (Cr_2O_3) scale. A formation of unstable chromium (IV) hydroxide as well as the volatilization of Cr_2O_3 led to the oxide spallation, thus reducing the corrosion resistance of the superalloy.

1. Introduction

Due to the tremendous increase in CO_2 level in the atmosphere that contributes to the global warming, the use of renewable energy has been gaining its importance nowadays. Renewable energy sources that claimed to be CO_2 neutral-like biomass are currently becoming the substitutes for fossil fuel in many power plants. Biomass resources are readily available which make it much more preferable resources compared to others. In real combustion of waste, some types of biomass release high level of salts compounds such as HCl, NaCl and KCl [1]. At high operating temperature in biomass processor, the presence of this complex alkali chloride is subsequently deposited on metallic surface or already-formed oxide layer led to degradation and corrosion damages of construction material which are required a constant repair [2].

Hastelloy C22 is a well-known nickel-based superalloy mostly studied by researchers as a material used for high temperature application [3–7]. The capability of Hastelloy C22 to offer an excellent performance under extreme condition, enables biomass boilers to operate at high temperature which in turns increase the efficiency of biomass fired plants. At elevated temperature, high nickel and chromium content of Hastelloy C22 helps it to resist oxidation, carburization sulfidation and halides attack [7-8]. Corrosion resistant of metal alloys depends on the chemical stability, arrangement, compactness, strength, plastic and thermal expansion coefficient of the corrosion product [10]. An oxide layer such as chromia (Cr_2O_3) is believed to have an ability to protect the alloys by a formation of a stable and dense oxide layer to avoid further corrosive media contact [11].

In this studies, Hastelloy C22 is proposed as a potential material for biomass boilers due to its excellent resistance to corrosion attack [12]. The experimental support is needed to justify the lifetimes and performance of Hastelloy C22. Therefore, the aims of this work are to investigate the influence of humidity on corrosion tendency of Hastelloy C22 and to study the corrosion mechanism of Hastelloy C22 with a presence of synthetic salt in normal and dry conditions.

2. Experimental study

2.1 Sample preparation

The compositions of materials used in this study are given in Table 1. As received Hastelloy C22 were cut into square approximately to 10x10x2 mm. The samples were then ground using silica carbide grinding paper from 300 grit up to 1000 grit, followed by polishing using alumina powder from 1 μm, 0.3 μm and 0.05 μm. Then, all samples were further cleaned in ethanol using ultrasonic bath for 10 minutes and dried to avoid any contamination.

Table 1. Chemical composition of test material Hastelloy C22 (ASTM B-575-97, UNS06022)

Element	Ni	Cr	Mo	W	Fe	Co	Si	C	V	Mn	P	S
wt %	57.36	21.20	13.50	2.90	3.80	0.93	0.03	0.003	0.014	0.25	0.007	0.004

2.2 High temperature corrosion test in salt

A mixture of 80% potassium chloride (KCl) and 20% of potassium sulphates (K_2SO_4) were being used in this research to represent the main corrosive species in superheater tubes for biomass combustion. Experiments were run at 650°C with an exposure time 218 hours for both dry and normal air conditions. Air for dry condition was dried by passing it through an alumina combustion boat containing silica gel. Then, the samples were placed inside a quartz glass tube positioned in the furnace to ensure that the atmosphere remained as dry as possible.

A small amount of synthetic salt was deposited on the surface of each samples by taping it gently with a glass rod. Exposure was carried out in temperature-controlled horizontal quartz glass tube furnace. Samples were raised to the final exposure temperature at a rate of 5°C min^{-1}. Under both condition, air flow was fixed at 4 dm^3 min^{-1} positioned horizontally, parallel to flow direction. After 28 hours, specimen was withdrawn from the furnace, cooled in air, cleaned with soft brush and washed in boiling water to remove the salt left on the surface. Specimen was then dried in hot air and placed in a desiccator to be ready to undergo characterization.

2.3 Characterization

The surface morphology of corroded samples after exposure in synthetic salt was observed by Scanning Electron Microscope (SEM) in Secondary Electron (SE) mode. Chemical compositions were determined by Energy Dispersive X-ray Spectroscopy (EDX). Phase identification of the corrosion product was conducted by X-ray Diffraction (XRD) meter with Cu Kα incident radiation (λ = 0.15406) operating at 2θ from 10° - 90°.

3. Results and discussion

3.1 Oxidation behaviour of Hastelloy C22 in normal and dry atmosphere

The oxidation behavior of Hastelloy C22 at temperature 650˚C has been discussed according to the type of atmosphere they were being exposed. Under the corrosive condition, Hastelloy C22 was commonly coated with oxide films which formed as a result of interaction between Hastelloy and molten salt mixture. Fig 1 shows the SEM micrograph of corroded Hastelloy C22 specimen in chloride salt under dry and normal atmospheres at different magnifications. Corrosion product of the specimen under normal air appears to be much colloidal-like tiny structure and granular-like structure on the surface area (Fig 1 (a)). While the corrosion product appears under dry air specimen seems to be an agglomeration of granular-like structure (Fig 1 (c)). The corrosion products formed under dry air condition is relatively more uniform and spread evently across the specimen surface as compared to normal air condition.

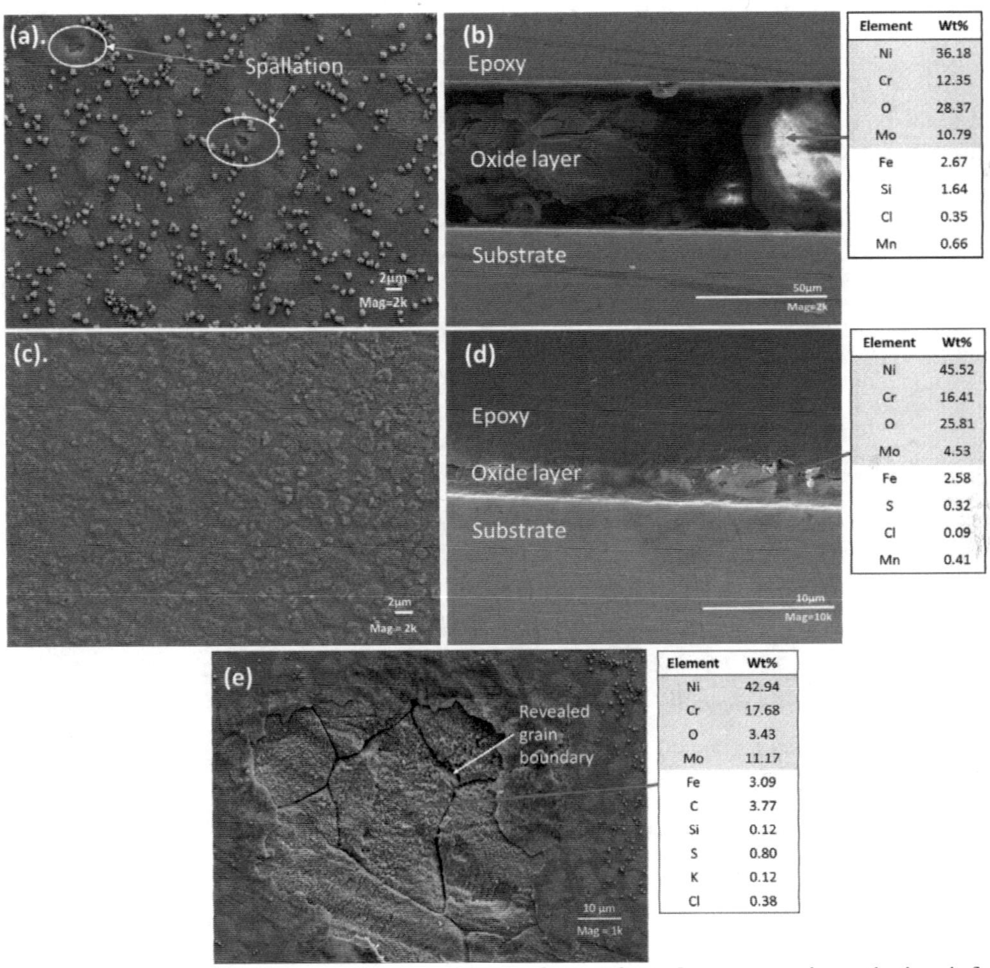

Element	Wt%
Ni	36.18
Cr	12.35
O	28.37
Mo	10.79
Fe	2.67
Si	1.64
Cl	0.35
Mn	0.66

Element	Wt%
Ni	45.52
Cr	16.41
O	25.81
Mo	4.53
Fe	2.58
S	0.32
Cl	0.09
Mn	0.41

Element	Wt%
Ni	42.94
Cr	17.68
O	3.43
Mo	11.17
Fe	3.09
C	3.77
Si	0.12
S	0.80
K	0.12
Cl	0.38

Fig. 1 SEM micrograph with elemental analysis of corrosion product after exposure in synthetic salt for 218 h with normal atmosphere at magnification of (a) 2000x (surface area), (b)2000x (cross section), dry atmosphere at magnification of (c) 2000x (surface area) (d) 10000x (cross section) and (e) Spallation of the sample under normal air at 1000x magnification

At certain area, specimen under normal air shows a spallation (Fig 1 (a) and (e)) due to bad adhesion between the oxide layer and the substrate. This scenario mostly happened when the corrosion products on the surface are not thermodynamically stable. Corrosion layer are fragile and alloy will be further exposed to the corrosion media and oxide started to form again. The corrosion occurs and the cycle continues until the grains being exposed.

Thickness of the oxides formed were measured by cross sectional morphology of the specimen. Corrosion layer of a specimen under normal air is thick (61.13 µm) and porous (Fig 1 (b)) compared to the specimen under dry air which is very thin (2.87 µm) and dense (Fig 1 (d)). This is due to the presence of air moisture and synthetic salt at high testing temperature that accelerates the corrosion reaction [13]. Thus, the oxide growth thicker for a specimen under normal air condition compared to dry air condition. However, a porous structure formed under normal air condition is unstable and cannot provide a protection to the alloy.

EDX result reveals that the corrosion product formed is constituted of Ni, Cr, Mo-rich oxides (Fig (1)). However, the results from XRD data only reveals the main peaks of NiCr and protective Cr_2O_3 peaks (Fig 2). The phases might be difficult to be detected due to the low crystallinity and low amount of oxide formed on the specimens. The peaks of protective Cr_2O_3 is slightly lower for the normal atmosphere due to the dissolution and volatilization of some Cr-rich oxides in the outer layer under presence of air moisture [12]. Hence, the oxide spallation occurs and XRD detected mostly the newly exposed substrate on the surface layer of the normal specimen. It has been proved by EDX data of the exposed grain boundary (Fig 1 (e)), which composed of lack amount of oxygen and high amount of nickel followed by chromium, molybdenum and others alloying elements.

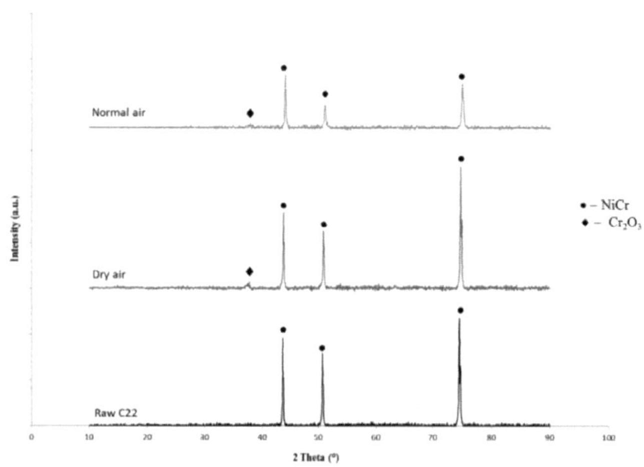

Specimen	Phase
Normal air	NiCr, Cr_2O_3
Dry air	NiCr, Cr_2O_3
Raw C22	NiCr

Fig 2 XRD pattern of corroded Hastelloy C22 in comparison with raw Hastelloy

3.2 Corrosion mechanism

The reaction between Cr_2O_3 layer with water vapour and oxygen at temperature above 400°C formed a volatile species of Cr(VI) hydroxide as shown in Eq. 1 which causes the loss of protective properties of the specimen under normal condition [6,11]. This phenomenon led to the chromium depletion of the oxide and in the alloy substrate, hence causing the breakdown of protective chromium-rich corundum type oxide. The present study is in accordance with the previous work [13-14], indicating that vaporization of Cr(VI) hydroxide does take place in oxygen with air and water environment, resulting in the depletion of the oxide in chromium.

The important features that contributes to the corrosion resistance of a certain material is its ability to form a stable protective oxide scales [7]. However, under chlorine-containing environment and a presence of water vapour, corrosion process is very likely to be accelerated. This resulting in increased oxidation and loose non-adherent scales that weaken and destroy the protective oxide scales. The corrosion scales are dominated by KCl due to its high corrosivity. Meanwhile, K_2SO_4 was believed to preserve the protective character of oxide scale, by avoiding the diffusion of gaseous chloride species to the metal surface and preventing the depletion of chromium within the oxide layer [15-16].

At high temperature around 650°C - 750°C, the synthetic salt are partially melted. The corrosion process is dominated by electrochemical corrosion as the liquid phase formed acts as a pathway for ionic transport. Furthermore, corrosion may be further enhanced when the protective surface oxide scale is dissolved by molten salt that are partially in liquid. Acceleration corrosion happened for oxygen as well as other aggressive corrosion species can get in and diffuse to the alloy substrate. Destruction of oxides at splat boundaries by chloride and the diffusion of chlorine along grain boundaries became the main factors that contributes to the failures of the Hastelloy. Consider the reactions in Eq. 2:

$$2Cr_2O_3(s) + 3O_2 + 4H_2O\ (g) \rightarrow 4CrO_2(OH)_2 \qquad \textbf{Eq. 1}$$

$$2RCl\ (s, l) + \frac{1}{2}Cr_2O_3(s) + \frac{5}{4}O_2\ (g) \rightarrow R_2CrO_4(s, l) + Cl_2 \qquad \textbf{Eq. 2}$$

Where RCl corresponds to KCl in this studies. Alkali chloride also can react with other chlorides to form low temperature melting eutectic compounds. For instance, $KCl\text{-}CrCl_2$ = 462-475°, $KCl\text{-}FeCl_2$ = 340-393°C, $KCl\text{-}FeCl_3$=202-220°C [19]. The compounds tend to volatile easily and help to transfer the alloy outwards which in turns accelerates the corrosion. Cr is the main protector for high temperature corrosion by forming a protective Cr_2O_3 layer. However it has lost its advantages when facing the water vapour and molten chloride salts. Meanwhile, NiO is not liable to be dissolved as compared to Cr_2O_3. This is mainly associated with the preferential dissolution of Cr_2O_3 that consumes oxygen and inhibits the dissolution of NiO. This preferential dissolution happens to MoO_2 as well which leave a NiO skeleton in the corrosion product [18]. This explains the occurrence of high amount of nickel and oxygen after the corrosion test.

4. Conclusion

The high temperature corrosion of Hastelloy C22 was studied at temperature 650°C in synthetic salt under two different atmospheres for 218 hours. The corrosion under normal air conditions was more severe than dry air. High test temperature with a presence of air moisture and synthetic salt aggravated the corrosion via preferential dissolution and formation of volatile hydroxide species. This volatile species converts the protective chromium oxide into a poorly protective oxide, hence reducing its protective capability.

The comparison of corrosion behaviour between normal and dry air conditions reveals that the presence of air moisture at high temperature increases the corrosion attack. However, a dense oxide structure that remains on dry air specimen proves that the Hastelloy C22 has a tendency to form a protective oxide layer. In future studies, attention should be drawn to the development of protective oxide layer prior to the corrosion test in order to improve the corrosion resistance of the alloy in molten salt at high temperature.

References

[1] Y. S. Li *et al.* 2014 *Corros. Sci.* 2014 64
[2] F. Abe, H. Kutsumi, H. Haruyama, and H. Okubo 2016 *Corros. Sci.* 114 1
[3] A. C. Lloyd, J. J. Noël, N. S. Mcintyre, and D. W. Shoesmith 2005 *JOM* 31
[4] M. A. Rodríguez and R. M. Carranza 2011 *J. Electrochem. Soc.* 158 C221
[5] O. Hedayati, N. Korei, M. Adeli, and M. Etminanbakhsh 2017 *J. Alloys Compd.* 712 172

[6] Q. Wang, Y. Zhang, S. Bai, and Z. Liu 2013 *J. Alloys Compd.* 553 253
[7] A. Fattah-alhosseini, A. Jalali, and S. Felegari 2015 *Arab. J. Sci. Eng.* 40 2985
[8] M. A. Rodríguez, R. M. Carranza, and R. B. Rebak 2005 *Metall. Mater. Trans. A* 36 1179
[9] W. Luo, Z. Liu, Y. Wang, and R. Yang 2012 36 212
[10] X. Tang, S. Wang, D. Xu, Y. Gong, J. Zhang, and Y. Wang 2013 *Ind. Eng. Chem. Res.* 52 18241
[11] S. Tuzi, H. Lai, K. Goransson, M. Thuvander, and K. Stiller 2017 *J. Nucl. Mater.* 486 350
[12] S. Liu, Z. Liu, Y. Wang, and J. Tang 2014 *Corros. Sci.* 83 396
[13] M. P. Brady *et al.* 2017 *Energy Sustain. Dev.* 37 20
[14] N. K. Othman, J. Zhang, and D. J. Young 2010 *Corros. Sci.* 52 2827
[15] E. J. Opila *et al.* 2007 2 1971
[16] X. Zhang and D. W. Shoesmith 2013 *Corros. Sci.* 76 424
[17] R. Altobelli, M. Cristina, and L. De Oliveira 2013 *Corros. Sci.* 76 6
[18] J. Pettersson, H. Asteman, J. E. Svensson, and L. G. Johansson 2005 *Oxid. Met.* 64 23
[19] G. J. Janz, C. B. Allen, J. R. J. Downey, And R. P. T. Tomkins 1978 *Chem. Informationsd.* 9 16

The Influence of Nitrogen on The Wear Resistance of Ferritic Stainless Steel

Nsikan Etim Dan, Siti Noor Syahirah Mohd Sabri, *Patthi Hussain and Hamed Mohebbi

Mechanical Engineering Department, Universiti Teknologi PETRONAS, 32610 Bandar Seri Iskandar, Perak Darul Ridzuan, Malaysia

*E-mail: patthi_hussain@utp.edu.my

Abstract. Ferritic stainless steel has low properties of wear resistance and hardness. Normal heat treatment on AISI 430 Ferritic Stainless Steel cannot be treated utilised to achieve this task. However, it can be treated through nitriding method. Nitrogen is well known to improve mechanical, wear and corrosion resistance of stainless steels. Normal heat treatment on AISI Ferritic Stainless Steel is not suitable to achieve this task. Thus, the aim of this work is to investigate the influence of nitriding on wear resistance properties of AISI 430 Ferritic Stainless Steel. The influence of nitriding on the steel were observed. Microstructures, hardness test and scratch resistance analysis for wear resistance were conducted and analysed. Nitriding was carried out at 1200°C for 1 h, 3 h and 5h using nitrogen gas as a medium. Martensitic transformation from ferrite was observed on the as-received samples. From microstructure observations, the influence of nitriding also could be seen. Martensite structure shows higher hardness compare to ferrite structure. Therefore, the hardness for 5 h nitrided sample is higher than 1 and 3 h nitrided sample as more martensite structures was formed at 5 h nitrided sample. Scratch resistance was conducted to analyse the influence of nitriding for wear resistance properties. The depth of penetration obtained shows how much improvement on tribology properties–are made. Results from testing conducted shows that the improvement of wear resistance properties is dependent on nitriding time. The longer the nitriding time, the more nitrogen is diffused into the sample. As more nitrogen atoms diffused into the steel, improvement on wear resistance properties are higher.

1. Introduction

High-temperature gas nitriding or solution nitriding was introduced as a method of adding nitrogen to stainless steels [1-3]. This method involves a diffusion process for nitrogen to permeate the surface of stainless steel through heat treatment in N_2 atmosphere at elevated temperature. There are three types of nitriding; gas nitriding, liquid/salt bath nitriding and ion/plasma nitriding. In this case study, the gas nitriding process to nitride the stainless steel was selected due to its simplicity, low cost and suitability to nitride the sample at any size.

IOP Conf. Series: Journal of Physics: Conf. Series **1082** (2018) 012109 doi:10.1088/1742-6596/1082/1/012109

Gas nitriding is a case-hardening process whereby nitrogen is introduced into the surface of a solid ferrous alloy by holding the metal at a suitable temperature in contact with a nitrogenous gas; usually ammonia resulting in the formation of hardened surface with in fatigue, wear and seizing resistance on machine components or tool. The depth, structure and properties of the surface layer depend on the particular alloying elements contained in the steel and on the processing variables [4].

In this case study, AISI 430 Ferritic Stainless Steel was used. Grade 430 contains chromium that has good corrosion resistance and formability characteristics with useful mechanical properties. Its ability to resist nitric acid attack permits its use in specific chemical applications however automotive trim and appliance components represents its largest fields of applications. These grades can be processed to develop an aesthetically pleasing, bright finish and, hence, are sometimes used for automotive trim and appliance molding. Ferritic stainless steels are used in the manufacture of engine mufflers, nuts, bolts and heat resistant tools. In some cases, such steels require high surface hardness. These grades can be hardened by cold rolling, but cannot be hardened as much as the austenitic alloys [5].

Tribology refers to the science and technology concerned with interacting surfaces in relative motion, including friction, lubrication, wear and corrosion. In many instances, low friction is desirable. The satisfactory operation of joints, for example, whether hinges on doors or human hip joints, demands a low friction force. Work done in overcoming friction in bearing and other mechanical component of machines is dissipated as heat, and its reduction will lead to an overall increase in efficiency [6]. Whenever surfaces move over each other, wear will occur: damage to one or both surfaces, generally involving progressive loss of material. In most cases wear is detrimental, leading to increased clearances between the moving components, unwanted freedom of movement and loss of precision - often vibration, increased mechanical loading and yet more rapid wear, and sometimes fatigue failure. As in the case of friction, though high wear rates are sometimes desirable; grinding and polishing, for example, employ wear processes to remove material rapidly and in a controlled manner, and a small amount of wear is often anticipated and even welcomed during the 'running-in' process in particular kinds of machinery [7].

One of the method of reducing friction, and often wear is to lubricate the surface. However in this case study, the method of nitriding was used to reduce friction and wear on AISI 430 Ferritic Stainless Steel. Thus, the wear resistance properties for as-received sample and nitrided sample were compared to investigate the influence of nitriding for this type of stainless steel. Scratch resistance test was used in this case study.

2. Materials and method

Material used in the study was AISI 430 steel with the composition of 0.04 wt% C, 0.28 wt%Si, 16.3 wt%Cr, 0.36 wt%Mn and Fe balance. Nitriding was carried in the tube furnace at 1200°C for 1 h, 3 h and 5 h.

Metallography was prepared by mounting using bakelite. The samples were observed under the Zeiss optical microscope Wear resistance was performed using scratch resistance equipment with SEM model. Hardness Testing was conducted using LECO LM247 AT micro-hardness Vickers.

3. Results and discussions

3.1 Microstructure observation

Fig. 1 shows the microstructure of as-received samples with mainly the presence of ferrite phases.

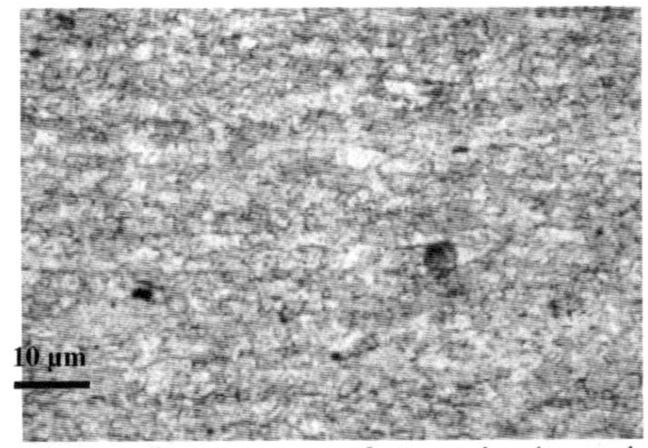

Fig. 1: Microstructure of as-received sample

After nitriding for 1ah, the transformation of ferrite to martensite on the surface was present and depicted as in Fig. 2. The diffusion of nitrogen was initiated from the surface to the inner core of the steel. The formation of martensites were developed with the formation of needle like structure and concentrated near the surface. It was estimated that one quarter of nitrided layer was formed i.e. 2.5 mm.

After 3 h of nitriding, more transformation of martensites from ferrites were observed and this is shown in Fig. 3. The figure illustrates thicker formation of martensites compare to 1 h nitrided sample. Some part was almost covered by transformed martensite.

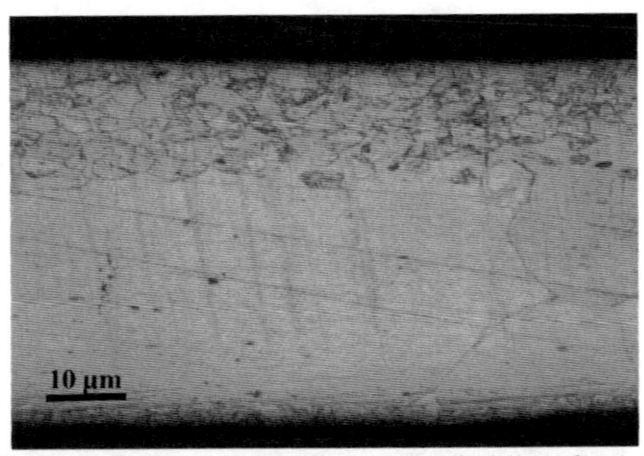

Fig. 2. Microstructure of sample nitriding for 1 h

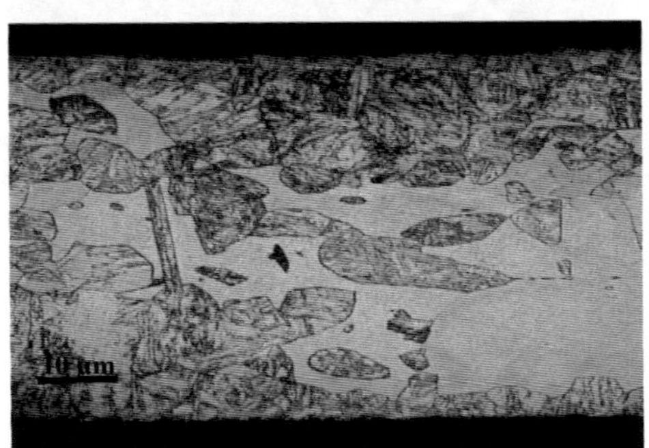

Fig. 3: Microstructure of sample nitriding for 3 h

When nitriding the sample for 5 h, the whole thickness of the steel was transformed from ferrites to martensites and this is shown in Fig. 4. This indicates that high concentration of nitrogen was diffused into 5 h sample when compared to nitriding the samples for 1 h and 3 h.

Fig. 4: Microstructure of sample nitriding for 5 h

3.2 Scratch resistance analysis

Scratch analysis was conducted to obtain the wear resistance properties of the nitride samples. Nitrided samples have low penetrations depth when compared to as-received sample. The normal force was set at 50N.

Fig. 5 shows the depth of penetration for as-received sample which is in the range of -350 to -550 μm. For the 1 h nitrided sample as in Fig. 6, the depth of penetration is in the range of -200 to -300 μm which is less than the as-received sample. The depth of 3 h nitrided sample as in Fig. 7 is in the range of -49 to -60 μm. This sample has less depth of penetration compared to the later samples. The 5 h nitrided samples as in Fig. 8 has the least depth of penetration of -50 μm which is a consistent depth of penetration throughout the sample thickness. This shows that the concentration of nitrogen is homogeneous throughout the layer of steel.

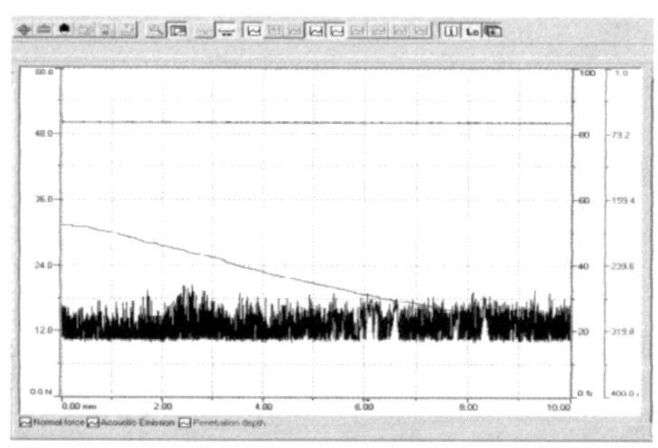

Fig.5: Scratch resistance- as received sample

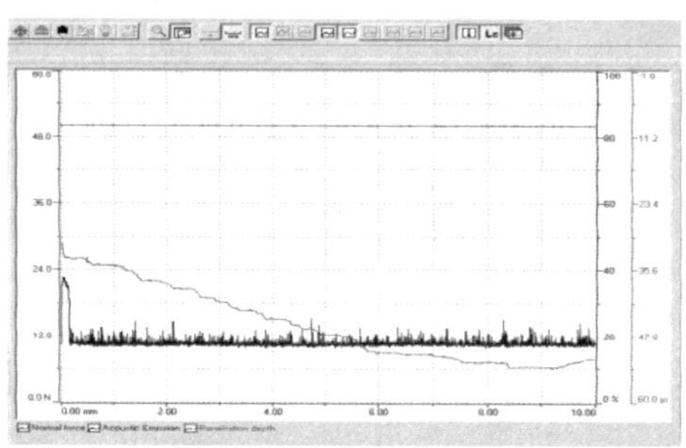

Fig. 6: Scratch resistance – 1 hour nitrided sample

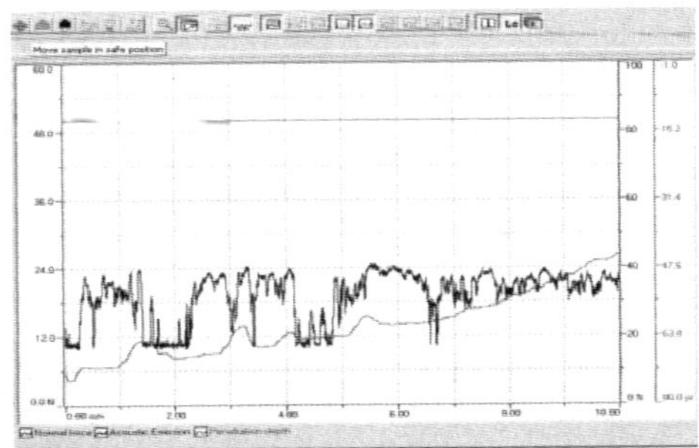

Fig. 7: scratch resistance - 3 hours nitriding

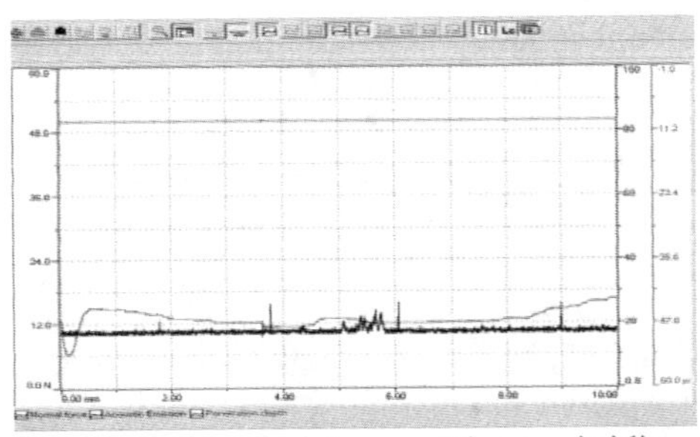

Fig. 8: scratch resistance - 5 hours nitriding

The overall depth of penetration of the scratch test is summarized in Fig. 9. The figure shows that 5 h nitrided sample has the lowest wear resistant of -50 μm and this is followed by the 3ah and the 1 h nitrided samples. The as-received has the lowest wear resistant of -550 μm.

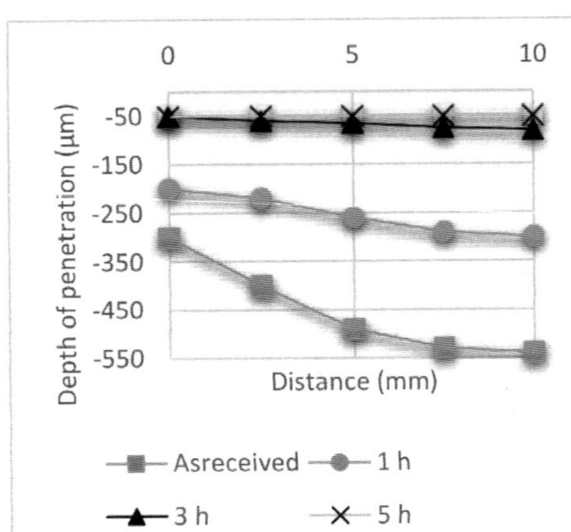

Fig. 9: Overall scratch tests

3.3 Hardness Test

The hardness values at the surface for the as-received, 1 h, 3 h and 5 h nitrided samples are 200, 400, 420 and 430 Hv, respectively (Fig. 10). It shows that the highest concentration of nitrogen contributes to the highest hardness. In this case, it reveals that 5 h nitrided sample has the highest hardness and the as-received sample has the lowest hardness value. The samples revealed that the core location have the lower hardness compared to near the surface of the nitride samples. This can be inferred that higher concertation of nitrogen with a longer holding time of nitriding produces the highest hardness compared to a lower holding time for nitriding.

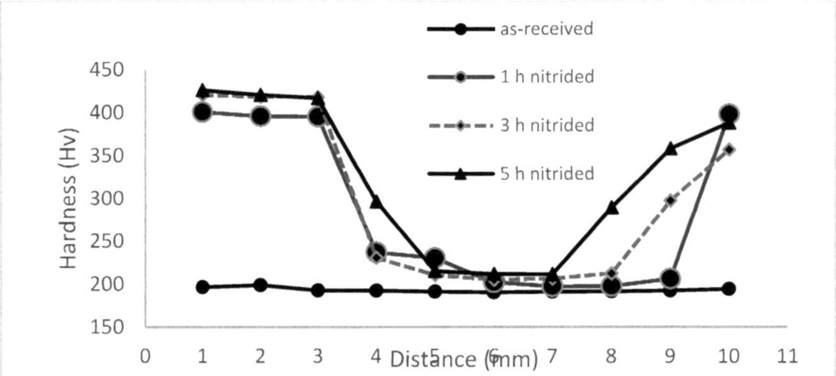

Fig. 10: Hardness Test - all samples

4. Conclusions

The influence of nitrogen on AISI 430 Ferritic Stainless Steel by nitriding affect the transformation of ferrite to martensite, high wear resistance and high hardness. The highest wear resistance is for 5 h nitrided sample with the depth of -50 μm depth of penetration and the least wear resistant is the as-received unnitrided sample with a depth of penetration -550 μm. The results reveal that sample nitrided for 5 h achieved the highest transformation of martensite from ferrite, the highest wear resistance and hardness value. Longer holding time, more nitrogen diffused and more improvement was made on the tribological properties of the steel. Wear resistance properties are improved by nitriding. As the time for nitriding increased, more nitrogen diffused into the samples. Higher nitrogen concentration was found near the surface when compared to the inner core of the sample. Hence, the longer the nitriding time, more improvement is made to tribological properties, in particular.

Acknowledgements

The authors would like to thank Universiti Teknologi PETRONAS under the URIF project 0153AA-B93 for providing the fund and facilities in the research work.

References

[1] JJ.H. Sung J.H.Kong D.K.Yoo H.Y.On D.J.Lee H.W.Lee 2008 *Materials Science and Engineering A* 489 38
[2] Nor Nurulhuda Md. Ibrahima, Patthi Hussain and Mohktar Awang 2014 *Web of Conferences* 4 0401 13 13 M 1-5
[3] Kamal Ariff Zainal Abidin, Elya Atikah Ismail1, Azman Zainuddin, Patthi Hussain 2014 *Web Conference* 13 04029 1
[4] George Krauss, 2000 Steels: Heat Treatment & Processing Principles, ASM Interantional
[5] T.Bell,A.Boyce and J.Lanagan. 1988 *ASM International*
[6] Speidel,M.O, in Proc.Stainless Steel world 2001 Conference, The Hague, the Netherlands, 13 98
[7] George Krauss, Steels: Heat Treatment & Processing Principles, ASM Interantional, 2000
[8] MJ Naele,TA Polak,M priest, Handbook of Surface Treatment and Coating, Professional Engineering Publishing, 2003
[6] Micheal D.Bryant, Journal of Tribology, University of Texas, April 2006
[7] D.T Hawkins,ASM Source Book on Nitriding

IOP Conf. Series: Journal of Physics: Conf. Series **1082** (2018) 012110 doi:10.1088/1742-6596/1082/1/012110

Corrosion Improvement and Microstructure Evaluation of SEM-Solid A356 Alloy by Ecap Process

M.A. Gebril[1], M.Z. Omar[1], I.F. Mohamed[1,2], N.K. Othman[3] and M.A.H. Abdelgnei[1]

[1]Department of mechanical and materials engineering, Faculty of Engineering and Built Enviroment, Universiti Kebangasaan Malaysia, 43600 UKM Bangi, Selangor, Malaysia

[2]Fuel Cell institute Universiti Kebangasaan Malaysia, 43600 UKM Bangi, Selangor, Malaysia

[3]Department of Applied Physics, Faculty of Science and Technology, Universiti Kebangasaan Malaysia, 43600 UKM Bangi, Selangor, Malaysia

Email: magmalki@yahoo.com

Abstract. Al-Si is a casting alloy developed mostly for automobile and aircraft engine industries. This research presents a study related to the semi solid casting of Al-Si A356 alloy produced by cooling slope known as thixoforming process. The sample has then undergone T6 heat treatment. The grain refinement was achieved suitable for thixoforming by Severe Plastic Deformation (SPD) technique through Equal-Channel Angular Pressing (ECAP). The ECAP was conducted at room temperature in which a 10 mm rod was pressed through a constrained channel of a die that was bent abruptly close to 120°. The coarse α-Al phase and silicon particles changes into a globular shape after thixoforming process and T6 heat treatment. The ECAP processing through route A leads to particle refinement after one pass and further refined with an average particle size of 41.2μm after two pass pressing. The potentiodynamic test showed that the corrosion rate reduced to 0.0024 mmpy after two passes of ECAP processing. It confirms that particle refinement through ECAP processing leads to significant enhancement of corrosion resistance in A356 alloy.

1. Introduction

A356 alloy is widely used in producing automotive components such as chassis and engine parts that are usually fabricated through various casting processes. The formation of casting defects such as micropores and microvoids is unavoidable during conventional casting processes and plays a decisive role in restricting design freedom because such defects degrade the mechanical and electrochemical properties. Thixoforming has less solidification shrinkage than traditional liquid casting. Thixoforming alloy is carried out in the semi-solid state to manufacture near net shaped products [1,2]. The significant feature of semi-solid microstructure its globular and none dendritic shape. The thixocasting process improves the component integrity and performance [3-5]. The microstructure need to be globular in the semisolid state prior to the application of thixoforming process [6]. Various methods have been proposed

IOP Conf. Series: Journal of Physics: Conf. Series **1082** (2018) 012110 doi:10.1088/1742-6596/1082/1/012110

to produce globular microstructures of α-Al. Cooling slope (C.S) is among the most popular method to produce globular microstructure in Al alloy. The major challenge for Al-Si alloy is precipitate of silicon particles in flake shape due to solubility limitation of Si in Al that affect the mechanical and electrochemical properties. Nevertheless, the refinement of hypoeutectic Al-Si alloy shows an improvement of mechanical properties of the alloy. SPD is most effective technique that able to refine grain structure to nano size or ultra-fine micrometers size [7]. Through SPD technique, it shows that more refine grain structure can be achieved in various metals and metallic alloys including brittle and ductile alloy [7]. ECAP is most effective mothed in severe plastic deformation process for fabricating bulk ultrafine grained [8]. However, decreasing the grain size and Si particles by deformation process my alter the corrosion properties of the alloy due to generation of dislocation sites that acting as a highway for diffusion and thus, the alloy becomes more susceptible to corrosion activities. The objective of this research is to refine Si particles by using ECAP process after T6 heat treatment and thus to investigate the effect of microstructure change to corrosion resistance of thixoformed A356 alloy after ECAP processing at room temperature.

2. Experimental study

A commercial A356 alloy was cast in ingot having initial dimensions of 30 (thickness) X 140 (length) X 60 (width) mm (as-cast). In semisolid processes, the material was melted in a graphite crucible set at 750°C in an argon atmosphere to prevent oxidation. The cooling slope was made with stainless steel, and its tilt angle, slope length, and slope temperature were 60°, 400 mm, and room temperature, respectively. A pouring temperature of 620°C was selected to limit the superheating of the melt. For thixoforming, samples were sectioned to dimensions of Ø 30 X 120 mm. Rapid reheating to 585 °C for 5 minutes was applied to allow for spheriodisation of the grains. The samples were compressed using a hydraulic cylinder press that provided a load of 20 kN and had a maximum compression velocity of 85 mm/s. The solution heat treatment was carried out over a period of 8 hours at 535 °C. The solution heat-treated samples were quenched in warm water at 60 °C and then aged at 180 °C for 3 hours [9]. Samples lubricated with MoS_2 were deformed by continuous ECAP pass route A (that is the sample is not rotated between each pass), in a die with two equal cross-section channels (10 mm in diameter). The angle between the two inner channels was 120 degrees, while the outer corner angle was 20 deg. A 50-ton hydraulic press was used, and the process was carried out at room temperature. ECAPed extremities billets were cutout to avoid any significant end effects as a result of inhomogeneous deformation. Cylindrical samples of 10 mm diameter and 30 mm height were taken. The microstructure of ECAPed samples were observed using Olympus optical microscope (OM) and ZEISS field emission scanning electron microscope (FESEM). Prior to microstructure characterisation, the samples were grounded with 400-1200 grit SiC abrasive papers, and then polished with (3 μm followed by 1 μm) diamond compound and then etched for approximately 10 s with Keller's reagent. Quantitative metallography analysis was carried out to measure the grain size according to ASTM E 112 standard. To measure the silicon particles size (particle width and length), Smart Tiffv2 software was carried out for at least 200 particles in each case. The electrochemical experiments were carried out in naturally-aerated 3.5% NaCl solution at room temperature and pH 6.5. Corrosion rate of the samples was accomplished with potentiostat GAMRY 3.2, connected to a three-electrode cell, the working electrode was the test material, graphite and silver/silver chloride (Ag/AgCl) electrodes were used as the counter and reference electrodes, respectively. Prior to the test, the as-cast and thixoforming samples were sequentially grounded with 400-1200 grit SiC abrasive papers and washed using distilled water and acetone, and then blown dried with compressed air.

IOP Conf. Series: Journal of Physics: Conf. Series **1082** (2018) 012110 doi:10.1088/1742-6596/1082/1/012110

3. Results and discussion

3.1 Microstructural prior to ECAP

Figure 1a illustrates the optical micrographs of the as-received samples with typical dendritic bright phase hypoeutectic structure surrounded by dark phase of Si eutectic. Figure 1b shows the thixoformed microstructure with primary α-Al phase which is transformed to nearly globular shape distributed uniformly and surrounded by coarse eutectic phase. Figure 1c and d shows the microstructure of eutectic Si after solution heat treatment T6 obviously transformed into spheroidized form in both as-casting and thixoformed alloys. The same behavior was also reported in [10] by the earlier study on A356 alloy after solution heat treatment. The transformation was explained by the formation of small fragmentations from needle-like Si platelets that gradually becomes more globular through the heating [11]. The average size of Si particles before and after T6 are shown in Table 1.

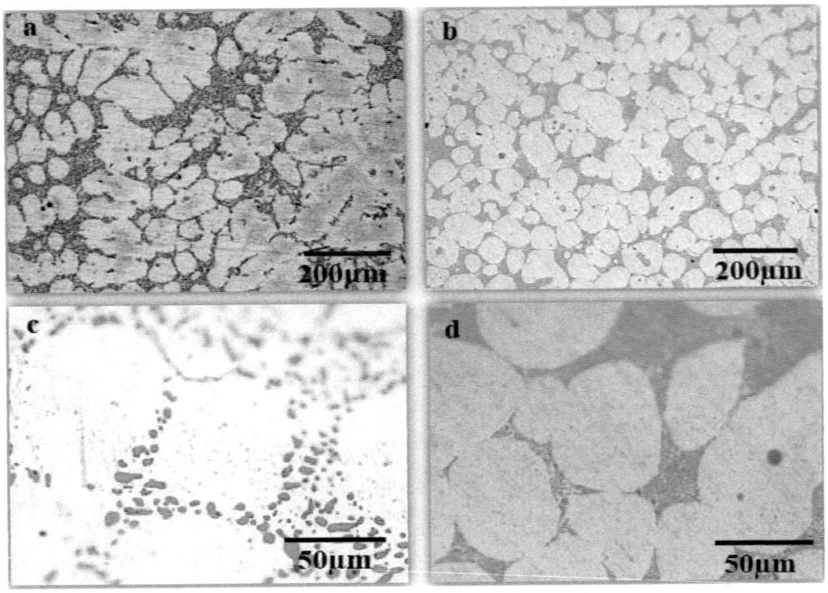

Figure 1. Optical micrographs of A356 alloy with (a) as-cast, (b) As-cast, T6, (c) thixoformed and (d) Thixoformed, T6.

Table 1. Average size of Si particles

Samples	Si size (μm)
As-cast	4.22
As-cast, T6	2.46
Thixoformed	3.10
Thixoformed T6	1.82
Thixoformed, T6 after 2 pass	1.61

3.2 Microstructural after ECAP process

Figure 2 a shows the microstructures of thixoforming alloy after ECAP processing at one and two presses, the microstructures show that a more compact primary α-Al grains surrounded by Si eutectic phase after ECAP process at one pass. Furthermore, the alloy showed more elongated shapes of α-Al when the number of press increases as in Figure 2b. Close observation shows some fragmentation of Si particles among elongated α-Al grains as enlarged in Figure (c and d).

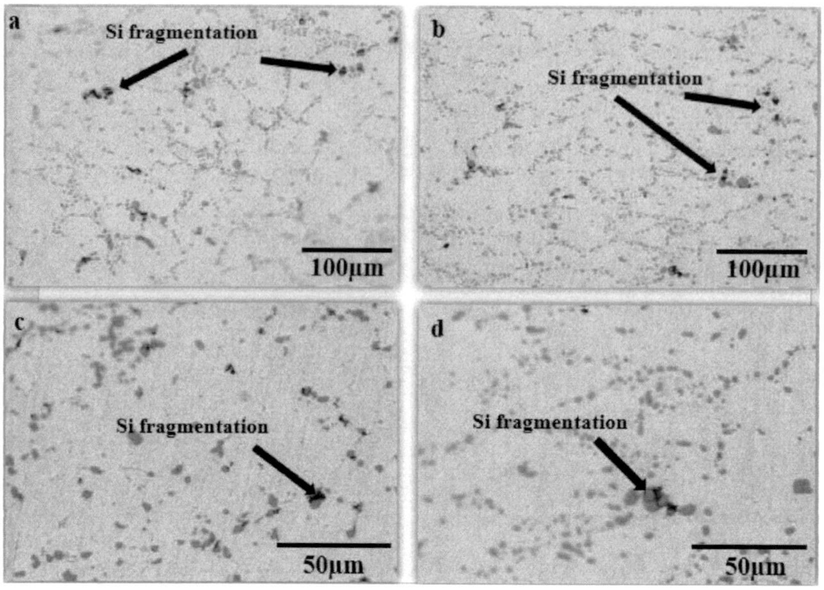

Figure 2. Optical micrographs of A3546 alloy with (a and b) T6 thixoformed after 1 pass, (c and d) T6 thixoformed after 2 passes with different magnification.

3.3 Corrosion evaluation

Figure 3(a-b) show micrographs of the surface of thixoformed and ECAPed thixoformed, T6 after 2 passes route A, respectively after immersion in 3.5% NaCl solution for 10 days. ECAPed thixoformed samples showed relatively smaller corrosion pit size than of thixoformed samples. The existence of pits surrounded by large rings product indicates that the pits are in a stable state [12]. From surface inspection, size of pits and size of rings around pits in thixoformed alloy were greater than ECAPed 2 pass thixoformed alloy which attributed to localized galvanic attack of the more active matrix eutectic mixture surrounded Si particles and the more noble particles [13]. The general absence of corrosion products within the rings surrounded the pits indicated that the cathodic reaction occurring on ring area while the anodic reaction occurred inside the pit itself due to concentration cell action [14]. Existing of big Si particles size accelerated the corrosion rate due to galvanic action with area ratio of cathode to anode because it is one of the key factors for determining the severity of corrosion with galvanic coupling. Figure 4 shows side view of thixoformed and ECAPed thixoformed sample corrosion revealed primary α-Al remained unattacked [15-17] and localized corrosion attacked and penetrated the eutectic mixture phase in both thixo (a) and (b) ECAPed, T6 after 2 pass route A. As shown in (b) after ECAP the eutectic mixture phase was distributed surrounding the elongated shape of α-Al and this is an important factor in galvanic corrosion, i.e. the distribution of the noble phases and less-noble eutectic mixture phase. The decreases in corrosion rate after 2 pass can be attributed to the reduction of Si particles and the reduction of shrinkage and microprosity due to thixoforming process as shows in Figure 4c and Table 2. It shows that the corrosion resistance improved from 0.054 mmy^{-1} to 0.0024 mmy^{-1} after 2 passes.

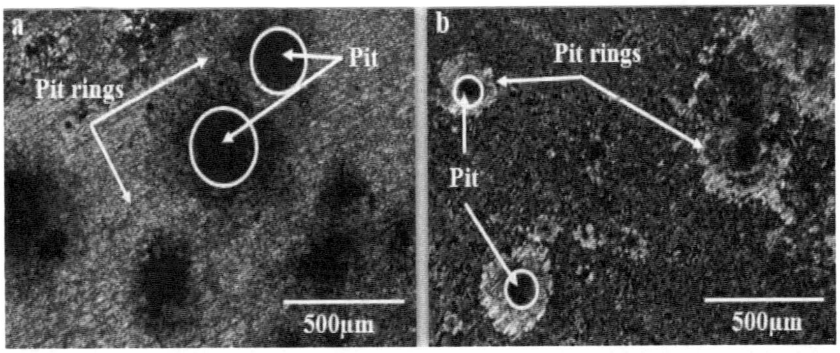

Figure 3. Surface morphology of A345 alloy (a) thixoformed and (b) ECAPed heat treated thixoformed, 2 passes after immersed in 3.5% wt solution for 10 days.

Table 2. Average values of polarization resistance (R*p*) and corrosion rate (C.r) of thixoformed A356 alloy evaluated from Fig. 4.

Pass No.	**R*p*** **(Ωcm^2)**	**C. r (mmpy)**
0 pass	$6.905x10^3$	0.0359
0 pass, thixoformed T6	$2.522x10^4$	0.0092
1 pass, thixoformed	$2.791x10^4$	0.0090
1 pass, thixoformed T6	$5.015x10^4$	0.0062
2 pass, thixoformed T6	$9.80x10^4$	0.0024

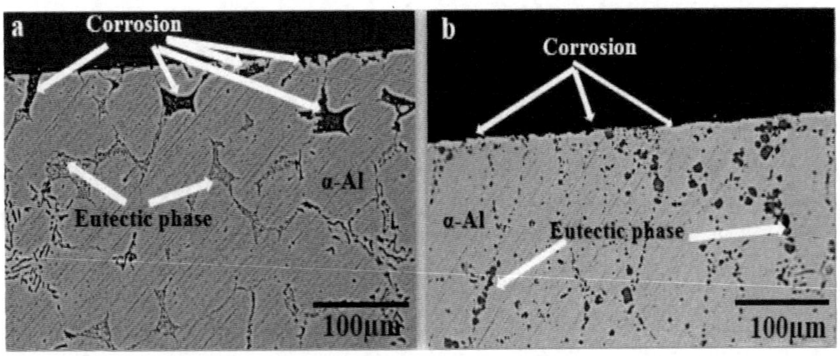

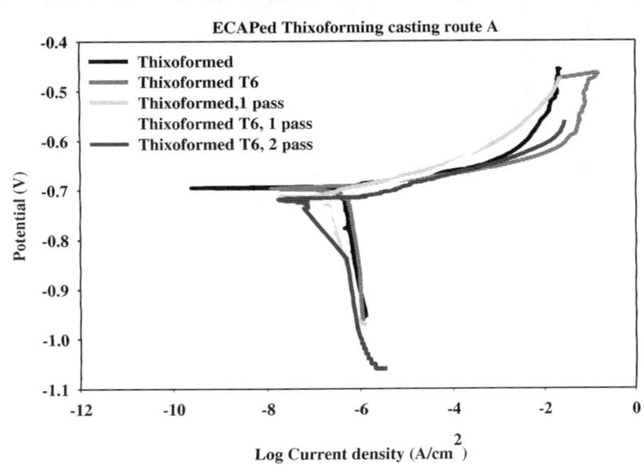

Figure 4. Side view of pitting corrosion of A356 alloy after 10 days' immersion in 3.5 wt.% NaCl solution (a) thixoformed, (b) ECAPed thixoformed T6, 2pass A and (c) Polarization curves of Thixoformed, thixoformed T6 and ECAPed thixoformed.

4. Conclusion

The brief conclusion of this work are presented below:

- The ECAP process had reduced the Si particles size from 4.22 to 1.82μm.
- The fragmentation of Si particles through ECAP processing as well as the reduction of shrinkage and microprosity after thixoforming reduced the microgalvanic effect due to the reduction of the ratio of anode to cathode.
- The corrosion rate has decreased to 0.0024 mmy^{-1} through application of ECAP processing.

Acknowledgement

The authors acknowledge the Ministry of Higher Education Malaysia for the Fundamental Research Grant Schemes (FRGS/1/2014/TK01/UKM/01/2) and (FRGS/1/2016/TK03/UKM/02/4).

References

[1] Flemings M C 1991 *Metall Trans A* 22A 957
[2] Fan Z 2002 *Int Mater Rev* 2002 47 49
[3] Birol Y 2009 *J Alloys Compd* 473 133
[4]Wang J G, Lin H Q, Li Y Q, and Jiang Q C 2008 *J Alloys Compd* 457 251
[5] Lee K, Kwon Y N, and Lee S 2008 *J Alloys Compd* 461 532
[6] Tzimas E, Zavaliangos 1999 *A. Acta Mater* 47 517
[7] Mohamed I F, Masuda T, Lee S, Edalati K, Horita Z, Hirosawa S, Matsuda K, Terada D, Omar M Z 2017 *Material science and engineering* A704 112
[8] Shin D H, Kim B C, Park K T 2000 *Acta Materialia* 48 3245
[9] ASM Handbook. Properties and selection: nonferrous alloy and special purposed materials, Vol.2. ASTM International: 1992.
[10] Möller H, Govender G & Stump W E 2007 *International Journal of Cast Metals Research* 20 340
[11] Juan Francisco Hernandez Paz, McGill University, Montreal, Canada, PhD thesis, September 2003.
[12] Son I, Nakano H, Oue S, Kobayashi S, Fukushim H, Horita Z 2007 *Materials Transactions* 48 21
[13] Song D, Ma A, Jiang J, Lin P, Yang D, Fan 2010 *J. Corrosion Science* 52 481
[14] William D. Callister, Jr. David G. Rethwisch, Material science and engineering an introduction, seventh edition, John Wiley & Sons, Inc. 2007.
[15] ASM Handbook. Properties and selection: nonferrous alloy and special purposed materials, vol. 2. ASM International; 1992.
[16] Goodarzy M H, Arabi H, Boutorabi M A, Seyedein S H, Hasani Najafabadi S H 2014 *J Alloys Compd* 585 753
[17] Barbucci A, Bruzzone G, Delucchi M, Panizza M, Cerisola G 2000 *Intermetallics* 8 305
[18] Yurt A, Bereket G, Ogretir C 2005 *J Mol Struct* 25 215

Institute of Physics
Dirac House, Temple Back
Bristol BS1 6BE UK

ISSN: 1742-6588
ISBN 978-1-5108-7299-8

CURRAN ASSOCIATES INC.
proceedings.com